Fundamental Theories of Physics

Volume 181

The international monograph series "Fundamental Theories of Physics" aims to stretch the boundaries of mainstream physics by clarifying and developing the theoretical and conceptual framework of physics and by applying it to a wide range of interdisciplinary scientific fields. Original contributions in well-established fields such as Quantum Physics, Relativity Theory, Cosmology, Quantum Field Theory, Statistical Mechanics and Nonlinear Dynamics are welcome. The series also provides a forum for non-conventional approaches to these fields. Publications should present new and promising ideas, with prospects for their further development, and carefully show how they connect to conventional views of the topic. Although the aim of this series is to go beyond established mainstream physics, a high profile and open-minded Editorial Board will evaluate all contributions carefully to ensure a high scientific standard.

More information about this series at http://www.springer.com/series/6001

Giulio Chiribella · Robert W. Spekkens
Editors

Quantum Theory: Informational Foundations and Foils

Springer

Editors
Giulio Chiribella
Department of Computer Science
The University of Hong Kong
Hong Kong
People's Republic of China

Robert W. Spekkens
Perimeter Institute for Theoretical Physics
Waterloo, Ontario
Canada

ISSN 0168-1222 ISSN 2365-6425 (electronic)
Fundamental Theories of Physics
ISBN 978-94-017-7302-7 ISBN 978-94-017-7303-4 (eBook)
DOI 10.1007/978-94-017-7303-4

Library of Congress Control Number: 2015945141

Springer Dordrecht Heidelberg New York London

Printed on acid-free paper

Springer Science+Business Media B.V. Dordrecht is part of Springer Science+Business Media (www.springer.com)

Contents

Part III Categories and Convex Sets

Part IV Quantum Versus Super-Quantum Correlations

Introduction

Giulio Chiribella and Robert W. Spekkens

The foundations of Quantum Mechanics are experiencing a golden age. In a timespan of less than two decades, an astonishing number of new results, ideas, and frameworks have revolutionized the way we think about the subject. A new research community is emerging worldwide, attracting scientists from a diverse spectrum of disciplines including physics, computer science, and mathematics. The keyword "foundations" is now included in the strategic priorities of many research institutions and funding agencies, and it regularly features as one of the hot topics in international conferences.

The abundance of ideas, approaches, and resources that have emerged poses some challenges however. For one, having a global vision of the field and reflecting on its high level goals is becoming increasingly difficult. For another, the sheer number of different frameworks that have been put forward risks creating a tower of Babel effect, fragmenting the community into smaller cliques that are unable to talk to one another. In addition, researchers who are joining the field have to cope with a fast-moving landscape where it can be hard to identify stable reference points.

These considerations led us to the project of this book, which aims to showcase the state of the art in quantum foundations. The book provides a collection of articles that deal with influential ideas in the field today, revealing the diversity of approaches on the one hand, and highlighting the common threads among them on the other.

G. Chiribella (✉)
Department of Computer Science, The University of Hong Kong, Pokfulam Road, Hong Kong, People's Republic of China
e-mail: giulio@cs.hku.hk

R.W. Spekkens
Perimeter Institute for Theoretical Physics, 31 Caroline Street North, Waterloo, Ontario N2L 2Y5, Canada
e-mail: rspekkens@perimeterinstitute.ca

G. Chiribella and R.W. Spekkens (eds.), *Quantum Theory: Informational Foundations and Foils*, Fundamental Theories of Physics 181,
DOI 10.1007/978-94-017-7303-4_1

1 Characteristics of the New Wave of Quantum Foundations

We start by outlining what is distinctive about the foundational research that this book aims to portray.

1.1 A Pragmatic Perspective

It is useful to distinguish between what one might call *dynamicist* and *pragmatist* traditions in physics. Within the dynamicist tradition, the physicist's job is to describe the natural dynamical behaviour of a system, without reference to human agents or their purposes. In the pragmatic approach, on the other hand, the laws of physics are characterized in terms of the extent to which we can learn and control the behaviour of physical systems. The distinction between the dynamicist and pragmatist points of view is nicely represented in competing formulations of the second law of thermodynamics. One that is clearly in the dynamicist tradition is Clausius's original statement:

> Heat can never pass from a colder to a warmer body without some other change, connected therewith, occurring at the same time [1].

On the other hand, the version of the Kelvin-Planck statement that is found in most textbooks is clearly pragmatic:

> It is impossible to devise a cyclically operating device, the sole effect of which is to absorb energy in the form of heat from a single thermal reservoir and to deliver an equivalent amount of work [2].

Quantum theory has always partaken in both traditions. Indeed, Schrödinger's wave mechanics and Heisenberg's matrix mechanics were distinguished in part by the fact that Schrödinger, following de Broglie's lead, sought to provide a description of the motion of particles, while Heisenberg, following Bohr's lead, espoused an operational philosophy and took his formalism to merely describe what would be observed in certain experimental circumstances. The new foundational work represents a renewed interest in exploring quantum theory within the pragmatic tradition.

1.2 Quantum Foundations in the Light of Quantum Information

The newfound popularity of the pragmatic tradition is tightly connected with the rise of quantum information theory. The real innovation of the recent foundational work is in the way researchers conceive the difference between quantum and classical theories [3]. Historically, quantum theory was taken to consist entirely of *restrictions* on our information-gathering ability; think for instance of the restriction imposed by the uncertainty principle. The quantum information revolution overturned this notion: a quantum world in fact holds new possibilities for information-processing tasks—in particular, communication tasks, cryptographic tasks and computational tasks—that could not be accomplished in classical physics.

Milestone applications of quantum information, such as secure quantum key distribution [4, 5], ultrafast quantum algorithms [6, 7], teleportation [8], and dense coding [9], stimulated the imagination of quantum theorists, and led them to ask questions that moved beyond the usual topics of foundational discussions: *Which principles of quantum theory can account for its information-processing advantages? Does the possibility of achieving one kind of information-processing advantage imply the possibility of achieving others? Is quantum theory the* only *theory where these advantages arise?* These questions were at the center of an influential research programme, launched by Fuchs [10, 11] and Brassard [12], that aimed to understand quantum theory in the light of quantum information. More specifically, the idea was to take certain facts about the information-processing features of a quantum world, for instance, the possibility of secure key distribution and the impossibility of secure bit commitment, and derive the quantum formalism from these. This line of inquiry gave birth to a new breed of foundational research with more pragmatic ambitions, with practitioners that split their time between developing novel practical applications of quantum information and achieving a deeper foundational understanding of quantum theory, with each activity informing the other.

1.3 The Shift from Interpretation to Reconstruction

Traditionally, the focus of many quantum foundations researchers was the *interpretation* of quantum theory. In most such works, the formalism of quantum theory was taken as given, and the goal was to infer from this formalism the correct story to tell about the nature of reality—typically, a story of dynamicist flavour. The Everett interpretation [13] and the deBroglie-Bohm interpretation [14] are examples. Models incorporating physical collapses [15, 16] are also proposed in an effort to secure a dynamicist story about quantum theory.

By contrast, the focus of the new wave is the *reconstruction* of quantum theory from physical principles. Contemporary researchers are looking for an answer to Wheeler's famous question *"Why the quantum?"* [17] and are driven to understand the origin of the formalism itself. Textbook postulates such as "a physical system is described by a complex Hilbert space", "pure states are described by unit vectors", "outcome probabilities are given by the Born rule", and "systems combine by the tensor product rule" are now regarded as abstract mathematical statements in need of a more fundamental explanation. Such an explanation would be akin in spirit to Einstein's derivation of the Lorentz transformations from the light postulate and the principle of relativity.

The goal is to find a compelling set of axioms that singles out quantum theory from among all possible theories. Finding an appealing axiomatization is a problem that has a long tradition, starting with the work of Birkhoff and von Neumann [18] and continuing through the works of Mackey [19], Ludwig [20], and Piron [21] and the tradition of quantum logic [22, 23]. What distinguishes the axiomatic work being pursued today is the use of notions inspired by quantum information theory, the emphasis on composite systems, the focus on finite-dimensional Hilbert spaces, and an insistence on axioms that are operationally meaningful.

1.4 The Operational Framework

Any question of the form "why *this*?" is implicitly asking "why not *that*?". Therefore, to tackle Wheeler's question, one first of all needs to be able to conceive of alternatives to quantum theory, ways the world *might have been.* In short, one requires a framework for describing a broad range of physical theories, including quantum and classical theories, but allowing more exotic alternatives as well.

One way to achieve such a framework is to focus on a strictly operational formulation of physical theories. An operational formulation is one wherein the primitive concepts are preparation procedures, transformation procedures, and measurement procedures, each understood as a specification of a list of instructions for an experimentalist, spelled out in sufficient detail that they could be implemented by any technician, as with a good recipe. The theory specifies a mathematical algorithm that fixes the probability distribution over outcomes for every possible measurement given every possible preparation and intervening transformation. When phyical theories are operationally formulated, therefore, the only relevant differences between them are differences in the sorts of experimental statistics that they allow.

The operational approach encourages one to focus on a characterization of quantum theory in terms of experimental facts, and to consequently avoid, as much as is possible, making claims that go beyond what is strictly required to describe these facts. This sort of exercise can be very useful for freeing the mind from all the baggage of classical preconceptions and previous attempts to interpret the quantum formalism. For many researchers, adopting this approach is not a rejection of the need for providing a dynamicist account of quantum theory, nor is it necessarily an endorsement of the notion that a physical theory is *nothing more* than an algorithm for predicting experimental statistics. Rather, it is considered an effective methodological tool for making progress on questions about the origin of the quantum formalism.

1.5 Foil Theories

A distinctive characteristic of contemporary foundations is the exploration of alternatives to quantum theory, that is, *foil theories*. A foil to X is something that helps to highlight the distinctive characteristics of X by contrasting with it.[1] Given a framework of possible theories that includes quantum theory, every nonquantum point in the landscape is a foil theory. Each such theory specifies a way the world might have been had it not been quantum.

We use the term 'foil' to highlight the attitude that is taken towards these theories: they are *not* being proposed as empirical competitors to quantum theory, with grand ambitions of usurping its throne. Rather, they serve to clarify what is distinctive about quantum theory. For instance, if one can identify a foil theory that shares some set of features with quantum theory, then that set of features cannot possibly be a complete set of axioms for quantum theory. Likewise, constructing foil theories is an essential step for proving the independence of a set of axioms: if one axiom is independent from another, then one should be able to devise a foil theory that satisfies the former but violates the latter.

1.6 Goals

One of the ambitions of researchers in quantum foundations is that the insights coming from their work will help with some of the big challenges of contemporary physics, such as the formulation of a quantum theory of gravity. Another ambition is to find alternatives to quantum theory that *could* eventually become empirical competitors. Given an axiomatic derivation of quantum theory, it suffices to modify a single axiom in order to get a consistent alternative. Furthermore, this approach can be used to avoid an important pitfall of more ad hoc approaches to developing alternatives to quantum theory, namely, that the latter may inadvertently violate fundamental principles that one would prefer not to abandon. A good example is the nonlinear modification of quantum theory proposed by Weinberg [24] which was subsequently shown to allow for superluminal signalling [25] and also to violate the second law of thermodynamics for the normal definition of entropy [26]. In the axiomatic approach, the fundamental principles that one wants to uphold can be built in from the outset.

A more practical application of this foundational work is to advance quantum technologies. Indeed, such work is beginning to clarify how information-processing capabilities can arise from foundational principles. For instance, cryptography based on Bell-inequality violations [5, 27] can be shown to be secure even if the devices used in the protocol are supplied by the adversary, as long as it is presumed that the adversary cannot signal superluminally [28, 29]. This idea, which originated from foundational works, led to an entire field of *device-independent cryptography* [28–32].

[1] "Whenever I marry," she continued after a pause which none interrupted, "I am resolved my husband shall not be a rival, but a foil to me."—from Jane Eyre, by Charlotte Brontë.

2 Frameworks for Operational Theories

It is worth spending a few words on the specific frameworks that have been developed in an attempt to achieve the aims described above. Because existing frameworks were found insufficient, many researchers opted to construct a new canvas for their portrait of quantum theory, with quantum information processing serving as their muse. The emphasis posed on the development of such frameworks is itself a distinctive trait of the new wave of foundational research.

To the outsider, it is hard to appreciate the importance of constructing the framework. But it is in fact a highly non-trivial task, where one is forced to make fundamental choices as to what is considered "general" (i.e. part of the notion of a physical theory) and what is considered "specific" (and hence a possible candidate for an axiom that identifies quantum theory). In a sense, what is a stake in the choice of a framework is the very definition of a physical theory.

Note that having a framework for operational theories is not only useful as an instrument for axiomatizations, but also as a playground for experimenting with alternative models of information processing. Such frameworks are increasingly being used to attempt to describe nonclassical phenomena in a language that does not presume the correctness of quantum theory. Not only is this pursued for the question of Bell inequality violations [33–36], but also for a number of applications to computer science and physics, including the study of communication complexity [37, 38], nonlocal computation [39], measurement-based computation [40–44], games and interactive proof systems [45–50], randomness amplification [51–54], causal networks [55–57], computability [58], complexity [59], key distribution [60], bit commitment [61–63], complementarity [64, 65], no cloning [63, 66, 67], teleportation [63, 68, 69], state discrimination [70–72], entropy [73–75], thermodynamics [76–78], general resource theories [79], and spacetime physics [80, 81]. This long list provides a good illustration of how fertile the development of new frameworks has been. In the following we identify the main directions along which the framework-building activity has developed so far.

2.1 The Framework of Convex Operational Theories

A particularly popular framework is that of *convex operational theories*, where preparations, transformations, and measurements are represented by elements of suitable convex sets, the dimension of which is fixed by the nature of the physical systems involved in the experiment.

The framework of convex operational theories is the contemporary descendant of the frameworks used in the tradition of operational quantum logic, in particular those introduced by Mackey [19], Ludwig [20], and Davis and Lewis [82]. In the new wave of quantum foundations, the first elaboration of this framework appeared in Hardy's 2001 axiomatization of quantum theory [83]. With respect to earlier works in quantum logic, Hardy's framework distinguishes itself by being more manageable and intuitive, partly because of its focus on finite-dimensional systems. This approach was brought to completion through a series of works by a number of other authors [66, 69, 84, 85].

2.2 The Category-Theoretic Framework

Due to the long tradition of using convex sets to represent the state spaces of physical systems, there is a strong temptation to identify the operational approach with the framework of convex operational theories. However, a substantial part of what defines a physical theory has nothing to do with convex sets, or even with probabilities. For example, operational notions such as composing two systems in parallel (this *and* that) and composing two physical processes in a sequence (do this *and then* do that) are more primitive than the notion of probability. Such notions of composition are the focus of the *category-theoretic framework* initiated by Abramsky and Coecke [68, 86–88]. In this framework, the mathematical structure describing a general physical theory, in particular the two notions of composition and how they interact, is that of a strict symmetric monoidal category. One of the characteristic features of the category-theoretic framework is that all the relations of interest can be encoded in diagrams, similar to those used in the representation of quantum circuits.

2.3 The Framework of Operational-Probabilistic Theories

The lesson of the category-theoretic framework is that the composition of systems and processes is fundamental to the operational structure of a theory and that one can talk about information processing without even having to mention probabilities. On the other hand, the precise probabilistic predictions of an operational theory are sometimes a feature of interest. If one is interested in *both* the compositional and the probabilistic features of a theory, then the framework of *operational-probabilistic theories*, recently developed by Chiribella et al. [63, 89, 90] and Hardy [91, 92], provides a supplementation of the category-theoretic framework with probabilistic structure.

In this framework, the category-theoretic notions are used to define circuits of physical processes. An experiment is represented by a closed circuit, starting from the preparation of a system and ending with a measurement having a particular outcome. The probabilistic structure is added on top of the circuit framework by introducing a rule that assigns probabilities to these closed circuits. The result of this construction is that states, transformations, and measurements are represented by elements of suitable vector spaces, as they are in the framework of convex-operational theories. However, the framework of operational-probabilistic theories allows one to describe also theories where the state space is not convex, such as Spekkens' toy theory [93].

In addition, it allows one to treat causality as an emergent feature in a broader class of physical theories where causality is not assumed as part of the framework [63].

When we wish to refer to a framework that can describe features of experimental probabilities, while remaining noncommital about whether it is the framework of convex operational theories or the more general framework of operational-probabilistic theories, we shall speak simply of the framework of *generalized probabilistic theories* (GPTs).

2.4 The Device-Independent Framework

Another popular framework is the *device-independent framework* [28, 29, 94, 95]. Here an experiment is not parsed into preparations, transformations and measurements, with a physical system of a particular dimension acting as a causal mediary between these. Rather, the experiment is treated as a black box, characterized completely by how it maps classical inputs to classical outputs. The roots of this approach can also be traced back to the quantum information revolution: considering input-output black boxes is a natural approach to the design of cryptographic protocols that are secure even if the functioning of the devices is not trusted. In this context, proving the security of a protocol independently of the inner workings of its black box components is desirable because the components may have been designed by one's adversary.

The device-independent framework is apt to capture the *device-independent features* of quantum theory. The paradigmatic example of a device-independent quantum feature is the Tsirelson bound [96], which can be viewed as an upper bound on the probability that two cooperating players win a game, known as the *CHSH* game after the seminal work of Clauser et al. [97]. In the CHSH game, the inputs are the questions asked by a referee to the two players, and the outputs are their answers. While playing the game, the players are allowed to share arbitrary entangled states and are allowed to perform arbitrary local measurements on their systems. Still, their winning probability is upper bounded, independently of the states they prepare and of the measurements they perform. The bound is device-independent, in that it depends only on the validity of quantum theory.

The CHSH game is the problem that got the device-independent approach started, when Popescu and Rohrlich [94] and Rastall [98] came up with a foil theory that is *more nonlocal than quantum theory*, i.e., it guarantees to the players a higher winning probability in the CHSH game. Nevertheless, any other game would define a device-independent feature of quantum theory. The ultimate device-independent feature is the specification of the full set of correlations (i.e. the conditional probability of the outputs given the inputs) that are achievable by local quantum measurements on a bipartite quantum state. This is known as *the quantum set*.

A particularly active line of research in recent years has been the problem of *deriving* device-independent features of quantum theory from information-theoretic principles. The ultimate dream of researchers working in this area is to derive the specific shape of the quantum set by using only device-independent axioms, that is, axioms that refer only to the conditional input-output probabilities. Although the study of information processing in generalized probabilistic theories and the study of device-independent features have developed on separate tracks until now, the time is ripe for uncovering connections between them. On the one hand, the tools developed in the study of axioms for generalized probabilistic theories may help to achieve a characterization of the quantum set, a project that is notoriously difficult. On the other hand, device-independent features may provide candidates for new axioms. A detailed discussion of the connections between the device-independent framework and the framework of general probabilistic theory can be found in Ref. [99].

3 Book Synopsis

The information-theoretic characterization of quantum theory is a general direction that unites the efforts of the new quantum foundationalists, although below this umbrella there is an exceptional variety of different approaches and goals. The book aims to provide a panoramic view of the field, including some of the most promising directions that have emerged in the past decade. It is divided into four sections, corresponding to the following themes:

1. Foil theories (Chaps. 1–3)
2. Axiomatizations (Chaps. 4–8)
3. Categories and convex sets (Chaps. 9–10)
4. Quantum versus super-quantum correlations (Chaps. 11–15)

This subdivision is meant as an aid for readers who are approaching the field for the first time and want to have an idea of the big picture. Many other organizational schemes would have worked just as well, and we therefore encourage readers to explore other paths through the various contributions. In the following, we provide a synopsis of the book through its four sections.

3.1 Foil Theories

We open the book with three examples of foil theories.

Wootters (Chap. 1) considers *real quantum theory* [100–102], which is the foil theory that results from replacing the complex field with the real field in the standard formalism of quantum theory. He considers the information transfer from a preparation to a measurement and shows that for certain natural ways of quantifying this transfer—for instance, the mutual information between the angle of a polarizer that prepares a photon's polarization and the relative frequency of outcomes in a measurement of polarization—the information transfer is optimized for real quantum theory and not for complex quantum theory. He further considers the question of whether some *other* notion of information transfer might pick out complex quantum theory rather than its real counterpart.

Schumacher and Westmoreland (Chap. 2) present and develop *modal quantum theory* [103], which replaces the complex field with a finite field. This necessitates a more dramatic modification of the quantum formalism than is required to replace the complex field with the real field. The foil theory that they construct is *possibilistic* rather than *probabilistic*: it does not specify the probabilities of different measurement outcomes, but only which outcomes are possible and which are impossible. Despite the fact that modal quantum theory is rather minimalist in the scope of states and measurements that it permits, it nonetheless reproduces a surprising number of qualitative features of quantum theory.

Spekkens (Chap. 3) considers a family of foil theories that arise from taking a classical statistical theory and imposing an epistemic restriction, that is, a restriction on the amount of knowledge any observer can have about the physical state of a classical system [93]. Depending on the type of degree of freedom being considered, the resulting foil theory either describes a subset of the preparations, transformations and measurements allowed in the full quantum theory for that type of degree of freedom, or it describes a distortion of such a subset that is inequivalent in its predictions to quantum theory. Both types are shown to reproduce a large number of phenomena that are usually taken to be distinctively quantum, but to lack others, thereby suggesting a distinction between weak and strong notions of nonclassicality.

3.2 Axiomatizations

This part of the book presents three different axiomatizations of quantum theory (Chaps. 4–6) along with two contributions on themes that are closely related to the axiomatic endeavour (Chaps. 7–8). For reasons of space, all of the axiomatization chapters confine themselves to presenting an outline of the main ideas behind the derivation of the Hilbert space formalism, while omitting the technicalities that go into the mathematical derivations (these can be found, of course, by referring to the original research articles).

Masanes and Müller (Chap. 4) present their 2011 axiomatization of quantum theory [104]. We start our lineup of axiomatization here because this work is a direct descendant of Hardy's seminal 2001 axiomatization, from which it inherits some of its axioms. With respect to Hardy 2001, the main progress here is in the elimination of one axiom, called the "Simplicity Axiom", which, compared to the others, seemed to be less motivated. Within both the Hardy 2001 and the Masanes-Müller 2011 axiomatizations, the feature that distinguishes quantum from classical theory is the fact that every two pure states are connected by a *continuous* path of pure states.

Chiribella, D'Ariano and Perinotti (Chap. 5) present their axiomatization [89] next. The central axiom here is the Purification Postulate, stating that every mixed state of a given physical system can be modelled as the marginal of a pure state of a larger composite system. This requirement directly implies many quantum features, such as no-cloning, teleportation, and the fact that every irreversible process can be modelled as the result of a reversible interaction between the system and an environment that is subsequently discarded [63]. A slogan for this axiomatization is that quantum theory is the only pure and reversible theory of information.

We conclude our lineup of axiomatizations with Hardy (Chap. 6) who presents his 2011 axiomatization [92]. In this axiomatic scheme, the emphasis is on the perfect distinguishability of states and on the possibility of performing computations reversibly. Hardy proves that there are only two theories compatible with his new set of axioms: classical and quantum. Once this result is established, therefore, one can identify quantum theory by choosing any feature that distinguishes it from classical theory. Insofar as this work constitutes a significant development of Hardy's influential 2001 axiomatization and incorporates tools and ideas introduced by other authors working on axiomatization, it is a good illustration of the progress of the field in the last decade.

Chapters 7 and 8 do not present new axiomatizations, but nonetheless concern themselves with the axiomatization project.

Dakić and Brukner (Chap. 7) note that within generalized probabilistic theories, experimental operations are described abstractly and do not make direct contact with more traditional concepts of physics, such as position in space, direction, and energy. Their work aims to bridge this gap to some extent. They show that, within a suitable class of theories, quantum theory embedded in a three-dimensional space is the only theory satisfying the consistency requirement that every possible transformation of a single elementary system can be generated by a symmetric interaction between the system and a macroscopic system which acts as a program for the desired transformation. Their work provides an example of the trend of applying the formalism of generalized probabilistic theories to a broader spectrum of topics in physics.

Fuchs and Stacey (Chap. 8) provide some critical remarks on existing axiomatizations of quantum theory and express some desiderata for future work. In addition to motivating the search for a more compelling picture, they review the QBist approach to the foundations of quantum theory [105–108], which aims to understand quantum theory within a subjective Bayesian approach to probability theory, in particular, as a modification to the manner in which experimental probabilities in different counterfactual scenarios are related to one another.[2]

3.3 *Categories and Convex Sets*

Chapters 9 and 10 expound the foundations of the category-theoretic framework and of the framework of convex operational theories, respectively. As we have noted,

[2]This chapter is a transcript of the talk given by Fuchs at the conference Conceptual Foundations and Foils for Quantum Information Processing, May 9–13, 2011

developing suitable frameworks is an essential step in the axiomatization of quantum theory and a subject of active research in its own right. The reader may well wonder why we chose to put the framework chapters *after* the axiomatizations chapters, rather than before. There are several reasons for our choice. First of all, the main message of the axiomatizations can be easily grasped without entering into the specific details of the framework. In fact, given the richness of nuances contained in the axiomatization works, too much attention to details could even hinder the first reading. On top of that, the frameworks used in the axiomatization chapters are often different from those presented in Chaps. 9 and 10. Finally, giving first a taste of what the study of operational theories can achieve is probably the best way to motivate the reader to a deeper excursion into the structural aspects of the framework.

Our excursion starts with Coecke, Duncan, Kissinger, and Wang (Chap. 9), who review the category-theoretic framework for describing operational theories [68, 86–88]. This chapter will take the reader through the quantum structures that are central to this approach, such as the tensor product structure, the compact structure associated to quantum teleportation, the dagger structure associated to the adjoint, and the Frobenius structure associated to orthonormal bases. These notions are expressed in terms of a diagrammatic calculus that allows mathematical proofs to be carried out entirely through the manipulation of diagrams. Using this framework, the authors provide a purely graphical treatment of complementarity and of the Greenberger-Horne-Zeilinger paradox at the end of the chapter.

Barnum and Wilce (Chap. 10) present the framework of convex operational theories [66, 69, 83–85]. Here, the structures of ordered vector spaces, geometry, and symmetry are the main protagonists. States of a given system are represented by points in a finite dimensional convex set, measurements by positive linear functionals, and physical transformations by positive linear maps. To illustrate some of these notions, the chapter presents many concrete examples of convex operational theories that are nonclassical but distinct from quantum theory. The treatment of tensor products and entanglement in convex operational theories is reviewed, as is the question of which information processing advantages of quantum theory are generic to convex operational theories that are nonclassical. Finally, the chapter discusses axioms for quantum theory based on considerations of symmetry and composition.

3.4 Quantum Versus Super-Quantum Correlations

In the final part of this book we present a number of important features of the set of quantum correlations, in particular, features that serve to distinguish it within a larger, hence "super-quantum", set of correlations compatible with the no-signalling principle.

Pawlowski and Scarani (Chap. 11) discuss the principle of Information Causality [109]. This is a device-independent principle that concerns the possibilities for communication within an operational theory, in particular, for a communication protocol known as a random access code, wherein the receiver only gets part of the data encoded by the sender, but is allowed to choose which part. An operational theory is said to be information causal if assisting a random access code with an arbitrary shared nonsignalling resource of correlations provides no advantage. They show that Information Causality implies the Tsirelson bound and many other features of the set of quantum correlations.

Navascués (Chap. 12) discusses the principle of Macroscopic Locality [110]. This approach makes use of the fact that the strength of nonsignalling correlations among microscopic systems has consequences for the strength of such correlations among macroscopic systems, that is, among collections of microscopic systems wherein one cannot address the constituents individually. To insist that an operational theory satisfy macroscopic locality is to insist that in the macroscopic limit it must look classical, in particular, it must look local in the sense of not violating a Bell inequality. This principle implies that the microscopic correlations must satisfy the Tsirelson bound and reproduces other features of the quantum set.

Acín, Almeida, Augusiak, and Brunner (Chap. 13) describe the foundational implications of a multipartite game called Guess Your Neighbor's Input (GYNI) [111]. The game of GYNI is one for which quantum does *not* provide an advantage over classical, but for which nonsignalling alternatives to quantum theory do provide an advantage. Thus, the game provides a natural separation between quantum correlations and super-quantum correlations. Various consequences for the project of deriving the quantum set are discussed: GYNI can be used to show that in the multipartite scenario, the no-signalling principle and the assumption that systems locally look quantum is not enough to recover the quantum set, unlike the bipartite case; and to derive Bell inequalities that do not admit of any quantum violation.

Finally, Colbeck and Renner (Chap. 14) consider the question of whether there might exist an extension of quantum theory, that is, an alternative theory that enables predictions that have less uncertainty than those of quantum theory, but which reproduces the quantum predictions when one averages over certain variables [112]. Using an assumption that seeks to formalize the notion that observers are free to choose the settings of their measurements, they prove a result that rules out such extensions.

4 Concluding Remarks

The goal of understanding what physical principles might underlie the formalism of quantum theory is an ambitious one. Nonetheless, this monograph testifies to the fact that real and sustained progress on the question has been achieved in recent years. We hope that readers will come away with a sense of the excitement and promise of contemporary research in the field of quantum foundations and that some may be inspired to contribute to the endeavour themselves in the years to come.

Acknowledgments The authors acknowledge the support of the Perimeter Institute for Theoretical Physics during the course of this project. Research at Perimeter Institute is supported by the Government of Canada through Industry Canada and by the Province of Ontario through the Ministry of Research and Innovation. GC also acknowledges the Foundational Questions Institute (FQXi-RFP3-1325) and the 1000 Youth Fellowship Program of China.

References

1. R. Clausius, *The Mechanical Theory of Heat: With Its Applications to the Steam-Engine and to the Physical Properties of Bodies* (J. van Voorst, London, 1867)
2. Y.V.C. Rao, *Chemical Engineering Thermodynamics* (Universities Press, Hyderabad, 1997)
3. L. Hardy, R.W. Spekkens, Why physics needs quantum foundations. Phys. Can. **66**(2), 73–76 (2010)
4. C.H. Bennett, G. Brassard, Quantum cryptography: public key distribution and coin tossing, in *Proceedings of IEEE International Conference on Computers, Systems and Signal Processing*, vol. 175, (New York, 1984), p. 8
5. A.K. Ekert, Quantum cryptography based on Bell's theorem. Phys. Rev. Lett. **67**(6), 661 (1991)
6. P. Shor, Polynomial-time algorithms for prime factorization and discrete logarithms on a quantum computer. SIAM J. Comput. **26**, 1484–1509 (1997)
7. L.K. Grover, A fast quantum mechanical algorithm for database search, in *Proceedings of the Twenty-eighth Annual ACM Symposium on Theory of Computing, STOC'96* (ACM, New York, 1996), pp. 212–219
8. C.H. Bennett, G. Brassard, C. Crépeau, R. Jozsa, A. Peres, W.K. Wootters, Teleporting an unknown quantum state via dual classical and Einstein-Podolsky-Rosen channels. Phys. Rev. Lett. **70**(13), 1895 (1993)
9. C. Bennett, S. Wiesner, Communication via one- and two-particle operators on Einstein-Podolsky-Rosen states. Phys. Rev. Lett. **69**, 2881 (1992)
10. C.A. Fuchs, Quantum foundations in the light of quantum information, *Nato Science Series, Subseries III: Computer and Systems Sciences*, vol. 182 (2001), pp. 38–82
11. C.A. Fuchs, Quantum mechanics as quantum information, mostly. J. Mod. Opt. **50**, 987–1023 (2003)
12. G. Brassard, Is information the key? Nat. Phys. **1**, 2–4 (2005)
13. H. Everett III, Relative state formulation of quantum mechanics. Rev. Mod. Phys. **29**, 454 (1957)
14. D. Bohm, A suggested interpretation of the quantum theory in terms of "Hidden" variables, I. Phys. Rev. **85**, 166 (1952)
15. G.C. Ghirardi, A. Rimini, T. Weber, Unified dynamics for microscopic and macroscopic systems. Phys. Rev. D **34**, 470 (1986)
16. G.C. Ghirardi, P. Pearle, A. Rimini, Markov processes in Hilbert space and continuous spontaneous localization of systems of identical particles. Phys. Rev. A **42**, 78 (1990)
17. J.A. Wheeler, Information, physics, quantum: the search for links, in *Complexity, Entropy, and the Physics of Information*, ed. by W. Zurek (Addison-Wesley, Redwood City, 1990)
18. G. Birkhoff, J. von Neumann, The logic of quantum mechanics. Ann. Math. **37**, 823 (1936)
19. G.W. Mackey, Quantum mechanics and Hilbert space. Am. Math. Mon. **64**, 45 (1957)
20. G. Ludwig, Versuch einer axiomatischen Grundlegung der Quantenmechanik und allgemeinerer physikaliseher Theorien. Zeitschrift für Physik **181**, 233–260 (1964)
21. C. Piron, Axiomatique quantique. Helv. Phys. Acta **37**, 439 (1964)
22. E.G. Beltrametti, G. Cassinelli, *The Logic of Quantum Mechanics* (Addison-Wesley, Reading, 1981)
23. B. Coecke, D. Moore, A. Wilce, Operational quantum logic: an overview, in *Current Research in Operational Quantum Logic: Algebras, Categories, Languages*, ed. by B. Coecke, D.J. Moore, A. Wilce (Kluwer Academic Publishers, Dordrecht, 2000), pp. 1–36

24. S. Weinberg, Precision tests of quantum mechanics. Phys. Rev. Lett. **62**, 485–488 (1989)
25. N. Gisin, Weinberg's non-linear quantum mechanics and supraluminal communications. Phys. Lett. A **143**, 12 (1990)
26. A. Peres, Nonlinear variants of Schrödinger's equation violate the second law of thermodynamics. Phys. Rev. Lett. **63**, 1114 (1989)
27. D. Mayers, A. Yao, Quantum cryptography with imperfect apparatus, in *Proceedings of the 39th Annual Symposium on Foundations of Computer Science* (1998), pp. 503–509
28. J. Barrett, L. Hardy, A. Kent, No signaling and quantum key distribution. Phys. Rev. Lett. **95**, 010503 (2005)
29. A. Acín, N. Gisin, Ll. Masanes, From Bell's theorem to secure quantum key distribution. Phys. Rev. Lett. **97**, 120405 (2006)
30. A. Acín, N. Brunner, N. Gisin, S. Massar, S. Pironio, V. Scarani, Device-independent security of quantum cryptography against collective attacks. Phys. Rev. Lett. **98**, 230501 (2007)
31. L. Masanes, S. Pironio, A. Acín, Secure device-independent quantum key distribution with causally independent measurement devices. Nat. Commun. **2**, 238 (2011)
32. U. Vazirani, T. Vidick, Fully device-independent quantum key distribution. Phys. Rev. Lett. **113**(14), 140501 (2014)
33. N. Brunner, D. Cavalcanti, S. Pironio, V. Scarani, S. Wehner, Bell nonlocality. Rev. Mod. Phys. **86**, 419 (2014)
34. J. Barrett, N. Linden, S. Massar, S. Pironio, S. Popescu, D. Roberts, Non-local correlations as an information theoretic resource. Phys. Rev. A **71**, 022101 (2005)
35. J. Barrett, S. Pironio, Popescu-Rohrlich correlations as a unit of nonlocality. Phys. Rev. Lett. **95**, 140401 (2005)
36. N. Brunner, P. Skrzypczyk, Non-locality distillation and post-quantum theories with trivial communication complexity. Phys. Rev. Lett. **102**, 160403 (2009)
37. W. van Dam, Implausible consequences of superstrong nonlocality. Nat. Comput. **12**(1), 9–12 (2013)
38. G. Brassard, H. Buhrman, N. Linden, A.A. Méthot, A. Tapp, F. Unger, Limit on nonlocality in any world in which communication complexity is not trivial. Phys. Rev. Lett. **96**, 250401 (2006)
39. N. Linden, S. Popescu, A.J. Short, A. Winter, Quantum nonlocality and beyond: limits from nonlocal computation. Phys. Rev. Lett. **99**, 180502 (2007)
40. J. Anders, D.E. Browne, Computational power of correlations. Phys. Rev. Lett. **102**, 50502 (2009)
41. R. Duncan, S. Perdrix, Rewriting measurement-based quantum computations with generalised flow, *Automata, Languages and Programming*. Lecture Notes in Computer Science, vol. 6199/2010 (2010), pp. 285–296
42. C. Horsman, Quantum picturalism for topological cluster-state computing. New J. Phys. **13**, 095011 (2011)
43. R. Duncan, A graphical approach to measurement-based quantum computing, in *Quantum Physics and Linguistics: A Compositional, Diagrammatic Discourse*, ed. by C. Heunen, M. Sadrzadeh, E. Grefenstette (2013). Also available as arXiv:1203.6242
44. R. Raussendorf, Contextuality in measurement-based quantum computation. Phys. Rev. A **88**, 22322 (2013)
45. R. Raz, A parallel repetition theorem. SIAM J. Comput. **27**(3), 763–803 (1998)
46. T. Holenstein, Parallel repetition: Simplifications and the no-signaling case, in *Proceedings of the Thirty-ninth Annual ACM Symposium on Theory of Computing, STOC'07* (ACM, New York, 2007), pp. 411–419
47. T. Ito, H. Kobayashi, K. Matsumoto, Oracularization and two-prover one-round interactive proofs against nonlocal strategies, in *24th Annual IEEE Conference on Computational Complexity, CCC'09* (2009), pp. 217–228
48. T. Ito, Polynomial-space approximation of no-signaling provers, in *Automata, Languages and Programming*, Lecture Notes in Computer Science, vol. 6198, ed. by S. Abramsky, C. Gavoille, C. Kirchner, F. Meyer auf der Heide, P. Spirakis (Springer, Berlin, 2010), pp. 140–151

49. Y.T. Kalai, R. Raz, R.D. Rothblum, How to delegate computations: the power of no-signaling proofs, in *Proceedings of the 46th Annual ACM Symposium on Theory of Computing, STOC'14* (ACM, New York, 2014), pp. 485–494
50. R. Arnon-Friedman, R. Renner, T. Vidick, Non-signalling parallel repetition using de Finetti reductions. arXiv:1411.1582
51. R. Colbeck, Quantum and relativistic protocols for secure multi-party computation. Ph.D. thesis, University of Cambridge, 2007. Also available as arXiv:0911.3814
52. S. Pironio et al., Random numbers certified by Bell's theorem. Nature **464**, 1021–1024 (2010)
53. R. Colbeck, A. Kent, A. Private randomness expansion with untrusted devices. J. Phys. A **44**, 095305 (2011)
54. R. Colbeck, R. Renner, Free randomness can be amplified. Nat. Phys. **8**, 450–454 (2012)
55. T. Fritz, Beyond Bell's theorem: correlation scenarios. New J. Phys. **14**(10) (2012)
56. J. Henson, R. Lal, M.F. Pusey, Theory-independent limits on correlations from generalized Bayesian networks. New J. Phys. **16**(11), 113043 (2014)
57. R. Chaves, C. Majenz, D. Gross, Information-theoretic implications of quantum causal structures. Nat. Commun. **6**, 5766 (2015)
58. T. Islam, S. Wehner, Computability limits non-local correlations. Phys. Rev. A **86**, 042109 (2012)
59. C.M. Lee, J. Barrett, Computation in generalised probabilistic theories, arXiv:1412.8671
60. L. Masanes, R. Renner, M. Christandl, A. Winter, J. Barrett, Full security of quantum key distribution from no-signaling constraints. IEEE Trans. Inf. Theory **60**(8), 4973–4986 (2014)
61. H. Buhrman, M. Christandl, F. Unger, S. Wehner, A. Winter, Implications of superstrong nonlocality for cryptography. Proc. R. Soc. A **462**(2071), 1919–1932 (2006)
62. H. Barnum, O.C.O. Dahlsten, M. Leifer, B. Toner, Nonclassicality without entanglement enables bit commitment, in *Proceedings of IEEE Information Theory Workshop, ITW'08* (2008), pp. 386–390
63. G. Chiribella, G.M. D'Ariano, P. Perinotti, Probabilistic theories with purification. Phys. Rev. A **81**, 062348 (2010)
64. B. Coecke, R. Duncan, Interacting quantum observables: categorical algebra and diagrammatics. New J. Phys. **13**, 043016 (2011)
65. B. Coecke, R. Duncan, A. Kissinger, Q. Wang, Strong complementarity and non-locality in categorical quantum mechanics, in *Proceedings of the 2012 27th Annual IEEE/ACM Symposium on Logic in Computer Science, LICS'12* (2012), pp. 245–254
66. H. Barnum, J. Barrett, M. Leifer, A. Wilce, Generalized no-broadcasting theorem. Phys. Rev. Lett. **99**, 240501 (2007)
67. S. Abramsky, No cloning in categorical quantum mechanics, in *Semantic Techniques in Quantum Computation*, ed. by S. Gay, I. Mackie (Cambridge University Press, Cambridge, 2010), pp. 1–28
68. S. Abramsky, B. Coecke, A categorical semantics of quantum protocols, in *Proceedings of the 19th Annual IEEE Symposium on Logic in Computer Science* (2004), pp. 415–425
69. H. Barnum, J. Barrett, M. Leifer, A. Wilce, Teleportation in general probabilistic theories, in *Proceedings of Symposia in Applied Mathematics*, vol. 71 (2012), pp. 25–48
70. G. Kimura, T. Miyadera, H. Imai, Optimal state discrimination in general probabilistic theories. Phys. Rev. A **79**, 062306 (2009)
71. K. Nuida, G. Kimura, T. Miyadera, Optimal observables for minimum-error state discrimination in general probabilistic theories. J. Math. Phys. **51**, 093505 (2010)
72. J. Bae, Distinguishability, ensemble steering, and the no-signaling principle. EPTCS **171**, 26–32 (2014). Also available as arXiv:1412.7917
73. H. Barnum, J. Barrett, L. Orloff Clark, M. Leifer, R. Spekkens, N. Stepanik, A. Wilce, R. Wilke, Entropy and information causality in general probabilistic theories. New J. Phys. **12**(3), 033024 (2010)
74. A.J. Short, S. Wehner, Entropy in general physical theories. New J. Phys. **12**, 033023 (2010)
75. G. Kimura, K. Nuida, H. Imai, Distinguishability measures and entropies for general probabilistic theories. Rep. Math. Phys. **66**, 175 (2010)

76. E. Hänggi, S. Wehner, A violation of the uncertainty principle implies a violation of the second law of thermodynamics. Nat. Commun. **4**, 1670 (2013)
77. N. Brunner, M. Kaplan, A. Leverrier, P. Skrzypczyk, Dimension of physical systems, information processing, and thermodynamics. New J. Phys. **16**, 123050 (2014)
78. G. Chiribella, C.M. Scandolo, Entanglement and thermodynamics in general probabilistic theories, arXiv:1504.07045
79. B. Coecke, T. Fritz, R.W. Spekkens, A mathematical theory of resources, arXiv:1409.5531
80. M.P. Müller, L. Masanes, Three-dimensionality of space and the quantum bit: an information-theoretic approach. New J. Phys. **15**, 053040 (2013)
81. M.P. Müller, J. Oppenheim, O.C.O. Dahlsten, The black hole information problem beyond quantum theory. J. High Energy Phys. **2012**(9), 116 (2012)
82. E.B. Davies, J.T. Lewis, An operational approach to quantum probability. Commun. Math. Phys. **17**, 239–260 (1970)
83. L. Hardy, Quantum theory from five reasonable axioms, arXiv:quant-ph/0101012
84. J. Barrett, Information processing in generalized probabilistic theories. Phys. Rev. A **75**, 032304 (2007)
85. H. Barnum, A. Wilce, Information processing in convex operational theories. Electron. Notes Theor. Comput. Sci. **270**, 3 (2011)
86. S. Abramsky, B. Coecke, Categorical quantum mechanics, in *Handbook of Quantum Logic and Quantum Structures: Quantum Logic*, ed. by K. Engesser, D.M. Gabbay, D. Lehmann (Elsevier, 2008), pp. 261–324
87. B. Coecke, Quantum picturalism. Contemp. Phys. **51**, 59 (2010)
88. B. Coecke, A universe of processes and some of its guises, in *Deep Beauty: Understanding the Quantum World Through Mathematical Innovation*, ed. by H. Halvorson (Cambridge University Press, Cambridge, 2010), pp. 129–186
89. G. Chiribella, G.M. D'Ariano, P. Perinotti, Informational derivation of quantum theory. Phys. Rev. A **84**, 012311 (2011)
90. G. Chiribella, Dilation of states and processes in operational-probabilistic theories, in *Proceedings 11th Workshop on Quantum Physics and Logic*, Electronic Proceedings in Theoretical Computer Science, vol. 172, ed. by B. Coecke, I. Hasuo, P. Panangaden (2014), pp. 1–14
91. L. Hardy, A formalism-local framework for general probabilistic theories including quantum theory. Math. Struct. Comput. Sci. **23**, 399–440 (2013)
92. L. Hardy, Reformulating and reconstructing quantum theory, arXiv:1104.2066v3
93. R.W. Spekkens, In defense of the epistemic view of quantum states: a toy theory. Phys. Rev. A **75**, 032110 (2007)
94. S. Popescu, D. Rohrlich, Causality and non-locality as axioms for quantum mechanics. Found. Phys. **24**, 379 (1994)
95. V. Scarani, The device-independent outlook on quantum physics. Acta Phys. Slovaca **62**, 347–409 (2012)
96. B.S. Tsirelson, Quantum generalizations of Bell's inequality. Lett. Math. Phys. **4**, 93 (1980)
97. J.F. Clauser, M.A. Horne, A. Shimony, R.A. Holt, Proposed experiment to test local hidden-variable theories. Phys. Rev. Lett. **23**(15), 880–884 (1969)
98. P. Rastall, Locality, Bell's theorem, and quantum mechanics. Found. Phys. **15**, 963 (1985)
99. G. Chiribella, X. Yuan, Bridging the gap between general probabilistic theories and the device-independent framework for nonlocality and contextuality (2015). arXiv:1504.02395
100. E.C. Stueckelberg, Quantum theory in real Hilbert space. Helv. Phys. Acta **33**(727), 458 (1960)
101. H. Araki On a characterization of the state space of quantum mechanics. Commun. Math. Phys. **75**, 1 (1980)
102. W.K. Wootters, Local accessibility of quantum states, in *Complexity, Entropy and the Physics of Information*, ed. by W.H. Zurek (Addison-Wesley, Redwood City, 1990), pp. 39–46
103. B. Schumacher, M.D. Westmoreland, Modal quantum theory. Found. Phys. **42**(7), 918–925 (2012)
104. L. Masanes, M. Müller, New J. Phys. **13**, 3001 (2011)

105. C.M. Caves, C.A. Fuchs, R. Schack, Quantum probabilities as Bayesian probabilities. Phys. Rev. A **65**, 022305 (2002)
106. C.A. Fuchs, R. Schack, A quantum-Bayesian route to quantum-state space. Found. Phys. **41**(3), 345–356 (2011)
107. C.A. Fuchs, R. Schack, Quantum-Bayesian coherence. Rev. Mod. Phys. **85**(4), 1693 (2013)
108. C.A. Fuchs, N.D. Mermin, R. Schack, An introduction to Qbism with an application to the locality of quantum mechanics. Am. J. Phys. **82**(8), 749–754 (2014)
109. M. Pawłowski, T. Paterek, D. Kaszlikowski, V. Scarani, A. Winter, M. Żukowski, Information causality as a physical principle. Nature **461**(7267), 1101–1104 (2009)
110. M. Navascués, H. Wunderlich, A glance beyond the quantum model. Proc. R. Soc. Lond. A: Math. Phys. Eng. Sci. **466**(2115), 881–890 (2010)
111. M. Almeida, J.-D. Bancal, N. Brunner, A. Acn, N. Gisin, S. Pironio, Guess your neighbor's input: a multipartite nonlocal game with no quantum advantage. Phys. Rev. Lett. **104**(23), 230404 (2010)
112. R. Colbeck, R. Renner, No extension of quantum theory can have improved predictive power. Nat. Commun. **2**, 411 (2011)

Part I
Foil Theories

Optimal Information Transfer and Real-Vector-Space Quantum Theory

William K. Wootters

1 Introduction

In 1936 Birkhoff and von Neumann initiated an axiomatic approach to the foundations of quantum mechanics, taking as their starting point postulates inspired by classical logic but adapted to the peculiar features of quantum theory [1]. Though they showed that many characteristics of quantum theory could be captured in this way, they could also see that their logical approach would not lead uniquely to standard quantum theory. In particular they noted that along with standard complex-vector-space quantum theory, the postulates could just as well be satisfied by a theory based on a real or quaternionic Hilbert space.[1]

Over the years other authors have taken other approaches to axiomatization and have found reasonable assumptions that favor the complex theory over the real and quaternionic models. One successful strategy along these lines has been to insist on the existence of an uncertainty principle of a specific form [5–7]. Another approach put forward by several authors relies on the fact that in standard quantum theory, it is possible to carry out a complete tomographic reconstruction of the state of a multipartite system entirely by means of local measurements on the individual components (taking into account correlations), with no need for global measurements on pairs of subsystems [8–14]. The real-vector-space theory does not have this property; so by adopting local tomography as an axiom, one rules out the real case. Surely, though, much of the appeal of these arguments comes from the fact that they succeed in leading us to what we believe to be the correct answer. If we had found ourselves living

[1] In fact, from this logical starting point, it has turned out to be difficult even to narrow the set of possibilities to just these three theories—real, complex, and quaternionic. Major theorems along these lines can be found in Refs. [2, 3]; see also Ref. [4]. Still, nothing in these papers favors the complex theory over the real or quaternionic theory.

W.K. Wootters (✉)
Department of Physics, Williams College, Williamstown, MA 01267, USA
e-mail: William.K.Wootters@williams.edu

G. Chiribella and R.W. Spekkens (eds.), *Quantum Theory: Informational Foundations and Foils*, Fundamental Theories of Physics 181,
DOI 10.1007/978-94-017-7303-4_2

in a world that seemed to be well described by real-vector-space quantum theory, we would not have regarded it as a logical problem that tomography requires global measurements. It would simply be another peculiar feature of quantum theory, like entanglement.[2] (I admit, though, that the local tomographic property of the complex theory does feel as if it could be a clue to something deeper.)

In this paper I would like to point out a particular property of real-vector-space quantum theory that I find especially intriguing: the transfer of information from a preparation to a measurement is *optimal* (in a sense to be explained below). Standard quantum theory does not have this property. So if we were trying to find a simple set of axioms that would generate real-vector-space quantum theory, we might well find ourselves adopting optimal information transfer as one of our axioms. This property of the real theory has been known for years—it appears in my 1980 doctoral dissertation [16]—but I would like to give a somewhat simpler and more intuitive presentation of it here.

One motivation for studying real-vector-space quantum theory is simply to shed light on the standard theory by comparison. But I would also like to keep open the possibility that the real-vector-space theory might turn out to be of value in its own right for describing our world. Several authors have given us reasons for not discounting this possibility. In a series of papers published around 1960, Stueckelberg and his collaborators developed an alternative formulation of quantum field theory based on a real Hilbert space [6, 17, 18]. In order to allow the existence of an uncertainty principle, Stueckelberg imposes a specific restriction on all the observables of the real-vector-space theory: every observable is required to commute with a certain operator that we can write as $I \otimes J$, where J is the 2×2 matrix $\begin{pmatrix} 0 & -1 \\ 1 & 0 \end{pmatrix}$ and I is the identity operator. (In the context of Stueckelberg's papers I is the identity on an infinite-dimensional real Hilbert space.) In effect, this restriction forces the matrix representing any observable to be composed of 2×2 blocks of the form $\begin{pmatrix} a & -b \\ b & a \end{pmatrix}$. Such 2×2 blocks add and multiply like complex numbers; so the theory becomes equivalent to the usual complex theory. One of the points Stueckelberg and his collaborators make in these papers is that in this formulation the time-reversal operator becomes linear, rather than antilinear as in the complex formulation. Around the same time, Dyson made the same point and argued that by bringing the time-reversal operator into our formalism, we are in effect basing our quantum theory on the field of real numbers [19].

Recently Gibbons has argued that the complex structure in quantum theory is intimately related to the existence of an arrow of time, and that in spacetimes in which such an arrow cannot be defined, one must fall back on a real-vector-space version of quantum theory [20]. In other work, Myrheim has pointed out that if one wants a version of the canonical commutation relation $[x, p] = i\hbar$ in a discrete system with finitely many values of position and momentum, one cannot use standard

[2] At least in real-vector-space quantum theory, one never needs to make global measurements involving more than *two* subsystems [15].

complex quantum theory: the trace of any commutator is zero in a finite-dimensional space, but the trace of $i\hbar$ is not zero. On the other hand, if we replace $i\hbar$ with $J\hbar$ (the same J as above), both sides of the equation have zero trace and there is no problem [21]. In the present paper I do not particularly build on any of these observations except insofar as they suggest that a real-vector-space version of quantum theory might be used to describe our actual world, and that the theory is worth studying for this reason as well as for whatever insights it might provide about standard quantum mechanics.

I begin in Sect. 2 by saying what I mean by "real-vector-space quantum theory." Then in Sects. 3 and 4 I present the property of optimal information transfer, first for a two-dimensional state space and then in d dimensions. As I have said, standard complex quantum theory does not have this property, and it is interesting to ask whether a revised statement of the problem might yield a positive answer even in the complex case. This is the subject of Sect. 5. Section 6 then summarizes our findings.

2 Real-Vector-Space Quantum Theory

One can summarize the basic structure of standard quantum theory in the following four statements:

1. A pure state is represented by a unit vector in a Hilbert space over the complex numbers.
2. An ideal repeatable measurement is represented by a set of orthogonal projection operators whose supports span the vector space. When a state $|s\rangle$ is subjected to the measurement $\{P_1, \cdots, P_m\}$, the probability of the ith outcome is $\langle s|P_i|s\rangle$. When the ith outcome occurs, the system is left in a state proportional to $P_i|s\rangle$.
3. A reversible transformation is represented by a unitary operator U. That is, for any initial state $|s\rangle$, the operation takes $|s\rangle$ to $U|s\rangle$.
4. A composite system has as its state space the tensor product of the state spaces of its components.

Of course other states, measurements and transformations are possible. Mixed states are averages of projection operators on pure states, and there also exist non-orthogonal measurements and irreversible transformations. But all such generalizations can be obtained from the cases listed above by applying them to a larger system and possibly discarding part of the system. I have chosen the above formulation partly to keep the discussion simple, but also because I do tend to think of orthogonal measurements and pure states as being more fundamental than their generalizations.

The real-vector-space theory has essentially the same structure, except that all vectors and matrices are limited to real components. The only changes in the above list are that "complex" is to be replaced by "real" in item 1, and "unitary" is to be replaced by "orthogonal" in item 3.

One might wonder what the analogue of the Schrödinger equation is in the real-vector-space theory. The Schrödinger equation generates a unitary transformation through a Hermitian operator, the Hamiltonian:

$$i\hbar\frac{d}{dt}|s\rangle = H|s\rangle. \tag{1}$$

If H is time independent, the unitary operator it generates over a time t is $U(t) = e^{-iHt/\hbar}$, since $|s(t)\rangle = U(t)|s(0)\rangle$ solves the differential equation. The analogous equation in the real-vector-space case should have an antisymmetric real matrix in place of $-iH$, since such a matrix generates orthogonal transformations. We can write the differential equation as

$$\frac{d}{dt}|s\rangle = S|s\rangle, \tag{2}$$

where S is an antisymmetric real operator. I like to call S the "Stueckelbergian" in honor of Ernst Stueckelberg (who of course did not use this term). If S is time independent, then the general solution of Eq. (2) is $|s(t)\rangle = e^{St}|s(0)\rangle$.

Another reasonable question is whether, for example, in a two-dimensional real space the operator

$$R = \begin{pmatrix} 1 & 0 \\ 0 & -1 \end{pmatrix} \tag{3}$$

should be allowed to count as a possible transformation [22]. It is an orthogonal matrix, so according to the above rules it does count. But there is no 2×2 Stueckelbergian that can generate this operator. This is because the operator R represents a reflection, not a rotation, and there is no continuous set of orthogonal transformations on a two-dimensional real space that takes us from the identity operator to a reflection operator.

Nevertheless, in a real-vector-space world it would still be possible to realize the operation R continuously by bringing in an ancillary two-dimensional system (that is, an ancillary "rebit"). To effect the transformation

$$\begin{pmatrix} s_1 \\ s_2 \end{pmatrix} \rightarrow \begin{pmatrix} s_1 \\ -s_2 \end{pmatrix}, \tag{4}$$

we can perform a controlled rotation on our ancillary rebit, conditioned on the state of the original rebit. By rotating the ancillary rebit by half a complete cycle, we can pick up the desired factor of -1. So it seems reasonable to allow orthogonal matrices with negative determinant to count as possible transformations.

3 Optimal Transfer of Information: The Two-Dimensional Case

Consider the following simple scenario. A stream of photons emerges from a linearly polarizing filter with its preferred axis oriented at an angle θ from the horizontal. Somewhere further along the photons' path there is a polarizing beam splitter and a pair of single-photon detectors, which together force each photon to yield either the horizontal outcome or the vertical outcome. The probability of "horizontal" is $p_0(\theta) = \cos^2\theta$. (The subscript "0" distinguishes this function from other hypothetical functions to be considered shortly.) This function allows someone observing the measurement results to gain information about the angle θ.

This scenario illustrates a typical feature of a quantum measurement: a measurement on a single instance of a system (in this case a single photon) cannot convey complete information about the system's preparation. But a large statistical sample of measurements on identically prepared copies can eventually home in on the values of the preparation parameters (in this case the single parameter θ). One does not encounter this limitation in classical physics, at least not for pure states: if a particle is placed at position x with momentum p, a measurement can directly reveal those values. This difference between classical and quantum physics reflects the fact that quantum theory is inherently probabilistic.

In our specific example, one can ask how well the information about θ is conveyed through the observed results. Specifically, one can quantify the mutual information between the measurement results and the value of θ. As we will see shortly, given the limitation imposed by the probabilistic nature of the polarization measurement, the transfer of information is optimal in the limit of a large number of trials. That is, in this example anyway, quantum mechanics orchestrates the optimal conveyance of information from the preparation to the measurement outcome.

Before justifying this statement, I want to note that it works only because we are in effect working in the real-vector-space theory. By limiting the possibilities to linear polarizations, we are ruling out all the polarization states one would normally represent with vectors having a nonzero imaginary part (circular and elliptical polarizations). We will return to this point toward the end of this section.

Now let us make the statement precise. We do so by comparing our actual world, in which the probability of "horizontal" is $p_0(\theta) = \cos^2\theta$, to a fictitious world in which the probability is given by some arbitrary function $p(\theta)$. In such a world, let N photons, each prepared with linear polarization angle θ, be subjected to a horizontal-vs-vertical polarization measurement. Let n be the number of these photons that yield the outcome "horizontal." The mutual information between the measurement results and the value of θ is based on the Shannon entropy H and is defined to be

$$I(n:\theta) = H(n) - H(n|\theta) = -\sum_{n=0}^{N} P(n)\ln P(n) + \frac{1}{2\pi}\int_0^{2\pi}\left(\sum_{n=0}^{N} P(n|\theta)\ln P(n|\theta)\right)d\theta. \tag{5}$$

Here we have assumed a uniform *a priori* distribution of θ over the interval $[0, 2\pi]$. (This is a crucial assumption that we discuss further below.) $P(n|\theta)$ is the probability of getting the horizontal outcome exactly n times if the photons are prepared in the state θ, and $P(n)$ is the probability of getting the horizontal outcome exactly n times in the absence of any information about θ (that is, when θ is uniformly distributed). Both $P(n|\theta)$ and $P(n)$ depend on the function $p(\theta)$. Note that in Eq. (5) we have written the mutual information as the average amount of information gained about the integer n upon learning the value of θ. It does have this interpretation, but it can alternatively be interpreted as the average amount of information one gains about the value of θ upon learning the value of n. (Mutual information is symmetric in its two arguments.) This latter interpretation is more descriptive of the scenario we are imagining, in which an observer at the polarizing beam splitter is trying to learn about the value of θ.

It turns out that for large N, $I(n:\theta)$ grows as $(1/2)\ln N$. We therefore consider the following limit, which has a finite upper bound:

$$\tilde{I} = \lim_{N\to\infty}\left[I(n:\theta) - \frac{1}{2}\ln N\right]. \tag{6}$$

We want to show that of all conceivable probability functions $p(\theta)$, the quantum mechanical function $p_0(\theta) = \cos^2\theta$ gives $\tilde{I}$ its largest possible value.

At this point we could proceed to compute $\tilde{I}$ starting from Eq. (5), but the calculation will be simpler, and I hope clearer, if we abstract the problem away from its quantum mechanical setting. The important point to notice is that $I(n:\theta)$ depends on the probability function $p(\theta)$ only through the *measure* it induces on the binary probability space. That is, before we have any knowledge of θ, we can use $p(\theta)$ and the assumed uniform distribution of θ to figure out how likely it is that the probability of "horizontal" lies in any given interval, and it is this weighting function that figures into $I(n:\theta)$.

The more abstract problem, then, can be stated as follows. Consider a two-outcome probabilistic experiment, and let (p_1, p_2) denote a point in the binary probability space with p_1 corresponding to outcome #1. The experiment is run N times, and outcome #1 is observed to occur n times. This observation gives the experimenter information about (p_1, p_2).[3] The mutual information I between the value of n and the value of (p_1, p_2) depends on the experimenter's *a priori* measure on probability space. Our problem is to find the *a priori* measure that maximizes the limit

$$\tilde{I} = \lim_{N\to\infty}\left[I - \frac{1}{2}\ln N\right]. \tag{7}$$

(The optimal measure will turn out to be unique.) We want to show that this optimal measure is the one induced by the quantum probability function $p_0(\theta) = \cos^2\theta$ when θ is uniformly distributed.

[3] The experimenter is trying to determine the value of an unknown probability. It may seem that this problem cannot be framed except in the context of an objective interpretation of the concept of probability, but this is not the case. The representation theorem of de Finetti shows how to express this kind of question within a subjective interpretation [23]. We note that the *quantum* de Finetti theorem does not hold in real-vector-space quantum theory [24], but this fact does not preclude a subjective interpretation of probability in our problem. In our problem the experimenter is trying to refine a distribution over ordinary probability space, to which the classical de Finetti theorem applies.

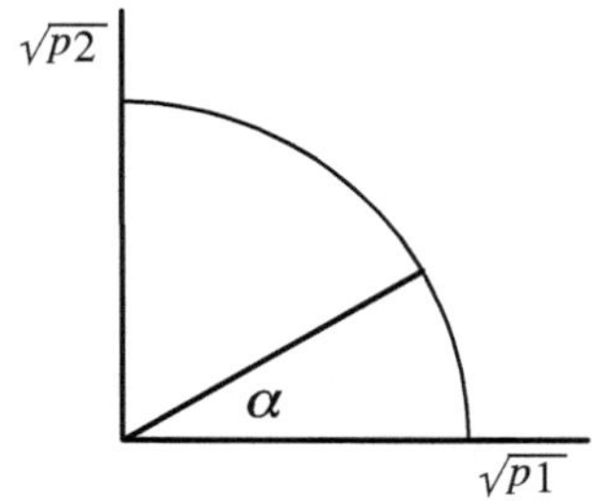

Fig. 1 The relation between (p_1, p_2) and α

In order to tackle this problem we need to choose a parameterization of the binary probability space. We could use p_1 or p_2 as our parameter, but it turns out to be more convenient to use a different parameter α defined by $(p_1, p_2) = (\cos^2\alpha, \sin^2\alpha)$, where $0 \leq \alpha \leq \pi/2$. The relation between α and (p_1, p_2) is illustrated in Fig. 1. (One might object that we seem to be smuggling some quantum mechanics into the calculation here, but we are not. The results will be entirely independent of our choice of parameter. Our choice merely simplifies the calculation.) Let $K(\alpha)d\alpha$ be the *a priori* measure on the set of values of α, normalized so that $\int_0^{\pi/2} K(\alpha)d\alpha = 1$. The mutual information between α and n can be written as

$$I(\alpha : n) = h(\alpha) - h(\alpha|n) = -\int_0^{\pi/2} K(\alpha)\ln K(\alpha)d\alpha + \sum_{n=0}^{N} P(n)\int_0^{\pi/2} P(\alpha|n)\ln P(\alpha|n)d\alpha, \tag{8}$$

where $h(\alpha)$ and $h(\alpha|n)$ are differential entropies.[4] Here $P(\alpha|n)$ is the probability distribution the experimenter assigns to α after seeing the value n, and $P(n)$ is the *a priori* probability of the value n as computed from the distribution $K(\alpha)$. (Note that if $K(\alpha)$ is derived from the probability function $p(\theta)$ under the assumption that θ is uniformly distributed, then $I(\alpha : n)$ is exactly equal to the quantity $I(n : \theta)$ given in Eq. (5)). The point of the next paragraph is to show that under modest assumptions about the function $K(\alpha)$, in the limit of very large N the second term on the right-hand side of Eq. (8) becomes *independent* of $K(\alpha)$. So we will only have to think about maximizing the first term.

To evaluate this second term, we need to write down expressions for $P(n)$ and $P(\alpha|n)$. We have

$$P(n) = \int_0^{\pi/2} P(n|\alpha)K(\alpha)d\alpha \tag{9}$$

and

$$P(\alpha|n) = \frac{P(n|\alpha)K(\alpha)}{P(n)}, \tag{10}$$

[4]The differential entropy is not the limit of the entropy of a discretized version of the continuous variable. However, a *mutual information* involving a continuous variable, being the *difference* between two differential entropies, is indeed the limit of the discretized mutual information [25].

where $P(n|\alpha)$ is given by the binomial distribution:

$$P(n|\alpha) = \frac{N!}{n!(N-n)!} p_1^n p_2^{N-n} \tag{11}$$

with $p_1 = \cos^2 \alpha$ and $p_2 = \sin^2 \alpha$. For any value of p_1 strictly between 0 and 1, it is possible to choose N large enough that the binomial distribution is well approximated by a Gaussian:

$$P(n|\alpha) \approx \frac{1}{\sqrt{2\pi N p_1 p_2}} \exp\left[-\frac{N}{2p_1 p_2}(n/N - p_1)^2\right]. \tag{12}$$

As N gets very large this distribution, regarded as a function of n/N, becomes arbitrarily highly peaked around $n/N = p_1$. Let $\alpha^{(n)}$ be defined so that $n/N = \cos^2 \alpha^{(n)}$. That is, $\alpha^{(n)}$ is the value of α corresponding to the observed outcome n. Then in the above exponent, we can approximate the quantity $(n/N - p_1)$ as

$$(n/N - p_1) = \cos^2 \alpha^{(n)} - \cos^2 \alpha \approx \frac{d\left(\cos^2 \alpha\right)}{d\alpha}\Delta\alpha = (-2\cos\alpha \sin\alpha)\Delta\alpha = -2\sqrt{p_1 p_2}\,\Delta\alpha, \tag{13}$$

where $\Delta\alpha = \alpha^{(n)} - \alpha$. This gives us

$$P(n|\alpha) \approx \frac{1}{\sqrt{2\pi N p_1 p_2}} \exp\left[-2N(\Delta\alpha)^2\right]. \tag{14}$$

Inserting this expression into Eq. (9), we again use the fact that the Gaussian is very highly peaked so that we can (i) extend the integral from $-\infty$ to ∞ without changing its value appreciably and (ii) evaluate everything outside the exponential at $\alpha = \alpha^{(n)}$. Then we get

$$P(n) \approx \frac{K\big(\alpha^{(n)}\big)}{2N\cos\alpha^{(n)}\sin\alpha^{(n)}}. \tag{15}$$

We now use Eqs. (10), (14) and (15) to approximate $P(\alpha|n)$:

$$P(\alpha|n) \approx \sqrt{\frac{2N}{\pi}} \exp\left[-2N(\Delta\alpha)^2\right]. \tag{16}$$

Using this expression and again relying on the narrowness of the Gaussian, we get

$$\int_0^{\pi/2} P(\alpha|n) \ln P(\alpha|n) d\alpha \approx \frac{1}{2} \ln\left(\frac{2N}{\pi e}\right). \tag{17}$$

Since this expression does not depend at all on n, it factors out of the sum in Eq. (8), so that the only sum we have to do is $\sum_n P(n)$, which is unity by definition. Putting the pieces together, we arrive at

$$I(\alpha : n) \approx -\int_0^{\pi/2} K(\alpha) \ln K(\alpha) d\alpha + \frac{1}{2} \ln \left(\frac{2N}{\pi e} \right). \tag{18}$$

And then subtracting $(1/2) \ln N$ as in Eq. (7) gives us

$$\tilde{I} = -\int_0^{\pi/2} K(\alpha) \ln K(\alpha) d\alpha + \frac{1}{2} \ln \left(\frac{2}{\pi e} \right). \tag{19}$$

The equality holds as long as our approximations become arbitrarily good as N gets larger. This will indeed be the case if the function $K(\alpha)$ is reasonably well behaved. A sufficient set of conditions on $K(\alpha)$ is that it be positive and differentiable on the interval $[0, \pi/2]$. Then when many trials are run, the range of likely values of α narrows to such a degree that the final distribution $P(\alpha|n)$ does not depend appreciably on the *a priori* distribution $K(\alpha)$.

The problem has now been reduced to finding out what distribution or distributions $K(\alpha)$ maximize the quantity $-\int_0^{\pi/2} K(\alpha) \ln K(\alpha) d\alpha$. The answer to this question is well known: the unique maximizing distribution is the *uniform* distribution $K(\alpha) = 2/\pi$. This result follows from the fact that the function $\phi(x) = -x \ln x$ is a strictly concave function of x for all positive values of x. Jensen's inequality then tells us that

$$(2/\pi) \int_0^{\pi/2} \phi[K(\alpha)] d\alpha \le \phi \left[(2/\pi) \int_0^{\pi/2} K(\alpha) d\alpha \right] = \phi(2/\pi), \tag{20}$$

with equality holding only for the constant function $K(\alpha) = 2/\pi$.[5]

Now we compare our result to quantum mechanics. Is this uniform distribution over α the one induced by the quantum probability law $p_0(\theta) = \cos^2 \theta$, when θ is uniformly distributed? First consider the values of θ from 0 to $\pi/2$. In that range the law $p_0(\theta) = \cos^2 \theta$ mirrors the definition of α and we have $\alpha = \theta$ (I am taking p_1 to correspond to the horizontal outcome.). So a uniform distribution of θ over this range would induce the uniform distribution of α. In the other three quadrants of the circle, that is, in the rest of the range of θ, the parameter α is not equal to θ but we still have $|d\alpha/d\theta| = 1$ (except at a finite number of points where α "bounces" off one of the endpoints of its range). Thus when θ is uniformly distributed, so is α. This completes our demonstration that the quantum probability function $p_0(\theta) = \cos^2 \theta$ is optimal.

Is the function $p_0(\theta) = \cos^2 \theta$ unique in this respect? The answer is no. Any function $p(\theta)$ that yields the same *a priori* measure on the binary probability space will be equally good. For example, any function of the form $p(\theta) = \cos^2(m\theta/2)$ where m is a positive integer yields the same distribution $K(\alpha) = 2/\pi$. And there

[5] Alternatively, instead of using differential entropies as in Eq. (8), we could have expressed the mutual information $I(\alpha : n)$ in terms of the Kullback-Leibler distances of both $K(\alpha)$ and $P(\alpha|n)$ from the uniform distribution over α. The calculation in Sect. 3 then tells us that $\tilde{I}$ is maximized when the Kullback-Leibler distance between $K(\alpha)$ and the uniform distribution is minimized, that is, when $K(\alpha)$ is itself the uniform distribution.

are many other, less physically interesting examples. Still, a typical function $p(\theta)$ will not have this optimization property.

Looking back over the above argument, one can see that the crucial feature is the exponent in Eq. (14): the coefficient of $(\Delta\alpha)^2$ depends only on N and not on α itself. In other words, the spread in the value of $\alpha^{(n)}$ depends only on the number of trials (when this number is large), and not on the probabilities (p_1, p_2). This is what is special about parameterizing probability space with the parameter α: it makes the statistical spread uniform. Once we have this fact, it is guaranteed that the final differential entropy $h(\alpha|n)$ will not depend on $K(\alpha)$. Therefore to maximize the mutual information, we want to maximize the initial differential entropy $h(\alpha)$ and we are thereby led to the uniform distribution.

There is perhaps a more direct way of seeing what is special about the function $p_0(\theta) = \cos^2\theta$. First note that the spread in n itself is *not* uniform over probability space. If one performs the binary experiment N times, the standard deviation in n/N is given by

$$\Delta(n/N) = \sqrt{\frac{p_1 p_2}{N}}, \tag{21}$$

which is smaller near the ends of probability space than near the middle. One can see this dependence in the exponent in Eq. (12). In our polarization experiment, an observer recording the frequency of occurrence of "horizontal" will therefore be more certain of the *probability* of "horizontal" when that probability is close to zero or one. (Again I am assuming that the experimenter's *a priori* distribution over probability space is reasonably smooth and that the number of trials is large). On the other hand, upon translating the uncertainty in probability to an uncertainty in θ, the observer must use the function $p(\theta)$. For the special case of $p_0(\theta) = \cos^2\theta$, the slope of this function exactly compensates for the varying size of $\Delta(n/N)$, so that the size of the resulting "region of uncertainty" of θ is independent of the value of n/N. Specifically,

$$\left|\frac{d}{d\theta}\cos^2\theta\right| = 2\,|\cos\theta\sin\theta| = 2\sqrt{p_0(\theta)[1 - p_0(\theta)]}, \tag{22}$$

which perfectly matches the dependence seen in Eq. (21). This compensation is illustrated in Fig. 2. Equalizing the uncertainties in θ also equalizes the uncertainties in α, which, as we have just seen, maximizes the mutual information in the limit of a large number of trials.

We now consider the case in which *all* pure polarization states are possible. The full set of pure states is the Bloch sphere—it includes the circular and elliptical polarizations—and the natural *a priori* measure is the uniform measure on the sphere, since this is the only probability measure invariant under all unitary transformations. We imagine a device that prepares a beam of photons in one of these polarization states, and further along the photons' path we imagine a person making the horizontal-vs-vertical measurement on each photon. The polarization is now determined by two parameters; for definiteness let us take them to be the polar angle β and the

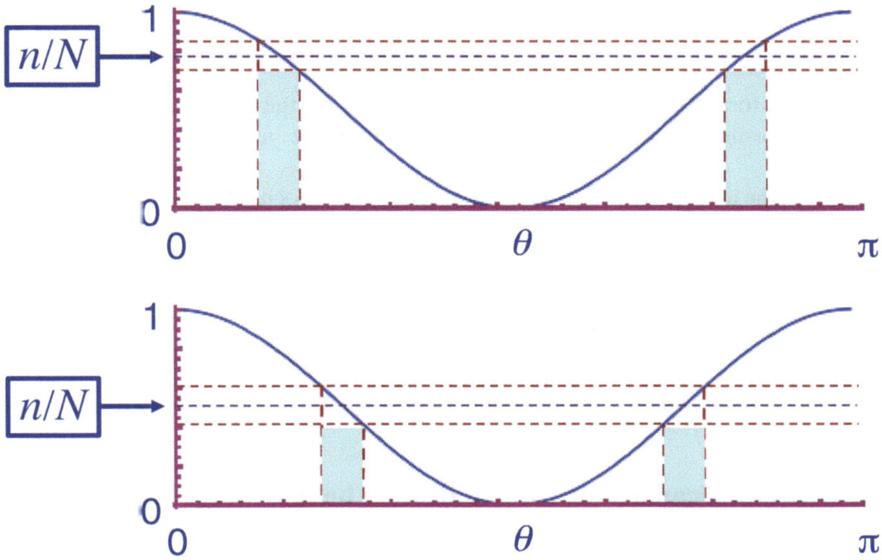

Fig. 2 The uncertainty in θ for two different values of n/N. Notice that the slope of the cosine-squared curve exactly compensates for the varying size of $\Delta(n/N)$, so that the "region of uncertainty" in θ has the same size for all values of n/N. (Here θ is plotted only up to π to make the diagram simpler)

azimuthal angle ϕ, and let the north and south poles of the sphere correspond to horizontal and vertical polarization. It is still possible to define the mutual information between the photons' preparation and the measurement outcomes; it could be written as $I(n : \beta, \varphi)$. Again this mutual information is the same as the quantity $I(\alpha : n)$ given in Eq. (8) and it is maximized only if the *a priori* distribution of α is the uniform distribution $K(\alpha) = 2/\pi$. But now the quantum mechanical law does *not* yield this distribution over the values of α. With the uniform distribution over the Bloch sphere, the parameter $\cos\beta$ is uniformly distributed over the interval $[-1, 1]$, and the quantum mechanical probability of "horizontal," $p(\theta) = (1/2)(1 + \cos\beta)$, is therefore uniformly distributed over the interval $[0, 1]$. To get the corresponding distribution of α, we use the relation $p_1 = \cos^2\alpha$ and the assumption that p_1 is uniformly distributed:

$$K(\alpha) = \left|\frac{dp_1}{d\alpha}\right| = 2\cos\alpha\sin\alpha. \tag{23}$$

Thus, rather than giving us the uniform distribution of α, the full Bloch sphere gives us a distribution that has a maximum in the middle of α's range.

We can see directly that this distribution does not allow as much information transfer as the optimal distribution. The relevant quantity is the integral in Eq. (19):

$$-\int_0^{\pi/2} K(\alpha) \ln K(\alpha) d\alpha. \tag{24}$$

For the uniform distribution over α, this quantity has the value $\ln(\pi/2) = 0.452$, whereas for the distribution $K(\alpha) = 2\cos\alpha\sin\alpha$, we get $1 - \ln 2 = \ln(e/2) = 0.307$.

Just as the uniform measure over the surface of the Bloch sphere is natural because it is invariant under all unitaries, in the real-vector-space theory where the set of pure states in two dimensions traces out a circle rather than a sphere, the uniform distribution over the circle is natural because it is invariant under all orthogonal transformations (rotations and reflections). That is, in the real-vector-space theory, we can use this invariance to justify our original assumption that the angle θ is uniformly distributed over the interval $[0, 2\pi]$.

4 Optimal Transfer of Information: The *d*-Dimensional Case

The above argument extends to a d-dimensional real vector space. Let a "redit" be a hypothetical quantum object whose pure states are vectors in a d-dimensional vector space over the real numbers. We now imagine an experiment in which a beam of N redits is prepared in a specific pure state $|s\rangle$. At some point further along the beam, an observer makes a fixed complete orthogonal measurement on each redit. The observer records the integers $n_1, \ldots, n_d$, where n_i is the number of times the ith measurement outcome occurs. We ask how much information the observer learns on average about the preparation $|s\rangle$, assuming (crucially) that the vector $|s\rangle$ is initially distributed uniformly over the unit sphere in the d-dimensional space. Again this average information gain is given by the mutual information, which we will write down shortly. The mutual information depends on the law that specifies the probability of the ith outcome given the preparation $|s\rangle$. In real-vector-space quantum theory, this law can be expressed as

$$p_i(|s\rangle) = s_i^2, \qquad i = 1, \ldots, d, \tag{25}$$

where $s_1, \ldots, s_d$ are the components of $|s\rangle$ in the basis defined by the measurement.

As before, what really matters in computing the mutual information is the *a priori* measure on probability space. The uniform measure over the unit sphere in d dimensions, together with Eq. (25), defines some specific *a priori* measure on probability space. We also want to consider other *a priori* measures, in order to show that the one induced by Eq. (25) is optimal. The probability space is now a $(d-1)$-dimensional set, since the probabilities must add to one. We could parameterize this set by the probabilities $p_1, \ldots, p_{d-1}$ of the first $d-1$ outcomes, but we instead choose to label the points of probability space by a unit vector $\vec{\gamma} = (\sqrt{p_1}, \ldots, \sqrt{p_d})$.

(Note that each $\sqrt{p_i}$ is non-negative; so $\vec{\gamma}$ is confined to the positive part of the unit sphere). We could go further and choose $d-1$ specific angular coordinates to locate this vector on the sphere (like the α of the preceding section), but we will not need to do so. Let $K(\vec{\gamma})d\vec{\gamma}$ be a generic *a priori* probability measure on the set of vectors $\vec{\gamma}$, where $d\vec{\gamma}$ is an infinitesimal $(d-1)$-dimensional surface element on the positive section of the unit sphere. Our goal is to find the distribution $K(\vec{\gamma})$ that maximizes the mutual information.

That mutual information can be written as follows:

$$I(\vec{\gamma}:\vec{n}) = h(\vec{\gamma}) - h(\vec{\gamma}|\vec{n}) = -\int K(\vec{\gamma}) \ln K(\vec{\gamma})d\vec{\gamma} + \sum P(\vec{n}) \int P(\vec{\gamma}|\vec{n}) \ln P(\vec{\gamma}|\vec{n})d\vec{\gamma}, \tag{26}$$

where $\vec{n} = (n_1, \ldots, n_d)$ specifies the number of times each outcome occurs. The sum is over all vectors $\vec{n}$ for which each n_i is a non-negative integer and $n_1 + \cdots + n_d = N$. The mutual information is now based on the multinomial distribution:

$$P(\vec{n}|\vec{\gamma}) = \frac{N!}{n_1! \cdots n_d!} p_1^{n_1} \cdots p_d^{n_d}. \tag{27}$$

(Here $p_j = \gamma_j^2$.) The functions appearing in Eq. (26) can be obtained from $P(\vec{n}|\vec{\gamma})$ as follows:

$$P(\vec{n}) = \int P(\vec{n}|\vec{\gamma})K(\vec{\gamma})d\vec{\gamma} \tag{28}$$

and

$$P(\vec{\gamma}|\vec{n}) = \frac{P(\vec{n}|\vec{\gamma})K(\vec{\gamma})}{P(\vec{n})}. \tag{29}$$

As N gets large, it will turn out that $I(\vec{\gamma}:\vec{n})$ grows as $[(d-1)/2]\ln N$. So we will compute the limiting value

$$\tilde{I} = \lim_{n\to\infty} \left[I(\vec{\gamma}:\vec{n}) - \left(\frac{d-1}{2}\right) \ln N \right]. \tag{30}$$

At this point the calculation is very similar to the one in the preceding section. As we did for the analagous equation in that case, we now show that the second term on the right-hand side of Eq. (26) becomes independent of $K(\vec{\gamma})$ as N approaches infinity.

For any fixed positive values of $p_1, \ldots, p_d$ and for large enough N, Eq. (27) can be approximated arbitrarily well by a Gaussian function:

$$P(\vec{n}|\vec{\gamma}) \approx \left[(2\pi N)^{d-1} p_1 p_2 \cdots p_d\right]^{-1/2} \exp\left[-\frac{N}{2} \sum_{i=1}^{d} \frac{(n_i/N - p_i)^2}{p_i} \right]. \tag{31}$$

It will be helpful to define the vectors

$$\vec{\gamma}^{(n)} = \left(\sqrt{\frac{n_1}{N}}, \ldots, \sqrt{\frac{n_d}{N}}\right) \qquad \text{and} \qquad \vec{\Delta\gamma} = \vec{\gamma}^{(n)} - \vec{\gamma}. \tag{32}$$

That is, $\vec{\gamma}^{(n)}$ is the vector of square roots of the observed frequencies of occurrence, whereas $\vec{\gamma}$ is the vector of square roots of the probabilities. The difference $\vec{\Delta\gamma}$ between these two vectors is likely to be small when N is large; so we will keep just the lowest-order term in this quantity. We can then rewrite the sum in the exponent of Eq. (31) as follows:

$$\sum_{i=1}^{d} \frac{(n_i/N - p_i)^2}{p_i} = \sum_{i=1}^{d} \frac{(\Delta p_i)^2}{p_i} \approx 4\left|\vec{\Delta\gamma}\right|^2, \tag{33}$$

where Δp_i is defined to be $(n_i/N) - p_i$ and the last step comes from

$$\Delta\gamma_i = \Delta\left(p_i^{1/2}\right) \approx \frac{1}{2} p_i^{-1/2} \Delta p_i. \tag{34}$$

We can therefore approximate Eq. (31) as

$$P(\vec{n}|\vec{\gamma}) \approx (2\pi N)^{-\left(\frac{d-1}{2}\right)} \frac{1}{\gamma_1\gamma_2\cdots\gamma_d} \exp\left(-2N\left|\vec{\Delta\gamma}\right|^2\right). \tag{35}$$

That is, $P(\vec{n}|\vec{\gamma})$, regarded as function of $\vec{\gamma}$, falls off as a Gaussian around the point $\vec{\gamma}^{(n)}$, with a spread that is isotropic and independent of the value of $\vec{\gamma}^{(n)}$. (This function is not, however, a *normalized* distribution of $\vec{\gamma}$; rather, it is normalized with respect to a sum over $\vec{n}$.)

In approximating the integral in Eq. (28), we rely on the narrowness of the Gaussian: the integral is over a section of a sphere, but we can treat it as being over an infinite flat space having $d - 1$ dimensions—the "plane" tangent to the sphere at the point $\vec{\gamma}^{(n)}$. We also evaluate everything outside the exponential at the point $\vec{\gamma}^{(n)}$. These approximations give us

$$P(\vec{n}) \approx (2N)^{-(d-1)} \frac{K(\vec{\gamma}^{(n)})}{\gamma_1^{(n)}\gamma_2^{(n)}\cdots\gamma_d^{(n)}}. \tag{36}$$

Inserting this expression and Eq. (35) into Eq. (29), we get

$$P(\vec{\gamma}|\vec{n}) \approx \left(\frac{2N}{\pi}\right)^{\frac{d-1}{2}} \exp\left(-2N\left|\vec{\Delta\gamma}\right|^2\right). \tag{37}$$

We can now do the second integral in Eq. (26), again treating the integral as if it were over an infinite $(d - 1)$-dimensional flat space. The result is

$$I(\vec{\gamma} : \vec{n}) \approx -\int K(\vec{\gamma}) \ln K(\vec{\gamma}) d\vec{\gamma} + \left(\frac{d-1}{2}\right) \ln\left(\frac{2N}{\pi e}\right). \tag{38}$$

Finally we subtract $[(d-1)/2]\ln N$ as in Eq. (30) to get

$$\tilde{I} = -\int K(\vec{\gamma}) \ln K(\vec{\gamma}) d\vec{\gamma} + \left(\frac{d-1}{2}\right) \ln\left(\frac{2}{\pi e}\right). \tag{39}$$

Note that Eq. (19) is a special case of this equation, with $d = 2$. As in that case, the expression is maximized by choosing $K(\vec{\gamma})$ to correspond to the uniform distribution:

$$K_{opt}(\vec{\gamma}) = \frac{2^d\, \Gamma(\frac{d}{2}+1)}{d\, \pi^{d/2}}. \tag{40}$$

The constant on the right-hand side of Eq. (40) is the reciprocal of the "surface area" of the positive section of the unit sphere.

The question now is whether the probability rule in real-vector-space quantum theory, $p_i = s_i^2$, induces the measure on probability space given by $K_{opt}(\vec{\gamma})$. The answer is yes, as is easily seen. The state vector $|s\rangle$ ranges over the full unit sphere in $\mathbb{R}^d$, but consider for now just the section of the sphere in which each s_i is positive. In that section the vector $\vec{\gamma}$ is equal to the vector $|s\rangle$, since $p_i = \gamma_i^2 = s_i^2$. So the uniform distribution of $|s\rangle$ over this section of the sphere induces the uniform distribution of $\vec{\gamma}$. The whole unit sphere in $\mathbb{R}^d$ consists, in effect, of 2^d copies of the positive section. So indeed, the uniform distribution of $|s\rangle$ over the whole sphere does correspond to the uniform distribution of $\vec{\gamma}$ expressed in Eq. (40). That is, the transfer of information from preparation to measurement is optimized in this d-dimensional case, just as it was in the two-dimensional case.

The complications involved in the information-theoretic calculation may obscure what is really a simple underlying fact. Imagine probability space as a $(d-1)$-dimensional flat "surface" in a d-dimensional space with orthogonal axes labeled $p_1, \ldots, p_d$. The surface consists of all points $(p_1, \ldots, p_d)$ such that each p_i is non-negative and the sum $p_1 + \cdots + p_d$ is equal to 1. Around each point on this surface one can imagine a small "region of uncertainty," representing the spread in the actual frequencies of occurrence of the d possible outcomes in N trials. These regions of uncertainty can be derived from the multinomial distribution, Eq. (27), and for large N their sizes and shapes can be read off the exponent in Eq. (31). (We could, for example, define "region of uncertainty" to be the range of values of $(n_1/N, \ldots, n_d/N)$ for which the exponent has magnitude less than 1.) One can see that these regions of uncertainty will have different sizes and shapes, depending on the location in probability space. For example, the largest is at the exact center, as shown in Fig. 3. But if we change the axes of probability space from p_i to $\gamma_i = \sqrt{p_i}$, probability space then looks like a section of a sphere. Again one can speak of a region of uncertainty around each point on this spherical surface, but now it happens that all the regions of uncertainty have the same size and shape—in fact they are all spherical—as we can see in the exponent of Eq. (35) and as is illustrated in Fig. 4. (The issue gets tricky near the edges. The closer one gets to the edge, the higher the value of N must be in order to see this uniformity. But no matter how close one is to the edge—as long as one is not *at* the edge—there is always such a value of N.)

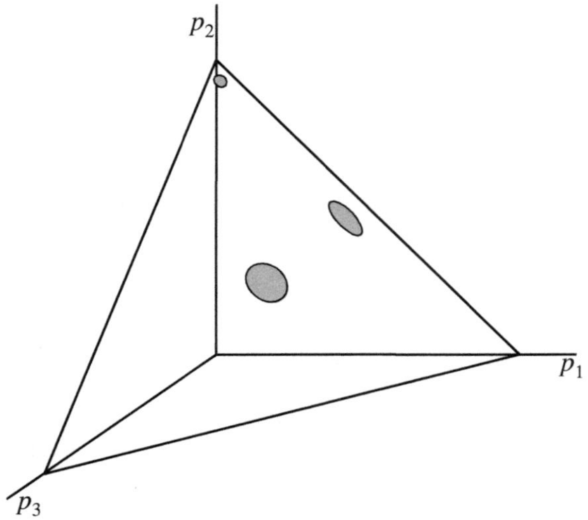

Fig. 3 Regions of uncertainty at different locations in the flat probability space. As one approaches an edge, the uncertainty shrinks along the direction perpendicular to the edge

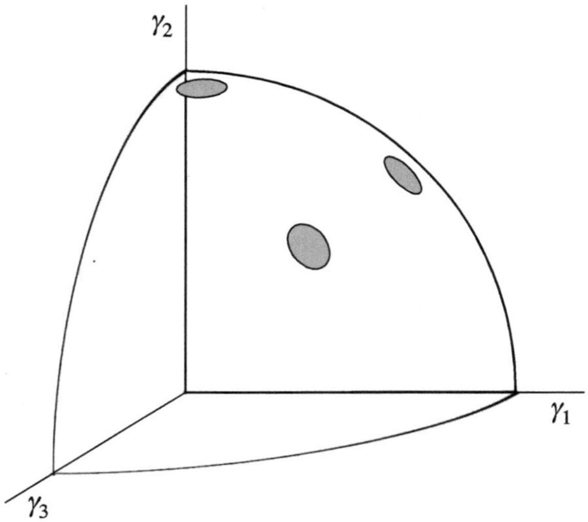

Fig. 4 Regions of uncertainty at different locations in the spherical probability space, with axes corresponding to the square roots of probability. Now all regions of uncertainty are isotropic and of the same size

In this sense, there is something special about representing probability space as a section of a sphere: it captures geometrically the statistical fluctuations in a large sample. What is special about real-vector-space quantum theory is that its set of pure states mirrors this representation of probability space.

As one would expect, the mathematical fact illustrated in Fig. 4 has been well noted in the statistics literature. Bhattacharyya in the 1940s proposed a distance measure between two probability distributions based on the *angle* between their $\vec{\gamma}$ vectors [26]. The square-root construction has been particularly explicit in the genetics literature. One can see diagrams of the positive section of the unit sphere in papers by Cavalli-

Sforza and collaborators from the 1960s, and these authors give credit for the idea to R.A. Fisher [27, 28] (as do Mosteller and Tukey [29]). In the present paper, I have used the square-root construction only to identify a special *measure* on probability space—the uniform measure on the spherical surface traced out by $\vec{\gamma}$. But one can also use it to define a special *metric* on the space, and this is what Bhattacharyya, Cavalli-Sforza and others have done. (One can find in Ref. [30] a review a various "genetic distances," some of which are based on the square-root construction.) Such a metric has also been used in work on quantum foundations [16, 31–33]. However, I want to emphasize that this special feature of the representation of probability space in terms of square roots of probability arises without any reference to quantum theory. It is simply a matter of statistics.

What about ordinary complex-vector-space quantum theory? In that theory each pure state is represented by a vector $|s\rangle$ in $\mathbb{C}^d$. The natural *a priori* distribution over pure states is the uniform distribution over the unit sphere in $\mathbb{C}^d$, that is, the unique distribution invariant under all unitary transformations. (We could just as well speak of a distribution over projection operators $|s\rangle\langle s|$ so as not to have to worry about the irrelevant overall phase factor in the vector $|s\rangle$, but for our purposes either picture leads to the same result.). For a complete orthogonal measurement, the probabilities of the outcomes are given by $p_i = |s_i|^2$, where the s_i's are the components of $|s\rangle$ in the basis defined by the measurement. We can ask what measure this probability rule, together with the *a priori* distribution of state vectors $|s\rangle$, induces on probability space. That question was answered by Sykora in 1974: it induces the *uniform* distribution, not on the spherical surface defined by $\vec{\gamma}$, but on the flat surface defined by $(p_1, \ldots, p_d)$ [34]. This is a remarkably simple and intriguing result, but this distribution is not the one that optimizes the transfer of information from preparation to measurement.

5 Optimal Information Transfer in Standard Quantum Mechanics?

The real-vector-space theory thus has a certain elegance to it, in that there is an optimal correspondence between the set of pure states and the set of probability distributions over the outcomes of a complete orthogonal measurement. The complex theory does not have this property, but one might wonder whether this is because we are not asking the question in the right way. That is, by somehow reframing the problem, might it be possible to see that the usual complex theory does exhibit the property of optimal information transfer in some altered sense?

For example, perhaps we are making a mistake to consider a complete orthogonal measurement. Such a measurement will never reveal the relative phases between the components of the state vector when it is written in the measurement basis. Instead we could consider a special case of a non-orthogonal measurement, namely, a symmetric informationally-complete measurement (a SIC) [35–37]. In d complex dimensions, such a measurement has d^2 possible outcomes. Each outcome corresponds to a pure state $|m_i\rangle\langle m_i|$, and the inner product between any two of these pure states has the same magnitude: $|\langle m_i|m_j\rangle|^2 = 1/(d+1)$. Numerical evidence strongly indicates that such symmetric measurements exist for all values of d up to 67 [38], and it would be reasonable to guess that they exist for all d. Such symmetric measurements figure prominently in the quantum Bayesian approach to understanding quantum mechanics [39, 40]. Is there a kind of optimal transfer of information from preparation to measurement that occurs when the measurement is a SIC?

With d^2 possible outcomes, one can estimate $d^2 - 1$ independent parameters by repeating the measurement on many identically prepared copies. This is exactly the number of parameters needed to specify a $d \times d$ density matrix, and indeed, any density matrix can be reconstructed with arbitrary precision from a fixed SIC applied to many copies. (This is the meaning of "informationally complete".). To state the question of optimal information transfer, we would need to specify an *a priori* measure on the set of all $d \times d$ density matrices. The measure should be unitarily invariant, but there are many unitarily invariant measures on this set. Is there at least one such measure for which the mutual information between the preparation (of a general mixed state) and the measurement outcomes is optimal?

One can see quickly that there is no such measure. The optimal *a priori* measure on probability space has already been determined in the preceding section. For d^2 possible outcomes, the optimal measure is the uniform measure over the (d^2-1)-dimensional spherical surface of probability space, when the axes correspond to the square roots of the probabilities. This measure clearly assigns nonzero weight to every nonzero volume of probability space. But if one performs a SIC on any state, the largest possible value of any probability is $1/d$. Thus the SIC does not make use of the whole probability space; so it is not providing information optimally in our sense, no matter what weighting function we place on the set of density matrices.

Let us try another version of the problem. Suppose we are given a specific entangled state of a pair of qubits, namely, the state

$$|\Phi^+\rangle = \frac{1}{\sqrt{2}}\left(|00\rangle + |11\rangle\right). \tag{41}$$

We imagine the first qubit is held by Alice and the second by Bob. Now Alice applies a unit-determinant unitary transformation U to her qubit—an element of $SU(2)$. She then sends her qubit to Bob, who performs a Bell measurement on the two qubits. That is, he distinguishes the four orthogonal states

$$\begin{aligned} |\Phi^+\rangle &= \frac{1}{\sqrt{2}}\left(|00\rangle + |11\rangle\right) \qquad & |\Phi^-\rangle &= \frac{1}{\sqrt{2}}\left(|00\rangle - |11\rangle\right) \\ |\Psi^+\rangle &= \frac{1}{\sqrt{2}}\left(|01\rangle + |10\rangle\right) \qquad & |\Psi^-\rangle &= \frac{1}{\sqrt{2}}\left(|01\rangle - |10\rangle\right) \end{aligned} \tag{42}$$

We imagine this whole procedure is repeated over and over—always with the same initial state, the same U, and the same Bell measurement—so that Bob can try to gain information about U from the outcomes of his measurements. We assume he already knows the initial state $|\Phi^+\rangle$. (This scenario is like superdense coding [41], except that we are not restricting U to a discrete set. Really what Bob is doing here is a restricted kind of process tomography [42, 43]—trying to infer the process U from the outcomes of measurements.). We can ask whether the transfer of information is optimal between Alice's choice of unitary transformation and the outcomes of Bob's measurements.

A general element of $SU(2)$ can be represented as

$$U = \exp\left[i(\theta/2)\hat{n}\cdot\vec{\sigma}\right], \tag{43}$$

where $\hat{n}$ is the unit vector defining the axis of the Bloch sphere around which Alice is rotating her qubit, θ is the angle of rotation—it runs from zero to 2π—and $\vec{\sigma}$ is the vector of Pauli matrices. The transformations U are in one-to-one correspondence with the points of a three-dimensional spherical surface, which we can imagine embedded in four dimensions. Specifically, we can label the point corresponding to the above U by the unit vector

$$v_U = (\cos(\theta/2), n_x\sin(\theta/2), n_y\sin(\theta/2), n_z\sin(\theta/2)). \tag{44}$$

The natural measure on $SU(2)$ is the uniform measure over this sphere—it is the unique measure that is invariant under left-multiplication (or right-multiplication) by any group element.

To determine whether the information is transferred optimally from Alice to Bob, we need to compute the probabilities of Bob's outcomes. It is straightforward to do so, and one finds that the probabilities are, in an arrangement parallel to that given in Eq. (42),

$$\begin{array}{cc} \cos^2(\theta/2) & n_z^2\sin^2(\theta/2) \\ n_x^2\sin^2(\theta/2) & n_y^2\sin^2(\theta/2) \end{array} \tag{45}$$

These probabilities are the squared components of the unit vector v_U given in Eq. (44). Thus the problem is equivalent to the case of real-vector-space quantum mechanics in four dimensions. So indeed, the information is transmitted optimally from Alice to Bob!

Does this example generalize to higher dimensions? The answer is no, at least not in any way that I can see. For example, in three dimensions, we would probably want Alice and Bob to start with the state $|\Phi\rangle = (|00\rangle + |11\rangle + |22\rangle)/\sqrt{3}$. Alice will perform a general unit-determinant unitary transformation U, and then Bob will measure both particles in the generalized Bell basis, which consists of the nine states

$$\begin{array}{lll} |00\rangle + |11\rangle + |22\rangle & |00\rangle + \omega|11\rangle + \omega^2|22\rangle & |00\rangle + \omega^2|11\rangle + \omega|22\rangle \\ |01\rangle + |12\rangle + |20\rangle & |01\rangle + \omega|12\rangle + \omega^2|20\rangle & |01\rangle + \omega^2|12\rangle + \omega|20\rangle \\ |02\rangle + |10\rangle + |21\rangle & |02\rangle + \omega|10\rangle + \omega^2|21\rangle & |02\rangle + \omega^2|10\rangle + \omega|21\rangle. \end{array} \tag{46}$$

Here $\omega = e^{2\pi i/3}$ and I have suppressed the normalization factor $1/\sqrt{3}$. A counting of parameters is initially encouraging: it takes eight real numbers to specify an element of $SU(3)$, and Bob's measurement yields eight independent probabilities. However, one quickly discovers that, as in the case of the SIC, the measurement does not make use of the whole probability space.

Consider specifically the probabilities of the second and third outcomes listed on the first row of Eq. (46); let us call these probabilities p_2 and p_3 (we imagine a list of nine probabilities $p_1, \ldots, p_9$, of which these are the second and third). In terms of the components u_{ij} of Alice's unitary matrix U, we have

$$p_2 = \frac{1}{9}\left|u_{00} + \omega^2 u_{11} + \omega u_{22}\right|^2 \qquad \text{and} \qquad p_3 = \frac{1}{9}\left|u_{00} + \omega u_{11} + \omega^2 u_{22}\right|^2, \tag{47}$$

so that the product has the value

$$p_2 p_3 = \frac{1}{81}\left|u_{00}^2 + u_{11}^2 + u_{22}^2 - u_{00}u_{11} - u_{00}u_{22} - u_{11}u_{22}\right|^2. \tag{48}$$

Now, in the whole probability space the maximum value of $p_2 p_3$ is 1/4, attained when $p_2 = p_3 = 1/2$. But given that each u_{ij} can have a magnitude no larger than 1, one can show that the expression in Eq. (48) cannot exceed the value $16/81 < 1/4$.[6] Thus a certain region of probability space is inaccessible in the scenario we are imagining. It follows that the information about U is not conveyed optimally to Bob through his measurement outcomes.

Thus, as far as I can tell, the property of optimal information transfer does not easily carry over from the real-vector-space theory to ordinary quantum mechanics.

---

[6] In proving this inequality, we are free to set u_{00} equal to 1. Then let $u_{11} = -a$ and $u_{22} = -b$ and the desired inequality becomes

$$|1 + a + b + a^2 + b^2 - ab| \leq 4$$

under the assumption that $|a| \leq 1$ and $|b| \leq 1$. (One can see that equality is achieved when $a = b = 1$.) This inequality is equivalent to

$$\left|A^2 + B^2 + (A - B)^2\right| \leq 8,$$

where $A = 1 + a$ and $B = 1 + b$. This last inequality can be proved by first noting that

$$\left|A^2 + B^2 + (A - B)^2\right| \leq |A^2 + B^2| + |A - B|^2$$

and then writing out the absolute values explicitly. One has to use the fact that A and B are both confined to a circle of unit radius in the complex plane, centered at the value 1.

6 Conclusions

In real-vector-space quantum theory, the number of parameters needed to specify a pure state is equal to the number of independent probabilities in a complete orthogonal measurement: both are equal to $d - 1$ for a state in d dimensions. So by measuring many copies prepared in an unknown pure state, one can hope to pin the state down to a finite number of small regions in state space. In this paper we have seen that this pinning down is in fact optimal, in the sense that the observer gains as much information about the state as could possibly be gained in any probabilistic theory, at least when the number of trials is very large.

Standard quantum theory, based on a complex vector space, does not have this property, and we have not been able to find a restatement of the problem for which the complex theory does achieve such an optimization (except for the case of a unitary transformation applied to a qubit). For our original statement of the problem, one can say that this lack of optimization comes from the fact that for any specific orthogonal measurement a complete specification of a pure state includes phase factors that have no effect on the probabilities of the outcomes. The presence of these phase factors changes the natural *a priori* measure on probability space, and the mutual information is no longer maximized.

Note that in the complex theory the number of real parameters needed to specify a pure state in d dimensions is $2d - 2$ if we do not count an irrelevant overall phase factor. This number is exactly *twice* the number of independent probabilities an orthogonal measurement can access, and it seems that this doubling of the number of parameters is what spoils the optimization. It is reasonable to ask whether there can be some deeper understanding of this factor of two, but at this point it is hard to have confidence in any particular answer.

In a sense, any axiomatization of quantum mechanics offers a potential answer to this question: whatever assumptions give rise to the structure of quantum theory also give rise to the factor of two. In his axiomatization, Goyal addresses the factor of two directly, formalizing it in his principle of complementarity: for a measurement that at some level generates $2d$ possible events, only d distinguishable outcomes can be observed, each corresponding to a *pair* of fundamental events. This principle, together with a principle of global gauge invariance, leads him to the basic structure of quantum mechanics [33]. One achieves a similar result by assuming that the underlying theory is real-vector-space quantum theory, but that because our knowledge is limited in some fundamental way we do not see the real-vector-space structure; we have access only to those observables that satisfy Stueckelberg's rule. (Goyal in fact relates his work to Stueckelberg's.) Imposing Stueckelberg's rule on a real vector space of $2d$ dimensions reduces the maximum number of orthogonal states from $2d$ to d, and it cuts in half the number of parameters required to specify a maximally pure state.[7]

[7] When standard quantum mechanics is expressed in real-vector-space terms, what we normally call a pure state is represented by a density matrix of rank 2.

While such an interpretation would give an important role to the real-vector-space theory, it raises a difficult question about the status of the main result in this paper. If the limitation on our knowledge is fundamental, then who are the observers for whom the transfer of information from preparation to measurement is optimal? Evidently it is not optimal for us, because whatever the underlying theory may be, the *effective* theory within which we live seems to be complex-vector-space quantum theory.

References

1. G. Birkhoff, J. von Neumann, Ann. Math., second series **37**, 823–843 (1936)
2. C. Piron, Helv. Phys. Acta **37**, 439–468 (1964)
3. M.P. Solèr, Commun. Algebra **23**, 219–243 (1995)
4. S.S. Holland, Bull. Am. Math. Soc. **32**, 205–234 (1995)
5. E.C.G. Stueckelberg, Helv. Phys. Acta **32**, 254 (1959)
6. E.C.G. Stueckelberg, Helv. Phys. Acta **33**, 727 (1960)
7. P. Lahti, M.J. Maczynski, J. Math. Phys. **28**, 1764–1769 (1987)
8. H. Araki, Commun. Math. Phys. **75**, 1 (1980)
9. S. Bergia, F. Cannata, A. Cornia, R. Livi, Found. Phys. **10**, 723 (1980)
10. W.K. Wootters, in *Complexity, Entropy and the Physics of Information*, ed. by W.H. Zurek (Addison-Wesley, Redwood City, 1990)
11. N.D. Mermin, Am. J. Phys. **66**, 753 (1998)
12. L. Hardy, arxiv:quant-ph/0101012 (2001)
13. J. Barrett, Phys. Rev. A **75**, 032304 (2007)
14. G. Chiribella, G.M. D'Ariano, P. Perinotti, Phys. Rev. A **84**, 012311 (2011)
15. L. Hardy, W.K. Wootters, Found. Phys. **42**, 454–473 (2012)
16. W.K. Wootters, The acquisition of information from quantum measurements. Ph.D. dissertation, University of Texas at Austin (1980)
17. E.C.G. Stueckelberg, M. Guenin, Helv. Phys. Acta **34**, 621 (1961)
18. E.C.G. Stueckelberg, M. Guenin, C. Piron, Helv. Phys. Acta **34**, 675 (1961)
19. F.J. Dyson, J. Math. Phys. **3**, 1199 (1962)
20. G.W. Gibbons, arxiv:1111.0457 (2011)
21. J. Myrheim, arxiv:quant-ph/9905037 (1999)
22. S. Aaronson, arxiv:quant-ph/0401062 (2004)
23. B. de Finetti, *Theory of Probability* (Wiley, New York, 1990)
24. C.M. Caves, C.A. Fuchs, R. Schack, J. Math. Phys. **43**, 4537 (2002)
25. T.M. Cover, J.A. Thomas, *Elements of Information Theory* (Wiley, New York, 1991), p. 231
26. A. Bhattacharyya, Sankhya **7**, 401–407 (1946)
27. L.L. Cavalli-Sforza, F. Conterio, Atti. Assoc. Genet. Ital. **5**, 333 (1960)
28. L.L. Cavalli-Sforza, A.W.F. Edwards, Am. J. Hum. Genet. **19**, 233 (1967)
29. F. Mosteller, J.W. Tukey, J. Am. Stat. Assoc. **44**, 174 (1949)
30. N. Takezaki, M. Nei, Genetics **144**, 389 (1996)
31. W.K. Wootters, Phys. Rev. D **23**, 357 (1981)
32. S.L. Braunstein, C.M. Caves, Phys. Rev. Lett. **72**, 3439 (1994)
33. P. Goyal, New J. Phys. **12**, 023012 (2010)
34. S. Sykora, J. Stat. Phys. **11**, 17–27 (1974)
35. G. Zauner, Quantum designs—foundations of a non-commutative theory of designs. Ph.D. dissertation, University of Vienna (1999)
36. C.M. Caves, Symmetric Informationally Complete POVMs, report 9 September 1999, http://info.phys.unm.edu/~caves/reports/infopovm.pdf
37. J.M. Renes, R. Blume-Kohout, A.J. Scott, C.M. Caves, J. Math. Phys. **45**, 2171 (2004)

38. A.J. Scott, M. Grassl, J. Math. Phys. **51**, 042203 (2010)
39. D.M. Appleby, A. Ericsson, C.A. Fuchs, Found. Phys. **41**, 564 (2011)
40. C.M. Fuchs, arxiv:1003.5209 (2010)
41. C. Bennett, S.J. Wiesner, Phys. Rev. Lett. **69**, 2881 (1992)
42. I.L. Chuang, M.A. Nielsen, J. Mod. Opt. **44**, 2455 (1997)
43. J.F. Poyatos, J.I. Cirac, P. Zoller, Phys. Rev. Lett. **78**, 390 (1997)

Almost Quantum Theory

Benjamin Schumacher and Michael D. Westmoreland

1 Introductory Remarks

1.1 Motivation

The remarkable features of quantum theory are best appreciated by comparing the theory to other possible theories—what Spekkens calls "foil" theories [1]. The most celebrated example of this approach was Bell's analysis [2], which showed that entangled quantum systems have statistical properties unlike any hypothetical local hidden variable model. More recently, there have been several efforts to give quantum theory an operational axiomatic foundation [3–5]. In these efforts, a general abstract framework is posited to describe system preparations, choices of measurement, observed results of measurement, and probabilities. Many possible theories can be expressed in the framework. The axioms embody fundamental aspects of quantum theory that uniquely identify it among them. A striking lesson of this work is that familiar quantum theory can be characterized by axioms that seem to have little to do with the traditional quantum machinery of states and observables in Hilbert space. The Hilbert space structure is "derived" from the operational axioms.

These approaches are based on two distinct concepts of generalization. First, we generalize within quantum theory to give the theory its most general form. For example, we generalize state vectors to density operators as a description of the quantum state of a system. Second, we generalize beyond quantum theory so that

B. Schumacher (✉)
Department of Physics, Kenyon College, Gambier, OH, USA
e-mail: schumacherb@kenyon.edu

M.D. Westmoreland (✉)
Department of Mathematical Sciences, Denison University, Granville, OH, USA
e-mail: westmoreland@denison.edu

G. Chiribella and R.W. Spekkens (eds.), *Quantum Theory: Informational Foundations and Foils*, Fundamental Theories of Physics 181,
DOI 10.1007/978-94-017-7303-4_3

we can embed it within a wider universe of possible theories. To be clear, we refer to these two processes as *development* within a theoretical framework and *extension* beyond that framework.

In this paper, we undertake these processes of development and extension, not for actual quantum theory (AQT), but for a close mathematical cousin of that theory. Modal quantum theory (MQT) [6] is a simplified toy model that reproduces many of the structural features of actual quantum theory. The underlying state space of MQT is a vector space $\mathcal{V}$ over an arbitrary field $\mathcal{F}$, which may be finite. MQT predicts, not the probabilities of the results of a measurement, but only which of those results are possible. This motivates the use of the term "modal", which in formal logic refers to operators asserting the possibility or necessity of a proposition [7]. Modal theories themselves can therefore be viewed as generalizations (extensions) of probabilistic theories.

1.2 Generalization

What is "generalization"? We begin our answer to this question with a simple example. Suppose we are devising simple substitution ciphers for English text. Each letter in the alphabet $\mathcal{A} = \{A, B, \ldots, Z\}$ is to be represented by some letter in $\mathcal{A}$. To begin with, we consider only extremely simple "transposition ciphers" in which exactly two letters are exchanged. For instance, we might exchange E and R, leaving all other letters alone.[1] An enciphered message can be decoded by applying the same transposition a second time.

To generalize this and make better ciphers, we now form compound ciphers by applying two or more transposition ciphers successively. Enciphered messages are decoded by applying the same transpositions in reverse order. Any cipher constructed out of transpositions can be described by a *permutation* of $\mathcal{A}$, an element of $S_{\mathcal{A}}$. Furthermore, any "permutation cipher" in $S_{\mathcal{A}}$ can be constructed in exactly this way, as a compound of pairwise transpositions. Thus, the concept of a permutation cipher is really a development of the original idea of a transposition.

Is further development possible? Consider the essential requirements for a "reasonable" cipher. A general substitution cipher c is a function $c : \mathcal{A} \rightarrow \mathcal{A}$. Since we need to be able to recover our plaintext correctly, it is appropriate to require as an axiom that c be a one-to-one function, so that each letter in the ciphertext can be decoded in only one way. Since $\mathcal{A}$ is finite, all one-to-one functions on $\mathcal{A}$ are permutations. This means that the generalization from transpositions to permutations already encompasses all reasonable substitution ciphers (as characterized by our axiom).

To generalize further, we must extend the idea of a cipher beyond simple letter substitutions. We might apply different substitution maps to different letters, or

[1] Such a cipher is not very hard to read.

encipher messages word-by-word. These more general ciphers will have some characteristics in common with substitution ciphers (such as the unique decipherability of an enciphered message), but will constitute a larger universe within which the substitution ciphers form a special class.

In the cipher story we can identify some general features. We begin with a basic theory based on a concept X. The process of development can involve several stages:

- *Construction.* In this stage, we devise a situation within the existing theory—that is, a situation that can be described using X—and show how this situation can be given a simpler or more natural description using X'. (*Compositions of transposition ciphers can be described as permutation ciphers.*)
- *Feasibility.* Often we are able to show that, if a situation is described by X', then it can always be given a more cumbersome description in terms of X. Informally, every instance of X' is feasible to construct from X. (*Every permutation cipher can be described as a composition of transposition ciphers.*)
- *Axiomatic characterization.* We may be led to impose one or more reasonable axioms that any situation ought to satisfy. Our development is most successful if we can establish that any "reasonable" situation (according to our axioms) can be encompassed by our generalized concept X'. (*Any uniquely decodable substitution cipher must be a permutation cipher.*)

If X' is feasible, then every instance of X' could be given a more cumbersome description in terms of X. In this case, the theory including X' is simply a development of the original one based on X. An axiomatic characterization tells us that the development is *complete*—that no further reasonable generalization of X is possible within the basic theory.

Once we have a complete development from X to X', further generalization must be an extension of the original theory.

- *Extension.* We can devise a broader framework Y within which the theory based on X is a special case. (*We can consider ciphers that are not based on letter-by-letter substitutions.*)

Once we have an extended framework Y, it is useful to ask what special properties the original theory may possess. Thus, we might investigate what distinguishing characteristics quantum theory has within the wider universe of probabilistic theories.

1.3 Scope of the Present Paper

The elementary features of modal quantum theory have been presented elsewhere [6, 8]. In the next section, we will briefly discuss the axioms for MQT, drawing the analogies between this theory and AQT. We will also discuss some of the properties of entangled states of simple systems in MQT.

Following this, we turn to a development of MQT analogous to the standard generalizations of states, measurements and dynamical evolution in AQT. Systems whose preparations are incompletely known, or which are entangled with other systems, require a more general description of their states. Measurement procedures and dynamical evolution for open systems require additional generalizations, which we will also explore. As in AQT, we can give axiomatic characterizations for these new concepts within the theory, showing that our development is, in the sense given above, complete.

To generalize further, we must embed MQT within a larger class of modal theories. We do this by analogy to the general probabilistic theories that have been used to analyze AQT. As in those theories, our modal theories are assumed to satisfy a version of the *no-signalling principle* [9], which states that the choice of measurement on one system cannot have a observable effect on the measurement results of a distinct system.

Finally, we note that any probabilistic theory can be viewed through "modal glasses", simply interpreting probabilities $p > 0$ as "possible" and $p = 0$ as "impossible". Thus, modal theories are generalizations of probabilistic theories. This generalization is actually an extension, since we will find modal theories that cannot be "resolved' to probabilistic ones. For situations that arise from systems in MQT, however, the situation is more complex. In the bipartite case we will show that a weak probabilistic resolution (which may assign $p = 0$ for a "possible" measurement result) can always be found.

2 Modal Quantum Theory

2.1 A Modal World

The world of modal quantum theory is a world without probabilities. Probabilities are so familiar that it is worthwhile to consider more carefully what their absence entails.

In a probabilistic world, an event x is assigned a numerical probability $p(x)$ such that $0 \leq p(x) \leq 1$. The probabilities are normalized, so that

$$\sum_x p(x) = 1 \tag{1}$$

where the sum extends over a set of mutually exclusive and exhaustive events. Probabilities are related to statistical frequencies. Suppose we perform N independent trials of an experiment and observe event x to occur N_x times. Then with high probability,[2]

$$p(x) \approx \frac{N_x}{N}. \tag{2}$$

[2]Note that the connection between probabilities and statistical frequencies is itself probabilistic! This highlights the difficulty in giving a non-circular operational interpretation of probability.

The possible results of an experiment may be labeled by numerical values v. The mean of such a random variable is given by

$$\langle v \rangle = \sum_v p(v)\, v. \tag{3}$$

In a *possibilistic* or *modal* world, we can only distinguish between possible and impossible events, but we do not assign any measure of likelihood to them. That is, we can identify a possible set

$$\mathcal{P} = \{x, x', \ldots\}. \tag{4}$$

The only "normalization" condition is the requirement that $\mathcal{P} \neq \emptyset$. If we perform an experiment many times, the set $\mathcal{R}$ of results that we see satisfies $\mathcal{R} \subseteq \mathcal{P}$. That is, every result we have actually seen is surely possible, but we can draw no definite conclusions about the possibility or impossibility of other results. Also, without any assignment of "weights" to the numerical results v of an experiment, we cannot compute a mean value $\langle v \rangle$.

The naive connection between probabilistic and modal pictures is that $x \in \mathcal{P}$ if and only if $p(x) \neq 0$. There are, however, some subtleties to be recognized. If we are assigning probabilities based on observed statistical frequencies, we cannot distinguish between a very rare event x (which may not have happened yet in our large but finite set of trials) and an impossible one. That is, we may be able to conclude that $p(x) \approx 0$ but not that x is impossible.

2.2 Basic Axioms

The axioms for modal quantum theory are closely related to those of actual quantum theory, as we can see in Table 1. The axioms presented are for the most elementary versions of each theory; we will develop them further below. Even without this development, however, we can identify some interesting features of MQT. Consider, for instance, the case where $\mathcal{F}$ is a finite field. A system with finite-dimensional $\mathcal{V}$ has only a finite set of possible state vectors. There are only finitely many distinct measurements or time evolution maps for the system, and time evolution must proceed in discrete steps.

The simplest possible system in MQT is a "modal bit" or *mobit* [6], for which $\dim \mathcal{V} = 2$. If we also choose the smallest field $\mathcal{F} = \mathbb{Z}_2$, then there are just three non-zero vectors in $\mathcal{V}$, which we can denote $|0)$, $|1)$ and $|\sigma) = |0) + |1)$. The dual space $\mathcal{V}^*$ also has three vectors, so that

$$\begin{array}{lll} (a\,|0) = 1 & (a\,|1) = 0 & (a\,|\sigma) = 1 \\ (b\,|0) = 0 & (b\,|1) = 1 & (b\,|\sigma) = 1. \\ (c\,|0) = 1 & (c\,|1) = 1 & (c\,|\sigma) = 0 \end{array} \tag{5}$$

Table 1 Elementary axioms for AQT and MQT

Actual quantum theory	Modal quantum theory
States. A system is described by a Hilbert space $\mathcal{H}$ over the field $\mathbb{C}$ of complex numbers. A state is a normalized $\lvert\psi\rangle \in \mathcal{H}$	*States*. A system is described by a vector space $\mathcal{V}$ over a field $\mathcal{F}$. A state is a non-zero $\lvert\psi) \in \mathcal{V}$
Measurements. A measurement is an orthonormal basis $\{\lvert k\rangle\}$ for $\mathcal{H}$. Each basis element represents a measurement outcome. For state $\lvert\psi\rangle$, the probability outcome k is $p(k) = \lvert\langle k \lvert \psi\rangle\rvert^2$	*Measurements*. A measurement is a basis $\{(k\rvert\}$ for $\mathcal{V}^*$. Each basis element represents a measurement outcome. For state $\lvert\psi)$, outcome k is possible if and only if $(k \lvert \psi) \neq 0$
Evolution. Over a given time interval, an isolated system evolves via a unitary operator U: $\lvert\psi\rangle \rightarrow U \lvert\psi\rangle$	*Evolution*. Over a given time interval, an isolated system evolves via an invertible operator T: $\lvert\psi) \rightarrow T \lvert\psi)$
Composite systems. The state space for a composite system is the tensor product of subsystem spaces: $\mathcal{H}^{(AB)} = \mathcal{H}^{(A)} \otimes \mathcal{H}^{(B)}$	*Composite systems*. The state space for a composite system is the tensor product of subsystem spaces: $\mathcal{V}^{(AB)} = \mathcal{V}^{(A)} \otimes \mathcal{V}^{(B)}$

Any pair of these dual vectors yields a basic measurement. There are thus three basic mobit measurements corresponding to the bases $X = \{(c\rvert, (a\rvert\}$, $Y = \{(b\rvert, (c\rvert\}$ and $Z = \{(a\rvert, (b\rvert\}$. The individual dual vectors in a measurement basis, which correspond to the results of the measurement, are called *effects*. It will sometimes be convenient to label the measurement results by $+$ and $-$, so we may write $(a\rvert = (+_z\rvert = (-_x\rvert$, etc. If a mobit is in the state $\lvert\sigma)$ and a Z-measurement is made, both outcomes $+_z$ and $-_z$ are possible.

As in AQT, we can compare MQT to a hypothetical hidden variable theory. Such a theory supposes that the system possesses some unknown variable λ such that, for a given value of λ, the result of any measurement is determined. As in AQT, we cannot completely exclude all hidden variable theories, though we can show that some kinds of are inconsistent with MQT. For instance, consider a *non-contextual* hidden variable theory [10], in which (given λ) the question of whether a given effect $(e\rvert$ will occur is independent of which other effects are present in the measurement basis. For a given value of λ, the theory would have to assign "yes" or "no" values to each of the dual vectors $(a\rvert$, $(b\rvert$ and $(c\rvert$, such that any pair of them includes exactly one "yes". This is plainly impossible. We conclude that the pattern of possibilities for a mobit system in MQT cannot be reproduced by any non-contextual hidden variable theory.

This is essentially an MQT version of the famous Kochen-Specker theorem of AQT [10]. The MQT argument has a similar structure to the original (both can be cast as graph-coloring problems) but is radically simpler. Furthermore, the AQT version of the Kochen-Specker theorem only applies for dim $\mathcal{H} \geq 3$, while the MQT version applies to any system of any dimension [11].

2.3 Entanglement

Composite systems in MQT may be in either product or entangled states. For instance, a pair of $\mathbb{Z}_2$-mobits has 15 possible states, of which 9 are product states and 6 are entangled. (For more complicated systems, the entangled states greatly outnumber the product states.)

Entangled states are marked by correlated measurement results. For example, consider the modal "singlet" state of two mobits:

$$|S) = |0, 1) - |1, 0) . \tag{6}$$

(The minus sign here allows us to generalize the state for any field $\mathcal{F}$. For $\mathcal{F} = \mathbb{Z}_2$, $-1 = +1$ and so $|S) = |0, 1) + |1, 0)$.) Note that, for any effect $(e|$,

$$(e, e\,|S) = (e\,|0)\,(e\,|1) - (e\,|1)\,(e\,|0) = 0. \tag{7}$$

Therefore, if we make the same measurement on both mobit subsystems, it is impossible that we obtain identical results.

The mobit measurements X, Y and Z yield nine possible joint measurements of a pair of mobits.[3] Let $(u, v|U, V)$ denote the situation in which measurements of U and V on two systems yield respective results u and v. Then we can summarize the measurement results for $|S)$ as follows:

- If the same measurement is made on each mobit, the results must disagree. Thus $(+, +|X, X)$ is impossible, and so on.
- If different measurements are made on the two mobits, all but one of the joint results are possible. Thus, $(+, -|X, Y)$ is impossible, but $(+, +|X, Y)$, $(-, +|X, Y)$ and $(-, -|X, Y)$ are all possible.

For AQT, Bell showed that the correlations between entangled quantum systems were incompatible with any local hidden variable theory [2]. He did this by devising a statistical inequality that holds for any local hidden variable theory, but which can be violated by entangled quantum systems. Unfortunately, a similar approach based on probabilities and expectation values is not available in MQT.

Hardy devised an alternate argument for AQT based only on possibility and impossibility [3]. He constructed a non-maximally entangled state $|\Psi\rangle$ for a pair of qubits together with a set of measurements having the following properties:

- $(+, +|A, D)$ and $(+, +|B, C)$ are both impossible—that is, they have quantum probability $p = 0$.
- $(+, +|B, D)$ is possible ($p > 0$).
- $(-, -|A, C)$ is impossible ($p = 0$).

How might a local hidden variable theory account for this situation? Since $(+, +|B, D)$ is possible, we restrict our attention to the set H of hidden variable values that

[3]There are also many measurements involving entangled effects.

yield this result. The result of a measurement on one qubit is unaffected by the choice of measurement on the other (locality). Furthermore, no allowed values of the hidden variables can lead to $(+, +|A, D)$ or $(+, +|B, C)$. Thus, for values in H, we must obtain the results $(-, +|A, D)$ and $(+, -|B, C)$. These jointly imply that the result $(-, -|A, C)$ would be obtained for values in H. But this contradicts AQT, in which $(-, -|A, C)$ is impossible.

The very same argument applies to the state $|S)$ in MQT, if we identify $A = X$, $B = Y$, $C = \bar{Z}$ (the negation of Z) and $D = \bar{Y}$. Thus we can conclude that no local hidden variable theory can account for the pattern of possible measurement outcomes generated by the entangled state $|S)$.

However, this argument has a weakness, because it only applies to those situations in which the joint outcome $(+, +|B, D) = (+, -|Y, Y)$ actually occurs. In AQT, we can assign a finite probability $p > 0$ to this result, so we can confidently expect it to arise in a large enough sample. But in MQT, the statement that the joint result is possible does not allow us to draw any such conclusion. The MQT version of the Hardy argument therefore applies only to a situation that may not, in fact, ever occur.

A stronger argument may be constructed along the following lines [6]. We imagine that the MQT state $|S)$ corresponds to a set H_S of possible values of a hidden variable. The variable controls the outcomes of possible measurements in a completely local way. For any particular value $h \in H_S$, the set of possible results of a measurement on one mobit depends only on the measurement choice for that mobit, not on the choice for the other mobit. Let $\mathcal{P}_h(E)$ be the set of possible results of a measurement of E for the hidden variable value h. Our locality assumption means that, given $V^{(1)}$ and $W^{(2)}$ measurements for the two mobits,

$$\mathcal{P}_h\big(V^{(A)}, W^{(B)}\big) = \mathcal{P}_h\big(V^{(A)}\big) \times \mathcal{P}_h\big(W^{(B)}\big)\,, \tag{8}$$

the simple Cartesian product of separate sets $\mathcal{P}_h(V^{(1)})$ and $\mathcal{P}_h(W^{(2)})$. The MQT set of possible results arising from $|S)$ should therefore be

$$\mathcal{P}\big(V^{(1)}, W^{(2)}|S\big) = \bigcup_{h \in H_S} \mathcal{P}_h\big(V^{(1)}\big) \times \mathcal{P}_h\big(W^{(2)}\big)\,. \tag{9}$$

The individual sets $\mathcal{P}_h(V^{(1)})$, etc., are simultaneously defined for all of the measurements that can be made on either mobit. Therefore, we may consider the set

$$\begin{aligned} \mathcal{J} = \bigcup_{h \in H_S} \; & \mathcal{P}_h\big(X^{(1)}\big) \times \mathcal{P}_h\big(Y^{(1)}\big) \times \mathcal{P}_h\big(Z^{(1)}\big) \\ & \times \mathcal{P}_h\big(X^{(2)}\big) \times \mathcal{P}_h\big(Y^{(2)}\big) \times \mathcal{P}_h\big(Z^{(2)}\big)\,. \end{aligned} \tag{10}$$

There might be up to $2^6 = 64$ elements in $\mathcal{J}$. However, since $\mathcal{J}$ can only contain elements that agree with the properties of $|S)$, we can eliminate many elements. For instance, the fact that corresponding measurements on the two mobits must give opposite results tells us that $(+, +, +, +, +, +)$ cannot be in $\mathcal{J}$, though

$(+, +, +, -, -, -)$ might be. However, when we apply all of the properties of $|S\rangle$ in this way, we find the surprising result that *all* of the elements of $\mathcal{J}$ are eliminated. *No* assignment of definite results to all six possible measurements can possibly agree with the correspondences obtained from the entangled MQT state $|S\rangle$. We therefore conclude that these correspondences are incompatible with any local hidden variable theory.

This argument can be recast in terms of a *pseudo-telepathy game* [12]. Two players, Alice and Bob, are separately asked questions drawn from a finite set. Their goal is to give answers that satisfy some joint criterion. The game is a pseudo-telepathy game if the goal could only be satisfied by classical players if they could communicate with each other. However, Alice and Bob may have a winning strategy if they share quantum entanglement. In our pseudo-telepathy game, Alice and Bob are each asked one of three questions (X, Y or Z), and their goal is to provide a joint answer consistent with the possible measurement outcomes of the entangled mobit state $|S\rangle$ described above. If Alice and Bob answer separately based on shared classical information, they cannot always win the game. If they share a mobit pair in $|S\rangle$, they can. (However, as we will see below in Sect. 5.3, this game has no perfect strategy in AQT.)

3 States and Measurements

3.1 Generalized States and Measurements in AQT

The axioms for MQT presented in Table 1 describe a "basic" version of the theory. In this section and the next, we will develop the theory to include more general kinds of states, measurements, and time evolution. Our development parallels the standard one in AQT [13], but also has many important differences.

In AQT, there are situations in which we cannot ascribe a definite state vector $|\psi\rangle$ to a system, either because its preparation is not completely known or because we have a subsystem of a larger composite system in an entangled state. In either case, we can construct a description of the situation from which we can make probabilistic predictions about the behavior of the system. We describe such *mixed states* by means of *density operators*.

Suppose, for instance that the system is prepared in one of several possible pure states, so that $|\psi_\alpha\rangle$ occurs with probability p_α. This mixture of states is described by the density operator

$$\rho = \sum_\alpha p_\alpha \, |\psi_\alpha\rangle\langle\psi_\alpha| \,. \tag{11}$$

If we make a measurement on the system corresponding to an orthonormal basis $\{|k\rangle\}$, then the overall probability of the result k is

$$p(k) = \sum_{\alpha} p_\alpha \, |\langle k \, | \psi_\alpha \rangle|^2 = \langle k | \, \rho \, | k \rangle \, . \tag{12}$$

Thus, the density operator ρ is sufficient to predict the probability of any basic measurement result, given the probabilistic mixture of states.

Different mixtures of states can yield the same ρ, and thus yield the same statistical predictions. We therefore say that the different mixtures correspond to the same mixed state. Conversely, different density operators ρ and ρ' will lead to different statistical predictions for at least some measurements.

Density operators can also be used to describe a system that is part of a composite system. Given a joint state $|\Psi\rangle$ of RQ, we can construct a density operator for Q via the partial trace operation:

$$\rho = \mathrm{Tr}_{(\mathrm{R})} \, |\Psi\rangle\langle\Psi| \, . \tag{13}$$

Again, this density operator predicts the probabilities for any basic measurement on subsystem Q itself, according to the rule in Eq. 12.

Every density operator arising from a mixture or a partial trace is a positive semidefinite operator of trace 1, and any such operator could arise in these ways. The set of positive semidefinite operators of trace 1 therefore constitutes our set of generalized states for a system.

We can also develop the concept of measurement in AQT. As a first step, we can "coarse-grain" a basic measurement, so that each outcome a corresponds to a projection operator Π_a (associated with the subspace of $\mathcal{H}$ spanned by the basis vectors $|k\rangle$ included in a). We can generalize further by supposing that we apply our measurement to a composite system and an *ancilla* system, which is regarded as part of the experimental apparatus. Then we find that each outcome a is associated with a positive semidefinite operator E_a, and that the probability of this outcome is

$$p(a) = \mathrm{Tr} \, \rho E_a. \tag{14}$$

The outcome operators E_a, sometimes called *effect operators*, sum to the identity:

$$\sum_a E_a = \mathbf{1}. \tag{15}$$

Our generalized model of measurement is thus a set $\{E_a\}$ of positive operators that satisfy Eq. 15. It can be further shown that any such set can be realized as a coarse-grained basic measurement on an extended system (i.e., they are feasible).

Finally, it is possible to give an axiomatic characterization of this development. The probability $p(a)$ of measurement result a is a functional of ρ, more completely written as $p(a) = p(a|\rho)$. The state ρ may arise as a mixture of two other states: $\rho = p_1\rho_1 + p_2\rho_2$. We infer that the probability of a for ρ should itself be a probabilistic combination:

$$p(a|\rho) = p_1 p(a|\rho_1) + p_2 p(a|\rho_2). \tag{16}$$

This motivates the axiom that $p(a|\rho)$ is a linear functional of ρ. Every such linear functional has the form of Eq. 14 for some operator E_a. Since the probability of a must be real and non-negative for any state ρ, the operator E_a is positive semidefinite; and since the probabilities must always sum to 1, Eq. 15 must also hold.

The developments of mixed states and generalized measurements in AQT thus illustrate the ideas of construction, feasibility and axiomatic characterization outlined in Sect. 1.2 above. We may regard this as a "complete" development within the theory. We are now ready to sketch the corresponding development in MQT.

3.2 Annihilators and Mixed States

Modal quantum theory is based on a vector space $\mathcal{V}$ of states and its dual $\mathcal{V}^*$ containing effects. It is convenient to summarize here a few definitions and elementary results about the subspaces of $\mathcal{V}$ and $\mathcal{V}^*$ [14].

Subspaces of $\mathcal{V}$ form a lattice under the "meet" and "join" operations $\wedge$ and $\vee$, where $\mathsf{A} \wedge \mathsf{B} = \mathsf{A} \cap \mathsf{B}$ and $\mathsf{A} \vee \mathsf{B} = \langle \mathsf{A} \cup \mathsf{B} \rangle$. (Here $\langle X \rangle$ is the linear span of a set X.) The minimal subspace in this lattice is $\langle 0 \rangle$, the 0-dimensional null subspace of $\mathcal{V}$.

Given a set A of vectors in $\mathcal{V}$, the annihilator A° is the set of dual vectors in $\mathcal{V}^*$ that "annihilate" vectors in A. That is,

$$A^\circ = \left\{ (e| \in \mathcal{V}^* : (e\,|a) = 0 \text{ for all } |a) \in A \right\}. \tag{17}$$

This can be easily turned around to define the annihilator of a subset of the dual space $\mathcal{V}^*$. In this case, the annihilator would be a subset of $\mathcal{V}$.

Within modal quantum theory, if A is a set of states, then A° includes all effects that are impossible for every state in A. Dually, if A is a set of effects, then A° includes all states for which every one of the effects in A is impossible. In spaces of finite dimension, the annihilator of a set has several straightforward properties.

- The annihilator A° is a subspace.
- If $A \subseteq B$, then $B^\circ \subseteq A^\circ$.
- The set A and its span $\langle A \rangle$ have the same annihilator: $A^\circ = \langle A \rangle^\circ$.
- A and B have the same annihilator if and only if $\langle A \rangle = \langle B \rangle$.
- $(A \cup B)^\circ = A^\circ \wedge B^\circ$.

Finally, we note that the annihilator of the annihilator is a subspace of the original space, and in fact $A^{\circ\circ} = \langle A \rangle$. If A is a subspace, then $\mathsf{A}^{\circ\circ} = \mathsf{A}$.

3.3 Mixed States in MQT

In modal quantum theory, a pure state of a system is represented by a state vector $|\psi)$ in $\mathcal{V}$. How should we represent a mixed state? We approach this question first by considering mixtures of pure states. Since MQT does not involve probabilities, a mixture is merely a set of possible state vectors: $M = \{|\psi_1\rangle, |\psi_2\rangle, \ldots\}$. A particular measurement outcome is possible provided it is possible for at least one of the states in the mixture. That is, effect $(e|$ is possible provided it is *not* in the annihilator M°.

Two different mixtures M_1 and M_2 will thus predict exactly the same possible effects if and only if $M_1^\circ = M_2^\circ$, so that $\langle M_1 \rangle = \langle M_2 \rangle$. We say that two such mixtures yield the same *mixed state*, and we identify that state with the *subspace* $\mathsf{M} \subseteq \mathcal{V}$ spanned by the elements of the mixture.[4]

We can also consider mixtures of two or more mixed states. If M_1 and M_2 are two subspaces of $\mathcal{V}$ associated with two states, then $\mathsf{M}_1 \vee \mathsf{M}_2$ is the subspace associated with a mixture of the two. Since any non-null subspace of $\mathcal{V}$ can be written as the span of a set of state vectors, it is an allowed mixed state.

As in AQT, mixed states in MQT can also arise when a composite system is in an entangled pure state. Suppose that the composite system RQ is in a state $|\Psi^{(\mathrm{RQ})}\rangle$, and consider the joint effect $(r^{(\mathrm{R})}, q^{(\mathrm{Q})}| = (r^{(\mathrm{R})}| \otimes (q^{(\mathrm{Q})}|$, which is possible provided $(r^{(\mathrm{R})}, q^{(\mathrm{Q})} |\Psi^{(\mathrm{RQ})}) \neq 0$. We can make sense of this by defining $\left|\psi_r^{(\mathrm{Q})}\right) = (r^{(\mathrm{R})} |\Psi^{(\mathrm{RQ})})$. That is, if we expand $|\Psi^{(\mathrm{RQ})})$ in a product basis $\{|a^{(\mathrm{R})}, b^{(\mathrm{Q})})\}$ we can write

$$\left|\psi_r^{(\mathrm{Q})}\right) = \left(r^{(\mathrm{R})}\right| \left(\sum_{a,b} \Psi_{ab} \left|a^{(\mathrm{R})}, b^{(\mathrm{Q})}\right) \right) = \sum_b \left(\sum_a \Psi_{ab} \left(r^{(\mathrm{R})} \middle| a^{(\mathrm{R})}\right) \right) \left|b^{(\mathrm{Q})}\right). \tag{18}$$

(The vector $\left|\psi_r^{(\mathrm{RQ})}\right)$ is independent of the choice of the $\{|a^{(\mathrm{R})}, b^{(\mathrm{Q})})\}$ basis chosen for this computation.) The joint effect $(r^{(\mathrm{R})}, q^{(\mathrm{Q})}|$ is possible provided $\left(q^{(\mathrm{Q})} \middle| \psi_r^{(\mathrm{Q})}\right) \neq 0$. Thus, it makes sense to interpret $\left|\psi_r^{(\mathrm{Q})}\right)$ as the conditional state of Q given the R-effect $(r^{(\mathrm{R})}|$ for the overall state $|\Psi^{(\mathrm{RQ})})$.[5]

To define the unconditional subsystem state for Q, we just define the Q-subspace of conditional states for all conceivable R-effects:

$$\mathsf{M}^{(\mathrm{Q})} = \left\{ \left(r^{(\mathrm{R})} \middle| \Psi^{(\mathrm{RQ})}\right) : \left(r^{(\mathrm{R})}\right| \in \mathcal{V}^{(\mathrm{R})*} \right\}. \tag{19}$$

An R-measurement is a basis $\{(k^{(\mathrm{R})}|\}$ of R-effects. Since these span $\mathcal{V}^{(\mathrm{R})*}$, we can see that

$$\mathsf{M}^{(\mathrm{Q})} = \left\langle \left\{ \left|\psi_k^{(\mathrm{Q})}\right) \right\} \right\rangle, \tag{20}$$

where $\left|\psi_k^{(\mathrm{Q})}\right) = (k^{(\mathrm{R})} |\Psi^{(\mathrm{RQ})})$.

This has the following important consequence. Whatever measurement is made on subsystem R, the mixture of conditional Q-states is exactly $\mathsf{M}^{(\mathrm{Q})}$ from Eq. 19.

---

[4]This clarifies a point about pure states, that $|\psi)$ and $c\,|\psi)$ are operationally equivalent for any $c \neq 0$. The two vectors span the same one-dimensional subspace of $\mathcal{V}$.

[5]Of course, if $\left|\psi_r^{(\mathrm{Q})}\right) = 0$, then it is not a legitimate state vector; but in this case, the R-effect $|r)$ is impossible. The formal inclusion of such phantom conditional states makes no difference to our analysis.

Thus, the choice of R-measurement by itself makes no observable difference in the observable properties of Q.

This analysis can be extended to the case where the composite system itself is in a mixed state. However, it is more convenient to delay that discussion until we have also generalized the concept of measurement in MQT.

3.4 Effects and Measurements

To generalize effects and measurements in modal quantum theory, it is instructive to begin with an axiomatic characterization.

In the abstract, an effect is simply a map E that assigns each subspace M of $\mathcal{V}$ an element of {possible, impossible}. But suppose $\mathsf{M} = \mathsf{M}_1 \vee \mathsf{M}_2$, the mixture of states M_1 and M_2. Then $E(\mathsf{M})$ should be possible if E is possible for either M_1 or M_2. We therefore adopt this requirement as an axiom for any reasonable effect map E. To make this a consistent rule, we will have to adopt the sensible convention that $E(\langle 0 \rangle)$ is always impossible.

Our axiom is equivalent to the statement that $E(\mathsf{M})$ is impossible if and only if both $E(\mathsf{M}_1)$ and $E(\mathsf{M}_2)$ are impossible. Therefore, for a given E we can consider the subspace $\mathsf{Z}_E \subseteq \mathcal{V}$

$$\mathsf{Z}_E = \bigvee \{\mathsf{M} : E(\mathsf{M}) \text{ is impossible}\}. \tag{21}$$

We see that $E(\mathsf{M})$ is impossible if and only if M is a subspace of Z_E. The map E can therefore be completely characterized by the annihilator subspace $\mathsf{Z}_E^\circ \subseteq \mathcal{V}^*$.

A generalized effect in MQT is thus defined to be a subspace $\mathsf{E} \subseteq \mathcal{V}^*$. For a generalized mixed state M, we say that $\mathsf{E}(\mathsf{M})$ is impossible if $\mathsf{M} \subseteq \mathsf{E}^\circ$ and possible otherwise. That is,

$$\mathsf{E}(\mathsf{M}) = \begin{cases} \text{possible} & (e\,|m) \neq 0 \quad \text{for some } (e| \in \mathsf{E}, |m) \in \mathsf{M} \\ \text{impossible} & (e\,|m) = 0 \quad \text{for all } (e| \in \mathsf{E}, |m) \in \mathsf{M} \end{cases} \tag{22}$$

This subspace characterization of generalized effects in MQT is exactly what we expect from a constructive approach. Beginning with an ordinary measurement given by the basis $\{(k|\}$ for $\mathcal{V}^*$, we can construct a coarse-grained effect E from a subset of the $(k|$ dual vectors. This effect is associated with a subspace of $\mathcal{V}^*$ (the one spanned by the relevant basis vectors). Both axiomatic and constructive approaches yield the same mathematical representation for generalized effects.

A generalized measurement will be a collection $\{\mathsf{E}_a\}$ of generalized effects (subspaces of $\mathcal{V}^*$) associated with the potential results of the measurement process. Some result must always occur, so we impose the requirement that, for any state M, at least one effect must be possible—that is, M cannot lie in the annihilator of all the generalized effects. Thus,

$$\bigcap_a \mathsf{E}_a^\circ = \langle 0 \rangle \,, \tag{23}$$

and so the generalized effects must satisfy

$$\bigvee_a \mathsf{E}_a = \mathcal{V}^*. \tag{24}$$

This is our "normalization" condition for a generalized measurement in MQT.

In actual quantum theory, a theorem due to Neumark [15] guarantees that any generalized positive operator measurement on a system Q can be realized by a basic measurement on a larger system. It is not difficult to confirm that an exactly analogous result holds in MQT. Thus, our generalized measurements are all feasible, in the sense discussed in Sect. 1.2. The constructive and axiomatic approaches coincide, and so our development is once again "complete".

3.5 Conditional States

In AQT, any pure entangled state $|\Psi\rangle$ of system RQ can be written in a special form, the *Schmidt decomposition* [13], as follows:

$$|\Psi\rangle = \sum_k \sqrt{\lambda_k}\, \left|k^{(\mathrm{R})}\right\rangle \otimes \left|k^{(\mathrm{Q})}\right\rangle \tag{25}$$

where $\{|k^{(\mathrm{R})}\rangle\}$ and $\{|k^{(\mathrm{Q})}\rangle\}$ are orthonormal bases for the two systems. It is easy to see that these are the eigenbases for the subsystem states $\rho^{(\mathrm{R})}$ and $\rho^{(\mathrm{Q})}$, and that the coefficients λ_k are the eigenvalues. The Schmidt decomposition is unique except for degeneracy among the λ_k values and some choices of relative phases among the two bases.

The MQT analogue of the Schmidt decomposition can be found as follows. Suppose $|\Psi^{(\mathrm{RQ})})$ is a joint state for RQ with subsystem mixed states $\mathsf{M}^{(\mathrm{R})}$ and $\mathsf{M}^{(\mathrm{Q})}$. Let $d^{(\mathrm{R})} = \dim \mathsf{M}^{(\mathrm{R})}$ and $d^{(\mathrm{Q})} = \dim \mathsf{M}^{(\mathrm{Q})}$. We introduce an R-basis $\{|k^{(\mathrm{R})})\}$ for which the first $d^{(\mathrm{R})}$ elements form a basis for $\mathsf{M}^{(\mathrm{R})}$. This means we can write

$$\left|\Psi^{(\mathrm{RQ})}\right) = \sum_k \left|k^{(\mathrm{R})}\right) \otimes \left|\psi_k^{(\mathrm{Q})}\right), \tag{26}$$

where the sum only requires the first $d^{(\mathrm{R})}$ terms. The state of Q is thus $\mathsf{M}^{(\mathrm{Q})} = \left\langle \{ \left|\psi_k^{(\mathrm{Q})}\right) \} \right\rangle$. Since $\mathsf{M}^{(\mathrm{Q})}$ is spanned by $d^{(\mathrm{R})}$ vectors, we conclude that $d^{(\mathrm{R})} \geq d^{(\mathrm{Q})}$. A symmetric argument establishes that $d^{(\mathrm{Q})} \geq d^{(\mathrm{R})}$, so the dimensions are equal. We therefore identify that $s = d^{(\mathrm{R})} = d^{(\mathrm{Q})}$ as the *Schmidt number* of the state $|\Psi^{(\mathrm{RQ})})$.

The s vectors $\{\left|\psi_k^{(\mathrm{Q})}\right)\}$ span a space of dimension s, so they must be linearly independent. We can thus construct a Q-basis $\{|k)^{(\mathrm{Q})}\}$ in which $|k^{(\mathrm{Q})}) = \left|\psi_k^{(\mathrm{Q})}\right)$ for $k \leq s$. Then

$$\left|\Psi^{(RQ)}\right) = \sum_{k} \left|k^{(R)}\right) \otimes \left|k^{(Q)}\right), \tag{27}$$

where the sum only includes s terms. This is a Schmidt decomposition for $|\Psi^{(RQ)})$. It is not unique, since we had the freedom to choose any basis for the mixed state (subspace) of one of the systems.

This has a useful consequence. Given a mixed state $\mathsf{M}^{(Q)}$ of Q, an entangled state $|\Psi^{(RQ)})$ of RQ that leads to this mixed state is called a *purification* of $\mathsf{M}^{(Q)}$ in RQ. Now consider two different purifications $\left|\Psi_1^{(RQ)}\right)$ and $\left|\Psi_2^{(RQ)}\right)$ for the same $\mathsf{M}^{(Q)}$. Fixing a common Q-basis $\{|k^{(Q)})\}$, we can write Schmidt decompositions for both purifications:

$$\left|\Psi_{1,2}^{(RQ)}\right) = \sum_{k} \left|k^{(R)}\right) \otimes \left|k_{1,2}^{(Q)}\right). \tag{28}$$

The two R-bases are connected by an invertible operator on $\mathcal{V}^{(R)}$: $T\left|k_1^{(R)}\right) = \left|k_2^{(R)}\right)$. Thus, two purifications of $\mathsf{M}^{(Q)}$ in RQ are connected via

$$\left|\Psi_2^{(RQ)}\right) = \left(T^{(R)} \otimes \mathbf{1}^{(Q)}\right)\left|\Psi_1^{(RQ)}\right); \tag{29}$$

that is, by an invertible transformation on R alone.

A theorem of Hughston et al. [16] (though earlier discussed by Schrödinger [17] and also by Jaynes [18]) relates mixtures to entangled states in AQT and characterizes those mixtures that can give rise to a given density operator ρ. An exactly analogous result holds in MQT. First, any mixture for $\mathsf{M}^{(Q)}$ can be realized as a mixture of conditional states arising from a purification of $\mathsf{M}^{(Q)}$. (The ability to realize different mixtures by a choice of measurement on the purifying subsystem is the MQT analogue of the familiar "steering" property of actual quantum theory [19].) Second, the elements of any two mixtures for a given mixed state are linear combinations of each other, with coefficients given by an invertible matrix of scalars. That is, if $\mathsf{M} = \langle\{|\psi_{k,1})\}\rangle = \langle\{|\psi_{k,2})\}\rangle$, then

$$\left|\psi_{l,2}\right) = \sum_{k} T_{lk} \left|\psi_{k,1}\right), \tag{30}$$

where the T_{lk} form an invertible matrix.

We now return to the question of conditional states and subsystem states for composite systems in MQT. How do these ideas work out in the context of generalized states and effects?

Suppose the composite system RQ is in the joint state $\mathsf{M}^{(RQ)}$, and the effect subspace $\mathsf{E}^{(R)}$ is part of some measurement on R. The conditional state of Q given this effect, which we can denote $\mathsf{M}_E^{(Q)}$, is defined via a map $\mathsf{C}(\cdot|\cdot)$:

$$\mathsf{M}_E^{(\mathrm{Q})} = \mathsf{C}\big(\mathsf{M}^{(\mathrm{RQ})}|\mathsf{E}^{(\mathrm{R})}\big) = \left\langle \left\{ \big(e^{(\mathrm{R})}\,\big|m^{(\mathrm{RQ})}\big) : \big(e^{(\mathrm{R})}\big| \in \mathsf{E}^{(\mathrm{R})}, \big|m^{(\mathrm{RQ})}\big) \in \mathsf{M}^{(\mathrm{RQ})} \right\} \right\rangle \tag{31}$$

If $\mathsf{M}_E^{(\mathrm{Q})} = \langle 0 \rangle$, then the effect $\mathsf{E}^{(\mathrm{R})}$ is impossible.

The map $\mathsf{C}(\cdot|\cdot)$ respects mixtures in both the joint state and the effect. That is,

$$\mathsf{C}\big(\mathsf{M}_1^{(\mathrm{RQ})} \vee \mathsf{M}_2^{(\mathrm{RQ})}|\mathsf{E}^{(\mathrm{R})}\big) = \mathsf{C}\big(\mathsf{M}_1^{(\mathrm{RQ})}|\mathsf{E}^{(\mathrm{R})}\big) \vee \mathsf{C}\big(\mathsf{M}_2^{(\mathrm{RQ})}|\mathsf{E}^{(\mathrm{R})}\big) \tag{32}$$

$$\mathsf{C}\big(\mathsf{M}^{(\mathrm{RQ})}|\mathsf{E}_1^{(\mathrm{R})} \vee \mathsf{E}_2^{(\mathrm{R})}\big) = \mathsf{C}\big(\mathsf{M}^{(\mathrm{RQ})}|\mathsf{E}_1^{(\mathrm{R})}\big) \vee \mathsf{C}\big(\mathsf{M}^{(\mathrm{RQ})}|\mathsf{E}_2^{(\mathrm{R})}\big) \tag{33}$$

Equation 31 generalizes the expression in Eq. 18 for conditional states of a composite system. It can also be used to define the unconditional subsystem state $\mathsf{M}^{(\mathrm{Q})}$, which we denote like so:

$$\mathsf{M}^{(\mathrm{Q})} = \mathsf{R}_{(\mathrm{R})}\big(\mathsf{M}^{(\mathrm{RQ})}\big) = \mathsf{C}\big(\mathsf{M}^{(\mathrm{RQ})}|\mathcal{V}^{(\mathrm{R})*}\big). \tag{34}$$

Since we take the linear span in Eq. 31, we only need to consider spanning sets for $\mathsf{M}^{(\mathrm{RQ})}$ and $\mathsf{E}^{(\mathrm{R})}$. That is, if $\mathsf{M}^{(\mathrm{RQ})} = \langle\{|\mu^{(\mathrm{RQ})})\}\rangle$ and $\mathsf{E}^{(\mathrm{R})} = \langle\{(\eta^{(\mathrm{R})}|\}\rangle$, then $\mathsf{M}_E^{(\mathrm{Q})} = \langle\{(\eta^{(\mathrm{R})}\,|\mu^{(\mathrm{RQ})})\}\rangle$. This is useful in calculations.

4 Open System Evolution

4.1 Type M Maps

According to the evolution postulates given in Table 1, the time evolution of state vectors in either actual or modal quantum theory can be described by a linear operator—unitary in the case of AQT ($|\psi\rangle \to U\,|\psi\rangle$), invertible in the case of MQT ($|\phi) \to T\,|\phi)$). In either case it is straightforward to generalize this to mixed states. The density operator in AQT evolves via $\rho \to U\rho U^\dagger$, and in MQT a subspace evolves according to

$$\mathsf{M} \to T\mathsf{M} = \{T\,|\phi) : |\phi) \in \mathsf{M}\}. \tag{35}$$

These postulates apply when the system in question is isolated. When a system is subject to noise or interaction with its environment, a more general description of time evolution is needed. In this section we trace this development.

Generalized operations can be of two types. *Conditional* operations do not take place with certainty but only happen when some objective condition (e.g., a measurement result) is observed. *Unconditional* operations are those that take place with certainty.

In actual quantum theory, a general operation is a map on density operators: $\rho \to \mathcal{E}(\rho)$. For an input density operator ρ, the output $\mathcal{E}(\rho)$ of an unconditional operation must also be a density operator—that is, a positive semidefinite operator of trace 1. For conditional operations, the output is subnormalized so that $p = \mathrm{Tr}\,\mathcal{E}(\rho)$ is the probability that the operation occurs. The map $\mathcal{E}$ must respect mixtures; that is,

$$\mathcal{E}\,(p_1\rho_1 + p_2\rho_2) = p_1\mathcal{E}(\rho_1) + p_2\mathcal{E}(\rho_2). \tag{36}$$

Thus $\mathcal{E}$ is a linear map on density operators. This is a powerful condition, since it allows us to extend $\mathcal{E}$ to a linear map on the space of all operators—that is, to a *superoperator*.

In MQT, a general operation $\mathcal{E}$ on a system is a map on the subspaces of $\mathcal{V}$: $\mathsf{M} \rightarrow \mathsf{M}' = \mathcal{E}\,(\mathsf{M})$. For an unconditional operation, the output of the map must always be a legitimate state, a non-null subspace. This means that $\mathcal{E}(\mathsf{M}) \neq \langle 0 \rangle$ for $\mathsf{M} \neq \langle 0 \rangle$. This requirement is relaxed for conditional operations. In that case, $\mathcal{E}(\mathsf{M}) = \langle 0 \rangle$ merely signifies that the condition of the operation cannot arise for the input state M.

General operations in MQT must also respect mixtures, meaning that

$$\mathcal{E}\,(\mathsf{M}_1 \vee \mathsf{M}_2) = \mathcal{E}(\mathsf{M}_1) \vee \mathcal{E}(\mathsf{M}_2). \tag{37}$$

(To maintain consistency, we adopt the convention that $\mathcal{E}(\langle 0 \rangle) = \langle 0 \rangle$.) The map $\mathcal{E}$ is not simply a linear superoperator, so we cannot easily extend it to inputs other than subspaces. However, Eq. 37 is still an important requirement. We will call subspace maps that respect mixtures in this way *Type M* maps.

Throughout the rest of this section, we will only consider unconditional operations in both AQT and MQT. The generalization to conditional operations is not difficult and is left as an exercise.

4.2 Constructive Approach

Consider a situation in which the system of interest S interacts with an external "environment" system E, where E is initially in some fixed state. In AQT, the dynamics of just such an *open* system S is described by a map $\mathcal{E}$ on density operators:

$$\rho \rightarrow \mathcal{E}(\rho) = \mathrm{Tr}_{(\mathrm{E})} U\,(\rho \otimes |0\rangle\langle 0|)\, U^{\dagger} \tag{38}$$

where $|0\rangle$ is the initial standard state of E and U is a unitary operator on the composite system SE.

By analogy, in MQT the evolution of an open system that interacts with an environment (initial state $\mathsf{M}_0^{(\mathrm{E})}$) can be described by the map $\mathcal{E}^{(\mathrm{S})}$ such that

$$\mathcal{E}^{(\mathrm{S})}(\mathsf{M}^{(\mathrm{S})}) = \mathsf{R}_{(\mathrm{E})}(T^{(\mathrm{SE})}(\mathsf{M}^{(\mathrm{S})} \otimes \mathsf{M}_0{}^{(\mathrm{E})})) \tag{39}$$

where $\mathsf{R}_{(\mathrm{E})}$ is the subsystem state reduction defined by Eq. 34. Without loss of generality, we may suppose that $\mathsf{M}_0^{(\mathrm{E})}$ is one-dimensional, since any mixed environment state can have a purification in a larger environment. We refer to these maps defined by invertible linear evolution on a larger system as *Type I* maps.

In AQT there is a way of representing the map $\mathcal{E}$ without the explicit involvement of the environment E in Eq. 38. Consider a particular basis $\{|e_k\rangle\}$ for the Hilbert space of the environment E. For each k define the operator A_k by

$$A_k |\phi\rangle = \langle e_k| U |\phi, 0\rangle \tag{40}$$

for any $|\phi\rangle$ in $\mathcal{H}^{(S)}$. Even though we have used the environment E and the interaction U in this definition, the A_k operators act on $\mathcal{H}^{(S)}$ alone. We may use the $\{|e_k\rangle\}$ basis to do the partial trace in Eq. 38. Given a pure state input $|\phi\rangle$,

$$\begin{aligned} \mathcal{E}(|\phi\rangle\langle\phi|) &= \sum_k \langle e_k| U (|\phi\rangle\langle\phi| \otimes |0\rangle\langle 0|) U^\dagger |e_k\rangle \\ &= \sum_k A_k |\phi\rangle\langle\phi| A_k^\dagger. \end{aligned} \tag{41}$$

And in general,

$$\mathcal{E}(\rho) = \sum_k A_k \rho A_k^\dagger. \tag{42}$$

This is called an *operator-sum representation* or *Kraus representation* of the map $\mathcal{E}$, and the operators A_k are called *Kraus operators* [13].

For an unconditional operation, the Kraus operators satisfy a normalization condition. If $\rho' = \mathcal{E}(|\phi\rangle\langle\phi|)$ for a normalized pure state $|\phi\rangle$, then

$$\begin{aligned} \mathrm{Tr}\,\rho' &= \sum_k \langle\phi| A_k^\dagger A_k |\phi\rangle \\ &= \langle\phi| \left(\sum_k A_k^\dagger A_k \right) |\phi\rangle . \end{aligned} \tag{43}$$

Since $\mathrm{Tr}\,\rho' = 1$ for every normalized input state $|\phi\rangle$,

$$\sum_k A_k^\dagger A_k = \mathbf{1}. \tag{44}$$

We can describe the map $\mathcal{E}$ entirely in terms of Kraus operators. It is often much more convenient to describe $\mathcal{E}$ in this way without considering the actual environment E, which might be very large and complex. An operator-sum representation is a compact description of how E affects the evolution of the state of S.

We can make the analogous construction in MQT. The system S interacts with an environment E, initially in the state $\mathsf{M}_0^{(E)} = \langle |0^{(E)}\rangle\rangle$, via the invertible operator $T^{(SE)}$. Let $\{(e_k^{(E)}|\}$ be a basis for $\mathcal{V}^{(E)*}$ and define the operator A_k on $\mathcal{V}^{(S)}$ by

$$A_k \left|\phi^{(S)}\right) = \left(e_k^{(E)}\right| T^{(SE)} \left|\phi^{(S)}, 0^{(E)}\right). \tag{45}$$

Given an initial S-state $\mathsf{M}^{(S)}$ spanned by state vectors $|m^{(S)})$, we have

$$\begin{aligned}
\mathcal{E}^{(S)}\left(\mathsf{M}^{(S)}\right) &= \mathsf{R}_{(E)}(T^{(SE)}(\mathsf{M}^{(S)} \otimes \mathsf{M}_0{}^{(E)})) \\
&= \left\langle \left(e_k^{(E)}\right| T^{(SE)} \left|m^{(S)}, 0^{(E)}\right) \right\rangle \\
&= \left\langle A_k \left|m^{(S)}\right) \right\rangle \\
\mathcal{E}^{(S)}\left(\mathsf{M}^{(S)}\right) &= \bigvee_k A_k \mathsf{M}^{(E)}.
\end{aligned} \tag{46}$$

The output of $\mathcal{E}^{(S)}$ acting on $\mathsf{M}^{(S)}$ is a mixture of images of $\mathsf{M}^{(S)}$ under the linear operators A_k. Equation 46 is the MQT analogue of the Kraus representation in Eq. 42. If a map $\mathcal{E}^{(S)}$ has a representation of this type, we say that it is a *Type L* map. (Note that we have shown that all Type I maps are also Type L.)

The individual operators A_k are not necessarily invertible. However, if $\mathcal{E}$ represents an unconditional operation on the MQT system S, then any non-null subspace M must evolve to a non-null subspace $\mathcal{E}(\mathsf{M})$. Thus, the A_k operators must satisfy

$$\bigcap_k \ker A_k = \langle 0 \rangle, \tag{47}$$

the MQT analogue of the normalization condition in Eq. 44.

4.3 Axiomatic Characterization

In AQT, every "physically reasonable" dynamical evolution map for an open system has both a unitary and a Kraus representation. Similarly, every "physically reasonable" evolution map in MQT is both Type I and Type L. To make sense of this claim, we must explain what is meant by a "physically reasonable" map.

Let us begin by reviewing the argument in AQT. A physically reasonable map $\mathcal{E}$ must be linear in the input density operator, making it a superoperator (an element of $\mathcal{B}(\mathcal{B}(\mathcal{H}))$). Furthermore, the output of $\mathcal{E}$ must be a valid density operator for any valid input state. This immediately implies two properties:

- $\mathcal{E}$ must be a *positive* map, in the sense that it maps positive operators to positive operators.
- $\mathcal{E}$ must be a *trace-preserving* map, so that $\mathrm{Tr}\,\mathcal{E}(A) = \mathrm{Tr}\,A$ for all operators A.

(Both properties are easy to prove in general, since any operator can be written as a linear combination of density operators.)

These two conditions are not sufficient to characterize "physically reasonable" linear maps, because there are positive, trace-preserving maps that cannot correspond to the time evolution of an quantum system. The easiest example arises for a simple qubit system. The Pauli operators X, Y and Z, together with the identity $\mathbf{1}$, form an operator basis. The following map $\mathcal{T}$ is positive:

$$\begin{array}{ll} \mathcal{T}(\mathbf{1}) = \mathbf{1} & \mathcal{T}(X) = X \\ \mathcal{T}(Y) = Y & \mathcal{T}(Z) = -Z \end{array}. \tag{48}$$

However, $\mathcal{T}$ does not describe the possible evolution of the state of an open qubit system. The reason is that the qubit is not necessarily alone in the universe. We may consider a second independent qubit whose state evolves according to the identity map $\mathcal{I}$. The composite system evolves according to the map $\mathcal{T} \otimes \mathcal{I}$. However, this extended map is *not* positive: entangled input states may map to operators having some negative eigenvalues. (See [13] for details.)

We need stronger property to characterize "physically reasonable" evolution maps for an open quantum system. The structure of the example just described provides a clue to what this stronger property looks like.

The map $\mathcal{E}$ for a system is said to be *completely positive* if $\mathcal{E} \otimes \mathcal{I}$ is positive whenever we append an independent quantum system. This means that, for any initial pure state $|\Psi\rangle$ of the composite system, the operator $\mathcal{E} \otimes \mathcal{I}(|\Psi\rangle\langle\Psi|)$ is positive. Since any system may be part of a composite system, we require that every "physically reasonable" map describing open system evolution must be linear, trace-preserving and completely positive. The importance of this requirement is shown by the following theorem.

Representation theorem for generalized dynamics in AQT. Let Q be a quantum system and $\mathcal{E}$ be a map on Q-operators. The following conditions are equivalent.

(a) $\mathcal{E}$ is a linear, trace-preserving, completely positive map.

(b) $\mathcal{E}$ has a "unitary representation". That is, we can introduce an environment system E, an initial environment state $|0\rangle$ and a joint unitary evolution U on QE so that

$$\mathcal{E}(G) = \mathrm{Tr}_{(E)} U \left(G \otimes |0\rangle\langle 0| \right) U^{\dagger}. \tag{49}$$

(c) $\mathcal{E}$ has a Kraus representation. That is, we can find operators A_k such that

$$\mathcal{E}(G) = \sum_k A_k G A_k^{\dagger}. \tag{50}$$

The Kraus operators satisfy the normalization condition of Eq. 44.

Once again, the constructive approach (unitary dynamics on a larger system) and the axiomatic characterization (linear, trace-preserving, completely positive maps) lead us to the same generalized unconditional operations in AQT. The representation theorem is a powerful and fundamental result in AQT. Details of its proof are given in Appendix D of [13].

So much for AQT. Can we find an analogous axiomatic characterization for "reasonable" state evolution in MQT? This is a tricky question, and not just because we lack access to an actual MQT world. A proposed evolution map $\mathcal{E}$ will map subspaces to subspaces, rather than operators to operators. Furthermore, the underlying field $\mathcal{F}$ may not include the notion of "positive elements". In a field of non-zero characteristic, any element added to itself sufficiently many times yields 0. Thus, in MQT there may be no analogue to the notion of a "positive map".

Nevertheless, there is a close analogue to the property of complete positivity that does make sense in MQT. As in AQT, this condition governs how a map $\mathcal{E}$ extends to one that applies to a larger composite system. Briefly, we require that an extension exists that commutes with the conditioning operation described in Sect. 3.5. Here is a more precise definition: We say that the subspace map $\mathcal{E}^{(S)}$ for modal quantum system S is *Type E* if for any other system R there exists a joint subspace map $\mathcal{E}^{(RS)}$ such that

$$\mathcal{E}^{(S)}\left(\mathsf{C}(\mathsf{M}^{(RS)}|\mathsf{E}^{(R)})\right) = \mathsf{C}\left(\mathcal{E}^{(RS)}(\mathsf{M}^{(RS)})|\mathsf{E}^{(R)}\right) \tag{51}$$

for any RQ-state $\mathsf{M}^{(RQ)}$ and R-effect $\mathsf{E}^{(R)}$. In other words, for a Type E map $\mathcal{E}^{(S)}$ and a system R, we can find a map $\mathcal{E}^{(RS)}$ so that the following diagram always commutes:

$$\begin{array}{ccc} \mathsf{M}^{(RS)} & \xrightarrow{\mathsf{C}(\cdot|\mathsf{E}^{(R)})} & \mathsf{M}^{(S)} \\ \mathcal{E}^{(RS)}\big\downarrow & & \big\downarrow\mathcal{E}^{(S)} \\ \mathcal{E}^{(RS)}(\mathsf{M}^{(RS)}) & \xrightarrow{\mathsf{C}(\cdot|\mathsf{E}^{(R)})} & \mathcal{E}^{(S)}(\mathsf{M}^{(S)}) \end{array} \tag{52}$$

We will require that any "reasonable" state evolution in MQT must be Type E. What is the motivation for such a condition? Suppose the state of system S evolves according to $\mathcal{E}^{(S)}$. It is always reasonable to suppose that another system R exists in the MQT universe. We imagine that R and S can be "independent" of one another—they might, for instance, be very far apart in space. The joint system RS is initially in the state $\mathsf{M}^{(RS)}$. The evolution of RS is described by some joint map $\mathcal{E}^{(RS)}$ that reflects the independence of the subsystems. We now imagine two experimental procedures.

- A measurement is made on system R, with the objective result corresponding to effect $\mathsf{E}^{(R)}$. Under this condition, S is in the state $\mathsf{C}(\mathsf{M}^{(RS)}|\mathsf{E}^{(R)})$. Now the dynamical evolution acts, so that the final state of S is $\mathcal{E}^{(S)}(\mathsf{C}(\mathsf{M}^{(RS)}|\mathsf{E}^{(R)}))$.
- The dynamical evolution acts, leading to the joint final state $\mathcal{E}^{(RS)}(\mathsf{M}^{(RS)})$. Now the measurement is made on system R, with the objective result corresponding to the effect $\mathsf{E}^{(R)}$ The conditional state of S is $\mathsf{C}(\mathcal{E}^{(RS)}(\mathsf{M}^{(RS)})|\mathsf{E}^{(R)})$.

Intuitively, if R and S are completely independent (and perhaps widely separated), the final S state should be independent of whether the measurement on R is performed before or after the evolution of S. This is exactly the requirement for a Type E map.

4.4 Representation Theorem for MQT

We now prove a result analogous to the representation theorem for AQT. Formally, we will show that

Representation theorem for generalized dynamics in MQT. Let S be a modal quantum system and $\mathcal{E}$ be a Type M map on subspaces for F. Then the following conditions are equivalent.

(a) $\mathcal{E}$ is Type E; that is, it can be extended in a way that commutes with the conditional operation.
(b) $\mathcal{E}$ is Type I; that is, it can be expressed as invertible linear evolution on a larger system.
(c) $\mathcal{E}$ is Type L; that is, it can be expressed as a mixture of linear maps satisfying Eq. 47.

As with the AQT result, this theorem is a strong characterization of the "reasonable" evolution maps in modal quantum theory. We have argued that all reasonable maps are Type E, and any Type I map is realizable by familiar linear evolution. Constructive and axiomatic approaches coincide.

We will prove the equivalence of the three conditions by establishing the cyclic implication L $\Rightarrow$ I $\Rightarrow$ E $\Rightarrow$ L. The first two implications are straightforward; the last requires a bit more work. Throughout, we take $\mathcal{E}^{(S)}$ to be a Type M map on S.

L $\Rightarrow$ I: First, assume that $\mathcal{E}^{(S)}$ is Type L. This means that there is a set of linear operators $\{A_k\}$ that yield $\mathcal{E}^{(S)}$ according to Eq. 46, and that these operators satisfy the normalization requirement (Eq. 47). Now we introduce an environment system E whose dimension is equal to the number of A_k operators. We fix an initial E-state $|0^{(E)})$ and a basis $\{|k^{(E)})\}$.

The set of SE states of the form $|\phi^{(S)}, 0^{(E)})$ constitute a subspace. Define the operator $T^{(SE)}$ on this subspace by

$$T^{(SE)} \left|\phi^{(S)}, 0^{(E)}\right) = \sum_k A_k \left|\phi^{(E)}\right) \otimes \left|k^{(E)}\right). \tag{53}$$

Because of the normalization requirement on the A_k operators, the right-hand side is never zero. Thus, the operator $T^{(SE)}$ is one-to-one on the subspace, and so we may extend it to an invertible operator on the whole of $\mathcal{V}^{(SE)}$. Given a mixed state $\mathsf{M}^{(S)}$, it is straightforward to show that

$$\mathsf{R}_{(E)} \left(T^{(SE)}(\mathsf{M}^{(S)} \otimes \mathsf{M}_0^{(E)})\right) = \left\langle\{A_k \mathsf{M}^{(S)}\}\right\rangle = \mathcal{E}^{(S)}(\mathsf{M}^{(S)}), \tag{54}$$

where $\mathsf{M}_0^{(E)} = \langle |0^{(E)})\rangle$. The map $\mathcal{E}^{(S)}$ is therefore Type I.

I $\Rightarrow$ E: Now assume that $\mathcal{E}^{(S)}$ is Type I, so that it is given by invertible evolution on the extended system SE as above. For any additional system R we define the map

$$\mathcal{E}^{(RS)}(\mathsf{M}^{(RS)}) = \mathsf{R}_{(E)} \left[(1^{(R)} \otimes T^{(SE)})(\mathsf{M}^{(RS)} \otimes \mathsf{M}_0^{(E)})\right]. \tag{55}$$

As we have already remarked, the reduction operation $\mathsf{R}_{(E)}$ is an "unconditional" conditioning operation. A direct application of the definition in Eq. 31 shows that iterated reduction with respect to independent subsystems (in our case, R and E) can be done in any order. This establishes that every Type I map must also be Type E.

E ⇒ L: It only remains to prove that Type E implies Type L. Let $\mathcal{E}^{(S)}$ be a Type E map for a modal quantum system S, which is represented by a vector space of finite dimension $\dim \mathcal{V}^{(S)} = d$. We append an identical quantum system R and consider the maximally entangled state

$$\left|\Phi^{(RS)}\right) = \sum_k \left|k^{(R)}, k^{(S)}\right). \tag{56}$$

Any initial state $|\psi^{(S)})$ of S could arise in the following way. The system RS is initially in the entangled state $|\Phi^{(RS)})$, and then a measurement is performed in R. The resulting state of S, conditional on the particular measurement outcome for R, happens to be $|\psi^{(S)})$.

We can do this more explicitly. Given $|\psi^{(S)}) = \sum_k g_k \, |k^{(S)})$, we can construct the R-effect

$$\left(\tilde{\psi}^{(R)}\right| = \sum_k g_k \left(k^{(R)}\right|. \tag{57}$$

If $\left(\tilde{\psi}^{(R)}\right|$ corresponds to one outcome of a basic measurement on R, then the associated conditional state of Q is $|\psi^{(Q)})$.

This reasoning can be generalized to mixed states. First, note that $\mathcal{M}^{(RS)} = \langle |\Phi^{(RS)})\rangle$ is the one-dimensional mixed state that corresponds to the "fully entangled" state $|\Phi^{(RS)})$. Now consider a general mixed Q-state

$$\mathsf{G}^{(Q)} = \left\langle \left\{ \left|g^{(Q)}\right) = \sum_k g_k \left|k^{(S)}\right) : (g_k) \in G \right\} \right\rangle, \tag{58}$$

where G is a set of d-tuples (g_k) of elements of $\mathcal{F}$. Now define the R-effect

$$\Gamma^{(R)} = \left\langle \left\{ \left(g^{(R)}\right| = \sum_k g_k \left(k^{(R)}\right| : (g_k) \in G \right\} \right\rangle. \tag{59}$$

Then

$$\mathsf{G}^{(Q)} = \mathsf{C}\left(\mathcal{M}^{(RS)} | \Gamma^{(R)}\right), \tag{60}$$

since $|g^{(Q)}) = (g^{(R)}\,|\Phi^{(RS)})$.

We use this machinery to construct a Type L representation the Type E map $\mathcal{E}^{(S)}$. Let $\mathcal{E}^{(RS)}$ be an extension of $\mathcal{E}^{(S)}$. Let $\left\{\left|m_\lambda^{(RS)}\right)\right\}$ be a set of RS states such that $\mathcal{E}^{(RS)}(\mathcal{M}^{(RS)}) = \left\langle\left\{\left|m_\lambda^{(RS)}\right)\right\}\right\rangle$ (where λ runs over some index set Λ). Given states $|g^{(S)})$ and associated effects $(g^{(R)}|$ as described above, we define A_λ as follows:

$$A_\lambda^{(S)} \left|g^{(S)}\right) = \left(g^{(R)} \left|m_\lambda^{(RS)}\right)\right.. \tag{61}$$

We now have

$$\begin{aligned}
\mathcal{E}^{(S)}(\mathsf{G}^{(S)}) &= \mathcal{E}^{(S)}\left(\mathsf{C}\left(\mathcal{M}^{(RS)}|\Gamma^{(R)}\right)\right) \\
&= \mathsf{C}\left(\mathcal{E}^{(RS)}(\mathcal{M}^{(RS)})|\Gamma^{(R)}\right) \\
&= \mathsf{C}\left(\langle\{m_\lambda^{(RS)}\}\rangle\,|\Gamma^{(R)}\right) \\
&= \langle\{(g^{(R)}\,|m_\lambda^{(RS)})\}\rangle \\
&= \langle\{A_\lambda{}^{(S)}\,|g^{(S)})\}\rangle \\
\mathcal{E}^{(S)}(\mathsf{G}^{(S)}) &= \bigvee_\lambda A_\lambda{}^{(S)}(\mathsf{G}^{(S)}).
\end{aligned} \tag{62}$$

Thus, any Type E map is also Type L.

We see that Type E maps in MQT play a role parallel to CP maps in actual quantum theory. In fact, the connection is stronger than this. We can adapt the definition of Type E maps to AQT: A linear map $\mathcal{E}^{(Q)}$ on density operators for Q is Type E provided there exists an extended map $\mathcal{E}^{(RQ)}$ on density operators of RQ that commutes with the formation of conditional Q states from effects on R. It is not hard to show that this condition is *equivalent* to complete positivity of $\mathcal{E}^{(Q)}$ and thus implies the existence of unitary ("Type L") and Kraus ("Type L") representations in AQT.

With the (now finished) proof of the representation theorem, we have completed our development of modal quantum theory to include generalized states, measurements, and dynamical evolution. This development has included both constructive approaches (based on the basic axioms in Table 1) and axiomatic characterizations. The two routes lead to the same place, a fact that gives us confidence that we have arrived at a "complete" development of the theory. Further generalization will necessarily involve an extension of MQT to a more general type of theory.

5 Generalized Modal Theories

5.1 Possibility Tables for Two Systems

In the study of the conceptual foundations of actual quantum theory, it is useful to consider AQT as an example of a more general class of probabilistic theories. Consider a system comprising two subsystems, designated 1 and 2. We can choose to make any of several possible measurements on each system and obtain various joint results with various probabilities. That is, our theory allows us to compute probabilities of the form $p(x, y|X^{(1)}, Y^{(2)})$, the probability of the joint outcome (x, y) given the choice of measurement $X^{(1)}$ on system 1 and $Y^{(2)}$ on system 2.

The state of the composite system can thus be described by a collection of joint probability distributions. These may be organized as a table. The rows and columns of the table correspond to the possible measurements on systems 1 and 2, respectively, like so:

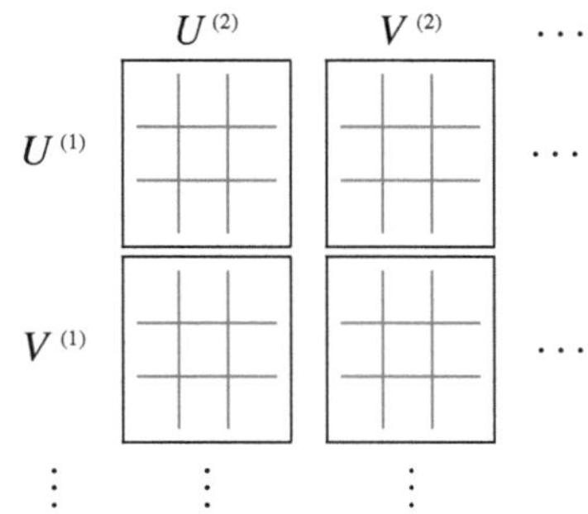

(63)

The theory is characterized by the set of possible states—that is, the possible collections of distributions in the table.[6]

All of the tables we consider satisfy the *no-signalling principle* which can be stated as follows [9]. For any choice of measurements $A^{(1)}$, $B^{(1)}$, $C^{(2)}$ and $D^{(2)}$,

$$\begin{aligned} p\,(a|A^{(1)}) &= \textstyle\sum_c p\,(a, c|A^{(1)}, C^{(2)}) = \sum_d p\,(a, d|A^{(1)}, D^{(2)}) \\ p\,(c|C^{(1)}) &= \textstyle\sum_a p\,(a, c|A^{(1)}, C^{(2)}) = \sum_b p\,(b, c|B^{(1)}, C^{(2)})\,. \end{aligned} \tag{64}$$

That is, the choice of system 2 measurement does not affect the overall probability of a system 1 outcome, and *vice versa.* In the table of joint distributions in Eq. 63, this means that any two distributions in the same row are connected, in that their sub-rows sum to the same values. A similar connection exists within each column as well.

Collections of distributions arising from a composite system in AQT satisfy the no-signalling principle. However, there are tables satisfying this principle that could not arise in AQT. These include examples of "states" of a composite system that are "more entangled" than quantum mechanics allows [20].

We can adapt this approach to construct generalized modal theories that extend MQT. The state of a composite system in a general modal theory would be a table similar to the one in Eq. 63, except that the individual "distributions" only indicate which joint outcomes are possible. We use the symbol X to denote a possible outcome, and a blank space for an impossible outcome. For instance, the table for a pair of mobits in $\mathbb{Z}_2$-MQT has just three rows and columns, corresponding to the three possible basic measurements for each mobit. For the entangled modal state $|S) = |0, 1) - |1, 0)$, we have

[6] AQT also allows "entangled" measurements on composite systems, measurements which cannot be reduced to separate measurements on the subsystems. Probabilities for non-entangled measurements, however, are sufficient to characterize the joint state of the system, so we restrict our attention to those.

$$\mathcal{S} = \left[\begin{array}{cccc} & X^{(2)} & Y^{(2)} & Z^{(2)} \\ X^{(1)} & \boxed{\begin{array}{c|c} & \mathsf{X} \\ \hline \mathsf{X} & \end{array}} & \boxed{\begin{array}{c|c} \mathsf{X} & \\ \hline \mathsf{X} & \mathsf{X} \end{array}} & \boxed{\begin{array}{c|c} \mathsf{X} & \mathsf{X} \\ \hline & \mathsf{X} \end{array}} \\ Y^{(1)} & \boxed{\begin{array}{c|c} \mathsf{X} & \mathsf{X} \\ \hline & \mathsf{X} \end{array}} & \boxed{\begin{array}{c|c} & \mathsf{X} \\ \hline \mathsf{X} & \end{array}} & \boxed{\begin{array}{c|c} \mathsf{X} & \\ \hline \mathsf{X} & \mathsf{X} \end{array}} \\ Z^{(1)} & \boxed{\begin{array}{c|c} \mathsf{X} & \\ \hline \mathsf{X} & \mathsf{X} \end{array}} & \boxed{\begin{array}{c|c} \mathsf{X} & \mathsf{X} \\ \hline & \mathsf{X} \end{array}} & \boxed{\begin{array}{c|c} & \mathsf{X} \\ \hline \mathsf{X} & \end{array}} \end{array} \right] \tag{65}$$

This table, like all tables arising from MQT systems, satisfies a modal version of the no-signalling principle. The question of whether a particular subsystem result is possible does not depend on what measurement is chosen for the other subsystem. Thus, if an X occurs in the table, at least one X must occur in the corresponding sub-rows to the right and left, and in the corresponding sub-columns above and below. We will only consider general modal theories satisfying the modal no-signalling principle.

In a general probabilistic theory, we can take the convex combination of two states and derive a "mixed" state. The set of allowed states is therefore a convex set. In a general modal theory, the mixture of two tables $\mathcal{R}$ and $\mathcal{T}$ is simply $\mathcal{R} \vee \mathcal{T}$, the table in which a joint outcome is possible if it is possible in either $\mathcal{R}$ or $\mathcal{T}$. (This corresponds to the usual mixture of states in MQT.) Finally, there is a natural partial ordering on states in a general modal theory. We say that $\mathcal{R} \preceq \mathcal{T}$ provided every possible result in $\mathcal{R}$ is also possible in $\mathcal{T}$.

5.2 Popescu-Rohrlich Boxes

Every generalized probabilistic table can be converted into a generalized modal table by replacing non-zero probabilities with X and zero probabilities with blanks. A table obeying the probabilistic no-signalling principle automatically yields one that obeys the modal version.

We use this idea to create a modal version of an important example of a generalized probabilistic model, the "nonlocal box" proposed by Popescu and Rohrlich [20]. This *PR box* satisfies the no-signalling principle but is in a sense more entangled than allowed by AQT. The modal version $\mathcal{P}$ looks like this:

$$\mathcal{P} = \left[\begin{array}{ccc} & C^{(2)} & D^{(2)} \\ A^{(1)} & \boxed{\begin{array}{c|c} \mathsf{X} & \\ \hline & \mathsf{X} \end{array}} & \boxed{\begin{array}{c|c} \mathsf{X} & \\ \hline & \mathsf{X} \end{array}} \\ B^{(1)} & \boxed{\begin{array}{c|c} \mathsf{X} & \\ \hline & \mathsf{X} \end{array}} & \boxed{\begin{array}{c|c} & \mathsf{X} \\ \hline \mathsf{X} & \end{array}} \end{array} \right] \tag{66}$$

We can summarize this pattern of possibilities in a simple way: For the measurement combinations (A, C), (A, D) and (B, C) the joint measurement results must always agree, but for (B, D) they always disagree. (The probabilistic PR box replaces X with probability 1/2 in Eq. 66.)

The PR box table $\mathcal{P}$ is minimal. That is, if any table $\mathcal{R}$ of similar dimensions satisfies the no-signalling principle, and if $\mathcal{R} \preceq \mathcal{P}$, then $\mathcal{R} = \mathcal{P}$.

Could the PR box table $\mathcal{P}$ in Eq. 66 arise from a composite system described by MQT? In fact, it cannot. Since $\mathcal{P}$ is minimal, it suffices to consider only pure states for system 12 together with measurements having non-overlapping effects. That is, the measurement $A^{(1)}$ consists of two effects (subspaces of $\mathcal{V}^{(1)*}$) A_+ and A_- such that $\mathsf{A}_+ \cap \mathsf{A}_- = \langle 0 \rangle$, and so on.

Suppose $|\Psi)$ is a modal quantum state that leads to the PR box table $\mathcal{P}$ in Eq. 66. As shown in the Appendix, the upper-left quarter of the table tells us that

$$|\Psi) = |\Psi_+) + |\Psi_-)\,, \tag{67}$$

where these two non-zero parts of $|\Psi)$ satisfy

$$|\Psi_+) \in \mathsf{A}_-^\circ \otimes \mathsf{C}_-^\circ \text{ and } |\Psi_-) \in \mathsf{A}_+^\circ \otimes \mathsf{C}_+^\circ. \tag{68}$$

The same state vector $|\Psi)$ gives rise to the possibilities in the upper-right quarter of $\mathcal{P}$ also. From this we can conclude that

$$|\Psi_+) \in \mathsf{A}_-^\circ \otimes \mathsf{D}_-^\circ \text{ and } |\Psi_-) \in \mathsf{A}_+^\circ \otimes \mathsf{D}_+^\circ. \tag{69}$$

We can continue around the table $\mathcal{P}$, arriving at the following facts:

$$|\Psi_+) \in \mathsf{B}_+^\circ \otimes \mathsf{D}_-^\circ \text{ and } |\Psi_-) \in \mathsf{B}_-^\circ \otimes \mathsf{D}_+^\circ. \tag{70}$$

$$|\Psi_+) \in \mathsf{B}_+^\circ \otimes \mathsf{C}_+^\circ \text{ and } |\Psi_-) \in \mathsf{B}_-^\circ \otimes \mathsf{C}_-^\circ. \tag{71}$$

This last pair of statements allows us to return to the upper-left corner, concluding that

$$|\Psi_+) \in \mathsf{A}_+^\circ \otimes \mathsf{C}_+^\circ \text{ and } |\Psi_-) \in \mathsf{A}_-^\circ \otimes \mathsf{C}_-^\circ. \tag{72}$$

Since the annihilator subspaces are non-overlapping, this contradicts Eq. 68. Thus, no such $|\Psi)$ exists for which the set of possible measurement results is described by the PR box pattern $\mathcal{P}$.

5.3 Probabilistic Resolutions

We have already noted that we can derive a generalized modal table from a generalized probability table, while respecting the no-signalling principle. Is it possible to do the reverse? That is, if we have a table of possibilities for a modal system, can we find a corresponding table of probabilities? We call this a *probabilistic resolution* of the modal table, and distinguish two different types.

- A *strong probabilistic resolution* assigns zero probability to every impossible result and non-zero probability to every possible result (X).
- A *weak probabilistic resolution* assigns zero probability to every impossible result. However, a "possible" result (X) may be assigned any probability, zero or non-zero. (See the discussion in Sect. 2.1.)

In either case, we require that the resulting table of distributions must satisfy the probabilistic no-signalling principle.

As an example, consider the PR box table $\mathcal{P}$ of Eq. 66. It is not difficult to show that this table has only one allowed probabilistic resolution, which is of the strong type:

$$
\begin{array}{ccc}
 & A^{(2)} & B^{(2)} \\
A^{(1)} & \begin{array}{|c|c|}\hline 1/2 & 0 \\ \hline 0 & 1/2 \\ \hline\end{array} & \begin{array}{|c|c|}\hline 1/2 & 0 \\ \hline 0 & 1/2 \\ \hline\end{array} \\
B^{(1)} & \begin{array}{|c|c|}\hline 1/2 & 0 \\ \hline 0 & 1/2 \\ \hline\end{array} & \begin{array}{|c|c|}\hline 0 & 1/2 \\ \hline 1/2 & 0 \\ \hline\end{array}
\end{array}
\qquad (73)
$$

Not all general modal tables actually have probabilistic resolutions of either type. Consider the following table (of which we have only shown the relevant parts):

$$
\mathcal{N} = \left[
\begin{array}{cccc}
 & U^{(2)} & V^{(2)} & W^{(2)} \\
U^{(1)} & \begin{array}{|c|c|c|}\hline & \mathsf{X} & \\ \hline & & \mathsf{X} \\ \hline \mathsf{X} & & \\ \hline\end{array} & \begin{array}{|c|c|c|}\hline \mathsf{X} & & \\ \hline & \mathsf{X} & \\ \hline & & \mathsf{X} \\ \hline\end{array} & \\
V^{(1)} & \begin{array}{|c|c|c|}\hline \mathsf{X} & & \\ \hline & \mathsf{X} & \\ \hline & & \mathsf{X} \\ \hline\end{array} & \begin{array}{|c|c|c|}\hline \mathsf{X} & & \\ \hline & \mathsf{X} & \\ \hline & & \mathsf{X} \\ \hline\end{array} & \begin{array}{|c|c|c|}\hline \mathsf{X} & & \\ \hline & \mathsf{X} & \\ \hline & \mathsf{X} & \\ \hline\end{array} \\
W^{(1)} & & \begin{array}{|c|c|c|}\hline \mathsf{X} & \mathsf{X} & \\ \hline & & \mathsf{X} \\ \hline & & \\ \hline\end{array} & \begin{array}{|c|c|c|}\hline \mathsf{X} & & \\ \hline & \mathsf{X} & \\ \hline & & \\ \hline\end{array}
\end{array}
\right]
\qquad (74)
$$

By inspection, $\mathcal{N}$ satisfies the modal no-signalling principle. When we attempt a probabilistic resolution, we quickly discover that all of the possibilities in the

$(U^{(1)}, U^{(2)})$, $(U^{(1)}, V^{(2)})$, $(V^{(1)}, U^{(2)})$ and $(V^{(1)}, V^{(2)})$ sub-tables must be assigned probability 1/3. We obtain

$$
\begin{array}{cccc}
 & U^{(2)} & V^{(2)} & W^{(2)} \\
U^{(1)} &
\begin{array}{|c|c|c|}\hline & 1/3 & \\ \hline & & 1/3 \\ \hline 1/3 & & \\ \hline\end{array} &
\begin{array}{|c|c|c|}\hline 1/3 & & \\ \hline & 1/3 & \\ \hline & & 1/3 \\ \hline\end{array} &
 \\
V^{(1)} &
\begin{array}{|c|c|c|}\hline 1/3 & & \\ \hline & 1/3 & \\ \hline & & 1/3 \\ \hline\end{array} &
\begin{array}{|c|c|c|}\hline 1/3 & & \\ \hline & 1/3 & \\ \hline & & 1/3 \\ \hline\end{array} &
\begin{array}{|c|c|c|}\hline 1/3 & & \\ \hline & 1/3 & \\ \hline & 1/3 & \\ \hline\end{array} \\
W^{(1)} &
 &
\begin{array}{|c|c|c|}\hline 1/3 & 1/3 & \\ \hline & & 1/3 \\ \hline & & \\ \hline\end{array} &
\begin{array}{|c|c|c|}\hline p & & \\ \hline & q & \\ \hline & & \\ \hline\end{array}
\end{array}
\tag{75}
$$

where we have for clarity omitted zero entries. The trouble arises in the lower-right corner $(W^{(1)}, W^{(2)})$. The probabilistic no-signalling principle imposes two sets of constraints on the probabilities p and q. Comparing to the $(W^{(1)}, V^{(2)})$ sub-table, we require $p = 2/3$ and $q = 1/3$. Comparing to the $(V^{(1)}, W^{(2)})$ sub-table, we require $p = 1/3$ and $q = 2/3$. We therefore conclude that no probabilistic resolution exists for modal table $\mathcal{N}$.

Under some circumstances, we can guarantee that a probabilistic resolution must exist. Suppose that a general modal table $\mathcal{R}$ arises from a local hidden variable theory. For each particular value h of the hidden variables, the outcomes of all joint measurements are determined. The resulting table $\mathcal{D}_h$ is thus deterministic—that is, each sub-table contains only a single X. The overall table $\mathcal{R}$ is thus a mixture of different $\mathcal{D}_h$ tables. Locality of the hidden variable theory means that each deterministic table $\mathcal{D}_h$ individually satisfies the no-signalling principle.

Suppose there are N distinct deterministic tables $\mathcal{D}_h$. Each deterministic table has an obvious probabilistic resolution in which each X entry is given probability 1. Now we assign each distinct $\mathcal{D}_h$ a probability of $1/N$, and take a mixture of their probabilistic resolutions with these weights. That is, if a particular outcome of a particular joint measurement is possible in M of the deterministic tables, it is assigned an overall probability M/N. The resulting table satisfies the probabilistic no-signalling principle, since it is a convex combination of no-signalling tables. Furthermore, it is a strong probabilistic resolution of $\mathcal{R}$, since it assigns a probability at least $1/N$ to each possible measurement outcome. Therefore, every general modal table arising from a local hidden variable theory has a strong probabilistic resolution.

The converse is certainly false. The PR box table $\mathcal{P}$ of Eq. 66 has a strong probabilistic resolution (Eq. 73). However, $\mathcal{P}$ is a minimal table, which means it cannot arise as a mixture of deterministic tables that satisfy the no-signalling principle. Therefore $\mathcal{P}$ cannot arise from any local hidden variable theory.

Now consider the modal table $\mathcal{S}$ arising from the $\mathbb{Z}_2$-MQT singlet state, as shown in Eq. 65. This has a unique probabilistic resolution, which we display below. Note that some of the possible outcomes have to be assigned probability zero—that is, only a weak probabilistic resolution can be given for this table:

$$
\begin{array}{cccc}
 & X^{(2)} & Y^{(2)} & Z^{(2)} \\
X^{(1)} & \begin{array}{c|c} & 1/2 \\ \hline 1/2 & \end{array} & \begin{array}{c|c} 1/2 & \\ \hline 0 & 1/2 \end{array} & \begin{array}{c|c} 1/2 & 0 \\ \hline & 1/2 \end{array} \\
Y^{(1)} & \begin{array}{c|c} 1/2 & 0 \\ \hline & 1/2 \end{array} & \begin{array}{c|c} & 1/2 \\ \hline 1/2 & \end{array} & \begin{array}{c|c} 1/2 & \\ \hline 0 & 1/2 \end{array} \\
Z^{(1)} & \begin{array}{c|c} 1/2 & \\ \hline 0 & 1/2 \end{array} & \begin{array}{c|c} 1/2 & 0 \\ \hline & 1/2 \end{array} & \begin{array}{c|c} & 1/2 \\ \hline 1/2 & \end{array}
\end{array}
\tag{76}
$$

There are a number of things to remark about the probabilistic resolution in Eq. 76. The MQT singlet state $|S)$ gives us an example of a table with a weak probabilistic resolution but not a strong probabilistic resolution. This gives us another proof that the modal properties of $|S)$ (represented in table $\mathcal{S}$) cannot be derived from any local hidden variable theory: if such a theory existed, the table would certainly have a strong probabilistic resolution.

As we have seen, the modal PR box table $\mathcal{P}$ of Eq. 66 cannot arise from an entangled composite system in MQT. Nevertheless, the weak probabilistic resolution of Eq. 76 does contain a *probabilistic* PR box! Consider the following section of the table:

$$
\begin{array}{ccc}
 & Z^{(2)} & Y^{(2)} \\
X^{(1)} & \begin{array}{c|c} 1/2 & 0 \\ \hline 0 & 1/2 \end{array} & \begin{array}{c|c} 1/2 & 0 \\ \hline 0 & 1/2 \end{array} \\
Y^{(1)} & \begin{array}{c|c} 1/2 & 0 \\ \hline 0 & 1/2 \end{array} & \begin{array}{c|c} 0 & 1/2 \\ \hline 1/2 & 0 \end{array}
\end{array}
\tag{77}
$$

This apparent paradox arises because a weak probabilistic resolution allows probability zero to be assigned to a possible measurement outcome.

A probabilistic PR box cannot arise in actual quantum theory. It follows that the behavior of an entangled composite system in MQT cannot be "simulated" by an entangled composite system in AQT. (This is why the pseudo-telepathy game for $|S)$ described in Sect. 2.3 has no winning strategy if the players can only share entangled states from AQT.)

5.4 A Hierarchy of Modal Theories

We have considered several distinct types of two-system modal tables.

- NSP is the set of tables satisfying the no-signalling principle. (This is our "universe" of tables.)
- SPR is the set of tables that have a strong probabilistic resolution.
- WPR is the set of tables that have a weak probabilistic resolution.
- LHV is the set of tables that have a local hidden variable model.
- MQT is the set of tables that can arise from a bipartite system in modal quantum theory.

As we have seen there are several relations between these classes:

$$\text{LHV} \subset \text{SPR} \subset \text{WPR} \subset \text{NSP}. \tag{78}$$

The inclusion relation is strict in each case. The PR box table $\mathcal{P}$ in Eq. 66 is in SPR but not LHV; the $\mathbb{Z}_2$ modal singlet table $\mathcal{S}$ in Eq. 65 is in WPR but not SPR; and the table $\mathcal{N}$ in Eq. 74 is in NSP but not WPR.

What about the set MQT? It is not hard to see that every table in LHV is also in MQT. We also know there are tables that are in MQT but not in LHV or SPR. Conversely, the PR box $\mathcal{P}$ (Eq. 66) is in SPR and WPR but not MQT. It remains to pin down the relation between MQT and WPR. We will prove that MQT $\subset$ WPR—that is, that every table that arises from the state of a bipartite system in MQT must have a weak probabilistic resolution.

To establish this, we will take advantage of several simplifications. Since a weak probabilistic resolution allows us to assign $p = 0$ for some possible outcomes, the addition of possibilities (X entries) to a modal table can never frustrate a weak probabilistic resolution. Therefore, we need only consider minimal modal tables in MQT, those that arise from pure bipartite states.

Every pure bipartite state $|\Psi)$ has a Schmidt decomposition (as in Eq. 27) with an integer Schmidt number s. The state vector therefore lies in a subspace we may denote $\mathcal{V} \otimes \mathcal{V}$, with $\dim \mathcal{V} = s$. The space $\mathcal{V}$ is a subspace of the state spaces for the two systems; but we can regard it as the effective state space for the particular situation described by $|\Psi)$. Any measurement on either subsystem can hence be regarded as a generalized measurement on $\mathcal{V}$. Therefore, we can suppose that $|\Psi)$ is a state of maximum Schmidt number for a pair of identical systems with state spaces $\mathcal{V}$ of dimension s. (The case where $s = 1$ is trivial, so we will assume that $s \geq 2$ and $|\Psi)$ is entangled.)

Generalized measurements whose effect subspaces have $\dim \mathsf{E}_a > 1$ can be viewed as "coarse-grained" versions of measurements with one-dimensional ("fine-grained") effects. If we can construct a weak probabilistic resolution for the fine-grained measurements, this will automatically give a resolution for the coarse-grained version. Therefore, we need only consider fine-grained measurements—that is, those whose effect subspaces are one-dimensional.

A fine-grained measurement can be viewed as a spanning set for $\mathcal{V}^*$. Every such spanning set contains a basis, and at least one of these basis effects must be possible for a given state. The "extra" effects can always be assigned probability zero. Therefore, we need only consider basic measurements, those that correspond to basis sets for $\mathcal{V}^*$.

Armed with all of these simplifications, let us consider a pair of identical systems in a pure entangled state $|\Psi^{(12)})$ of maximum Schmidt number. For each pair of basic measurements, we arrive at an $s \times s$ sub-table of possibilities. Let us focus our attention on one such sub-table, with measurement bases $\{\left(e_j^{(1)}\right|\}$ (the rows) and $\{(f_k^{(2)}|\}$ (the columns).

For each $\left(e_j^{(1)}\right|$, define the set

$$F_j = \{(f_k^{(2)}| : \left(e_j^{(1)} f_k^{(2)} |\Psi^{(12)}\right) \neq 0\}. \tag{79}$$

That is, for each system 1 effect, we consider the set of system 2 effects that are jointly possible given state $|\Psi^{(12)})$. Consider next a set E containing d system 1 effects $\left(e_j^{(1)}\right|$. Each $\left(e_j^{(1)}\right|$ corresponds to a conditional state $\left|\psi_j^{(2)}\right) = \left(e_j^{(1)} |\Psi^{(12)}\right)$. Since $|\Psi^{(12)})$ is maximally entangled, these are non-zero and linearly independent. Hence, the effects in E correspond to a set of system 2 states that span a subspace $\mathsf{M}_E^{(2)}$ of dimension d.

A basic system 2 measurement on M_E must have at least d possible outcomes. These correspond to the system 2 effects in the set $\bigcup_E F_j$. We have shown that the collection $F = \{F_j\}$ of sets has the property that, for any set E of basic system 1 effects,

$$\#\left(\bigcup_E F_j\right) \geq \#(E), \tag{80}$$

where $\#(K)$ is the number of elements in finite set K. By Hall's Marriage Theorem [21], we can conclude that the collection F has a set of distinct representatives. That is, for each $\left(e_j^{(1)}\right|$ we can identify a corresponding $\left(f_j^{(2)}\right|$ such that

- $\left(e_j^{(1)} f_j^{(2)} |\Psi^{(12)}\right) \neq 0$ for all j, and
- $(f_i^{(2)}| \neq \left(f_j^{(2)}\right|$ when $i \neq j$.

In our sub-table, this means we can identify a set of the possible joint outcomes (the X's) such that each row and each column contains exactly one of them.

We therefore make the following probability assignment. Each impossible joint outcome, of course, is assigned $p = 0$. We also assign $p = 0$ to all of the possible joint outcomes except for those we have identified above, one in each row and column. These are assigned $p = 1/s$.

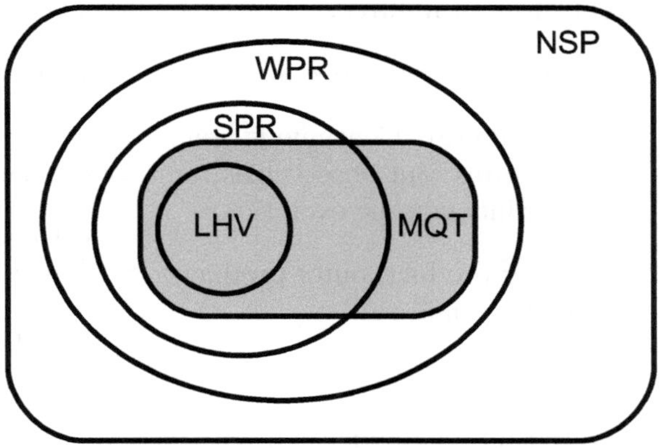

Fig. 1 The hierarchy of bipartite states in modal theories

The same procedure can be applied for each sub-table independently. In every case, the total probability for each row and for each column is $1/s$. Therefore, the probabilistic no-signalling principle is automatically satisfied. Our construction (via Hall's Marriage Theorem) yields a weak probabilistic resolution for the modal table associated with the entangled state $|\Psi^{(12)}\rangle$. Every table that arises from a bipartite state in MQT has a weak probabilistic resolution.

In terms of our hierarchy of modal theories, we have shown that MQT $\subset$ WPR. Our conclusions are summarized in Fig. 1. It is worth noting that all of the six distinct regions in this diagram are non-empty. Thus, for example, table $\mathcal{S}$ of Eq. 65 is in MQT but not SPR; table $\mathcal{P}$ of Eq. 66 is in SPR but not MQT; and table $\mathcal{N}$ of Eq. 74 is within NSP but not WPR. Other examples are easy to construct.

6 Concluding Remarks

6.1 What MQT Has, and What It Does not Have

As diverting an exercise as MQT is, its real purpose is to shed light on the structure of actual quantum theory. It is remarkable how many of the features of AQT are retained, at least in some form, even in such a primitive theory. An incomplete summary can be found in Fig. 2. In the left-hand column we have listed aspects of AQT that are not found in MQT; in the right-hand column, we have listed aspects of AQT that do have analogies in MQT. The key point is that *nothing in the right-hand column logically depends on anything in the left-hand column.*

Furthermore, as we have seen, the process of generalization is very similar in AQT and MQT. In both theories we can develop more general concepts of state, measurement and time evolution, and these generalizations can be characterized in both constructive and axiomatic ways. Both theories can also be extended to more general (probabilistic or modal) theories. Within these more general types of theories,

MQT does not have:

- Probabilities, expectations
- ($\mathcal{F}$ finite) Continuous sets of states and observables, or continuous time evolution
- Inner product, outer product, orthogonality
- Convexity
- Hermitian conjugation (†)
- Density operators
- Effect operators
- CP maps
- Unextendable product bases

MQT does have:

- "Classical" versus "quantum" theories
- Superposition, interference
- Complementary measurements
- Entanglement
- No local hidden variables
- Kochen-Specker theorem, "free will" theorem
- Superdense coding, teleportation, "steering" of mixtures
- Mixed states, generalized effects, generalized evolution maps
- No cloning theorem
- Nonclassical models of computation

Fig. 2 Properties and structures of actual quantum theory that either are or are not present in MQT

the quantum theories have special properties—e.g., PR boxes are excluded in either theory, and every bipartite state in MQT has a weak probabilistic resolution.

This last point deserves further comment. We have imagined a modal world, one which supports the distinction between "possible" and "impossible" events without necessarily imposing any probability measure. As we have seen, it is not always possible to make a reasonable probability assignment in such a modal world. The table $\mathcal{N}$ of Eq. 74 provides an example that respects the modal no-signalling principle, but within which we cannot assign probabilities respecting the probabilistic NSP.

Under what circumstances, then, can we make reasonable probability assignments to a set of possibilities? In the bipartite case, we have shown that this can always be done for joint measurements on a modal quantum system. That is, the underlying structure of MQT somehow "makes room" for probabilities. It remains to be seen whether this sheds any light on the way in which probabilities arise in the real world.

6.2 Open Problems

Modal quantum theory is an exceptionally rich "toy model" of physics. Despite the known features of the theory summarized in Fig. 2, there remain many open questions.

- Although we have shown that bipartite systems in MQT support weak probabilistic resolutions, we do not know whether this is true for entangled states of three or more systems.
- We have established many properties of pure entangled states for MQT systems, but we know much less about mixed entangled states. For example, we do not know whether there are "bound" entangled states in MQT [22]. (The usual AQT construction cannot be adapted to MQT, since there are no unextendable product bases in MQT.)
- Many results and ideas of quantum information and quantum computation have direct analogues in MQT. For instance, MQT supports both superdense coding and teleportation [6]. It is straightforward to show that the Deutsch-Jozsa oracle algorithm (distinguishing constant and balanced functions with a single query) can be implemented without change on a modal quantum computer with $\mathcal{F} = \mathbb{Z}_3$ [23]. However, a great deal of work remains to be done along these lines.[7]
- It is possible to regard actual quantum theory as a special type of modal quantum theory in which $\mathcal{F} = \mathbb{C}$ and we have special restrictions on the allowed measurements and time evolution operators. What (if anything) can be gained by analyzing AQT in this way?

We believe that in the investigation of these and other open problems MQT will shed further light on the mathematical structure of quantum theory.

Acknowledgments We have benefitted from discussions of MQT with many colleagues. Howard Barnum and Alex Wilce helped us clarify the mathematical representation of measurement within the theory. Charles Bennett and John Smolin suggested several questions about entangled states. Gilles Brassard pointed out that the "no hidden variables" results in MQT are best described by pseudo-telepathy games. Our research students Arjun Singh (Denison) and Peter Johnson (Kenyon) participated in the early development of the MQT model. Rob Spekkens (no stranger to thought-provoking "foil" theories) has been particularly helpful at many stages of this project.

We would also like to thank the Perimeter Institute for its hospitality and the organizers of the workshop there on "Conceptual Foundations and Foils for Quantum Information Processing", May 9–13, 2011.

Appendix

Here we fill in the details of the argument in Sect. 5.2. For convenience, we will suppose that modal quantum systems 1 and 2 are both described by the state space

[7] Some observations are obvious. In a world without probabilities, we are interested in the *zero-error capacities* of communication channels and computer algorithms that reach *deterministic* results.

$\mathcal{V}$, and that the same two-outcome measurement is performed on each. The effect subspaces E and F in $\mathcal{V}^*$ are non-overlapping, so that $\mathsf{E} \cap \mathsf{F} = \langle 0 \rangle$. Finally, we assume that the joint possibility table for the state $|\Psi)$ is as follows:

	E	F
E	X	
F		X

(81)

(Each sub-table of Eq. 66 is of this form.)

We can find a basis for $\mathcal{V}^*$ of the form $\{(e_i|, (f_m|\}$, where the $\{(e_i|\}$ spans E and $\{(f_m|\}$ spans F. The dual basis $\{|e_i), |f_m)\}$ of $\mathcal{V}$ therefore has the property that $|e_i)$ is annihilated by every $(f_m|$ and $|f_m)$ is annihilated by every $(e_i|$. In fact, $\{|e_i)\}$ spans the annihilator F° and $\{|f_m)\}$ spans E°. We can expand the composite state $|\Psi)$ in this way:

$$|\Psi) = \sum_{ij} \alpha_{ij} \left|e_i e_j\right) + \sum_{in} \beta_{in} \left|e_i f_n\right) + \sum_{mj} \gamma_{mj} \left|f_m e_j\right) + \sum_{mn} \delta_{mn} \left|f_m f_n\right). \qquad (82)$$

From Eq. 81, we can see that the effect $\mathsf{E} \otimes \mathsf{F}$ is impossible, which implies that $(e_i f_n |\Psi) = \beta_{in} = 0$ for every i, n. In the same way, because $\mathsf{F} \otimes \mathsf{E}$ is impossible, $\gamma_{mj} = 0$ for every m, j. Therefore,

$$|\Psi) = |\Psi_{ee}) + \left|\Psi_{ff}\right), \qquad (83)$$

where $|\Psi_{ee}) \in \mathsf{F}^\circ \otimes \mathsf{F}^\circ$ and $\left|\Psi_{ff}\right) \in \mathsf{E}^\circ \otimes \mathsf{E}^\circ$.

Though we have supposed that the two systems are of the same type and that the same measurement is made on each, it is easy to adapt this argument to more general situations, provided the effect subspaces are non-overlapping.

References

1. R. Spekkens, Evidence for the epistemic view of quantum states: a toy theory. Phys. Rev. A **75**, 032110 (2007)
2. J.S. Bell, On the Einstein-Podolsky-Rosen paradox. Physics **1**, 195 (1964)
3. L. Hardy, Reformulating and reconstructing quantum theory, e-print arxiv:1104.2066
4. G. Chiribella, G.M. DAriano, P. Perinotti, Informational derivation of quantum theory. Phys. Rev. A **84**(1), 012311 (2011)
5. L. Masanes, M.P. Mller, A derivation of quantum theory from physical requirements. New J. Phys. **13**, 063001 (2011)
6. B. Schumacher, M.D. Westmoreland, Modal quantum theory, e-print arXiv:1010:2937
7. G. Hughes, M. Cresswell, *A New Introduction to Modal Logic* (Routledge, London, 1996)
8. B. Schumacher, M.D. Westmoreland, Non-contextuality and free will in modal quantum theory, e-print arXiv:1010:5452
9. C. Simon, V. Bužek, N. Gisin, The no-signaling condition and quantum dynamics. Phys. Rev. Lett. **87**, 170405 (2001)

10. S. Kochen, E. Specker, The problem of hidden variables in quantum mechanics. J. Math. Mech. **17**, 59–88 (1967)
11. B. Schumacher, M. Westmoreland, Locality non-contextuality and free will in modal quantum theory, Report on poster presented at Conceptual Foundations and Foils for QIP, Perimeter Institute (2011). http://www.perimeterinstitute.ca/Events/
12. G. Brassard, A. Broadbent, A. Tapp, Multi-party pseudo-telepathy, in *Proceedings of the 8th International Workshop on Algorithms and Data Structures*. Lecture Notes in Computer Science, vol. 2748 (2003), pp. 1–11
13. B. Schumacher, M.D. Westmoreland, *Quantum Processes, Systems and Information* (Cambridge University Press, Cambridge, 2010)
14. P. Halmos, *Naive Set Theory* (Springer, New York, 1974)
15. I.M. Gelfand, M.A. Neumark, On the embedding of normed rings into the ring of operators in Hilbert space. Rec. Math. [Mat. Sbornik] N.S. **54**, 197–213 (1943)
16. L.P. Hughston, R. Jozsa, W.K. Wootters, A complete classification of quantum ensembles having a given density matrix. Phys. Lett. A **183**, 14–18 (1993)
17. E. Schrödinger, Probability relations between separated systems. Proc. Camb. Philos. Soc. **32**, 446–452 (1936)
18. E.T. Jaynes, Information theory and statistical mechanics. II. Phys. Rev. **108**(2), 171–190 (1957)
19. H.M. Wiseman, S.J. Jones, A.C. Doherty, Steering, entanglement, nonlocality, and the Einstein-Podolsky-Rosen paradox. Phys. Rev. Lett. **98**, 140402 (2007)
20. S. Popescu, D. Rohrlich, Nonlocality as an axiom. Found. Phys. **24**, 379–385 (1994)
21. P. Hall, On representatives of subsets. J. Lond. Math. Soc. **10**, 26–30 (1935)
22. C.H. Bennett, D.P. DiVincenzo, T. Mor, P.W. Shor, J.A. Smolin, B.M. Terhal, Unextendible product bases and bound entanglement. Phys. Rev. Lett. **82**, 5385 (1999)
23. D. Deutsch, R. Jozsa, Rapid solutions of problems by quantum computation. Proc. R. Soc. Lond. A **439**, 553 (1992)

Quasi-Quantization: Classical Statistical Theories with an Epistemic Restriction

Robert W. Spekkens

1 Introduction

1.1 Epistricted Theories

Start with a classical theory for some degree of freedom and consider the statistical theory associated with it. This is the theory that describes the statistical distributions over the space of physical states and how they change over time. If one then postulates, as a fundamental principle, that there is a restriction on what kinds of statistical distributions can be prepared, then the resulting theory reproduces a large part of quantum theory, in the sense of reproducing its operational predictions. This article reviews recent work on such theories and their relevance for notions of nonclassicality, for the interpretation of the quantum state, and for the program of deriving the formalism of quantum theory from axioms.

Some clarifications are in order regarding statistical theories. Given a system whose physical state is drawn from some ensemble of possibilities, the statistical distribution associated to this ensemble can be taken either to describe relative frequencies of physical properties within the virtual ensemble or it can be taken to describe the knowledge that an agent has about an individual system when she knows that it was drawn from that ensemble. The latter sort of language is preferred by those who take a Bayesian approach to statistics, and we shall adopt it here. The distinction between the physical state of a system and an agent's state of knowledge of that physical state will be critical in what follows. As such, we will make use of some jargon to clearly distinguish the two sorts of states. Recalling the Greek terms for reality and for knowledge, *ontos* and *epistēmē*, we will henceforth refer to physical

R.W. Spekkens (✉)
Perimeter Institute for Theoretical Physics, 31 Caroline St. N, Waterloo,
Ontario N2L 2Y5, Canada
e-mail: rspekkens@perimeterinstitute.ca

G. Chiribella and R.W. Spekkens (eds.), *Quantum Theory: Informational Foundations and Foils*, Fundamental Theories of Physics 181,
DOI 10.1007/978-94-017-7303-4_4

Table 1 Theories arising from imposing certain epistemic restrictions on statistical theories for various classical theories, and the subtheories of quantum theory that they correspond to

Classical ontological theory	Statistical theory for the classical ontological theory	Epistemically-restricted statistical theory for the classical ontological theory
Mechanics	Liouville mechanics	Gaussian epistricted mechanics = Gaussian subtheory of quantum mechanics
		Quadrature epistricted mechanics = Quadrature subtheory of quantum mechanics
Trits	Statistical theory of trits	Quadrature epistricted theory of trits = Quadrature/Stabilizer subtheory for qutrits
Bits	Statistical theory of bits	Quadrature epistricted theory of bits $\simeq$ Quadrature/Stabilizer subtheory for qubits
Optics	Statistical optics	Gaussian epistricted optics = Gaussian subtheory of quantum optics
		Quadrature epistricted optics = quadrature subtheory of quantum optics

states as *ontic states* and states of knowledge as *epistemic states* [1]. The theory that governs the evolution of ontic states is an *ontological theory*, while the *statistical theory* describes the evolution of epistemic states. A restriction on knowledge is an *epistemic restriction*. The theories we are considering, therefore, are epistemically-restricted statistical theories of classical systems. Given that this is a rather unwieldy descriptor, we introduce the term *epistricted theory* as an abbreviation to it.

It is worth considering in a bit more detail the scheme by which one infers from a given classical theory the epistricted version thereof. One starts with a particular classical ontological theory (first column of Table 1). We are here considering the usual notion of an ontological theory: one which provides a kinematics and a dynamics, that is, a hypothesis about the possible physical states that a system can occupy at a given time and a law describing how each state can evolve over time.

One then constructs the statistical theory for the classical ontological theory under consideration (second column of Table 1). The fundamental object here is a statistical distribution over the physical state space rather than a point in the physical state space, that is, an epistemic state rather than an ontic state. The statistical theory answers questions such as: If the physical state of the system undergoes deterministic dynamics, how does the statistical distribution change over time? Or, more precisely, if an agent assigns a statistical distribution over physical states at one time and she knows the dynamics, what statistical distribution should she assign at a later time? If an agent implements a measurement on the system and takes note of the outcome, how should she update her statistical distribution?

In the final and most significant step of the theory-construction scheme, one postulates a fundamental restriction on the sorts of statistical distributions that can describe an agent's knowledge of the system (third column of Table 1).

As a first example, consider the classical ontological theory of particle mechanics. The associated statistical theory is what is sometimes called *Liouville mechanics*. If one then postulates a classical version of the uncertainty principle as the epistemic restriction [2], then one obtains a theory which we shall refer to here as *Gaussian epistricted mechanics* (it was called *epistemically-restricted Liouville mechanics* in Ref. [2]). This theory is equivalent to a subtheory of quantum mechanics, the Gaussian subtheory, which is defined in Ref. [2].

The case of optics is a straightforward extension of the case of mechanics because each optical mode is a scalar field and the phase spaces of a field mode and of a particle are both Euclidean. The canonically conjugate variables, which are position and momentum in the mechanical case, are field quadratures in the optical case. The statistical theory of optics is well-known [3]. Upon postulating an epistemic restriction in the form of an uncertainty principle, one obtains the optical analogue of the Gaussian subtheory of quantum mechanics, namely, the Gaussian subtheory of quantum optics, which is sometimes referred to as *linear quantum optics*. The latter theory includes a wide variety of quantum optical experiments.

One can apply the same strategy for a classical ontological theory wherein the fundamental degrees of freedom are discrete, so that every system has an integer number d of ontic states. It is unusual for physicists to discuss discrete degrees of freedom in a classical context. Nonetheless, this is done when considering the possibility of models that are cellular automata. It is also common when describing the physics of digital computers. The language of computation, therefore, is a natural one for describing such a theory.

The simplest case to consider is $d = 2$, in which case the fundamental degree of freedom is a bit. A collection of such fundamental degrees of freedom corresponds to a string of bits. An interaction between two distinct degrees of freedom can be understood as a gate acting on two bits. Similarly for interactions between n systems. General dynamics, which corresponds to an arbitrary sequence of interactions, can be understood as a circuit. The statistical theory of bits is just a theory of the statistical distributions over the possible bit-strings, how these evolve under gates, and how these are updated as a result of registering the outcome of measurements performed on the bits. One then imposes a restriction on what kinds of statistical distributions can characterize an agent's knowledge of the value of the bit-string.

This is the arena in which the first epistricted theory was constructed [1]. The restriction on knowledge was implemented through a principle that asserted that any agent could have the answers to at most half of a set of questions that would specify the ontic state of the system. Consequently, when one has maximal knowledge, then the number of independent questions that are answered is equal to the number of independent questions that are unanswered; in this case, one's measure of knowledge is equal to one's measure of ignorance. This epistemic restriction was dubbed the *knowledge-balance principle*, and the epistricted theory of bits that resulted was called a *toy theory* in Ref. [1]. This theory mirrors very closely a subtheory of the quantum theory of qubits, namely, the one which is known to quantum information theorists as the *stabilizer formalism* and which we will term the *stabilizer subtheory* of qubits. It will be presented in Sect. 3. Stabilizer states are defined to be the eigenstates of products of Pauli operators, stabilizer measurements are measurements of commuting sets of products of Pauli operators, and stabilizer transformations are unitary transformations which take stabilizer states to stabilizer states. Although the toy theory is not operationally equivalent to the stabilizer subtheory, it reproduces qualitatively the same phenomenology. Also, the toy theory can be cast in the same sort of language as the stabilizer theory, as noted in Ref. [4].

Subsequent work sought to develop an epistricted theory for discrete systems with d ontic states, where $d > 2$. There were two natural avenues to pursue: generalize the knowledge-balance principle used in Ref. [1] or devise a discrete version of the classical uncertainty principle used in Ref. [2]. The former approach was pursued by van Enk [5].[1] However, some important work by Gross [6] established that it is possible to define a discrete phase space for a d-level systems where d is an odd prime and it is possible to define a Wigner representation based on this phase space such that the stabilizer theory for these qudits admits of a nonnegative Wigner representation. Gross's Wigner representation can be understood as a hidden variable model for the stabilizer subtheory. This suggests that one should be able to define a classical theory of d-level systems using this discrete phase space and then to find an epistemic restriction that yields precisely this hidden variable model. In other words, Gross's work strongly suggests that one should look for an epistemic restriction that appeals to the phase-space structure, analogously to the epistemic restriction that was used in the Gaussian epistricted mechanics [2].

Such an epistemic restriction was subsequently identified [7]. Using the phase-space structure, one can define *quadrature variables* for the classical system. The epistemic restriction then asserts that one can have joint knowledge of a set of quadrature variables if and only if they commute relative to a discrete analogue of the Poisson bracket. The epistemic restriction is dubbed *classical complementarity* and the theory that results is called the *quadrature epistricted theory* of d-level classical systems.

If we apply the complementarity-based epistemic restriction in the case of $d = 2$, the resulting theory—the quadrature epistricted theory of bits—turns out to be equivalent to the toy theory of Ref. [1], and as mentioned previously, this is operationally a close phenomenological cousin of the stabilizer theory of qubits.

On the other hand, for d an *odd* prime, i.e., any prime besides 2, the quadrature epistricted theory reproduces *precisely* the stabilizer theory for qudits. For such

[1] We are here refering to the first part of Ref. [5]. In the second part, the author proposes a theory wherein there is a restriction on what can be known about the outcome of measurements, rather than a restriction on what can be known about some underlying ontic state. As such, the latter theory is not an epistricted theory.

values of d, the epistemic restriction of classical complementarity turns out to be inequivalent to the knowledge-balance principle. The latter specifies only that at most half of the full set of variables can be known, whereas the former picks out particular halves of the full set of variables, namely, the halves wherein all the variables Poisson-commute. Because the restriction of classical complementarity actually reproduces the stabilizer theory for qudits while the knowledge-balance principle does not [7], epistemic restrictions based on the symplectic structure seem to be preferable to those based on a principle of knowledge balance.

We will also show that on the quantum side, one can define the notion of a quadrature *observable*, a quantum analogue of a classical quadrature variable. In $d = 2$, the Pauli operators are both unitary and Hermitian; as unitaries, they constitute the quantum analogue of classical phase-space displacements, while as observables, they correspond to our quadrature observables. In $d > 2$, on the other hand, the generalized Pauli operators are unitary but not always Hermitian and therefore cannot always be interpreted as observables. Consequently, the stabilizer of a state in $d > 2$ specifies the unitaries that leave the state invariant, not the observables for which the state is an eigenstate. In $d > 2$, the quadrature observables are the ones that are defined in terms of the eigenbases of the generalized Pauli operators. They provide a means for acheiving a characterization of stabilizer states for any d as joint eigenstates of a commuting set of quadrature observables. This characterization is more analogous to our characterization, in epistricted theories, of the valid epistemic states as states wherein one has joint knowledge of a Poisson-commuting set of quadrature variables.

Finally, the epistemic restriction of classical complementarity can also be applied to particle mechanics, where it is different from the restriction based on the uncertainty principle that is used in Ref. [2]. In particular, a smaller set of statistical distributions are considered valid epistemic states. Using the principle of classical complementarity, one obtains a different theory at the end, which we call *quadrature epistricted mechanics*. We prove that this is equivalent to a subtheory of quantum mechanics that we will call the *quadrature subtheory of quantum mechanics* and which we will describe in detail in Sect. 3. The latter stands to the Gaussian subtheory of quantum mechanics as the quadrature epistricted theory of mechanics stands to the Gaussian epistricted theory of mechanics. One can similarly define analogous theories for optics.

It follows that the epistemic restriction of classical complementarity provides the beginning of a unification of all known epistricted theories. It can be applied for both continuous and discrete degrees of freedom, and the formalism can be made to look precisely the same in each case.

It remains an open question whether one can find a form of the epistemic restriction that is applicable to an arbitrary degree of freedom and that when applied in the case of a d-level system yields the Stabilizer/quadrature subtheory of qudits while when applied in the case of continuous variable systems yields the Gaussian subtheory of quantum mechanics/optics rather than merely the quadrature subtheory.

Guided by the bridge between the epistricted theories and the quantum subtheories, we present the formalism of the associated quantum subtheories in a unified manner for continuous and discrete degrees of freedom. This presentation focusses on quadrature observables rather than stabilizer groups and helps to reveal the analogies between the subtheories for the different degrees of freedom.

For any epistemic restriction that is applicable to many different degrees of freedom, such as the principle of classical complementarity described here, one can think of the process of applying this restriction to the corresponding classical statistical theories as a kind of quantization scheme, or more precisely, a *quasi-quantization* scheme. It is "quasi" because it does not succeed at obtaining the full quantum theory from its classical counterpart and because in certain cases, such as binary variables, it does not even yield a subtheory of quantum theory.[2] Unlike normal quantization schemes, which are mathematically inspired, the quasi-quantization scheme of this approach is *conceptually* inspired. There is no ambiguity about how to interpret the formalism that results.

Although our quasi-quantization scheme has already been applied to a few different sorts of degrees of freedom, it is clear that one could apply it to others. Vector fields are a good example, one which promises the possibility of a quasi-quantization of classical electrodynamics. By finding the appropriate epistemic restriction on a statistical theory of electrodynamics, one can imagine deriving a theory that might be equivalent to—or perhaps, as for the case of bits, merely analogous to—some subtheory of quantum electrodynamics.[3] At present, it is not obvious how to do this because the epistemic restrictions that have worked best for the degrees of freedom considered thus far have made reference to canonically conjugate degrees of freedom. One therefore expects to encounter precisely the same difficulties that were faced by those who attempted a canonical quantization of classical electrodynamics. Presumably, therefore, it would be useful to develop a Lagrangian, or least-action quasi-quantization scheme in addition to the canonical one. If one could succeed at devising an epistricted theory of electrodynamics, then it would of course be very interesting to attempt to apply quasi-quantization to classical theories of gravity. This would not yield a full quantum theory of gravity, but it might reconstruct some subtheory, or a distorted version of such a subtheory.

The rest of the introduction makes explicit what can and cannot be explained in epistricted theories, together with their significance for interpretation and axiomatization. We have put this material up front rather than at the end of the paper for

[2]Note that for the purposes of this article, the term "quantum theory" refers to a theory schema that can be applied to many different degrees of freedom: particles, fields and discrete systems.

[3]Note that the theory of *stochastic electrodynamics* has some significant similarities to an epistricted theory of electrodynamics, but there are also significant differences. Many authors who describe themselves as working on stochastic electrodynamics posit a *nondeterministic* dynamical law for the fields, whereas an epistricted theory of electrodynamics is one wherein agents merely lack knowledge of the electrodynamic fields, which continue to evolve deterministically. That being said, Boyer's version of stochastic electrodynamics [8] does not posit any modification of the dynamical law and so is closer to what we are imagining here. A second difference is that in stochastic electrodynamics, there is no epistemic restriction on the matter degrees of freedom. However, if one degree of freedom can interact with another, then to enforce an epistemic restriction on one, it is necessary to enforce a similar epistemic restriction on the other. In other words, the assumptions of stochastic electrodynamics were inconsistent. The sort of epistricted theory of electrodynamics we propose here is one that would apply the epistemic restriction to the matter and to the fields.

the benefit of those readers who are reluctant to engage with the detailed development until they have had certain questions answered, in particular, questions about the precise explanatory scope of these epistricted theories, and the question of why one should care about a quantization scheme that does not recover the full quantum theory.

1.2 Explanatory Scope

We return now to the claim that epistricted theories reproduce a "large part" of quantum theory. At this stage, a sceptic might be unconvinced on the grounds that for each classical ontological theory, the subtheory of the corresponding quantum theory that has been derived via this quantization scheme is *far* from the full quantum theory. For instance, Gaussian epistricted mechanics yields a part of quantum mechanics wherein the dynamics include only those Hamiltonians that are at most quadratic in position and momentum observables [2]. Clearly, this is a small subset of all possible Hamiltonians. Nonetheless, we argue that the relative size of the space of Hamiltonians is not the correct metric by which to assess this project. The primary object of the exercise is to achieve conceptual clarity on the principles that might underly quantum theory. As such, it is better to ask: how many distinctively quantum phenomena are reproduced within these subtheories? In particular, how many of the phenomena that are usually taken to defy classical explanation? In terms of the phenomena they include, the subtheories of quantum theory one obtains by an epistemic restriction *do* subsume a large part of the full theory. In support of this claim, Table 2 provides a categorization of some prominent quantum phenomena into those that arise in epistricted theories (on the left), and those that do not (on the right). As one can easily see, for this particular list, the lion's share are found on the left, and this set includes many of the phenomena that are typically taken to provide the greatest challenge to the classical worldview.[4]

Note that it is typically the case that if one looks hard enough at a given quantum phenomenon that appears on the left list, one can usually find *some* feature of it that cannot be explained within an epistricted theory. When we place a given phenomenon on the left, therefore, what we are claiming is that an epistricted theory can reproduce *the features of this phenomenon that are most frequently cited as making it difficult to understand classically*. Consider the example of quantum teleportation. What is most frequently taken to be mysterious about teleportation from a classical perspective is that the amount of information that is required to describe the quantum state exceeds the amount of information that is communicated in the protocol. This is just as true, however, if one seeks to teleport a quantum state within the stabilizer theory of qubits: for a single qubit, this subtheory includes only *six* distinct quantum states (rather than an infinite number), but the teleportation protocol still succeeds while communicating

[4]It should be noted that many researchers had previously recognized the possibility of recovering many of these quantum phenomena if one compared quantum states to probability distributions in a classical statistical theory [9–12].

Table 2 Categorization of quantum phenomena

Phenomena arising in epistricted theories	Phenomena not arising in epistricted theories
Noncommutativity	Bell inequality violations
Coherent superposition	Noncontextuality inequality violations
Collapse	Computational speed-up (if it exists)
Complementarity	Certain aspects of items on the left
No-cloning	
No-broadcasting	
Interference	
Teleportation	
Remote steering	
Key distribution	
Dense coding	
Entanglement	
Monogamy of entanglement	
Choi-Jamiolkowski isomorphism	
Naimark extension	
Stinespring dilation	
Ambiguity of mixtures	
Locally immeasurable product bases	
Unextendible product bases	
Pre and post-selection effects	
Quantum eraser	
And many others...	

only *two* bits of classical information, which is less than $\log_2 6$ and hence not enough to describe a state drawn from this set. As such, we judge the teleportation protocol in the stabilizer formalism to include the essential mystery of teleportation, and, because an epistricted theory can reproduce this notion of teleportation, we put teleportation on the left-hand list. One can always point to features of the teleportation protocol in the full quantum theory that *do not* arise in the stabilizer theory of qubits, for instance, the fact that it works for an infinite number of quantum states rather than just six. However, this is not the feature of teleportation that is typically cited as "the mystery". Such incidental features are the sorts of things that we mean to include on the right-hand side of our classification under "certain aspects of items on the left". Of course, one of the lessons of this categorization exercise is that the usual story about what is mysterious about a given quantum phenomenon should be supplanted by one that highlights these more subtle features, and these should henceforth be our focus when puzzling about quantum theory.

The left-hand list includes basic quantum phenomena, such as non-commutativity of observables, interference, coherent superposition, collapse of the wave function, the existence of complementary bases, and a no-cloning theorem [13]. It also includes many quantum information processing tasks, such as teleportation [14] and key distribution [15]. A large part of entanglement theory [16] is there, as are more exotic phenomena, such as locally indistinguishable product bases [17] and unextendible product bases [18]. One also gets many of the distinctive relations that hold between (and within) the sets of quantum states, quantum measurements, and quantum transformations, such as the Choi-Jamiołkowski isomorphism between bipartite states and unipartite operations [19, 20], the Naimark extension of positive-operator valued measures into projector-valued measures [21], the Stinespring dilation of irreversible operations into reversible (unitary) operations [22], and the fact that there are many convex decompositions and many purifications of a mixed state.

On the right-hand side, we find Bell-inequality violations [23] and noncontextuality-inequality violations [24–26]. This is expected, as these phenomena are the operational signatures of the impossiblity of locally causal and noncontextual ontological models respectively, whereas the particular sorts of epistemic restrictions that have been considered to date yield theories that satisfy local causality and noncontextuality (even the generalized sense of noncontextuality of Ref. [25]). The right-hand side also includes quantum computational speedup, with the caveat that the claim of an exponential speed-up is predicated on certain unproven conjectures, such as the factoring problem being outside the complexity class P.

There are also some phenomena that have not yet been conclusively categorized. Two examples are: the quantization of quantities such as energy and angular momentum and the statistics of indistinguishable particles.

There is always some satisfaction in adding a quantum phenomenon to the left-hand list: it suggests that the idea of an epistemic restriction captures much of the innovation of quantum theory, that the phenomena in question is not so mysterious after all. However, the right-hand list is the one that we would most like to see grow, because the phenomena that appear there are the ones that still seem surprising, and it is by focusing on these that one can best develop the research program wherein quantum states are understood as states of knowledge.

Our quasi-quantization scheme sheds light on the old question "what is the conceptual innovation of quantum theory relative to classical theories?" In particular, it implies that the frontier between what *can* and what *cannot* be explained classically extends much deeper into quantum territory than previously thought. This is because in order to pronounce a phenomenon nonclassical, one should be maximally permissive in *how* a classical theory manages to reproduce the phenomenon. For instance, when considering whether certain operational statistics admit of a local or a noncontextual hidden variable model, one must allow an arbitrary space of ontic states, i.e., arbitrary hidden variables. With the benefit of hindsight, one sees that previous assessments of the scope of classical explanations were overly pessimistic because they did not consider the possibility that some phenomenon exhibited by quantum states was reproduced by classical *epistemic states* rather than by classical *ontic states*.

Phenomena arising in an epistricted theory might still be considered to exhibit a type of nonclassicality insofar as an epistemic restriction is, strictly speaking, an assumption that goes beyond classical physics. But it is a weak type of nonclassicality, as it ultimately can be understood via a relatively modest addendum to the classical worldview. By contrast, quantum phenomena that do not arise in epistricted theories constitute a strong type of nonclassicality, one which marks a significant departure from the classical worldview. Table 2 may therefore be understood as sorting quantum phenomena into categories of weak and strong nonclassicality.

1.3 Interpretational Significance

Epistricted theories serve to highlight the existence (and the appeal) of a type of ontological model that has previously received almost no attention. With the exception of a model of a qubit proposed by Kochen and Specker in 1967, previous ontological models have been such that the ontic state included a description of the quantum state, and therefore any two distinct pure quantum states necessarily described different ontic states. This was true whether the model considered the space of ontic states to be precisely the space of pure quantum states, or whether the quantum state was supplemented by additional variables, such as occurs, for instance, in Bohmian mechanics. By contrast, in epistricted theories, two distinct pure quantum states that are nonorthogonal correspond to two probability distributions that overlap on one or more ontic states. Indeed, it was the work on epistricted theories that led to the articulation of the distinction between a ψ-ontic model, wherein the ontic state encodes the quantum state, and a ψ-epistemic model, wherein it does not [1, 25, 27].

For the subtheories of quantum theory described above (Gaussian and quadrature quantum mechanics and the stabilizer theory of qudits for d an odd prime), ψ-epistemic models exist and provide a compelling causal explanation of the operational predictions of those theories. The breadth of quantum phenomenology that is reproduced within epistricted theories suggests that something about the principles underlying these theories must be correct. The assumption that quantum states should be interpreted as epistemic states rather than ontic states seems a good candidate.

This in no way implies, however, that the innovation of quantum theory is *merely* to impose an epistemic restriction on some underlying classical physics. This is clearly *not* the only innovation of quantum theory. As we have noted, epistricted theories, considered as ontological models, are by construction both local and noncontextual, and because of Bell's theorem and the Kochen-Specker theorem, we know that the full quantum theory cannot be explained by such models. If quantum computers really do allow an exponential speed-up over their classical counterparts, then this too cannot be reproduced by such models. What the success of epistricted theories suggests, rather, is that pure quantum states and statistical distributions over classical ontic states are *the same category of thing*, namely, an epistemic thing. This is a point of view that has been central also to the "QBist" research program [28–30].

A question that naturally arises is whether one can construct a ψ-epistemic model of the *full* quantum theory, or whether one can find natural assumptions under which such models are ruled out, This question was first posed by Lucien Hardy and was formalized in Ref. [27]. It has become the subject of much debate in recent years [31–33]. It is worth noting, however, that the standard framework for ontological models contains many implicit assumptions, including the idea that the correct formalism for describing epistemic states is classical probability theory. This assumption can be questioned, and indeed, the nonlocality and contextuality of quantum theory already suggest that it should be abandoned, as argued in [34, 35].

The investigation of epistricted theories, therefore, need not—and indeed *should not*—be considered as the first step in a research program that seeks to find a ψ-epistemic ontological model of the full quantum theory. Even though such a model could always circumvent any no-go theorems by violating their assumptions, it would be just as unsatisfying as a ψ-ontic ontological model insofar as it would need to be explicitly nonlocal and contextual. Rather, the investigation of epistricted theories is best considered as a first step in a larger research program wherein the framework of ontological models—in particular the use of classical probability theory for describing an agent's incomplete knowledge—is ultimately rejected, but where one holds fast to the notion that a quantum state is epistemic.

1.4 Significance for the Axiomatic Program

Most reconstruction efforts are focussed on recovering the formalism of the full quantum theory. However, it may be that there are particularly elegant axiomatic schemes that are not currently in our reach and that the road to progress involves temporarily setting one's sights a bit lower. The quasi-quantization scheme described here only recovers certain subtheories of the full quantum theory, which include only a subset of the preparations, transformations and measurements that the latter allows. However, such derivations may provide clues for more ambitious axiomatization schemes. In particular, it provides further evidence for the usefulness of foundational principles asserting a fundamental restriction on what can be known [28, 36, 37]. It also seems to highlight the importance of symplectic structure, which is not currently a feature of any reconstruction program.

Furthermore, epistricted theories constitute a foil to the full quantum theory in two senses. When they are operationally equivalent to a subtheory of the full quantum theory, they are still a foil in the sense that the universe might have been governed by this subtheory rather than the full theory. Why did nature choose the full theory rather than the subtheory? When the epistricted theory is not operationally equivalent to the corresponding quantum subtheory, as in the case of bits, it is a foil not only to the full quantum theory, but to the quantum subtheory as well. In this case, the epistricted theory describes a set of operational predictions that are not instantiated in our universe and again the question is: why did nature not avail itself of this option?

Because epistricted theories share so much of the operational phenomenology of quantum theory, they constitute points in the landscape of possible operational theories that are particularly close to quantum theory. They are therefore particularly helpful in the project of determining what is unique about quantum theory. For any purported attempt to derive the formalism of quantum theory from axioms, it is useful to ask which of the axioms rule out epistricted theories. Axiom sets that may seem promising at first can often be ruled out immediately if they fail to pass this simple test.

Finally, epistricted theories have significance for the problem of developing mathematical frameworks for describing the landscape of operational theories. In particular, they provide a test of the *scope* of any given framework. Broadness of scope is the key virtue of any framework because an axiomatic derivation of quantum theory is only as impressive as the size of the landscape within which it is derived. For instance, the formalism of C^*-algebras essentially includes only operational theories that are fully quantum or fully classical, or that are quantum within each of a set of superselection sectors and classical between these. This is a rather limited scope, and consequently axiomatizations within this framework are less impressive than those formulated within broader frameworks.

On this front, epistricted theories serve to highlight two deficiencies in the prevailing framework of convex operational theories.

First, quadrature epistemic theories are an example of a *possibilistic* or *modal* theory, wherein one does not specify the *probabilities* of measurement outcomes, but only which outcomes are *possible* and which are *impossible*. This perspective on the toy theory of Ref. [1], for instance, is emphasized in Ref. [38]. Possibilistic theories have recently been receiving renewed attention [39–41] because in the context of discussions of Bell's theorem they highlight the fact that quantum theory cannot merely imply an innovation to probability theory but must also imply an innovation to logic.

Second, these epistricted theories have operational state spaces that are not convex. They only allow certain mixtures of operational states. As such, they cannot be captured by the prevailing framework of *convex* operational theories [42, 43] because these assume from the outset a convex state space. On the other hand, epistricted theories can be captured by the category-theoretic framework for process theories [44], as shown in [45], or by the framework of operational-probabilistic theories described in Ref. [46]. Epistricted theories therefore provide a concrete example of how the category-theoretic framework necessarily describes some real estate in the landscape of foil theories that is not on the map of the convex operational framework.

2 Quadrature Epistricted Theories

2.1 Classical Complementarity as an Epistemic Restriction

The criterion on the joint knowability of classical variables that is used here is inspired by the criterion on the *joint measurability* of quantum observables.

Guiding analogy:
A set of observables is *jointly measurable* if and only if it is commuting relative to the matrix commutator.
A set of variables is *jointly knowable* if and only if it is commuting relative to the Poisson bracket.

The full epistemic restriction that we adopt is a combination of this notion of joint knowability together with the restriction that the only variables that can be known by the agent are *linear* combinations of the position and momentum variables. We refer to such variables as *quadrature variables*.[5] We term the full epistemic restriction *classical complementarity*.

Classical complementarity: The valid epistemic states are those wherein an agent knows the values of a set of quadrature variables that commute relative to the Poisson bracket and is maximally ignorant otherwise.

It is presumed that maximal ignorance corresponds to a probability distribution that is uniform over the region of phase-space consistent with the known values of the quadrature variables. In the case of a phase-space associated to a continuous field, uniformity is evaluated relative to the measure that is invariant under phase-space displacements. Hence a valid epistemic state is a uniform distribution over the ontic states that is consistent with a given valuation of some Poisson-commuting set of quadrature variables. It is because of the uniformity of these distributions that they can be understood as merely specifying, for the given constraints, which ontic states are possible and which are impossible. Consequently, the epistemic state in this case is aptly described as a *possibilistic* state.

There is a subtlety here. The epistemic restriction is assumed to apply *only* to what an agent can know about a set of variables based on information acquired entirely to the past or entirely to the future of those variables. It is not assumed to apply to what an agent can know about a set of variables based on pre- and post-selection. The same caveats on applicability hold for the quantum uncertainty principle, so this constraint on applicability is not unexpected.

To describe the epistemic restriction in more detail, we introduce some formalism. The continuous and discrete cases are considered in turn.

Continuous degrees of freedom. Assume n classical continuous degrees of freedom. The configuration space is $\mathbb{R}^n$ and a particular configuration is denoted

$$(\mathsf{q}_1, \mathsf{q}_2, \ldots, \mathsf{q}_n) \in \mathbb{R}^n. \tag{1}$$

These could describe the positions of n particles in a 1-dimensional space, or the positions of $n/3$ particles in a 3-dimensional space, or the amplitudes of n scalar fields, etcetera. The associated phase space is

$$\Omega \equiv \mathbb{R}^{2n} \tag{2}$$

and we denote a point in this space by

$$\mathbf{m} \equiv (\mathsf{q}_1, \mathsf{p}_1, \mathsf{q}_2, \mathsf{p}_2, \ldots, \mathsf{q}_n, \mathsf{p}_n) \in \Omega. \tag{3}$$

[5]This terminology comes from optics, where it was originally used to describe a pair of variables that are canonically conjugate to one another. It was inherited from the use of the expression in astronomy, where it applies to a pair of celestial bodies and describes the configuration in which they have an angular separation of 90° as seen from the earth.

We consider real-valued functionals over this phase space

$$f : \Omega \to \mathbb{R}. \tag{4}$$

In particular, the functionals associated with the position and momentum of the ith degree of freedom are defined respectively by

$$q_i(\mathbf{m}) = \mathtt{q}_i, \;\; p_i(\mathbf{m}) = \mathtt{p}_i. \tag{5}$$

The Poisson bracket is a binary operation on a pair of functionals, defined by

$$[f, g]_{\mathrm{PB}}(\mathbf{m}) \equiv \sum_{i=1}^{n} \left(\frac{\partial f}{\partial q_i} \frac{\partial g}{\partial p_i} - \frac{\partial f}{\partial p_i} \frac{\partial g}{\partial q_i} \right)(\mathbf{m}). \tag{6}$$

In particular, we have

$$\left[q_i, p_j\right]_{\mathrm{PB}}(\mathbf{m}) = \delta_{i,j}. \tag{7}$$

The assumption of classical complementarity incorporates a restriction on the sorts of functionals that an agent can know. Specifically, an agent can only know the value of a functional that is *linear* in the position and momentum functionals, that is, those of the form

$$f = \mathtt{a}_1 q_1 + \mathtt{b}_1 p_1 + \cdots + \mathtt{a}_n q_n + \mathtt{b}_n p_n + \mathtt{c}, \tag{8}$$

where $\mathtt{a}_1, \mathtt{b}_1, \ldots, \mathtt{a}_n, \mathtt{b}_n, \mathtt{c} \in \mathbb{R}$. (Note that functionals that differ only by addition of a scalar or by a multiplicative factor ultimately describe the same property.) We will call these *quadrature functionals* or *quadrature variables*. The vector of coefficients of the position and momentum functionals for a given quadrature functional will be denoted by the boldface of the notation used for the functional itself. The vector $\mathbf{f}$ specifying the position and momentum dependence of the quadrature functional f defined in Eq. (8) is

$$\mathbf{f} \equiv (\mathtt{a}_1, \mathtt{b}_1, \ldots, \mathtt{a}_n, \mathtt{b}_n), \tag{9}$$

such that if we define the vector of position and momentum functionals

$$\mathbf{z} \equiv (q_1, p_1, \ldots, q_n, p_n), \tag{10}$$

we can express f as

$$f = \mathbf{f}^T \mathbf{z} + \mathtt{c}. \tag{11}$$

Similarly, the action of the functional f on a phase space vector $\mathbf{m}$ is given by

$$f(\mathbf{m}) = \mathbf{f}^T \mathbf{m} + \mathtt{c}. \tag{12}$$

In other words, the space of quadrature functionals is the dual of the phase space Ω, but each functional f is associated with a vector in the phase space, $\mathbf{f} \in \Omega$. Note that the vectors associated with the position and momentum functionals q_i and p_i are $\mathbf{q}_i \equiv (0, 0, \ldots, 1, 0, \ldots, 0, 0)$ where the only nonzero component is $\mathtt{a}_i$ and $\mathbf{p}_i \equiv (0, 0, \ldots, 0, 1, \ldots, 0, 0)$, where the only nonzero component is $\mathtt{b}_i$.

It is not difficult to see that the Poisson bracket of two quadrature functionals always evaluates to a functional that is uniform over the phase space. Its value is equal to the symplectic inner product of the associated vectors,

$$[f, g]_{PB}(\mathbf{m}) = \langle \mathbf{f}, \mathbf{g} \rangle, \tag{13}$$

where

$$\langle \mathbf{f}, \mathbf{g} \rangle \equiv \mathbf{f}^T J \mathbf{g}, \tag{14}$$

with T denoting transpose and J denoting the skew-symmetric $2n \times 2n$ matrix with components $J_{ij} \equiv \delta_{i,j+1} - \delta_{i+1,j}$, that is,

$$J \equiv \begin{pmatrix} 0 & 1 & 0 & 0 & \ldots \\ -1 & 0 & 0 & 0 & \\ 0 & 0 & 0 & 1 & \\ 0 & 0 & -1 & 0 & \\ \vdots & & & & \ddots \end{pmatrix}. \tag{15}$$

(Note that J squares to the negative of the $2n \times 2n$ identity matrix, $J^2 = -I$, it is an orthogonal matrix, $J^T J = I$, it has determinant $+1$, and it has an inverse given by $J^{-1} = J^T = -J$.) For instance, for $\Omega = \mathbb{R}^2$, if $\mathbf{f} = (\mathtt{a}, \mathtt{b})$ and $\mathbf{g} = (\mathtt{a}', \mathtt{b}')$, then $\langle \mathbf{f}, \mathbf{g} \rangle = \mathtt{a}\mathtt{b}' - \mathtt{b}\mathtt{a}'$.

The symplectic inner product on a phase space Ω is a bilinear form $\langle \cdot, \cdot \rangle : \Omega \times \Omega \to \mathbb{R}$ that is skew-symmetric ($\langle \mathbf{f}, \mathbf{g} \rangle = -\langle \mathbf{g}, \mathbf{f} \rangle$ for all $\mathbf{f}, \mathbf{g} \in \Omega$) and non-degenerate (if $\langle \mathbf{f}, \mathbf{g} \rangle = 0$ for all $\mathbf{g} \in \Omega$, then $\mathbf{f} = 0$). By equipping the vector space Ω with the symplectic inner product, it becomes a symplectic vector space. This connection to symplectic geometry allows us to provide a simple geometric interpretation of the Poisson-commuting sets of quadrature functionals, which we will present in Sect. 2.2.1.

Discrete degrees of freedom. For discrete degrees of freedom, the formalism is precisely the same, except that variables are no longer valued in the real field $\mathbb{R}$, but a finite field instead. Recall that all finite fields have order equal to the power of a prime. We shall consider here only the case where the order is itself a prime, denoted d, in which case the field is isomorphic to the integers modulo d, which we will denote by $\mathbb{Z}_d$. Therefore, the configuration space is $(\mathbb{Z}_d)^n$, the associated phase space is

$$\Omega \equiv (\mathbb{Z}_d)^{2n},$$

the functionals have the form

$$f : \Omega \rightarrow \mathbb{Z}_d,$$

and the linear functionals are of the form of Eq. (8),

$$f = \mathtt{a}_1 q_1 + \mathtt{b}_1 p_1 + \cdots + \mathtt{a}_n q_n + \mathtt{b}_n p_n + \mathtt{c}, \tag{16}$$

but where $\mathtt{a}_1, \mathtt{b}_1, \ldots, \mathtt{a}_n, \mathtt{b}_n, \mathtt{c} \in \mathbb{Z}_d$ and the sum denotes addition modulo d. It follows that the vector $\mathbf{f} \equiv (\mathtt{a}_1, \mathtt{b}_1, \ldots, \mathtt{a}_n, \mathtt{b}_n)$ associated with the functional f lives in the phase space $\Omega \equiv (\mathbb{Z}_d)^n$ as well.

The Poisson bracket, however, cannot be defined in the conventional way, because without continuous variables we do not have a notion of derivative. Nonetheless, one can define a discrete version of the Poisson bracket in terms of finite differences. For any functionals $f : \Omega \rightarrow \mathbb{Z}_d$ and $g : \Omega \rightarrow \mathbb{Z}_d$, their Poisson bracket, denoted $[f, g]_{PB}$, is also such a functional, the one defined by

$$[f, g]_{PB}(\mathbf{m}) \equiv \sum_{i=1}^{n} \begin{bmatrix} (f(\mathbf{m} + \mathbf{q}_i) - f(\mathbf{m}))(g(\mathbf{m} + \mathbf{p}_i) - g(\mathbf{m})) \\ -(f(\mathbf{m} + \mathbf{p}_i) - f(\mathbf{m}))(g(\mathbf{m} + \mathbf{q}_i) - g(\mathbf{m})) \end{bmatrix}, \tag{17}$$

where the differences in this expression are evaluated with modular arithmetic. The requirement that

$$\left[q_i, p_j\right]_{\mathrm{PB}}(\mathbf{m}) = \delta_{i,j}, \tag{18}$$

is clearly satisfied. Furthermore, it is straightforward to verify that Eq. (13) also holds under this definition, so that one can relate the Poisson bracket in the discrete setting to the symplectic inner product on the discrete phase space, $\langle \cdot, \cdot \rangle : \Omega \times \Omega \rightarrow \mathbb{Z}_d$.

Simple examples. It is useful to consider some simple examples of commuting pairs of quadrature variables, that is, some examples of 2-element sets $\{f, g\}$ such that $[f, g]_{PB} = 0$. Any quadrature variable defined for system 1 commutes with any quadrature variable for system 2, e.g., the pair

$$\mathtt{a}_1 q_1 + \mathtt{b}_1 p_1, \quad \mathtt{a}_2 q_2 + \mathtt{b}_2 p_2 \tag{19}$$

is a commuting pair for any values of $\mathtt{a}_1, \mathtt{b}_1, \mathtt{a}_2, \mathtt{b}_2 \in \mathbb{R}$ (or $\mathtt{a}_1, \mathtt{b}_1, \mathtt{a}_2, \mathtt{b}_2 \in \mathbb{Z}_d$). Additionally, there are commuting pairs of quadrature variables describing joint properties of the two systems, for instance

$$q_1 - q_2, \quad p_1 + p_2 \tag{20}$$

(when the field is $\mathbb{Z}_d$, the coefficient -1 is equivalent to $d - 1$, so that $q_1 - q_2 = q_1 + (d - 1)q_2$).

Another useful concept in the following will be that of *canonically conjugate* variables. A pair of variables are said to be canonically conjugate if $[f, g]_{PB} = 1$. On a single system, the pair of quadrature variables

$$\mathtt{a}q + \mathtt{b}p, \quad -\mathtt{b}q + \mathtt{a}p \tag{21}$$

are canonically conjugate for any values $\mathtt{a}, \mathtt{b} \in \mathbb{R}$ (or $\mathtt{a}, \mathtt{b} \in \mathbb{Z}_d$) such that $\mathtt{a}^2 + \mathtt{b}^2 = 1$; in particular $\{q, p\}$ is such a pair.

Note that we were able to present these examples without specifying the nature of the field. We will follow this convention of presenting results in a unified field-independent manner for the next few sections.

2.2 *Characterization of Quadrature Epistricted Theories*

2.2.1 The Set of Valid Epistemic States

Using the connection between the Poisson bracket for quadrature functionals and the symplectic inner product, one obtains a geometric interpretation of the epistemic restriction and the valid epistemic states.

To specify an epistemic state one must specify: (i) the set of quadrature variables that are known and (ii) the values of these variables. We will consider each aspect in turn.

The epistemic restriction asserts that the only sets of variables that are jointly knowable are those that are Poisson-commuting (which is to say that every pair of elements in the set Poisson-commutes). Note, however, that if every variable in a set has a known value, then any function of those variables also has a known value, in particular any linear combinations of those variables has a known value. It follows that for any Poisson-commuting set of variables, we can close the set under linear combination and preserve the property of being Poisson-commuting. In terms of the vectors representing these variables, this implies that we can take their *linear span* while preserving the property of having vanishing symplectic inner product for every pair of vectors. In terms of the symplectic geometry, a subspace all of whose vectors have vanishing symplectic inner product with one another is called an *isotropic* subspace of the phase space. Formally, a subspace $V \subseteq \Omega$ is isotropic if

$$\forall \mathbf{f}, \mathbf{g} \in V : \langle \mathbf{f}, \mathbf{g} \rangle = 0. \tag{22}$$

It follows that we can parametrize the different possible sets of known variables in terms of the isotropic subspaces of the phase space Ω.

For a $2n$-dimensional phase space, the maximum possible dimension of an isotropic subspace is n. These are called *maximally isotropic* or *Lagrangian* subspaces. This case corresponds to the maximal possible knowledge an agent can have according to the epistemic restriction. The agent then knows a *complete* set of Poisson-commuting variables, which is the analogue of measuring a complete set of commuting observables in quantum theory.

For a given Poisson-commuting set of variables, define a basis of that set to be any subset containing linearly independent elements and from which the entire set can be obtained by linear combinations. In the symplectic geometry, this corresponds to a vector basis for the associated isotropic subspace. There are, of course, many choices of bases for a given isotropic subspace or Poisson-commuting set.

Next, we must characterize the possible value assignments to a Poisson-commuting set of quadrature variables. That is, we must specify a linear functional v acting on a quadrature functional f and taking values in the appropriate field (continuous or discrete) such that $v(f)$ is the value assigned to f. Denote the isotropic subspace of Ω that is associated to this Poisson-commuting set by V, such that $\mathbf{f} \in V$ is the vector associated with the quadrature function f. The set of value assignments corresponds precisely to the set of vectors in V. In other words, for every vector $\mathbf{v} \in V$, which we call a *valuation vector*, we obtain a distinct value assignment v, via

$$v(f) = \mathbf{f}^T \mathbf{v}.$$

To see this, it suffices to note that the ontic state of the system determines the values of all functionals and therefore the set of possible value assignments is given by the set of possible ontic states. Specifically, each ontic state $\mathbf{m} \in \Omega$ defines the value assignment

$$v_{\mathbf{m}}(f) = \mathbf{f}^T \mathbf{m}.$$

However, many different ontic states yield the same value assignment. Denoting the projector onto V by P_V, we can express the relevant equivalence relation thus: the ontic states $\mathbf{m}$ and $\mathbf{m}'$ yield the same value assignment to the quadrature functionals associated to V if and only if $P_V \mathbf{m} = P_V \mathbf{m}'$. It follows that the set of possible value assignments to the functionals associated to V can be parametrized by the set of projections of all ontic states $\mathbf{m} \in \Omega$ into V, which is simply the set of ontic states in V. This establishes what we set out to prove.

As an example, consider the case where we have two degrees of freedom, so that Ω is 4-dimensional, and suppose that the set of quadrature variables that are jointly known are the position variables, $\{q_1, q_2\}$, and that these are known to each take the value 1. In this case, the associated isotropic subspace $V \subseteq \Omega$, and the valuation vector $\mathbf{v} \in V$ are, respectively,

$$V = \text{span}\{\mathbf{q}_1, \mathbf{q}_2\} \tag{23}$$

$$= \text{span}\{(1, 0, 0, 0), (0, 0, 1, 0)\} \tag{24}$$

$$= \{(\mathtt{s}, 0, \mathtt{t}, 0) : \mathtt{s}, \mathtt{t} \in \mathbb{R}/\mathbb{Z}_d\} \tag{25}$$

$$\mathbf{v} = (1, 0, 1, 0). \tag{26}$$

These are depicted in green in Fig. 1.

Next, we consider a given epistemic state, where the known quadrature variables are specified by an isotropic subspace $V \subseteq \Omega$, and their values are specified by $\mathbf{v} \in V$, and ask: what probability distribution over the phase space Ω does it correspond to? Recalling that this probability distribution should be maximally uninformative

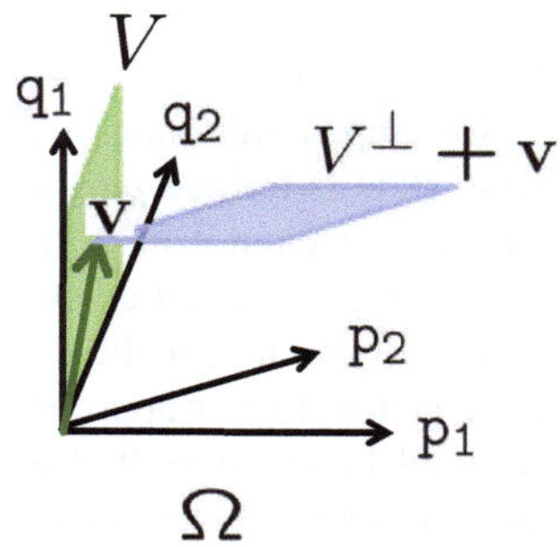

Fig. 1 A schematic of the 4-dimensional phase space of ontic states, Ω, the isotropic subspace $V \subseteq \Omega$ associated with the known quadrature variables, the valuation vector $\mathbf{v} \in V$ specifying the values of the known variables and the manifold $V^{\perp}+\mathbf{v}$ corresponding to the ontic support of the associated epistemic state

relative to the given constraint, the answer is simply a uniform distribution on the set of all ontic states that yield this value assignment, that is, on the set

$$\begin{aligned}&\{\mathbf{m} \in \Omega : \mathbf{f}^T\mathbf{m} = \mathbf{f}^T\mathbf{v}\ \forall \mathbf{f} \in V\}\\ &= \{\mathbf{m} \in \Omega : P_V\mathbf{m} = \mathbf{v}\}.\end{aligned} \tag{27}$$

If we denote the subspace of Ω that is orthogonal to V (relative to the Euclidean inner product) by $V^{\perp}$, and we denote the translation of a subspace W by a vector $\mathbf{v}$ as $W+\mathbf{v} \equiv \{\mathbf{m} : \mathbf{m} = \mathbf{w}+\mathbf{v},\ \mathbf{w} \in W\}$, then it is clear that the set of ontic states of (27) is simply

$$V^{\perp}+\mathbf{v}.$$

For instance, in the example described above,

$$V^{\perp} = \{(0, \mathtt{s}, 0, \mathtt{t}) : \mathtt{s}, \mathtt{t} \in \mathbb{R}/\mathbb{Z}_d\}, \tag{28}$$

and consequently the set of ontic states consistent with the agent's knowledge is

$$V^{\perp}+\mathbf{v} \equiv \{(1, \mathtt{s}, 1, \mathtt{t}) : \mathtt{s}, \mathtt{t} \in \mathbb{R}/\mathbb{Z}_d\}, \tag{29}$$

which is depicted in blue in Fig. 1.

As a probability distribution over Ω, the epistemic state associated to $(V, \mathbf{v})$ has the following form

$$\mu_{V,\mathbf{v}}(\mathbf{m}) = \frac{1}{\mathcal{N}_V}\delta_{V^{\perp}+\mathbf{v}}(\mathbf{m}) \tag{30}$$

where we have introduced the notation

$$\delta_{V^{\perp}+\mathbf{v}}(\mathbf{m}) \equiv \prod_{\mathbf{f}^{(i)}:\mathrm{span}\{\mathbf{f}^{(i)}\}=V} \delta(\mathbf{f}^{(i)T}\mathbf{m} - \mathbf{f}^{(i)T}\mathbf{v}), \tag{31}$$

where in the discrete case $\delta(\mathtt{c}) = 1$ if $\mathtt{c} = 0$ and $\delta(\mathtt{c}) = 0$ otherwise, while in the continuous case δ denotes a Dirac delta function. In this expression, $\{\mathbf{f}^{(i)}\}$ can be any basis of V. Geometrically, $\mu_{V,\mathbf{v}}$ is simply the uniform distribution over the ontic states in $V^{\perp} + \mathbf{v}$.

Some epistemic states are seen to be mixtures of others in this theory. A valid epistemic state is termed *pure* if it is convexly extremal among valid epistemic states, that is, if it cannot be formed as a convex combination of other valid epistemic states. Non-extremal epistemic states are termed *mixed.* Note that we are judging extremality relative to the set of valid epistemic states, not relative to the set of all epistemic states. In our approach, the pure epistemic states are those corresponding to maximal knowledge, that is, knowledge associated to a complete set of Poisson-commuting quadrature variables. Note, however, that because of the epistemic restriction, maximal knowledge is always incomplete knowledge.

2.2.2 The Set of Valid Transformations

In addition to specifying the valid epistemic states, we must also specify what transformations of the epistemic states are allowed in our theory. To begin with, we consider the reversible transformations on an isolated system.

Suppose an agent knows the precise ontological dynamics of a system over some period of time. This transformation is represented by a bijective map on the ontic state space, and this induces a bijective map on the space of epistemic states.

Because we assume that the underlying ontological theory has symplectic structure, it follows that the allowed transformations must be within the set of *symplectic transformations* (sometimes called *symplectomorphisms*). The requirement that the epistemic restriction must be preserved under the transformation implies that the valid transformations are a subset of the symplectic transformations, namely, those that map the set of quadrature variables to itself. Each such transformation can be represented in terms of its action on the phase space vector $\mathbf{m} \in \Omega$ as

$$\mathbf{m} \mapsto S\mathbf{m} + \mathbf{a} \tag{32}$$

where $\mathbf{a} \in \Omega$ is a phase-space displacement vector and where S is a $2n \times 2n$ *symplectic matrix*, that is, one which preserves the symplectic form J defined in Eq. (15),

$$S^T J S = J, \tag{33}$$

or equivalently, one which preserves symplectic inner products, i.e., $(S\mathbf{m})^T \, J \, (S\mathbf{m}') = \mathbf{m}^T J \mathbf{m}' \; \forall \mathbf{m}, \mathbf{m}' \in \Omega$. These are combinations of phase-space rotations and phase-space displacements.

Equation (32) describes an affine transformation, but it does not include all such transformations because S is not a general linear matrix. Following [6], we call transformations of the form of Eq. (32) *symplectic affine transformations*. Two such transformations, $S \cdot + \mathbf{a}$ and $S' \cdot + \mathbf{a}'$ compose as

$$S\left(S'\cdot+\mathbf{b}'\right)+\mathbf{b}=SS'\cdot+(S\mathbf{b}'+\mathbf{b}). \tag{34}$$

The inverse of a symplectic matrix S is $S^{-1}=J^TS^TJ$, and the inverse of the phase-space displacement $\mathbf{a}$ is of course $-\mathbf{a}$. We call the resulting group of transformations the *symplectic affine group*.

If the epistemic state is described by a probability distribution/density over ontic states, $\mu:\Omega\to\mathbb{R}_+$, then under the ontological transformation $\mathbf{m}\mapsto S\mathbf{m}+\mathbf{a}$, the transformation induced on the epistemic state is

$$\mu(\mathbf{m})\mapsto\mu'(\mathbf{m})=\mu(S^{-1}\mathbf{m}-\mathbf{a}), \tag{35}$$

We can equivalently represent this transformation by a conditional probability distribution $\Gamma_{S,\mathbf{a}}:\Omega\times\Omega\to\mathbb{R}_+$, that is,

$$\mu'(\mathbf{m})=\int d\mathbf{m}'\Gamma_{S,\mathbf{a}}(\mathbf{m}|\mathbf{m}')\mu(\mathbf{m}'), \tag{36}$$

where

$$\Gamma_{S,\mathbf{a}}(\mathbf{m}|\mathbf{m}')=\delta(\mathbf{m}-(S\mathbf{m}'+\mathbf{a})). \tag{37}$$

There is a subtlety worth noting at this point. The map $\mu\mapsto\mu'$ on the space of probability distributions, which is induced by the map $\mathbf{m}\mapsto S\mathbf{m}+\mathbf{a}$ on the space of ontic states, has the following property: it maps the set of valid epistemic states (those satisfying the classical complementarity principle) to itself. However, not every map from the set of valid epistemic states to itself can be induced by some map on the space of ontic states. A simple counterexample is provided by the map corresponding to time reversal. For a single degree of freedom, time reversal is represented by the map $\mathbf{m}=(\mathtt{q},\mathtt{p})\mapsto\mathbf{m}'=(\mathtt{q},-\mathtt{p})$, which obviously fails to preserve the symplectic form. In terms of symplectic geometry, it is a reflection rather than a rotation in the phase space. Nonetheless, it maps isotropic subspaces to isotropic subspaces and therefore it also maps valid epistemic states to valid epistemic states. Therefore, in considering a given map on the space of distributions over phase space, it is not sufficient to ensure that it takes valid epistemic states to valid epistemic states, one must also ensure that it arises from a possible ontological dynamics. We say that the map must *supervene* upon a valid ontological transformation [2].

Note that if the phase space is over a *discrete* field, then the transformations must be discrete in time. Only in the case of continuous variables can the transformations be continuous in time and only in this case can they be generated by a Hamiltonian.

In addition to transformations corresponding to reversible maps over the epistemic states, there are also transformations corresponding to irreversible maps. These correspond to the case where information about the system is lost. The most general such transformation corresponds to adjoining the system to an ancilla that is prepared in a valid epistemic state, evolving the pair by some symplectic affine transformation that involves a nontrivial coupling of the two, and finally marginalizing over the ancilla.

The reason this leads to a loss of information about the ontic state of the system is that the transformation of the system depends on the initial ontic state of the ancilla, and the latter is never completely known, by virtue of the epistemic restriction.

2.2.3 The Set of Valid Measurements

We must finally address the question of which *measurements* are consistent with our epistemic restriction. We will distinguish sharp and unsharp measurements. The sharp measurements are the analogues of those associated with projector-valued measures in quantum theory and can be defined as those for which the outcome is deterministic given the ontic state. The unsharp measurements are the analogues of those in quantum theory that cannot be represented by a projector-valued measure but instead require a positive operator-valued measure; they can be defined as those for which the outcome is not deterministic given the ontic state.

We begin by considering the valid *sharp* measurements. Without the epistemic restriction, one could imagine the possibility of a sharp measurement that would determine the values of *all* quadrature variables, and hence also determine what the ontic state of the system was prior to the measurement. Given classical complementarity, however, one can only jointly retrodict the values of a set of quadrature variables if these are a Poisson-commuting set, and therefore the only sets of quadrature variables that can be jointly measured are the Poisson-commuting sets.

Given that every Poisson-commuting set of quadrature variables defines an isotropic subspace, the valid sharp measurements are parametrized by the isotropic subspaces. Furthermore, the possible joint value-assignments to a Poisson-commuting set of variables associated with isotropic subspace V are parametrized by the vectors in V, so that the outcomes of the measurement associated with V are indexed by $\mathbf{v} \in V$.

Such measurements can be represented as a conditional probability, specifying the probability of each outcome $\mathbf{v}$ given the ontic state $\mathbf{m}$, namely,

$$\xi_V(\mathbf{v}|\mathbf{m}) = \delta_{V^\perp+\mathbf{v}}(\mathbf{m}), \tag{38}$$

where $\delta_{V^\perp+\mathbf{v}}(\mathbf{m})$ is defined in Eq. (31). We refer to the set $\{\xi_V(\mathbf{v}|\mathbf{m}) : \mathbf{v} \in V\}$, considered as functions over Ω, as the *response functions* associated with the measurement.

The set of all valid *unsharp* measurements can then be defined in terms of the valid sharp measurements as follows. An unsharp measurement on a system is valid if it can be implemented by adjoining to the system an ancilla that is described by a valid epistemic state, coupling the two by a symplectic affine transformation, and finally implementing a valid sharp measurement on the system+ancilla. Note that this construction of unsharp measurements from sharp measurements on a larger system is the analogue of the Naimark dilation in quantum theory.

A full treatment of measurements would include a discussion of how the epistemic state is updated when the system survives the measurement procedure, but we will not discuss the transformative aspect of measurements in this article.

2.2.4 Operational Statistics

Suppose that one prepares a system with phase space Ω in the epistemic state $\mu_{V,\mathbf{v}}(\mathbf{m})$ associated with isotropic subspace V and valuation vector $\mathbf{v}$, and one subsequently implements the sharp measurement associated with the isotropic subspace V'. What is the probability of obtaining a given outcome $\mathbf{v}' \in V'$? The answer follows from an application of the law of total probability. The probability is simply

$$\sum_{\mathbf{m}\in\Omega} \xi_{V'}(\mathbf{v}'|\mathbf{m})\mu_{V,\mathbf{v}}(\mathbf{m}). \tag{39}$$

If a symplectic affine transformation $\mathbf{m} \mapsto S\mathbf{m} + \mathbf{a}$ is applied between the preparation and the measurement, the probability of outcome $\mathbf{v}'$ becomes

$$\sum_{\mathbf{m}\in\Omega} \xi_{V'}(\mathbf{v}'|\mathbf{m}) \sum_{\mathbf{m}'\in\Omega} \Gamma_{S,\mathbf{a}}(\mathbf{m}|\mathbf{m}')\mu_{V,\mathbf{v}}(\mathbf{m}'). \tag{40}$$

These statistics constitute the operational content of the quadrature epistricted theory.

2.3 *Quadrature Epistricted Theory of Continuous Variables*

We now turn to concrete examples of quadrature epistricted theories for particular choices of the field. In this section, we consider the case of a phase space of n *real* degrees of freedom, $\Omega = \mathbb{R}^{2n}$. We begin by discussing the valid epistemic states for a single degree of freedom, $n = 1$. In this case, the phase space is 2-dimensional and the isotropic subspaces are the set of 1-dimensional subspaces. We have depicted a few examples in Fig. 2. The isotropic subspace V is depicted in light green, the valuation vector $\mathbf{v}$ is depicted as a dark green arrow, and the set $V^{\perp}+\mathbf{v}$ of ontic states in the support of the associated epistemic state is depicted in blue. Figure 2a depicts a state of knowledge wherein position is known (and hence momentum is unknown). Figure 2b depicts the vice-versa. Figure 2c corresponds to knowing the value of a quadrature $(\cos\theta)\,q + (\sin\theta)\,p$ (and hence having no knowledge of the canonically conjugate quadrature $-(\sin\theta)\,q + (\cos\theta)\,p$). Finally, an agent could know nothing at all, in which case the epistemic state is just the uniform distribution over the whole phase space, as depicted in Fig. 2d.

If one considers a pair of continuous degrees of freedom, then it becomes harder to visualize the epistemic states because the phase space is 4-dimensional. Nonetheless, we present 3-dimensional projections as a visualization tool. We know that for every pair of isotropic subspace and valuation vector, $(V, \mathbf{v})$, there is a distinct epistemic state. In Fig. 3a, we depict the example where q_1 and q_2 are the known variables and

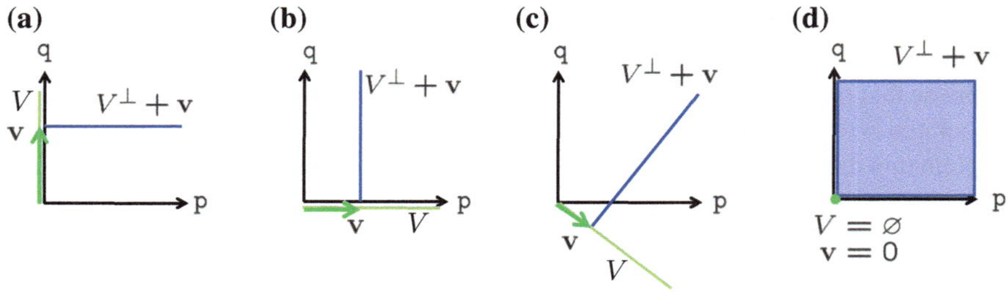

Fig. 2 Examples of valid epistemic states for a single continuous variable system

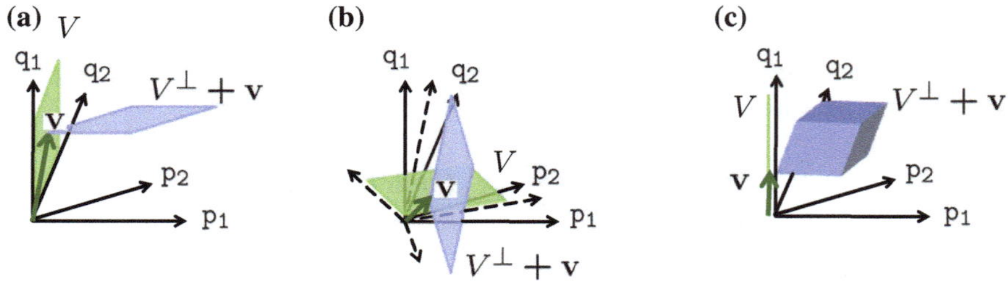

Fig. 3 Examples of valid epistemic states for a pair of continuous variable systems

both take the value 1, so that $V = \text{span}\{\mathbf{q}_1, \mathbf{q}_2\} = \text{span}\{(1, 0, 0, 0), (0, 0, 1, 0)\}$ and $\mathbf{v} = (1, 0, 1, 0)$, while in Fig. 3b, it is $q_1 - q_2$ and $p_1 + p_2$ that are the known variables and both take the value 1, so that $V = \text{span}\{\mathbf{q}_1 - \mathbf{q}_2, \mathbf{p}_1 + \mathbf{p}_2\} = \text{span}\{(1, 0, -1, 0), (0, 1, 0, 1)\}$ and $\mathbf{v} = \left(\frac{1}{2}, \frac{1}{2}, -\frac{1}{2}, \frac{1}{2}\right)$. In the example of Fig. 3c, only a single variable, q_1, is known and takes the value 1, so that $V = \text{span}\{\mathbf{q}_1\} = \text{span}\{(1, 0, 0, 0)\}$ and $\mathbf{v} = (1, 0, 0, 0)$.

2.4 Quadrature Epistricted Theory of Trits

We turn now to discrete systems. We begin with the case where the configuration space of every degree of freedom is three-valued, i.e., a *trit*, and represented therefore by $\mathbb{Z}_3$, the integers modulo 3. The configuration space of *n* degrees of freedom is $\mathbb{Z}_3^n$ and the phase space is $\Omega = (\mathbb{Z}_3)^{2n}$.

For a single system ($n = 1$), we can depict Ω as a 3×3 grid. Consider all of the quadrature functionals that can be defined on such a system. They are of the form $f = \mathtt{a}q + \mathtt{b}p + \mathtt{c}$ where $\mathtt{a}, \mathtt{b}, \mathtt{c} \in \mathbb{Z}_3$. Some of these functionals partition the phase-space in equivalent ways. It suffices to look at the inequivalent quadrature functionals. There are four of these:

$$q,\ p,\ q + p,\ q + 2p. \tag{41}$$

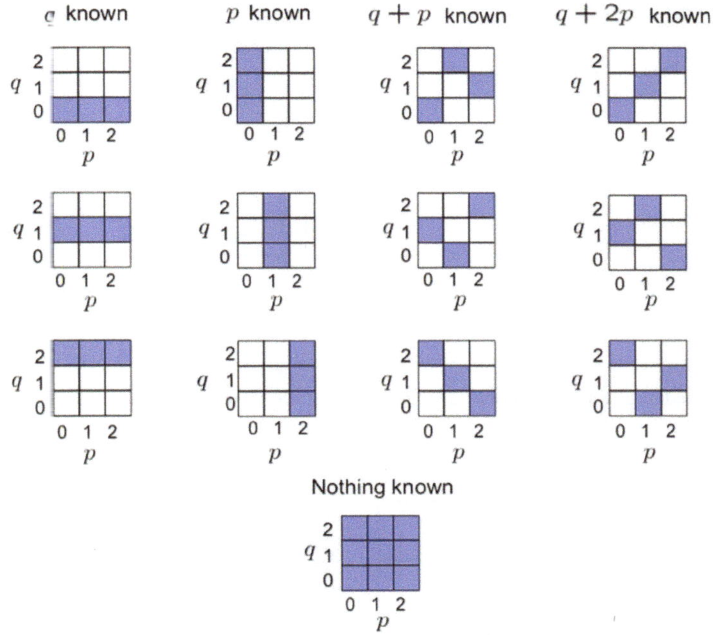

Fig. 4 All the valid epistemic states for a single trit. There are twelve states of maximal knowledge (one variable known) and a single state of nonmaximal knowledge (no variable known)

Note that because addition is modulo 3, $q + 2p$ could equally well be written $q - p$.

Because no two of these functionals Poisson-commute, the principle of classical complementarity implies that an agent can know the value of at most one of these variables. It follows that there are twelve pure epistemic states, depicted in Fig. 4. The only mixed state is the state of complete ignorance. Here we depict in blue the ontic states in the support of the epistemic state. We have not explicitly depicted the isotropic subspace and valuation vector, but these are analogous to what we had in the continuous variable case.

Next, we can consider *pairs* of trits ($n = 2$). The quadrature variables are linear combinations of the positions and momentum of each, with coefficients drawn from $\mathbb{Z}_3$. Just as in the continuous case, one now has quadrature variables that describe joint properties of the pair of systems. The complete sets of Poisson-commuting variables now contain a pair of variables. Rather than attempting to portray the 4-dimensional phase space, as we did in the continuous case, we can depict each 2-dimensional symplectic subspace along a line, as in Fig. 5. This is the "Sudoku puzzle" depiction of the two-trit phase space.

89

Figures 6a, b, and 7a, b each depict a mixed epistemic state, wherein the value of a single quadrature variable is known. Figures 6c and 7c depict pure epistemic states, wherein the values of a pair of Poisson-commuting variables are known. If one of

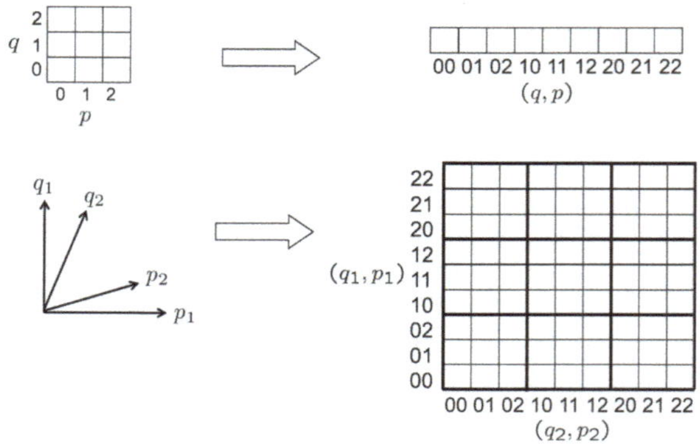

Fig. 5 Embedding a 4-dimensional discrete phase space in a 2-dimensional grid resembling a Sudoku puzzle

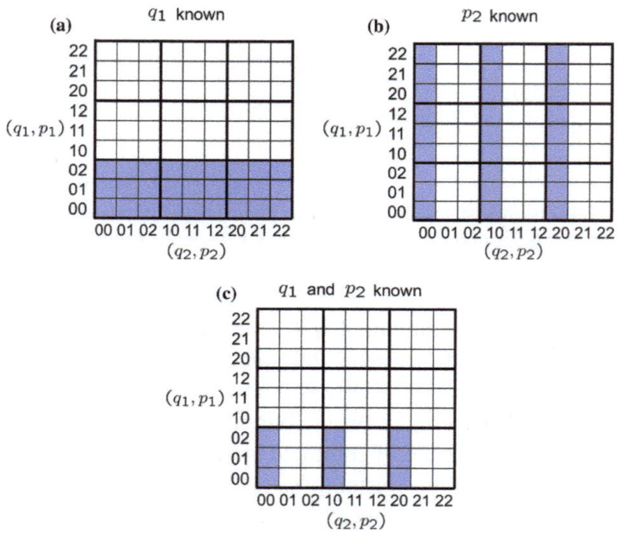

Fig. 6 Examples of epistemic states for two trits. **a** and **b** are examples where one variable is known. **c** is an example where two variables are known, corresponding to a product quantum state

the pair of known variables refers to the first subsystem and the other refers to the second subsystem, as in Fig. 6c, the epistemic state corresponds to a product state in quantum theory. If both of the known variables describe joint properties of the pair of trits, as in Fig. 7c, the epistemic state corresponds to an entangled state.

The valid reversible transformations are the affine symplectic maps on the phase-space. These correspond to a particular subset of the permutations. Some examples are depicted in Fig. 8.

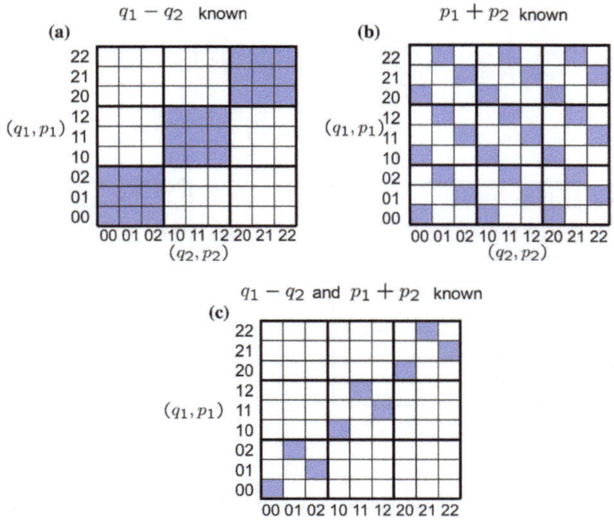

Fig. 7 Examples of epistemic states for two trits. **a** and **b** are examples where one variable is known. **c** is an example where two variables are known, corresponding to an entangled quantum state

Just as in the continuous case, the valid measurements are those that determine the values of a set of Poisson-commuting quadrature variables. For instance, for a single trit, there are only four inequivalent measurements: of q, of p, of $q + p$ and

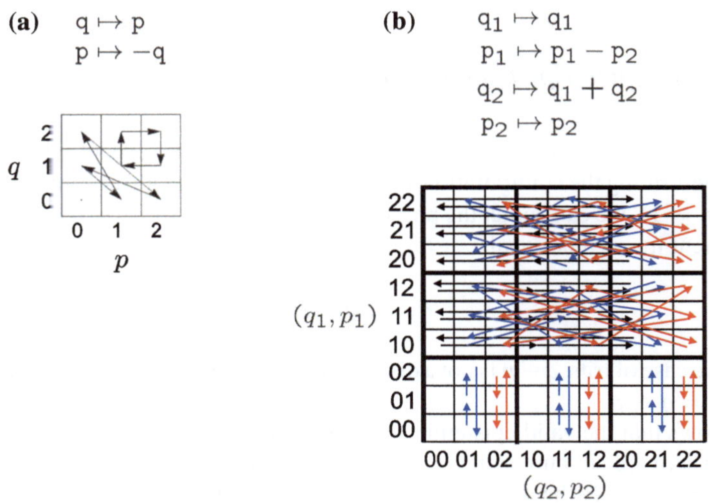

Fig. 8 Examples of valid reversible transformations for **a** one trit, and **b** two trits

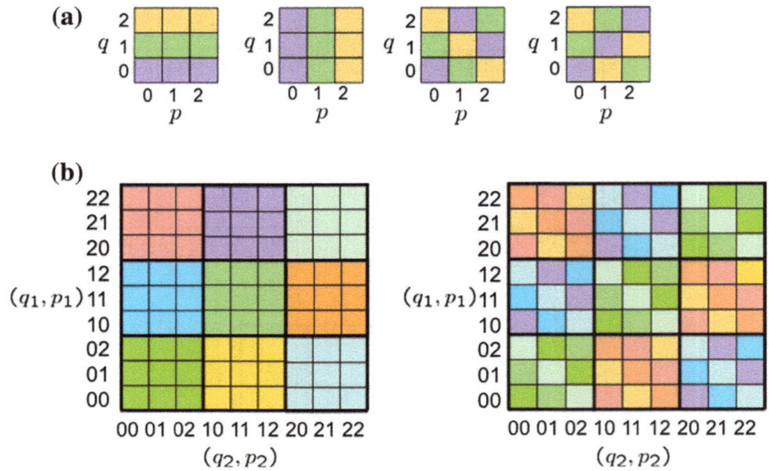

Fig. 9 **a** The set of valid sharp measurements on a single trit. **b** Examples of valid sharp measurements on a pair of trits. The support of the response function corresponding to a particular outcome is coloured uniformly

of $q + 2p$, depicted in Fig. 9a, with different colours denoting different outcomes. Figure 9b depicts some valid measurements on a pair of trits. The left depicts a joint measurement of q_1 and q_2, which corresponds to a product basis in quantum theory. The right depicts a joint measurement of $q_1 - q_2$ and $p_1 + p_2$, which corresponds to a basis of entangled states.

2.5 Quadrature Epistricted Theory of Bits

The epistricted theory of bits is very similar to that of trits, except with $\mathbb{Z}_2$ rather than $\mathbb{Z}_3$ describing the configuration space of a single degree of freedom. For a single system ($n = 1$), we can depict the phase space Ω as a 2×2 grid. There are only three inequivalent linear functionals:

$$q, \ p, \ q + p. \tag{42}$$

Unlike the case of trits, $q - p$ is not a distinct functional because in arithmetic modulo 2, $q - p = q + p$.

It follows that the valid epistemic states for a single system are those depicted in Fig. 10. There are six pure states and one mixed state. We adopt a similar graphical convention to depict the 4-dimensional phase space of a pair of bits as we did for a pair of trits, presented in Fig. 11. Because the combinatorics are not so bad for

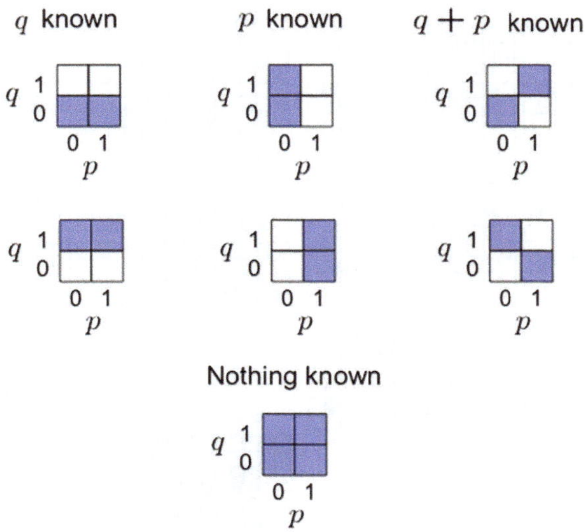

Fig. 10 All the valid epistemic states for a single bit. There are six states of maximal knowledge (one variable known) and a single state of nonmaximal knowledge (no variable known)

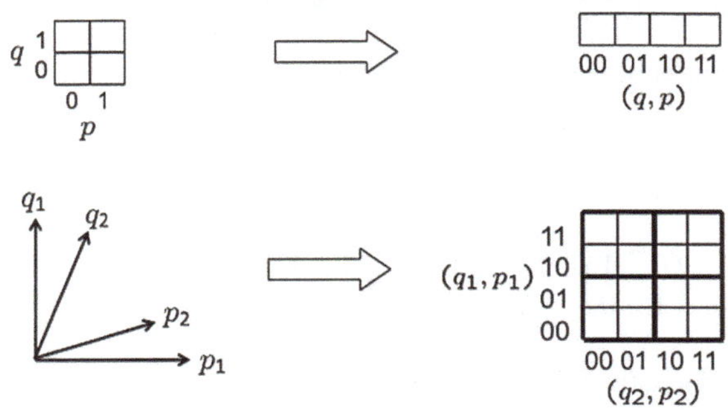

Fig. 11 Embedding a 4-dimensional discrete phase space in a 2-dimensional grid

the case of bits, we depict *all* of the valid epistemic states for a pair of bits in Fig. 12. We categorize these into those for which two variables are known (the pure states) and those for which only one or no variable is known (the mixed states). We also categorize these according to whether they exhibit correlation between the two subsystems or not. The pure correlated states correspond to the entangled states.

The reversible transformations for the case of a single system ($n = 1$) are particularly simple. In this case, $\Omega = (\mathbb{Z}_d)^2$, and the symplectic form is simply $J = \begin{pmatrix} 0 & 1 \\ 1 & 0 \end{pmatrix}$, because $-1 = 1$ in arithmetic modulo 2. As such, the symplectic matrices in this case are those with elements in $\mathbb{Z}_2$ and satisfying $S^T \begin{pmatrix} 0 & 1 \\ 1 & 0 \end{pmatrix} S = \begin{pmatrix} 0 & 1 \\ 1 & 0 \end{pmatrix}$. These are

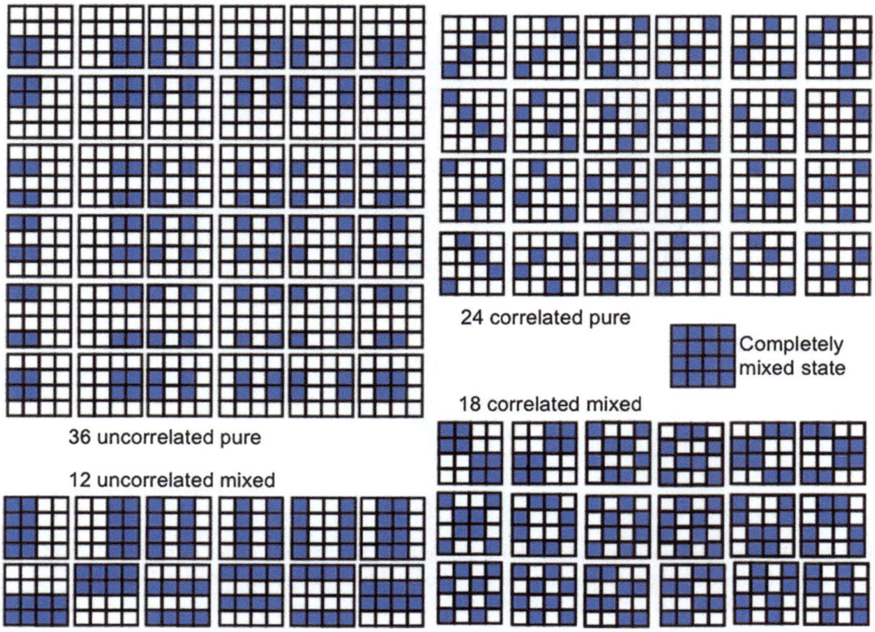

Fig. 12 All the valid epistemic states for a pair of bits

all the 2 × 2 matrices having at least one column containing a 0, that is,

$$\begin{pmatrix}1 & 0\\0 & 1\end{pmatrix}, \begin{pmatrix}0 & 1\\1 & 0\end{pmatrix}, \begin{pmatrix}1 & 0\\1 & 1\end{pmatrix}, \begin{pmatrix}1 & 1\\0 & 1\end{pmatrix}, \begin{pmatrix}0 & 1\\1 & 1\end{pmatrix}, \begin{pmatrix}1 & 1\\1 & 0\end{pmatrix}, \tag{43}$$

corresponding respectively to the transformations

$$\begin{matrix}\mathtt{q}\mapsto\mathtt{q}\\\mathtt{p}\mapsto\mathtt{p}\end{matrix}, \begin{matrix}\mathtt{q}\mapsto\mathtt{p}\\\mathtt{p}\mapsto\mathtt{q}\end{matrix}, \begin{matrix}\mathtt{q}\mapsto\mathtt{q}\\\mathtt{p}\mapsto\mathtt{q}+\mathtt{p}\end{matrix}, \begin{matrix}\mathtt{q}\mapsto\mathtt{q}+\mathtt{p}\\\mathtt{p}\mapsto\mathtt{p}\end{matrix}, \begin{matrix}\mathtt{q}\mapsto\mathtt{p}\\\mathtt{p}\mapsto\mathtt{q}+\mathtt{p}\end{matrix}, \begin{matrix}\mathtt{q}\mapsto\mathtt{q}+\mathtt{p}\\\mathtt{p}\mapsto\mathtt{q}\end{matrix}. \tag{44}$$

Each of these symplectic transformations can be composed with the four possible phase-space displacements,

$$\begin{matrix}\mathtt{q}\mapsto\mathtt{q}\\\mathtt{p}\mapsto\mathtt{p}\end{matrix}, \begin{matrix}\mathtt{q}\mapsto\mathtt{q}+1\\\mathtt{p}\mapsto\mathtt{p}\end{matrix}, \begin{matrix}\mathtt{q}\mapsto\mathtt{q}\\\mathtt{p}\mapsto\mathtt{p}+1\end{matrix}, \begin{matrix}\mathtt{q}\mapsto\mathtt{q}+1\\\mathtt{p}\mapsto\mathtt{p}+1\end{matrix}. \tag{45}$$

In all, this leads to 24 reversible symplectic affine transformations, which are depicted in Fig. 13. Given that there are only 24 permutations on the discrete phase space, we see that *every* reversible ontic transformation is physically allowed in this case.

On the other hand, for a pair of systems ($n = 2$), only a subset of the permutations of the ontic states correspond to valid sympectic affine transformations.

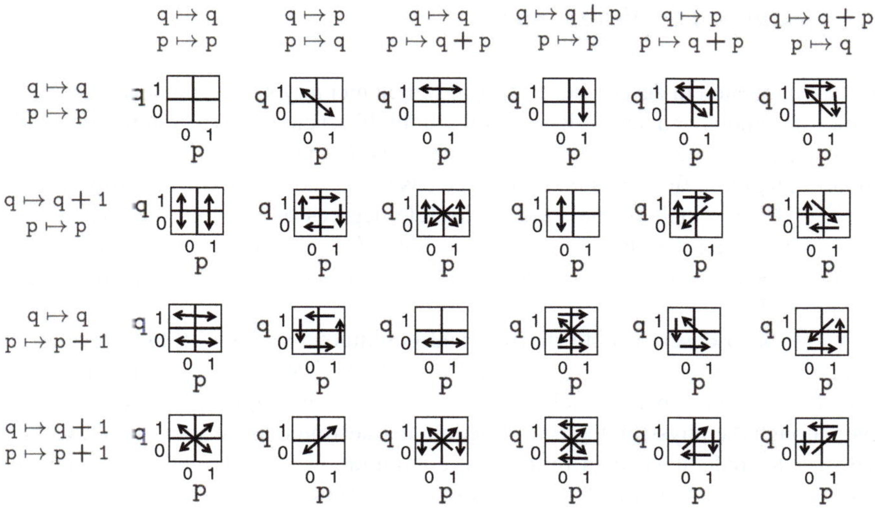

Fig. 13 All the valid transformations for a single bit

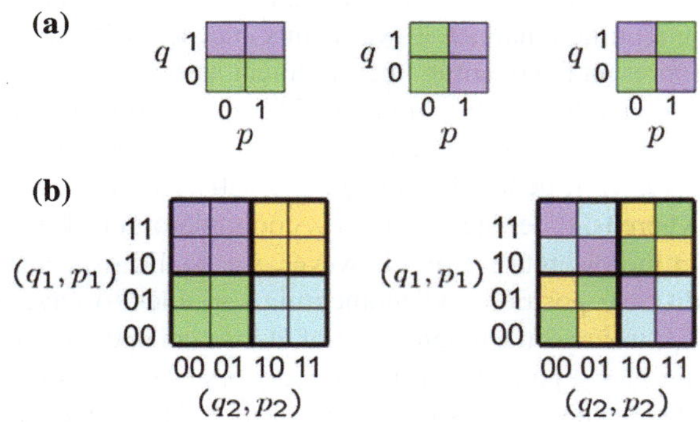

Fig. 14 a The set of valid measurements on a single bit. b Examples of valid measurements on a pair of bits

In Fig. 14a, we present the valid reproducible measurements on a single bit, and in Fig. 14b we present some examples of such measurements on a pair of bits, one corresponding to a product basis and the other an entangled basis.

3 Quadrature Quantum Subtheories

We now shift our attention to quantum theory, and build up to a definition of the subtheories of quantum theory that our epistricted theories will ultimately be shown to reproduce.

3.1 Quadrature Observables

We are interested in describing collections of elementary systems that each describe some continuous or discrete degree of freedom. If the elementary system is a continuous degree of freedom, it is associated with the Hilbert space $\mathcal{L}^2(\mathbb{R})$, the space of square-integrable functions on $\mathbb{R}$. For the case of n such systems, the Hilbert space is $\mathcal{L}^2(\mathbb{R})^{\otimes n} = \mathcal{L}^2(\mathbb{R}^n)$. The sorts of discrete degrees of freedom we consider are those wherein all the elementary systems have d levels where d is a prime. These are described by the Hilbert space $\mathbb{C}^d$. For n such systems, the Hilbert space is $(\mathbb{C}^d)^{\otimes n} = \mathbb{C}^{dn}$.

We seek to describe both discrete and continuous systems in the language of symplectic structure. For a scalar field, for instance, we describe each mode of the field in terms of a pair of field quadratures. In the example of a 2-level system, even though the physical degree of freedom in question may be spin or polarization, we seek to understand it in terms of a configuration variable and its canonically conjugate momentum. In all of these cases, we will conventionally refer to the pair of conjugate variables, regardless of the degrees of freedom they describe, as 'position' and 'momentum'.

We wish to present the quadrature subtheories for the continuous and discrete cases in a unified manner. Towards this end, we will avoid using a Hermitian operator to represent the quantum measurement associated to a quadrature variable. The reason is that although this works well for the continuous case, it fails to make sense in the discrete case. Recall that in the continuous case, we can define Hermitian operators on $\mathcal{L}^2(\mathbb{R})$, denoted $\hat{q}$ and $\hat{p}$, and satisfying the commutation relation $[\hat{q}, \hat{p}] = \hat{\mathbb{1}}$, where $[\cdot, \cdot]$ denotes the matrix commutator and $\hat{\mathbb{1}}$ is the identity operator on $\mathcal{L}^2(\mathbb{R})$. In the discrete setting, however, we would expect the operators associated to the discrete position and momentum variables to have eigenvalues in the finite field $\mathbb{Z}_d$, whereas the eigenvalues of Hermitian operators are necessarily real. Even if we did pick a pair of Hermitian operators to serve as discrete position and momentum observables, these would necessarily fail to provide an analogue of the commutation relation $[\hat{q}, \hat{p}] = \hat{\mathbb{1}}$, because in a finite-dimensional Hilbert space, the commutator of any two Hermitian operators has vanishing trace and therefore cannot be proportional to the identity operator on that space.

In any case, within the fields of quantum foundations and quantum information, there has been a move away from representing measurements by Hermitian operators because the eigenvalues of these operators are merely arbitrary labels of the measurement outcomes and have no operational significance. It is only the projectors in the spectral resolution of such a Hermitian operator that appear in the Born rule and hence only these that are relevant to the operational statistics. Therefore, a measurement with outcome set K is associated with a set of projectors $\{\Pi_k : k \in K\}$ such that $\Pi_k^2 = \Pi_k, \ \forall k \in K$ and $\sum_{k \in K} \Pi_k = \mathbb{1}$ (integral in the case of a continuum of outcomes). Such a set is called a *projector-valued measure* (PVM).

In the continuous variable case, we define the position observable, denoted $\mathcal{O}_q$, to be the PVM consisting of projectors onto position eigenstates,

$$\mathcal{O}_q \equiv \{\hat{\Pi}_q(\mathtt{q}) : \mathtt{q} \in \mathbb{R}\},$$

where

$$\hat{\Pi}_q(\mathtt{q}) \equiv |\mathtt{q}\rangle_q \langle \mathtt{q}|.$$

The momentum observable, denoted $\mathcal{O}_p$, is defined to be the PVM of projectors onto momentum eigenstates

$$\mathcal{O}_p \equiv \{\hat{\Pi}_p(\mathtt{p}) : \mathtt{p} \in \mathbb{R}\},$$

where

$$\hat{\Pi}_p(\mathtt{p}) \equiv |\mathtt{p}\rangle_p \langle \mathtt{p}|,$$

and the momentum eigenstates are related to the position eigenstates by a Fourier transform,

$$|\mathtt{p}\rangle_p \equiv \frac{1}{2\pi\hbar} \int_{\mathbb{R}} \mathrm{d}\mathtt{q} e^{i\frac{\mathtt{qp}}{\hbar}} |\mathtt{q}\rangle_q. \tag{46}$$

Strictly speaking, one needs to make use of rigged Hilbert space to define position and momentum eigenstates rigorously but we will adopt the standard informal treatment of such states here.

In the discrete case, we can also define position and momentum observables in this way. A discrete position basis for $\mathbb{C}^d$ (which one can think of as the *computational basis* in a quantum information setting) can be chosen arbitrarily. Denoting this basis by $\{|\mathtt{q}\rangle_q : \mathtt{q} \in \mathbb{Z}_d\}$, the PVM defining the position observable, denoted $\mathcal{O}_q$, is

$$\mathcal{O}_q \equiv \{\hat{\Pi}_q(\mathtt{q}) : \mathtt{q} \in \mathbb{Z}_d\},$$

where $\hat{\Pi}_q(\mathtt{q}) \equiv |\mathtt{q}\rangle_q \langle \mathtt{q}|$. We can define a discrete momentum basis, denoted $\{|\mathtt{p}\rangle_p : \mathtt{p} \in \mathbb{Z}_d\}$, via a discrete Fourier transform,

$$|\mathtt{p}\rangle_p \equiv \frac{1}{\sqrt{d}} \sum_{\mathtt{q} \in \mathbb{Z}_d} e^{i2\pi\frac{\mathtt{qp}}{d}} |\mathtt{q}\rangle. \tag{47}$$

and in terms of it, the PVM defining the momentum observable,

$$\mathcal{O}_p \equiv \{\hat{\Pi}_p(\mathtt{p}) : \mathtt{p} \in \mathbb{Z}_d\},$$

where $\hat{\Pi}_p(\mathtt{p}) \equiv |\mathtt{p}\rangle_p \langle \mathtt{p}|$. If one does not associate a Hermitian operator to each observable, then joint measurability of two observables can no longer be decided by the commutation of the associated Hermitian operators. Rather, it is determined by whether the associated PVMs commute or not, where two PVMs are said to commute if every projector in one commutes with every projector in the other.

To define the rest of the quadrature observables (and the commuting sets of these), we must first define a unitary representation of the symplectic affine transformations.

We begin by specifying the unitaries that correspond to phase-space displacements. To do this in a uniform manner for discrete and continuous degrees of freedom, we define functions $\chi : \mathbb{R} \to \mathbb{C}$ and $\chi : \mathbb{Z}_d \to \mathbb{C}$ as

$$\begin{aligned}\chi(\mathtt{c}) &= e^{i\frac{\mathtt{c}}{\hbar}} \text{ for } \mathtt{c} \in \mathbb{R}\\ \chi(\mathtt{c}) &= e^{i\frac{2\pi}{d}\mathtt{c}} \text{ for } \mathtt{c} \in \mathbb{Z}_d, \text{ when } d \text{ is an odd prime}\\ \chi(\mathtt{c}) &= e^{i\frac{\pi}{2}\mathtt{c}} \text{ for } \mathtt{c} \in \mathbb{Z}_d, \text{ when } d = 2.\end{aligned} \tag{48}$$

In the continuous case, this is the standard exponential function; in the discrete case where d is an odd prime, $\chi(\mathtt{a})$ is the $\mathtt{a}$th power of the dth root of unity; in the discrete case where $d = 2$, $\chi(\mathtt{a})$ is the $\mathtt{a}$th power of the fourth (not the second) root of unity. In terms of this function, we can define a unitary that shifts the position by $\mathtt{q}$, where $\mathtt{q} \in \mathbb{R}$ in the continuous case and $\mathtt{q} \in \mathbb{Z}_d$ in the discrete case, as

$$\begin{aligned}\hat{S}(\mathtt{q}) &= \sum_{\mathtt{p}\in\mathbb{R}/\mathbb{Z}_d} \chi(\mathtt{qp})|\mathtt{p}\rangle_p\langle\mathtt{p}|\\ &= \sum_{\mathtt{q}'\in\mathbb{R}/\mathbb{Z}_d} |\mathtt{q}'-\mathtt{q}\rangle_q\langle\mathtt{q}'|\end{aligned} \tag{49}$$

and a unitary that boosts the momentum by $\mathtt{p}$, where $\mathtt{p} \in \mathbb{R}$ in the continuous case and $\mathtt{p} \in \mathbb{Z}_d$ in the discrete case, as

$$\begin{aligned}\hat{B}(\mathtt{p}) &= \sum_{\mathtt{q}\in\mathbb{R}/\mathbb{Z}_d} \chi(\mathtt{qp})|\mathtt{q}\rangle_q\langle\mathtt{q}|\\ &= \sum_{\mathtt{p}'\in\mathbb{R}/\mathbb{Z}_d} |\mathtt{p}'-\mathtt{p}\rangle_p\langle\mathtt{p}'|\end{aligned} \tag{50}$$

Note that the shift unitaries do not commute with the boost unitaries. The unitaries corresponding to phase-space displacements—typically called the *Weyl operators*—are proportional to products of these. In particular, the Weyl operator associated with the phase-space displacement vector $\mathbf{a} = (\mathtt{q}, \mathtt{p}) \in \mathbb{R}^2/(\mathbb{Z}_d)^2$ is defined to be

$$\hat{W}(\mathbf{a}) = \chi(2\mathtt{pq})\hat{S}(\mathtt{q})\hat{B}(\mathtt{p}). \tag{51}$$

This is easily generalized to the case of a phase-space displacement for n degrees of freedom, $\mathbf{a} = (\mathtt{q}_1, \mathtt{p}_1, \ldots, \mathtt{q}_n, \mathtt{p}_n) \in \mathbb{R}^{2n}/(\mathbb{Z}_d)^{2n}$ via the tensor product,

$$\hat{W}(\mathbf{a}) = \bigotimes_{i=1}^{n} \chi(2\mathtt{p}_i\mathtt{q}_i)\hat{S}(\mathtt{q}_i)\hat{B}(\mathtt{p}_i). \tag{52}$$

For $\mathbf{a}, \mathbf{a}' \in \Omega$, the product of the corresponding Weyl operators is

$$\hat{W}(\mathbf{a})\hat{W}(\mathbf{a}') = \chi(2\langle\mathbf{a}, \mathbf{a}'\rangle)\hat{W}(\mathbf{a} + \mathbf{a}'). \tag{53}$$

Thus it is clear that the Weyl operators constitute a projective unitary representation of the group of phase-space displacements $\mathbf{m} \rightarrow \mathbf{m} + \mathbf{a}$, where the composition law is

$$(\cdot + \mathbf{a}) + \mathbf{a}' = \cdot + (\mathbf{a} + \mathbf{a}'). \tag{54}$$

Next, we define a projective unitary representation $\hat{V}$ of the symplectic group acting on a $2n$-dimensional phase space Ω. For every $2n \times 2n$ symplectic matrix $S : \Omega \rightarrow \Omega$ there is a unitary $\hat{V}(S)$ acting on the Hilbert space $\mathcal{L}(\mathbb{R}^n)/\mathbb{C}^{dn}$, such that

$$\hat{V}(S)\hat{V}(S') = e^{i\phi}\hat{V}(SS') \tag{55}$$

for some phase factor $e^{i\phi}$. These can be defined via their action on the Weyl operators. Specifically, $\forall \mathbf{a} \in \Omega$,

$$\hat{V}(S)\hat{W}(\mathbf{a})\hat{V}^{\dagger}(S) \propto \hat{W}(S\mathbf{a}). \tag{56}$$

In the following, we will often consider the action of these unitaries under conjugation, therefore, we define the superoperators associated to phase-space displacement $\mathbf{a}$ and symplectic matrix S,

$$\begin{aligned} \mathcal{W}(\mathbf{a})(\cdot) &\equiv \hat{W}(\mathbf{a}) \cdot \hat{W}(\mathbf{a})^{\dagger}, \\ \mathcal{V}(S)(\cdot) &\equiv \hat{V}(S) \cdot \hat{V}(S)^{\dagger}. \end{aligned} \tag{57}$$

Note that Eq. (56) implies that

$$\mathcal{W}(\mathbf{a}) \circ \mathcal{V}(S)(\cdot) = \mathcal{V}(S) \circ \mathcal{W}(S^{-1}\mathbf{a})(\cdot). \tag{58}$$

In the classical theory, every Poisson-commuting set of quadrature functionals $\{f^{(1)}, f^{(2)}, \ldots, f^{(k)}\}$ can be obtained from every other such set by a symplectic linear transformation (here, $k \leq n$). The proof is as follows. If $f^{(i)} = \mathbf{f^{(i)}}^T\mathbf{z}$ is a quadrature functional, then so is $\tilde{f}^{(i)} = (S\mathbf{f}^{(i)})^T\mathbf{z}$ for all $i \in \{1, \ldots, k\}$ when S is a symplectic matrix. Furthermore, if the initial set is Poisson-commuting, then $\langle \mathbf{f}^{(i)}, \mathbf{f}^{(j)} \rangle = 0$ for all $i \neq j \in \{1, \ldots, k\}$, and then because

$$\begin{aligned} \langle \tilde{\mathbf{f}}^{(i)}, \tilde{\mathbf{f}}^{(j)} \rangle &= \langle S\mathbf{f}^{(i)}, S\mathbf{f}^{(j)} \rangle \\ &= \langle \mathbf{f}^{(i)}, \mathbf{f}^{(j)} \rangle, \end{aligned} \tag{59}$$

it follows that $\langle \tilde{\mathbf{f}}^{(i)}, \tilde{\mathbf{f}}^{(j)} \rangle = 0$ for all $i \neq j \in \{1, \ldots, k\}$ so the final set is Poisson-commuting as well. Here, we have used the fact that the symplectic inner product is invariant under the action of a symplectic matrix.

We can define commuting sets of quantum quadrature *observables* similarly. Consider a single degree of freedom, $\Omega = \mathbb{R}^2/\mathbb{Z}_d^2$. Denote by S_f the symplectic matrix that takes the position functional q to a quadrature functional f, so that $S_f\mathbf{q} = \mathbf{f}$. (Given that $\mathbf{q} \equiv (1, 0)$, we see that $\mathbf{f}$ is the first column of S_f.) We define the quadrature *observable* associated with f, denoted $\mathcal{O}_f$, to be the image under the action of the unitary $\hat{V}(S_f)$ of the position observable, that is,

$$\mathcal{O}_f \equiv \{\hat{\Pi}_f(\mathtt{f}) : \mathtt{f} \in \mathbb{Z}_d\},$$

where

$$\hat{\Pi}_f(\mathtt{f}) \equiv \mathcal{V}(S_f)(\hat{\Pi}_q(\mathtt{f})). \tag{60}$$

It is useful to note how these projectors transform under phase-space displacements and symplectic matrices. By definition of the quadrature observables, we infer that for a symplectic matrix S,

$$\mathcal{V}(S)(\hat{\Pi}_f(\mathtt{f})) = \hat{\Pi}_{Sf}(\mathtt{f}), \tag{61}$$

where Sf denotes the quadrature functional associated to the vector $S\mathbf{f} \in \Omega$. Now consider the action of a Weyl superoperator. First note that the projectors onto position eigenstates transform as

$$\mathcal{W}(\mathbf{a})(\hat{\Pi}_q(\mathtt{q})) = \hat{\Pi}_q(\mathtt{q} + q(\mathbf{a})).$$

It follows that if $f = S_f q$, then

$$\begin{aligned} \mathcal{W}(\mathbf{a})(\hat{\Pi}_f(\mathtt{f})) &= \mathcal{W}(\mathbf{a})(\hat{\Pi}_{S_f q}(\mathtt{f})), \\ &= \mathcal{W}(\mathbf{a})\mathcal{V}(S_f)(\hat{\Pi}_q(\mathtt{f})), \\ &= \mathcal{V}(S)\mathcal{W}(S_f^{-1}\mathbf{a})(\hat{\Pi}_q(\mathtt{f})), \\ &= \mathcal{V}(S)(\hat{\Pi}_q(\mathtt{f} + q(S_f^{-1}\mathbf{a}))), \\ &= \hat{\Pi}_f(\mathtt{f} + f(\mathbf{a})). \end{aligned} \tag{62}$$

In all,

$$\mathcal{V}(S)\mathcal{W}(\mathbf{a})(\hat{\Pi}_f(\mathtt{f})) = \hat{\Pi}_{Sf}(\mathtt{f} + f(\mathbf{a})). \tag{63}$$

The case of n degrees of freedom, $\Omega = \mathbb{R}^2/\mathbb{Z}_d^2$, is treated similarly. In this case, our quadrature observables need not be rank-1. Our fiducial quadrature can be taken to be q_1, the position functional for system 1. The associated quadrature observable is

$$\mathcal{O}_{q_1} \equiv \{\hat{\Pi}_{q_1}(\mathtt{q}_1) \otimes \mathbb{1}_2 \otimes \cdots \otimes \mathbb{1}_n : \mathtt{q}_1 \in \mathbb{R}/\mathbb{Z}_d\}.$$

For an arbitrary functional on the n systems, $f : \Omega \to \mathbb{R}/\mathbb{Z}_d$, we find the symplectic matrix S_f such that $S_f \mathbf{q}_1 = \mathbf{f}$, and we define the quadrature observable associated with f to be

$$\mathcal{O}_f \equiv \{\hat{\Pi}_f(\mathtt{f}) : \mathtt{f} \in \mathbb{R}/\mathbb{Z}_d\}.$$

where

$$\hat{\Pi}_f(\mathtt{f}) \equiv \hat{V}(S_f)\left(\hat{\Pi}_{q_1}(\mathtt{f}) \otimes \mathbb{1}_2 \otimes \cdots \otimes \mathbb{1}_n\right)\hat{V}(S_f)^{\dagger}.$$

It follows that for every classical quadrature *functional* f, there is a corresponding quadrature *observable* $\mathcal{O}_f$, which stands in relation to the position and momentum observables as f stands to the position and momentum functionals.

As an aside, one may note that in the continuous variable case, the quadrature observables are simply the spectral resolutions of those Hermitian operators that are linear combinations of position and momentum operators. In particular, for a quadrature observable $\mathcal{O}_f$ associated to a vector $\mathbf{f} \in \Omega$, the associated Hermitian operator is simply

$$\hat{f} = \mathbf{f}^T \hat{\mathbf{z}},$$

where

$$\hat{\mathbf{z}} \equiv (\hat{q}_1, \hat{p}_1, \ldots, \hat{q}_n, \hat{p}_n), \tag{64}$$

is the vector of position and momentum operators. Hence for every classical quadrature *variable* $f = \mathbf{f}^T \mathbf{z}$, as defined in Eq. (11), there is a corresponding quadrature *operator* $\hat{f} = \mathbf{f}^T \hat{\mathbf{z}}$, where we have simply replaced the position and momentum functionals with their corresponding Hermitian operators.

We are now in a position to describe the commuting sets of quadrature observables. A set of quadrature observables $\{\mathcal{O}_{f^{(1)}}, \ldots, \mathcal{O}_{f^{(k)}}\}$ is a commuting set if and only if the corresponding quadrature functionals $\{f^{(1)}, \ldots, f^{(k)}\}$ are Poisson-commuting. The proof is as follows. The functionals $\{f^{(1)}, \ldots, f^{(k)}\}$ are Poisson-commuting if and only if they can be obtained by some symplectic transformation from any other such set, in particular, the set of position functionals for the first k systems, $\{q_1, \ldots, q_k\}$. In other words, $\{f^{(1)}, \ldots, f^{(k)}\}$ are Poisson-commuting if and only if there is a sympectic matrix S such that $\mathbf{f}^{(i)} = S\mathbf{q}_i$ for all $i \in \{1, \ldots, k\}$ (which implies that the vectors $\mathbf{f}^{(i)}$ are the first k columns of S). Given the definition of quadrature observables, this condition is equivalent to the statement that there exists a symplectic matrix S such that $\mathcal{O}_{f^{(i)}} = \hat{V}(S)\mathcal{O}_{q_i}\hat{V}(S)^{\dagger}$ for all $i \in \{1, \ldots, k\}$. But given that the elements of the set $\{\mathcal{O}_{q_1}, \ldots, \mathcal{O}_{q_k}\}$ (the position observables for the first k systems) commute, and commutation relations are preserved under a unitary, it follows that the elements of the set $\{\mathcal{O}_{f^{(1)}}, \ldots, \mathcal{O}_{f^{(k)}}\}$ commute if and only if there exists such an S, hence they commute if and only if the corresponding quadrature functionals $\{f^{(1)}, \ldots, f^{(k)}\}$ Poisson-commute.

Again, this has a simple interpretation in the continuous variable case. There, it is easy to verify that the matrix commutator of two quadratures operators is equal to the symplectic inner product of the corresponding vectors, that is, $[\hat{f}, \hat{g}] = \langle \mathbf{f}, \mathbf{g} \rangle$. In particular, it follows that $[\hat{f}, \hat{g}] = 0$ if and only if $\langle \mathbf{f}, \mathbf{g} \rangle = 0$, which provides another proof of the fact that a commuting set of quadrature observables is associated with an isotropic subspace of the phase space.

As described in Sect. 2.2.1, every set of Poisson-commuting quadrature functionals defines an isotropic subspace $V \subseteq \Omega$ and therefore the sets of commuting quadrature observables are also parameterized by the isotropic subspaces of Ω. If a commuting set of quadrature observables is such that the corresponding quadrature functionals are associated with an isotropic subspace V, then this set defines a single quadrature observable, denoted $\mathcal{O}_V$, by

$$\mathcal{O}_V = \{\hat{\Pi}_V(\mathbf{v}) : \mathbf{v} \in V\}$$

where

$$\hat{\Pi}_V(\mathbf{v}) \equiv \prod_{\mathbf{f}^{(i)}:\mathrm{span}(\mathbf{f}^{(i)})=V} \hat{\Pi}_{f^{(i)}}\left(f^{(i)}(\mathbf{v})\right). \tag{65}$$

For instance, the quadrature functionals $f = q_1 - q_2$ and $g = p_1 + p_2$ are Poisson-commuting and therefore the associated quadrature observables, $\mathcal{O}_f$ and $\mathcal{O}_g$, commute, which is to say that the projectors $\{\hat{\Pi}_f(\mathtt{f}) : \mathtt{f} \in \mathbb{R}/\mathbb{Z}_d\}$ all commute with the projectors $\{\hat{\Pi}_g(\mathtt{g}) : \mathtt{g} \in \mathbb{R}/\mathbb{Z}_d\}$. If $V = \mathrm{span}\{\mathbf{f}, \mathbf{g}\}$, then the possible pairs of values for the two observables can be expressed as the possible components of a vector $\mathbf{v} \in V$ along the basis vectors $\mathbf{f}$ and $\mathbf{g}$ respectively. These are the pairs $\{(\mathtt{f}, \mathtt{g})\}$ such that $\mathtt{f} = \mathbf{f}^T\mathbf{v} = f(\mathbf{v})$ and $\mathtt{g} = \mathbf{g}^T\mathbf{v} = g(\mathbf{v})$ for some $\mathbf{v} \in V$. It follows that we can parametrize the possible values of this commuting set by vectors $\mathbf{v} \in V$.

In the continuous variable case, $\hat{\Pi}_f(\mathtt{f})$ is the projector onto the eigenspace of $\hat{q}_1 - \hat{q}_2$ with eigenvalue $\mathtt{f}$, $\hat{\Pi}_g(\mathtt{g})$ is the projector onto the eigenspace of $\hat{p}_1 + \hat{p}_2$ with eigenvalue $\mathtt{g}$, and $\hat{\Pi}_V(\mathbf{v})$ is the projector onto the joint eigenspace of $\hat{q}_1 - \hat{q}_2$ and $\hat{p}_1 + \hat{p}_2$ with eigenvalues $(\mathtt{f}, \mathtt{g})$, which corresponds to an Einstein-Podolsky-Rosen entangled state.

With this background established, we are in a position to define the quadrature quantum subtheories.

3.2 *Characterization of Quadrature Quantum Subtheories*

In this section, we define a quadrature subtheory of the quantum theory for a given system (discrete or continuous). In the discrete case, this subtheory is closely connected to the stabilizer formalism, a connection that we make precise in the appendix.

3.2.1 The Set of Valid Quantum States

In order to define the valid quantum states in the quadrature quantum subtheory, we use the guiding analogy of Sect. 2.1, together with the isomorphism between quadrature functionals and quadrature observables noted above.

As we have just seen, for both the discrete and continuous cases, every commuting set of quadrature observables is associated to an isotropic subspace $V \subset \Omega$. Furthermore, every set of values that these observables can jointly take is associated to a vector $\mathbf{v} \in V$. We have denoted the projector that yields these values by $\hat{\Pi}_V(\mathbf{v})$. The quantum states that are part of the quadrature subtheory, termed *quadrature states*, are simply the density operators that are proportional to such projectors. It follows that the quadrature states are parameterized by pairs consisting of an isotropic subspace V and a valuation vector $\mathbf{v} \in V$ (in precisely the same way as one parametrizes the set of valid epistemic states in the epistricted classical theory). Specifically, it is the set of states of the form

$$\rho_{V,\mathbf{v}} = \frac{1}{\mathcal{N}_V}\hat{\Pi}_V(\mathbf{v}), \tag{66}$$

where $V \subseteq \Omega$ is isotropic, $\mathbf{v} \in V$, and $\mathcal{N}_V$ is a normalization factor. Equivalently, if $\{\mathbf{f}^{(i)}\}$ is a basis of V, then

$$\rho_{V,\mathbf{v}} \equiv \frac{1}{\mathcal{N}_V} \prod_{\{\mathbf{f}^{(i)}:\text{span}\{\mathbf{f}^{(i)}\}=V\}} \hat{\Pi}_{f^{(i)}}\left(f^{(i)}(\mathbf{v})\right). \tag{67}$$

3.2.2 The Set of Valid Transformations

Because the overall phase of a Hilbert space vector is physically irrelevant, physical states are properly represented by density operators, and consequently a reversible physical transformation is not represented by a unitary operator but rather by the superoperator corresponding to conjugation by that unitary.

When a Weyl operator $\hat{W}(\mathbf{a})$ acts by conjugation, it defines what we will call the *Weyl superoperator*,

$$\mathcal{W}(\mathbf{a})(\cdot) \equiv \hat{W}(\mathbf{a})(\cdot)\hat{W}(\mathbf{a})^{\dagger}.$$

Unlike the Weyl operators of two phase-space displacements, which, by Eq. (53), commute if and only if the corresponding phase-space displacement vectors have vanishing symplectic inner product,

$$[\hat{W}(\mathbf{a}), \hat{W}(\mathbf{a}')] = 0 \text{ if and only if } \langle \mathbf{a}, \mathbf{a}' \rangle = 0,$$

the Weyl superoperators of any two phase-space displacements necessarily commute,

$$[\mathcal{W}(\mathbf{a}), \mathcal{W}(\mathbf{a}')] = 0 \; \forall \mathbf{a}, \mathbf{a}' \in \Omega.$$

This follows from Eq. (53) and the skew-symmetry of the symplectic inner product. It follows that

$$\mathcal{W}(\mathbf{a})\mathcal{W}(\mathbf{a}') = \mathcal{W}(\mathbf{a}+\mathbf{a}') \; \forall \mathbf{a}, \mathbf{a}' \in \Omega.$$

As such, the Weyl superoperators constitute a nonprojective representation of the group of phase-space displacements, Eq. (54).

Next, we consider the projective unitary representation $\hat{V}$ of the symplectic group acting by conjugation. This defines a superoperator representation of the symplectic group which is nonprojective, that is, for

$$\mathcal{V}(S)(\cdot) \equiv \hat{V}(S)(\cdot)\hat{V}(S)^{\dagger},$$

we have

$$\mathcal{V}(S)\mathcal{V}(S') = \mathcal{V}(SS').$$

The Clifford group of unitaries is defined as those which, when acting by conjugation, take the set of Weyl operators to itself. That is, a unitary $\hat{U}$ is in the Clifford group if $\forall \mathbf{b} \in \Omega$,

$$\hat{U}\hat{W}(\mathbf{b})\hat{U}^{\dagger} = c(\mathbf{b})\hat{W}(S\mathbf{b}), \tag{68}$$

for some maps $c : \Omega \to \mathbb{C}$ and $S : \Omega \to \Omega$.

It turns out that every such unitary can be written as a product of a Weyl operator and an element of the unitary projective representation of the symplectic group, that is,

$$\hat{U}(S, \mathbf{a}) = \hat{W}(\mathbf{a})\hat{V}(S),$$

for some symplectic matrix $S : \Omega \to \Omega$ and phase-space vector $\mathbf{a} \in \Omega$. From Eqs. (53) and (55) we infer that a product of such unitaries is

$$\hat{U}(S, \mathbf{a})\hat{U}(S', \mathbf{a}') = e^{i\phi}\hat{U}(SS', S\mathbf{a}' + \mathbf{a}). \tag{69}$$

for some phase factor ϕ. Recalling Eq. (34), it is clear that the Clifford group of unitaries $\hat{U}(S, \mathbf{a})$ constitutes a projective representation of the symplectic affine group.

When a Clifford unitary $\hat{U}(S, \mathbf{a})$ acts by conjugation, it defines what we will call a *Clifford superoperator* $\mathcal{U}(S, \mathbf{a})(\cdot) \equiv \hat{U}(S, \mathbf{a})(\cdot)\hat{U}(S, \mathbf{a})^{\dagger}$. It follows that

$$\mathcal{U}(S, \mathbf{b})\mathcal{U}(S', \mathbf{b}') = \mathcal{U}(SS', \mathbf{b} + S\mathbf{b}'),$$

and therefore, recalling Eq. (34), these form a nonprojective representation of the symplectic affine group.

The reversible transformations that are included in quadrature quantum mechanics are precisely those associated with Clifford superoperators. These map every quadrature state to another quadrature state.

The valid *irreversible* transformations in the quadrature subtheory are those that admit of a Stinespring dilation of the following form: the system is coupled to an ancilla of arbitrary dimension that is prepared in a quadrature state, the system and ancilla undergo a reversible transformation associated with a Clifford superoperator, and a partial trace operation is performed on the ancilla.

3.2.3 The Set of Valid Measurements

Finally, the reproducible measurements included in quadrature quantum subtheories are simply those associated with a commuting set of quadrature observables. Recall that these are parametrized by the isotropic subspaces $V \subset \Omega$, and correspond to PVMs of the form $\{\hat{\Pi}_V(\mathbf{v}) : \mathbf{v} \in V\}$, as defined in Eq. (65).

The most general measurement allowed is one whose Naimark extension can be achieved by preparing an ancilla in a quadrature state, coupling to the system via a Clifford superoperator, and finally measuring a commuting set of quadrature observables on system+ancilla.

4 Comparing Quantum Subtheories to Epistricted Theories

4.1 Equivalence for Continuous and Odd-Prime Discrete Cases

The operational equivalence result is proven using the Wigner representation. The latter is a quasi-probability representation of quantum mechanics, wherein Hermitian operators on the Hilbert space are represented by real-valued functions on the corresponding classical phase space.

For the case of n continuous degrees of freedom, where the Hilbert space is $\mathcal{L}^2(\mathbb{R}^n)$ and the phase space is $\mathbb{R}^{2n}$, the Wigner representation is a well-known formulation of quantum theory, particularly in the field of quantum optics [47, 48]. For the case of *discrete* degrees of freedom, there are many proposals for how to define a quasi-probability representation that is analogous to Wigner's but for a discrete phase space. We here make use of a proposal due to Gross [6], which is built on (but distinct from) a proposal by Wootters [49]. For n d-level systems (qudits), where d is a prime, the phase space is taken to be $(\mathbb{Z}_d)^{2n}$.

We shall attempt to present the proof for the continuous case and for the odd-prime discrete case in a unified notation. Towards this end, we will provide a definition of the Wigner representation that is independent of the nature of the phase space. In the case of $\Omega = \mathbb{R}^{2n}$ and $\Omega = (\mathbb{Z}_d)^{2n}$ for d an odd prime, our definition will reduce, respectively, to the standard Wigner representation and the discrete Wigner representation proposed by Gross [6]. Marginalizing over the entire phase space Ω will be denoted by a sum over Ω in all of our expressions, which will be taken to represent a discrete sum in the discrete case and an integral with a phase-space invariant measure in the continuous case.

4.1.1 Wigner Representation of Quantum Theory

The Wigner representation of an operator $\hat{O}$, denoted $\hat{W}_{\hat{O}}(\mathbf{m})$, can be understood as the components of that operator in a particular basis for the vector space of Hermitian operators where the inner product is the Hilbert-Schmidt inner product, $\langle \hat{O}, \hat{O}' \rangle \equiv \mathrm{tr}(\hat{O}\hat{O}')$. The elements of this operator basis are indexed by the elements of the phase space and termed the *phase-space point operators*. Denoting this operator basis by $\{\hat{A}(\mathbf{m}) : \mathbf{m} \in \Omega\}$, we have

$$\hat{W}_{\hat{O}}(\mathbf{m}) = \mathrm{Tr}[\hat{O}\hat{A}(\mathbf{m})]. \tag{70}$$

The phase-space point operators can be defined as the symplectic Fourier transform of the Weyl operators (which in turn are defined for both continuous and discrete degrees of freedom in Eq. (52)),

$$\hat{A}(\mathbf{m}) \equiv \frac{1}{\mathcal{N}_\Omega} \sum_{\mathbf{m}' \in \Omega} \chi(\langle \mathbf{m}, \mathbf{m}' \rangle) \hat{W}(\mathbf{m}'). \tag{71}$$

where $\mathcal{N}_\Omega$ is a normalization factor chosen to ensure that

$$\mathrm{Tr}[\hat{A}(\mathbf{m})] = 1.$$

The key property of the phase-space point operators is that they transform covariantly under symplectic affine transformations,

$$\mathcal{U}(S, \mathbf{a}) \left[\hat{A}(\mathbf{m}) \right] \propto \hat{A}(S\mathbf{m} + \mathbf{a}), \tag{72}$$

which can be inferred from Eq. (71) and the manner in which the Weyl operators transform under the action of the Clifford superoperators, Eq. (56). This in turn implies that the Wigner representation of an operator also transforms covariantly under symmplectic affine transformations,

$$\begin{aligned} \hat{W}_{\mathcal{U}(S,\mathbf{a})(\hat{O})}(\mathbf{m}) &= \mathrm{tr}\left(\mathcal{U}(S, \mathbf{a})(\hat{O}) \hat{A}(\mathbf{m}) \right) \\ &= \mathrm{tr}\left(\hat{O}\, \mathcal{U}(S^{-1}, -\mathbf{a})(\hat{A}(\mathbf{m})) \right) \\ &= \hat{W}_{\hat{O}}(S^{-1}\mathbf{m} - \mathbf{a}). \end{aligned} \tag{73}$$

In both the discrete and continuous cases, we have

$$\frac{1}{\mathcal{N}_\Omega} \sum_{\mathbf{m} \in \Omega} \chi(\langle \mathbf{m}, \mathbf{m}' \rangle) = \delta_{\mathbf{0}}(\mathbf{m}'),$$

where $\delta_0(\mathbf{m}') = \prod_{i=1}^{n} \delta(q_i')\delta(p_i')$ for $\mathbf{m}' \equiv (q_1', p_1', \ldots, q_n', p_n')$ and where δ denotes the Dirac-delta function in the continuous case and a Kronecker-delta in the discrete case. It then follows from Eq. (71) that

$$\begin{aligned}\sum_{\mathbf{m}\in\Omega} \hat{A}(\mathbf{m}) &= \sum_{\mathbf{m}'\in\Omega} \delta(\mathbf{m}')\hat{W}(\mathbf{m}'),\\ &= \hat{W}(\mathbf{0}),\\ &= \mathbb{1}.\end{aligned} \tag{74}$$

Consequently the trace of an arbitrary operator is given by the normalization of the corresponding Wigner representation on the phase-space,

$$\mathrm{Tr}(\hat{O}) = \sum_{\mathbf{m}\in\Omega} W_{\hat{O}}(\mathbf{m}).$$

The phase-space point operators are Hermitian, and therefore the Wigner representation of any Hermitian operator is real-valued. They are orthogonal,

$$\mathrm{Tr}\left(\hat{A}(\mathbf{m})\hat{A}(\mathbf{m}')\right) \propto \delta(\mathbf{m}-\mathbf{m}'), \tag{75}$$

and form a complete basis for the operator space relative to the Hilbert-Schmidt inner product, that is, for arbitrary $\hat{O}$,

$$\sum_{\mathbf{m}\in\Omega} \hat{A}(\mathbf{m})\mathrm{Tr}\left(\hat{A}(\mathbf{m})\hat{O}\right) = \hat{O}.$$

It follows from this completeness that for any pair of Hermitian operators $\hat{O}$ and $\hat{O}'$,

$$\mathrm{Tr}\left(\hat{O}\hat{O}'\right) = \sum_{\mathbf{m}\in\Omega} W_{\hat{O}}(\mathbf{m})W_{\hat{O}'}(\mathbf{m}). \tag{76}$$

The Wigner representation of a quantum state ρ is the function $W_\rho : \Omega \to \mathbb{R}$ defined by

$$W_\rho(\mathbf{m}) = \mathrm{Tr}[\rho\hat{A}(\mathbf{m})], \tag{77}$$

where the fact that $\mathrm{Tr}(\rho) = 1$ implies

$$\sum_{\mathbf{m}\in\Omega} W_\rho(\mathbf{m}) = 1. \tag{78}$$

A superoperator $\mathcal{E}$ corresponding to the transformation $\rho \mapsto \mathcal{E}(\rho)$ can be modelled in the Wigner representation by a conditional quasiprobability function $W_{\mathcal{E}}(\mathbf{m}'|\mathbf{m})$

such that

$$W_\rho(\mathbf{m}) \mapsto \sum_{\mathbf{m}'\in\Omega} W_{\mathcal{E}}(\mathbf{m}|\mathbf{m}')W_\rho(\mathbf{m}').$$

Specifically, the function $W_{\mathcal{E}} : \Omega \times \Omega \to \mathbb{R}$ is defined as

$$W_{\mathcal{E}}(\mathbf{m}|\mathbf{m}') = \mathrm{Tr}\left[\hat{A}(\mathbf{m})\mathcal{E}\left(\hat{A}(\mathbf{m}')\right)\right] \tag{79}$$

If $\mathcal{E}$ is trace-preserving, then

$$\sum_{\mathbf{m}\in\Omega} W_{\mathcal{E}}(\mathbf{m}|\mathbf{m}') = 1. \tag{80}$$

A sharp measurement with outcome set K, associated with a projector-valued measure $\mathcal{O} \equiv \{\hat{\Pi}_{\mathbf{k}} : \mathbf{k} \in K\}$ is represented by a conditional quasi-probability function $W_{\mathcal{O}} : K \times \Omega \to \mathbb{R}$ defined by

$$\begin{aligned} W_{\mathcal{O}}(\mathbf{k}|\mathbf{m}) &= W_{\hat{\Pi}_{\mathbf{k}}}(\mathbf{m}), \\ &= \mathrm{Tr}[\hat{\Pi}_{\mathbf{k}}\hat{A}(\mathbf{m})], \end{aligned} \tag{81}$$

where the fact that $\sum_{\mathbf{k}\in K} \hat{\Pi}_{\mathbf{k}} = \hat{\mathbb{1}}$ implies that

$$\sum_{\mathbf{k}\in K} W_{\mathcal{O}}(\mathbf{k}|\mathbf{m}) = 1. \tag{82}$$

Finally, we can infer from Eq. (76) that the probability of obtaining outcome $\mathbf{k}$ in a measurement of $\{\hat{\Pi}_{\mathbf{k}} : \mathbf{k} \in K\}$ on the state ρ is expressed in the Wigner representation as

$$\mathrm{Tr}\left(\hat{\Pi}_{\mathbf{k}}\rho\right) = \sum_{\mathbf{m}\in\Omega} W_{\mathcal{O}}(\mathbf{k}|\mathbf{m})W_\rho(\mathbf{m}). \tag{83}$$

Similarly, if a transformation associated with the completely-positive trace-preserving map $\mathcal{E}$ acts between the preparation and the measurement, then the probability of obtaining outcome k is expressed in the Wigner representation as

$$\mathrm{Tr}\left(\hat{\Pi}_{\mathbf{k}}\mathcal{E}(\rho)\right) = \sum_{\mathbf{m}\in\Omega} W_{\mathcal{O}}(\mathbf{k}|\mathbf{m}) \sum_{\mathbf{m}'\in\Omega} W_{\mathcal{E}}(\mathbf{m}|\mathbf{m}')W_\rho(\mathbf{m}'). \tag{84}$$

Note that if $W_\rho(\mathbf{m})$ is nonnegative, it can be interpreted as a probability distribution on phase-space. Similarly, if $W_{\mathcal{O}}(\mathbf{k}|\mathbf{m})$ and $W_{\mathcal{E}}(\mathbf{m}|\mathbf{m}')$ are nonnegative, then can be interpreted as conditional probability distributions. In this case, Eqs. (83) and (84) for the probability of a measurement outcome can be understood as an application of the law of total probability, in analogy with Eqs. (39) and (40). This sort of interpretation is indeed possible for the quadrature quantum subtheories and yields precisely the operational predictions of the quadrature epistricted theory. To show this, it remains only to show that the Wigner representation of the preparations, transformations and measurements of the quadrature subtheory are precisely equal to the epistemic states, transition probabilities and response functions of the quadrature epistricted theory.

4.1.2 Wigner Representation of the Quadrature Quantum Subtheory

Our proof of equivalence relies on two of the defining features of the Wigner representation. First, the fact that the Wigner representation transforms covariantly under the symplectic affine transformations, Second, the fact that the Wigner representation of the projectors defining the position and momentum observables are the response functions associated to the position and momentum functionals in the classical theory, that is,

$$\begin{aligned} W_{\hat{\Pi}_{q_i}(\mathtt{q}_i)}(\mathbf{m}) &= \delta(q_i(\mathbf{m}) - \mathtt{q}_i), \\ W_{\hat{\Pi}_{p_j}(\mathtt{p}_j)}(\mathbf{m}) &= \delta(p_j(\mathbf{m}) - \mathtt{p}_j). \end{aligned} \tag{85}$$

It follows from these facts that the Wigner representation of the projectors in the quadrature observable $\mathcal{O}_f$ are equal to the response functions associated with the corresponding quadrature functional f in the classical theory,

$$\begin{aligned} W_{\hat{\Pi}_f(\mathtt{f})}(\mathbf{m}) &= W_{\hat{\Pi}_{S_f q_1}(\mathtt{f})}(\mathbf{m}), \\ &= W_{\mathcal{V}(S_f)(\hat{\Pi}_{q_1}(\mathtt{f}))}(\mathbf{m}), \\ &= W_{\hat{\Pi}_{q_1}(\mathtt{f})}(S_f^{-1}\mathbf{m}), \\ &= \delta(q_1(S_f^{-1}\mathbf{m}) - \mathtt{f}), \\ &= \delta((S_f q_1)(\mathbf{m}) - \mathtt{f}), \\ &= \delta(f(\mathbf{m}) - \mathtt{f}) \\ &= \delta(\mathbf{f}^T\mathbf{m} - \mathtt{f}). \end{aligned} \tag{86}$$

As noted previously, the sharp measurements that are included in the quadrature quantum subtheory are those associated to a set of commuting quadrature observables $\{\mathcal{O}_{f^{(i)}}\}$ which in turn is associated with a PVM $\mathcal{O}_{V'} \equiv \{\Pi_{V',\mathbf{v}'} : \mathbf{v}' \in V'\}$ where $V' = \text{span}\{\mathbf{f}^{(i)}\}$. Given that $\hat{\Pi}_V(\mathbf{v}) \equiv \prod_{\{\mathbf{f}^{(i)}:\text{span}(\mathbf{f}^{(i)})=V\}} \hat{\Pi}_{f^{(i)}}\left(\mathbf{f}^{(i)T}\mathbf{v}\right)$ (Eq. (92)), and using Eq. (86), we conclude that

$$W_{\hat{\Pi}_{V'}(\mathbf{v}')}(\mathbf{m}) = \prod_{\{\mathbf{f}^{(i)}:\text{span}(\mathbf{f}^{(i)})=V\}} \delta(\mathbf{f}^{(i)T}\mathbf{m} - \mathbf{f}^{(i)T}\mathbf{v}). \tag{87}$$

Recalling Eq. (38), we see that the Wigner representation of the projector valued measure associated with $(V', \mathbf{v}')$ is the set of response functions associated with $(V', \mathbf{v}')$ in the quadrature epistricted theory, that is,

$$W_{\mathcal{O}_{V'}}(\mathbf{v}'|\mathbf{m}) = \xi_{V'}(\mathbf{v}'|\mathbf{m}).$$

The Wigner representation of the quadrature state associated with $(V, \mathbf{v})$ is

$$\begin{aligned} W_{\rho_{V,\mathbf{v}}}(\mathbf{m}) &= \mathrm{Tr}\left(\rho_{V,\mathbf{v}}\hat{A}(\mathbf{m})\right) \\ &= \frac{1}{\mathcal{N}_V} \prod_{\mathbf{f}^{(i)}:\mathrm{span}\{\mathbf{f}^{(i)}\}=V} \delta(\mathbf{f}^{(i)T}\mathbf{m} - \mathbf{f}^{(i)T}\mathbf{v}) \end{aligned} \tag{88}$$

where we have used Eqs. (66) and (87). Recalling Eqs. (30) and (31), we conclude that the Wigner representation of the quadrature state associated with $(V, \mathbf{v})$ is the epistemic state associated with $(V, \mathbf{v})$ in the quadrature epistricted theory, that is,

$$W_{\rho_{V,\mathbf{v}}}(\mathbf{m}) = \mu_{V,\mathbf{v}}(\mathbf{m}).$$

The Wigner representation of the Clifford superoperator $\mathcal{U}(S, \mathbf{a})$ is

$$\begin{aligned} W_{\mathcal{U}(S,\mathbf{a})}(\mathbf{m}|\mathbf{m}') &= \mathrm{Tr}\left[\hat{A}(\mathbf{m})\mathcal{U}(S,\mathbf{a})\left(\hat{A}(\mathbf{m}')\right)\right] \\ &= \mathrm{Tr}\left[\hat{A}(\mathbf{m})\hat{A}(S\mathbf{m}' + \mathbf{a})\right] \\ &= \delta(\mathbf{m} - (S\mathbf{m}' + \mathbf{a})). \end{aligned} \tag{89}$$

Here, the first equality follows from the form of the Wigner representation of superoperators, Eq. (79). The second equality follows from the fact that the phase-space point operators transform covariantly under the action of the Clifford superoperators, Eq. (72). The third equality in Eq. (90) follows from the orthogonality of the phase-space point operators, Eq. (75).

Recalling Eq. (37), we see that this is precisely the transition probability associated with the symplectic affine transformation, $\Gamma_{S,\mathbf{a}}(\mathbf{m}|\mathbf{m}')$, in the quadrature epistricted theory,

$$W_{\mathcal{U}(S,\mathbf{a})}(\mathbf{m}|\mathbf{m}') = \Gamma_{S,\mathbf{a}}(\mathbf{m}|\mathbf{m}'). \tag{90}$$

This concludes the proof of equivalence.

4.2 *Inequivalence for Bits/Qubits*

In the case where $d = 2$, the only even prime, the situation is more complicated. We have shown that in *both* the quadrature epistricted theory of bits and in the quadrature subtheory of qubits, we have: (i) the set of possible operational states is isomorphic to the set of pairs $(V, \mathbf{v})$ where V is an isotropic subspace of the phase-space $\Omega = (\mathbb{Z}_2)^{2n}$ and $\mathbf{v} \in V$; (ii) the set of possible sharp measurements is isomorphic to the set of isotropic subspaces V' (with the different outcomes associated to the elements

$\mathbf{v}' \in V'$); (iii) the set of possible reversible transformations is isomorphic to the elements $(S, \mathbf{a})$ of the symplectic affine group acting on $\Omega = (\mathbb{Z}_2)^{2n}$. It follows that the operational states, measurements and transformations of one theory are respectively isomorphic to those of the other. The valid *unsharp* measurements and *irreversible* transformations are defined in terms of the sharp and reversible ones respectively, and they are defined *in the same way* in the quadrature epistricted theory and the quadrature quantum subtheory. It follows that we also have the unsharp measurements and irreversible transformations of one theory isomorphic to those of the other.

Despite this strong structural similiarly, the two theories nonetheless make different predictions. The particular algorithm that takes as input a triple of preparation, measurement and transformation and yields as output a probability distribution over measurement outcomes, is not equivalent in the two theories. More precisely,

$$\mathrm{Tr}\left(\hat{\Pi}_{V'}(\mathbf{v}')\mathcal{U}_{S,\mathbf{a}}(\rho_{V,\mathbf{v}})\right) \neq \sum_{\mathbf{m}\in\Omega} \xi_{V'}(\mathbf{v}'|\mathbf{m}) \sum_{\mathbf{m}'\in\Omega} \Gamma_{S,\mathbf{a}}(\mathbf{m}|\mathbf{m}')\mu_{V,\mathbf{v}}(\mathbf{m}').$$

Gross's Wigner representation for discrete systems only works for systems of dimension d for d a power of an odd prime. It therefore does not work for $d = 2$. Nonetheless, a Wigner representation can be constructed for the quadrature subtheory of qubits. One can define it in terms of tensor products of the phase-space point operators for a single qubit as proposed by Gibbons, Hoffman and Wootters [49]. In this representation, we have

$$\mathrm{Tr}\left(\hat{\Pi}_{V'}(\mathbf{v}')\,\mathcal{U}(S,\mathbf{a})\left(\rho_{V,\mathbf{v}}\right)\right) = \sum_{\mathbf{m}\in\Omega} W_{\mathcal{O}_{V'}}(\mathbf{v}'|\mathbf{m}) \sum_{\mathbf{m}'\in\Omega} W_{\mathcal{U}(S,\mathbf{a})}(\mathbf{m}|\mathbf{m}')W_{\rho_{V,\mathbf{v}}}(\mathbf{m}').$$

So why can't we identify the Wigner representations with the corresponding objects in the epistricted theory, just as we did for d an odd prime and in the continuous case? The problem is that in the qubit case, the Wigner functions representing quadrature states sometimes go negative. It follows that these cannot be interpreted as probability distributions over the phase space. Similarly, the Wigner representations of quadrature observables and Clifford superoperators also sometimes go negative and hence cannot always be interpreted as conditional probability distributions.

It is also straightforward to prove that no alternative definition of the Wigner representation can achieve positivity. First, we make use of a fact shown in Ref. [50], that if a set of preparations and measurements supports a proof of contextuality in the sense of Ref. [25], then *all* quasiprobability representations must necessarily involve negativity. It then suffices to note that the quadrature subtheory is contextual. There are many ways of seeing this. For instance, Mermin's magic square proof of contextuality using two qubits [51] uses only the resources of the stabilizer theory of qubits. The same is true of the Greenberger-Horne-Zeilinger proof of nonlocality using three qubits [52], which is also a proof of contextuality.

The quadrature subtheory of qubits simply makes different operational predictions than the quadrature epistricted theory of bits. It admits of contextuality and nonlocality proofs while the quadrature epistricted theory is local and noncontextual by construction.[6]

By contrast, the quadrature subtheory of odd-prime qudits and the quadrature subtheory of mechanics make precisely the same predictions as the corresponding epistricted theories. They are consequently devoid of any contextuality or nonlocality because they admit of hidden variable models that are both local and noncontextual—the quadrature epistricted theory *is* the hidden variable model. Other differences between the two theories are discussed in Ref. [1].

The conceptual significance of the difference between the quadrature subtheory of qubits and the epistricted theory of bits remains a puzzle, despite various formalizations of the difference [4, 38]. This puzzle is perhaps the most interesting product of these investigations.

It shows in particular that whatever conceptual innovation over the classical world-view is required to achieve the phenomenology of contextuality and nonlocality, it must be possible to make sense of this innovation even in the thin air of the quadrature subtheory of qubits. This is an advantage because the latter theory uses a more meager palette of concepts than full quantum theory. For instance, we can conclude that it must be possible to describe the innovation of quantum over classical in terms of possibilistic inferences rather than probabilistic inferences.

Acknowledgments I acknowledge Stephen Bartlett and Terry Rudolph for discussions on the quadrature subtheory of quantum mechanics, Jonathan Barrett for suggesting to define the Poisson bracket in the discrete case in terms of finite differences, and Giulio Chiribella, Joel Wallman and Blake Stacey for comments on a draft of this article. Much of the work presented here summarizes unpublished results obtained in collaboration with Olaf Schreiber. Research at Perimeter Institute is supported by the Government of Canada through Industry Canada and by the Province of Ontario through the Ministry of Research and Innovation.

Appendix A: Quadrature Quantum Subtheories and the Stabilizer Formalism

In quantum information theory, there has been a great deal of work on a particular quantum subtheory for discrete systems of prime dimension (qubits and qutrits in particular) which is known as the *stabilizer formalism* [6, 53].

A stabilizer state is defined as a joint eigenstate of a set of commuting Weyl operators. By Eq. (53), two Weyl operators commute if and only if the corresponding phase-space displacement vectors have vanishing symplectic inner product,

$$[\hat{W}(\mathbf{a}), \hat{W}(\mathbf{a}')] = 0 \text{ if and only if } \langle \mathbf{a}, \mathbf{a}' \rangle = 0.$$

[6]It seems that the quadrature epistricted theory of bits is about as close as one can get to the stabilizer theory for qubits while still being local and noncontextual.

Consequently, the sets of commuting Weyl operators, and therefore the stabilizer states, are parametrized by the isotropic subspaces of Ω. Specifically, for each isotropic subspace M of Ω and each vector $\mathbf{v} \in JM \equiv \{J\mathbf{u} : \mathbf{u} \in M\}$, we can define a stabilizer state $\rho^{(\text{stab})}_{M,\mathbf{v}}$ as the projector onto the joint eigenspace of $\{\hat{W}(\mathbf{a}) : \mathbf{a} \in M\}$ where $\hat{W}(\mathbf{a})$ has eigenvalue $\chi(\langle \mathbf{v}, \mathbf{a} \rangle)$.

We will show here that the set of stabilizer states is precisely equivalent to the set of quadrature states.

To describe the connection, it is convenient to introduce some additional notions from symplectic geometry. The *symplectic complement* of a subspace V, which we will denote as V^C, is the set of vectors that have vanishing symplectic inner product with every vector in V,

$$V^C \equiv \{\mathbf{m}' \in \Omega : \mathbf{m}'^T J \mathbf{m} = 0, \ \forall \mathbf{m} \in V\},$$

where J is the symplectic form, defined in Eq. (15). This is not equivalent to the Euclidean complement of a subspace V, which is the set of vectors that have vanishing Euclidean inner product with every vector in V,

$$V^\perp \equiv \{\mathbf{m}' \in \Omega : \mathbf{m}'^T \mathbf{m} = 0, \ \forall \mathbf{m} \in V\},$$

The composition of the two complements will be relevant in what follows. It turns out that the latter is related to V by an isomorphism; it is simply the image of V under left-multiplication by the symplectic form J,

$$(V^\perp)^C \equiv JV = \{J\mathbf{u} : \mathbf{u} \in V\}.$$

Note that if V is isotropic, then $(V^\perp)^C$ is as well.

Proposition 1 *Consider the quadrature state $\rho_{V,\mathbf{v}}$, with V an isotropic subspace of Ω and $\mathbf{v} \in V$ a valuation vector, which is the joint eigenstate of the commuting set of quadrature observables $\{\mathcal{O}_f : \mathbf{f} \in V\}$, where the eigenvalue of $\mathcal{O}_f$ is $f(\mathbf{v})$. This is equivalent to the stabilizer state $\rho^{(\text{stab})}_{M,\mathbf{v}}$, which is the joint eigenstate of the commuting set of Weyl operators $\{\hat{W}(\mathbf{a}) : \mathbf{a} \in M\}$ where $M \equiv (V^\perp)^C$ is the isotropic subspace that is the symplectic complement of the Euclidean complement of V, and where the eigenvalue of $\hat{W}(\mathbf{a})$ is $\chi(\langle \mathbf{v}, \mathbf{a} \rangle)$.*

Proof Consider first a single degree of freedom. Every quadrature observable $\mathcal{O}_f$ can be expressed in terms of the position observable $\mathcal{O}_q$ as follows: if S_f is the symplectic matrix such that $\mathbf{f} = S_f \mathbf{q}$, then $\mathcal{O}_f = \hat{V}(S_f)\mathcal{O}_q \hat{V}(S_f)^\dagger$. Now note that the position basis can equally well be characterized as the eigenstates of the boost operators. Specifically, $\hat{B}(\mathtt{p})|\mathtt{q}\rangle_q = \chi(\mathtt{qp})|\mathtt{q}\rangle_q$, that is, an element $|\mathtt{q}\rangle_q$ of the position basis is an eigenstate of the set of operators $\{\hat{B}(\mathtt{p}) : \mathtt{p} \in \mathbb{R}/\mathbb{Z}_d\}$ where the eigenvalue of $\hat{B}(\mathtt{p})$ is $\chi(\mathtt{qp})$. The element $|\mathtt{f}\rangle_f$ of the basis associated to the quadrature operator $\mathcal{O}_f$ is defined as $|\mathtt{f}\rangle_f \equiv \hat{V}(S_f)|\mathtt{f}\rangle_q$ and consequently can be characterized as an eigenstate of the set of operators $\{\hat{V}(S_f)\hat{B}(\mathtt{g})\hat{V}(S_f)^\dagger : \mathtt{g} \in \mathbb{R}/\mathbb{Z}_d\}$ where the eigenvalue of $\hat{V}(S_f)\hat{B}(\mathtt{g})\hat{V}(S_f)^\dagger$ is $\chi(\mathtt{f}\mathtt{g})$. This can be stated equivalently as follows: the element $|\mathtt{f}\rangle_f$ of the basis associated to the quadrature operator $\mathcal{O}_f$ is the eigenstate of the set of Weyl operators $\{\hat{W}(\mathbf{a}) : \mathbf{a} \in \text{span}(S_f\mathbf{p})\}$ where the eigenvalue of $\hat{W}(\mathbf{a})$ is $\chi(\mathtt{f}\langle\mathbf{f}, \mathbf{a}\rangle)$. Noting that

$$\begin{aligned} \text{span}(S_f\mathbf{p}) &= \text{span}(S_f J\mathbf{q}), \\ &= \text{span}(JS_f\mathbf{q}), \\ &= \text{span}(J\mathbf{f}), \end{aligned} \tag{91}$$

we can just as well characterize $\hat{\Pi}_f(\mathtt{f})$ as the projector onto the joint eigenspace of the Weyl operators $\{\hat{W}(\mathbf{a}) : \mathbf{a} \in \text{span}(J\mathbf{f})\}$.

Now consider n degrees of freedom. The quadrature state associated with $(V, \mathbf{v})$ has the form

$$\rho_{V,\mathbf{v}} = \frac{1}{\mathcal{N}} \prod_{\mathbf{f}^{(i)}:\text{span}(\mathbf{f}^{(i)})=V} \hat{\Pi}_{f^{(i)}}\left(f^{(i)}(\mathbf{v})\right). \tag{92}$$

By an argument similar to that used for a single degree of freedom, this is an eigenstate of the Weyl operators $\{\hat{W}(\mathbf{a}) : \mathbf{a} \in \text{span}(J\mathbf{f}^{(i)})\}$ where the eigenvalue of $\hat{W}(\mathbf{a})$ is $\chi(\langle\mathbf{v}, \mathbf{a}\rangle)$. Noting that $\text{span}(J\mathbf{f}^{(i)}) = JV = (V^{\perp})^C$, we have our desired isomorphism. ■

The stabilizer formalism allows all and only the Clifford superoperators as reversible transformations. The sharp measurements that are included in the stabilizer formalism are the ones associated with PVMs corresponding to the joint eigenspaces of a set of commuting Weyl operators, which, by Proposition 1, are precisely those corresponding to the joint eigenspaces of a set of commuting quadrature observables. It follows that the stabilizer formalism coincides precisely with the quadrature subtheory.

Gross has argued that the discrete analogue of the Gaussian quantum subtheory for continuous variable systems is the stabilizer formalism [6]. Our results show that the connection between the discrete and continuous variable cases is a bit more subtle than this. In the continuous variable case, there is a distinction between the Gaussian subtheory and the quadrature subtheory, with the latter being contained within the former. In the discrete case, there is no distinction, so the stabilizer formalism can be usefully viewed as either the discrete analogue of the Gaussian subtheory or as the discrete analogue of the quadrature subtheory. While Gross's work showed that the stabilizer formalism for discrete systems could be defined similarly to how one defines Gaussian quantum mechanics, our work has shown that it can also be defined in the same way that one defines quadrature quantum mechanics.

To our knowledge, quadrature quantum mechanics has not previously received much attention. However, given that it is a natural continuous variable analogue of the stabilizer formalism for discrete systems, it may provide an interesting paradigm for exploring quantum information processing with continuous variable systems.

References

1. R.W. Spekkens, Evidence for the epistemic view of quantum states: a toy theory. Phys. Rev. A **75**(3), 032110 (2007)
2. S.D. Bartlett, T. Rudolph, R.W. Spekkens, Reconstruction of Gaussian quantum mechanics from Liouville mechanics with an epistemic restriction. Phys. Rev. A **86**(1), 012103 (2012)
3. M. Born, E. Wolf, *Principles of Optics: Electromagnetic Theory of Propagation, Interference and Diffraction of Light* (Cambridge University Press, Cambridge, 1999)
4. M.F. Pusey, Stabilizer notation for Spekkens' toy theory. Found. Phys. **42**(5), 688–708 (2012)
5. S.J. Van Enk, A toy model for quantum mechanics. Found. Phys. **37**(10), 1447–1460 (2007)
6. D. Gross, Hudson's theorem for finite-dimensional quantum systems. J. Math. Phys. **47**, 122107 (2006)
7. O. Schreiber, R.W. Spekkens, The power of epistemic restrictions in reconstructing quantum theory: from trits to qutrits, unpublished, 2008. R.W. Spekkens, The power of epistemic restrictions in reconstructing quantum theory, Talk, Perimeter Institute, http://pirsa.org/09080009/, 10 August 2008
8. T.H. Boyer, *Foundations of Radiation Theory and Quantum Electrodynamics, Chapter A Brief Survey of Stochastic Electrodynamics* (Plenum, New York, 1980)
9. C.M. Caves, C.A. Fuchs, Quantum information: how much Information in a state vector? (1996). arXiv:quant-ph/9601025
10. J.V. Emerson, Quantum chaos and quantum-classical correspondence. Ph.D. thesis, Simon Fraser University, Vancouver, Canada (2001)
11. L. Hardy, Disentangling nonlocality and teleportation (1999). arXiv:quant-ph/9906123
12. K.A. Kirkpatrick, Quantal behavior in classical probability. Found. Phys. Lett. **16**(3), 199–224 (2003)
13. W.K. Wootters, W.H. Zurek, A single quantum cannot be cloned. Nature **299**(5886), 802–803 (1982)
14. C.H. Bennett, G. Brassard, C. Crépeau, R. Jozsa, A. Peres, W.K. Wootters, Teleporting an unknown quantum state via dual classical and Einstein-Podolsky-Rosen channels. Phys. Rev. Lett. **70**(13), 1895 (1993)
15. C.H. Bennett, G. Brassard et al., Quantum cryptography: public key distribution and coin tossing, in *Proceedings of IEEE International Conference on Computers, Systems and Signal Processing*, vol. 175, (New York 1984)
16. R. Horodecki, P. Horodecki, M. Horodecki, K. Horodecki, Quantum entanglement. Rev. Mod. Phys. **81**(2), 865 (2009)
17. C.H. Bennett, D.P. DiVincenzo, C.A. Fuchs, T. Mor, E. Rains, P.W. Shor, J.A. Smolin, W.K. Wootters, Quantum nonlocality without entanglement. Phys. Rev. A **59**(2), 1070 (1999)
18. C.H. Bennett, D.P. DiVincenzo, T. Mor, P.W. Shor, J.A. Smolin, B.M. Terhal, Unextendible product bases and bound entanglement. Phys. Rev. Lett. **82**(26), 5385 (1999)
19. M.D. Choi, Completely positive linear maps on complex matrices. Linear Algebra Appl. **10**, 285–290 (1975)
20. A. Jamiołkowski, Linear transformations which preserve trace and positive semidefiniteness of operators. Rep. Math. Phys. **3**, 275–278 (1972)
21. M.A. Naimark. Izv. Akad. Nauk SSSR, Ser. Mat. **4**:277–318 (1940)
22. W.F. Stinespring, Positive functions on C*-algebras. Proc. Am. Math. Soc. **6**(2), 211–216 (1955)
23. J.S. Bell, On the Einstein Podolsky Rosen paradox. Physics **1**(3), 195–200 (1964)

24. S. Kochen, E.P. Specker, The problem of hidden variables in quantum mechanics. J. Math. Mech. **17**, 59 (1967)
25. R.W. Spekkens, Contextuality for preparations, transformations, and unsharp measurements. Phys. Rev. A **71**(5), 052108 (2005)
26. Y.C. Liang, R.W. Spekkens, H.M. Wiseman, Specker's parable of the overprotective seer: a road to contextuality, nonlocality and complementarity. Phys. Rep. **506**(1), 1–39 (2011)
27. N. Harrigan, R.W. Spekkens, Einstein, incompleteness, and the epistemic view of quantum states. Found. Phys. **40**, 125 (2010)
28. C.M. Caves, C.A. Fuchs, R. Schack, Quantum probabilities as Bayesian probabilities. Phys. Rev. A **65**(2), 022305 (2002)
29. C.A. Fuchs, Quantum mechanics as quantum information, mostly. J. Mod. Opt. **50**(6–7), 987–1023 (2003)
30. C.A. Fuchs, R. Schack, Quantum-Bayesian coherence. Rev. Mod. Phys. **85**(4), 1693 (2013)
31. M.F. Pusey, J. Barrett, T. Rudolph, On the reality of the quantum state. Nat. Phys. **8**(6), 475–478 (2012)
32. P.G. Lewis, D. Jennings, J. Barrett, T. Rudolph, Distinct quantum states can be compatible with a single state of reality. Phys. Rev. Lett. **109**(15), 150404 (2012)
33. R. Colbeck, R. Renner, Is a system's wave function in one-to-one correspondence with its elements of reality? Phys. Rev. Lett. **108**(15), 150402 (2012)
34. M.S. Leifer, R.W. Spekkens, Towards a formulation of quantum theory as a causally neutral theory of Bayesian inference. Phys. Rev. A **88**(5), 052130 (2013)
35. C.J. Wood, R.W. Spekkens, The lesson of causal discovery algorithms for quantum correlations: causal explanations of bell-inequality violations require fine-tuning (2012). arXiv:1208.4119
36. A. Zeilinger, A foundational principle for quantum mechanics. Found. Phys. **29**(4), 631–643 (1999)
37. T. Paterek, B. Dakić, Č. Brukner, Theories of systems with limited information content. New J. Phys. **12**(5), 053037 (2010)
38. B. Coecke, B. Edwards, R.W. Spekkens, Phase groups and the origin of non-locality for qubits. Electron. Notes Theor. Comput. Sci. **270**(2), 15–36 (2011)
39. S. Mansfield, T. Fritz, Hardy's non-locality paradox and possibilistic conditions for non-locality. Found. Phys. **42**(5), 709–719 (2012)
40. S. Abramsky, L. Hardy, Logical bell inequalities. Phys. Rev. A **85**(6), 062114 (2012)
41. B. Schumacher, M.D. Westmoreland, Modal quantum theory. Found. Phys. **42**(7), 918–925 (2012)
42. J. Barrett, Information processing in generalized probabilistic theories. Phys. Rev. A **75**(3), 032304 (2007)
43. L. Hardy, Quantum theory from five reasonable axioms (2001). arXiv:quant-ph/0101012
44. B. Coecke, E.O. Paquette, Categories for the practising physicist, in *New Structures for Physics*. Lecture Notes in Physics, ed. by B. Coecke (Springer, Berlin, 2009), pp. 173–286
45. B. Coecke, B. Edwards, Toy quantum categories. Electron. Notes Theor. Comput. Sci. **270**(1), 29–40 (2011)
46. G. Chiribella, G.M. DAriano, P. Perinotti, Probabilistic theories with purification. Phys. Rev. A **81**(6), 062348 (2010)
47. E. Wigner, On the quantum correction for thermodynamic equilibrium. Phys. Rev. **40**(5), 749 (1932)
48. C. Gardiner, P. Zoller, *Quantum Noise: A Handbook of Markovian and Non-Markovian Quantum Stochastic Methods with Applications to Quantum Optics*, vol. 56 (Springer, New York, 2004)
49. K.S. Gibbons, M.J. Hoffman, W.K. Wootters, Discrete phase space based on finite fields. Phys. Rev. A **70**(6), 062101 (2004)
50. R.W. Spekkens, Negativity and contextuality are equivalent notions of nonclassicality. Phys. Rev. Lett. **101**(2), 20401 (2008)
51. N.D. Mermin, Hidden variables and the two theorems of John Bell. Rev. Mod. Phys. **65**(3), 803 (1993)

52. D.M. Greenberger, M.A. Horne, A. Zeilinger, Going beyond Bell's theorem, *Bell's Theorem, Quantum Theory and Conceptions of the Universe* (Springer, New York, 1989), pp. 69–72
53. D. Gottesman, The Heisenberg representation of quantum computers (1998). arXiv:quant-ph/9807006

Part II
Axiomatizations

Information-Theoretic Postulates for Quantum Theory

Markus P. Müller and Lluís Masanes

1 Introduction

By all standards, quantum theory is one of the most successful theories of physics. It provides the basis of particle physics, chemistry, solid state physics, and it is of paramount importance for many technological achievements. So far, all experiments have confirmed its universal validity in all parts of our physical world. Unfortunately, quantum theory is also one of the most mysterious theories of physics.

In the text books, quantum theory is usually introduced by stating several abstract mathematical postulates: *States are unit vectors in a complex Hilbert space; probabilities are given by the Born rule; the Schrödinger equation describes time evolution in closed systems*, to name just some of them. As many students recognize—and experienced researchers over the years sometimes tend to forget—these postulates seem arbitrary and do not have a clear meaning. It is true that they work very well and are in accordance with experiments, but *why are they true?* Why is nature described by these counterintuitive laws of complex Hilbert spaces?

What at first sight seems to be a physically vacuous, philosophical question is in fact of high relevance to theoretical physics, in particular for *attempts to generalize quantum theory*. There have been several attempts in the past to construct natural modifications of quantum theory—either to set up experimental tests of quantum physics, or to adapt it in a way which allows for easier unification with general

M.P. Müller (✉)
Departments of Applied Mathematics and Philosophy, University of Western Ontario, 1151 Richmond Street, London ON N6A 5B7, Canada
e-mail: markus@mpmueller.net

L. Masanes (✉)
Department of Physics and Astronomy, University College London, London WC1E 6BT, UK
e-mail: l.masanes@ucl.ac.uk

G. Chiribella and R.W. Spekkens (eds.), *Quantum Theory: Informational Foundations and Foils*, Fundamental Theories of Physics 181,
DOI 10.1007/978-94-017-7303-4_5

relativity. However, modification of quantum theory turned out to be a surprisingly difficult task.

A historical example is given by Weinberg's [1] non-linear modification of quantum theory. Only a few months after his proposal was published, Gisin [2] demonstrated that the resulting theory has an unexpected poisonous property: it allows for superluminal signalling. It can be shown in general that other proposals of this kind must face a similar fate [3]. It seems as if the usual postulates of quantum theory are intricately intertwined, in a way such that modification of one postulate makes the combination of the others collapse into a physically meaningless–or at least problematic–theory.

One possible way to overcome this difficulty is to find alternative postulates for quantum theory that have a clear physical interpretation and do not refer to the mathematical structure of complex Hilbert spaces. The search for simple operational axioms dates back to Birkhoff and von Neumann [4], and includes work by Mackey [5], Ludwig [6], Alfsen and Shultz [7] and many others. The advent of quantum information theory initiated new ideas and methods to approach this problem, resulting in the pioneering work by Hardy [8], and a recent wave of axiomatizations of quantum theory, including Dakić and Brukner's work [9], our result [10], the reconstruction by the Pavia group [11], alternative formulations by Hardy [12, 13] and Zaopo [14].

In this paper, we give a self-contained summary of our results in [10], where we derive the formalism of quantum theory from four natural information-theoretic postulates. They can loosely be stated as follows:

1. The state of a composite system is characterized by the statistics of measurements on the individual components.
2. All systems that effectively carry the same amount of information have equivalent state spaces.
3. Every pure state of a system can be transformed into every other by continuous reversible time evolution.
4. In systems that carry one bit of information, all measurements which give non-negative probabilities are allowed by the theory.

Below, we show how to derive the usual formalism of quantum theory from these postulates. Surprisingly, the complex numbers and Hilbert spaces pop out even though they are not mentioned in the postulates. This is true for all the axiomatization approaches mentioned above, starting with Hardy's work [8]: these results allow us to gain a better understanding of the usual quantum formalism, and resolve some of the mystery around ad hoc postulates like the Born rule.

Every axiomatization has its own benefits. We think that the main advantage of our work [10]—as described in this paper—is its *parsimony*: our postulates are close to a *minimal* set of postulates for quantum theory. Accomplishing the goal of minimality would mean to have a set of axioms such that dropping or weakening any one of the axioms will always yield new solutions in addition to quantum theory. Currently, we do not know if we have actually achieved this goal, though we think that we are pretty close to it (this will be discussed in more detail in Sect. 6). Our attempt to have as few assumptions as possible is also reflected in the background assumptions: for example, we do not assume apriori that the composition of three systems into a joint system is associative, or that pairs of generalized bits admit an analogue of a "swap" operation.

Our result suggests an obvious method to obtain natural modifications of quantum theory: *drop or weaken one of the postulates, and work out mathematically what the resulting set of theories looks like*. It is clear that minimality of the axioms (in the sense just described) is crucial for this method. In contrast to the usual formulation of quantum theory, we know for sure that the corresponding alternative "post-quantum" theories are consistent and do not allow for superluminal signalling as in Weinberg's approach. This is due to the fact that the no-signalling principle is built in as a background assumption. In a way, those theories will be "quantum theory's closest cousins": they are not formulated in terms of Hilbert spaces, but share as many characteristic features with quantum theory as possible.

As the simplest possible modification, suppose we drop the word "continuous" from Postulate 3—that is, we allow for discrete reversible time evolution. Then another solution in addition to quantum theory appears: in this theory, states are probability distributions, and reversible time evolution is given by permutations of outcomes. This is exactly *classical probability theory* on discrete sample spaces. It turns out to be the unique additional solution in this case.

2 What Do We Mean by "Quantum Theory"?

When talking about axiomatizing quantum theory, there is sometimes confusion about what we actually mean by it. The term "quantum theory" arouses association with many different aspects of physics that are usually treated in quantum mechanics text books, such as particles, the hydrogen atom, three-dimensional position and momentum space and many more.

However, a more careful definition should apply here. As an analogy, consider the theory of statistical mechanics. This theory consists of an application of probability theory to mechanics, which means in particular that abstract probability theory can be studied detached from statistical physics—and this has been done in mathematics for a long time.

Similarly, we can consider quantum mechanics to be a combination of an abstract probabilistic theory—*quantum theory*—and classical mechanics. Abstract quantum theory can be studied detached from its mechanical realization; the main difference to the previous example lies in the historical fact that the development of quantum mechanics preceded that of abstract quantum theory. In this terminology, we understand by "quantum theory" the statement that

- states are vectors (resp. density matrices) in a complex Hilbert space,
- probabilities are computed by the Born rule resp. trace rule,
- the possible reversible transformations are the unitaries,
- measurements are described by projection operators, and thus observables are given by self-adjoint matrices.

The "classical mechanics" part, on the other hand, determines the type of Hilbert space to consider (such as $L^2(\mathbb{R}^3)$), the choice of "Hamiltonians" H which generate the time evolution, $U(t) = \exp(iHt)$, and the choice of initial states of that time evolution. This conceptual distinction has proven particularly useful in the development of quantum information theory. It seems that this distinction was always implicit when expressing the desire to "quantize" any classical physical theory, that is, to combine it with abstract quantum theory.

Thus, since we are aiming for a reconstruction of abstract quantum theory, we will not refer to position, momentum, or Hamiltonians in this paper. Instead, we only use the notions of abstract probability theory: of events, happening with certain probabilities, and of transformations modifying the probabilities. Furthermore, we restrict our analysis to finite-dimensional systems: we argue that the main mystery is *why to have a complex Hilbert space at all.* If this is understood in finite dimensions, it seems only a small conceptual (though possibly mathematically challenging) step to guess the correct infinite-dimensional generalizations.

Since we presuppose probabilities as given, we also do not address the question where these probabilities come from. Hence we also ignore the question about what happens in a quantum measurement, and all other interpretational mysteries encompassing the formulation of quantum theory. Instead, we restrict ourselves to ask how the mathematical formalism of quantum theory can be derived from simpler postulates, and what possible modifications of it we might hope to find in nature.

Questions that we would like to address:

- How can we understand (that is, derive) the complex Hilbert space formalism from simple operational assumptions on probabilities?
- What other probabilistic theories are operationally closest to quantum theory?

Questions/problems that we do *not* address:

- How should we interpret "probability", and where does it come from?
- The measurement problem.
- Interpretation of quantum mechanics.

In order to formulate our postulates, we work with a simple and general framework encompassing all conceivable ways to formulate physical theories of probability: this is the framework of *generalized probabilistic theories.*

3 Generalized Probabilistic Theories

Classical probability theory (abbreviated CPT henceforth) is used to describe processes which are not deterministic. This is achieved by assuming a particular mathematical structure: a probability space with a unique fixed probability

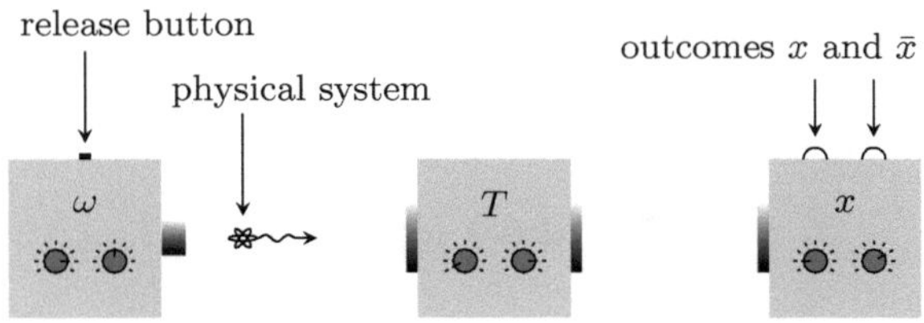

Fig. 1 General experimental set up. From *left* to *right* there are the preparation, transformation and measurement devices. As soon as the release button is pressed, the preparation device outputs a physical system in the state specified by the knobs. The next device performs the transformation specified by its knobs (which in particular can be "do nothing"). The device on the *right* performs the measurement specified by its knobs, and the outcome (x or $\bar{x}$) is indicated by the corresponding light

measure, which is used to assign probabilities to all random variables. The framework of generalized probabilistic theories [4, 5, 8, 15–18] generalizes this approach in a simple way. We will now give a brief introduction to this framework, built on general considerations of what constitutes an experiment in physics. For more detailed introductions, we refer the reader to [16, 17], and for nice presentations of the main ideas to [21, 22].

In order to set up a common picture, we consider Fig. 1 as the model for what constitutes a physical experiment. This is just an illustration: the events that we describe may as well be natural processes that happen without human or technological intervention.

The main idea (cf. Fig. 1) is that physical systems can cause objective events which we call "measurement outcomes"—for example clicks of detectors. We say that two systems are in the same state ω if all outcome probabilities of all possible measurements are the same. In order to test this empirically, we always assume that we can prepare a physical system in a given state as often as we want. That is, we may think of a *preparation device* which produces a physical system in a particular state.

3.1 States and Measurements

Single outcomes of measurements are called *effects*, and are denoted by uppercase letters such as E. The probability of obtaining outcome E, if measured on state ω, will be denoted $E(\omega)$. This way, effects become maps from states to probabilities in [0, 1].

What can we say about the set of all possible states ω in which a given system can be prepared? Suppose we have two preparation devices; one of them prepares the system in some state ω, the other one prepares it in some state φ. Then we can use these devices to construct a new device, which tosses a coin, and then prepares either state ω with probability $p \in [0, 1]$, or state φ with probability $1 - p$. We denote this new state by

$$\omega' := p\omega + (1-p)\varphi.$$

Clearly, if we apply a measurement on ω', we get outcome E with probability

$$E(\omega') = pE(\omega) + (1-p)E(\varphi).$$

Thus, by this construction, we see that states ω become elements of an affine space, and effects E are affine maps. The set of all possible states—called the *state space* $\mathcal{S}$— will be a subset of this affine space. We have just seen that $\omega \in \mathcal{S}$ and $\varphi \in \mathcal{S}$ imply $p\omega + (1-p)\varphi \in \mathcal{S}$ if $0 \leq p \leq 1$; that is, state spaces are convex sets (similar reasoning is given in [8, 17, 19]).

In principle, state spaces can be infinite-dimensional (and in fact, in many physical situations, they are). However, in this paper, we will only consider finite-dimensional state spaces. Then, states ω are determined by finitely many coordinates, and we may use this to construct a more concrete representation of states. Denote the dimension of a state space $\mathcal{S}$ by d. Then, by choosing d affinely independent effects $E_1, \ldots, E_d$, the probabilities $E_1(\omega), \ldots, E_d(\omega)$ determine ω uniquely. We now use the representation

$$\omega = \begin{bmatrix} 1 \\ E_1(\omega) \\ E_2(\omega) \\ \vdots \\ E_d(\omega) \end{bmatrix} =: \begin{bmatrix} 1 \\ \omega_1 \\ \omega_2 \\ \vdots \\ \omega_d \end{bmatrix} \in \mathcal{S} \subset \mathbb{R}^{d+1}. \tag{1}$$

The choice of $E_1, \ldots, E_d$ is arbitrary, subject only to the restriction that they are affinely independent. We call a set of effects with this property *fiducial*, and we refer to $E_1(\omega), \ldots, E_d(\omega)$ as *fiducial outcome probabilities* [8]. The component $\omega_0 := 1$ has been introduced for calculational convenience: it allows us to write the affine effects E as *linear* functionals on the larger space $\mathbb{R}^{d+1}$. It will also turn out to be particularly useful in calculations involving composite state spaces.

In the following, we will assume that state spaces $\mathcal{S}$ are topologically closed and bounded, i.e. compact (for a physical motivation see [10]). The extremal points of the convex set $\mathcal{S}$ will be called *pure states*; these are states ω which cannot be written as mixtures $p\varphi + (1-p)\varphi'$ of other states $\varphi \neq \varphi'$ with $0 < p < 1$. It follows from the compactness of $\mathcal{S}$ that every state can be written as a convex combination of at most $d+1$ pure states [20].

Measurements with n outcomes are described by a collection of n effects $E_1, E_2, \ldots, E_n$ with the property $E_1(\omega) + E_2(\omega) + \ldots + E_n(\omega) = 1$ for all states ω. This expresses the fact that outcome i happens with probability $E_i(\omega)$, and the total probability is one. Note that two effects E and F can only be part of the same measurement if $E(\omega) + F(\omega) \leq 1$ for all states ω. Sets of fiducial effects (as introduced above) do not necessarily have this property. A single effect E is always part of a measurement with two outcomes E and $\bar{E}$, where $\bar{E}(\omega) := 1 - E(\omega)$.

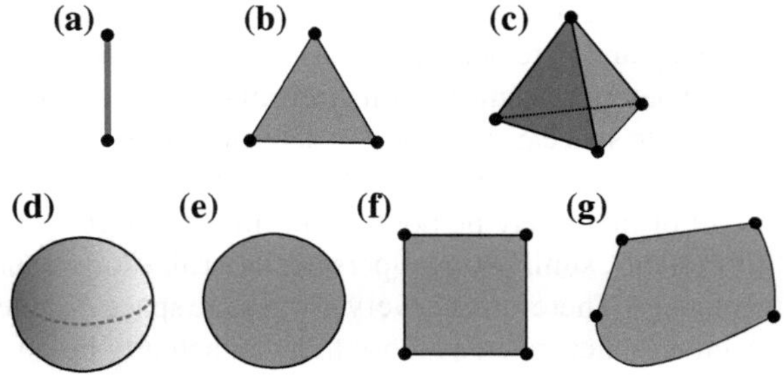

Fig. 2 Examples of convex state spaces: **a** is a classical bit, **b** and **c** are classical 3- and 4-level systems, **d** is a quantum bit, **e** is the projection of a qubit, **f** and **g** are neither classical nor quantum. Note that quantum n-level systems for $n \geq 3$ are *not* balls

Figure 2 gives some examples of convex state spaces. First, consider a classical bit, which is described within CPT. We can think of a coin which shows either heads or tails; in general, it can be in one of those configurations with some probability. The probability p of showing heads determines the state uniquely, since the tails probability must be $1 - p$. Thus, $p \in [0, 1]$ is a fiducial probability; recalling (1), we can represent states as $\omega = [1, p]^{\mathrm{T}}$. This yields a one-dimensional state space, with two pure states $[1, 0]^{\mathrm{T}}$ and $[1, 1]^{\mathrm{T}}$, corresponding to coins which deterministically show heads or tails. It is depicted in Fig. 2a.

Similarly, classical n-level systems have states which correspond to probability distributions $p_1, \ldots, p_n$. Since $p_n = 1 - (p_1 + \cdots + p_{n-1})$, the numbers $p_1, \ldots, p_{n-1}$ are fiducial outcome probabilities, yielding states $\omega = [1, p_1, \ldots, p_{n-1}]^{\mathrm{T}}$. Geometrically, the resulting state spaces are simplices. They are depicted in Fig. 2b, c for $n = 2$ and $n = 3$.

Quantum systems look very different: as it is well-known, states of quantum 2-level systems, i.e. qubits, can be parametrized by a vector $\vec{r} \in \mathbb{R}^3$ with $|\vec{r}| \leq 1$, such that every density matrix can be written $\rho = (\mathbf{1} + \vec{r} \cdot \vec{\sigma})/2$, with $\vec{\sigma} = (\sigma_x, \sigma_y, \sigma_z)$ the Pauli matrices. Thus, we can use the vector $[1, r'_x, r'_y, r'_z]^{\mathrm{T}}$ to represent states, where $r'_i := (1 + r_i)/2$ is the probability to measure "spin up" in i-direction. This state space is the famous (slightly reparametrized) Bloch ball, cf. Fig. 2d.

Figure 2e shows a state space which is a projection of the Bloch ball: it corresponds to the effective state space that we obtain if, for some reason, spin measurements in z-direction are physically impossible to implement, with states $\omega = [1, r'_x, r'_y]^{\mathrm{T}}$. The square state space in Fig. 2f describes a system for which there exist two independent effects, say X and Y, that can yield probabilities $X(\omega)$ and $Y(\omega)$ in $[0, 1]$ arbitrarily and independently from each other. States will be of the form $\omega = [1, \omega_x, \omega_y]^{\mathrm{T}}$, with $\omega_x = X(\omega)$ and $\omega_y = Y(\omega)$.

Consider the two yes-no-measurements which correspond to the effects X and Y; we can interpret these as spin measurements in two orthogonal directions, with "yes"-outcome X or Y for "spin up", and "no"-outcome $\bar{X}$ or $\bar{Y}$ for "spin down". If we perform either one of these measurements on the state $\omega = (1, 1, 1)$, then we will get the "yes"-outcome with unit probability – and this is true for both measurements. If we consider the analogous measurements on the circle state space, we see that the corresponding behavior becomes impossible: if one of the spin measurements yields outcome "yes" with certainty, then the other spin measurement must give outcome "yes" with probability 1/2. This follows from $r_x^2 + r_y^2 \leq 1$.

Thus, the circle state space shows a form of *complementarity*, which is not present in the square state space. As this example illustrates, the state space of a physical system determines many of its information-theoretic properties. Given a description of the state space $\mathcal{S}$, we can also determine the set of all linear functionals which map states to the unit interval [0, 1], that is, the candidates for possible effects. However, not all of them may be possible to implement in physics: maybe some of them are "forbidden", similarly as superselection rules forbid some superpositions in quantum mechanics. Therefore, to every given state space $\mathcal{S}_A$, there is a set of "allowed effects" which are interpreted as those that can actually be physically performed.

We introduce some notions which will be useful later: A set of states $\omega_1, \ldots, \omega_n$ is called *distinguishable* if there is a measurement with outcomes represented by effects $E_1, \ldots, E_n$, such that $E_i(\omega_j) = \delta_{ij}$, which is 1 if $i = j$ and 0 otherwise. The interpretation is that we can build a device which perfectly distinguishes the different states ω_j. Given a physical system A, we define the *capacity* N_A as the maximal size of any set of distinguishable states $\omega_1, \ldots, \omega_n \in \mathcal{S}_A$. A measurement which is able to distinguish N_A states (that is, as much as possible) will be called *complete*. For a quantum state space, N_A equals the dimension of the underlying complex Hilbert space.

We denote the real vector space which carries $\mathcal{S}_A$ by V_A. Then effects are elements of the dual space V_A^*. For a quantum N-level system, V_A is the real vector space of Hermitian $N \times N$-matrices with complex entries. Following Wootters and Hardy [8, 23], we also use the notation $K_A := \dim V_A = \dim(\mathcal{S}_A) + 1$, that is the number of degrees of freedom that is necessary to describe an unnormalized state. For a qubit, for example, we have $N_A = 2$, but $K_A = 4$. In quantum theory, $K_A = N_A^2$ equals the number of independent real parameters in a density matrix (dropping normalization). In classical probability theory, we always have $K_A = N_A$.

3.2 Transformations

A transformation is a map T which takes a state to another state. Which transformations are actually possible is a question of physics. However, there are certain minimal assumptions that every transformation must necessarily satisfy in order to be physically meaningful in the context of convex state spaces. First, transformations must respect probabilistic mixtures—that is,

$$T(p\omega + (1 - p)\varphi) = pT(\omega) + (1 - p)T(\varphi).$$

This is because both sides of the equation can be interpreted as the result of randomly preparing ω or φ (with probabilities p resp. $1-p$) and applying the transformation T. Thus, transformations (from one system to itself) are affine maps which map a state space $\mathcal{S}_A$ into itself; we can always assume that they are linear maps $T : V_A \to V_A$.

If both T and T^{-1} are physically allowed transformations, we call T *reversible*. The set of reversible transformations on a physical system A is a group $\mathcal{G}_A$. For physical reasons, we assume that $\mathcal{G}_A$ is topologically closed, hence a compact group [24] (it may be a finite group).

Reversible transformations map a state space bijectively onto itself—hence they are symmetries of the state space. For example, in quantum theory, reversible transformations are the unitary conjugations, $\rho \mapsto U\rho U^\dagger$. In the Bloch ball representation of the qubit (as in Fig. 2d), these maps are represented as rotations, such that the group of reversible transformations is isomorphic to $SO(3)$.

However, as this example also shows, not all symmetries are automatically allowed reversible transformations: a reflection in the Bloch ball is a symmetry, but it is not an allowed transformation (in the density matrix picture, it would correspond to an anti-unitary map).

In summary, for what follows, a physical system A is specified by three mathematical objects: the state space $\mathcal{S}_A$, the group of reversible transformations $\mathcal{G}_A$ (which is a compact subgroup of all symmetries of $\mathcal{S}_A$), and a set of physically allowed effects. The latter will not be given a particular notation, but we assume that the set of allowed effects is topologically closed. For obvious physical reasons, if E is an allowed effect and $T \in \mathcal{G}_A$, then $E \circ T$ is an allowed effect; similarly, convex combinations of allowed effects are allowed.

3.3 Composite Systems

If we are given two physical systems A and B, we would like to define a *composite system* AB which is also a physical system in the sense described above, with its own state space $\mathcal{S}_{AB}$, group of reversible transformations $\mathcal{G}_{AB}$, and set of allowed effects.

In contrast to quantum theory, the framework of general probabilistic theories allows many different possible composites for two given systems A and B. Every possible composite AB has a set of minimal physical assumptions that it must satisfy:

- If $\omega_A \in \mathcal{S}_A$ and $\omega_B \in \mathcal{S}_B$ are two local states, then there is a distinguished state $\omega_A\omega_B \in \mathcal{S}_{AB}$ which is interpreted as the result of *preparing* ω_A *and* ω_B *independently on the subsystems A and B*.
- If E_A and E_B are local allowed effects on A and B, then there is a distinguished allowed effect $E_A E_B$ on AB which is interpreted as *measuring* E_A *on A and* E_B *on B independently,* yielding the total probability that outcome E_A happens on system A, and outcome E_B happens on system B.
- This intuition is mathematically expressed by demanding that

$$E_A E_B(\omega_A\omega_B) = E_A(\omega_A)E_B(\omega_B)$$

where both $E_A E_B$ and $\omega_A\omega_B$ are affine in both arguments. This also formalizes the physical assumption that the temporal order of the local preparations resp. measurements is irrelevant.

From the previous point, we can infer that we can represent independent local preparations $\omega_A\omega_B$ and measurement outcomes $E_A E_B$ by tensor products:

$$E_A E_B \equiv E_A \otimes E_B, \quad \omega_A\omega_B \equiv \omega_A \otimes \omega_B.$$

The vector space V_{AB} that carries the composite state space must thus satisfy

$$V_A \otimes V_B \subseteq V_{AB}. \tag{2}$$

For the dimensions of these spaces, we obtain

$$K_A K_B \leq K_{AB}. \tag{3}$$

Now consider two different measurements (for simplicity with two outcomes) $E_B, \bar{E}_B := \mathbf{1}_B - E_B$ and $F_B, \bar{F}_B := \mathbf{1}_B - F_B$, where $\mathbf{1}_B$ denotes the trivial effect on system B which yields unit probability on every normalized state. We can think of an agent *Bob*, holding system B, who may decide freely (say, according to some local random variable) whether to perform measurement $E_B, \bar{E}_B$ or $F_B, \bar{F}_B$.

Suppose that Alice (holding system A) performs some measurement after Bob has chosen and performed his measurement on a bipartite state ω_{AB}. The marginal probability that she obtains (not knowing Bob's outcome) is the same in both cases:

$$\begin{aligned} E_A \otimes \mathbf{1}_B(\omega_{AB}) &= E_A \otimes E_B(\omega_{AB}) + E_A \otimes \bar{E}_B(\omega_{AB}) \\ &= E_A \otimes F_B(\omega_{AB}) + E_A \otimes \bar{F}_B(\omega_{AB}). \end{aligned}$$

The same holds with the roles of A and B reversed. This equation follows from our assumptions above on how to represent local measurements. We have proven that our assumptions imply the *no-signalling property*: Bob cannot send information to Alice merely by his choice of local measurement (and vice versa). Moreover, the previous equation shows that the outcome probabilities of all of Alice's measurements are described by the *reduced state* $\omega_A := \mathrm{Id}_A \otimes \mathbf{1}_B(\omega_{AB})$ (note that Id_A is the identity transformation, while $\mathbf{1}_B$ is a linear functional). This state corresponds to the marginal of ω_{AB} on A, and is uniquely characterized by the equation

$$E_A(\omega_A) = E_A \otimes \mathbf{1}_B(\omega_{AB})$$

for all functionals (in particular, all allowed effects) E_A.

For physically meaningful composites AB, we should demand that reduced states ω_A, ω_B of all bipartite states $\omega_{AB} \in \mathcal{S}_{AB}$ are valid local states themselves. In fact, we will demand something which is stronger and contains this as a special case. Suppose that Alice and Bob share ω_{AB} and Bob performs a measurement and obtains outcome E_B. Knowing this outcome leaves a *conditional* state $\omega_A^{E_B}$ at Alice's side, which by elementary probability theory satisfies

$$E_A(\omega_A^{E_B}) = \frac{E_A \otimes E_B(\omega_{AB})}{\mathbf{1}_A \otimes E_B(\omega_{AB})}. \tag{4}$$

We demand that $\omega_A^{E_B} \in \mathcal{S}_A$ for all allowed effects E_B and all $\omega_{AB} \in \mathcal{S}_{AB}$. The reduced state ω_A can be written

$$\omega_A = \lambda\omega_A^{E_B} + (1-\lambda)\omega_A^{\bar{E}_B}$$

with $\lambda = \mathbf{1}_A \otimes E_B(\omega_{AB})$; thus, $\omega_A \in \mathcal{S}_A$ by convexity.

In some situations, this condition is automatically satisfied, namely if all effects on A and B are allowed (recall that not all effects need to be physically possible to implement; above, we have discussed that only a subset of effects might be physically allowed). The proof will also illustrate that the *cone of unnormalized states* is a useful concept.

Lemma 1 *Suppose that A and B are state spaces such that all effects are allowed. Then, the inclusion of conditional states in the local state spaces follows directly from the fact that the composite state space AB contains all product states and effects.*

Proof Define the *cone of unnormalized states* A_+ on A by

$$A_+ := \{\lambda\omega_A \mid \omega_A \in \mathcal{S}_A, \lambda \geq 0\}.$$

Since $\mathbf{1}_A(\lambda\omega) = \lambda$ for $\omega \in \mathcal{S}_A$, a vector $\omega \in A_+$ is a normalized state, i.e. $\omega \in \mathcal{S}_A$, if and only if $\mathbf{1}_A(\omega_A) = 1$.

The *cone of unnormalized effects* is

$$A^+ := \{\lambda E_A \mid E_A(\omega_A) \in [0, 1] \text{ for all } \omega_A \in \mathcal{S}_A, \lambda \geq 0\}.$$

Since we have said that all effects are allowed, every linear map $E_A : V_A \to \mathbb{R}$ with $E_A(\omega) \in [0, 1]$ is an allowed effect. The set A^+ contains all non-negative multiples of those. Both sets A_+ and A^+ are *closed convex cones* [25], where "cones" refers to the fact that if x is in the set, then λx is also in the set for all $\lambda \geq 0$.

It is now easy to see that A^+ is the "dual cone" $(A_+)^*$ of A_+, where

$$(A_+)^* \equiv \{E : V_A \to \mathbb{R} \mid E(\omega) \geq 0 \text{ for all } \omega \in A_+\}.$$

Since $(A_+)^{**} = A_+$, we get also that A_+ is the dual cone of A^+; in other words,

$$A_+ = \{\omega \in V_A \mid E(\omega) \geq 0 \text{ for all } E \in A^+\}.$$

Recall the definition of the conditional state in (4). It follows directly from this definition that $E_A(\omega_A^{E_B}) \geq 0$ for all allowed effects E_A, hence for all $E_A \in A^+$. But then, we must have $\omega_A^{E_B} \in A_+$. Since $\mathbf{1}_A(\omega_A^{E_B}) = 1$, we get $\omega_A^{E_B} \in \mathcal{S}_A$. The same reasoning holds for B instead of A. □

Our state spaces also carry a group of reversible transformations. If $G_A \in \mathcal{G}_A$ is a reversible transformation on A, and $G_B \in \mathcal{G}_B$ one on B, it is physically clear that we should be able to accomplish both transformations locally independently; i.e., $G_A \otimes G_B \in \mathcal{G}_{AB}$. We will assume that composite state spaces satisfy this condition.

One of our postulates below will be the postulate of *local tomography*. This is an additional condition on composites AB which is sometimes, but not always imposed in the framework of general probabilistic theories: It states that

global states are uniquely determined by the statistics of local measurement outcomes.

Local measurement outcomes correspond to effects of the form $E_A \otimes E_B$. Thus, the postulate of local tomography states that $E_A \otimes E_B(\omega_{AB}) = E_A \otimes E_B(\varphi_{AB})$ for all E_A, E_B implies that $\omega_{AB} = \varphi_{AB}$.

Since the E_A span the dual space V_A^*, and the E_B span V_B^*, the local measurement outcomes span a $(K_A K_B)$-dimensional subspace of V_{AB}^*:

$$\dim \operatorname{span}\{E_A \otimes E_B\} = (\dim V_A^*)(\dim V_B^*) = K_A K_B.$$

Any state $\omega_{AB} \in \mathcal{S}_{AB}$ can thus be uniquely specified by $K_A K_B$ linear coordinates

$$E_A^{(i)} \otimes E_B^{(j)}(\omega_{AB}), \qquad i = 1, \ldots, K_A; \;\; j = 1, \ldots, K_B;$$

in fact, one of these coordinates is redundant, since $\mathbf{1}_A \otimes \mathbf{1}_B(\omega_{AB}) = 1$, so $K_A K_B - 1$ coordinates are sufficient. Thus, we obtain an injective affine map from the $(K_{AB}-1)$-dimensional convex set $\mathcal{S}_{AB}$ into $\mathbb{R}^{K_A K_B - 1}$, which proves that

$$K_{AB} - 1 = \dim \mathcal{S}_{AB} \leq K_A K_B - 1.$$

Due to Eq. (3), we obtain

$$K_{AB} = K_A K_B.$$

Reading the argumentation backwards shows that this equation is in fact *equivalent* to local tomography, as pointed out by Hardy [8]. It also follows from Eq. (2) that

$$V_{AB} = V_A \otimes V_B.$$

Thus, we get a certain type of tensor product rule for composite state spaces, including $\mathbf{1}_{AB} = \mathbf{1}_A \otimes \mathbf{1}_B$. Note that this is *not* as strong as the tensor product rule of quantum theory, which in addition uniquely specifies the set of global states on composite systems. In contrast, our tensor product rule only says that the surrounding vector spaces satisfy $V_{AB} = V_A \otimes V_B$, but does not uniquely specify $\mathcal{S}_{AB}$ in terms of $\mathcal{S}_A$ and $\mathcal{S}_B$. In particular, classical probability theory satisfies this tensor product rule as well. Suppose that A is a classical bit, and B is a classical 3-level system.

Then the composite AB is classical 6-level system, i.e. $K_{AB} = 6$, while $K_A = 2$ and $K_B = 3$. We get $K_{AB} = K_A K_B$, which is equivalent to local tomography.

To see that this framework allows for state spaces that are physically very different from quantum theory, suppose that A and B are both the square state space from Fig. 2f. Then, define the global state space $\mathcal{S}_{AB}$ as the set of all vectors $x \in AB$ with $E_A \otimes E_B(x) \in [0, 1]$ for all effects E_A and E_B, and $\mathbf{1}_A \otimes \mathbf{1}_B(x) = 1$ (normalization). It turns out that this state space contains so-called *PR-box states* that violate the Bell-CHSH inequality by more than any quantum states [17]. The set of states $\mathcal{S}_{AB}$ itself turns out to be the eight-dimensional *no-signalling polytope* for two parties with two measurements and two outcomes each. The fact that these state spaces can have stronger non-locality than quantum theory has been extensively studied [16, 17, 27–31] and is a main reason for the popularity of general probabilistic theories in quantum information.

It is important to keep in mind that the conditions above do not determine the composite state space $\mathcal{S}_{AB}$ uniquely, even if $\mathcal{S}_A$ and $\mathcal{S}_B$ are given. For example, if $\mathcal{S}_A$ and $\mathcal{S}_B$ are quantum state spaces, then the usual quantum tensor product is a possible composite $\mathcal{S}_{AB}$, but there are infinitely many other possibilities: one of them is to define $\mathcal{S}_{AB}$ as the set of unentangled global states. It satisfies all conditions mentioned above.

3.4 Equivalent State Spaces

In classical physics, choosing a different inertial coordinate system does not alter the physical predictions of Newtonian mechanics. A similar statement is true for convex states spaces.

Consider a system A, given by a state space $\mathcal{S}_A$, a group of transformations $\mathcal{G}_A$, and some allowed effects. Suppose that B is another system, and suppose that there is an invertible linear map $L : V_A \to V_B$ such that

- $\mathcal{S}_B = L(\mathcal{S}_A)$,
- E_A is an allowed effect on A if and only if $E_A \circ L^{-1}$ is an allowed effect on B,
- $\mathcal{G}_B = L \circ \mathcal{G}_A \circ L^{-1}$.

We will then call A and B *equivalent*. Physically, this means that the systems A and B are of the same type in the following sense. Suppose that we prepare a state ω_A, perform a transformation T_A, and finally ask for the occurrence of an effect E_A. The total probability of this is then the same as if we prepare the state $\omega_B = L\omega_A$, perform a transformation $T_B = L \circ T_A \circ L^{-1}$, and ask for the occurrence of the effect $E_B := E_A \circ L^{-1}$. In this sense, all physical scenarios on A can be "translated" into physical scenarios on B, and vice versa. One may then argue that the linear map L just mediates between two different ways of describing exactly the same type of physical system. As an example, we may describe the state space of a qubit either as a set of 2×2 density matrices, or as a set of three-dimensional real vectors, i.e. Bloch vectors. These are two different descriptions for exactly the same physics.

Thus, in our endeavor to derive quantum theory, we have to prove that all state spaces satisfying our postulates are equivalent to quantum state spaces.

4 The Postulates

In this section, we describe our postulates and explain their physical meaning. We start with an axiom on composite state spaces that has already been mentioned in Sect. 3.3 above:

Postulate 1 (Local tomography) *The state of a composite system AB is completely characterized by the statistics of measurements on the subsystems A, B.*

The name "local tomography" comes from the interpretation that state tomography on composite systems can be done by performing local measurements and subsequently comparing the outcomes to uncover correlations. As already mentioned, this postulate is equivalent to $K_{AB} = K_A K_B$, where K_A denotes the number of degrees of freedom needed to specify an unnormalized state on A.

Our second postulate formalizes a property of physics that physicists intuitively take for granted, and that is in fact used very often in performing real experiments. Imagine some physical three-level system (that is, with three perfectly distinguishable states and no more: $N = 3$) that we can access in the lab (it might be quantum, classical, or describable within another theory). Now suppose that, for some reason, we have a situation where we *never* find the system in the third of the three distinguishable configurations on performing a measurement.

To have a concrete example, consider a quantum system that consists of three energy levels which can be occupied by a single particle. Suppose the system is constructed such that the third energy level is actually never occupied (maybe because the corresponding energy is too high).

The consequence that we expect is the following: *We effectively have a two-level system.* This is definitely true for quantum theory, and classical probability theory, but it is not necessarily true for other generalized probabilistic theories. In general, for any number of levels (perfectly distinguishable states) N, we expect to have a corresponding state space $\mathcal{S}_N$. And the collection of states $\omega \in \mathcal{S}_N$ which has probability zero to be found in the N th level upon measurement should be equivalent to $\mathcal{S}_{N-1}$.

In actual physics, this property is used all the time: We apply "effective descriptions" of physical systems, by ignoring impossible configurations. Qubits manufactured in the lab usually actually correspond to two levels of a system with many more energy levels, set up in a way such that the additional energy levels have probability close to zero to be occupied.

One may argue that practicing physics would be very difficult if this property did not hold: we would then possibly have to take into account unobservable potential configurations even if they are never seen. Their presence or absence would affect the resulting state space that we actually observe. The following "subspace postulate", first introduced by Hardy [8], formalizes this idea. It is actually somewhat stronger than our discussion motivates: it also implies that, for every N, there is a *unique* type of N-level system $\mathcal{S}_N$.

The notions of complete measurements and equivalent state spaces were defined in Sects. 3.1 and 3.4.

Postulate 2 (Equivalence of subspaces) *Let $\mathcal{S}_N$ and $\mathcal{S}_{N-1}$ be systems with capacities N and $N-1$, respectively. If $E_1, \ldots, E_N$ is a complete measurement on $\mathcal{S}_N$, then the set of states $\omega \in \mathcal{S}_N$ with $E_N(\omega) = 0$ is equivalent to $\mathcal{S}_{N-1}$.*

The notion of equivalence needs some discussion. Postulate 2 states the equivalence of $\mathcal{S}_{N-1}$ and

$$\mathcal{S}'_{N-1} := \{\omega \in \mathcal{S}_N \mid E_N(\omega) = 0\}. \tag{5}$$

Denote the real linear space which contains $\mathcal{S}_N$ by V_N; define V_{N-1} analogously, and set $V'_{N-1} := \mathrm{span}(\mathcal{S}'_{N-1})$. Equivalence means first of all that there is an invertible linear map $L : V_{N-1} \to V'_{N-1}$ such that $L(\mathcal{S}_{N-1}) = \mathcal{S}'_{N-1}$. But it also means that transformations and measurements on one of them can be implemented on the other. We now describe in more detail what this means.

Every effect E on $\mathcal{S}_N$ defines an effect on $\mathcal{S}'_{N-1}$ by restricting it to the corresponding linear space, resulting in $E \restriction V'_{N-1}$. Equivalence implies that the resulting set of effects is in one-to-one correspondence with the set of effects on $\mathcal{S}_{N-1}$, as described in Sect. 3.4.

The transformations on $\mathcal{S}'_{N-1}$ are defined analogously. To be more specific, define $\bar{\mathcal{G}}'_{N-1}$ as the set of transformations in $\mathcal{S}_N$ that preserve $\mathcal{S}'_{N-1}$ (or, equivalently, V'_{N-1}):

$$\bar{\mathcal{G}}'_{N-1} := \{T \in \mathcal{G}_N \mid T\mathcal{S}'_{N-1} = \mathcal{S}'_{N-1}\}.$$

The set of reversible transformations $\mathcal{G}'_{N-1}$ is defined as the restriction of all these transformations to $\mathcal{S}'_{N-1}$ (or rather, as linear maps, to V'_{N-1}):

$$\mathcal{G}'_{N-1} = \left\{T \restriction V'_{N-1} \mid T \in \bar{\mathcal{G}}'_{N-1}\right\}.$$

Equivalence means that

$$\mathcal{G}'_{N-1} = L \circ \mathcal{G}_{N-1} \circ L^{-1}.$$

Concretely, if $U \in \mathcal{G}_{N-1}$ is any reversible transformation on a state space of capacity $N-1$, then the transformation $\tilde{U} := L \circ U \circ L^{-1}$ is a reversible transformation on $\mathcal{S}'_{N-1}$, i.e. $\tilde{U} \in \mathcal{G}'_{N-1}$. As such, it can be written $\tilde{U} = T \restriction \mathcal{S}'_{N-1}$ for some reversible transformation $T \in \mathcal{G}_N$.

It is important to note that we *don't have full information on T*—that is, our postulate does not specify T uniquely, given $\tilde{U}$. By definition, T preserves $\mathcal{S}'_{N-1}$ and therefore the subspace V'_{N-1}, but we do not know how it acts on the complement of that subspace—it might act as the identity there, or it might have a non-trivial action. Postulate 2 does not specify this. In general, there may (and will) be different T which implement the same $\tilde{U}$ on the subspace.

Using Postulate 2 iteratively, we see that state spaces of smaller capacity are included (in the sense described above) in those of larger capacity; symbolically,

$$\mathcal{S}_1 \subsetneq \mathcal{S}_2 \subsetneq \mathcal{S}_3 \subsetneq \ldots$$

Our next postulate describes the idea that any actual physical theory of probabilities must allow for ample possibilities of reversible time evolution. In situations where "no information is lost"—assuming that this situation applies to closed systems—, these systems A must evolve reversibly, that is, according to some subgroup of the group of reversible transformation $\mathcal{G}_A$. Clearly, if this group is trivial (contains only the identity), physics becomes "frozen": no reversible time evolution is possible at all.

Postulate 3 proclaims a minimal amount of transformational richness for reversible time evolution: as a minimal requirement, it states that the group of reversible transformations should act transitively on the pure states. That is, if we prepare a pure state ω, and φ is another (desired) pure state on the same state space, then there should be a reversible transformation T which maps ω to φ:

Postulate 3 (Symmetry) *For every pair of pure states $\omega, \varphi \in \mathcal{S}_A$, there is a reversible transformation $T \in \mathcal{G}_A$ such that $T\omega = \varphi$.*

It is easy to see that Postulate 3 is true for quantum theory: every pure state can be mapped to every other by some unitary. This example also shows that Postulate 3 is rather weak: in quantum theory, even tuples of perfectly distinguishable pure states $\omega_1, \ldots, \omega_n$ can be mapped to other tuples $\varphi_1, \ldots, \varphi_n$ by suitable unitaries. This is a much higher degree of symmetry than what is demanded by Postulate 3.

There is one postulate remaining. As we discussed in Sect. 3.1, given some state space $\mathcal{S}_A$, not all effects (i.e. linear functionals on A which are non-negative on $\mathcal{S}_A$) may be physically allowed. Similarly as for superselection rules, it might be true that some effects are impossible to implement (an example would be a state space that allows only noisy measurements, and no outcome whatsoever occurs with probability zero).

In order for our axiomatization to work, we need to exclude this possibility: we postulate that all mathematically well-defined effects correspond to allowed measurement outcomes. As it turns out, it is sufficient to postulate this for a 2-level system $\mathcal{S}_2$ (i.e. a generalized bit). In combination with the other postulates, it follows for all other state spaces.

Postulate 4 (All measurements allowed) *All effects on $\mathcal{S}_2$ are outcome probabilities of possible measurements.*

From a mathematical point of view, this postulate could also be regarded as a background assumption: structurally, it says that the class of considered theories is the class of models where the effects are automatically taken as the dual of the states. In other words, it means that whenever we refer to "measurements" in the other postulates, we actually refer to collections of effects without considering the possibility that additional physical conditions might prevent their implementation.

It is interesting to note that Postulate 4 can be replaced by a different formulation, which has first been suggested in the axiomatization by G. Chiribella et al. [11]. It refers to "completely mixed states", which are states that are in the relative interior of the convex set of states:

Postulate 4' [11] If a state is not completely mixed, then there exists at least one state that can be perfectly distinguished from it.

5 How Quantum Theory Follows from the Postulates

We are now ready to carry out the reconstruction of quantum theory (QT) from the postulates. As it turns out, there will be another solution to Postulates 1–4, which is classical probability theory (CPT). By this we mean the theory where the states are finite probability distributions, and the reversible transformations are the permutations. Figure 2a–c shows what classical probability distributions look like in terms of convex sets: they are simplices.

Therefore, we will now prove the following theorem:

Theorem 1 (Main Result) *The only general probabilistic theories, satisfying Postulates 1–4 above, are equivalent to one of the following two theories:*

- ***Classical probability theory (CPT)**: The state space is the set of probability distributions,*

$$\mathcal{S}_N = \{(p_1, \ldots, p_N) \mid p_i \geq 0, \sum_i p_i = 1\},$$

and the reversible transformations $\mathcal{G}_N$ are the permutations on $\{1, \ldots, N\}$.

- ***Quantum theory (QT):** The state space $\mathcal{S}_N$ is the set of density matrices on N-dimensional complex Hilbert space,*

$$\mathcal{S}_N = \left\{\rho \in \mathbb{C}^{N \times N} \mid \rho \geq 0,\ \mathrm{Tr}\rho = 1\right\},$$

and the group of reversible transformations $\mathcal{G}_N$ is the projective unitary group, that is, the set of maps $\rho \mapsto U\rho U^\dagger$ with $U^\dagger U = \mathbf{1}$.

In both cases, all effects must be allowed. Working out the set of effects (that is, linear functionals on states yielding values between 0 and 1), one easily recovers the usual measurements of CPT and QT.

In this paper, we will not give the full reconstruction in all details; the full proof can be found in our more technical paper [10]. Instead, we will try to give a self-contained summary of the reconstruction, its main ideas, and some interesting observations in the course of the argument.

Before starting to do this, let us discuss a simple observation regarding Theorem 1. In order to rule out CPT—and hence to single out QT uniquely—we can tighten Postulate 3 by replacing it with the following modification:

Postulate 3C (Continuous symmetry) *For every pair of pure states* $\omega, \varphi \in \mathcal{S}_A$, *there is a continuous family of reversible transformations* $\{G_t\}_{t\in[0,1]}$ *such that* $G_0\omega = \omega$ *and* $G_1\omega = \varphi$.

In other words, every pure state can be "continuously moved" into every other pure state. A statement like this is expected to be true in physical systems with continuous reversible time evolution—which is the case that seems to be true, to good approximation, in our universe. The consequence is:

$$\begin{array}{l} The\ only\ general\ probabilistic\ theory\ that\ satisfies \\ Postulates\ 1, 2, 3C, and\ 4,\ is\ quantum\ theory\ (QT). \end{array}$$

5.1 Why Bits are Balls

In QT, the state space of a 2-level system (that is, a generalized bit, or qubit, $\mathcal{S}_2$) is a three-dimensional ball, the Bloch ball. In CPT, the (classical) bit instead is a line segment, as shown in Fig. 2. In fact, this is a ball, too: it is a one-dimensional unit ball. However, quantum N-level systems with $N \geq 3$ are not balls: they contain mixed states in their topological boundary [46].

We will now show that all theories satisfying our postulates must have Euclidean ball states spaces as generalized bits. The dimension of this ball will not be determined yet; this will be done later on.

Our argument proceeds in two steps: first, we show that the state space $\mathcal{S}_2$ cannot have lines in its boundary; that is, we exclude the fact that $\mathcal{S}_2$ has proper faces as in the left picture of Fig. 3. Using convex geometry language, we prove that $\mathcal{S}_2$ is *strictly convex.*

As a second step, we show that the symmetry property, Postulate 3, enforces $\mathcal{S}_2$ to be a Euclidean ball. The reason for this comes from group representation theory: since the group of transformations acts linearly, there is an inner product such that all transformations are orthogonal with respect to it.

Lemma 2 *The state space of the generalized bit* $\mathcal{S}_2$ *is strictly convex.*

Proof Consider any effect E with $0 \leq E(\omega) \leq 1$ for all states $\omega \in \mathcal{S}_2$. Then this effect belongs to a two-outcome measurement (as defined in Sect. 3.1), consisting of the two effects E and $\mathbf{1} - E$. It is important to understand that the level sets $\{x \mid E(x) = c\}$ are hyperplanes of codimension 1, due to linearity of E. This is true for all state spaces $\mathcal{S}$. On the other hand, given some hyperplane, we can construct a corresponding effect E (with some freedom of offset and scaling) that has this hyperplane as its level set.

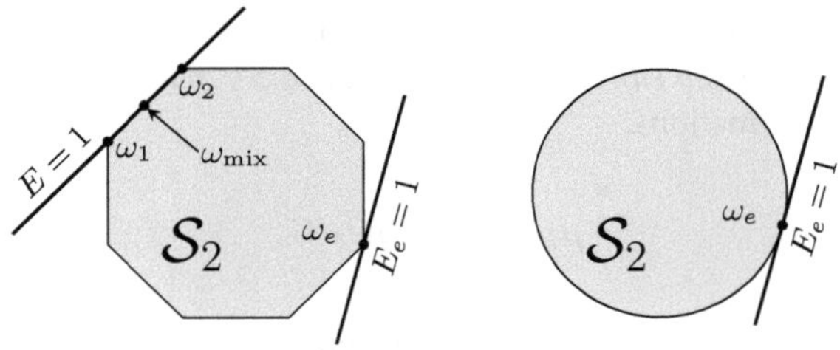

Fig. 3 Like every compact convex set, the bit state space $\mathcal{S}_2$ contains pure states ω_e that are exposed—that is, there is an effect E_e such that ω_e is the unique state where this effects attains value 1. Due to Postulate 2, this proves that $\mathcal{S}_1$ contains a single state only. Now suppose $\mathcal{S}_2$ had lines in its boundary, as in the *left picture*. Then we would analogously find another effect E that attains value 1 on a non-trivial face. Consequently, Postulate 2 would tell us that $\mathcal{S}_1$ contains infinitely many states—a contradiction. Thus, $\mathcal{S}_2$ must be strictly convex as in the *right picture*. Euclidean ballness follows from group representation theory

Like every compact convex set, $\mathcal{S}_2$ has at least one pure state ω_e which is exposed [26]—that is, there is a hyperplane which touches the convex set only in ω_e and in no other point. Thus, we can construct an effect E_e such that the corresponding hyperplane is $\{x \mid E_e(x) = 1\}$, i.e. $E_e(\omega_e) = 1$, and $\min_{\omega\in\mathcal{S}_2} E_e(\omega) = 0$. But then, $(E_e, \mathbf{1} - E_e)$ distinguishes two states perfectly, which is the maximal number for a bit – in other words, this is a *complete measurement*.

Now Postulate 2 says that

$$\{\omega \in \mathcal{S}_2 \mid (\mathbf{1} - E_e)(\omega) = 0\} = \{\omega \in \mathcal{S}_2 \mid E_e(\omega) = 1\} \\ = \{\omega_e\} \simeq \mathcal{S}_1 .$$

In other words, $\mathcal{S}_1$ is a trivial state space which contains only a single state. Now suppose that $\mathcal{S}_2$ has lines in its boundary, and therefore non-trivial faces, as depicted on the left-hand side of Fig. 3. Then we find a supporting hyperplane that touches $\mathcal{S}_2$ in infinitely many states. Constructing a corresponding effect E and repeating the argument from above, we analogously argue that $\mathcal{S}_1$ must contain infinitely many states. This is a contradiction. □

Balls do not have lines in their boundary, but there are many other strictly convex sets—for example, imagine a droplet-like figure. However, Postulate 3 says that there is lots of symmetry in the state space $\mathcal{S}_2$: all pure states (which we now know means all states in the topological boundary due to Lemma 2) are connected by reversible transformations.

From this, one can prove that.

Lemma 3 *The state space $\mathcal{S}_2$ is equivalent to a Euclidean ball (of some dimension $d_2 := K_2 - 1$).*

Recall that we denote the dimension of the set of *unnormalized* states by K_N; therefore, the set of normalized states $\mathcal{S}_N$ has dimension $K_N - 1$. We will not prove

Lemma 3 here, but only sketch where it comes from. An important notion turns out to be the *maximally mixed state*. On any state space $\mathcal{S}_N$, define μ_N as a mixture over the group of transformations,

$$\mu_N := \int_{\mathcal{G}_N} G\omega\, dG,$$

where $\omega \in \mathcal{S}_N$ is any pure state. This is an integral over the invariant measure of the group; see [32, 33] for details of its definition. It follows from the connectedness of all pure states (Postulate 3) that μ_N does not depend on the choice of the pure state ω. Moreover, μ_N turns out to be the unique state which is invariant with respect to all reversible transformations,

$$G\mu_N = \mu_N \qquad \text{for all } G \in \mathcal{G}_N.$$

All states $\omega \in \mathcal{S}_N$ span an affine space of dimension $K_N - 1$. We can now consider μ_N to be the origin of that affine space, turning it into a linear space. Then reversible transformations $G \in \mathcal{G}_N$ act linearly; they preserve the origin. States ω are represented by their difference vectors $\hat{\omega} := \omega - \mu_N$ that live in this linear space. If a reversible transformation T maps ω to φ, then it also maps $\hat{\omega}$ to $\hat{\varphi}$. By group representation theory, there is an inner product on this linear space which is invariant with respect to all reversible transformations. As a consequence, if ω and φ are arbitrary pure states, then there is a reversible transformation T such that $T\hat{\omega} = \hat{\varphi}$ due to Postulate 3, and so $\|\hat{\omega}\| = \|\hat{\varphi}\|$ for the norm corresponding to this inner product. In the case of a bit, i.e. $N = 2$, strict convexity implies that we obtain the full Euclidean ball, with the pure states on the surface and the maximally mixed state μ_N in the center.

5.2 *The Multiplicativity of Capacity*

So far, we know that if we combine two state space A and B, the joint state space has dimension $K_{AB} = K_A K_B$ – this is due to Postulate 1, local tomography, as discussed in Sect. 3.3. However, we do not yet know whether the same equality is true for capacity N. An important step in the derivation of quantum theory is to prove this. As it turns out, a key insight is that the maximally mixed state must be multiplicative: if we have two state spaces A and B, then the maximally mixed state on the composite system AB (assuming our postulates) is

$$\mu_{AB} = \mu_A \otimes \mu_B.$$

This is easily proved from the fact that μ_{AB} must in particular be invariant with respect to all *local* reversible transformations, leaving $\mu_A \otimes \mu_B$ as the only possibility. A further key lemma is the following:

Lemma 4 *If there are n perfectly distinguishable pure states* $\omega_1, \ldots, \omega_n \in \mathcal{S}_N$ *that average to the maximally mixed state, i.e.*

$$\mu_N = \frac{1}{n}\sum_{i=1}^{n} \omega_i,$$

then $n = N$.

Proof Clearly, $N \geq n$, since N is the maximal number of perfectly distinguishable states. On the other hand, let $\varphi_1, \ldots, \varphi_N$ be a set of perfectly distinguishable pure states on $\mathcal{S}_N$, and $E_1, \ldots, E_N$ the corresponding effects, i.e. $E_i(\varphi_j) = \delta_{ij}$. Since $1 = \sum_{i=1}^{N} E_i(\mu_N)$, there must be some k such that $E_k(\mu_N) \leq 1/N$. By Postulate 3, there is a reversible transformation $G \in \mathcal{G}_N$ with $G\omega_1 = \varphi_k$. Thus

$$\frac{1}{N} \geq E_k(\mu_N) = E_k \circ G(\mu_N) = \frac{1}{n}\sum_{i=1}^{n} E_k \circ G(\omega_i)$$
$$\geq \frac{1}{n} E_k \circ G(\omega_1) = \frac{1}{n}.$$

Thus, we also have $N \leq n$, proving the claim. □

In quantum theory, the maximally mixed state on an N-dimensional Hilbert space is the density matrix

$$\mu_N = \frac{\mathbf{1}_N}{N} = \frac{1}{N}\sum_{i=1}^{N} |\psi_i\rangle\langle\psi_i|,$$

if $|\psi_1\rangle, \ldots, |\psi_N\rangle$ denotes an orthonormal basis of $\mathbb{C}^N$—that is, if these are pure states that are perfectly distinguishable. This is in agreement with Lemma 4. Moreover, we can prove that an analogous formula holds for every theory satisfying our Postulates 1–4:

Lemma 5 *For every N, there are N pure perfectly distinguishable states* $\omega_1, \ldots, \omega_N \in \mathcal{S}_N$ *such that*

$$\mu_N = \frac{1}{N}\sum_{i=1}^{N} \omega_i.$$

We only sketch the proof here: For $N = 1$, the statement is trivially true, since $\mathcal{S}_1$ contains only a single state. For $N = 2$, we know that $\mathcal{S}_N$ is a Euclidean ball, with the maximally mixed state in the center. Thus, taking ω_1 and ω_2 as two antipodal points on the ball (say, north and south pole), we get

$$\mu_2 = \frac{1}{2}(\omega_1 + \omega_2),$$

and these states are perfectly distinguishable by an analogue of a quantum spin measurement. Now consider a generalized bit A, and k copies of this physical system denoted $A_1, \ldots, A_k$. We can form a joint system $A^{(k)} := A_1 A_2 \ldots A_k$; since we do not yet know that we have associativity of composition, we mean by this $((A_1 A_2) A_3) A_4 \ldots$. Then the maximally mixed state on the resulting state space is

$$\mu_{A^{(k)}} = \mu_2 \otimes \ldots \otimes \mu_2 = \frac{1}{2^k} \sum_{i_1,\ldots,i_k=1,2} \omega_{i_1} \otimes \ldots \otimes \omega_{i_k}.$$

Since in locally tomographic composites, products of pure states are pure, the $\omega_{i_1} \otimes \ldots \otimes \omega_{i_k}$ are all pure states, and they are perfectly distinguishable by product measurements. Thus, Lemma 4 shows that the capacity of $A^{(k)}$ must be $N_{A^{(k)}} = 2^k$. This proves Lemma 5 for all N which are a power of two. For all other N, the lemma is proved by using the fact that $\mathcal{S}_N$ is embedded in some $A^{(k)}$ for some k large enough due to Postulate 2, and then constructing the maximally mixed state on $\mathcal{S}_N$ in a clever way from that on $A^{(k)}$.

Now we can form the tensor product of the equations

$$\mu_{N_A} = \frac{1}{N_A} \sum_{i=1}^{N_A} \omega_i^A \quad \text{and} \quad \mu_{N_B} = \frac{1}{N_B} \sum_{j=1}^{N_B} \omega_j^B,$$

and we obtain

$$\mu_{N_{AB}} = \mu_{N_A} \otimes \mu_{N_B} = \frac{1}{N_A N_B} \sum_{i=1}^{N_A} \sum_{j=1}^{N_B} \omega_i^A \otimes \omega_j^B,$$

and Lemma 4 tells us that capacity must be multiplicative:

Lemma 6 $N_{AB} = N_A N_B$.

Why is this equation so important? As noticed by Hardy [8], it allows us to draw a surprising conclusion. Every state space $\mathcal{S}_N$ has unnormalized dimension K_N. Since $K_{AB} = K_A K_B$ and $N_{AB} = N_A N_B$ for all state spaces A and B due to our postulates, we get the following facts:

- The function $N \mapsto K_N$ maps natural numbers to natural numbers, and is strictly increasing due to Postulate 2.
- It satisfies $K_{N_1 N_2} = K_{N_1} K_{N_2}$, and $K_1 = 1$.

As shown in [8], these simple conditions imply that there must be an integer $r \geq 1$ such that

$$K_N = N^r. \tag{6}$$

Now recall that the dimension of the bit state space (which is a Euclidean ball) is $d_2 := K_2 - 1$. It follows that

$$d_2 \in \{1, 3, 7, 15, 31, \ldots\}$$

since $K_2 = 2^r$ for some $r \in \mathbb{N}$. Thus, we see in particular that the bit state space is an *odd*-dimensional Euclidean ball. The next subsection will deal with the case $d_2 = 1$; as we will see, this case corresponds to classical probability theory.

5.3 How to Get Classical Probability Theory (CPT)

Suppose that $d_2 = K_2 - 1 = 1$; that is, the generalized bit is a one-dimensional ball, as shown in Fig. 2. A line segment like this describes a classical bit. What can we say about N-level systems for $N \geq 3$ in this case? Equation (6) tells us that the parameter r must be $r = 1$, and thus

$$K_N = N$$

for all N, not only for $N = 2$.

Choose N perfectly distinguishable pure states $\omega_1, \ldots, \omega_N \in \mathcal{S}_N$, and $E_1, \ldots, E_N$ the corresponding effects with $E_i(\omega_j) = \delta_{ij}$ as well as $\sum_i E_i = \mathbf{1}$. It is easy to see that the states must be linearly independent; since $K = N$, they span the full unnormalized state space.

Thus, every state ω can be written $\omega = \sum_{i=1}^N \alpha_i \omega_i$, with $\alpha_i \in \mathbb{R}$ and $\sum_i \alpha_i = \mathbf{1}(\omega) = 1$. But then, $E_j(\omega) = \alpha_j \geq 0$, and so this decomposition of ω is in fact a convex decomposition.

In other words, the full state space $\mathcal{S}_N$ is a convex combination of $\omega_1, \ldots, \omega_N$—that is, a classical simplex as in Fig. 2a–c. These are exactly the state spaces of CPT. Moreover, since for $N = 2$, we can permute the two pure states due to Postulate 3, we can use the subspace postulate to conclude that every pair of pure states on $\mathcal{S}_N$ can be interchanged. These transpositions generate the full permutation group, which must thus be the group of reversible transformations $\mathcal{G}_N$. We have therefore proven the following:

In the case $d_2 = 1$, we get classical probability theory as the unique solution of Postulates 1–4.

5.4 The Curious 7-Dimensional Case

Let us now consider the remaining cases, i.e. the cases where the dimension of the Euclidean bit ball is $d_2 = K_2 - 1 \in \{3, 7, 15, 31 \ldots\}$. The generalized bit carries a group of reversible transformations $\mathcal{G}_2$; by our background assumptions mentioned in Sect. 3.2, this must be a topologically closed matrix group. Since it maps the unit ball into itself, it must be a subgroup of the orthogonal group. Closed subgroups of Lie groups are Lie groups; therefore, $\mathcal{G}_2$ is itself a Lie group.

Denote by $\mathcal{G}_2^0$ the connected component of $\mathcal{G}_2$ containing the identity matrix. We have

$$\mathcal{G}_2^0 \subseteq SO(d_2).$$

We know from Postulate 3 that for every pair of pure states $\omega, \varphi \in \mathcal{S}_2$, there is a reversible transformation $T \in \mathcal{G}_2$ with $T\omega = \varphi$. In other words, $\mathcal{G}_2$ acts *transitively* on the unit sphere, that is, the surface of the unit ball. It can be shown that this implies that $\mathcal{G}_2^0$ is itself transitive on the unit sphere.

At first sight, it seems that this enforces $\mathcal{G}_2^0$ to be the full special orthogonal group $SO(d_2)$, but this intuition is wrong. For example, the group of 4×4-matrices

$$\left\{ \begin{pmatrix} \mathrm{re}\, U & \mathrm{im}\, U \\ -\mathrm{im}\, U & \mathrm{re}\, U \end{pmatrix} \;\middle|\; U \in SU(2) \right\}$$

acts transitively on the surface of the 4-dimensional unit ball, even though it is a proper subgroup of $SO(4)$. The set of all compact connected Lie matrix groups which act transitively on the unit sphere has been classified in [34–37]. In general, there are many possibilities. Fortunately, however, we have additional information: we know that the bit ball has *odd dimension* $d_2 := K_2 - 1$. It turns out that there remain only two possibilities:

- If $d_2 \neq 7$, then $\mathcal{G}_2^0 = SO(d_2)$.
- If $d_2 = 7$, then $\mathcal{G}_2^0$ is either $SO(7)$ or of the form MG_2M^{-1}, where M is a fixed orthogonal matrix, and $\mathcal{G}_2$ is the fundamental representation of the exceptional Lie group G_2.

In fact, $d_2 = 7$ appears in our list of possible dimensions of the bit ball, because $7 = 2^3 - 1$. In our endeavor to derive quantum theory from Postulates 1–4, we will have to show that all the cases $d_2 \in \{7, 15, 31, \ldots\}$ violate at least one postulate. Thus, we see that the case $d_2 = 7$ has to be (and is) treated separately.

The appearance of $d_2 = 7$ as a special case seems like an almost unbelievable coincidence. Is there some deeper significance to this case? Might there be some interesting unknown theory waiting to be discovered which has 7-dimensional balls as bits and the exceptional Lie group G_2 as the analogue of local unitaries? We do not know.

5.5 Subspace Structure and 3-Dimensionality

Having discussed the case of classical probability theory with bit ball dimension $d_2 = 1$, the remaining cases are

$$d_2 \in \{3, 7, 15, 31, \ldots\}.$$

We will now show that all dimensions $d_2 \geq 7$ are incompatible with the postulates, leaving only the case $d_2 = 3$—that is, the Bloch ball of quantum theory. For the rest of this chapter, we ignore the special case $d_2 = 7$ with $\mathcal{G}_2^0 = MG_2M^{-1}$ and G_2 the exceptional Lie group; it can be ruled out by an analogous argument.

In the following, we will parametrize the single bit state space as

$$\mathcal{S}_2 = \left\{ \begin{pmatrix} 1 \\ \hat{\omega} \end{pmatrix} \mid \hat{\omega} \in \mathbb{R}^{d_2}, \|\hat{\omega}\| \leq 1 \right\}.$$

The maximally mixed state becomes $\mu = (1, \mathbf{0})^T$, where $\mathbf{0} \in \mathbb{R}^{d_2}$ denotes the zero vector. Let $n := (1, 0, \ldots, 0)^T \in \mathbb{R}^{d_2}$, then we have two pure states $\omega_1 := (1, n)^T \in \mathcal{S}_2$ and $\omega_2 := (1, -n)^T \in \mathcal{S}_2$, corresponding to the north and south pole of the ball. These states are pure, and they are perfectly distinguished by the measurement consisting of the two effects (for $\omega \in \mathcal{S}_2$)

$$E_1(\omega) := (1 + \langle \hat{\omega}, n \rangle)/2,$$
$$E_2(\omega) := (1 - \langle \hat{\omega}, n \rangle)/2.$$

We know that if we combine two bits into a joint state space, we obtain a state space of capacity four that we call $\mathcal{S}_{2,2}$. It is equivalent to $\mathcal{S}_4$. Thus, the product states $\omega_i \otimes \omega_j$ with $i, j = 1, 2$ represent four perfectly distinguishable states in $\mathcal{S}_{2,2}$, and the corresponding product effects $E_i \otimes E_j$ constitute a complete measurement. Recall, however, that the joint state space $\mathcal{S}_{2,2}$ is not fully known so far—all we know is that the surrounding linear space is the tensor product of the local spaces. At this stage, we do not yet have a complete description of the set of all states in $\mathcal{S}_{2,2}$ or $\mathcal{S}_4$.

Using the subspace postulate twice, i.e. Postulate 2, we obtain that the set of states ω with $(E_1 \otimes E_1 + E_2 \otimes E_2)(\omega) = 1$ is again equivalent to a single bit. This turns out to be a surprisingly restrictive requirement that we are now going to exploit. Denote this set of states by F (it is a face of the state space $\mathcal{S}_{2,2}$), then

$$F = \{\omega \in \mathcal{S}_{2,2} \mid (E_1 \otimes E_1 + E_2 \otimes E_2)(\omega) = 1\} \simeq \mathcal{S}_2.$$

In the following, we will label the two bits by indices A and B for convenience. The group $\mathcal{G}_2 = SO(d_2)$ contains a subgroup $\mathcal{G}_2^s$ which leaves the axis containing north and south pole invariant, i.e.

$$\mathcal{G}_2^s := \{G \in \mathcal{G}_2 \mid G\omega_1 = \omega_1 \text{ and } G\omega_2 = \omega_2\} \simeq SO(d_2 - 1).$$

If $R \in SO(d_2 - 1)$, then its action as an element of $\mathcal{G}_2^s$ is

$$\left(1, \omega^{(1)}, \ldots, \omega^{(d_2)}\right)^T \mapsto \left(1, \omega^{(1)}, R(\omega^{(2)}, \ldots, \omega^{(d_2)})\right)^T.$$

Suppose we apply one transformation of this kind on each part of a bipartite state ω locally; that is, a transformation $G_A \otimes G_B$ with $G_A, G_B \in \mathcal{G}_2^s$. Then we have

$(E_1 \otimes E_1 + E_2 \otimes E_2)(\omega) = 1$ if and only if $(E_1 \otimes E_1 + E_2 \otimes E_2)(G_A \otimes G_B(\omega)) = 1$. Thus, this transformation leaves the face F invariant:

$$(G_A \otimes G_B)F = F.$$

We know that the dimension of the linear span of F is $d_2 + 1$, since it is equivalent to $\mathcal{S}_2$. We will now explore in more detail how the transformations $G_A \otimes G_B$ act on the face F. In particular, we are interested in the structure of invariant subspaces.

First, consider a single bit. Its unnormalized states are carried by a real vector space $V_A = \mathbb{R}^{d_2+1}$ that we can decompose in the following way:

$$V_A = \mathbb{R} \cdot \begin{pmatrix} 1 \\ 0 \\ \vdots \\ 0 \end{pmatrix} \oplus \mathbb{R} \cdot \begin{pmatrix} 0 \\ 1 \\ \vdots \\ 0 \end{pmatrix} \oplus A',$$

where A' denotes the set of all vectors with first two components zero. Since $\mu = (1, 0, \ldots, 0)^T$ and $G\mu = \mu$, as well as $\omega_1 = (1, 1, 0, \ldots, 0)^T$ and $G\omega_1 = \omega_1$ for all $G \in \mathcal{G}_2^s$, these three subspaces are all invariant.

Consequently, the vector space which carries two bits, $V_{AB} \equiv V_A \otimes V_B$, contains the subspace $A' \otimes B'$ which is invariant with respect to all transformations $G_A \otimes G_B$ for $G_A, G_B \in \mathcal{G}_2^s$. This defines an action of $SO(d_2 - 1) \times SO(d_2 - 1)$ on the subspace $A' \otimes B'$.

With a bit of work, one can show that the face F contains at least one state ω which has non-zero overlap with $A' \otimes B'$. Denote the projection of that vector onto this subspace by $\omega_{A' \otimes B'}$. We know that every $(G_A \otimes G_B)(\omega)$ is a valid state in the face F, and its component in the aforementioned subspace is $(G_A \otimes G_B)(\omega_{A' \otimes B'})$. Now imagine we apply all the local transformations $G_A \otimes G_B$ to the vector $\omega_{A' \otimes B'}$, and we are interested in the orbit—that is, in the set of all vectors that we can generate this way.

If $d_2 \geq 4$, then the group $SO(d_2 - 1)$ has a nice property in terms of group representation theory [32]: it is irreducible. That is, its action on $\mathbb{C}^{d_2-1}$ does not leave any non-trivial subspaces invariant. This allows us to draw an important conclusion: it implies [32] that the product group $SO(d_2 - 1) \times SO(d_2 - 1)$ is also irreducible. But then, the orbit $(G_A \otimes G_B)(\omega_{A' \otimes B'})$ must span the full space $A' \otimes B'$, which has dimension $(d_2 - 1)^2$—this is a very large orbit.

In fact, it is too large for the subspace postulate: above, we have concluded from Postulate 2 that the span of the face F (which is preserved by those local transformations) must have dimension $d_2 + 1$, which is less than $(d_2 - 1)^2$ if $d_2 > 3$. Thus, we obtain a contradiction: if the bit ball has dimension $d_2 \in \{7, 15, 31, \ldots\}$, it is impossible to combine two bits into a joint state space which satisfies all our postulates.

As it turns out, this is not true if $d_2 = 3$: the group $SO(d_2 - 1) = SO(2)$ leaves the span of $(1, i)^{\mathrm{T}}$ invariant; that is, $SO(2)$ is reducible. Thus, this case is not ruled out by the reasoning above. In group-theoretic terms, this reducibility is related to the fact that $SO(2)$ is Abelian. In other words, *the fact that rotations commute in 3–1 dimensions can be seen as a possible reason of the fact that the Bloch ball is 3-dimensional.*

Lemma 7 *The dimension of the bit ball must be* $d_2 = 3$.

We have thus uncovered a group-theoretic explanation why the smallest non-trivial quantum systems have three mutually incompatible, independent components and not more. Due to Postulate 4, we can find all possible measurements on this state space: all effects (that is, linear functionals) which yield probabilities in the interval $[0, 1]$ correspond to outcome probabilities of possible measurements. It is easy to see that these effects are in one-to-one correspondence with the quantum measurements (POVMs) on a single qubit.

Furthermore, we know that the group of reversible transformations contains $SO(3)$, the rotations of the Bloch ball, which correspond to the unitary transformations on a qubit. At this point, however, we do not yet know whether $\mathcal{G}_2 = SO(3)$ or $\mathcal{G}_2 = O(3)$.

5.6 Quantum Theory on N-level Systems for $N \geq 3$

In the previous section, we have derived quantum theory for single bits. It remains to show that our postulates also predict quantum theory for all N-level systems with $N \geq 3$. As before, we only sketch the main proof ideas, and refer the reader to [10] for proof details.

For a single bit in state $\omega = (1, \hat{\omega})^T$, we can obtain the usual representation as a density matrix by applying a linear map $L : \mathbb{R}^4 \to \mathbb{C}^{2\times 2}_{sa}$, where the latter symbol denotes the real vector space of self-adjoint complex 2×2-matrices. This map L is defined by linear extension of

$$L(\omega) := (\mathbf{1} + \hat{\omega} \cdot \vec{\sigma})/2,$$

where $\vec{\sigma} = (\sigma_x, \sigma_y, \sigma_z)$ denotes the Pauli matrices. The representation that we obtain (applying L in a suitable way to effects and transformations as well) is equivalent in the sense of Sect. 3.4 to the Bloch ball representation.

If we have the state space $\mathcal{S}_{2,2}$ of two bits, we can use the map $L \otimes L$ to represent states $\omega \in \mathcal{S}_{2,2}$ by self-adjoint 4×4-matrices $L \otimes L(\omega)$. Recall that we have constructed a face F of $\mathcal{S}_{2,2}$ in the previous subsection. Analyzing F in a bit more detail, one can show that it contains a family of pure states ω_u, where $u \in [0, \pi)$, which are mapped by $L \otimes L$ onto

$$L \otimes L(\omega_u) = |\psi_u\rangle\langle\psi_u|,$$

where

$$|\psi_u\rangle = \cos\frac{u}{2}|0\rangle \otimes |0\rangle + \sin\frac{u}{2}|1\rangle \otimes |1\rangle$$

for some orthonormal basis $\{|0\rangle, |1\rangle\}$. This is an entangled quantum state with Schmidt coefficients $\cos(u/2)$ and $\sin(u/2)$. Choosing u appropriately, it can attain any value between 0 and 1. Thus, by applying local unitaries (which corresponds to the $SO(3)$-rotations of the local balls), we can generate all pure quantum states.

Denoting $\mathcal{S}'_{2,2} := L \otimes L(\mathcal{S}_{2,2})$, we have proven the following:

Lemma 8 *$\mathcal{S}'_{2,2}$ contains all pure 2-qubit quantum states as pure states.*

The next step is somewhat tricky: we have to show that there are no further (non-quantum) states in $\mathcal{S}'_{2,2}$. The idea is to show that *all quantum effects* are allowed effects on $\mathcal{S}'_{2,2}$. Then, if there were additional non-quantum states in $\mathcal{S}'_{2,2}$, some of these effects would give negative probabilities, which is impossible.

We know that the product effects are allowed on $\mathcal{S}_{2,2}$. Applying the transformation $L \otimes L$, some of the corresponding effects in $\mathcal{S}'_{2,2}$ are the maps

$$\rho \mapsto \mathrm{Tr}\,(P_1 \otimes P_2 \rho)\,,$$

where P_1 and P_2 are one-dimensional projectors. If $T \in \mathcal{G}_{2,2} \simeq \mathcal{G}_4$ is any reversible transformation on $\mathcal{S}_{2,2}$, denote the corresponding transformation on $\mathcal{S}'_{2,2}$ by $T' \in \mathcal{G}'_{2,2}$. It maps states ρ to $T'(\rho)$. Suppose we could show the equation

$$\mathrm{Tr}(P_1 \otimes P_2 T'(\rho)) = \mathrm{Tr}((T')^{-1}(P_1 \otimes P_2)\rho). \tag{7}$$

Then we would be done: due to Postulate 3, transformations $T' \in \mathcal{G}'_{2,2}$ can map every pure product state to every other pure state, in particular, to every pure entangled quantum state. This way, $(T')^{-1}$ in the equation above would generate all entangled quantum effects from the product effect $P_1 \otimes P_2$. This is exactly what we want.

Why does Eq. (7) hold? Up to a factor 1/4, the map $L^{\otimes 2}$ is an isometry: for all $x, y \in \mathbb{R}^4 \otimes \mathbb{R}^4$, we have

$$\mathrm{Tr}\left(L^{\otimes 2}(x) L^{\otimes 2}(y)\right) = \frac{1}{4}\langle x, y\rangle.$$

Thus, translating Eq. (7) from $\mathcal{S}'_{2,2}$ back to $\mathcal{S}_{2,2}$, we have to prove that

$$\langle E_1 \otimes E_2, T\omega\rangle = \langle T^{-1}(E_1 \otimes E_2), \omega\rangle.$$

This is satisfied if $T^T = T^{-1}$ for all $T \in \mathcal{G}_{2,2}$. In fact, we have

Lemma 9 *All reversible transformations $T \in \mathcal{G}_{2,2}$ act as orthogonal matrices on $\mathbb{R}^4 \otimes \mathbb{R}^4$.*

The proof of this lemma is non-trivial and somewhat surprising: it uses Schur's Lemma from group representation theory, together with the fact that there exist certain kinds of SWAP and CNOT operations on two bits. These operations are constructed by using Postulate 2.

Due to Lemma 9, all the above argumentation becomes solid: Eq. (7) is valid, and we get

Lemma 10 *$\mathcal{S}'_{2,2}$ is the set of 2-qubit quantum states, and the allowed effects are the quantum effects.*

So what about the transformations? First of all, we know that that the transformation group of a *single* bit must be $SO(3)$—it cannot be $O(3)$, because local reflections would correspond to partial transpositions which generate negative eigenvalues on entangled states. Furthermore, every transformation $T \in \mathcal{G}_{2,2}$ is a linear isometry on the set of self-adjoint 4×4-matrices that maps the set of density matrices into itself.

According to Wigner's Theorem [38, 39], only unitary and anti-unitary maps satisfy this. However, due to Wigner's normal form, anti-unitary maps generate reflections in some Bloch ball faces of the state space, which is impossible due to Postulate 2.

So $\mathcal{G}_{2,2}$ is a subgroup of the unitary group. Due to Postulate 3, it maps some pure product state to an entangled state. In other words, $\mathcal{G}_{2,2}$ contains an entangling unitary, and also all local unitaries. It is a well-known fact from quantum computation [40] that these transformations generate the full unitary group.

We have thus shown

Lemma 11 *The group of reversible transformations $\mathcal{G}'_{2,2}$ on two bits corresponds to the unitary conjugations, i.e. the maps $\rho \mapsto U\rho U^\dagger$ with $U \in SU(4)$.*

It is now clear that what we did for two bits can also be done for n bits. Since every $\mathcal{S}_N$ is contained in some $\mathcal{S}_{2^n}$ for n large enough, we can use the subspace postulate to conclude that every state space $\mathcal{S}_N$ is equivalent to the quantum N-level state space.

6 Conclusions and Outlook

We have shown that the Hilbert space formalism of quantum theory can be reconstructed from four natural, information-theoretic postulates. We hope that this reconstruction—together with other recent axiomatizations [8, 9, 11–14]—contributes to a better understanding of quantum theory, and sheds light on some of the mysterious aspects of its formalism, such as the appearance of complex numbers or unitaries.

One of the main motivations for this work, as mentioned in the introduction, was to find a "minimal" set of postulates, in the sense that removing or weakening any one of the postulates yields new solutions in addition to quantum theory. Classifying these additional solutions means to analyze "quantum theory's closest cousins": these are theories that are operationally close to quantum theory, but not described by the Hilbert space (or C^*-algebra) formalism. These theories make physical predictions that differ from quantum theory [41] and that can be tested experimentally [42].

Have we achieved the goal of minimality? The postulate which seems to be the strongest is Postulate 2, which was called "Subspace Axiom" by Hardy [8]. In fact, in follow-up work [43, 44], we show that Postulate 2 can be significantly weakened: it can be replaced by the requirements that generalized bits carry exactly one bit of information and not more, and that the state of any system can be reversibly encoded in a sufficiently large number of generalized bits. As a further benefit, quantum theory with superselection rules appears as an additional solution. In particular, continuous reversible interaction is sufficient to single out $d_2 = 3$ as the dimensionality of the Bloch ball [43]. On the other hand, Postulate 1 seems crucial: removing it yields at least quantum theory over the real numbers as an additional solution.

It is currently an open problem whether classical probability theory and quantum theory are the unique theories satisfying Postulates 1, 3 and 4. It seems unlikely that Postulate 4 can be dropped: adding restrictions to the possible measurements in quantum theory may allow to construct a counterexample. Furthermore, all current axiomatizations seem to indicate that some assumption on the group of reversible transformations, as in Postulate 3, is crucial, since this gives the power of group representation theory and the Euclidean structure of the Bloch ball. Interesting progress has been made recently by Hardy [12], where the corresponding axiom only postulates the existence of suitable *permutations*.

Thus, we have not yet fully achieved the goal of minimality, but we think that our set of postulates is very close to it. In particular, having as few background assumptions as possible may yield interesting new state spaces that are overlooked if the full pictorial background framework of quantum circuits is assumed. For example, one might consider the following weaker version of Postulate 1.

Postulate 1' For every triple (but not necessarily for every pair) of state spaces A, B and C, there is a tomographically-local composite ABC which satisfies all other postulates.

It remains an interesting open problem to find a minimal set of axioms, prove its minimality, and systematically characterize all theories which satisfy some of these axioms, but not all of them. Besides being of interest in its own right, thorough understanding of alternative routes that nature might have taken may be of crucial importance for experimental tests of quantum theory, such as tests for higher-order interference [47].

Acknowledgments Research at Perimeter Institute is supported by the Government of Canada through Industry Canada and by the Province of Ontario through the Ministry of Research and Innovation. LM acknowledges support from CatalunyaCaixa.

References

1. S. Weinberg, Ann. Phys. NY **194**, 336 (1989)
2. N. Gisin, Weinberg's non-linear quantum mechanics and supraluminal communications. Phys. Lett. A **143**, 1–2 (1990)
3. C. Simon, V. Bužek, N. Gisin, No-signaling condition and quantum dynamics. Phys. Rev. Lett. **87**, 170405 (2001)
4. G. Birkhoff, J. von Neumann, The logic of quantum mechanics. Ann. Math. **37**, 823 (1936)
5. G.W. Mackey, *The Mathematical Foundations of Quantum Mechanics* (W. A. Benjamin Inc., New York, 1963)
6. G. Ludwig, *Foundations of Quantum Mechanics I and II* (Springer, New York, 1985)
7. E.M. Alfsen, F.W. Shultz, *Geometry of State Spaces of Operator Algebras* (Birkhäuser, Boston, 2003)
8. L. Hardy, Quantum theory from five reasonable axioms, arxiv:quant-ph/0101012v4
9. B. Dakić, C. Brukner, Quantum theory and beyond: is entanglement special?, in *Deep beauty*, ed. by H. Halvorson (Cambridge Press, 2011), arXiv:0911.0695v1
10. Ll. Masanes, M.P. Müller, A derivation of quantum theory from physical requirements. New J. Phys. **13**, 063001 (2011)
11. G. Chiribella, G.M. D'Ariano, P. Perinotti, Informational derivation of quantum theory. Phys. Rev. A **84**, 012311 (2011)
12. L. Hardy, Reformulating and Reconstructing Quantum Theory, arXiv:1104.2066v1
13. L. Hardy, The Operator Tensor Formulation of Quantum Theory. Phil. Trans. R. Soc. A **370**, 3385–417 (2012), arXiv:1201.4390v1
14. M. Zaopo, Information Theoretic Axioms for Quantum Theory, arXiv:1205.2306
15. L. Hardy, *Foliable Operational Structures for General Probabilistic Theories*, in "Deep beauty", Editor Hans Halvorson (Cambridge Press, 2011), arXiv:0912.4740v1
16. H. Barnum, A. Wilce, *Information processing in convex operational theories*, DCM/QPL (Oxford University 2008), arXiv:0908.2352v1
17. J. Barrett, *Information processing in generalized probabilistic theories*, Phys. Rev. A **75**, 032304 (2007), arXiv:quant-ph/0508211v3
18. G. Chiribella, G. M. D'Ariano, P. Perinotti; *Probabilistic theories with purification*; Phys. Rev. A **81**, 062348 (2010), arXiv:0908.1583v5
19. A.S. Holevo, *Statistical Structure of Quantum Theory* (Springer, Berlin, 2001)
20. R. T. Rockafellar, *Convex Analysis*, (Princeton University Press, 1970)
21. G. Brassard, Is information the key? Nat. Phys. **1**, 2 (2005)
22. Č. Brukner, Questioning the rules of the game. Physics **4**, 55 (2011)
23. W.K. Wootters, Quantum mechanics without probability amplitudes. Found. Phys. **16**, 391–405 (1986)
24. A. Baker, *Matrix Groups, An Introduction to Lie Group Theory* (Springer, London, 2006)
25. C.D. Aliprantis, R. Tourky, *Cones and Duality*, (American Mathematical Society, 2007)
26. S. Straszewicz, Über exponierte Punkte abgeschlossener Punktmengen. Fund. Math. **24**, 139–143 (1935)
27. M. Navascues, H. Wunderlich, A glance beyond the quantum model. Proc. R. Soc. Lond. A **466**, 881–890 (2009). arXiv:0907.0372v1
28. W. van Dam, Implausible Consequences of Superstrong Nonlocality. Nat. Comput. **12**(1), 9–12 (2012), arXiv:quant-ph/0501159v1
29. D. Gross, M. Müller, R. Colbeck, O.C.O. Dahlsten, All reversible dynamics in maximally non-local theories are trivial. Phys. Rev. Lett. **104**, 080402 (2010), arXiv:0910.1840v2
30. M. Pawlowski, T. Paterek, D. Kaszlikowski, V. Scarani, A. Winter, M. Zukowski, A new physical principle: information causality. Nature **461**, 1101 (2009), arXiv:0905.2292v3
31. S. Popescu, D. Rohrlich, *Causality and Nonlocality as Axioms for Quantum Mechanics*, Proceedings of the Symposium on Causality and Locality in Modern Physics and Astronomy (York University, Toronto, 1997), arXiv:quant-ph/9709026v2

32. W. Fulton, J. Harris, *Graduate texts in mathematics*, Representation Theory (Springer, Berlin, 2004)
33. B. Simon, *Representations of Finite and Compact Groups*. Graduate Studies in Mathematics, vol. 10 (American Mathematical Society, Providence, 1996)
34. D. Montgomery, H. Samelson, Transformation groups of spheres. Ann. Math. **44**, 454–470 (1943)
35. A. Borel, Some remarks about Lie groups transitive on spheres and tori. Bull. A.M.S. **55**, 580–587 (1949)
36. A.L. Onishchik, V.V. Gorbatsevich, *Lie Groups and Lie Algebras I, Encyclopedia of Mathematical Sciences 20* (Springer, Berlin, 1993)
37. A. L. Onishchik, *Transitive compact transformation groups*, Mat. Sb. (N.S.) **60**(102):4 447–485 (1963); English translation: Amer. Math. Soc. Transl. (2) **55**, 153–194 (1966)
38. V. Bargmann, Note on Wigner's theorem on symmetry operations. J. Math. Phys. **5**, 862–868 (1964)
39. E.P. Wigner, Normal form of antiunitary operators. J. Math. Phys. **1**, 409–413 (1960)
40. M.J. Bremner, C.M. Dawson, J.L. Dodd, A. Gilchrist, A.W. Harrow, D. Mortimer, M.A. Nielsen, T.J. Osborne, Practical scheme for quantum computation with any two-qubit entangling gate. Phys. Rev. Lett. **89**, 247902 (2002), arXiv:quant-ph/0207072v1
41. T. Paterek, B. Dakić, Č. Brukner, Theories of systems with limited information content. New J. Phys. **12**, 053037 (2010)
42. C. Ududec, H. Barnum, J. Emerson, Three slit experiments and the structure of quantum theory. Found. Phys. **41**, 396–405 (2010)
43. Ll. Masanes, M. P. Müller, D. Pérez-García, R. Augusiak, Entanglement and the three-dimensionality of the Bloch ball. J. Math. Phys. **55**, 122203 (2014), arXiv:1111.4060
44. Ll. Masanes, M. P. Müller, R. Augusiak, D. Pérez-García, Existence of an information unit as a postulate of quantum theory. Proc. Natl. Acad. Sci. USA **110**(41), 16373 (2013), arXiv:1208.0493
45. H. Barnum, A. Wilce, Local tomography and the Jordan structure of quantum theory. Found. Phys. 44, 192–212 (2014), arXiv:1202.4513
46. I. Bengtsson, K. Życzkowski, *Geometry of Quantum States* (University Press, Cambridge, 2006)
47. U. Sinha, C. Couteau, T. Jennewein, R. Laflamme, G. Weihs, Ruling out multi-order interference in quantum mechanics. Science **329**, 418–421 (2010)

Quantum from Principles

Giulio Chiribella, Giacomo Mauro D'Ariano and Paolo Perinotti

1 Introduction

Quantum foundations is an old field—as old as quantum mechanics itself. Among the early works stand out the seminal papers by Einstein, Podolski, and Rosen [1] and Schrödinger [2], who addressed quantum entanglement for the first time, exploring quantum mechanics within the Hilbert space formulation. Almost at the same time, Birkhoff and von Neumann [3] looked at the theory in a wider framework allowing for alternative theories. From that angle, it was natural to ask what is special about quantum mechanics and why Nature obeys its peculiar laws. The take of Birkhoff and von Neumann was that quantum theory should be regarded as a new form of logic, whose laws could be derived from physically motivated axioms. This programme gave rise to the tradition of quantum logic [4–8], whose ramifications are still the object of active research [9].

Researchers in quantum logic managed to derive a significant part of the quantum framework from logical axioms. However, there is a general consensus that the axioms put forward in this context are not as insightful as one would have hoped. For both experts and non-experts, it is hard to figure out what is the moral of the quantum-logic axiomatizations. What is special about quantum theory after all? Why

G. Chiribella
Department of Computer Science, The University of Hong Kong, Pokfulam road, Hong Kong, People's Republic of China
e-mail: giulio@cs.hku.hk

G.M. D'Ariano (✉) · P. Perinotti (✉)
QUIT Group, Dipartimento di Fisica, Università di Pavia, via Bassi 6, 27100 Pavia, Italy
e-mail: dariano@unipv.it

P. Perinotti
e-mail: paolo.perinotti@unipv.it

G.M. D'Ariano · P. Perinotti
INFN Sezione di Pavia, via Bassi 6, 27100 Pavia, Italy

G. Chiribella and R.W. Spekkens (eds.), *Quantum Theory: Informational Foundations and Foils*, Fundamental Theories of Physics 181,
DOI 10.1007/978-94-017-7303-4_6

should quantum theory be preferred to alternative theories? Not many answers can be found in the popular accounts of quantum logic (see e.g. the Wikipedia entry [10]) and even understanding what the axioms are requires delving into a highly specialized literature.

The ambition to find a more insightful axiomatization reemerged with the rise of quantum information. The new field showed that the mathematical axioms of quantum theory imply striking operational consequences, such as quantum key distribution [11, 12], quantum algorithms [13, 14], no-cloning [15, 16], quantum teleportation [17] and dense coding [18]. A natural question is: Can we reverse the implication and *derive* the mathematics of quantum theory from some of its operational consequences? This question is at the core of a research programme launched by Fuchs [19] and Brassard [20], which can be synthesized by the motto *"quantum foundations in the light of quantum information"* [21].[1] The ultimate goal of the programme is to reconstruct the whole structure of quantum theory from few simple principles of information-theoretic nature.

One may wonder why quantum information theorists should be more successful than their predecessors in the axiomatic endeavour. A good reason is the following: In the pre-quantum information era, quantum theory was viewed like an impoverished version of classical theory, lacking the ability to make deterministic predictions about the outcomes of experiments. Clearly, this perspective offered no vantage point for explaining why the world should be quantum. Contrarily, quantum information provided plenty of positive reasons for preferring quantum theory to its classical counterpart. Turning some of these reasons into axioms then appeared as a promising route towards a compelling axiomatization. Pioneering works along this route are those by Hardy [23] and D'Ariano [24, 25]. More recently, the programme flourished, leading to an explosion of new axiomatizations [26–33].

Here we review the axiomatization of Ref. [26]. In this work, quantum theory is derived from six principles, formulated in a general framework of theories of information. The first five principles—Causality, Purity of Composition, Local Discriminability, Perfect State Discrimination, and Ideal Compression—express ordinary properties that are shared by quantum and classical information theory: such principles define what we call a *standard* theory of information. Among all standard theories of information, the sixth principle—Purification—identifies quantum theory uniquely. Purification states that every random preparation can be simulated via non-random preparation procedure, in which the system is prepared together with an environment. An agent that has access to both the system and the environment would then have maximal control of the preparation—*maximal* in the sense that no other agent could conceivably have higher control. The moral of our work is that *Quantum Theory is the theory that allows maximal control of randomness*, giving us—at least in principle—the power to control all possible preparations and all possible dynamics.

[1]This was also the title of one influential conference, held in May 2000 at the Université de Montréal [22], which kickstarted the new wave of quantum axiomatizations.

The chapter is structured as follows: in Sect. 2 we provide an introduction to the framework of *operational-probabilistic theories*—general theories of information arising from the combination of the circuit framework with probability theory. Then, Sect. 3 presents the background to the reconstruction, discussing the main standing assumptions—finite-dimensionality, non-determinism, and closure under limits—and introducing a few basic operational tasks: signalling, collecting side information, doing state tomography, distinguishing states, compressing information, and simulating preparations. The principles are then analyzed in Sect. 4. Section 5 provides a guided tour through the main results in our reconstruction, showing how the main features of quantum theory can be derived directly from the principles. Finally, the conclusions are drawn in Sect. 6.

2 Operational-Probabilistic Theories

In order to reconstruct quantum theory and the features of quantum information, one needs a framework capable to describe a variety of alternative theories. Different frameworks have been proposed for this scope, under the broad name of *general probabilistic theories* [23–27, 34–39]. Our reconstruction is based on a specific variant of general probabilistic theories, which we call *operational-probabilistic theories (OPTs)* [26, 34]. OPTs are an extension of probability theory, in which events can be connected into circuits. Technically, OPTs arise from the combination of the categorical framework of Abramsky and Coecke [40–42] with the toolbox of elementary probability theory. We regard such a combination as the natural mathematical object describing a "general theory of information". In the following we present a concise summary of the OPT framework.

2.1 Operational Structure

2.1.1 Systems

Systems are labels, which determine how different events can be connected to one another. We denote systems by capital letters, such as A, B, C, and so on. The letter I will be reserved for the *trivial system*, representing "nothing".[2] The set of all systems under consideration will be denoted by Sys.

Every two systems A and B can be considered together as a composite system, denoted by $\mathrm{A} \otimes \mathrm{B}$. The composition of systems is associative, namely

$$\mathrm{A} \otimes (\mathrm{B} \otimes \mathrm{C}) = (\mathrm{A} \otimes \mathrm{B}) \otimes \mathrm{C} \qquad \forall \mathrm{A}, \mathrm{B}, \mathrm{C} \tag{1}$$

[2]More precisely, "nothing that the theory cares to describe".

and has the trivial system as identity element, namely

$$\mathrm{A} \otimes \mathrm{I} = \mathrm{I} \otimes \mathrm{A} = \mathrm{A} \qquad \forall \mathrm{A}. \tag{2}$$

The second condition means that considering system A together with "nothing" is the same as considering system A alone.

2.1.2 Events

An event of type A $\to$ B represents the occurrence of a transformation that converts the input system A into the output system B. An event $\mathcal{E}$ of type A $\to$ B will be represented graphically as

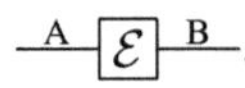

The set of all events of type A $\to$ B will be denoted by $\mathsf{Transf}(\mathrm{A} \to \mathrm{B})$, identifying events with the corresponding transformations.

When the input and output systems are composite systems, we draw boxes with multiple wires. For example, the box

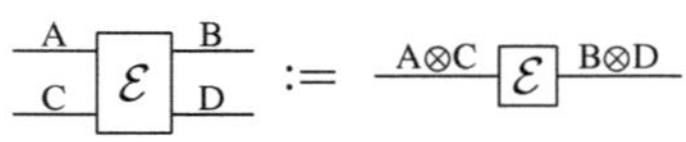

represents an event of type $(\mathrm{A} \otimes \mathrm{C}) \to (\mathrm{B} \otimes \mathrm{D})$.

Some types of events are particularly important and deserve a name of their own. An event of type I $\to$ A is a *preparation-event* (or simply, a *preparation*), that is, an event that makes system A available to further processing. An event of type A $\to$ I is an *observation-event* (or simply, an *observation*), after which system A is no longer available. Preparation- and observation-events will be represented as

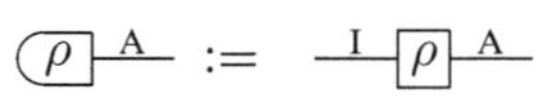

and

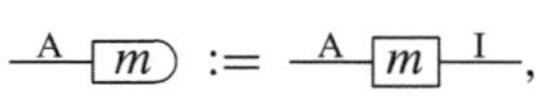

respectively. We will often use the Dirac-like notation $(a|$ and $|\rho)$ for the observation a and the preparation ρ, respectively.

Events of type I $\to$ I will be called *scalars* [40]. Scalars will be represented "out of the box", as

$s :=$ 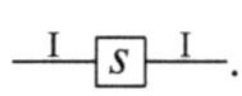

Later, scalars will be associated to probabilities. For the moment, however, they are just a special type of events.

2.1.3 Composition of Events

Events can be connected into networks through the following operations

1. *Sequential composition*: an event of type $\mathrm{A} \to \mathrm{B}$ can be connected to an event of type $\mathrm{B} \to \mathrm{C}$, yielding an event of type $\mathrm{A} \to \mathrm{C}$.
2. *Parallel composition*: an event of type $\mathrm{A} \to \mathrm{A}'$ can be composed with an event of type $\mathrm{B} \to \mathrm{B}'$, yielding an event of type $(\mathrm{A} \otimes \mathrm{B}) \to (\mathrm{A}' \otimes \mathrm{B}')$.

Intuitively, the sequential composition represents two events happening at "subsequent time steps".[3] The sequential composition of two events $\mathcal{E}$ and $\mathcal{F}$ of matching types is denoted by $\mathcal{F} \circ \mathcal{E}$ and is represented graphically as

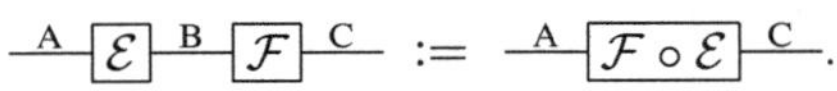

This graphical notation is justified by the requirement that sequential composition be associative, namely

$$\mathcal{G} \circ (\mathcal{F} \circ \mathcal{E}) = (\mathcal{G} \circ \mathcal{F}) \circ \mathcal{E}, \tag{3}$$

for arbitrary events $\mathcal{E}$, $\mathcal{F}$ and $\mathcal{G}$ of matching types. In addition to associativity, sequential composition is required to have an identity element for every system. The *identity on system* A, denoted by $\mathcal{I}_{\mathrm{A}}$, is the special event of type $\mathrm{A} \to \mathrm{A}$ identified by the conditions

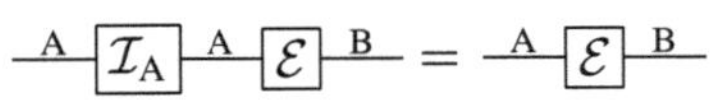 (4)

and

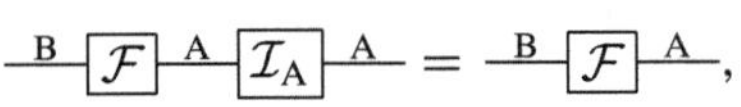 (5)

required to be valid for arbitrary systems A, B and arbitrary events $\mathcal{E}$ and $\mathcal{F}$ of types $\mathrm{A} \to \mathrm{B}$ and $\mathrm{B} \to \mathrm{A}$, respectively. The intuitive content of the above equations is that $\mathcal{I}_{\mathrm{A}}$ represents the process that "does nothing on the system". Consistently, we use the graphical notation

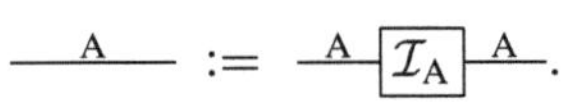

[3] *Per se*, the mathematical formalism does not force us to interpret the order of sequential composition as an order in time. Nevertheless, composition in time is the reference situation that we will have in mind when phrasing our axioms.

Mathematically, conditions (3)–(5) impose that the events form a *category* [43, 44], in which the systems are the objects and the events are the arrows. For the sequential composition of a preparation and an observation we will often use the Dirac-like notation,

$$(a|\rho) :=$$ 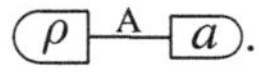 (6)

Let us consider parallel composition. The parallel composition of two events $\mathcal{E}$ and $\mathcal{F}$ is denoted as $\mathcal{E} \otimes \mathcal{F}$ and is represented graphically as

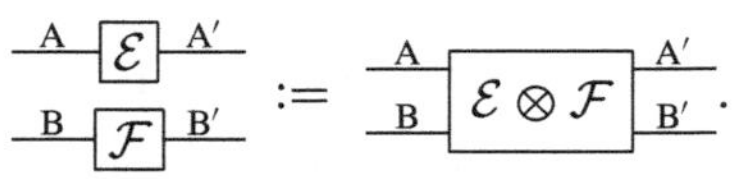

The graphical notation is justified by the requirement of the following condition

$$(\mathcal{E} \otimes \mathcal{F}) \circ (\mathcal{G} \otimes \mathcal{H}) = (\mathcal{E} \circ \mathcal{G}) \otimes (\mathcal{F} \circ \mathcal{H}), \tag{7}$$

where $\mathcal{E}, \mathcal{F}, \mathcal{G}$, and $\mathcal{H}$ are arbitrary events of matching types. Such condition is necessary for the graphical notation to make sense, since in graphical notation the two sides of Eq. (7) look exactly the same. In addition to Eq. (7), parallel composition is required to satisfy the condition

$$\mathcal{I}_{\mathrm{A}\otimes\mathrm{B}} = \mathcal{I}_{\mathrm{A}} \otimes \mathcal{I}_{\mathrm{B}}. \tag{8}$$

Mathematically, the presence of parallel composition turns the category of events into a strict monoidal category, whose key properties are summarized by Eqs. (1), (2), (7), and (8). We denote such category by Transf.

2.1.4 Reversible Events

An event $\mathcal{E}$ of type $\mathrm{A} \to \mathrm{B}$ is *reversible* iff there exists another event $\mathcal{F}$, of type $\mathrm{B} \to \mathrm{A}$, such that

 (9)

and

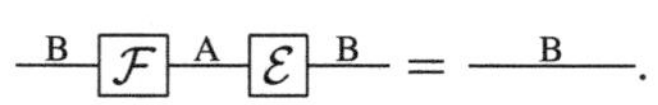 (10)

When this is the case, we write $\mathcal{F} = \mathcal{E}^{-1}$ and we say that systems A and B are *operationally equivalent* (or simply *equivalent*).

We denote by RevTransf($\mathrm{A} \to \mathrm{B}$) the set of reversible events of type $\mathrm{A} \to \mathrm{B}$. Such set (which may be empty) depends on the specific theory. In general, we require the existence of a reversible event that swaps pairs of systems. Given two systems A and B, the *swap of* A *with* B—denoted by $\mathcal{S}_{A,B}$—is a reversible event of type $(A \otimes B) \to (B \otimes A)$ satisfying the condition

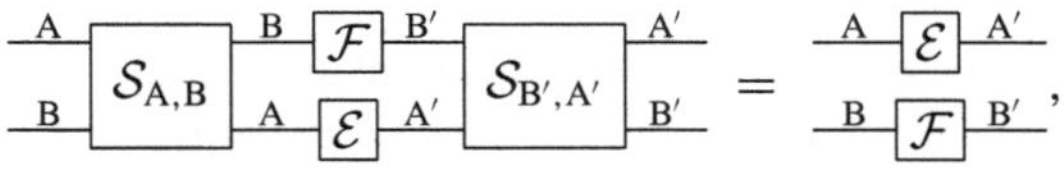

(11)

for arbitrary systems A, B, A′, B′ and arbitrary events $\mathcal{E}$, $\mathcal{F}$, as well as the conditions

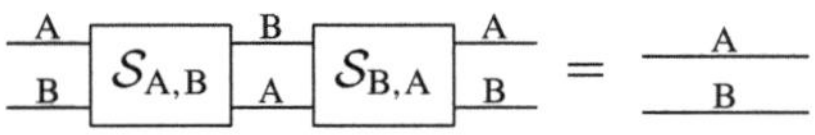

(12)

and

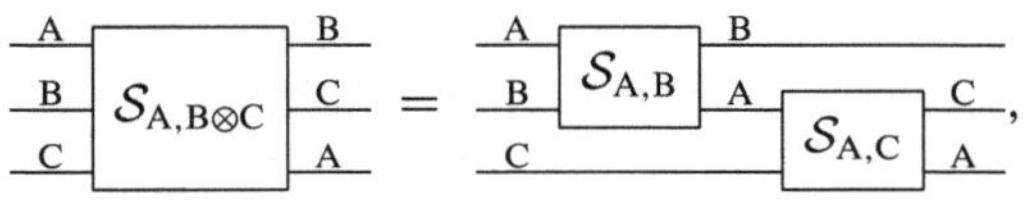

(13)

The presence of the swap, with the related Eqs. (11)–(13), turns the strict monoidal category into a strict *symmetric* monoidal category [45, 46] (strict SMC, for short).

2.1.5 Tests

A test represents a process, which can generally be non-deterministic—i.e. it can result in multiple alternative events. Specifically, a test of type $A \to B$ is collection of events of type $A \to B$, labelled by outcomes in a suitable outcome set X. The test $\boldsymbol{\mathcal{E}} := \{\mathcal{E}_x\}_{x\in X}$ is represented graphically as

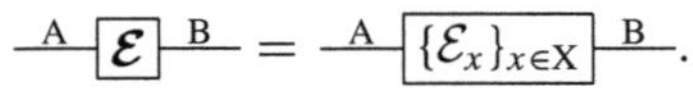

When two events/transformations belong to the same test, we say that they are *coexisting*.

The set of tests of type $A \to B$ with outcomes in X will be denoted by $\mathsf{Tests}(A \to B, X)$. We will restrict our attention to tests with a *finite* outcome set.

Tests with $|X| = 1$ are called *deterministic*, because only one event can take place. We will often identify a deterministic test $\{\mathcal{E}_{x_0}\}$ with the corresponding event $\mathcal{E}_{x_0}$, saying that $\mathcal{E}_{x_0}$ is a *deterministic event* (or a *deterministic transformation*). The deterministic transformations form a strict symmetric monoidal subcategory of Transf, denoted by $\mathsf{DetTransf}$.

Some types of tests are particularly important and deserve a name of their own. A test of type $I \to A$ is a *preparation-test* (or an *ensemble*), which prepares system A in a non-deterministic way, with the possible preparations labelled by different outcomes. A test of type $A \to I$ is an *observation-test*, corresponding to a demolition measurement that absorbs system A while producing an outcome.

2.1.6 Composition of Tests

Not all collections of events are "tests". Whether or not a specific collection is a test is determined by the theory, compatibly with two basic requirements:

1. the set of tests must be closed under sequential and parallel composition
2. the set of tests must contain deterministic tests corresponding to reversible events.

Let us discuss these requirements in more detail:

1. The sequential composition of two tests $\boldsymbol{\mathcal{E}} = \{\mathcal{E}_x\}_{x\in\mathrm{X}}$ and $\boldsymbol{\mathcal{F}} = \left\{\mathcal{F}_y\right\}_{y\in\mathrm{Y}}$ of matching types is defined as

$$\boldsymbol{\mathcal{F}} \circ \boldsymbol{\mathcal{E}} := \left\{\mathcal{F}_y \circ \mathcal{E}_x\right\}_{(x,y)\in\mathrm{X}\times\mathrm{Y}} .$$

   The test $\boldsymbol{\mathcal{F}} \circ \boldsymbol{\mathcal{E}}$ represents a cascade of two (generally non-deterministic) processes, wherein each process can result in a number of alternative events. Similarly, the parallel composition of two tests is defined as

$$\boldsymbol{\mathcal{E}} \otimes \boldsymbol{\mathcal{F}} := \left\{\mathcal{E}_x \otimes \mathcal{F}_y\right\}_{(x,y)\in\mathrm{X}\times\mathrm{Y}}$$

   and represents two non-deterministic processes occurring in parallel. The composition of tests induces a composition of their outcome spaces via the Cartesian product. As a consequence, the set of all outcome spaces must be closed under this operation. We will denote such a set by **Outcomes**.
2. If $\mathcal{U}$ is a reversible event of type $\mathrm{A} \to \mathrm{B}$, we require that there exists a deterministic test $\boldsymbol{\mathcal{U}} := \{\mathcal{U}\}$. In particular, there must be a deterministic test $\boldsymbol{\mathcal{I}}_\mathrm{A} := \{\mathcal{I}_\mathrm{A}\}$ corresponding to the identity on system A and a deterministic test $\boldsymbol{\mathcal{S}}_{\mathrm{A,B}} := \{\mathcal{S}_{\mathrm{A,B}}\}$ corresponding to the swap of systems A and B.

Note that all the basic equations valid for events can be lifted to tests: for example, the identity test acts as identity element with respect to sequential composition, that is, one has

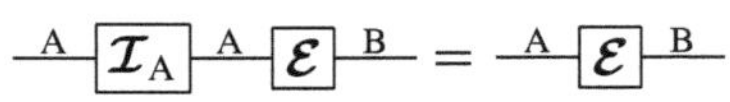

(14)

and

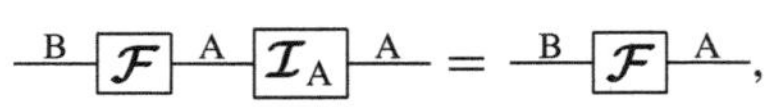

(15)

for arbitrary systems A, B and for arbitrary tests $\boldsymbol{\mathcal{E}}$ and $\boldsymbol{\mathcal{F}}$ of types $\mathrm{A} \to \mathrm{B}$ and $\mathrm{B} \to \mathrm{A}$, respectively. Since events form a strict SMC, also the tests form a strict SMC, which we denote by **Tests**.

2.1.7 Summary About the Operational Structure

Summarizing the ideas introduced so far, an *operational structure* consists of a triple

$$\mathsf{Op} = (\mathsf{Transf}, \mathsf{Outcomes}, \mathsf{Tests}),$$

where Transf is a strict symmetric monoidal category, Outcomes is a collection of sets closed under Cartesian product, and Tests is a strict symmetric monoidal category, related to Transf and Outcomes as described in the previous paragraph. Intuitively, the operational structure describes

1. what can be done (connecting tests)
2. what can be observed (outcomes), and
3. what can happen (events).

2.2 *Probabilistic Structure*

The goal of a physical theory is not only to *describe* a class of experiments, but also to *make predictions* about the outcomes of such experiments. In the following we show how this can be accomplished by adding a probabilistic structure on top of the operational structure.

2.2.1 Assigning Probabilities

An *experiment* consists in sequence of tests that starts from a preparation-test and ends with an observation-test, leaving no open wires, as in the following example

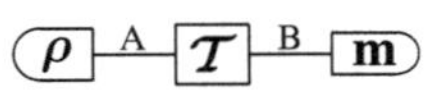

. (16)

If we compose all the tests involved in an experiment, we obtain a single test, which transforms the trivial system into itself. In order to make predictions on the outcomes of the experiment, we need a rule assigning a probability to the events of such test. The rule is provided by the probabilistic structure of the theory:

Definition 1 (*Probabilistic structure*) Let Op be an operational structure. A *probabilistic structure* for Op is a map $\mathsf{Prob} : \mathsf{Transf}(\mathrm{I} \to \mathrm{I}) \to [0, 1]$, which associates a given scalar s to a probability $\mathsf{Prob}(s)$, in accordance to the following two requirements:

1. *Consistency*: $\sum_{x \in \mathrm{X}} \mathsf{Prob}(s_x) = 1$ for every outcome set $\mathrm{X} \in \mathsf{Outcomes}$ and for every test $s \in \mathsf{Tests}(\mathrm{I} \to \mathrm{I}, \mathrm{X})$
2. *Independence*: $\mathsf{Prob}(s \otimes t) = \mathsf{Prob}(s)\, \mathsf{Prob}(t)$ for every pair of scalars s and t.

The consistency requirement guarantees that we can interpret $\mathsf{Prob}(s_x)$ as the probability of the outcome $x \in \mathrm{X}$. The independence requirement guarantees that experiments that involve independent tests on two systems give rise to uncorrelated outcomes. As observed by Hardy [27, 38], independence is equivalent to the requirement that probabilities can assigned to the outcomes of an experiment in a way that is independent of the context in which the experiment is performed. Note that the map Prob needs not be surjective: for example, in a *deterministic* theory the range of Prob are only the values 0 and 1.

We are now ready to give the formal definition of OPT:

Definition 2 An *operational-probabilistic theory* Θ is a pair $(\mathsf{Op}, \mathsf{Prob})$ consisting of an operational structure Op and of a probabilistic structure for Op.

2.2.2 Statistically Equivalent Events

Once probabilities are introduced, it is natural to identify events that give rise to the same probabilities in all possible circuits. Precisely, we say that two events of type $\mathrm{A} \to \mathrm{B}$, say $\mathcal{E}$ and $\mathcal{E}'$, are *statistically equivalent* iff

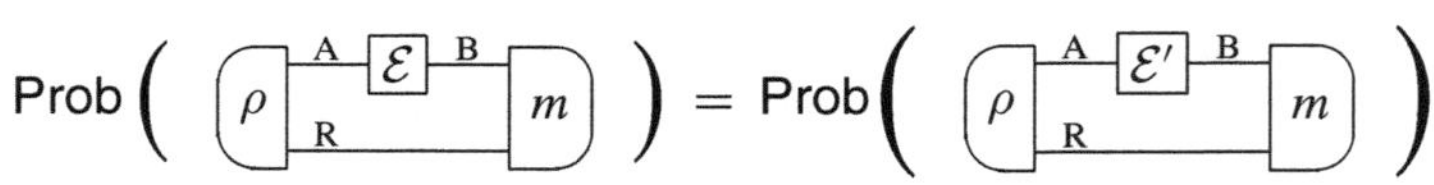

for every system R, every preparation-event $\rho \in \mathsf{Transf}(\mathrm{I} \to \mathrm{A} \otimes \mathrm{R})$ and every observation-event $m \in \mathsf{Transf}(\mathrm{B} \otimes \mathrm{R} \to \mathrm{I})$. We denote by $[\mathcal{E}]$ the equivalence class of the event $\mathcal{E}$.

Equivalence classes can be composed in sequence and parallel in the obvious way

$$[\mathcal{F}] \circ [\mathcal{E}] := [\mathcal{F} \circ \mathcal{E}], \qquad [\mathcal{E}] \otimes [\mathcal{F}] := [\mathcal{E} \otimes \mathcal{F}]$$

and it is easily verified that both definitions are well-posed. Furthermore, $[\mathcal{I}_{\mathrm{A}}]$ and $[\mathcal{S}_{\mathrm{A,B}}]$ behave like the identity on A and the swap between A and B, respectively. As a result, the equivalence classes of events form a strict SMC, which we denote by $[\mathsf{Transf}]$.

Similar considerations apply to tests: the equivalence class of a test $\boldsymbol{\mathcal{E}} = \{\mathcal{E}_x\}_{x \in \mathrm{X}}$ is defined as $[\boldsymbol{\mathcal{E}}] := \{[\mathcal{E}_x]\}_{x \in \mathrm{X}}$ and the sequential/parallel composition of equivalence classes of tests are induced by the sequential/parallel composition of events:

$$[\boldsymbol{\mathcal{F}}] \circ [\boldsymbol{\mathcal{E}}] := [\boldsymbol{\mathcal{F}} \circ \boldsymbol{\mathcal{E}}], \qquad [\boldsymbol{\mathcal{E}}] \otimes [\boldsymbol{\mathcal{F}}] := [\boldsymbol{\mathcal{E}} \otimes \boldsymbol{\mathcal{F}}].$$

Again, the equivalence class of $[\boldsymbol{\mathcal{I}}_{\mathrm{A}}]$ and $[\boldsymbol{\mathcal{S}}_{\mathrm{A,B}}]$ behave like the identity and the swap. As a result, the equivalence classes of tests form a strict SMC, which we denote by $[\mathsf{Tests}]$.

2.2.3 The Quotient OPT

The notion of statistical equivalence allowed us to transform the original operational structure $\mathsf{Op} = (\mathsf{Transf}, \mathsf{Outcomes}, \mathsf{Tests})$ into a new operational structure $[\mathsf{Op}] := ([\mathsf{Transf}], \mathsf{Outcomes}, [\mathsf{Tests}])$, which we call the *quotient operational structure*. The operational structure $[\mathsf{Op}]$ comes with an obvious probabilistic structure $[\mathsf{Prob}]$, defined as

$$[\mathsf{Prob}]([s]) := \mathsf{Prob}(s) \qquad \forall s \in \mathsf{Transf}(\mathrm{I} \to \mathrm{I}).$$

It is indeed immediate to verify that the consistency and independence conditions in Definition 1 are satisfied. As a result, the original OPT $\Theta = (\mathsf{Op}, \mathsf{Prob})$ has been turned into a new OPT $[\Theta] := ([\mathsf{Op}], [\mathsf{Prob}])$, which we call the *quotient OPT*. Intuitively, the quotient OPT contains all the information that is statistically relevant, disregarding those distinctions that have no consequences for the purpose of making probabilistic predictions.

In the following we will focus on quotient OPTs: by default, an OPT will be a *quotient* OPT. Accordingly, we will omit the symbol of equivalence class everywhere and write $\Theta = (\mathsf{Op}, \mathsf{Prob})$, assuming that equivalence classes have been already taken from the start. This is equivalent to requiring the following *separation property* [47]:

Definition 3 An OPT satisfies the *separation property* iff for every pair of systems A and B and every pair of events $\mathcal{E}$ and $\mathcal{E}'$ of type $\mathrm{A} \to \mathrm{B}$ the condition

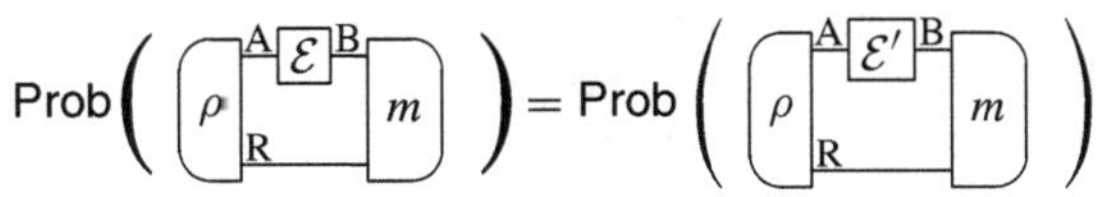

$$\forall \mathrm{R} \in \mathsf{Sys} \qquad \forall \rho \in \mathsf{Transf}(\mathrm{I} \to \mathrm{A} \otimes \mathrm{R}) \qquad \forall m \in \mathsf{Transf}(\mathrm{B} \otimes \mathrm{R} \to \mathrm{I})$$

implies $\mathcal{E} = \mathcal{E}'$.

In a quotient OPT preparation-events (respectively, observation-events) will be called *states* (respectively, *effects*) and we will use the notation $\mathsf{St}(\mathrm{A}) := \mathsf{Transf}(\mathrm{I} \to \mathrm{A})$ (respectively, $\mathsf{Eff}(\mathrm{A}) := \mathsf{Transf}(\mathrm{A} \to \mathrm{I})$).

2.2.4 Vector Space Representation of an OPT

OPTs satisfying the separation property have a convenient representation in terms of ordered vector spaces and positive maps. The construction proceeds in four steps:

1. The separation property guarantees that a scalar s can be identified with its probability $\mathsf{Prob}(s)$. Hence, from now on we will omit Prob and will identify the set of scalars $\mathsf{Transf}(\mathrm{I} \to \mathrm{I})$ with a subset of the real interval $[0, 1]$.
2. By the separation property, a state $\rho \in \mathsf{St}(\mathrm{A})$ can be identified with the real-valued function $\hat{\rho} : \mathsf{Eff}(\mathrm{A}) \to \mathbb{R}$ defined by

$$\widehat{\rho}(m) := \quad (\rho|\overset{\mathrm{A}}{\text{---}}|m)$$

(indeed, one has $\rho = \sigma$ if and only if $\widehat{\rho} = \widehat{\sigma}$). Since real-valued functions form a vector space, we can define the vector (sub)space spanned by the states of system A as

$$\mathsf{St}_{\mathbb{R}}(\mathrm{A}) := \mathsf{Span}_{\mathbb{R}}\,\{\rho \mid \rho \in \mathsf{St}(\mathrm{A})\}\,.$$

Limiting ourselves to linear combination with positive coefficients we obtain the proper cone $\mathsf{St}_{+}(\mathrm{A})$, which turns $\mathsf{St}_{\mathbb{R}}(\mathrm{A})$ into an ordered vector space.

3. Every effect $m \in \mathsf{Eff}(\mathrm{A})$ defines a *linear* function $\widehat{m} : \mathsf{St}_{\mathbb{R}}(\mathrm{A}) \to \mathbb{R}$, via the relation

$$\widehat{m}\left(\sum_i c_i\,\rho_i\right) := \sum_i c_i \quad (\rho|\overset{\mathrm{A}}{\text{---}}|m_i), \qquad \forall\{c_i\} \subset \mathbb{R}\,, \quad \forall\{\rho_i\} \subset \mathsf{St}(\mathrm{A}).$$

It is immediate to see that the definition is well-posed, namely $\widehat{m}\left(\sum_i c_i\,\rho_i\right) = \widehat{m}\left(\sum_j c'_j\,\rho'_j\right)$ whenever $\sum_i c_i\,\rho_i = \sum_j c'_j\,\rho'_j$. Again, the effect m can be identified with the linear function $\widehat{m}$ thanks to the separation property. Taking linear combinations of effects we obtain the vector space

$$\mathsf{Eff}_{\mathbb{R}}(\mathrm{A}) := \mathsf{Span}_{\mathbb{R}}\,\{m \mid m \in \mathsf{Eff}(\mathrm{A})\}\,,$$

while restricting to positive linear combinations we obtain the proper cone $\mathsf{Eff}_{+}(\mathrm{A})$. As a result, also $\mathsf{Eff}_{\mathbb{R}}(\mathrm{A})$ is an ordered vector space.

4. Every event $\mathcal{E}$ of type $\mathrm{A} \to \mathrm{B}$ induces a linear map $\widehat{\mathcal{E}} : \mathsf{St}_{\mathbb{R}}(\mathrm{A}) \to \mathsf{St}_{\mathbb{R}}(\mathrm{B})$, via the definition

$$\widehat{\mathcal{E}}\left(\sum_i c_i\,\rho_i\right) := \sum_i c_i\;(\mathcal{E} \circ \rho_i)\,, \qquad \forall\{c_i\} \subset \mathbb{R}\,, \quad \forall\{\rho_i\} \subset \mathsf{St}(\mathrm{A}).$$

Again, it is not hard to see that the definition is well-posed, namely that $\widehat{\mathcal{E}}\left(\sum_i c_i\,\rho_i\right) = \widehat{\mathcal{E}}\left(\sum_j c'_j\,\rho'_j\right)$ whenever $\sum_i c_i\,\rho_i = \sum_j c'_j\,\rho'_j$. Note that the map $\widehat{\mathcal{E}}$ is not only linear, but also *positive*: indeed, it sends elements of the cone $\mathsf{St}_{+}(\mathrm{A})$ to elements of the cone $\mathsf{St}_{+}(\mathrm{B})$. We call $\widehat{\mathcal{E}}$ the *state change* associated to $\mathcal{E}$.

At this point, a few remarks are in order:

1. *Linearity versus convexity*. Traditionally, the linearity of state changes has been argued from the assumption that the state space $\mathsf{St}(\mathrm{A})$ is convex. However, our argument shows that such assumption is *not* needed: the probabilistic structure alone suffices to define the linear map $\widehat{\mathcal{E}}$.
2. *Finite versus infinite dimensional systems*. For a given system A, we define D_{A} to be the dimension of the vector space $\mathsf{St}_{\mathbb{R}}(\mathrm{A})$ and we say that system A is *finite*

dimensional if D_{A} is finite. For finite systems, one has the equality $\mathsf{Eff}_{\mathbb{R}}(\mathrm{A}) = \mathsf{St}_{\mathbb{R}}(\mathrm{A})^*$ where $\mathsf{St}_{\mathbb{R}}(\mathrm{A})^*$ is the vector space of all linear functionals on $\mathsf{St}_{\mathbb{R}}(\mathrm{A})$. For infinite dimensional systems, such an equality may not hold.

3. *The no-restriction hypothesis.* Since effects are identified with positive linear functions, one has the inclusion $\mathsf{Eff}_+(\mathrm{A}) \subseteq \mathsf{St}_+(\mathrm{A})^*$, where $\mathsf{St}_+(\mathrm{A})^*$ denotes the dual cone of $\mathsf{St}_+(\mathrm{A})$

$$\mathsf{St}_+(\mathrm{A})^* := \{m \in \mathsf{St}_{\mathbb{R}}(\mathrm{A})^* \mid m(\rho) \geq 0 \quad \forall \rho \in \mathsf{St}_+(\mathrm{A})\}. \tag{17}$$

Even for finite dimensional systems, the inclusion $\mathsf{Eff}_+(\mathrm{A}) \subseteq \mathsf{St}_+(\mathrm{A})^*$ may not be an equality. The assumption $\mathsf{Eff}_+(\mathrm{A}) = \mathsf{St}_+(\mathrm{A})^*$ is known as *No-Restriction Hypothesis* [34]. We stress that such an assumption is *not* made in our derivation.

4. *Transformations versus linear maps.* Unlike in the case of states and effects, the correspondence between the transformation $\mathcal{E}$ and the linear map $\widehat{\mathcal{E}}$ may not be one-to-one. The reason for this is that the difference between two transformations $\mathcal{E}$ and $\mathcal{E}'$ may show up when one applies them locally on a part of a composite system: one can have $\widehat{\mathcal{E} \otimes \mathcal{I}_{\mathrm{R}}} \neq \widehat{\mathcal{E}' \otimes \mathcal{I}_{\mathrm{R}}}$ for some $\mathrm{R} \in \mathsf{Sys}$ even if $\widehat{\mathcal{E}} = \widehat{\mathcal{E}'}$. This problem disappears if one assumes the axiom of Local Tomography, as we will see later in this chapter. In the lack of Local Tomography, however, the transformation $\mathcal{E}$ can still be identified with a linear map: for this purpose, one can choose the linear map $\widehat{\mathcal{E}}_\oplus$ defined by [47]

$$\widehat{\mathcal{E}}_\oplus := \bigoplus_{\mathrm{R} \in \mathsf{Sys}} \widehat{\mathcal{E} \otimes \mathcal{I}_{\mathrm{R}}}. \tag{18}$$

The map $\widehat{\mathcal{E}}_\oplus$ transforms elements of the (infinite-dimensional) vector space $\mathsf{St}_{\mathbb{R},\oplus}(\mathrm{A}) := \bigoplus_{\mathrm{R} \in \mathsf{Sys}} \mathsf{St}_{\mathbb{R}}(\mathrm{A} \otimes \mathrm{R})$ into elements of the (infinite-dimensional) vector space $\mathsf{St}_{\mathbb{R},\oplus}(\mathrm{B}) := \bigoplus_{\mathrm{R} \in \mathsf{Sys}} \mathsf{St}_{\mathbb{R}}(\mathrm{B} \otimes \mathrm{R})$. Then, the separation property guarantees that the correspondence between $\mathcal{E}$ and $\widehat{\mathcal{E}}_\oplus$ is one-to-one.

5. *The vector space of transformations.* So far we have defined the vector spaces of states and effects. A vector space of transformations can be defined using the one-to-one correspondence with the linear maps in Eq. (18) and setting

$$\mathsf{Transf}_{\mathbb{R}}(\mathrm{A} \to \mathrm{B}) := \mathsf{Span}_{\mathbb{R}}\{\mathsf{Transf}(\mathrm{A} \to \mathrm{B})\}. \tag{19}$$

Again, a proper cone $\mathsf{Transf}_+(\mathrm{A} \to \mathrm{B})$ can be defined by restricting the attention to linear combinations with positive coefficients. Note that, in general, the vector space $\mathsf{Transf}_{\mathbb{R}}(\mathrm{A} \to \mathrm{B})$ and the cone $\mathsf{Transf}_+(\mathrm{A} \to \mathrm{B})$ can be infinite-dimensional *even if both systems* A *and* B *are finite dimensional*. However, this is not the case when the theory satisfies the Local Tomography.

2.2.5 Closure Under Coarse-Graining

A key notion that comes with the probabilistic structure is the notion of *coarse-graining*: given a test $\mathcal{T} = \{\mathcal{T}_x\}_{x\in\mathrm{X}}$, one can decide to identify some outcomes, thus obtaining another, coarse-grained test. Mathematically, a coarse-graining is defined by partitioning the outcome set X into mutually disjoint subsets $\{\mathrm{X}_y\}_{y\in\mathrm{Y}}$. Relative to such partition, the coarse-graining of the test $\mathcal{T}$ is the test $\mathcal{T}' = \{\mathcal{T}'_y\}_{y\in\mathrm{Y}}$ defined by[4]

$$\mathcal{T}'_y := \sum_{x\in\mathrm{X}_y} \mathcal{T}_x, \tag{20}$$

setting $\mathcal{T}'_y = 0$ for $\mathrm{X}_y = \emptyset$, where 0 is the zero element of the vector space $\mathsf{Transf}_{\mathbb{R}}(\mathrm{A} \to \mathrm{B})$. Note that, by calling $\mathcal{T}'$ a *test* we have implicitly made two assumptions, namely that

1. the set Y belongs to $\mathsf{Outcomes}$
2. the collection $\{\mathcal{T}'_y\}_{y\in\mathrm{Y}} \subset \mathsf{Transf}_{\mathbb{R}}(\mathrm{A} \to \mathrm{B})$ belongs to $\mathsf{Tests}(\mathrm{A} \to \mathrm{B}, \mathrm{Y})$.

From now on, we will require that our OPT is *closed under coarse-graining*, meaning that the above conditions are satisfied.

By coarse-graining over all outcomes of a test $\mathcal{T} \in \mathsf{Tests}(\mathrm{A} \to \mathrm{B}, \mathrm{X})$ one obtains a deterministic test, identified with the deterministic transformation $\mathcal{T} := \sum_{x\in\mathrm{X}} \mathcal{T}_x \in \mathsf{DetTransf}(\mathrm{A} \to \mathrm{B})$. In particular, when a preparation test $\boldsymbol{\rho} \in \mathsf{Tests}(\mathrm{I} \to \mathrm{A}, \mathrm{X})$ satisfies $\sum_{x\in\mathrm{X}} \rho_x = \rho$ we say that the test $\boldsymbol{\rho}$ is an *ensemble decomposition of* ρ.

2.2.6 Summary of the OPT Framework

Let us sum up the main points discussed so far. We defined an OPT as a pair $\Theta = (\mathsf{Op}, \mathsf{Prob})$, consisting of an operational structure $\mathsf{Op} = (\mathsf{Transf}, \mathsf{Outcomes}, \mathsf{Tests})$ and of a probabilistic structure Prob that assigns probabilities to scalars. We restricted our attention to OPTs that satisfy the Separation Property (Definition 3), which implies that one can identify scalars with probabilities, states with elements of suitable vector spaces, and effects with linear functionals over them. Transformations with nontrivial input and output induce linear maps on the corresponding state spaces. Finally, in agreement with the probabilistic interpretation, we demanded that the theory Θ be closed under coarse-graining.

[4]Note that the summation is well-defined thanks to the vector space structure of $\mathsf{Transf}_{\mathbb{R}}(\mathrm{A} \to \mathrm{B})$.

3 Background of the Quantum Reconstruction

In this section we provide some background that will be useful for our reconstruction of quantum theory. We start by reviewing three standing assumptions: finite-dimensionality, non-determinism, and closure under operational limits. We will then review the operational tasks that motivate our axioms.

3.1 *Standing Assumptions*

Here we introduce three standing assumptions that will be made in the rest of the chapter. These assumptions are common to all recent axiomatizations of quantum theory, and could be even incorporated in the OPT framework. We keep them separate from the rest, both for clarity of presentation and for the sake of maintaining the OPT framework as flexible as possible. The assumptions are the following:

1. *Finite dimensionality*. We restrict our attention to finite systems, i.e. systems with finite dimensional state spaces. Operationally, this means that the state of every system can be identified from the statistics of a *finite number of finite-outcome measurements*. Of course, the implicit assumption here is that finite systems exist and form a sub-theory of our theory, meaning that the operational structure Op contains a non-trivial substructure $\mathsf{FiniteOp}$, consisting of transformations, outcome sets, and tests involving only finite systems.
2. *Non-determinism*. While the OPT framework accommodates a variety of theories, here we focus on OPTs that are non-deterministic, meaning that there exists at least one experiment for which the outcome is not determined a priori. Mathematically, this means that the range of the probability function Prob is not just $\{0, 1\}$. Note that non-determinism is a weaker assumption than convexity of the state spaces: there exist indeed examples of theories—such as Spekkens' toy theory [48]—that are non-deterministic and yet do not have convex state spaces.
3. *Closure under operational limits*. Suppose that $(\mathcal{T}_n)_{n\in\mathbb{N}}$ is a sequence of transformations of type $\mathrm{A} \to \mathrm{B}$ and that $\mathcal{T}$ is an element of the vector space $\mathsf{Transf}_{\mathbb{R}}(\mathrm{A} \to \mathrm{B})$ such that

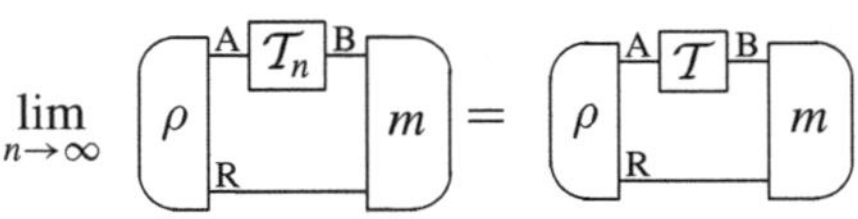

$$\forall \mathrm{R} \in \mathsf{Sys} \quad \forall \rho \in \mathsf{Transf}(\mathrm{I} \to \mathrm{A} \otimes \mathrm{R}) \quad \forall m \in \mathsf{Transf}(\mathrm{B} \otimes \mathrm{R} \to \mathrm{I}),$$

   meaning that the probability of every experiment involving $\mathcal{T}_n$ converges to the probability of an hypothetical experiment involving $\mathcal{T}$. When this is the case, we assume that $\mathcal{T}$ belongs to $\mathsf{Transf}(\mathrm{A} \to \mathrm{B})$. Operationally, one can think of the sequence $(\mathcal{T}_n)_{n\in\mathbb{N}}$ as a *limit procedure* to implement the transformation $\mathcal{T}$.

3.2 Basic Operational Tasks

We now give a brief list of the operational notions on which our axioms are based.

3.2.1 Signalling

When a number of devices are connected into a network, it is natural to ask whether one node of the network can signal to another. For example, given the experiment

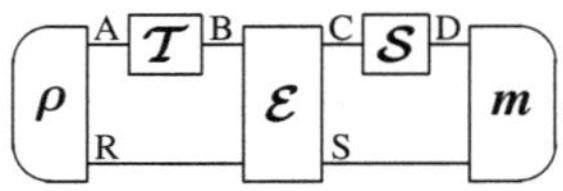

(21)

one can ask whether the choice of the test $\mathcal{T}$ can influence the outcome of the test $\mathcal{S}$. Precisely, the question is whether or not the marginal probability distribution for the outcomes of $\mathcal{S}$ (obtained by summing over the outcomes of all the other tests in the network) depends on $\mathcal{T}$. Denoting the marginal probability distribution by $p(x|\mathcal{T})$, $x \in \mathrm{X}$, we say that the node occupied by the test $\mathcal{T}$ *does not signal* to the node occupied by the test $\mathcal{S}$ iff

$$p(x|\mathcal{T}_0) = p(x|\mathcal{T}_1) \qquad \forall x \in \mathrm{X},$$

for every possible choice of tests $\mathcal{T}_1$ and $\mathcal{T}_2$. Similarly, one can ask whether the node occupied by the test $\mathcal{S}$ can signal to the node occupied by the test $\mathcal{T}$. Now, note that the test $\mathcal{S}$ is performed *after* the test $\mathcal{T}$: if the node occupied by $\mathcal{S}$ can signal to the node occupied by $\mathcal{T}$ we say that the circuit of Eq. (21) allows for *signalling from the future to the past.*

3.2.2 Collecting Side Information

Suppose that the test $\mathcal{T} = \{\mathcal{T}_x\}_{x\in\mathrm{X}}$ is obtained from the test $\mathcal{T}' = \{\mathcal{T}_z\}_{z\in\mathrm{Z}}$ via coarse-graining, namely

$$\mathcal{T}_x = \sum_{z\in\mathrm{Z}_x} \mathcal{T}'_z \qquad \forall x \in \mathrm{X},$$

where $\{\mathrm{Z}_x\}_{x\in\mathrm{X}}$ is a partition of Z into disjoint subsets. In this case we say that $\mathcal{T}'$ *refines* $\mathcal{T}$. Now, it is convenient to relabel the outcomes of $\mathcal{T}'$ as $z = (x, y)$, with $x \in \mathrm{X}$ and $y \in \mathrm{Z}_x$, and to write $\mathcal{T}'_{x,y}$ in place of $\mathcal{T}'_z$. In this way, we can think of the random variable y as a *side information*, which is not accessible to the agent Alice performing the test $\mathcal{T}$, but may be accessible to some other agent Eve. This picture is particularly relevant to cryptographic scenarios, wherein Eve could be an eavesdropper attempting to collect as much information as possible. In all such scenarios, a special role is played by those transformations that do not leak any useful side information. We call such transformations *pure*:

Definition 4 We say that a transformation $\mathcal{E}$ is *pure*[5] iff for every test $\boldsymbol{\mathcal{T}}$ containing $\mathcal{E}$ and for every test $\boldsymbol{\mathcal{T}}'$ refining $\boldsymbol{\mathcal{T}}$ one has

$$\mathcal{T}'_{x_0,y} = p_y\, \mathcal{T}_{x_0}, \tag{22}$$

where x_0 is the outcome such that $\mathcal{T}_{x_0} = \mathcal{E}$ and $\{p_y\}$ is a probability distribution.

Informally, the purity condition (22) states that the side information possessed by Eve is uncorrelated with the transformation $\mathcal{E}$ taking place in Alice's laboratory. We denote the set of pure transformations of type A $\to$ B by $\mathsf{PurTransf}(\mathrm{A} \to \mathrm{B})$. In the special case of transformations with trivial input we will use the notation $\mathsf{PurSt}(\mathrm{A})$ (respectively, $\mathsf{PurEff}(\mathrm{A})$), referring to *pure states* (respectively, *pure effects*). An *pure test* is a test consisting of pure transformations.

Transformations that are not necessarily pure will be called *mixed*. Among the mixed transformations, the ones that are in the interior of the cone $\mathsf{Transf}_+(\mathrm{A} \to \mathrm{B})$ play an important role. They are defined as follows:

Definition 5 A transformation $\mathcal{E} \in \mathsf{Transf}(\mathrm{A} \to \mathrm{B})$ is called *internal* iff for every transformation $\mathcal{F} \in \mathsf{Transf}(\mathrm{A} \to \mathrm{B})$ there exists a transformation $\mathcal{G}$ and a scaling constant $\lambda > 0$ such that

1. $\mathcal{E} = \lambda \mathcal{F} + \mathcal{G}$
2. $\lambda \mathcal{F}$ and $\mathcal{G}$ coexist in a test.[6]

Roughly speaking, an internal transformation is compatible with the occurrence of any other transformation of the same type. Internal transformations with trivial input (output) will be called *internal states* (*internal effects*).

---

[5]In previous works, we used different names for transformations that do not allow for side information: in Refs. [26, 34] they were called *atomic*, while in the popularized version of Ref. [49] they were called *fine-grained*. We apologize with our readers for the changes of terminology, due to an ongoing search for the word that best captures this operational concept. In this chapter, we adopted the word *pure* because (i) this term is the standard one in the case of states and (ii) using the same term for transformations should hopefully ease the reading. Still, a warning is in order: when the set of transformations $\mathsf{Transf}(\mathrm{A} \to \mathrm{B})$ is convex, the pure transformations $\mathsf{PurTransf}(\mathrm{A} \to \mathrm{B})$ *may not coincide* with the extreme points of $\mathsf{Transf}(\mathrm{A} \to \mathrm{B})$. For example, in quantum theory the identity effect I_A is an extreme point of the set of effects, but is not pure in the sense of our definition because it can be decomposed e.g. as $I_\mathrm{A} = \sum_{n=1}^{d_\mathrm{A}} P_n$, where the effects $\{P_n = |n\rangle\langle n| \mid n = 1, \dots, d_\mathrm{A}\}$ represent a projective measurement on some orthonormal basis $\{|n\rangle \mid n = 1, \dots, d_\mathrm{A}\}$.

[6]Note that, in principle, our definition of "internal transformations" may not include all the transformations in the interior of the cone, because the $\lambda \mathcal{F}$ and $\mathcal{G}$ may fail to coexist in a test. However, this annoying discrepancy disappears under the mild assumption that the set of transformations is convex. Later, we will justify this assumption on the basis of the Causality axiom.

3.2.3 State Tomography

The task of state tomography consists in identifying the state of a system from the statistics of a restricted set of observations. Suppose that an experimenter is able to perform a set of observation-tests and let M be the set of all effects appearing in such tests.

Definition 6 We say that the effects in M are *tomographically complete* for system A iff, for every pair of states ρ and ρ' of system A, one has the implication

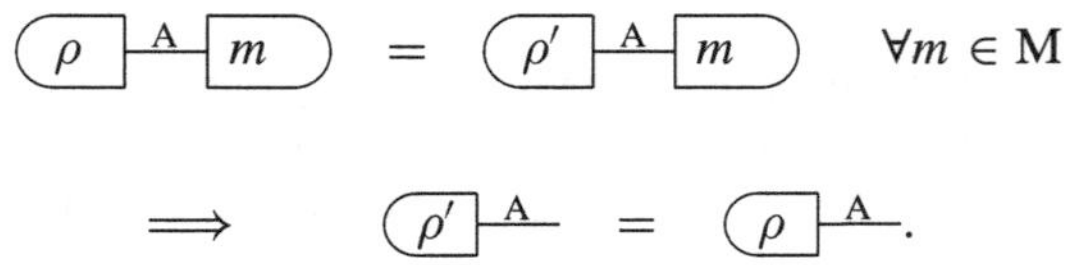

In the contrapositive: if two states are different, then the difference can be detected from the statistics of some effect in M.

Let us consider state tomography for composite systems. Suppose that two experimenters Alice and Bob perform measurements on two systems A and B, respectively, and that Alice (Bob) is able to perform the set of measurements with effects M (N). Then, by coordinating their choices of measurements and by communicating the outcomes to each other, Alice and Bob can observe the statistics of all product measurements. Hence, their set of measurement effects will be

$$\mathrm{M} \otimes \mathrm{N} := \{m \otimes n \mid m \in \mathrm{M}, \quad n \in \mathrm{N}\}.$$

Now the question is: is there a choice of measurement effects M and N such that the set M ⊗ N tomographically complete? In the affirmative case, we say that system A ⊗ B *allows for local tomography*:

Definition 7 System A ⊗ B *allows for local tomography* iff, for every pair of states $\rho, \rho' \in \mathsf{St}(\mathrm{A} \otimes \mathrm{B})$, one has the implication

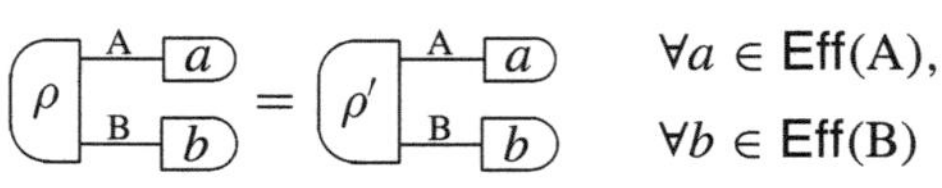

(23)

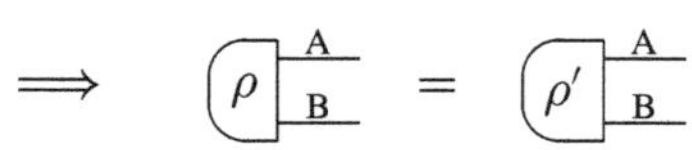

(24)

More generally, we have the following

Definition 8 An K-partite system $\mathrm{A} = \bigotimes_{k=1}^{K} \mathrm{A}_k$ *allows for local tomography* iff for every $k \in \{1, \ldots, K\}$ there exists a set of measurement effects M_k on system A_k such that the set $\bigotimes_{k=1}^{K} \mathrm{M}_k$ is tomographically complete.

For a given OPT, it is easy to see that the following conditions are equivalent:

1. every multipartite system allows for local tomography
2. every bipartite system allows for local tomography.

In other words, the possibility of local tomography for arbitrary composite systems can be established by just checking bipartite systems.

3.2.4 State Discrimination

The task of state discrimination can be presented as a game featuring a player and a referee. The referee prepares a physical system A in a state ρ_x, belonging to some set $\{\rho_x \mid x \in \mathrm{X}\}$ known to the player. The player is asked to guess the label x. In order to do that, she performs a measurement **m** with outcomes in X: upon finding the outcome x', she will guess that the state was $\rho_{x'}$. If the player guesses right all the times, we say that the states are perfectly distinguishable:

Definition 9 The states $\{\rho_x \mid x \in \mathrm{X}\}$ are *perfectly distinguishable* iff there exists a measurement **m** such that

$$(m_x|\rho_{x'}) = \delta_{x,x'} \qquad \forall x, x' \in \mathrm{X}.$$

When this is the case, we say that ***m*** is a *discriminating measurement*.

Note that, in order to be perfectly distinguishable, the states must be

1. *normalized*, namely $\|\rho_x\| = 1 \, \forall x \in \mathrm{X}$, where $\|\cdot\|$ is the operational norm [34] given by $\|\rho\| = \sup_{a \in \mathsf{Eff}(\mathrm{A})} (a|\rho)$
2. *non-internal*: indeed, if a state $\rho_{x'}$ is internal, then $(m_x|\rho_{x'}) = 0$ implies $m_x = 0$, in contradiction with the condition $(m_x|\rho_x) = 1$.

Note that *a priori* an OPT may not have any distinguishable states at all. However, the existence of distinguishable states is essential if we want our theory to include classical computation and classical information theory.

3.2.5 Ideal Compression

A preparation-test $\boldsymbol{\rho} \in \mathsf{Tests}(\mathrm{I} \to \mathrm{A}, \mathrm{X})$ can be thought as describing a *source of information*. An interesting question is how well such information can be transferred from the original system to another physical support, say system B. An *encoding* of the preparation-test $\boldsymbol{\rho}$ is a deterministic transformation $\mathcal{E} \in \mathsf{DetTransf}(\mathrm{A} \to \mathrm{B})$, which transforms $\boldsymbol{\rho}$ into a new preparation-test $\boldsymbol{\rho}' := \{\mathcal{E} \circ \rho_x\}_{x \in \mathrm{X}}$. The states $\{\mathcal{E} \circ \rho_x \mid x \in \mathrm{X}\}$ are called *codewords*.

The ideal property of an encoding is to be lossless, in the following sense:

Definition 10 An encoding $\mathcal{E} \in \mathsf{DetTransf}(\mathrm{A} \to \mathrm{B})$ is *lossless for the preparation-test* $\boldsymbol{\rho} \in \mathsf{Tests}(\mathrm{I} \to \mathrm{A}, \mathrm{X})$ iff there exists a deterministic transformation $\mathcal{D} \in \mathsf{DetTransf}(\mathrm{B} \to \mathrm{A})$, called the *decoding*, such that

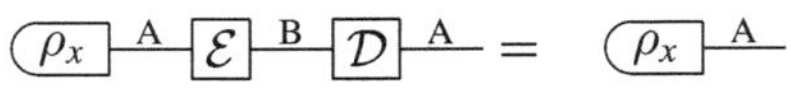

$\forall x \in \mathrm{X}.$ (25)

We say that

- $\mathcal{E}$ is a *lossless encoding for the state* $\rho \in \mathsf{DetSt}(\mathrm{A})$ iff $\mathcal{E}$ is a lossless encoding for every ensemble decomposition of ρ.
- $\mathcal{E}$ a *lossless encoding of system* A *into system* B iff $\mathcal{E}$ is a lossless encoding for all states $\rho \in \mathsf{DetSt}(\mathrm{A})$.

The notion of encoding offers an operational way to compare the size of different systems: naturally, we can say that system A is *no larger* than system B iff there exists a lossless encoding of A into B.

Among all possible encodings, we now consider the compressions:

Definition 11 A *compression* of system A into system B is an encoding $\mathcal{E} \in \mathsf{DetTransf}(\mathrm{A} \to \mathrm{B})$ where B is no larger than A.

How much can we compress a given state? The ultimate limit to compression is when *every* state of system B is proportional to a codeword, i.e. when every state $\sigma \in \mathsf{St}(\mathrm{B})$ can be written as $\sigma = \lambda \mathcal{E} \rho_{x_0}$, for some scaling constant $\lambda \geq 0$ and some state ρ_{x_0} belonging to some ensemble decomposition of ρ. When this is the case, we say that the compression $\mathcal{E}$ is *maximally efficient*. Summing up, we have the following

Definition 12 A transformation $\mathcal{E} \in \mathsf{DetTransf}(\mathrm{A} \to \mathrm{B})$ is an *ideal compression of the state* $\rho \in \mathsf{DetSt}(\mathrm{A})$ iff it is lossless and maximally efficient.

3.2.6 Simulating Preparations

A state can be prepared in many different ways. For example, a state ρ_{A} could be prepared by a circuit that involves many auxiliary systems, which interact with A and are finally discarded. We refer to these systems as the *environment* and describe them collectively as a single system E. Assuming that the system and the environment are initially uncorrelated, the fact that the circuit prepares the state ρ_{A} is expressed by the diagram

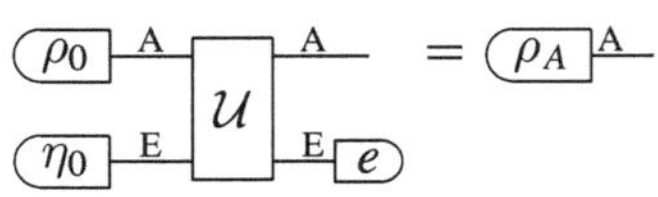

(26)

where ρ_0 and η_0 are the initial states of system and environment, respectively, $\mathcal{U}$ is a transformation representing all the system-environment interaction, and e is a some effect. By defining the state $\rho_{AE} := \mathcal{U}(\rho_0 \otimes \eta_0)$ the circuit of Eq. (26) can be simplified to

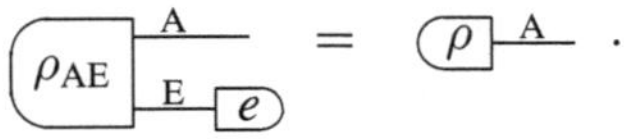

(27)

To capture the idea that the environment is discarded, we require the effect e to be *deterministic*:

Definition 13 A *simulation* of the preparation ρ_A is a triple (E, ρ_{AE}, e) where E is a system, ρ_{AE} is a state of A $\otimes$ E, and e is a deterministic effect satisfying Eq. (27). If the state ρ_{AE} is pure, we say that (E, ρ_{AE}, e) is a *pure simulation*—or, more concisely, a *purification*—of ρ_A.

Purifications arise, for example, when we start from a *pure* product state $\alpha_0 \otimes \eta_0 \in \mathsf{PurSt}(A \otimes E)$ and evolve it through a *reversible* transformation $\mathcal{U}$. A purification gives the agent maximal control over the process of preparation: indeed, an agent possessing systems A and E can be sure that no side information can hide outside her laboratory.

Given the importance of purifications, it is important to ask how many of them can be found for a given state. From a purification there are two trivial ways to generate new ones:

1. by transforming the environment with a reversible transformation $\mathcal{U}_E$ such that $(e|\mathcal{U} = (e|$, and
2. by appending a dummy system D to the environment, prepared in a pure deterministic state δ_D such that $\rho_{AE} \otimes \delta_D$ is pure.

We say that a pure simulation is *essentially unique* if it is unique up to trivial transformations:

Definition 14 A state ρ_A has an *essentially unique purification* iff for every two purifications (E, Ψ, e) and (E′, Ψ', e') with E = E′ one has

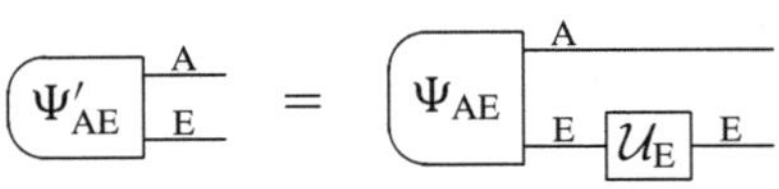

(28)

and[7]

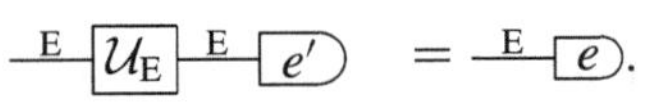

(29)

for some reversible transformation $\mathcal{U}_E$.

[7] It turns out that the second condition is automatically satisfied if the theory satisfies the Causality axiom—see the next section.

4 The Principles

We are now ready to state our principles for quantum theory. We divide them into five *Axioms* and one *Postulate*.[8] The five axioms are

A1 **Causality**. No signal can be sent from the future to the past.
A2 **Purity of Composition**. No side information can hide in the composition of two pure transformations.
A3 **Local Tomography**. State tomography can be performed with only local measurements.
A4 **Perfect State Discrimination**. Every normalized non-internal state can be perfectly distinguished from some other state.
A5 **Ideal Compression**. Every state can be compressed in an ideal way.

The five Axioms express generic and rather unsurprising features, which are common to classical and quantum theory. We regard the theories satisfying these axioms as *standard*. The Postulate is

P6 **Purification**. Every preparation can be simulated via a pure preparation in an essentially unique way.

Purification brings in a radically non-classical feature: the idea that randomness can be simulated through the preparation of pure states. We will see that this feature singles out quantum theory uniquely among all standard OPTs.

4.1 Causality

Causality states that signals cannot be sent from the future to the past. To check this condition, it is sufficient to look at a special class of circuits, consisting of a single preparation-test, followed by a single observation-test. Precisely, we have the following

Proposition 1 *An OPT satisfies Causality if and only if for every system* $\mathrm{A} \in \mathsf{Sys}$, *every preparation-test* $\boldsymbol{\rho} \in \mathsf{Tests}(\mathrm{I} \to \mathrm{A}, \mathrm{X})$ *and every pair of observation-tests* $\mathbf{m_0} \in \mathsf{Tests}(\mathrm{A} \to \mathrm{I}, \mathrm{Y}_0)$ *and* $\mathbf{m_1} \in \mathsf{Tests}(\mathrm{A} \to \mathrm{I}, \mathrm{Y}_1)$ *one has*

$$p(x|\boldsymbol{m}_0) = p(x|\boldsymbol{m}_1) \qquad \forall x \in \mathrm{X},$$

with $p(x|\boldsymbol{m}_i) := \sum_{y_i \in \mathrm{Y}_i} (m_{y_i}|\rho_x)$.

An even simpler condition for causality is given by

[8] We differentiate the names in order to highlight the different roles of these principles in our reconstruction. Mathematically, there is no difference between axioms, postulates, background assumptions, and requirements in the OPT framework (all of them are "axioms"). The point of using different names is just to provide a more intuitive picture.

Proposition 2 *A theory satisfies Causality if and only if every system* A *has a unique deterministic effect* $e_{\mathrm{A}} \in \mathsf{DetEff}(\mathrm{A})$.

In categorical terms, the uniqueness of the deterministic effect can be phrased as "terminality of the tensor unit" in the category of deterministic transformations DetTransf. Categories where the tensor unit is terminal have been introduced by Coecke and Lal [50, 51], who named them *causal categories*.

Recall that deterministic effects can be used to describe "discarding operations", whereby a physical system is eliminated from the description. Now, Causality is equivalent to the statement that every physical system can be discarded in a unique way. Thanks to Causality, we can define the marginals of a bipartite state in a canonical way.

Definition 15 Let ρ_{AB} be a state of system $\mathrm{A} \otimes \mathrm{B}$. The *marginal* of ρ_{AB} on system A is the state ρ_{A} defined as

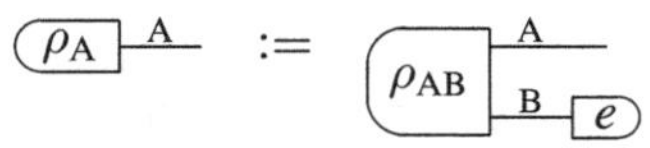

4.1.1 Causality and No-Signalling

An important consequence of Causality is the impossibility to signal without interaction: in the lack of any interaction between system A an system B, it is impossible to influence the probability distribution of a test on system A by performing tests on system B. The precise statement is the following

Proposition 3 *For every pair of systems* A *and* B, *for every state* ρ_{AB}, *and every triple of tests* $\boldsymbol{\mathcal{A}} \in \mathsf{Tests}(\mathrm{A} \to \mathrm{A}', \mathrm{X})$, $\boldsymbol{\mathcal{B}}_0 \in \mathsf{Tests}(\mathrm{B} \to \mathrm{B}'_0, \mathrm{Y}_0)$ *and* $\boldsymbol{\mathcal{B}}_1 \in \mathsf{Tests}(\mathrm{B} \to \mathrm{B}'_1, \mathrm{Y}_1)$ *one has*

$$p\,(x|\boldsymbol{\mathcal{B}}_0) = p\,(x|\boldsymbol{\mathcal{B}}_1) \qquad \forall x \in \mathrm{X},$$

with $p(x|\boldsymbol{\mathcal{B}}_i) := \sum_{y_i \in \mathrm{Y}_i} (e_{\mathrm{B}_i}|\, \mathcal{A}_x \otimes \mathcal{B}_{i,y_i}\, |\rho_{\mathrm{AB}})$, $i \in \{0, 1\}$.

4.1.2 Causality and Conditional Tests

We introduced Causality as a negative statement:

C: the choice of tests performed in the future *cannot* affect the outcome probabilities of tests performed in the past.

The axiom can be reformulated in a positive, and slightly stronger way:

C′: the outcomes of tests performed in the past *can* affect the choice of tests performed in the future.

Technically, Condition $\mathbf{C}'$ establishes the possibility of performing *conditional tests*, defined as follows:

Definition 16 Given a test $\boldsymbol{\mathcal{T}} \in \mathsf{Tests}(\mathrm{A} \to \mathrm{B}, \mathrm{X})$ and a collection of tests $\{\boldsymbol{\mathcal{S}}_x \in \mathsf{Tests}(\mathrm{B} \to \mathrm{C}, \mathrm{Y}_x) \mid x \in \mathrm{X}\}$, the *conditional test* associated to them is the collection of transformations

$$\{\boldsymbol{\mathcal{S}}_x\} \odot \boldsymbol{\mathcal{T}} := \left\{ \overset{A}{\text{—}}\boxed{\mathcal{T}_x}\overset{B}{\text{—}}\boxed{\mathcal{S}^x_{y_x}}\overset{C}{\text{—}} \;\middle|\; x \in \mathrm{X}, y_x \in \mathrm{Y}_x \right\}.$$

Condition $\mathbf{C}'$ states that such collection is actually a *test*, meaning that

1. the set $\mathrm{Z} = \bigcup_{x \in \mathrm{X}} \{x\} \times \mathrm{Y}_x$ belongs to $\mathsf{Outcomes}$, and
2. the collection $\{\boldsymbol{\mathcal{S}}_x\} \odot \boldsymbol{\mathcal{T}}$ belongs to $\mathsf{Tests}(\mathrm{A} \to \mathrm{C}, \mathrm{Z})$.

The relation between $\mathbf{C}$ and $\mathbf{C}'$ is the following:

1. $\mathbf{C}'$ *implies* $\mathbf{C}$,
2. $\mathbf{C}$ implies that the theory can be enlarged to another theory satisfying $\mathbf{C}'$: thanks to $\mathbf{C}$, all conditional tests can be included without losing the consistency of the probabilistic structure [34].

Since conditional tests can be included, we will always assume that they *are* included, i.e. we will take the validity of $\mathbf{C}'$ as part of the Causality package.

4.1.3 Convexity

The ability to perform conditional tests brings naturally to convexity of the sets of physical transformations. This result can be obtained in two steps:

1. Under the standing assumptions that the theory is not deterministic and that the set $\mathsf{Transf}(\mathrm{I} \to \mathrm{I})$ is closed, we obtain that $\mathsf{Transf}(\mathrm{I} \to \mathrm{I})$ is the whole interval $[0, 1]$. In other words, every number in the interval $[0, 1]$ can be interpreted as the probability of some outcome in some test allowed by the theory.
2. Given two transformations $\mathcal{T}_0, \mathcal{T}_1 \in \mathsf{Transf}(\mathrm{A} \to \mathrm{B})$, the convex combination $p\mathcal{T} + (1-p)\mathcal{T}'$ can be generated by
   (a) performing a binary test with the outcomes 0 and 1 generated with probabilities $p_0 = p$ and $p_1 = 1 - p$
   (b) conditionally on the occurrence of the outcome i, performing a test $\boldsymbol{\mathcal{T}}_i$ containing the transformation $\mathcal{T}_i$
   (c) coarse-graining over the appropriate outcomes of the conditional test.

The above observations show that convexity needs not be assumed from the start, but can be derived from non-determinism and Causality (in the positive formulation $\mathbf{C}'$), under the standard assumption that the set of probabilities generated by tests in the theory is closed.

4.1.4 Rescaling

In addition to convexity, conditional tests guarantee that every state is proportional to a *normalized* state. Specifically, given a state ρ of a generic system A, one can define the normalized state $\widetilde{\rho} := \rho/(e_{\rm A}|\rho)$. An approximate way to prepare the state $\widetilde{\rho}$ is to

1. pick a binary test $\{\rho_0, \rho_1\}$ such that $\rho_1 = \rho$
2. perform it N times, generating a string of outcomes $(x_1, x_2, \ldots, x_N)$
3. perform a conditional test that discards $N-1$ systems, keeping only a system i such that $x_i = 1$, if such a system exists, or otherwise keeping only the first system
4. coarse-grain over all outcomes, thus obtaining the deterministic state

$$\rho_N := (1 - p_N)\,\widetilde{\rho} + p_N\,\widetilde{\rho}_0 \qquad\qquad p_N = (e_{\rm A}|\rho_0)^N .$$

Clearly, the state ρ_N converges to $\widetilde{\rho}$ when N goes to infinity. Hence, the standard assumption that the set of states is closed guarantees that $\widetilde{\rho}$ is a state allowed by the theory.

4.2 Purity of Composition

Purity of Composition is a very primitive rule about how information propagates in time. Mathematically, the axiom consists of the implication

$$\begin{aligned}\mathcal{A} \in \mathsf{PurTransf}(\mathrm{A}\to\mathrm{B})\,,\ \mathcal{B} \in \mathsf{PurTransf}(\mathrm{B}\to\mathrm{C}) \\ \Longrightarrow \qquad \mathcal{B}\circ\mathcal{A} \in \mathsf{PurTransf}(\mathrm{A}\to\mathrm{C}),\end{aligned}$$

required to be valid for all systems $\mathrm{A}, \mathrm{B}, \mathrm{C} \in \mathsf{Sys}$ and for all pure transformations $\mathcal{A}$ and $\mathcal{B}$.

Think of a world where this were not the case. In that world, an agent Alice could perform a test $\boldsymbol{\mathcal{A}} \in \mathsf{Tests}(\mathrm{A}\to\mathrm{B}\,,\mathrm{X})$ with such degree of control that, upon knowing the outcome, she could not possibly know better what happened to her system. Immediately after, another agent Bob could perform another test $\boldsymbol{\mathcal{B}} \in \mathsf{Tests}(\mathrm{B}\to\mathrm{C}\,,\mathrm{Y})$ also having maximal knowledge of the system's conditional evolution. Still, some of the resulting transformations $\mathcal{B}_y\mathcal{A}_x$ may not be pure. This means that $\mathcal{B}_y\mathcal{A}_x$ can be simulated by a third party—Charlie—by performing one test $\{\mathcal{C}_z\}_{z\in\mathrm{Z}}$ and joining together the outcomes in a suitable subset $\mathrm{S}_{xy} \subset \mathrm{Z}$

$$\xrightarrow{\ \mathrm{A}\ }\boxed{\mathcal{A}_x}\xrightarrow{\ \mathrm{A}\ }\boxed{\mathcal{B}_y}\xrightarrow{\ \mathrm{A}\ } = \sum_{z\in\mathrm{S}_{xy}} \xrightarrow{\ \mathrm{A}\ }\boxed{\mathcal{C}_z}\xrightarrow{\ \mathrm{A}\ } . \qquad (30)$$

Although this scenario is logically conceivable, it rises a puzzling question: What is the extra information about? Which physical parameters correspond to the outcome z? Surely the information is not about what happened in the first step, because Alice already had maximal knowledge about this. Nor it is about what happened in the second step, because Bob has maximal information about that. The outcome z has to specify a feature of how the two time steps interacted together—in a sense, a kind of information that is *non-local in time*. Quantum theory is non-local, but not in such an extreme way! Indeed, pure transformations in quantum theory are described by completely positive maps with a single Kraus operator, i.e. of the form $\mathcal{A}_x(\cdot) = A_x \cdot A_x^\dagger$ and $\mathcal{B}_y(\cdot) = B_y \cdot B_y^\dagger$, and clearly the composition of two pure transformations is still pure: $\mathcal{B}_y\mathcal{A}_x(\cdot) = (B_y A_x) \cdot (B_y A_x)^\dagger$. Purity of Composition guarantees this property at the level of first principles.

4.3 Local Tomography

Local Tomography implies that even if a state is entangled, the information it contains can be extracted by local measurements. This fact reconciles the holism of entanglement and the reductionist idea that the full information about a composite system can be obtained by studying its parts [25].

Mathematically, Local Tomography states that product effects form a separating set for the vector space $\mathsf{St}_{\mathbb{R}}(\mathrm{A} \otimes \mathrm{B})$. Equivalently,[9] they form a spanning set for the dual space $\mathsf{St}_{\mathbb{R}}(\mathrm{A} \otimes \mathrm{B})^* \equiv \mathsf{Eff}_{\mathbb{R}}(\mathrm{A} \otimes \mathrm{B})$. Hence, we must have the conditions

$$\mathsf{Eff}_{\mathbb{R}}(\mathrm{A} \otimes \mathrm{B}) = \mathsf{Eff}_{\mathbb{R}}(\mathrm{A}) \otimes \mathsf{Eff}_{\mathbb{R}}(\mathrm{B}) \quad \text{and} \quad \mathsf{St}_{\mathbb{R}}(\mathrm{A} \otimes \mathrm{B}) = \mathsf{St}_{\mathbb{R}}(\mathrm{A}) \otimes \mathsf{St}_{\mathbb{R}}(\mathrm{B}), \tag{31}$$

where $\otimes$ in the r.h.s. denote the tensor product of finite dimensional vector spaces. Equation (31) implies that the dimensions of the vector spaces in question satisfy the product relation [23]

$$D_{\mathrm{A}\otimes\mathrm{B}} = D_{\mathrm{A}}\, D_{\mathrm{B}}. \tag{32}$$

Moreover, a generic state $\rho \in \mathsf{St}(\mathrm{A} \otimes \mathrm{B})$ and a generic effect $m \in \mathsf{Eff}(\mathrm{A} \otimes \mathrm{B})$ can be expanded as

$$\rho = \sum_{i,j} \rho_{ij}\, \left(v_i \otimes w_j\right) \quad \text{and} \quad m = \sum_{i,j} m_{ij}\, \left(v_i^* \otimes w_j^*\right), \tag{33}$$

where $[\rho_{ij}]$ and $[m_{ij}]$ are real matrices, $\{v_i\}_{i=1}^{D_{\mathrm{A}}}$ and $\{w_j\}_{j=1}^{D_{\mathrm{B}}}$ are bases for the vector spaces $\mathsf{St}_{\mathbb{R}}(\mathrm{A})$ and $\mathsf{St}_{\mathbb{R}}(\mathrm{B})$, respectively, and $\{v_i^*\}_{i=1}^{D_{\mathrm{A}}}$ and $\{w_j^*\}_{j=1}^{D_{\mathrm{B}}}$ are the dual bases,

[9]Recall that we are assuming that the state spaces are finite-dimensional.

defined by the relations $(v_i^*|v_k) = \delta_{ik}$ and $(w_j^*|w_l) = \delta_{jl}$, respectively. As a result, the probability of the effect m on the state ρ can be expressed as

$$(m|\rho) = \mathrm{Tr}[m\,\rho], \tag{34}$$

having committed a little abuse of notation in using the letter m (respectively, ρ) both for the effect (respectively, state) and for the corresponding matrix $[m_{ij}]$ (respectively, $[\rho_{ij}]$).

Finally, the decomposition in Eq. (33) implies the following

Theorem 1 *In a theory satisfying Local Tomography, the correspondence between a transformation $\mathcal{E} \in \mathsf{Transf}(\mathrm{A} \to \mathrm{B})$ and the linear map $\widehat{\mathcal{E}} : \mathsf{St}_{\mathbb{R}}(\mathrm{A}) \to \mathsf{St}_{\mathbb{R}}(\mathrm{B})$ is invertible.*

In other words, Local Tomography guarantees that physical transformations can be characterized in the simplest possible way: by preparing a set of input states and performing a set of measurements on the output.

A remarkable example of a theory that does not satisfy Local Tomography is quantum theory on real Hilbert spaces [52], RQT for short. In this theory, states and effects are real symmetric matrices, and transformations are represented by completely positive maps mapping symmetric matrices into symmetric matrices. The failure of the relation $D_{\mathrm{A}\otimes\mathrm{B}} = D_\mathrm{A}\, D_\mathrm{B}$ was first noted by Araki [53]. More explicitly, Wootters [54] noted that two different bipartite states can be locally indistinguishable, as in the following extreme example:

$$\rho = \frac{1}{2}|\Phi_+\rangle\langle\Phi_+| + \frac{1}{2}|\Psi_-\rangle\langle\Psi_-| \qquad\qquad \rho' = \frac{1}{2}|\Phi_-\rangle\langle\Phi_-| + \frac{1}{2}|\Psi_+\rangle\langle\Psi_+| \tag{35}$$

with $|\Phi_\pm\rangle := (|0\rangle|0\rangle \pm |1\rangle|1\rangle)/\sqrt{2}$ and $|\Psi_\pm\rangle = (|0\rangle|1\rangle \pm |1\rangle|0\rangle)/\sqrt{2}$. Here the states ρ and ρ' have orthogonal support and therefore are perfectly distinguishable. However, it is easy to check that one has

$$\rho - \rho' = \frac{1}{2}\begin{pmatrix} 0 & -1 \\ 1 & 0 \end{pmatrix} \otimes \begin{pmatrix} 0 & -1 \\ 1 & 0 \end{pmatrix},$$

and, therefore, $\mathrm{Tr}[(\rho - \rho')(P_\mathrm{A} \otimes P_\mathrm{B})] = 0$ for every pair of real symmetric matrices P_A and P_B. In other words, ρ and ρ' give exactly the same statistics for all possible local measurements.

RQT has another, closely related quirk: two *different* transformations of system A can act in the same way on all states of A. For example, consider the qubit channels $\mathcal{C}$ and $\mathcal{C}'$, whose action on a generic 2×2 matrix is defined by

$$\mathcal{C}(M) := \frac{1}{2}M + \frac{1}{2}YMY \qquad \text{and} \qquad \mathcal{C}'(M) := \frac{1}{2}ZMZ + \frac{1}{2}XMX,$$

X, Y, and Z being the Pauli matrices. When acting on symmetric matrices, the two channels give exactly the same output: one has $\mathcal{C}(\tau) = \mathcal{C}'(\tau) = I/2$ for every symmetric matrix τ. On the other hand, one has

$$(\mathcal{C} \otimes \mathcal{I})(|\Phi_+\rangle\langle\Phi_+) = \rho \qquad (\mathcal{C}' \otimes \mathcal{I})(|\Phi_+\rangle\langle\Phi_+) = \rho',$$

where ρ and ρ' are the two perfectly distinguishable states defined in Eq. (35). This means that, in fact, the two transformations $\mathcal{C}$ and $\mathcal{C}'$ are perfectly distinguishable with the help of a reference system. For a more extensive discussion of tomography in RQT we refer the reader to subsection V.A of Ref. [34] and to the work of Hardy and Wootters [55].

4.4 Perfect State Discrimination

Perfect State Discrimination is an optimistic statement about the possibility to encode bits without error. It guarantees that every state that *could* be part of a set of perfectly distinguishable states *is* indeed perfectly distinguishable from some other state.

By virtue of Perfect State Discrimination, every normalized non-internal state ρ_0 can be perfectly distinguished from some state ρ_1. As a result, the two states ρ_0 and ρ_1 can be used to encode the value of a bit without errors. It is easy to see that Quantum theory satisfies the axiom. Indeed, a density matrix is internal if and only if it has full rank. Hence, a non-internal density matrix ρ_0 must have a kernel, so that every state ρ_1 with support in the kernel of ρ_0 will be perfectly distinguishable from ρ_0.

4.5 Ideal Compression

Ideal Compression expresses the idea that information is *fungible*, i.e. independent of the physical support in which it is encoded. The axiom implies non-trivial statements about the state spaces arising in the theory. For example, suppose that the theory contains a system whose space of deterministic states is a square. Then, the theory should contain also a system whose space of deterministic states is a segment—in other words, the theory should contain a classical bit. Indeed, only in this way one could encode a side of the square in a lossless and maximally efficient way. More generally, Ideal Compression imposes that the every face of the convex set of deterministic states be in one-to-one correspondence with the set of deterministic states of some smaller physical system.

Ideal Compression is clearly satisfied by quantum theory. Indeed, every density matrix of rank r can be compressed ideally to a density matrix of an r-dimensional quantum system. For example, the two-qubit density matrix

$$\rho = \begin{pmatrix} \rho_{00,00} & 0 & 0 & \rho_{00,11} \\ 0 & 0 & 0 & 0 \\ 0 & 0 & 0 & 0 \\ \rho_{11,00} & 0 & 0 & \rho_{11,11} \end{pmatrix} \tag{36}$$

can be compressed ideally to the one-qubit density matrix

$$\mathcal{E}(\rho) = \begin{pmatrix} \rho_{00,00} & \rho_{00,11} \\ \rho_{11,00} & \rho_{11,11} \end{pmatrix} \tag{37}$$

with encoding and decoding channels given by

$$\mathcal{E}(\cdot) := V^{\dagger}(\cdot)\, V + \mathrm{Tr}[(I - V\, V^{\dagger})\,(\cdot)\,]\, |0\rangle\langle 0| \qquad V := |0\rangle|0\rangle\langle 0| + |1\rangle|1\rangle\langle 1|$$
$$\mathcal{D}(\cdot) := V\,(\cdot)\, V^{\dagger}.$$

Note that Ideal Compression refers to a *single-shot, zero error scenario*, i.e. a scenario where the source is used only once and no decoding errors are allowed. Such a scenario is different from the asymptotic scenario considered in Shannon's [56] and Schumacher's [57] compression, wherein small decoding errors are allowed, under the condition that they vanish in the asymptotic limit of infinitely many uses of the same source.

4.6 Purification

While our first five axioms expressed standard requirements for information-processing, Purification brings in a radically new idea: at least in principle, every state can be prepared by an agent who has maximal control over all the systems involved in the preparation process. In short, Purification allows us to harness randomness by controlling the environment. The idea does not apply only to preparations, but also to arbitrary deterministic transformations: combining Purification with Causality and Local Tomography, we can prove the following

Theorem 2 ([34]) *For every deterministic transformation* $\mathcal{T} \in \mathsf{DetTransf}(\mathrm{A} \to \mathrm{A}')$, *there exist two systems* E *and* E′, *a pure state* $\eta \in \mathsf{PurSt}(\mathrm{E})$, *and a reversible transformation* $\mathcal{U} \in \mathsf{RevTransf}(\mathrm{A} \otimes \mathrm{E} \to \mathrm{A}' \otimes \mathrm{E}')$ *such that*

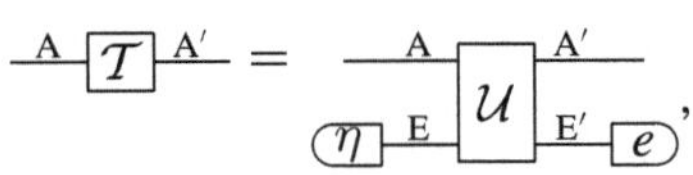

(38)

where e is the unique deterministic effect of system E′.

In other words, Purification implies that every irreversible process can be simulated through reversible interactions between the system and its environment, with the environment initialized in a pure state. This result is a necessary condition for the formulation of physical theories in which elementary processes are reversible at the fundamental level.

Purification is known to be satisfied by quantum mechanics. For example, consider a single-qubit mixed state, diagonalized as

$$\rho = p_0|0\rangle\langle 0| + p_1|1\rangle\langle 1|, \tag{39}$$

for some suitable orthonormal basis $\{|0\rangle, |1\rangle\}$. A purification of the state ρ can be obtained by adding a second qubit and by preparing the two qubits in the pure state

$$|\Psi\rangle := \sqrt{p_0}\,|0\rangle|0\rangle + \sqrt{p_1}\,|1\rangle|1\rangle. \tag{40}$$

Indeed, it is immediate to see that ρ is the marginal of the density matrix $|\Psi\rangle\langle\Psi|$ on the first qubit. In addition, any other purification $|\Psi'\rangle$—using a single qubit as the purifying system—must be of the form $|\Psi\rangle' = (I \otimes U)\,|\Psi\rangle$ for some unitary matrix U.

In the quantum information community, taking purifications is a standard approach to quantum communication, cryptography, and quantum error correction. The approach is familiarly known with the nickname of "going to the Church of the larger Hilbert space".[10] Purification is known among mathematicians as the *Gelfand-Naimark-Segal construction* [59, 60].

Two important remarks are in order:

1. *Purification, entanglement, and quantum information*. Purification is intimately linked with the phenomenon of *entanglement* [2], namely the existence of pure bipartite states Ψ_{AB} that are not of the product form $\psi_A \otimes \psi_B$. In the OPT framework, the link is made precise by the following

   Proposition 4 *Let Θ be a theory satisfying Causality, Local Tomography, and Purification. Then, there are only two alternatives: either Θ is deterministic, or Θ exhibits entanglement.*

   Under our standing assumption that the theory is non-deterministic, entanglement follows from Purification as a necessary consequence.

   Entanglement is a very peculiar feature—far from what we experience in our everyday life. How can we claim that we know A *and* B if we do not know A alone? This puzzling feature had been noted already in the early days of quantum theory, when Schrödinger famously wrote: "Another way of expressing the peculiar situation is: *the best possible knowledge of a whole does not necessarily include the best possible knowledge of all its parts*" [2]. And, in the same paper: "I would not call that *one* but rather *the* characteristic trait of quantum mechanics, the

[10]The expression is due to John Smolin, see e.g. the lecture notes [58].

one that enforces its entire departure from classical lines of thought". In a sense, our reconstruction can be considered as a mathematical proof of Schrödinger's intuition[11]: on the background of five standard axioms satisfied by both classical theory and quantum theory, Purification is the ingredient that allows to reconstruct the Hilbert space framework and the distinctive information-theoretic features of quantum theory. Combined with Causality and Local Tomography, Purification already reproduces an impressive list of quantum-like features, like no-cloning, no-programming, information-disturbance tradeoff, no bit commitment, conclusive teleportation and entanglement swapping, the reversible dilation of channels, the state-transformation isomorphism, the structure of error correction, and the structure of no-signalling channels [34].

2. *Purification and the Many Wolrd Interpretation.* Pondering about the meaning of Purification, one may tempted to conclude that it favours the Many Worlds Interpretation (MWI) of quantum mechanics [61]. In fact, Purification is feature of quantum theory, and, as such, it does not favour the MWI more than quantum theory itself does. Whether or not quantum theory provides any evidence for many worlds is a debatable point, but the validity of Purification is independent of such interpretative issue. Furthermore, we stress that we did not phrase Purification as an ontological statement about "how processes occur in nature", but rather an operational statement about the agent's ability to simulate physical processes with maximal control. Purification is *compatible* with the idea that processes are reversible at the fundamental level, and its validity is a *necessary condition* for building up a physical description of nature in terms of pure states and reversible processes. Still, here we do not make any commitment as to how processes are realized in nature, because this would unnecessarily limit the range of application of our results.

5 The Reconstruction of Quantum Theory

Here we provide a summary of the reconstruction of Refs. [26, 34], highlighting the key theorems and providing a guide to the original papers. The scope of the reconstruction is not just to derive the Hilbert space framework, but also to rebuild the key quantum features directly from first principles. Accordingly, we try to derive as much as possible of quantum theory directly from the axioms, leaving Hilbert spaces to the very end. We organize our results in six subsections:

1. Elementary facts.
2. Correlation structures.
3. Distinguishability structures.

[11] It is worth stressing that Schrödinger's paper was not just about the *existence* of entangled states, but also about how entanglement interacted with the reversible dynamics and with the process of measurement (cf. the notion of *steering*, which made its first appearance in the very same paper).

4. Interaction between correlation and distinguishability structures.
5. Qubit features.
6. The density matrix.

5.1 Elementary Facts

5.1.1 From Local Tomography

Local Tomography implies a few useful facts:

1. If $\alpha \in \mathsf{St}(\mathrm{A})$ and $\beta \in \mathsf{St}(\mathrm{B})$ are pure, then also $\alpha \otimes \beta$ is pure.
2. Let ρ_{AB} be a state of the composite system $\mathrm{A} \otimes \mathrm{B}$ and, assuming Causality, let ρ_{A} be its marginal on system A. If ρ_{A} is pure, then ρ_{AB} is a product state.
3. If $\rho_{\mathrm{A}} \in \mathsf{St}(\mathrm{A})$ and $\rho_{\mathrm{B}} \in \mathsf{St}(\mathrm{B})$ are internal states, then also $\rho_{\mathrm{A}} \otimes \rho_{\mathrm{B}}$ is an internal state.
4. Suppose that every system A has a unique *invariant state* χ_{A}, i.e. a unique state satisfying the condition $\mathcal{U}\chi_{\mathrm{A}} = \chi_{\mathrm{A}}$ for every reversible transformation $\mathcal{U}$. Then, $\chi_{\mathrm{A}\otimes\mathrm{B}} = \chi_{\mathrm{A}} \otimes \chi_{\mathrm{B}}$.

5.1.2 From Purification

Purification has a few immediate consequences. First, all pure states of a given system are connected to one another through reversible transformations:

Proposition 5 *For every system* $\mathrm{A} \in \mathsf{Sys}$ *and every pair of pure states* $\alpha, \alpha' \in \mathsf{PurSt}(\mathrm{A})$ *there exists a reversible transformation* $\mathcal{U}$ *such that* $\alpha' = \mathcal{U}\,\alpha$.

To prove this fact, it is enough to pick a system B and pure state $\beta \in \mathsf{PurSt}(\mathrm{B})$, consider the states $\Psi = \alpha \otimes \beta$ and $\Psi' = \alpha' \otimes \beta$ as purifications of β, and invoke the essential uniqueness of purification [Eq. (28)]. Mathematically, the above proposition expresses the fact that the action of the reversible transformations is *transitive* on the set of pure states—a requirement that played an important role in many recent reconstructions, see e.g. [23, 28, 29]. A byproduct of transitivity is

Proposition 6 *Every system* $\mathrm{A} \in \mathsf{Sys}$ *has a unique invariant state* χ_{A}.

Finally, combining Ideal Compression and Purification it is easy to see that every state has a *minimal purification*, in the following sense

Definition 17 Let $\Psi \in \mathsf{PurSt}(\mathrm{A} \otimes \mathrm{B})$ be a pure state with marginals ρ_{A} and ρ_{B} on systems A and B, respectively. We say that Ψ is a *minimal purification* of ρ_{A} iff ρ_{B} is internal.

To construct a minimal purification, it is enough to take an arbitrary purification and to compress the state of the purifying system.

5.2 *Correlation Structures*

5.2.1 Pure Steering

One of the most important consequences of our axioms is that pure bipartite states enable *steering*, namely the ability to remotely generate every desired ensemble decomposition of a marginal state [2, 62]:

Proposition 7 (Pure Steering) *Let Ψ be a pure state of the composite system $\mathrm{A}\otimes\mathrm{B}$, let ρ be the marginal of Ψ on system* A, *and let $\boldsymbol{\rho} = \{\rho_x\}_{x\in\mathrm{X}}$ be an ensemble decomposition of ρ. Then there exists a measurement $\boldsymbol{b} = \{b_x\}_{x\in\mathrm{X}}$ such that*

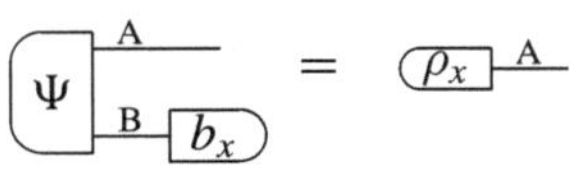

$$\forall x \in \mathrm{X}. \tag{41}$$

Pure steering is the essential ingredient for a number of major results. The first result is the existence of *pure, tomographically faithful states*. A state $\rho \in \mathsf{St}(\mathrm{A}\otimes\mathrm{B})$ is called *tomographically faithful* for system A iff the implication

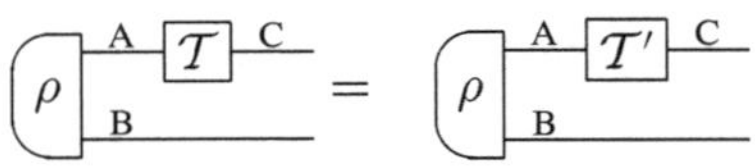

$$\Longrightarrow \quad \mathcal{T} = \mathcal{T}', \tag{42}$$

holds for every system C and every pair of transformations $\mathcal{T}$ and $\mathcal{T}'$ of type $\mathrm{A} \to \mathrm{C}$. Thanks to Pure Steering and Local Tomography, we are able to construct tomographically faithful pure states:

Proposition 8 *Let ρ_{A} be an internal state of system* A *and let $\Psi \in \mathsf{PurSt}(\mathrm{A}\otimes\mathrm{B})$ be a purification of ρ_{A}. Then, Ψ is tomographically faithful for system* A.

The result can be improved by choosing a *minimal* purification: in this way, the pure state Ψ is faithful on both systems A and B. We call such a state *doubly faithful*.

5.2.2 Conjugate Systems

A canonical choice of doubly faithful state is obtained by picking a minimal purification of the invariant state χ_{A}. We denote such purification by $\Phi \in \mathsf{PurSt}(\mathrm{A}\otimes\overline{\mathrm{A}})$ and call system $\overline{\mathrm{A}}$ the *conjugate of system* A. The name is motivated by the following facts:

1. system $\overline{\mathrm{A}}$ is uniquely defined, up to operational equivalence
2. the marginal of Φ on system $\overline{\mathrm{A}}$ is the invariant state $\chi_{\overline{\mathrm{A}}}$ (cf. Corollary 46 of [34]), meaning that we have $\overline{\overline{\mathrm{A}}} = \mathrm{A}$, up to operational equivalence.

Summarizing, the state Φ satisfies the relations

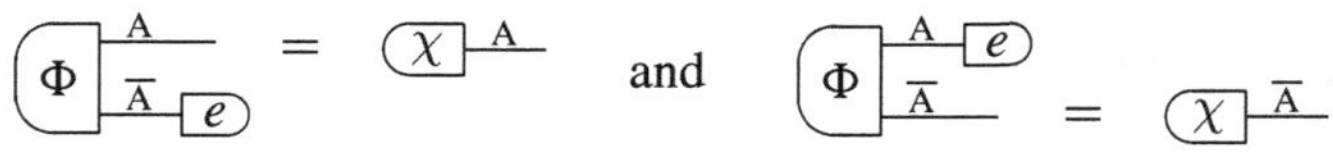

(43)

By analogy with quantum theory, we call Φ a *Bell state*.

5.2.3 The State-Transformation Isomorphism

For a given transformation $\mathcal{T}$, we define the (generally unnormalized) state

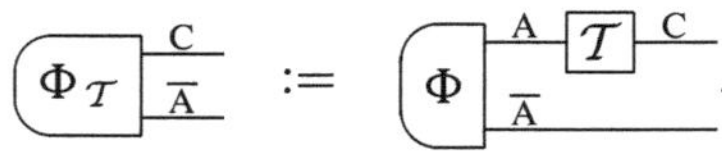

(44)

and call the correspondence $\mathcal{T} \mapsto \Phi_{\mathcal{T}}$ the *state-transformation isomorphism*. Since the Bell state Φ is doubly faithful, the correspondence is one-to-one. In quantum theory, the state-transformation isomorphism coincides with the Choi isomorphism [63]. By analogy, we call the state $\Phi_{\mathcal{T}}$ the *Choi state*.

A powerful byproduct of the state-transformation isomorphism is that the normalized states completely identify the theory:

Theorem 3 *Let Θ and Θ' be two theories with the same set of systems. If the sets of normalized states of Θ and Θ' coincide for all systems, then the two theories coincide.*

Thanks to this result, deriving the density matrix representation of normalized states is sufficient to derive the whole of quantum theory.

5.2.4 Conclusive Entanglement Swapping

An important consequence of Pure Steering is the possibility of entanglement swapping, namely the possibility to generate entanglement remotely by performing a joint measurement. Consider, as a prototype of entangled state, the Bell state Φ. Then, it is possible to show that there exists a pure effect $E \in \mathsf{PurEff}(\overline{\mathrm{A}} \otimes \mathrm{A})$ and a non-zero probability $p_{\mathrm{A}} > 0$ such that

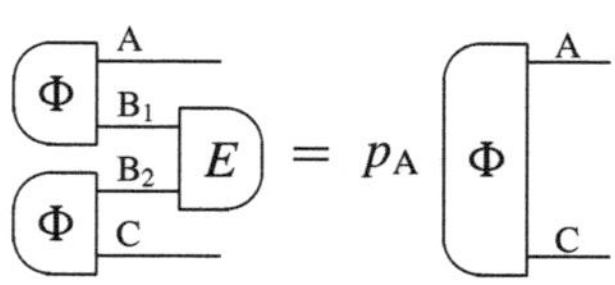

$$\mathrm{A} \equiv \mathrm{B}_2\,, \quad \mathrm{B}_1 \equiv \mathrm{C} \equiv \overline{\mathrm{A}}. \tag{45}$$

This diagram represents an instance of *conclusive entanglement swapping*: conditionally on the occurrence of the effect E, the two systems A and C are prepared in the Bell state, consuming the initial entanglement present in the composite systems $A \otimes B_1$ and $B_2 \otimes C$.

The possibility of entanglement swapping follows easily from Pure Steering: Since the states χ_A and $\chi_{\overline{A}}$ are internal, Local Tomography implies that their product $\chi_A \otimes \chi_{\overline{A}}$ is internal. Hence, there must exist a non-zero probability $p_A > 0$ such that

$$\chi_A \otimes \chi_{\overline{A}} = p_A \, \Phi + (1 - p_A) \, \tau, \tag{46}$$

for some state τ. Applying Pure Steering (Proposition 7) to the pure state $\Phi \otimes \Phi$ and to the ensemble $\{p_A \, \Phi \, , (1 - p_A) \, \tau\}$ one can find a binary measurement $\{E, e_{B_1} \otimes e_{B_2} - E\}$ such that the entanglement swapping condition (45) holds. Using the fact that the state $\Phi \otimes \Phi$ is doubly faithful, it is easy to see that the effect E must be pure.

5.2.5 Conclusive Teleportation

By the state-transformation isomorphism, conclusive entanglement swapping is equivalent to *conclusive teleportation* [17], expressed by the diagram

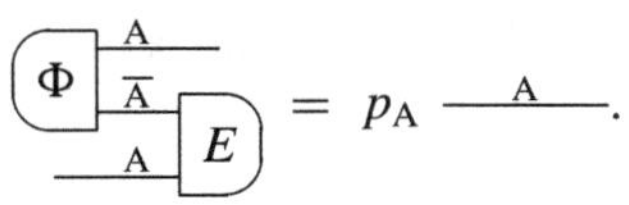

(47)

Indeed, the entanglement swapping diagram (45) is equivalent to the condition $\Phi_{\mathcal{T}} = \Phi_{\mathcal{T}'}$, with

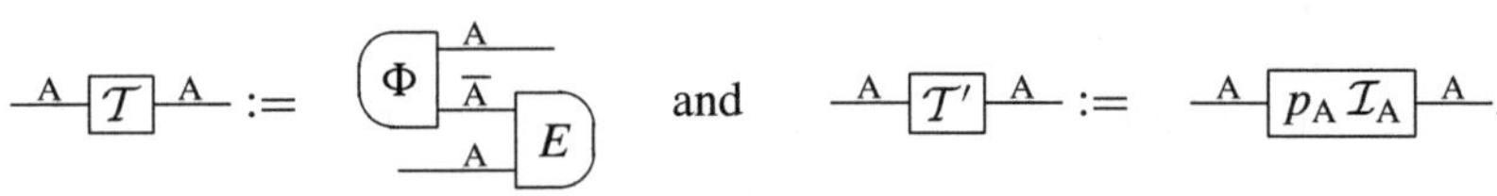

(48)

By the state-transformation isomorphism, $\Phi_{\mathcal{T}} = \Phi_{\mathcal{T}'}$ implies $\mathcal{T} = \mathcal{T}'$, which is nothing but the teleportation condition.

5.2.6 The Teleportation Upper Bound

Combined with Local Tomography, the teleportation diagram allows us to upper bound the dimension of the state space. The idea is to write the teleportation diagram in matrix elements, by expanding Φ and E as

$$\Phi = \sum_{ik} \Phi_{ik} \, (v_i \otimes w_k) \qquad \text{and} \qquad E = \sum_{jl} E_{jl} \left(w_j^* \otimes v_l^* \right), \tag{49}$$

with suitable bases $\{v_i\}_{i=1}^{D_{\mathrm{A}}}$ and $\{w_j\}_{j=1}^{D_{\overline{\mathrm{A}}}}$. In this representation, Eq. (47) becomes

$$[\,\Phi\, E\,]_{il} = p_{\mathrm{A}}\,\delta_{il}, \tag{50}$$

and, taking the trace,

$$\mathrm{Tr}[\Phi E] = p_{\mathrm{A}}\, D_{\mathrm{A}}. \tag{51}$$

On the other hand, we have

$$\mathrm{Tr}[\Phi E] = (E|\, \mathcal{S}_{\mathrm{A},\overline{\mathrm{A}}}\, |\Phi) \leq 1, \tag{52}$$

which combined with Eq. (51) leads to bound

$$D_{\mathrm{A}} \leq \frac{1}{p_{\mathrm{A}}}. \tag{53}$$

Clearly, in order to have the best bound we need to find the *maximum* probability of teleportation. To discover what the maximum is, we need to move our attention to the distinguishability structures implied by our axioms.

5.3 Distinguishability Structures

5.3.1 No Disturbance Without Information

Our first move is to derive a simple result about the structure of measurements: a measurement that extracts no information from a face of the state space can be implemented without disturbing that face. By *face of the state space* we mean a face of the convex set of deterministic states.[12] We say that the measurement $\boldsymbol{m} \in \mathsf{Tests}(\mathrm{A} \to \mathrm{I}, \mathrm{X})$ *does not extract information* from the face F iff there exists a set of probabilities $\{p_x\}_{x\in\mathrm{X}}$ such that

$$(m_x|\tau) = p_x \qquad \forall x \in \mathrm{X}, \quad \forall \tau \in F.$$

Also, we say that a test $\boldsymbol{\mathcal{T}} \in \mathsf{Tests}(\mathrm{A} \to \mathrm{A}, \mathrm{X})$ *does not disturb the face* F iff $\sum_{x\in\mathrm{X}} \mathcal{T}_x|\tau) = |\tau)$ for every state $\tau \in F$.

[12] We recall that a *face* of a convex set C is a convex subset $F \subseteq C$ satisfying the condition that, for every $x \in F$, if x is a non-trivial convex combination of x_1 and x_2 with $x_1, x_2 \in C$, then x_1 and x_2 belong to F.

With this terminology, our result is the following:

Proposition 9 *If a measurement* $\mathbf{m}$ *does not extract information from the face* F, *then there exists a test* $\mathcal{T}$ *that realizes the measurement—namely* $(e_{\mathrm{A}}|\mathcal{T}_x = (m_x|$, $\forall x \in \mathrm{X}$*—and does not disturb* F.

This result has two important consequences. First, it allows us to establish whether or not a set of perfectly distinguishable set can be extended:

Proposition 10 *Let* $\mathrm{S} = \{\rho_x \mid x \in \mathrm{X}\}$ *be a set of perfectly distinguishable states and let* ω_{S} *be its barycenter, defined as*

$$\omega_{\mathrm{S}} := \frac{1}{|\mathrm{X}|} \sum_{x \in \mathrm{X}} \rho_x .$$

Then, the following are equivalent:

1. *the set* S *is* maximal, *i.e. no other set* $\mathrm{S}' \supset \mathrm{S}$ *can consist of perfectly distinguishable states*
2. *the barycenter of* S *is internal.*

Another important consequence is that only the *pure* maximal sets can have maximum cardinality:

Proposition 11 *Let* S *be a maximal set of perfectly distinguishable states of system* A. *If one of the states in* S *is not pure, then there exists another maximal set* $\mathrm{S}' \subset \mathsf{St}(\mathrm{A})$, *consisting only of pure states and having strictly larger cardinality* $|\mathrm{S}'| > |\mathrm{S}|$.

Combining the above points we have that every pure state belongs to some maximal set of perfectly distinguishable pure states. For short, we call such sets *pure maximal sets*.

5.3.2 Duality Between Pure States and Pure Effects

For a pure maximal set S, we observe that the measurement that distinguishes the states in S must consist of *pure* effects. Hence, for every pure state $\alpha \in \mathsf{PurSt}(\mathrm{A})$ there exists an pure effect a such that $(a|\alpha) = 1$. Expanding on this observation, we establish a one-to-one correspondence between pure normalized states and pure normalized effects,[13] denoted by $\mathsf{PurSt}_1(\mathrm{A})$ and $\mathsf{PurEff}_1(\mathrm{A})$, respectively.

Theorem 4 *For every system* $\mathrm{A} \in \mathsf{Sys}$, *there exists a one-to-one map* $\dagger : \mathsf{PurSt}_1(\mathrm{A}) \to \mathsf{PurEff}_1(\mathrm{A})$, *sending pure normalized states to pure normalized effects and satisfying the condition*

$$(\alpha^\dagger|\alpha) = 1 \qquad \forall \alpha \in \mathsf{PurSt}_1(\mathrm{A}).$$

[13] We call an effect of system A *normalized* iff there exists an effect a state ρ such that $(a|\rho) = 1$.

The proof is rather elaborate. The two main steps are

1. proving that every pure normalized effect *a identifies* a pure state α, meaning that $(a|\rho) = 1$ if and only if $\rho = \alpha$.
2. proving that, if two pure effects identify the same state, then they must coincide.

The second step uses Pure Steering in an essential way, suggesting that the distinguishability features of quantum theory are deeply connected with its correlation features.

5.3.3 The Informational Dimension

An easy consequence of the state-effect duality is that every two pure normalized effects are connected by a reversible transformation, just like the pure states. In turn, this leads to a useful result.

Proposition 12 *For a given system* A $\in$ Sys, *all pure maximal sets have the same cardinality.*

The proof idea is simple: let $\boldsymbol{a} = \{a_x\}_{x\in \mathrm{X}}$ be the measurement that distinguishes among the states in a pure maximal set $\mathrm{S} = \{\alpha_x \mid x \in \mathrm{X}\}$. As we already observed, all the effects in $\boldsymbol{a}$ must be pure. Since every two pure normalized effects are connected by a reversible transformation, we must have $a_x = a \circ \mathcal{U}_x \; \forall x \in \mathrm{X}$, where a is fixed (but otherwise arbitrary) effect in $\mathsf{PurEff}_1(\mathrm{A})$ and $\mathcal{U}_x$ is a reversible transformation. Applying the effects to the invariant state χ we then obtain

$$(a_x|\chi) = (a|\chi) \qquad \forall x \in \mathrm{X},$$

and summing over x we get the equality $1 = |\mathrm{X}|\,(a|\chi)$. Hence, the cardinality of the maximal set S is $|\mathrm{S}| \equiv |\mathrm{X}| = 1/(a|\chi)$. Since S is a generic pure maximal set, we proved the desired result.

In the following, the cardinality of the pure maximal sets in A will be denoted by d_A. We call it the *informational dimension*, because it is the number of distinct classical messages that can be encoded in system A and decoded without error. In Quantum Theory, d_A is the dimension of the Hilbert space associated to system A.

For composite systems, the informational dimension has the product form:

Proposition 13 *For every pair of systems* A *and* B *one has* $d_{\mathrm{A}\otimes\mathrm{B}} = d_\mathrm{A}\, d_\mathrm{B}$.

The reason is simply that the product of two pure maximal sets for systems A and B is a pure maximal set for $\mathrm{A}\otimes\mathrm{B}$: it is pure, because the product of two pure states is pure (by Local Tomography) and it is maximal because the product of two internal states is internal (again, by Local Tomography)—hence, maximality follows by Proposition 10.

5.3.4 The Spectral Theorem

An important consequence of the state-effect duality is the ability to decompose every state as a mixture of perfectly distinguishable pure states. The crucial step is to prove such a decomposition for the invariant state:

Lemma 1 *For every pure maximal set* $\{\alpha_x\}_{x=1}^{d_\mathrm{A}} \subset \mathsf{PurSt}(\mathrm{A})$ *one has* $\chi = \frac{1}{d_\mathrm{A}} \sum_{x=1}^{d_\mathrm{A}} \alpha_x$.

This result is extremely important, because it helps us to cope with the existence of different maximal sets of pure states. To begin with, it allows us to prove the analogue of the spectral theorem:

Theorem 5 (Spectral Decomposition) *For every vector* $v \in \mathsf{St}_\mathbb{R}(\mathrm{A})$ *there exists a pure maximal set* $\{\alpha_x\}_{x=1}^{d_\mathrm{A}} \subset \mathsf{PurSt}(\mathrm{A})$ *and a set of real coefficients* $\{c_x\}_{x=1}^{d_\mathrm{A}}$ *such that*

$$v = \sum_{x=1}^{d_\mathrm{A}} c_x \, \alpha_x . \tag{54}$$

Similarly, for every vector $w \in \mathsf{Eff}_\mathbb{R}(\mathrm{A})$ *there exists a pure discriminating measurement* $\{a_x\}_{x=1}^{d_\mathrm{A}}$ *and a set of real coefficients* $\{d_x\}_{x=1}^{d_\mathrm{A}}$ *such that*

$$w = \sum_{x=1}^{d_\mathrm{A}} d_x \, a_x . \tag{55}$$

5.3.5 Orthogonal Faces

Thanks to the spectral theorem, it is easy to retrieve the basic structures of quantum logic. In general, the faces of a convex set C form a bounded lattice, with partial order $\preceq$ corresponding to set-theoretic inclusion and with meet and join operations defined as $F \wedge G := F \cap G$ and $F \vee G := \bigcap\{H \mid F \subseteq H\, ,\; G \subseteq H\}$, respectively. The lattice is bounded, with the convex set C being the top element and the empty set $\emptyset$ being the bottom element. Hence, the set of deterministic states $C_\mathrm{A} := \mathsf{DetSt}(\mathrm{A})$ in a convex theory can be seen as a lattice in the above way. However, our axioms imply much more: according to them, the faces of the state space form an *orthomodular lattice*, i.e. a lattice with an operation of orthogonal complement $\perp$ satisfying the orthomodularity condition $F \preceq G \Longrightarrow G = F \vee (G \wedge F^\perp)$.

Let us see why this is the case. For a given face $F \subseteq C_\mathrm{A}$ we can pick a set of perfectly distinguishable pure states $\mathrm{S}_F = \{\alpha_x\}_{x=1}^{d_F} \subset F$ that is *maximal in* F, meaning that no other state in F can be distinguished perfectly from the states in S_F. Then, we can define the *barycenter of* F as

$$\omega_F := \frac{1}{d_F} \sum_{x=1}^{d_F} \alpha_x . \tag{56}$$

Since the face F can be compressed into the state space of a smaller system, Lemma 1 guarantees that the definition of the state ω_F depends only on F, and not on the maximal set S_F. In other words, Eq. (56) sets up a one-to-one correspondence between faces and their barycenters.

Now, we can extend the set S_F to a pure maximal set for system A, say $\{\alpha_x\}_{x=1}^{d_\mathrm{A}}$. Let us define the set $\mathrm{S}_{F^\perp} := \{\alpha_x\}_{x=d_F+1}^{d_\mathrm{A}}$ and denote by $F^\perp$ the smallest face containing $\mathrm{S}_{F^\perp}$. By construction, it is easy to verify that the set $\mathrm{S}_{F^\perp}$ is maximal in $F^\perp$ and therefore

$$\omega_{F^\perp} = \frac{1}{d_\mathrm{A} - d_F} \sum_{x=d_F+1}^{d_\mathrm{A}} \alpha_x .$$

$F^\perp$ can be equivalently characterized as the face containing all the states that are perfectly distinguishable from F. Moreover, it is not hard to show that

1. $F \vee F^\perp = C_\mathrm{A}$
2. $F \wedge F^\perp = \emptyset$
3. $\left(F^\perp\right)^\perp \equiv F$
4. $F \preceq G \Longrightarrow G^\perp \preceq F^\perp$
5. $F \preceq G \Longrightarrow G = F \vee \left(G \wedge F^\perp\right)$,

where the last two properties are proven by picking a pure maximal set for F, extending it to a pure maximal set for G, and extending the latter to a pure maximal set for the whole convex set C_A. Properties 1–4 show that the operation $\perp$ is an *orthogonal complement*, while property 5 is the orthomodularity condition. Hence, we obtained that the set of faces must be an orthomodular lattice.

5.3.6 Orthogonal Effects

By the state-effect duality, we can associate every face F with an effect a_F, defined as

$$a_F := \sum_{x=1}^{d_F} \alpha_x^\dagger, \tag{57}$$

where $\mathrm{S}_F = \{\alpha_x\}_{x=1}^{d_F}$ is a pure maximal set in F. Again, it is easy to see that the definition of a_F is independent of the choice of maximal set S_F. Indeed, by definition one has $a_F + a_{F^\perp} = e_\mathrm{A}$ for every pure maximal set $S_{F^\perp}$. Varying S_F without varying $S_{F^\perp}$ shows that the definition of a_F depends only on F.

Thanks to the spectral theorem, a_F can be operationally characterized the only effect that happens with unit probability on F and with zero probability on $F^\perp$:

Proposition 14 *a_F is the unique effect $a \in \mathsf{Eff}(\mathrm{A})$ satisfying the conditions*

$$\begin{aligned}(a|\rho) &= 1 \qquad \forall \rho \in F\\ (a|\sigma) &= 0 \qquad \forall \sigma \in F^{\perp}.\end{aligned}$$

For this reason, we call a_F the *identifying effect* of the face F. The set of identifying effects inherits the structure of orthomodular lattice from the set of faces, via the following definitions

1. $a_F \preceq a_G$ iff $F \preceq G$,
2. $a_F \wedge a_G := a_{F\wedge G}$,
3. $a_F \vee a_G := a_{F\vee G}$, and
4. $a_F^{\perp} := a_{F^{\perp}}$.

In quantum theory, the lattice of identifying effects is the lattice of projectors on subspaces of the Hilbert space. It is easy to see that the partial order $\preceq$ coincides with the partial order $\leq$ induced by the probabilities, namely $a_F \preceq a_G$ if and only if $(a_F|\rho) \leq (a_G|\rho)$ for every state ρ.

5.3.7 Orthogonal Projections

Faces of the state space can also be associated with physical transformations, in the following way:

Definition 18 A transformation $\Pi_F \in \mathsf{Transf}(\mathrm{A} \to \mathrm{A})$ is an *orthogonal projection* on the face $F \subseteq C_{\mathrm{A}}$ iff the following conditions are satisfied[14]

$$(\rho \xrightarrow{\mathrm{A}} \boxed{\Pi_F} \xrightarrow{\mathrm{A}} \quad = \quad (\rho \xrightarrow{\mathrm{A}} \qquad \forall \rho \in F \tag{59}$$

$$(\sigma \xrightarrow{\mathrm{A}} \boxed{\Pi_F} \xrightarrow{\mathrm{A}} \quad = \quad 0 \qquad \forall \sigma \in F^{\perp}. \tag{60}$$

[14]In the original work [26], we also required that projections be *pure*. However, in the context of our axioms, purity is implied by the two conditions in the present definition. This follows from the fact that *i)* one can construct a pure projection, and *ii)* it is possible to prove that projections are unique. A sketch of proof is the following: First, one can prove that for every pure state $\alpha \in F$ one must have $(\alpha^{\dagger}|\,\Pi_F = (\alpha^{\dagger}|$ (this follows from the definition and from Proposition 14). As a consequence, one also has $(a_F|\,\Pi_F = (a_F|$. This implies that, for every state $\rho \in \mathsf{St}(\mathrm{A})$, the unnormalized state $\Pi_F\,|\rho)$ is proportional to a state in F. Now, for two projections Π_F and Π'_F one must have

$$(\alpha^{\dagger}|\,\Pi_F|\rho) = (\alpha^{\dagger}|\rho) = (\alpha^{\dagger}|\,\Pi'_F|\rho), \tag{58}$$

for every pure state $\alpha \in F$. Since the states $\Pi_F\,|\rho)$ and $\Pi'_F\,|\rho)$ are proportional to states in F and $\alpha \in F$ is a generic pure state, Ideal Compression implies $\Pi_F\,|\rho) = \Pi'_F\,|\rho)$, or equivalently, $\Pi_F = \Pi'_F$, because the state ρ is generic.

The definition is non-empty: thanks to Purification and Purity of Composition, we are able to construct a *pure* projection Π_F for every face F. Moreover, it follows from the definition that the projection Π_F is unique.

In addition to purity, projections have a number of properties, including

1. $(a_F^{\perp}|\,\Pi_F = 0$
2. $(a_G|\,\Pi_F = (a_G|$ whenever $G \preceq F$
3. for every input state ρ, the normalized output state $\tau := \Pi_F\,|\rho)/(e_{\mathrm{A}}|\,\Pi_F\,|\rho)$ belongs to F
4. $\Pi_G\,\Pi_F = \Pi_F\Pi_G = \Pi_G$ whenever $G \preceq F$.

5.4 Interaction Between Correlation and Distinguishability Structures

We have seen that our axioms imply peculiar features, both in the way systems correlate and in the way states can be distinguished. It is time to combine these two types of features and to explore the consequences.

5.4.1 The Schmidt Bases

Combining Pure Steering and Spectral Decomposition, we are now in position to give the operational version of the Schmidt bases in quantum theory. The result can be summarized as follows:

Proposition 15 *Let* Ψ *be a pure state of* A $\otimes$ B *and let* ρ_{A} *and* ρ_{B} *be its marginals on systems* A *and* B, *respectively. Then, for every spectral decomposition*

$$\rho_{\mathrm{A}} = \sum_{x=1}^{r} p_x\,\alpha_x,$$

there exists a set of perfectly distinguishable pure states $\{\beta_x\}_{x=1}^{r} \subset \mathsf{PurSt}(\mathrm{B})$ *such that*

$$\rho_{\mathrm{B}} = \sum_{x=1}^{r} p_x\,\beta_x. \tag{61}$$

Moreover, one has

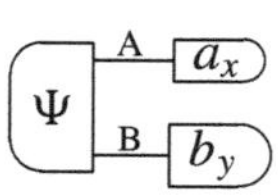

$$= \begin{cases} p_x\,\delta_{xy} & x, y \in \{1, \ldots, r\} \\ 0 & x, y \notin \{1, \ldots, r\} \end{cases} \tag{62}$$

for every two measurements $\boldsymbol{a} = \{a_x\}_{x=1}^{k_A}$ *and* $\boldsymbol{b} = \{b_y\}_{y=1}^{k_B}$ *satisfying* $a_x = \alpha_x^\dagger$ *and* $b_y = \beta_y^\dagger$ *for every* $x \leq r$ *and for every* $y \leq r$.

In particular, applying the result to the Bell state Φ, we obtain that the invariant state $\chi_{\overline{A}}$ can be decomposed as $\chi_{\overline{A}} = \frac{1}{d_A}\sum_{x=1}^{d_A} \overline{\alpha}_x$, for a suitable set of perfectly distinguishable pure states $\{\overline{\alpha}_x\}_{x=1}^{d_A}$. In particular, this implies that conjugate systems have the same informational dimension:

Corollary 1 *For every system* A, *one has* $d_{\overline{A}} = d_A$.

Combined with the fact that the informational dimension is multiplicative (Proposition 13), the above result implies that the composite system $A\otimes\overline{A}$ has informational dimension

$$d_{A\otimes\overline{A}} = d_A^2.$$

5.4.2 The Maximum Probability of Conclusive Teleportation

In our construction of conclusive teleportation, the teleportation probability was equal to the probability of the state Φ in an ensemble decomposition of the invariant state $\chi_A \otimes \chi_{\overline{A}}$, cf. Eq. (46). Now, since $\chi_A \otimes \chi_{\overline{A}}$ is the invariant state, it can be decomposed as

$$\chi_A \otimes \chi_{\overline{A}} = \frac{1}{d_A^2}\sum_{x=1}^{d_A^2} \Phi_x,$$

for every pure maximal set $\{\Phi_x\}_{x=1}^{d_A^2}$. The maximum probability of the Bell state in a convex decomposition of $\chi_A \otimes \chi_{\overline{A}}$ is then given by

$$p_A^{\max} = \frac{1}{d_A^2}. \tag{63}$$

Inserting the above equality into the teleportation upper bound (53) we obtain the relation

$$D_A \leq d_A^2. \tag{64}$$

In the next paragraph we will see how to obtain the converse inequality.

5.4.3 The Teleportation Lower Bound

Thanks to the state-effect duality, it is possible to establish a lower bound on the state space dimension. The proof is a little bit laborious and consists of two steps:

1. show that the effect $\Phi^\dagger$ that identifies the Bell state is of the form

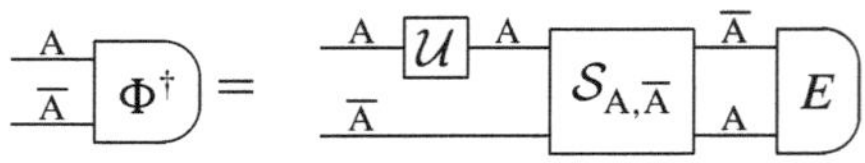

   where E is the effect achieving maximum teleportation probability, $\mathcal{S}_{\mathrm{A},\overline{\mathrm{A}}}$ is the swap, and $\mathcal{U}$ is some reversible transformation.
2. show that, with a suitable choice of basis for the vector space $\mathsf{St}_{\mathbb{R}}(\mathrm{A})$, every reversible transformation $\mathcal{U}$ is represented by an orthogonal matrix $M_{\mathcal{U}}$.

Once these two results are established, we can expand the Bell state Φ and the teleportation effect E as in Eq. (49), thus obtaining

$$1 = (\Phi^\dagger|\Phi) = \mathrm{Tr}[\Phi\, E\, M_{\mathcal{U}}] = p_{\mathrm{A}}^{\max}\ \mathrm{Tr}[M_{\mathcal{U}}] \le p_{\mathrm{A}}^{\max}\, D_{\mathrm{A}}, \tag{65}$$

having used the teleportation equality $\Phi\, E = p_{\mathrm{A}}^{\max}\, I_{D_{\mathrm{A}}}$ and the fact that the trace of an orthogonal matrix cannot be larger than the trace of the identity. Hence, we obtained the *teleportation lower bound*

$$D_{\mathrm{A}} \ge \frac{1}{p_{\mathrm{A}}^{\max}}. \tag{66}$$

Combining the teleportation lower bound with Eqs. (63) and (64), we obtain the equality

$$D_{\mathrm{A}} = d_{\mathrm{A}}^2. \tag{67}$$

5.5 *Qubit Structures*

So far, we avoided giving a concrete representation of our state spaces: all the quantum features that we have shown followed *directly* from the principles. We now proceed to analyze some features that are more closely related to the concrete geometrical shape of the quantum state spaces. We will first see that all two-dimensional systems in our theory have qubit state spaces. Leveraging on this fact, we will then derive two features of higher-dimensional systems: (i) an operational version of the superposition principle, and (ii) the fact that all systems of the same dimension are operationally equivalent.

5.5.1 Derivation of the Qubit

Showing that the states of a two-dimensional system can be described by density matrices is quite easy. This can be done geometrically, by showing that the deterministic states form a 3-dimensional Euclidean ball. The 3-dimensionality is obvious from the equality $D_A = d_A^2$, which for $d_A = 2$ implies that the convex set $C_A = \mathsf{DetSt}(A)$ is a three-dimensional manifold.[15] Then, we can make a simple geometrical reasoning:

1. all the pure states are generated from a fixed pure state by application of reversible transformations, and, by choosing a suitable basis for the state space, such transformations act in the 3-dimensional space as orthogonal matrices.
2. all states on the border of C_A are pure—otherwise, Perfect State Discrimination and Proposition 11 would imply $d_A > 2$. This means that, if we move away from the invariant state χ_A in an arbitrary direction, at some point we will hit a pure state.

In the ordinary 3-dimensional space, the sphere is the only (closed) 3-dimensional convex set generated by orthogonal matrices and with only pure states on the border.

Once we established that the convex set C_A is a sphere, we can represent every normalized state $\rho \in C_A$ with a density matrix S_ρ. In particular, the pure states will be of the form

$$S_\alpha = \begin{pmatrix} p & \sqrt{p(1-p)}\, e^{-i\theta} \\ \sqrt{p(1-p)}\, e^{i\theta} & 1-p \end{pmatrix} = |\alpha\rangle\langle\alpha| \tag{68}$$

$$|\alpha\rangle := \sqrt{p}\,|0\rangle + e^{i\theta}\sqrt{1-p}\,|1\rangle,$$

for some probability $p \in [0, 1]$ and some phase $\theta \in [0, 2\pi)$. Once we have chosen this representation, it is obvious that every effect $a \in \mathsf{Eff}(A)$ must be described by a positive semidefinite matrix E_a upper bounded by the identity and that probabilities are given by the Born rule

$$(a|\rho) = \mathrm{Tr}[E_a\, S_\rho]. \tag{69}$$

Moreover, the state-effect duality imposes that *all* such matrices represent valid effects.

5.5.2 The Superposition Principle

Pure states in quantum theory satisfy the so-called "superposition principle", which just means that they are in one-to-one correspondence with the rays of the underlying Hilbert space. *Per se*, this statement has hardly any operational meaning. However, one can formulate an operational version in general OPTs:

[15] In general, the dimension of the convex set C_A is given by $D_A - 1$.

Definition 19 (*Superposition Principle*) We say that system A *satisfies the superposition principle* iff for every pure maximal set $\mathrm{S} = \{\alpha_x \mid x \in \mathrm{X}\} \subset \mathsf{PurSt}(\mathrm{A})$ and for every probability distribution $\{p_x\}_{x\in\mathrm{X}}$ there exists one pure state ψ such that

$$\left(\psi\right|\overset{\mathrm{A}}{\text{—}}\left| a_x\right) = p_x \qquad \forall x \in \mathrm{X}, \tag{70}$$

for every measurement $\boldsymbol{a} = \{a_x\}_{x\in\mathrm{X}}$ that perfectly distinguishes among the states in the maximal set S.

Now, in a theory satisfying our principles we know that the two-dimensional systems are quantum—and therefore satisfy the superposition principle. Thanks to Ideal Compression, it is then easy to generalize the result to systems of arbitrary dimension: given two perfectly distinguishable pure states, one can encode them into a two-dimensional system, use the Bloch sphere representation to find the superposition state, and come back with the decoding operation. Iterating this procedure, we can superpose any number of perfectly distinguishable pure states.

As a simple application of the superposition principle, we obtain the following

Proposition 16 *A state ρ_{A} with spectral decomposition $\rho_{\mathrm{A}} = \sum_{x=1}^{r} p_x\,\alpha_x$ has a purification with purifying system* B *if and only if $d_{\mathrm{B}} \geq r$.*

The "only if" part was already clear from the Schmidt decomposition. For the "if" part, it is enough to pick r perfectly distinguishable pure states of B, say $\{\beta_x\}_{x=1}^{r}$, and to superpose the product states $\{\alpha_x \otimes \beta_x\}_{x=1}^{r}$ with probabilities $\{p_x\}_{x=1}^{r}$. The resulting pure state $\Psi \in \mathsf{PurSt}(\mathrm{A} \otimes \mathrm{B})$ is the desired purification.

5.5.3 The Superposition Principle for Transformations

The superposition principle allows us to glue distinguishable states in any way we like. Thanks to the state-transformation isomorphism, we can extend this idea to transformations. For example, consider a set of pure transformations $\{\mathcal{A}_x \mid x \in \mathrm{X}\} \subset \mathsf{PurTransf}(\mathrm{A} \to \mathrm{B})$ and suppose that they have *orthogonal support*, that is, that there exists a set of orthogonal faces $\{F_x \mid x \in \mathrm{X}\}$ such that

$$\mathcal{A}_x = \mathcal{A}_x\,\Pi_{F_x} \qquad \forall x \in \mathrm{X}. \tag{71}$$

Then, it is possible to find a pure transformation $\mathcal{A} \in \mathsf{PurTransf}(\mathrm{A} \to \mathrm{B})$ such that

$$\mathcal{A}\,\Pi_{F_x} = \mathcal{A}_x \quad \forall x \in \mathrm{X}. \tag{72}$$

The result follows by noticing that the Choi states $\{\Phi_{\mathcal{A}_x} \mid x \in \mathrm{X}\}$ are proportional to pure and perfectly distinguishable states and by applying the superposition principle to corresponding the normalized states.

5.5.4 Equivalence of Pure Maximal Sets up to Reversible Transformations

Using the superposition principle for transformations we can prove that all pure maximal sets of the same cardinality are equivalent:

Proposition 17 *Let* $\{\alpha_x\}_{x=1}^{d_A}$ *and* $\{\beta_y\}_{y=1}^{d_B}$ *be pure maximal sets for systems* A *and* B, *respectively. If* $d_A = d_B$, *then there exists a reversible transformation* $\mathcal{U} \in \mathsf{Transf}(A \to B)$ *such that*

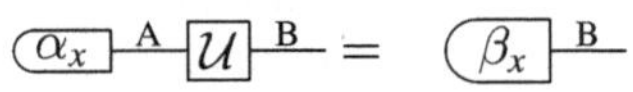

$\forall x \in X.$

The result follows immediately from the application of the superposition principle to the pure transformations $\mathcal{A}_x = |\beta_x)(\alpha_x^\dagger|$. As a corollary, we have that all systems of the same dimension are operationally equivalent.

5.6 *The Density Matrix*

We finally reached to the end of the reconstruction. It is now time to enter into the specific details of the Hilbert space formalism of quantum theory. Our strategy to reconstruct the Hilbert space formalism is to show that, for every system A, there exists a one-to-one linear map from the vector space $\mathsf{St}_{\mathbb{R}}(A)$ to the space of $d_A \times d_A$ Hermitian matrices, with the property that the convex set of deterministic states is mapped to the convex set of density matrices (non-negative matrices with unit trace).

Let us see how this can be proven. Since the dimension of the state space satisfies the relation $D_A = d_A^2$, every vector $v \in \mathsf{St}_{\mathbb{R}}(A)$ can be represented as square $d_A \times d_A$ real matrix M_v. In turn, the matrix M_v can be turned into a complex Hermitian matrix S_v, applying the linear transformation

$$S_v := \left(M_v + M_v^T\right) + i\left(M_v - M_v^T\right), \tag{73}$$

where M^T denotes the transpose of M. The problem is now to find a suitable representation in which normalized states $\rho \in C_A$ correspond to density matrices, that is $S_\rho \geq 0$ and $Tr[S_\rho]$. To find such a representation, we follow Hardy's method [23]: we pick a pure maximal set $\{\alpha_m\}_{m=1}^{d_A}$ and define the diagonal elements of the matrix S_ρ as

$$[S_\rho]_{mm} := (\alpha_m^\dagger|\rho),$$

In this way, we guarantee the unit-trace condition $\mathrm{Tr}[S_\rho] = 1$. To define the off-diagonal elements, we consider the two-dimensional faces $F_{mn} := \{\alpha_m\} \vee \{\alpha_n\}$, $n > m$. Projecting the state inside these faces, we obtain the states

$$|\rho^{mn}) = \frac{\Pi_{F_{mn}}|\rho)}{(e_A|\,\Pi_{F_{mn}}\,|\rho)} \qquad n > m.$$

Since every state ρ^{mn} is belongs to a two-dimensional face, it can be encoded into a qubit system and can be associated with a density matrix τ^{mn}. The off-diagonal elements $[S_\rho]_{mn}$ and $[S_\rho]_{nm}$ are defined in term of the qubit density matrix τ^{mn}, as

$$[S_\rho]_{mn} := [\tau^{mn}]_{01} \qquad \text{and} \qquad [S_\rho]_{nm} := [\tau^{mn}]_{10}.$$

The matrix S_ρ defined in this way is clearly Hermitian and, with a little bit of work, one can see that the linear map $\rho \mapsto S_\rho$ is one-to-one.

At this point the problem is to guarantee that the matrix S_ρ is positive. We consider first the case of pure states $\alpha \in \mathsf{PurSt}(A)$, for which one has

$$[S_\alpha]_{mn} = \sqrt{p_m\, p_n}\, e^{i\theta_{mn}}$$

where $\{p_m\}_{m=1}^{d_A}$ is a suitable probability distribution and $\{\theta_{mn}\}$ are phases satisfying the conditions $\theta_{mm} = 0$ for every m and $\theta_{nm} = -\theta_{mn}$ for every $n > m$. This expression follows from the fact that each state $|\alpha^{mn}) = \Pi_{F_{mn}}|\alpha)/(e_A|\Pi_{F_{mn}}|\alpha)$ is pure and, once encoded into a qubit, it has a density matrix of the form (68). In order to prove positivity, we need to show that the phases θ_{mn} are of the form $\theta_{mn} = \gamma_m - \gamma_n$, for some phases $\{\gamma_m\}$. The strategy is to prove the result first in dimension $d_A = 3$ and then to extend it to arbitrary dimensions.

Once we have proven that pure states correspond to rank-one projectors, it remains to show that *all* such projectors correspond to pure states. This can be done by using the superposition principle (both for states and for reversible transformations). Having proven that the set of pure states is in one-to-one correspondence with the set of rank-one projectors, it follows by convexity that the set of states is in one-to-one correspondence with the set of density matrices. In short, all state spaces are quantum.

To complete our reconstruction, we invoke Theorem 3, which guarantees that the tests in our theory are in one-to-one correspondence with the tests allowed by quantum theory.

6 Conclusions

Quantum theory can be rebuilt from bottom to top starting from six basic principles. The principles do not refer to specific physical systems such as particles or waves: instead, they are the rules that dictate how information can be processed. The first five principles—Causality, Purity of Composition, Local Tomography, Perfect State Discrimination, and Ideal Compression—can be thought of as requirements for a standard theory of information. On the background of these five principles, the sixth—Purification—stands out as *the* quantum principle, which brings in counterintuitive features like entanglement, no cloning, and teleportation. Purification gives the agent the power to harness randomness, by simulating the preparation of every state through the preparation of a pure bipartite state. When this is done, the agent has an intrinsic guarantee that no side information can hide outside her control. The moral of our reconstruction is quantum theory is the standard theory of information that allows for maximal control of randomness.

Acknowledgments The work is supported by the Templeton Foundation under the project ID# 43796 *A Quantum-Digital Universe*, by the Foundational Questions Institute through the large grant *The fundamental principles of information dynamics* (FQXi-RFP3-1325), by the 1000 Youth Fellowship Program of China, and by the National Natural Science Foundation of China through Grants 11450110096 and 11350110207. GC acknowledges the hospitality of the Simons Center for the Theory of Computation and of Perimeter Institute for Theoretical Physics. Research at Perimeter Institute for Theoretical Physics is supported in part by the Government of Canada through NSERC and by the Province of Ontario through MRI.

References

1. A. Einstein, B. Podolsky, N. Rosen, Can quantum-mechanical description of physical reality be considered complete? Phys. Rev. **47**, 777–780 (1935). doi:10.1103/PhysRev.47.777
2. E. Schrödinger, Discussion of probability relations between separated systems, in *Mathematical Proceedings of the Cambridge Philosophical Society*, vol. 31 (Cambridge University Press, Cambridge, 1935), pp. 555–563
3. G. Birkhoff, J.V. Neumann, The logic of quantum mechanics. Ann. Math. **37**(4), 823–843 (1936). http://www.jstor.org/stable/1968621
4. G.W. Mackey, Quantum mechanics and Hilbert space. Am. Math. Mon. **64**, 45–57 (1957)
5. G. Ludwig, Versuch einer axiomatischen grundlegung der quantenmechanik und allgemeinerer physikalischer theorien. Zeitschrift für Physik **181**(3), 233–260 (1964)
6. C. Piron, Axiomatique quantique. Helvetica Physica Acta **37**(4–5), 439 (1964)
7. J. Jauch, C. Piron, On the structure of quantal proposition systems, in *The Logico-Algebraic Approach to Quantum Mechanics* (Springer, Berlin, 1975), pp. 427–436
8. E.G. Beltrametti, G. Cassinelli, *The Logic of Quantum Mechanics*, vol. 15 (Cambridge University Press, Cambridge, 2010)
9. B. Coecke, D. Moore, A. Wilce, *Current Research in Operational Quantum Logic: Algebras, Categories, Languages*, vol. 111 (Springer, Berlin, 2000)
10. Quantum logic, http://en.wikipedia.org/wiki/Quantum_logic. Accessed: 2015-04-30
11. C.H. Bennett, G. Brassard, Quantum cryptography: public key distribution and coin tossing, in *Proceedings of IEEE International Conference on Computers, Systems and Signal Processing* (1984), pp. 175–179
12. A.K. Ekert, Quantum cryptography based on bell's theorem. Phys. Rev. Lett. **67**(6), 661 (1991)
13. P.W. Shor, Algorithms for quantum computation: discrete logarithms and factoring, in *IEEE 35th Annual Symposium on Foundations of Computer Science, 1994 Proceedings* (1994), pp. 124–134
14. L.K. Grover, A fast quantum mechanical algorithm for database search, in *Proceedings of the Twenty-eighth Annual ACM Symposium on Theory of Computing, STOC'96* (ACM, New York, 1996), pp. 212–219. doi:10.1145/237814.237866
15. W. Wootters, W. Zurek, A single quantum cannot be cloned. Nature **299**(5886), 802–803 (1982)
16. D. Dieks, Communication by EPR devices. Phys. Lett. A **92**(6), 271–272 (1982)
17. C.H. Bennett, G. Brassard, C. Crépeau, R. Jozsa, A. Peres, W.K. Wootters, Teleporting an unknown quantum state via dual classical and Einstein-Podolsky-Rosen channels. Phys. Rev. Lett. **70**(13), 1895 (1993)

18. C.H. Bennett, S.J. Wiesner, Communication via one-and two-particle operators on Einstein-Podolsky-Rosen states. Phys. Rev. Lett. **69**(20), 2881 (1992)
19. C.A. Fuchs, Quantum mechanics as quantum information, mostly. J. Mod. Opt. **50**(6–7), 987–1023 (2003)
20. G. Brassard, Is information the key? Nat. Phys. **1**(1), 2–4 (2005)
21. C.A. Fuchs et al., Quantum foundations in the light of quantum information. NATO Sci. Ser. Sub Ser. III Comput. Syst. Sci. **182**, 38–82 (2001)
22. C.A. Fuchs (ed.), *Coming of Age With Quantum Information* (Cambridge University Press, Cambridge, 2011)
23. L. Hardy, Quantum theory from five reasonable axioms, arXiv preprint arXiv:quant-ph/0101012
24. G.M. D'Ariano, How to derive the Hilbert-space formulation of quantum mechanics from purely operational axioms. AIP Conf. Proc. **844**, 101 (2006)
25. G.M. D'Ariano, Probabilistic theories: what is special about quantum mechanics, in *Philosophy of Quantum Information and Entanglement*, ed. by A. Bokulich, G. Jaeger (Cambridge University Press, Cambridge, 2010), pp. 85–126. doi:10.1017/CBO9780511676550.007
26. G. Chiribella, G.M. D'Ariano, P. Perinotti, Informational derivation of quantum theory. Phys. Rev. A **84**, 012311 (2011). doi:10.1103/PhysRevA.84.012311
27. L. Hardy, Reformulating and reconstructing quantum theory, arXiv:1104.2066
28. L. Masanes, M.P. Müller, A derivation of quantum theory from physical requirements. New J. Phys. **13**(6), 063001 (2011)
29. B. Dakic, C. Brukner, Quantum theory and beyond: is entanglement special?, in *Deep Beauty: Understanding the Quantum World Through Mathematical Innovation*, ed. by H. Halvorson (Cambridge University Press, Cambridge, 2011), pp. 365–392
30. P. Goyal, K.H. Knuth, J. Skilling, Origin of complex quantum amplitudes and Feynman's rules. Phys. Rev. A **81**, 022109 (2010). doi:10.1103/PhysRevA.81.022109
31. L. Masanes, M.P. Müller, R. Augusiak, D. Pérez-García, Existence of an information unit as a postulate of quantum theory. Proc. Natl. Acad. Sci. **110**(41), 16373–16377 (2013)
32. A. Wilce, Conjugates, correlation and quantum mechanics, arXiv:1206.2897
33. H. Barnum, M.P. Mueller, C. Ududec, Higher-order interference and single-system postulates characterizing quantum theory, arXiv:1403.4147
34. G. Chiribella, G.M. D'Ariano, P. Perinotti, Probabilistic theories with purification. Phys. Rev. A **81**, 062348 (2010). doi:10.1103/PhysRevA.81.062348
35. J. Barrett, Information processing in generalized probabilistic theories. Phys. Rev. A **75**, 032304 (2007). doi:10.1103/PhysRevA.75.032304
36. H. Barnum, J. Barrett, M. Leifer, A. Wilce, Generalized no-broadcasting theorem. Phys. Rev. Lett. **99**(24), 240501 (2007). doi:10.1103/PhysRevLett.99.240501
37. H. Barnum, J. Barrett, M. Leifer, A. Wilce, Teleportation in general probabilistic theories, ArXiv e-prints arXiv:0805.3553
38. L. Hardy, A formalism-local framework for general probabilistic theories, including quantum theory. Math. Struct. Comput. Sci. **23**(02), 399–440 (2013). doi:10.1017/S0960129512000163
39. H. Barnum, A. Wilce, Information processing in convex operational theories. Electron. Notes Theor. Comput. Sci. **270**(1), 3–15 (2011)
40. S. Abramsky, B. Coecke, A categorical semantics of quantum protocols, in *Proceedings of the 19th Annual IEEE Symposium on Logic in Computer Science* (IEEE, 2004), pp. 415–425. doi:10.1109/LICS.2004.1319636
41. S. Abramsky, B. Coecke, Categorical quantum mechanics, in *Handbook Of Quantum Logic and Quantum Structures: Quantum Logic*, ed. by K. Engesser, D.M. Gabbay, D. Lehmann (Elsevier, 2008), pp. 261–324. doi:10.1016/B978-0-444-52869-8.500141
42. B. Coecke, A universe of processes and some of its guises, in *Deep Beauty: Understanding the Quantum World Through Mathematical Innovation*, ed. by H. Halvorson (Cambridge University Press, Cambridge, 2010), pp. 129–186
43. B. Coecke, É.O. Paquette, Categories for the practising physicist, in *New Structures for Physics* (Springer, Berlin, 2011), pp. 173–286

44. S. Mac Lane, *Categories for the Working Mathematician*, vol. 5 (Springer, Berlin, 1978)
45. B. Coecke, Quantum picturalism. Contemp. Phys. **51**(1), 59–83 (2010). doi:10.1080/00107510903257624
46. P. Selinger, A survey of graphical languages for monoidal categories, in *New Structures for Physics*. Lecture Notes in Physics, vol. 813, ed. by B. Coecke (2011). doi:10.1007/978-3-642-12821-9_4
47. G. Chiribella, Dilation of states and processes in operational-probabilistic theories, in *Proceedings 11th Workshop on Quantum Physics and Logic*, Kyoto, Japan, 4–6th June 2014. Electronic Proceedings in Theoretical Computer Science, vol. 172, ed. by B. Coecke, I. Hasuo, P. Panangaden (Open Publishing Association, 2014), pp. 1–14. doi:10.4204/EPTCS.172.1
48. R.W. Spekkens, Evidence for the epistemic view of quantum states: a toy theory. Phys. Rev. A **75**, 032110 (2007). doi:10.1103/PhysRevA.75.032110
49. G. Chiribella, G.M. D'Ariano, P. Perinotti, Quantum theory, namely the pure and reversible theory of information. Entropy **14**(10), 1877–1893 (2012)
50. B. Coecke, R. Lal, Causal categories: relativistically interacting processes. Found. Phys. **43**(4), 458–501 (2013)
51. B. Coecke, Terminality implies non-signalling, arXiv preprint arXiv:1405.3681
52. E.C. Stueckelberg, Quantum theory in real Hilbert space. Helv. Phys. Acta **33**(727), 458 (1960)
53. H. Araki, On a characterization of the state space of quantum mechanics. Commun. Math. Phys. **75**(1), 1–24 (1980)
54. W.K. Wootters, Local accessibility of quantum states. Complex. Entropy Phys. Inf. **8**, 39–46 (1990)
55. L. Hardy, W.K. Wootters, Limited holism and real-vector-space quantum theory. Found. Phys. **42**(3), 454–473 (2012)
56. C. Shannon, A mathematical theory of communication. Bell Syst. Tech. J. **27**(3), 379–423 (1948). doi:10.1002/j.1538-7305.1948.tb01338.x
57. B. Schumacher, Quantum coding. Phys. Rev. A **51**(4), 2738 (1995)
58. C.H. Bennett, More about entanglement and cryptography (2007) http://www.lancaster.ac.uk/users/esqn/windsor07/Lectures/Bennett2.pdf. Accessed: 2014-11-14
59. I.M. Gelfand, M.A. Naimark, On the imbedding of normed rings into the ring of operators in Hilbert space. Mat. Sb. **54**(2), 197–217 (1943)
60. I.E. Segal, Irreducible representations of operator algebras. Bull. Am. Math. Soc. **53**(2), 73–88 (1947). doi:10.1090/S0002-9904-1947-08742-5
61. H. Everett III, Relative state formulation of quantum mechanics. Rev. Mod. Phys. **29**(3), 454 (1957)
62. H. Barnum, C.P. Gaebler, A. Wilce, Ensemble steering, weak self-duality, and the structure of probabilistic theories. Found. Phys. **43**(12), 1411–1427 (2013)
63. M.-D. Choi, Completely positive linear maps on complex matrices. Linear Algebra Appl. **10**(3), 285–290 (1975)

Reconstructing Quantum Theory

Lucien Hardy

1 Motivation

The standard axioms of QT are rather ad hoc. Where does this structure come from? Can we write down natural axioms, principles, laws, or postulates from which can derive this structure? Compare with the Lorentz transformations and Einstein's two postulates for special relativity. Or compare with Kepler's Laws and Newton's Laws. The standard axioms of quantum theory look rather ad hoc like the Lorentz transformations or Kepler's laws. Can we find a natural set of postulates for quantum theory that are akin to Einstein's or Newton's laws?

The real motivation for finding deeper postulates for quantum theory is that it may help us go beyond quantum theory to a theory of quantum gravity (just as Einstein's work helped him go beyond special relativity to his theory of General Relativity). It is in the finding of new physics that we can expect a real payoff of this program.

In [28] I showed how classical probability theory and quantum theory are the only two theories consistent with the set of postulates given above in the abstract. In this chapter I will explain the meaning of these postulates and indicate how the main steps of the proof work. The reconstruction takes place in the context of the circuit framework which I will describe.

L. Hardy (✉)
Perimeter Institute, 31 Caroline Street North,
Waterloo, ON N2L 2Y5, Canada
e-mail: lhardy@perimeterinstitute.ca

G. Chiribella and R.W. Spekkens (eds.), *Quantum Theory: Informational Foundations and Foils*, Fundamental Theories of Physics 181,
DOI 10.1007/978-94-017-7303-4_7

2 A Personal History of Reconstruction

A dozen years or so years ago Christopher Fuchs implored the community to "find an information theoretic reason" for axioms of QT (in multiple talks and a few papers [15, 17]). Further, Chris Fuchs and Gilles Brassard invited me to a workshop in Montreal in 2000 on this issue amongst others (see notes in [16]). I accepted their invitation but was, in the event, unable to attend. However, I was already hooked. The work I began preparing for that workshop led to my paper "Quantum theory from five reasonable axioms" [27]. In modern form (see [23]), the axioms given there can be stated in the following way:

Information Systems having, or constrained to have, a given information carrying capacity have the same properties.
Information locality Same as **P2** above.
Tomographic locality Same as **P3** above.
Continuous reversibility There exists a continuous reversible transformation between any pair of pure states.
Simplicity States are specified by the smallest number of probabilities consistent with the other axioms.

The simplicity axiom stands out as being less reasonable than the others. If we drop it then we may get a hierarchy of theories. This leads to two possibilities. Either there do exist higher theories in this hierarchy or there do not. For many years I tried to find such theories, and I tried to prove that such theories do not exist. I also tried to find other reasonable axioms that rule out higher theories in this hierarchy. It was not until 2009 that progress was made by others. In 2010 Chiribella, D'Ariano, and Perinotti (CDP) [6] showed how considerations concerning teleportation can be used to get rid of the need for a simplicity axiom in certain contexts. In 2010 [7] they found a set of postulates for quantum theory which, by virtue of the techniques developed in [6], did not require a simplicity axiom. Independently Dakić and Brukner [10] showed how one can replace the simplicity axiom with the assumption that any state for a two level system can be written as a mixture of perfectly distinguishable states (modulo some technical problems in their proof arising from the unfortunate existence of a subgroup of SO(7) that is transitive on the 6-sphere). In 2010 Masanes and Müller [34] showed how to replace the simplicity axiom with a different axiom saying that all mathematically well defined measurements for a two-level system are allowed. The axiom sets of Dakić and Brukner, and of Masanes and Müller are slight modifications on my original axiom set from 2001. The axioms of CDP are quite different (except for the assumption of tomographic locality). Masanes, Müller, Augusiak, and Pérez-García provide another set of axioms employing tomographic locality, continuous reversibility, and another axiom concerning the existence of an informational unit. Another set of axioms using tomographic locality was given by Marco Zaopo [45].

There has been much work recently by many people along less related lines (Fuchs [18], Goyal [19], Wilce [42], Rau [38, 39], Fivel [12], ...). Further, there is, in fact, a long history of attempts to reconstruct QT (von Neumann [41], Mackey [33], Birkoff and von Neumann [4], Zierler [46], Piron [37] Ludwig [32], Rovelli [40], and many others).

Many of these reconstruction attempts employ the so called "convex probabilities framework". This goes back to originally to Mackey and has been worked on (and sometimes rediscovered) by many others since including Ludwig [32], Davies and Lewis [11], Gunson [21], Mielnik [36], Araki [2], Gudder *et al.* [20], Foulis and Randall [14], Fivel [13] as well as more recent incarnations [3, 27].

The circuit framework used here [22, 28] (see also [23, 24]) might be regarded as a marriage of the convex probabilities framework and the pictorial (or categorical) approach of Abramsky and Coecke [1, 9]. A similar framework has been developed by Chiribella, D'Ariano, and Perinotti [6]. The pictorial approach of Abramsky and Coecke is important because of its emphasis on composition as a basic primitive.

In 2003 Clifton, Bub, and Halvorson [8] were inspired by a suggestion of Fuchs and Brassard to take a different approach to reconstructing quantum theory. They showed that some features of quantum theory follow if one imposes no-bit-commitment, no-broadcasting, and no-signalling within the C^* algebraic framework (rather than the convex probabilities framework).

3 The Circuit Framework

In this section we will present the circuit framework. The basic idea is that circuits can be built from operations. An operation corresponds to one use of an apparatus with some particular outcome, or subset of possible outcomes specified. Operations have some number of systems as inputs and some number of systems as outputs. We can wire together operations. If we have no open inputs or outputs left over then we have a circuit. We employ three background assumptions for this framework. The main one is that we can associate a probability with a circuit (the joint probability that the outcomes are in the associated outcome sets on each operation).

3.1 Operations

We can notate an operation diagrammatically or symbolically as follows.

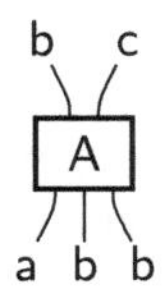

$$\Longleftrightarrow \qquad A^{b_4 c_5}_{a_1 b_2 b_3}$$

The integer subscripts in the symbolic representation will be used to show where the wires go and have no significance beyond this. An operation, A, corresponds to one use of an apparatus and has the following features.

- *Inputs and outputs.* Come in various types, a, b, ... The inputs correspond to wires going in the bottom of the box in the diagrammatic notation, or subscripts in the symbolic notation. The outputs correspond to wires coming out the top and to superscripts.
- *A setting*, s(A). We can think of this as corresponding to certain positions for knobs, buttons, and dials that may be on the apparatus.
- *An outcome set*, o(A). This is a subset of all the possible outcomes for this use of the apparatus.

If $x_A \in o(A)$ then we say operation A "happened". If we have a different setting, or a different outcome set, then we have a different operation and should notate this with a different letter (e.g. B rather than A).

3.2 Wires

Outputs can be connected to inputs by wires.

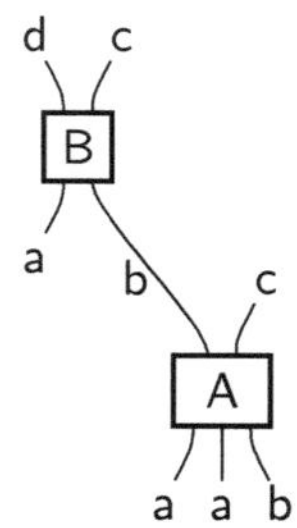

$$\iff \quad A^{b_4 c_5}_{a_1 a_2 b_3} B^{d_7 c_8}_{a_6 b_4}$$

Note how the wire linking the two boxes corresponds to the repeated integer (the 4 on b_4). These diagrams are interpreted graphically. In particular, vertical position has no meaning. We can distort the graph in any way we wish without changing the physical meaning so long as the wires remain attached to the same positions on the boxes and the boxes maintain their orientations.

There are certain wiring rules.

- *One wire:* At most one wire can be connected to any given input or output.
- *Type matching:* Wires can connect inputs and outputs of the same type.
- *No closed loops:* If we trace from output to input along wires through the operations then we cannot get back to the operation we started at.

The last rule is to rule out closed time like loops.

3.3 *Fragments*

The most general object we can consider is a collection of operations wired together. For example,

Fragment E

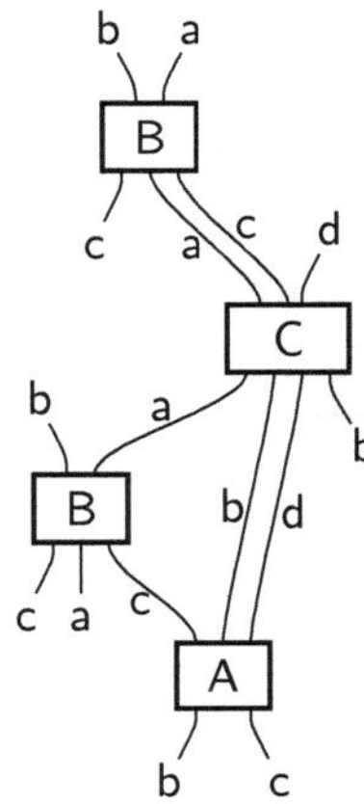

$$\Longleftrightarrow \quad A^{c_1 b_2 d_3}_{b_{15} c_{16}} B^{b_6 a_7}_{c_4 a_5 c_1} C^{a_9 c_{10} d_{11}}_{a_7 b_2 d_3 b_8} B^{b_{13} a_{14}}_{c_{12} a_9 c_{10}}$$

Such objects are called fragments (as they are fragments of circuits). In general fragments may have open inputs and outputs. Fragments have

- *A setting*, s(E), given by specifying the setting on each operation.
- *An outcome set*, o(E), (equals o(A) × o(B) × o(C) × o(A) in this case). We say the fragment "happened" if the outcome is in the outcome set.
- *A wiring*, w(E), given by specifying the input/output pairs which are wired together.

3.4 *Circuits*

Circuits have no open inputs or outputs. For example,

Circuit H

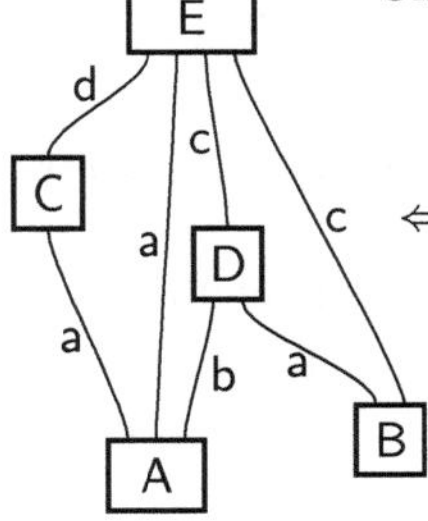

$$\Longleftrightarrow \quad A^{a_1 a_2 b_3} B^{a_4 c_7} C^{d_5}_{a_1} D^{c_6}_{b_3 a_4} E_{d_5 a_2 c_6 c_7}$$

Circuits are special cases of fragments. Circuits have

- *A setting*, s(H), given by specifying the setting on each operation.

- *An outcome set*, $o(H)$, given by specifying the outcome set at each operation (equals $o(A) \times o(B) \times o(C) \times o(D) \times oE$ in this case). We say the fragment "happened" if the outcome is in the outcome set.
- *A wiring*, $w(E)$, given by specifying which input/output pairs are wired together.

3.5 Preparations, Results, and Transformations

A **preparation** is a fragment having only outputs. Here are some examples:

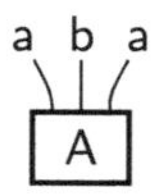

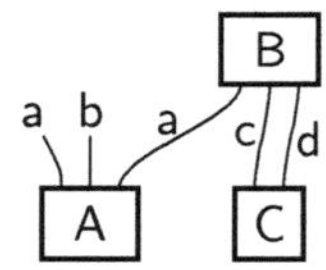

We will associate *states* with preparations.

A **result** is a fragment having only inputs. Here are some examples:

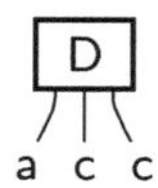

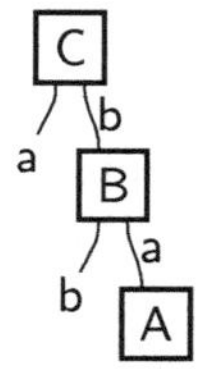

we will associate *effects* with results. A **measurement** is a collection of results corresponding to the same setup with disjoint outcome sets whose union is the set of all outcomes for this setup.

A **transformation** is a fragment that has inputs and outputs that is *used in transformation mode*. Here are some examples

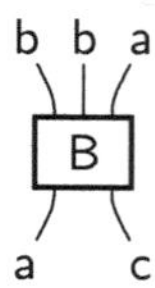

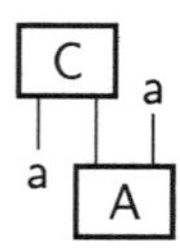

A fragment is used in transformation mode if we do not feed outputs into inputs on this fragment (either directly or indirectly).

A transformation, $B^{a_2}_{a_1}$, is reversible if there exists another transformation, $\tilde{B}^{a_3}_{a_2}$, such that $B^{a_2}_{a_1}\tilde{B}^{a_3}_{a_2}$ is the identity transformation. Note that the identity transformation is defined to have the property that, if it inserted on any wire in any circuit, then the probability for that circuit remains unchanged.

3.6 The First Background Assumption

We need three background assumptions in setting up the circuit framework. The first is the following.

Assump 1. We can associate a probability with any given circuit (the probability that the circuit "happens"), and this probability depends only on the specification of the given circuit (the knob settings and outcome sets at the operations, and the wiring).

For example.

Prob 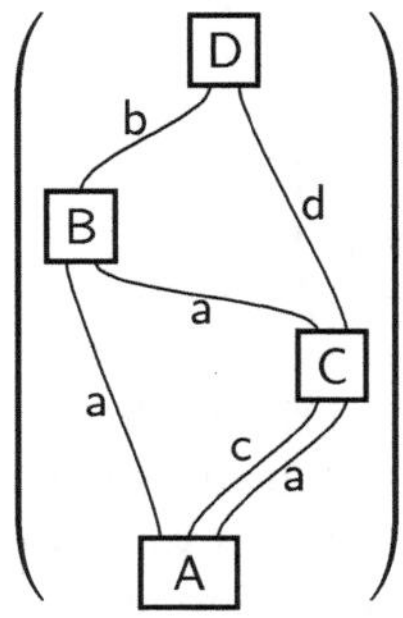 is well conditioned

Note that we make this assumption for circuits, not for fragments generally. Indeed, a fragment with open inputs and/or outputs cannot be expected to satisfy this since the probability may depend on what is done with these open ports.

3.7 The State

Want to associate a state with a preparation

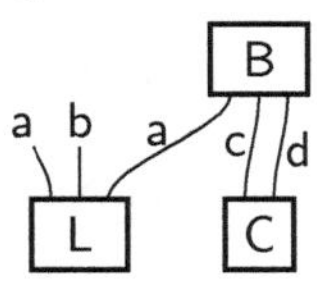

There exist many results which complete this into a circuit. Here are a few examples:

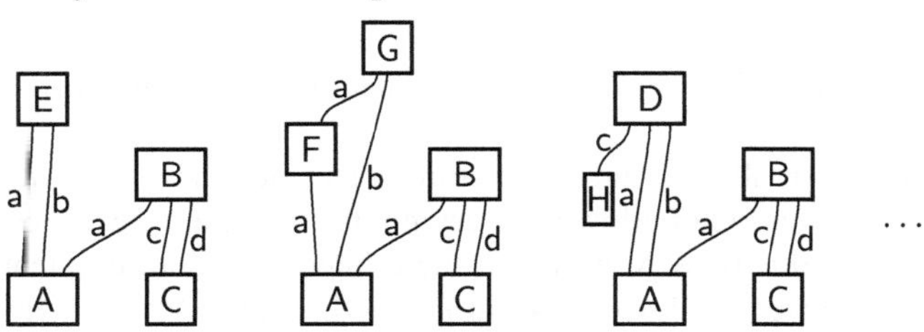

The state associated with a preparation should enable us to predict the probability for every circuit containing this preparation.

We could specify the state associated with a preparation by giving a list of probabilities for every circuit made with this preparation. This would be a very long list and rather cumbersome to work with. However, physical theories typically relate different quantities. Consequently it should be possible to pick out a subset of results such that specifying the probabilities for just the circuits containing the given preparation and results from this subset is sufficient to allow us to calculate the probability for any other circuit containing this preparation. For example, in Quantum Theory we can calculate all the probabilities for a spin-half particle from just the probabilities

$$\begin{pmatrix} p_{x+} \\ p_{y+} \\ p_{z+} \\ p_{z-} \end{pmatrix}$$

as these suffice to determine the density matrix for this system. In fact, these probabilities suffice to determine the elements of the density matrix by linear equations. We will insist on linearity in what follows. This is well justified when one considers taking mixtures of states (see Appendix B of [28] for example). We call the choice of results used to specify the state **fiducial results**. In general this choice is not unique.

3.8 Using Fiducial Results to Define States

It is worth paying attention to the font used in the notation below. Consider preparations of the form $\mathsf{A}^{\mathsf{a}_1}$. We choose a fiducial set of results

$$\mathsf{X}^{a_1}_{\mathsf{a}_1} \quad \text{for } a_1 = 1 \text{ to } K_{\mathsf{a}}$$

The state associated with preparation $\mathsf{A}^{\mathsf{a}_1}$ is given by

$$A^{a_1} := \text{Prob}(\mathsf{A}^{\mathsf{a}_1}\mathsf{X}^{a_1}_{\mathsf{a}_1})$$

A fiducial set is a minimal set such that, for any result, $\mathsf{B}_{\mathsf{a}_1}$ there exists an **effect** B_{a_1} such that

$$\text{Prob}(\mathsf{A}^{\mathsf{a}_1}\mathsf{B}_{\mathsf{a}_1}) = A^{a_1}B_{a_1} \quad \text{(summation over } a_1 \text{ implied)}$$

This is linear. If we allow arbitrary mixtures then must have linearity here [28]. However, even if we do not allow arbitrary mixtures, we are free to consider only linear relations of this type even though there may be a more efficient non-linear expression. Associated with **preparation** $\mathsf{A}^{\mathsf{a}_1}$ is the **state**, A^{a_1}. This is a list of K_{a} fiducial probabilities from which all other probabilities can be calculated. Associated with the **result** $\mathsf{B}_{\mathsf{a}_1}$ is the **effect** B_{a_1}. This is a list of K_{a} real coefficients (which can be negative).

3.9 *Pure States*

A **mixed state** is one that can be simulated by a mixture of preparations. i.e.

$$A^{a_1} = \lambda B^{a_1} + (1 - \lambda)C^{a_1}$$

where $0 < \lambda < 1$ and $B^{a_1} \neq C^{a_1}$. A **pure state** is one that cannot be simulated by a mixture of preparations. A transformation is **non-mixing** if it preserves purity (up to normalization).

3.10 *Maximal Sets*

A very important notion is that of a maximal set.

> A **maximal set of distinguishable states** is any set containing the maximum number, N_{a}, of states for which there exists some measurement, called a **maximal measurement**, which can identify which state from the set we have in a single shot.

We also need the following notion.

> A **maximal effect** is associated with each outcome of a maximal measurement.

We can notate these notions diagrammatically as

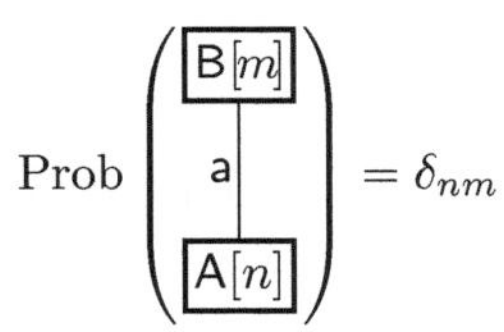

or symbolically as

$$A^{a_1}[n]B_{a_1}[m] = \delta_{nm}$$

where $m, n = 1$ to N_{a}.

In quantum theory maximal sets of distinguishable states are associated with an orthonormal basis. Then N_{a} is the dimension of the Hilbert space, maximal measurements correspond to non-degenerate observables, and maximal effects correspond to rank-one projectors.

In classical probability theory there is a unique maximal set of distinguishable states and it is usually understood to correspond to the underlying states of reality.

3.11 Two More Assumptions for Framework

We need two more background assumptions for the circuit framework.

Assump 2. There exists at least one type of system having $N_\mathsf{a} > 1$ and K_a finite.

Recall that N_a is the maximum number of states in a distinguishable set and K_a is the number of probabilities that must be provided to specify the state. In quantum theory we note that systems having finite N_a also have finite K_a.

We will give the third assumption without defining all the terms.

Assump 3. If, for any accuracy $\delta > 0$, there exists a fragment $A[\delta]$ that is operationally indiscernible from a given hypothetical fragment, Q, then there actually exists a fragment with the probabilistic properties of Q.

This assumption is used to obtain the property that the space of fragments is compact in an appropriate sense. The reader is referred to [28] for more details.

3.12 Permutation Transformations

We can define permutation transformations with respect to a given maximal set of distinguishable states

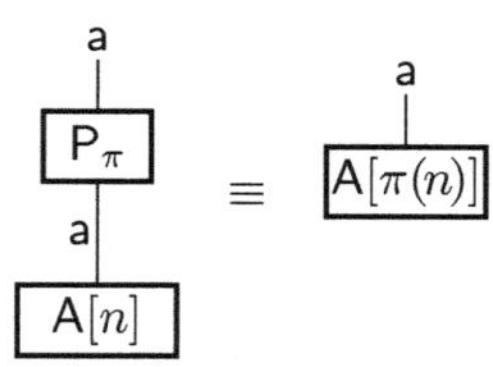

for some permutation π. That is, a permutation transformation permutes the elements of a maximal set of distinguishable states.

3.13 P1

For the purpose of clarity, it is worth discussing the first postulate at this stage.

P1 *Logical Sharpness.* There is a one-to-one map between pure states and maximal effects such that we get unit probability.

This means that for any given pure state there is a unique maximal effect giving unit probability, and that for any given maximal effect there is a unique pure state giving unit probability. In pictures, there is a one-to-one map between pure states and maximal effects:

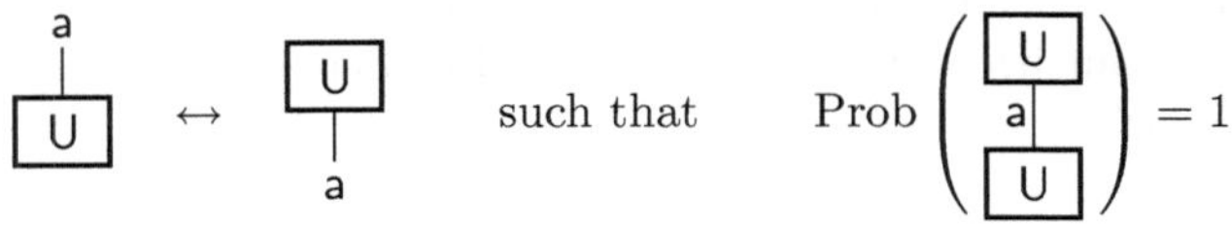

Interestingly, causality follows from this postulate. This is the property that choices in the future do not influence probabilities the past. The causality property was introduced by CDP as corresponding to the existence of a unique deterministic effect [6] and used as a postulate in their reconstruction [7].

3.14 Informational Faces and Non-flat Sets of States

An **informational face**, S, is the full set of states having support only on some subset, O_S, of the outcomes of some maximal measurement, $\{\mathsf{B}_{\mathsf{a}_1}[m]\}$. Basically, these are sets of states constrained to have a certain information carrying capacity. The states in $\bar{S}$ have support on the complement subset of outcomes, $\bar{O}_S$, for the same maximal measurement. In convex geometry a face is given by the intersection of the convex set in question and a supporting hyperplane. A supporting hyperplane is one which has no elements of the convex set on one side. The supporting hyperplane defining S is given by the equation

$$\left(\sum_{m\in\bar{O}_S} B_{a_1}[m]\right) A^{a_1} = 0 \tag{1}$$

Faces are, themselves, convex sets. In quantum theory all faces are, in fact, informational faces by virtue of the *spectrality* property (any state can be written as a convex combination of states in a maximal distinguishable set). However, this need not be the case and we do not assume this here.

A set of states is **non-flat** if it is a spanning subset of some informational face. It could be an over-complete spanning subset and consequently the informational face is, itself, non-flat. If **P1** holds then we can think of a non-flat set of states as a kind of generalization of the notion of a pure state. In fact it follows from **P1** that any single member non-flat set of states consists of a state proportional to a pure state. Thus, we can think of a set containing a single pure state as being the simplest type of non-flat set in a hierarchy of bigger and bigger non-flat sets.

We need the following notion to understand **P5**.

A transformation is said to be **non-flattening** if, for any non-flat set of states we send in, we get a non-flat set of states out.

It follows from **P1** that all non-flattening transformations are also non-mixing. Interestingly, in quantum theory the converse is true also: all non-mixing transformations are non-flattening.

3.15 Filters

A filter, F, is defined with respect to a given informational face, S.

> **A filter** is a transformation that
>
> - passes unchanged states in S
> - blocks states in $\bar{S}$

For a filter defined with respect to an informational face S given by maximal measurement $\{\mathsf{B}[m]\}$ and outcome set O_{S} we have

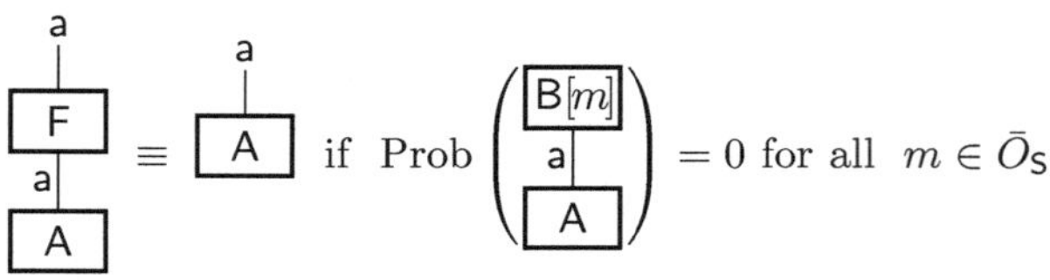

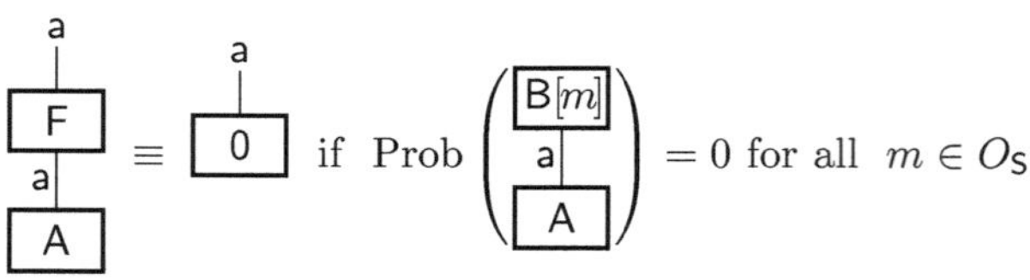

Here

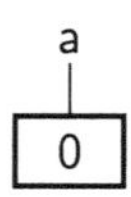

is the preparation corresponding to the null state. The null state is the state that gives probability zero for any circuit it is part of. The components of the null state are, therefore, all equal to zero.

4 The Postulates

Classical probability theory and quantum theory are only two theories consistent with the following postulates.

P1 *Logical sharpness.* There is a one-to-one map between pure states and maximal effects such that we get unit probability.

P2 *Information locality.* A maximal measurement on a composite system is effected if we perform maximal measurements on each of the components. Equivalently $N_{\mathsf{ab}} = N_{\mathsf{a}} N_{\mathsf{b}}$.

P3 *Tomographic locality.* The state of a composite system can be determined from the statistics collected by making measurements on the components. Equivalently $K_{\mathsf{ab}} = K_{\mathsf{a}} K_{\mathsf{b}}$.

P4′ *Permutability.* There exists a reversible transformation on any system effecting any given permutation of any given maximal set of distinguishable states for that system.

P5 *Sturdiness.* Filters non-flattening.

4.1 Ruling Out the Classical Case

To single out quantum theory it suffices to add *anything* that is inconsistent with classical probability and consistent with quantum theory. The key property of non-classical theories is that $K_{\mathsf{a}} > N_{\mathsf{a}}$ for non-trivial systems (i.e. systems having $N_{\mathsf{a}} > 1$). One way to ensure this is to replace **P4′** with

P4 *Compound permutability.* There exists a compound reversible transformation on any system effecting any given permutation of any given maximal set of distinguishable states for that system.

A compound transformation is one that can be made from two sequential transformations (neither equal to the identity). The advantage of this is that it requires only adding a single word (the word "compound") to one of the existing postulates. However, as we just mentioned, we could add any property inconsistent with classical probability theory so long as it is consistent with quantum theory. For example, we could simply demand that there are more pure states than there are states in any maximal distinguishable set of states for non-trivial systems.

4.2 P2

Our second postulate is the following.

P2 *Information locality.* A maximal measurement on a composite system is effected if we perform maximal measurements on each of the components.

This means the set of results (with $m = 1$ to N_{a} and $n = 1$ to N_{b})

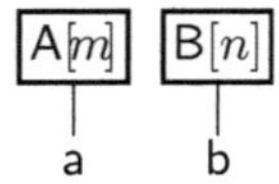

is a maximal measurement.

If N_a is the maximum number of distinguishable states then **P2** is equivalent to statement that

$$N_\mathsf{ab} = N_\mathsf{a} N_\mathsf{b}$$

This is very natural. For example, if we have a die ($N_\mathsf{a} = 6$) and a coin ($N_\mathsf{b} = 2$) then we have $N_\mathsf{ab} = 12$. We call this "information locality" since the total information capacity is given by adding together the local information capacities:

$$\log N_\mathsf{ab} = \log N_\mathsf{a} + \log N_\mathsf{b}$$

This postulate is looks innocent but it is actually very powerful. Certainly we can imagine situations in which this postulate is not true (see [30]).

4.3 P3

Our third postulate is used in many of the recent reconstructions of quantum theory. It can be stated in the following way.

P3 *Tomographic locality.* The state of a composite system can be determined from the statistics collected by making measurements on the components.

Pictorially this means we can determine the state associated with the preparation, $\mathsf{A}^{\mathsf{a}_1\mathsf{b}_2}$, by determining the probabilities for circuits of the form

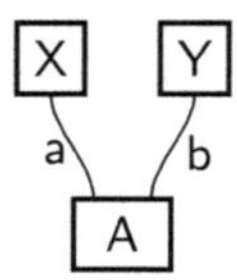

It follows from this that we can write the state as $A^{a_1 b_2}$ where this is a list of joint probabilities determined by putting separate fiducial results on system a_1 and b_2. In fact, more generally, it follows from tomographic locality that we can represent an arbitrary operation such as $\mathsf{B}_{\mathsf{a}_1\mathsf{b}_2\mathsf{c}_3}^{\mathsf{d}_4\mathsf{e}_5}$ by a tensor $B_{a_1 b_2 c_3}^{d_4 e_5}$. Actually, this fact is an equivalent statement of tomographic locality. In words the equivalent statement is that an arbitrary operation can be fully characterized by local process tomography. Then the probability for a circuit is given the scalar obtained by contracting over indices where there are wires in the circuit. For example,

$$\text{Prob}\left(\mathsf{A}^{\mathsf{a}_1\mathsf{b}_2\mathsf{c}_3\mathsf{f}_6}\mathsf{B}_{\mathsf{a}_1\mathsf{b}_2\mathsf{c}_3}^{\mathsf{d}_4\mathsf{e}_5}\mathsf{C}_{\mathsf{d}_4}\mathsf{D}_{\mathsf{e}_5\mathsf{f}_6}\right) = A^{a_1 b_2 c_3 f_6} B_{a_1 b_2 c_3}^{d_4 e_5} C_{d_4} D_{e_5 f_6} \tag{2}$$

In [22, 28] a tensor such as $B^{d_4e_5}_{a_1b_2c_3}$ correspond to putting a more general object called a duotensor into standard form. Duotensors play an important role in the full reconstruction. However, we will not discuss them further here.

Another equivalent statement of tomographic locality is that

$$K_{\mathsf{ab}} = K_{\mathsf{a}}K_{\mathsf{b}}$$

(where K_{a} is number of probabilities required to specify state). Hence we see that information locality and tomographic locality are very similar postulates (they were grouped together in [27]).

There exist other equivalent statements of the tomographic locality assumption (see [28] for some of them).

4.4 *P4′*

The forth postulate concerns the ability to permute the states in a maximal set of distinguishable states by means of a reversible transformation.

P4′ *Permutability.* There exists a reversible transformation on any system effecting any given permutation of any given maximal set of distinguishable states for that system.

In pictures we can say that exists reversible P_π

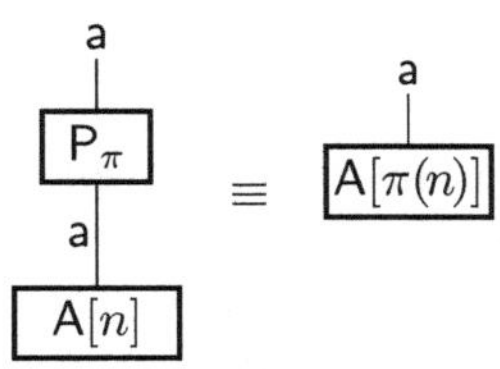

for any maximal set of distinguishable states, $A^{a_1}[n]$ and permutation, π. Reversibility means that this transformation is reversible when applied to any state (not just the members of the maximal set of distinguishable states).

This postulate implies we can perform lossless arbitrary translation of a message encoded with respect to any alphabet to one encoded with respect to any permutation of this same alphabet.

4.5 *P5*

The last postulate concerns filters.

P5 *Sturdiness.* Filters are non-flattening.

Recall that a set of states is said to be **non-flat** if it is a spanning subset some informational face. One way to think of this property is that sets of states resist being squashed (hence the name "sturdiness"). Quantum states are not as sensitive as we might have imagined. A filter is a pretty dramatic transformation. However, according to this postulates, sets of states remain as intact as they can under the circumstances.

5 Outline of Reconstruction

The full reconstruction, while only using elementary mathematics, is rather lengthy. Here we will only give an outline of some of the main steps.

First, using **P1**, **P2**, **P3**, and **P4**′ we

- show that there exists a reversible transformation between any pair of pure states,
- construct arbitrary filters,
- show there exist types with $N = 1, 2, 3, \ldots$,
- show that systems having same N are equivalent,
- show that $K_{\mathsf{a}} = N_{\mathsf{a}}^r$ where $r = 1, 2, 3, \ldots$ (the Wootters hierarchy [43, 44].

Using **P5** as well we

- show that gebits (generalized bits, i.e. systems having $N_{\mathsf{a}} = 2$) correspond to hyperspheres,
- show that all points on the hypersphere correspond to pure states,
- show how to do teleportation,
- prove that $K_{\mathsf{a}} = N_{\mathsf{a}}$ or $K_{\mathsf{a}} = N_{\mathsf{a}}^2$.

This gives us the bit or the qubit. Interestingly, getting the qubit is the most difficult part of this reconstruction as well as many others. Having got the qubit we get the appropriate constraints on quantum theory in general (not just the $N_{\mathsf{a}} = 2$ case) by showing that a certain "magic operation" can implement any complete set of superoperators (superoperators corresponding to a set of operations associated with a given apparatus with disjoint outcome sets where the union of these outcome sets is the full set of outcomes). To complete this last step and get quantum theory in general we employ the duotensor framework [22] and the operator tensor formulation of quantum theory [28, 29]. We will provide an outline of how some aspects of the reconstruction work in the following subsections. For the full details of the proofs (which are mostly omitted here) the reader is referred to [28].

5.1 Reversible Transformation Between Pure States

Let

$$\{\mathsf{U}^{\mathsf{a}_1}[n] : n = 1 \text{ to } N_{\mathsf{a}}\} \quad \text{and} \quad \{\mathsf{V}^{\mathsf{a}_1}[n] : n = 1 \text{ to } N_{\mathsf{a}}\}$$

be maximal sets of distinguishable preparations for a. Let

$$\{\mathsf{W}^{\mathsf{b}_2}[m] : m = 1 \text{ to } N_{\mathsf{b}}\}$$

be a maximal set of distinguishable preparations for b. We will denote the maximal measurement that distinguishes these maximal sets of distinguishable states by $\{\mathsf{U}_{\mathsf{a}_1}[n]\}$, $\{\mathsf{V}_{\mathsf{a}_1}[n]\}$, and $\{\mathsf{W}_{\mathsf{b}_2}[m]\}$.

It follows from **P2** that

$$\{\mathsf{U}^{\mathsf{a}_1}[n]\mathsf{W}^{\mathsf{b}_2}[m] : nm = 11, 12, \dots N_{\mathsf{a}}N_{\mathsf{b}}\}$$

is a maximal set for ab. Similarly,

$$\{\mathsf{V}^{\mathsf{a}_1}[n]\mathsf{W}^{\mathsf{b}_2}[m] : nm = 11, 12, \dots N_{\mathsf{a}}N_{\mathsf{b}}\}$$

is another maximal set for ab.

Let P be the reversible transformation that permutes $\mathsf{U}^{\mathsf{a}_1}[n]\mathsf{W}^{\mathsf{b}_2}[m]$ according to

$$\pi_{\mathsf{P}} = (nm \leftrightarrow mn)$$

Let Q be the reversible transformation that permutes $\mathsf{V}^{\mathsf{a}_1}[n]\mathsf{W}^{\mathsf{b}_2}[m]$ according to

$$\pi_{\mathsf{Q}} = (nm \leftrightarrow mn)$$

Note we choose b such that $N_{\mathsf{b}} = N_{\mathsf{a}}$. Then a little thought shows that

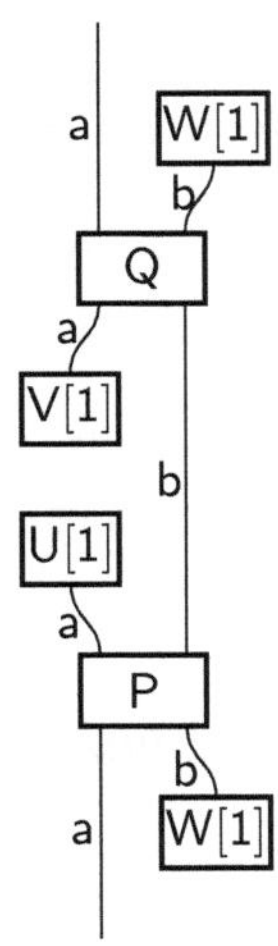

does the job. We can prove that this is a reversible transformation and it clearly takes $\mathsf{U}^{\mathsf{a}_1}[1]$ to $\mathsf{V}^{\mathsf{a}_3}[1]$. In fact it actually does a bit more. It takes $\mathsf{U}^{\mathsf{a}_1}[n]$ to $\mathsf{V}^{\mathsf{a}_3}[n]$ for $n = 1$ to N_{a}.

5.2 Arbitrary Filters are Possible

It can be shown that the transformation

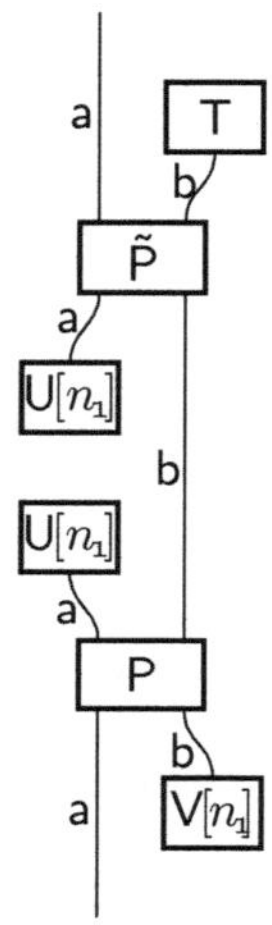

(3)

effects an arbitrary filter where $N_{\mathsf{a}} = N_{\mathsf{b}}$ and n_1 is any integer chosen from O_S. Here the transformation P is a permutation transformation with permutation

$$\pi = \begin{pmatrix} nm \leftrightarrow mn \text{ if } n \text{ and } m \in O(S) \\ nm \leftrightarrow nm \text{ otherwise} \end{pmatrix} \tag{4}$$

and $\tilde{\mathsf{P}}$ is the transformation that reverses P. The result T_{b} is a deterministic result (its outcome set is equal to the set of all outcomes).

5.3 Systems with Same are N Equivalent

These substitutions prove equivalence when $N_{\mathsf{a}} = N_{\mathsf{b}}$.

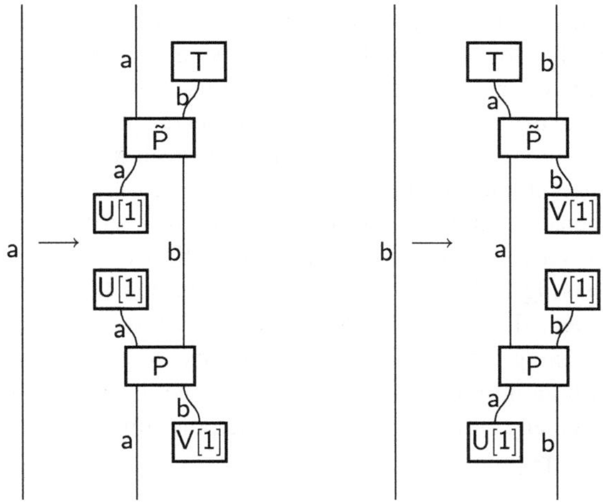

(5)

where

$$\pi = (nm \leftrightarrow mn) \tag{6}$$

With these substitutions we can replace any wire of type a by one of type b (and vice versa) without changing the probability for the given circuit.

5.4 *Proof that $K_{\mathsf{a}} = N_{\mathsf{a}}^r$*

It follows from the first four postulates that

- $K_{\mathsf{a}} = K(N_{\mathsf{a}})$ (since systems having the same N_{a} are equivalent).
- $K(N+1) > K(N)$ (since we can filter systems down).
- $K(N_{\mathsf{a}}N_{\mathsf{b}}) = K(N_{\mathsf{a}})K(N_{\mathsf{b}})$ (by **P2**).

It can be shown that

$$K_{\mathsf{a}} = N_{\mathsf{a}}^r \quad \text{where } r = 1, 2, 3, \ldots$$

follows (the proof of this uses the decomposition of N_{a} and N_{b} into prime numbers). This relationship was first suggested by Wootters [43, 44] and hence we term it the Wootters hierarchy. It was first proven that this relationship follows from the above more basic premises in [27]

5.5 *All Points on Hypersphere Correspond to Pure States*

It can be shown to follow from fact that there exists a reversible group of transformations between pure states that all pure states must live on surface of a hypersphere.

We then need to show that all points on this hypersphere correspond to pure states for the gebit. The proof of this starts with a getrit (a system having $N_a = 3$). We prepare a system constrained to an $N_a = 2$ informational face of this getrit (i.e. we consider states constrained to a gebit space). Next we filter on the getrit space but with respect to a maximal measurement that has one maximal result having full support on the afore mentioned gebit and one maximal result having only partial support on this gebit. What happens is that the states emerging out of this filter are also gebit states but they move closer to one of the poles of the gebit (the one associated with the maximal result having full support). If we keep filtering like this then an initially non-flat set of states will, by **P5** remain non-flat but will move closer to this pole. We can produce a spanning set of states that are as close to the pole as we like. It then follows that within an infinitesimal region of the pole there must be points lying in any direction. Since any state could serve as the pole, this proves that all points on the hypersphere are populated.

5.6 Getting the Qubit and $K_a = N_a^2$

The classical case corresponds to the 1-sphere with just two pure states. If we are in the non-classical case then we want to prove that this hypersphere must be the 2-sphere corresponding to the qubit of quantum theory. Consequently, we want to prove that for $N_a = 2$ we have $K_a = 4$ (in the non-classical case). This means that we get the Bloch ball (since one parameter counted in K_a corresponds to normalization). This proof is adopted from a beautiful proof due to Chiribella, D'Ariano, and Perinotti [6, 7] using teleportation.

We start by assuming that we are in the non-classical case. Consider the gebit preparation

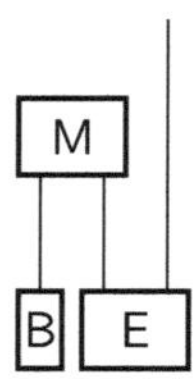

(7)

where B^{a_1} is a gebit preparation, $E^{a_2 a_3}$ is an entangled pure state in $S_{\{11,22\}}$ and $M_{a_1 a_2}$ is a certain maximal entangled effect (these do not exist in the classical case but must exist in the non-classical case). We can show that the transformation on B^{a_1} is non-flattening using **P5**. The pure states for preparation B^{a_1} lie on the surface of a hypersphere. Under the transformation in (7) this hypersphere is transformed to an hyper-ellipsoid. Hence we can use the preparation in (7) to prepare a state proportional to any pure state by making an appropriate choice of preparation B. The state that is prepared is not necessarily equal to B^{a_1} so we do not necessarily have faithful teleportation. However, we can use this result to obtain the following result.

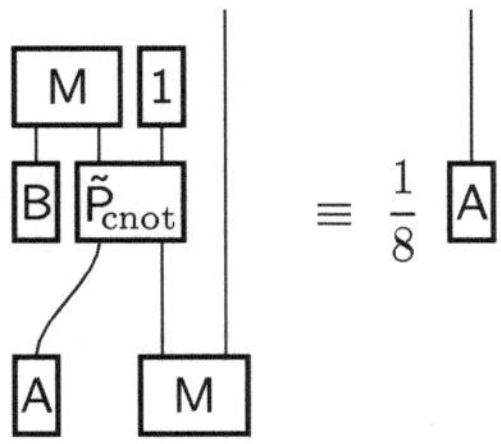

(8)

for any state A^{a_1} for the gebit. Now we do have faithful (probabilistic) teleportation. We work in a computational basis for the gebit denoted by 0 and 1. Here B^{a_2} is a special choice of state. In fact it must be an *equatorial state*—a pure state on the equator of the hypersphere between the two poles (these exist only in the non-classical case). Also, P_{cnot} is a reversible permutation transformation effecting the permutation associated with the $CNOT$ gate in the computational basis.

Since we now have faithful (albeit probabilistic) teleportation, we have

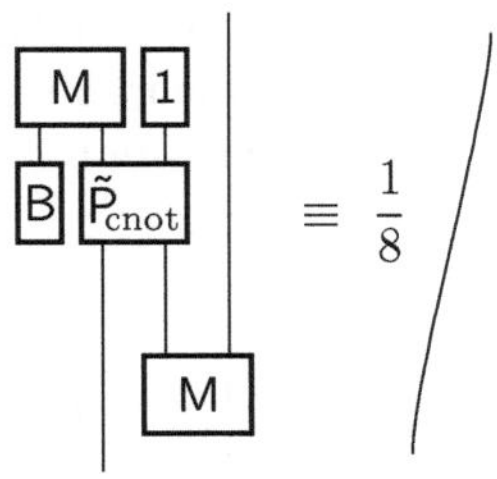

We can also prove that

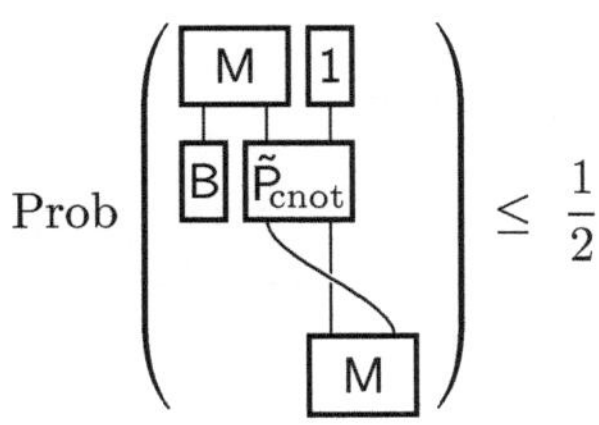

(9)

using the fact that B^{a_2} is equatorial. For convenience we put

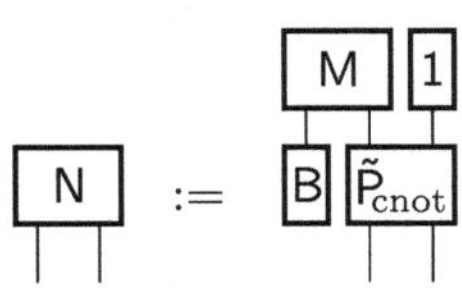

(10)

Then

$$N_{a_1 a_2} M^{a_2 a_3} = \frac{1}{8} I_{a_1}^{a_3} \tag{11}$$

where $I_{a_1}^{a_3}$ is the identity. Hence,

$$N_{a_1 a_2} M^{a_2 a_1} = \frac{1}{8} I_{a_1}^{a_1} = \frac{1}{8} K_{\mathsf{a}} \tag{12}$$

since the trace of the identity is equal to the dimension of the space on which it acts. But we also have

$$N_{a_1 a_2} M^{a_2 a_1} \leq \frac{1}{2} \tag{13}$$

It follows that, for a gebit, $K_{\mathsf{a}} \leq 4$. Hence, $K = N^2$ in general.

5.7 The Magic Operation

The last part of the proof shows how to use the fact that the gebit is equal to the qubit along with the postulates to get quantum theory in general. The key part of this is showing that the following set of operations (for different outcomes l of a maximal measurement)

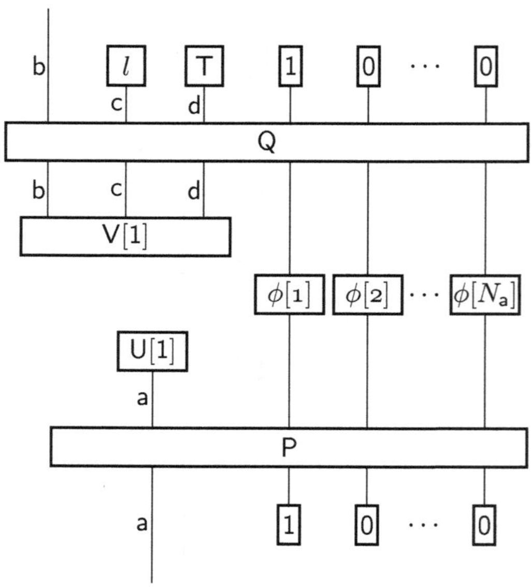

(14)

can generate any complete set of operations in quantum theory. Here P and Q are appropriately chosen reversible permutation transformations, $\{\phi[n] : n = 1 \text{ to } N_{\mathsf{a}}\}$ are appropriately chosen phases, c and d are ancillary systems having appropriate N_{c} and N_{d}, and $\mathsf{V}^{\mathsf{b}_2\mathsf{c}_3\mathsf{d}_4}[1]$ is an appropriately chosen preparation (for a pure state). The unlabeled wires represent qubits. T is the deterministic effect. This proof employs the duotensor and operator tensor frameworks and the reader is referred to [28] for details.

6 Conclusions

I have provided a set of operational postulates from which quantum theory can be reconstructed. This does not require a simplicity assumption as did my earlier work [27] from over a decade ago. This is one of a number of recent reconstructions [7, 10, 34, 35, 45] along similar lines which use the assumption of tomographic locality (like [27]) and do not need a simplicity assumption. There are strong connections between these different approaches and many of the proof techniques are similar. What appears as a postulate in one approach appears as a low level theorem in another and vice versa. One might think of a set of postulates as being a little akin to a choice of coordinate system used to represent some shape. If we find a good coordinate system then the shape appears simple. The fact that there are a number of good postulate sets that are fairly simply related to each other is similar to the fact that there are often a number of good choices of simply related coordinate system for viewing a shape. While one may have preferences for one or the other set of postulates, there is not really much to distinguish them.

However, I am left with the sense that some much deeper insights are still left to be had. One reason for this sense is that in the operator tensor formulation [28, 29] of quantum theory (and, similarly, in the quantum combs approach [5]) preparations, transformations, and results are all treated on a fairly equal footing. However, this is not true of the postulate set presented here or the others I have mentioned. Surely the postulates should also treat all kinds of operations on an equal footing (so far as this is possible). Further motivation for this comes from quantum gravity. We do not yet have a theory of quantum gravity. However, when we do, it seems likely that we will have to contend with indefinite causal structure. We will not be able to say whether some particular interval is space-like, time-like, or null, but rather can expect to have something like a quantum superposition of these different cases. Then we cannot be sure that some ports on an operation are inputs and others are outputs (since these notions assume definite causal structure). In this case we cannot distinguish preparations, transformations, and results. Quantum theory might reasonably be expected to be obtained as a limit of quantum gravity (in the limit as we have definite causal structure). In this limit distinct notions of preparations, transformations, and results might emerge. However, fundamentally (before the limit is taken), they are not distinct. It would, then, be great if a set of postulates for quantum theory treated them more-or-less on an equal footing. Some of these postulates may then go over to a theory of quantum gravity. In [25, 26, 31] I developed the "causaloid framework" for general probabilistic theories that can accommodate indefinite causal structure. One can put quantum theory into this framework and it does, indeed, have the feature that preparations, transformations, and results are then treated on a (more-or-less) equal footing (indeed, the causaloid formulation of quantum theory is the origin of the operator tensor formulation).

I think that the real test of this research program will be the progress that is made towards a theory of quantum gravity using these newly developed techniques. It is in constructing new physical theories that we can really test whether we are on the right path since then we have to make new predictions and account for new experimental data. Furthermore, the more fundamental our physical theory, the more natural we can expect our postulates to be.

Acknowledgments First and foremost I am grateful to Christopher Fuchs for getting me thinking about these questions in the first place. He is responsible for initiating the present wave of reconstruction work. I am also grateful to Giulio Chiribella, Mauro D'Ariano, and Paolo Perinotti both for discussions on these matters and for writing a number of inspirational papers. Over the years I have benefited by discussions with many people on the subject of reconstructing quantum theory to whom I am grateful.
Research at Perimeter Institute is supported by the Government of Canada through Industry Canada and by the Province of Ontario through the Ministry of Economic Development and Innovation.

References

1. S. Abramsky, B. Coecke, A categorical semantics of quantum protocols, in *Proceedings of the 19th Annual IEEE Symposium on Logic in Computer* (2004), pp. 415–425
2. H. Araki, On a characterization of the state space of quantum mechanics. Commun. Math. Phys. **75**(1), 1–24 (1980)
3. J. Barrett, Information processing in generalized probabilistic theories. Phys. Rev. A **75**(3), 032304 (2007)
4. G. Birkhoff, J. von Neumann, The logic of quantum mechanics. Ann. Math. **37**(4), 823–843 (1936)
5. G. Chiribella, G.M. D'Ariano, P. Perinotti, Theoretical framework for quantum networks. Phys. Rev. A **80**(2), 022339 (2009)
6. G. Chiribella, G.M. D'Ariano, P. Perinotti, Probabilistic theories with purification. Phys. Rev. A **81**(6), 062348 (2010). arXiv:0908.1583
7. G. Chiribella, G.M. D'Ariano, P. Perinotti, Informational derivation of quantum theory. Phys. Rev. A **84**, 012–111 (2011). arXiv:1011.6451
8. R. Clifton, J. Bub, H. Halvorson, Characterizing quantum theory in terms of information-theoretic constraints. Found. Phys. **33**(11), 1561–1591 (2003)
9. B. Coecke, Quantum picturalism. Contemp. Phys. **51**, 59–83 (2010)
10. B. Dakic, C. Brukner, Quantum theory and beyond: is entanglement special?, in *Deep Beauty: Understanding the Quantum World Through Mathematical Innovation*, ed. by H. Halvorson (Cambridge University Press, 2011), pp. 365–392 (2009). Arxiv reprint. arXiv:0911.0695
11. E.B. Davies, J.T. Lewis, An operational approach to quantum probability. Commun. Math. Phys. **17**(3), 239–260 (1970)
12. D.I. Fivel. Derivation of the rules of quantum mechanics from information-theoretic axioms. Found. Phys. pp. 1–28 (2010). Arxiv preprint. arXiv:1010.5300
13. D.I. Fivel, How interference effects in mixtures determine the rules of quantum mechanics. Phys. Rev. A **50**(3), 2108 (1994)
14. D.J. Foulis, C.H. Randall, Empirical logic and tensor products. Interpret. Found. Quantum Theory **5**, 9–20 (1979)
15. C.A. Fuchs et al. Quantum foundations in the light of quantum information. Nato Sci. Ser. Sub Ser. III Comput. Syst. Sci. **182**, 38–82 (2001). ArXiv preprint. arXiv:quant-ph/0106166

16. C.A. Fuchs, Coming of age with quantum information (2009)
17. C.A. Fuchs, Quantum mechanics as quantum information (and only a little more) (2002). Arxiv preprint. arxiv:quant-ph/0205039
18. C.A. Fuchs, R. Schack, A quantum-Bayesian route to quantum-state space. Found. Phys. 1–12 (2010)
19. P. Goyal, Information-geometric reconstruction of quantum theory. Phys. Rev. A **78**(5), 052120 (2008)
20. S. Gudder, S. Pulmannová, S. Bugajski, E. Beltrametti, Convex and linear effect algebras. Rep. Math. Phys. **44**(3), 359–379 (1999)
21. J. Gunson, On the algebraic structure of quantum mechanics. Commun. Math. Phys. **6**(4), 262–285 (1967)
22. L. Hardy, A formalism-local framework for general probabilistic theories including quantum theory (2010). Arxiv preprint. arXiv:1005.5164
23. L. Hardy, Foliable operational structures for general probabilistic theories. *Deep Beauty: Understanding the Quantum World Through Mathematical Innovation*, ed. by H. Halvorson. (2011), pp. 409–442. arXiv:0912.4740
24. L. Hardy, Operational structures as a foundation for probabilistic theories, PIRSA:09080011 (can be viewed at http://pirsa.org/09080011/) (2009)
25. L. Hardy, Probability theories with dynamic causal structure: a new framework for quantum gravity (2005). Arxiv preprint. arxiv:gr-qc/0509120
26. L. Hardy, Quantum gravity computers: on the theory of computation with indefinite causal structure. *Quantum Reality, Relativistic Causality, and Closing the Epistemic Circle: Essays in Honour of Abner Shimony*, (2009), p. 379
27. L. Hardy, Quantum theory from five reasonable axioms (2001). Arxiv preprint arxiv:quant-ph/0101012
28. L. Hardy, Reformulating and reconstructing quantum theory (2011). Arxiv preprint. arXiv:1104.2066
29. L. Hardy, The operator tensor formulation of quantum theory. Philos. Trans. R. Soc. A: Math., Phys. Eng. Sci., **370**(1971), 3385–3417 (2012). arXiv:1201.4390
30. L. Hardy, W.K. Wootters, Limited holism and real-vector-space quantum theory. Found. Phys. (2012), 1–20. arXiv:1005.4870
31. L. Hardy, Towards quantum gravity: a framework for probabilistic theories with non-fixed causal structure. J. Phys. A: Math. Theor. **40**, 3081 (2007)
32. G. Ludwig, *An Axiomatic Basis of Quantum Mechanics*. vols. I, II (Springer, Berlin 1985, 1987)
33. G.W. Mackey, *The Mathematical Foundations of Quantum Mechanics: A Lecture-note Volume* (Addison-Wesley, 1963)
34. L. Masanes, M.P. Müller, A derivation of quantum theory from physical requirements. New J. Phys. **13**(5), 063001 (2011). arXiv:1004.1483
35. L. Masanes, M.P. Müller, A. Augusiak, Pérez-García. A digital approach to quantum theory. (2012). Arxiv preprint. arXiv:1208.0493
36. B. Mielnik, Theory of filters. Commun. Math. Phys. **15**(1), 1–46 (1969)
37. C. Piron, Axiomatique quantique. Helv. Phys. Acta **37**, 439 (1964)
38. J. Rau, Measurement-based quantum foundations. Found. Phys., 1–9 (2010)
39. J. Rau, On quantum vs. classical probability. Ann. Phys. **324**(12), 2622–2637 (2009)
40. C. Rovelli, Relational quantum mechanics. Int. J. Theor. Phys. **35**, 1637 (1996)
41. J. von Neumann, *Mathematical Foundations of Quantum Mechanics* (Princeton University Press, 1996)
42. A. Wilce. Four and a half axioms for finite dimensional quantum mechanics. (2009). Arxiv preprint. arXiv:0912.5530
43. W.K. Wootters, Local accessibility of quantum states, in *Complexity, Entropy, and the Physics of Information: The Proceedings of the 1988 Workshop on Complexity, Entropy, and the Physics of Information held May–June, 1989, in Santa Fe, New Mexico* (Westview Press, 1990), p. 39

44. W.K. Wootters, Quantum mechanics without probability amplitudes. Found. Phys. **16**(4), 391–405 (1986)
45. M. Zaopo, Information theoretic axioms for quantum theory (2011) ArXiv preprint. arXiv:1205.2306
46. N. Zierler, Axioms for non-relativistic quantum mechanics. *The Logico-algebraic Approach to Quantum Mechanics* (1975), p.149

The Classical Limit of a Physical Theory and the Dimensionality of Space

Borivoje Dakić and Časlav Brukner

1 Introduction

"Physical space is not a space of states" writes Bengtsson in his article entitled "Why is space three dimensional?" [1]. Indeed, although the state space dimension for a macroscopic object is exponentially large (in the number of object's constituents), we still find ourselves organizing data into a three-dimensional manifold called "space". Why is this discrepancy? Can there be more dimensions? In past different approaches have been taken to show that the three-dimensional space is special, such as bio-topological argument [2], stability of planet orbits [2], stability of atoms [3] or elementary particle properties [4]. The existence of extra dimensions has been proposed as a possibility for physics beyond the standard model [5–10].

In this work we will address the questions given above within the operational approach to general probabilistic theories [11–13]. There the basic ingredients of the theory are primitive laboratory procedures by which physical systems are prepared, transformed and measured by laboratory devices, but the systems are *not* necessarily described by quantum theory. General probabilistic theories are shown to share many features that one previously have expected to be uniquely quantum, such as probabilistic predictions for individual outcomes, the impossibility of copying unknown states (no cloning) [14], or violation of Bell's inequalities [15, 16]. Why then nature

B. Dakić · Č. Brukner (✉)
Vienna Center for Quantum Science and Technology (VCQ), Faculty of Physics, University of Vienna, Boltzmanngasse 5, 1090 Vienna, Austria
e-mail: caslav.brukner@univie.ac.at

B. Dakić
Centre for Quantum Technologies, National University of Singapore, 117543
3 Science Drive 2, Singapore, Singapore

Č. Brukner
Institute of Quantum Optics and Quantum Information (IQOQI), Austrian Academy of Sciences, Boltzmanngasse 3, 1090 Vienna, Austria

G. Chiribella and R.W. Spekkens (eds.), *Quantum Theory: Informational Foundations and Foils*, Fundamental Theories of Physics 181,
DOI 10.1007/978-94-017-7303-4_8

obeys quantum mechanics rather than other probabilistic theory? Recently, there have been several approaches, answering this question by reconstructing quantum theory from a plausible set of axioms that demarcate phenomena that are exclusively quantum from those that are common to more general probabilistic theories [13, 17–31].

In probabilistic theories the macroscopic laboratory devices are standardly assumed to be classically describable, but are not further analyzed. The "position" of the switch at the transformation device or the record on the observation screen have only an abstract meaning and are not linked to the concepts of position, time, direction, or energy of "traditional" physics (or to use Barnum's words "the full, meaty physical theory" is still missing [32]). As a result of the reconstructions of quantum theory, one derives a finite-dimensional, or countable infinite-dimensional, Hilbert space as an operationally testable, abstract formalism concerned with predictions of frequency counts in future experiments with no appointment of concrete physical labels to physical states or measurement outcomes. In standard textbook approach to quantum mechanics this appointment is "inherited" from classical mechanics and is formalized through the first quantization—the set of explicit rules that relate classical phase variables with quantum-mechanical operators. However, these rules lack an immediate operational justification. This calls for a "completion" of operational approaches to quantum mechanics with the "meaty physics". Our work can be understood as a step in this direction.

In an operational approach one interprets parameters that describe physical states, transformations, and measurements, as the parameters that specify the configurations of macroscopic instruments in physical space by which the state is prepared, transformed, and measured. Within this approach it is natural to assume the state space and the physical space to be isomorphic to each other. The isomorphism of the two spaces is realized in quantum mechanics for the elementary directional degree of freedom (spin-1/2). The state space of the spin is a three-dimensional unit ball (the Bloch ball) and its dimension and the symmetry coincide with those of the Euclidian (non-relativistic) three-dimensional space in which classical macroscopic instruments are embedded. This was first pointed out by von Weizsäcker who writes [33]: "It [quantum theory of the simple alternative] contains a two-dimensional complex vector space with a unitary metric, a two-dimensional Hilbert space. This theory has a group of transformations which is surprisingly near-isomorphic with a group of rotations in the real three-dimensional Euclidian space. This has been known for a very long time. I propose to take this isomorphism seriously as being the real reason why ordinary space is three-dimensional." In a different vein, Penrose demonstrated that the angles of three-dimensional space can be modeled by spin networks in semi-classical states of "large spins" [34] and Wootters showed a relation between the statistical distinguishability in quantum mechanics and geometry [35].

Whereas von Weizsäcker based his proposal on a mathematical isomorphism between the two spaces, there are very compelling physical evidences that they indeed are related to each other. The Einstein-de Haas effect [36] as well as the Barnett effect [37] demonstrate a deep relationship between magnetism, angular momentum, and elementary quantum spin. In the Einstein-de Hass effect an external magnetic field, generated by electric current through the coil surrounding a ferromagnet, leads to the mechanical rotation of the ferromagnet (or reversely, in the Barnett effect, a spinning ferromagnet can change its magnetization). The two effects phenomenologically demonstrate that the quantum spin is indeed of the same nature as the angular momentum of macroscopic rotating bodies as perceived in classical mechanics. One can therefore associate mathematical properties to the elementary quantum spin that are typical for a vector (more precisely, pseudo-vector) in a three-dimensional space, such as three coordinates, orientation in space, or building the cross products with other vectors. For example, the precession of the spin in the external magnetic field (the Larmor precession) is due to torque on the spin, which is given by the cross product between the spin and the field.

If one assumes that quantum theory is universal [38], one should be able to arrive at an explanation of macroscopic devices (such as those for preparation, transformation and measurement of elementary spins) in terms of classical physics and three-dimensional space from within quantum theory. This would allow to invert the logic from the previous paragraph and argue that the symmetry of the classical angular momenta as embedded in the three-dimensional Euclidian space should follow from the symmetry of the elementary quantum spin. One could offer such an explanation in the "classical" or "macroscopic" limit of quantum theory. It is known that the spin coherent states [39, 40]—which are the states of a large number of identically prepared elementary spins—acquire an effective description of a classical spin embedded in the ordinary three-dimensional space under the restriction of coarse-grained measurements [41]. These (macroscopic) states are "robust": they are stable with respect to small perturbations, such as those caused by repeated observations, giving rise to "objective" properties in the classical limit. For example, if one flips only a few spins of a ferromagnet, the system will turn into an orthogonal state, but we will identify it as the very same magnet at the macroscopic level. The macroscopic distinguishability can be reached only if a sufficiently large number of spins (of the order of square-root of the total number of spins) are flipped in which case we perceive it as a new state of magnetization.

The spin coherent states can serve as "reference states" with respect to which one can define the notion of "direction". Preparation, rotation or measurement of the elementary quantum spin along some "direction" has then only relative meaning with respect to such quantum reference frames [42–46] which become classical ones in the limit of a large number of spins constituting the coherent state. In the limit the spin coherent states can be understood as representing the classical magnetic field in which other quantum spins may evolve. Importantly, the group of transformations of an individual quantum spin is then generated by a rotationally invariant interaction between the spin and the coherent state, i.e. by a pairwise invariant interaction between the spin and each of the constituting spins of the coherent state [45]. The invariance is required as there is no external reference frame. The spin coherent states define directions in terms of two (polar) angles in the three-dimensional Euclidian space, and thus give rise, through the relative angle, to the notion of "neighboring" orientations, without having such a notion from the very beginning.

We have seen that there are phenomenological and mathematical evidences for the isomorphism between the state space of elementary quantum spin and the physical space. Central for the argument are coherent states, which can be understood as representing macroscopic fields in the physical space, on one hand, are described in quantum theory as states in Hilbert space for which all the spins are prepared in the same quantum state, on the other hand.

The notion of coherent states is not exclusive for quantum theory but can be straightforwardly extended to general probabilistic theories as well, as a state of the collection of a large number of equally prepared elementary systems. It is legitimate to think that starting with the theory that differs from quantum theory and going into the limit of states with a large number of elementary systems and coarse-grained measurements one might arrive at "classical physics" embedded in a space of dimensions different than the one of our everyday life [47]. For example, quaternionic quantum theory describing non-relativistic spin requires the physical space to have five dimensions, and the octavic quantum theory requires the space of nine dimensions [48].

Here we investigate the possibility of having higher-dimensional physical spaces in the macroscopic ("classical") limit. Our analysis is restricted to non-relativistic geometry of space (not space-time and not curved spaces) and directional degrees of freedom (spin). It is clear that one can imagine a vast variety of manifolds as possible candidates for the space (for example, as odd as the donut shape). Our focus here is onto the most natural generalization of the experienced (three-dimensional) non-relativistic space: the Euclidean d-dimensional isotropic space. We have seen that the symmetry of the state space of the elementary quantum spin (three-dimensional Bloch sphere) has the symmetry of the three-dimensional physical space. This strongly suggests that one needs to go outside of quantum framework to explore possibilities of higher-dimensional physical spaces. The natural choice to start with are systems for which the state space is d-dimensional Bloch sphere and we call them "generalized spins". They can be derived from an information-theoretic analysis [49] and come as the most natural generalization of quantum spin. All such systems share fundamental features with the quantum spin, such as quadratic uncertainty relations for mutually unbiased (complementarity) observables, isotropic set of states, its rotational symmetry etc. They only differ in the dimension d of the state space.[1]

A large number of equally prepared generalized spins define a (generalized) spin coherent state. Under the restriction of coarse-grained measurements such spin coherent state acquires an effective description of a classical vector embedded in the d-dimensional space. One might think that the analogue with quantum theory can be developed further in that a spin coherent state can define the "field of the magnet"

---

[1]The argument for choosing to consider the generalized spins can be made more rigorous. In the spirit of operational theories we assume that every continuous reversible transformation (e.g. rotation) of macroscopic devices in the physical space generates a continuous reversible transformation of the system in the state space between two pure states. This excludes the "box-world" [12, 16] systems as they have discrete state spaces. The systems with relaxed uncertainty relations [50] are also excluded since they require non-linear transformations.

in which the elementary generalized spin can evolve, analogous to the Larmor precession but in a higher-dimensional physical space. With no preferred direction one would require the pairwise interaction between the generalized spin and each of the constituting spins of the coherent state to be invariant under the simultaneous group action on both (the rotational invariance). Here we show that no such interaction between the spin and the macroscopic field can generate the group of transformations of the spin unless its state space and the physical space in which the field acts are both three-dimensional—as in quantum theory and in our three-dimensional world.

In more precise terms we impose the following requirements on theory:

- **(Closeness)** *The dynamics of the elementary system of the theory can always be generated through the invariant interaction of the system with the macroscopic device that itself is obtained from within theory in the macroscopic limit.*
- **(Macroscopic states)** *The macroscopic transformation device ("magnetic field") is in a coherent state in which the constituting elementary systems are all equally prepared.*

The precondition for considering the two requirements is that ***the symmetry of the elementary system and of the macroscopic device by which the system is transformed are both those of the Euclidean d-dimensional space***. We show that if the elementary interactions between the elementary systems are *pairwise*, the two requirements can only be fulfilled if the underlying theory is quantum theory and $d = 3$.

An important restriction under which our result is obtained is that the generalized spin interacts *pairwise* with each single spin constituting the large spin in the coherent state. We show that if we relax this assumption, there are group invariant interactions between three or more generalized spins. This means that the spin under consideration could interact with several other spins, each one belonging to a different coherent state, and that such interaction could generate the group of transformation of the spin. The notion of the "field of the magnet" would then be extended such that it is represented not by a single but several coherent states. This opens up a possibility of having higher-dimensional Euclidian physical spaces compatible with underlying generalized probabilistic theory different from quantum theory. Nonetheless, we leave the question open of whether such a theory can be fully constructed in a mathematically consistent way.

In a recent work [51], Müller and Masanes gave an information-theoretic analysis of the relationship between the geometry of the state space of an elementary system (directional degree of freedom) and the classical space in which the macroscopic devices are embedded. In their work, they consider "spin" as an elementary directional degree of freedom to be measured by a macroscopic measurement device ("generalized Stern-Gerlach magnet") that can be oriented along arbitrary direction in d-dimensional physical space. Assuming that any spatial direction can be encoded in a physical state of the spin and no further information is encoded in the state, they derive that the state space is the d-dimensional Bloch sphere. In the next step, they show that such systems can exhibit continuous non-trivial dynamics only in three dimensions with the constraint to the *locally-tomographic* theories [52–54].

In the present approach, we take a different route. From the very beginning we consider the systems that have d-dimensional Bloch sphere as the state space and obtain the dimensionality of the physical space by the requirement that the theory is "closed". In a probabilistic theory the dynamics of a single system is assumed to be generated by an external field of macroscopic devices. In a closed probabilistic theory the fields are not notions from "outside" of the theory, but are obtained from within it in the macroscopic limit. Furthermore, we extend the study to the more general class of theories that are not in general *locally tomographical* [55]. The assumption of so called *local-tomography* states that the global state of a composite system can be learned trough local statistics. We allow for more general situations where the state of a composite system may include a set of *global parameters* that cannot be learned trough local statistics but trough the global (entangling) measurements on a whole system [56]. The prototype of the theory that involves global parameters is quantum mechanics based on real amplitudes [57]. This theory can be reconstructed within an information-theoretic approach [56]. Most of the previous information-theoretic reconstructions of quantum theory [13, 26, 29, 31], as well as the work of Ref. [51] adopt local tomography (e.g. directly eliminating real quantum mechanics), in contrast to our work here.

In conclusion, we reconstruct, the three-dimensional space and quantum mechanics trough the macroscopic limit under the constraint of pairwise elementary interactions. Interestingly, higher-dimensional space may arise in the limit if one allows for multipartite elementary interactions, i.e. ternary and more.

2 The Classical Limit of Quantum Theory and Three-Dimensionality of Space

In the operational approach to quantum mechanics the notion of quantum state refers to a well-defined configuration of the macroscopic instrument by which preparation of the state is defined. For example, the "horizontal" polarization of photon is specified by the "reference direction" of a classical object relative by which the polarization is prepared, such as the plane of the polarizing filter. On the other hand, if quantum mechanical laws are universal, then macroscopic, classical objects, such as polarizing filters, themselves should allow a description from within quantum mechanics.

One can consider a macroscopic object as a collection of large number of elementary quantum systems, which are in one of "macroscopically distinct states". The latter are defined as quantum states that can still be differentiated even if the measurement precision is poor and one performs coarse-grained measurements. The states can be repeatedly measured by different observers or copied with negligible disturbance. They are "robust" under disturbance or losses of a sufficiently small number of constituent quantum systems. These properties give rise to a level of "objectivity" of the macroscopically distinct states among the observers. A good example of such "classical" states are large spin-coherent states [39, 40] under the restriction of coarse-grained measurements. For the spin-J system the spin coherent states are defined as the eigenstates with the largest eigenvalue of spin projection along direction $\vec{n}$:

$$\hat{J}_{\vec{n}}|\vec{n}\rangle = J|\vec{n}\rangle, \tag{1}$$

where $\hat{J}_{\vec{n}} = \vec{n}\hat{\vec{J}}$ and $\vec{n} = \sin\theta\cos\phi\vec{e}_x + \sin\theta\sin\phi\vec{e}_y + \cos\theta\vec{e}_z$ (with no external reference frame assumed, $\vec{n}$ should be understood as a parametrization of the spin state with no further immediate physical interpretation). Their expansion in the eigenbasis of $\hat{J}_z$ reads:

$$|\vec{n}\rangle = \sum_{m=-J}^{J} \binom{2J}{J+m}^{\frac{1}{2}} \left(\cos\frac{\theta}{2}\right)^{J+m} \left(\sin\frac{\theta}{2}\right)^{J-m} e^{-im\phi}|m\rangle. \tag{2}$$

The spin-J particle can be considered as a composite system consisting of N spin-1/2 particles. The spin-J coherent state is then the product state of N equally prepared spin-1/2 particles ($J = N/2$)

$$|\vec{n}\rangle = |\vec{n}\rangle_1|\vec{n}\rangle_2 \ldots |\vec{n}\rangle_N \tag{3}$$

In the limit of large J (or large N) the spin coherent states acquire the properties of "classical" states. The probability of obtaining outcome m of J_z is given by the binomial distribution $p_m = |\langle m|\vec{n}\rangle|^2$. In the limit it reduces to the normal distribution:

$$p_m = \frac{1}{\sqrt{2\pi}\sigma} e^{\frac{(m-\mu)^2}{2\sigma^2}}, \tag{4}$$

where $\sigma = \sqrt{N}\sin\theta$ is the width of distribution and $\mu = N/2\cos\theta$ is the mean value. The overlap between two spin-coherent states

$$|\langle\vec{n}_1|\vec{n}_2\rangle|^2 = \left(\frac{1+\vec{n}_1\vec{n}_2}{2}\right)^N \longrightarrow \delta_{\vec{n}_1,\vec{n}_2}, \tag{5}$$

becomes exponentially small in the limit of large N.

The uncertainty of measuring J_z is given by the standard deviation σ. Under the restriction of coarse-grained measurements where the outcomes are merged into "slots" of size much larger than the standard deviation, the Gaussian cannot be distinguished anymore from the delta function [41] and the spin-coherent states become effectively "classical vectors" in three-dimensional space.

There are two independent ways in which large spin-coherent states can be said to induce the properties of the physical space. Firstly, they can be used to define the "reference direction" in a three-dimensional space, though one lacks this notion in the abstract Hilbert space formulation of quantum theory to start with. With no external reference frame only rotationally invariant observables can be measured, such as the total spin length. Consider a "large" spin of length J in a spin-coherent state $|\vec{n}\rangle$ and a "small" spin of length 1/2. It can be shown that the probability distribution for the outcomes $J + 1/2$ ("aligned") and $J - 1/2$ ("anti-aligned") of the total spin length approaches the probability distribution for the outcomes of spin projection of the spin-1/2 along the direction $\vec{n}$ in the classical limit ($N \rightarrow +\infty$) [42, 45]. In that way, the spin-coherent states define the complete *set of measurements* for the elementary spin. The set has the same dimensionality and the symmetry as the three-dimensional Euclidian physical space. We call these *static* properties of the space.

Secondly, spin-coherent states can generate non-trivial *dynamics* in three-dimensional space. A macroscopic spin in a coherent state can serve as an "external magnetic field" around which another spin can precess, i.e. it serves as a *transformation device* for the elementary spins [45]. Since there is no preferred direction beside the one defined by the large spin one requires the interaction between the elementary spin and the large spin to be *rotationally invariant*. To illustrate it consider the situation as given in Fig. 2 (left). A single spin-1/2 particle interacts with N spins prepared in a coherent state along direction $\vec{n}$. Total interaction Hamiltonian is the sum of all pairwise interactions $H = \sum_{n=1}^{N} H^{(0n)}$ where $H^{(0n)}$ labels the interaction between the single spin and nth spin of the macroscopic system. There is only one rotationally invariant Hamiltonian, that is the Heisenberg spin-spin interaction $H^{(0n)} = J_n \vec{\sigma}_0 \vec{\sigma}_n$, where J_n is the coupling constant. It can be shown that in the macroscopic limit the elementary spin only negligibly affects the state of a large spin and the dynamics of the elementary spin becomes unitary [45]:

$$e^{itH}|\psi\rangle|\vec{n}\rangle \approx (e^{itH_{eff}}|\psi\rangle)|\vec{n}\rangle, \tag{6}$$

where $H_{eff} = \vec{B}\vec{\sigma}$ is the effective Hamiltonian and $\vec{B}$ represents the strength of "macroscopic field" around which the "small spin" precesses (see Appendix 1 for details).

3 Generalized Spins and Higher-Dimensional Space

3.1 Single System

Is there a microscopic theory that in its macroscopic limit leads to classical physics embedded in a physical space of dimension higher than three? Following previous discussions one can expect that the elementary (two-level) system with the d-dimensional sphere $\mathcal{S}^{(d-1)}$ as state space gives rise to coherent states and "magnetic fields" embedded in a d-dimensional Euclidean space in the macroscopic limit. Such an elementary system is non-quantum because it represents a two-level system with more than three degrees of freedom. Within the information-theoretic framework of generalized theories, such generalized bit (here called "generalized spin") is derived as the most natural generalization of qubit—the system that is fundamentally limited to the content of one bit of information [26, 49]. Other information-theoretic approaches lead to the derivation of the same class of systems, e.g. by adopting *information causality* [58] or *continuous reversible dynamics* [27, 29].

The state of generalized spin is represented by a vector in a d-dimensional real space, $\mathbf{x} = (x_1, \ldots, x_d)$. The probability $P_1(\mathbf{x}, \mathbf{y})$ to obtain the spin along direction $\mathbf{y}$ when the state is prepared along direction $\mathbf{x}$ is expressed trough the generalized Born rule [26]:

$$P_1(\mathbf{x}, \mathbf{y}) = \frac{1}{2}(1 + \mathbf{x}^{\mathrm{T}}\mathbf{y}). \tag{7}$$

The set of pure states satisfy $P(\mathbf{x}, \mathbf{x}) = 1$ and is represented by a unit sphere $\mathcal{S}^{d-1}$ in d-dimensions (see Fig. 1). The characteristic feature differentiating between the theories is the number d of parameters required to describe the state completely. For example, classical probability has one parameter, real quantum mechanics has two, complex (standard) quantum mechanics has three and the one based on quaternions has five parameters. A lower-order theory of the single system can always be embedded in a higher-order ones in the same way in which classical theory of a bit can be embedded in qubit theory.

Following the operational approach we assume that the continuous reversible transformations of macroscopic devices acting upon the system generates the continuous reversible transformation of the state of the system. Therefore, the set of physical transformations is a continuous (Lie) group. Furthermore, if an arbitrary reversible transformation of the states can be realized manipulating the macroscopic device, then the group of physical transformations is transitive on a sphere [59, 60], i.e. any pure state can be transformed to any other in a continuous fashion. We will

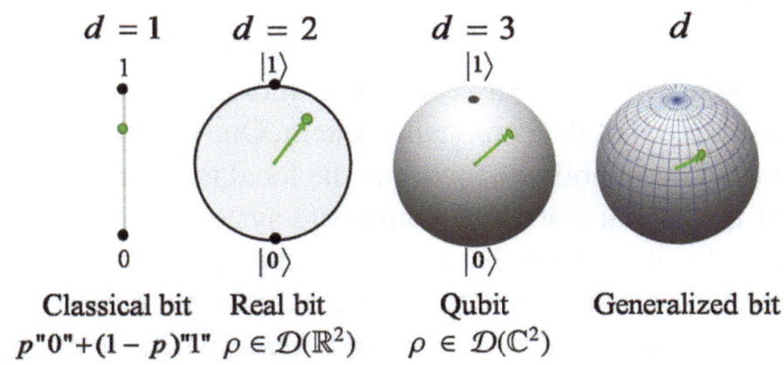

Fig. 1 Figure taken from Ref. [26]. State spaces of a generalized spin or generalized bit (two-level system). The minimal number of real parameters d is needed to specify the (mixed) state completely. From *left* to *right* A classical bit with one parameter (the weight p in the mixture of two bit values), a real bit with two real parameters (state $\rho \in \mathcal{D}(\mathbb{R}^2)$ is represented by 2×2 real density matrix), a qubit (quantum bit) with three real parameters (state $\rho \in \mathcal{D}(\mathbb{C}^2)$ is represented by 2×2 complex density matrix) and a and a generalized bit for which d real parameters are needed to specify the state. In the classical limit, a theory of elementary system with d parameters gives rise to physics of macroscopic, classical "fields" embedded in d-dimensional physical space (see main text)

consider minimal group transitive on $\mathcal{S}^{d-1}$, which is thus necessarily within the set of physical transformations (see Appendix 2). The existence of such "reversible transformations of macroscopic devices" is usually assumed *ad hoc*. The aim of this work is exactly to show that they do not always exist, if the macroscopic devices are not considered "outside" of the theory, but are required to be obtained from within it in the classical limit.

3.2 Generalized Spin-Coherent States

Generalized spin-coherent states can be straightforwardly introduced in generalized probabilistic theories. For every dimension d, they are collections of N equally prepared generalized spins. The preparation can be parameterized by a direction $\vec{n}$ in a d-dimensional space. Equations (4) and (5), derived in quantum theory, remain valid here as well. In the macroscopic limit of large N, the effective description of the coherent states is that of classical vectors embedded in a d-dimensional Euclidian space. We address here the question of whether generalized spin coherent states can generate non-trivial dynamics of individual spins in the space, similarly as the one given by Eq. (6). We will next show that with pairwise invariant interaction between elementary spins this is not possible except when $d = 3$. We then discuss possible generalizations of our approach to multi-spin invariant interactions that might give rise to non-trivial dynamics in higher-dimensional spaces.

4 Dynamics and Macroscopic Limit

4.1 The Composite System

In order to describe interactions between two or more generalized spins we need to introduce a representation of the composite system. One of the characteristics of both classical and quantum probabilistic theory is the local tomography [52–54], namely the property that the global state of a composite system is completely determined by the statistics of local measurements. For example, a state of two classical bits $\vec{p} = (p_{00}, p_{01}, p_{10}, p_{11})$, where e.g. p_{01} denotes the probability to obtain "spin up" on the first spin and "spin down" on the second one, can be equivalently represented by three numbers (x, y, t):

$$x = p_{00} + p_{01} - p_{10} - p_{11}, \tag{8}$$

$$y = p_{00} - p_{01} + p_{10} - p_{11}, \tag{9}$$

$$t = p_{00} - p_{01} - p_{10} + p_{11}. \tag{10}$$

The local statistics is given by mean values x and y of probabilities measured on the first and the second spin, respectively, whereas t is the mean value of correlation (difference between the probabilities that the two spins are the same and that they are different). Similarly, the density matrix ρ of two qubits can be decomposed as

$$\rho = \frac{1}{4}\left(\mathbb{1} \otimes \mathbb{1} + \sum_{i=1}^{3} x_i \sigma_i \otimes \mathbb{1} + \sum_{j=1}^{3} y_j \mathbb{1} \otimes \sigma_j + \sum_{i,j=1}^{3} T_{ij} \sigma_i \otimes \sigma_j\right), \tag{11}$$

where σ_i, $i = 1, 2, 3$, are Pauli operators. Vectors $\mathbf{x} = (x_1, x_2, x_3)$ and $\mathbf{y} = (y_1, y_2, y_3)$ are called local Bloch vectors and are the mean values of the Pauli operators and T is the 3×3 correlation matrix with elements $T_{ij} = \langle \sigma_i \sigma_j \rangle$.

Not all generalized probabilistic theories fulfill local tomography; an example is quantum mechanics based on real amplitudes. For the real bit only two Pauli matrices σ_1 and σ_3 correspond to physical observables, because σ_2 is a complex matrix. However, $\sigma_2 \otimes \sigma_2$ is a real matrix, and thus it corresponds to a physical observable, although it cannot be measured locally. In general, a real density matrix ρ can be represented in a form

$$\rho = \frac{1}{4}\left(\mathbb{1} \otimes \mathbb{1} + \sum_{i=1}^{3} x_i \sigma_i \otimes \mathbb{1} + \sum_{j=1}^{3} y_j \mathbb{1} \otimes \sigma_j + \sum_{i,j=1}^{3} T_{ij} \sigma_i \otimes \sigma_j + \lambda \sigma_2 \otimes \sigma_2\right), \tag{12}$$

where λ is a global parameter. Therefore, we can represent the state of a composite system by 4-tuple $(\mathbf{x}, \mathbf{y}, T, \lambda)$.

We now introduce a representation of the composite system of two generalized spins. Firstly, we assume that local measurements on individual spins are well defined (i.e. probabilities for local measurements are non-negative and they sum up to one). Secondly, if the subsystems of a composite system are emitted from two independent sources, we assume that the joint probability distribution is factorizable. Consequently, one can define properly the set of product states as triples $(\mathbf{x}, \mathbf{y}, T_p)$, where $T_p = \mathbf{x}\mathbf{y}^{\mathrm{T}}$. However, for general non-product states there might be some global parameters missing in the state description. Therefore, in the general case we associate a 4-tuple $\vec{\psi}_{12} = (\mathbf{x}, \mathbf{y}, T, \Lambda)$ to the state of a composite system, where $\mathbf{x}, \mathbf{y}$ are the local Bloch vectors, T is a $d \times d$ real matrix that represents correlations and $\Lambda = (\lambda_1, \lambda_2, \dots)$ is a collection of global parameters that can be present in the state description but are not accessible trough statistics of local measurements.

We define the probability distribution for obtaining the two local local spins "up" along measurement directions $\mathbf{a}, \mathbf{b}$ to be

$$P_{12}(\vec{\psi} \mid \mathbf{a}, \mathbf{b}) = \frac{1}{4}(1 + \mathbf{x}\mathbf{a} + \mathbf{y}\mathbf{b} + \mathbf{a}T\mathbf{b}), \tag{13}$$

where $\vec{\psi} = (\mathbf{x}, \mathbf{y}, T, \Lambda)$ is the state of the composite system. The formula can also be interpreted as the overlap between the state $\vec{\psi}$ and product state $\vec{\phi}_p = (\mathbf{a}, \mathbf{b}, \mathbf{a}\mathbf{b}^{\mathrm{T}}, 0)$:

$$P_{12}(\vec{\psi}, \vec{\phi}_p) = \frac{1}{4}(1 + \vec{\psi}^{\mathrm{T}}\vec{\phi}_p). \tag{14}$$

A general state of N spins is represented by $\vec{\psi}_N = (\mathbf{x}_1, \dots, \mathbf{x}_N, T_{12}, \dots, T_{123}, \dots, T_{1\dots N}, \Lambda)$, where $\mathbf{x}_i$ is the local Bloch vector of the ith spin, tensors $T_{i_1 i_2 \dots}$ represents correlations (two-spin, three-spin etc.) and $\Lambda = (\Lambda_{12}, \Lambda_{13}, \dots, \Lambda_{123}, \dots)$ is the set of all global parameters, where, for example, Λ_{123} is the global parameter related to subsystems 1, 2 and 3.

4.2 Dynamics in Macroscopic Limit

Dynamics of an individual generalized spin as generated by a transformation device is given by:

$$\frac{dx_i}{dt} = g_{ij}x_j, \tag{15}$$

where $[G]_{ij} = g_{ij}$ is the generator of evolution and t is the parameter of the transformation, usually taken to be time. Here and in the rest of the article the summation over repeated indices is always assumed. The integral version of the formula reads

$$\mathbf{x}(t) = U(t)\mathbf{x}(0), \tag{16}$$

where $U(t) = \exp(tG)$ is the reversible transformation that belongs to the group of transformation $\mathcal{G}$ of the generalized spin. Our main objective is to investigate if such a dynamics can be obtained as a mean field approximation of the theory. Note that in quantum mechanics this is the case and Eq. (16) is equivalent to Eq. (6). More precisely, we want to find out whether formula (16) for the dynamical evolution of an individual generalized spin can be seen as a consequence of its interaction with a system composed of a large number of generalized spins (e.g. in coherent state). A negative answer to this question would indicate that the theory is not *closed*.

We represent a single spin by its local Bloch vector $\mathbf{x}$ and the "large" system by a state $\vec{\psi}_N$. In the limit of large N, the following holds

$$W_N(t)\vec{\psi}_N \otimes \mathbf{x} = \vec{\psi}_N \otimes U(t)\mathbf{x} + \vec{O}(N, t), \tag{17}$$

where W_N represents the joint evolution of the system and of the field after duration t of the interaction. The state $\vec{\psi}_N \otimes \mathbf{x}$ represents the product state of a joint system (large + small system), in a sense that all the correlation tensors are factorized. If the dynamics of the small spin can be reproduced from the interaction, one has

$\vec{O}(N, t) \to 0$ in the limit when the number of spins N goes to infinity. Consequently, one recovers the Eqs. (15) and (16) exactly, the initial state $\vec{\psi}_N$ of the large system remains almost unchanged, and the dynamics factorizes.

4.3 Pairwise Interaction

Here we assume that all the interactions are pairwise at the elementary level. In Sect. 7 we will relax this assumption. The state of the composite system of two elementary generalized spins is represented by $\vec{\psi}_{12} = (\mathbf{x}, \mathbf{y}, T, \Lambda)$. The dynamical law reads $\vec{\psi}_{12}(t) = W_{12}(t)\vec{\psi}_{12}(0)$, where t is the duration of interaction. One has $W_{12}(t) = \exp(tH)$, where H is the generator of the interaction W_{12}. The differential version of the dynamical law reads:

$$\frac{dx_i}{dt} = a_{ij}x_j + b_{ij}y_j + \mu_{ijk}T_{jk} + L_{in}\lambda_n, \tag{18}$$

where a_{ij}, b_{ij}, μ_{ijk}, L_{in} are the components of the generator H. Similarly, one can write the differential equation for $\frac{dy_i}{dt}$, $\frac{dT_{ij}}{dt}$ and $\frac{d\lambda_n}{dt}$.

The small spin in the state $\mathbf{x}(t)$ is assumed to interact via pairwise interaction with each of N spins constituting the large spin. The dynamical equation for the small spin is given by:

$$\frac{dx_i}{dt} = a_{ij}x_j + \sum_{s=1}^{N} \left(b_{ij}^{(s)} y_j^{(s)} + \mu_{ijk}^{(s)} T_{jk}^{(s)} + L_{in}^{(s)} \lambda_n^{(s)} \right), \tag{19}$$

where $\mathbf{y}^{(s)}$ is the Bloch vector of the sth spin of the large system, $T^{(s)}$ is the correlation tensor of the small spin and the sth spin and $\Lambda^{(s)} = (\ldots, \lambda_n^{(s)}, \ldots)$ is the set of global parameters of the small spin and all spins of the large one.

We assume that each of the N constituents of the large system interacts with the small spin in a "same way", the only difference being in the strength of interaction (for example, because one spin is physically closer to the small spin than the other one). Thus, one has

$$b_{ij}^{(s)} = \beta_s b_{ij} \quad \mu_{ijk}^{(s)} = J_s \mu_{ijk}, \tag{20}$$

where β_s and J_s are the coupling constants defining the strength of interaction (they can be different due to the spatial distribution of particles that constitute the large system). Here b_{ij} and μ_{ijk} are constants that are characteristic of the pairwise interaction and they are assumed to be the same for all particles.

If we assume that in the macroscopic limit, the state of the large system changes negligibly during the interaction time, we obtain

$$T_{ij}^{(s)}(t) = x_i(t)y_j^{(s)}(0),\ \lambda_n^{(s)}(t) = \lambda_n^{(s)}(0) = 0,\ y_j^{(s)}(t) = y_j^{(s)}(0). \tag{21}$$

The Eq. (19) becomes:

$$\frac{dx_i(t)}{dt} = \left(a_{ij} + \mu_{ijk}\sum_{s=1}^{N} J_s y_k^{(s)}(0)\right)x_j(t) + b_{ij}\sum_{s=1}^{N}\beta_s y_j^{(s)}(0).$$

If $b_{ij} = 0$ (otherwise, the equation above does not represent unitary dynamics), this equation becomes equivalent to Eq. (15) in the limit of very large N, in which case one obtains:

$$g_{ij} = a_{ij} + \mu_{ijk}B_k. \tag{22}$$

Here $\mathbf{B} = \sum_{s=1}^{N} J_s\mathbf{y}^{(s)}(0)$ can be understood as an analog of macroscopic field or magnetization, resembling the field produced by a ferromagntic in quantum mechanics. Assuming that the large system is in a spin-coherent state $\vec{n}$, one obtains the same expression as in the case of quantum mechanics: $\vec{B} = N\langle J\rangle\vec{n}$, where $\langle J\rangle = \frac{1}{N}\sum_{i=1}^{N} J_n$.

5 Covariant Interaction

The dynamical equation that follows from (22) reads

$$\frac{dx_i}{dt} = (a_{ij} + \mu_{ijk}B_k)x_j, \tag{23}$$

where B_k is the component of the macroscopic field. Since we want the dynamics to be reversible (and therefore to transform pure states into pure states), the equation above should preserve the norm of $\mathbf{x}$. Therefore one has:

$$a_{ij} = -a_{ji} \text{ and } \mu_{ijk} = -\mu_{jik}. \tag{24}$$

The dynamics is solely generated by the field B_k, since a_{ij} and μ_{ijk} are constants that arise from the pairwise dynamics (18).

Let $\mathcal{G}$ be the group of transformations of a single spin (see Sect. 3.1). Since there is no external reference direction we assume that the dynamical law (23) is covariant, i.e. it has the same form in all frames of reference. More precisely, for any reversible transformation $R \in \mathcal{G}$ (note that R is a transformation on a sphere $\mathcal{S}^{(d-1)}$, therefore it is real and orthogonal $RR^T = \mathbb{1}$) that maps old coordinates of the spin and field into the new ones, $x_i' = R_{ii_1}x_{i_1}$ and $B_i' = R_{ii_1}B_{i_1}$, we assume that the dynamical law keeps the same form in new coordinates:

$$\frac{dx_i'}{dt} = (a_{ij} + \mu_{ijk}B_k')x_j'. \tag{25}$$

The tensors a_{ij} and μ_{ijk} do not change because they are constants of interaction. After the substitution one obtains:

$$R_{ii_1}\frac{dx_{i_1}}{dt} = (a_{ij} + \mu_{ijk}R_{kk_1}B_{k_1})R_{jj_1}x_{j_1}. \tag{26}$$

If we multiply the last equation with $R^{-1} = R^T$ we obtain

$$\frac{dx_{i_1}}{dt} = (a_{ij}R_{ii_1}R_{jj_1})x_{j_1} + (\mu_{ijk}R_{ii_1}R_{jj_1}R_{kk_1})B_{k_1}x_{j_1}. \tag{27}$$

Therefore, for all $R \in \mathcal{G}$ we require:

$$R_{ii_1}R_{jj_1}a_{i_1j_1} = a_{ij}, \tag{28}$$

$$R_{ii_1}R_{jj_1}R_{kk_1}\mu_{i_1j_1k_1} = \mu_{ijk}. \tag{29}$$

Under these conditions, the pairwise interaction (18) is invariant under simultaneous change of the local reference frames.

If $\mu_{ijk} = 0$, then the dynamics given by (23) becomes trivial as it does not depend on the internal state of the transformation device but only on the interaction constant a_{ij} (i.e. the set of transformations becomes one-parameter Lie group). We require that Eq. (29) has non-trivial solution $\mu_{ijk} \neq 0$.

6 Main Proofs

Here we show that only $d = 3$ gives non-trivial solution of the Eqs. (28) and (29). Recall that the group of physical transformations $\mathcal{G}$ contains the minimal group transitive on the sphere $\mathcal{S}^{(d-1)}$. All such groups are summarized in the Appendix 2.

6.1 *Hint to Representation Theory*

Our result is based on the group representation theory. We therefore first introduce some basic notions of the representation theory. For an abstract group $\mathcal{G}$ and element $g \in \mathcal{G}$ we say that a matrix $D(g) \in \mathrm{Mat}(\mathcal{H})$, where $\mathcal{H}$ is a vector space, defines a representation of $\mathcal{G}$ if $D(g_1g_2) = D(g_1)D(g_2)$ for every two group elements g_1 and g_2. In this work we consider only unitary (orthogonal) representations. Representation is called reducible if there exists a nontrivial invariant subspace for all the matrices $D(g)$. Otherwise it is irreducible (IR) representation. Therefore, the group induces a decomposition of the vector space $\mathcal{H} = \oplus_\mu \mathcal{H}^{(\mu)}$ into irreducible subspaces $\mathcal{H}^{(\mu)}$ and

$$D(g) = \oplus_\mu a_\mu \Delta^{(\mu)}(g), \tag{30}$$

where $\Delta^{(\mu)}(g)$ is an IR representation that appears with the frequency a_μ. The dimension of the IR subspace is $|\mathcal{H}^{(\mu)}| = |\mu| a_\mu$, where $|\mu|$ is the dimension of the IR representation $\Delta^{(\mu)}$. The frequency of some IR representation can be computed as

$$a_\mu = (\chi^{(\mu)}, \chi) = \frac{1}{|\mathcal{G}|} \sum_{g \in \mathcal{G}} \chi^{(\mu)}(g^{-1}) \chi(g), \tag{31}$$

where $\chi(g) = \mathrm{Tr}(D(g))$ and $\chi^{(\mu)}(g) = \mathrm{Tr}(\Delta^{(\mu)}(g))$ are the characters of the representations.

For two representations $D_1(\mathcal{G})$ and $D_2(\mathcal{G})$ one can define the tensor product $(D_1 \otimes D_2)(\mathcal{G})$ that is representation of $\mathcal{G}$ itself. If D_1 and D_2 are IR, then the decomposition of $D_1 \otimes D_2$ is called Clebsch–Gordan (CG) series. In this work, it will be of particular interest to compute the frequency of the trivial representation $\Delta^{(1)}(g) = 1$. The following lemma will be used (see Appendix for the proof):

Lemma 1 *CG series of the product $\Delta^{(\mu)} \otimes \Delta^{(\nu)}$, where $\Delta^{(\mu)}$, $\Delta^{(\nu)}$ are real and irreducible, contains the trivial representation if and only if $\mu = \nu$ and then the trivial representation appears once, only.*

The main purpose of introducing the tools of representation theory is to solve Eqs. (28) and (29). The left hand side of Eqs. (28) and (29) can be seen as an action of the Kronecker products $D(\mathcal{G}) \otimes D(\mathcal{G})$ and $D(\mathcal{G}) \otimes D(\mathcal{G}) \otimes D(\mathcal{G})$, respectively, with $D(\mathcal{G})$ being the representation of the group of transformation $\mathcal{G}$ and $D(R) = R \in \mathcal{G}$. The solutions a_{ij} and μ_{ijk} are invariant under the action of the group of transformations $\mathcal{G}$, hence they lie within the totally invariant IR subspace that belongs to the trivial representation. Therefore, we will need the CG decomposition of $D \otimes D$ and $D \otimes D \otimes D$ in order to solve Eqs. (28) and (29).

6.2 *d* Odd Case and *d* $\neq$ 7

If we assume d odd, $d > 1$ and $d \neq 7$, the set of physical transformations contains the special orthogonal group $\mathrm{SO}(d) \lhd \mathcal{G}$ (see Appendix 2).

The easiest way to solve Eq. (28) is as follows. We rewrite it into a matrix form: $RAR^{\mathrm{T}} = A$ for all $R \in \mathrm{SO}(d)$. This is possible only if A is a scalar matrix $A = a\mathbb{1}$. Taking into account Eq. (24) we conclude $A = 0$. Although this gives the solution in the particular case considered, we will proceed with full group-representation analysis of Eqs. (28) and (29) as we will need it later on.

The set of $d \times d$ orthogonal matrices of the unit determinant define a d-dimensional, real IR representation of $\mathrm{SO}(d)$ and we label it as Δ^d with $\Delta^d(R) = R$. The left-hand side of Eqs. (28) and (29) can be seen as the action of product representation $\Delta^d(R) \otimes \Delta^d(R)$ and $\Delta^d(R) \otimes \Delta^d(R) \otimes \Delta^d(R)$. The solutions a_{ij} and μ_{ijk} lie within the IR subspace that belongs to the trivial representation in CG series of $\Delta^d \otimes \Delta^d$ and $\Delta^d \otimes \Delta^d \otimes \Delta^d$, respectively.

Let us analyze the product $\Delta^d \otimes \Delta^d$. Note that this representation commutes with the permutation group S_2 (of two elements). Therefore $\Delta^d \otimes \Delta^d$ can be decomposed on invariant subspaces that are irreducible under the action of S_2. There are two of them and they define symmetric and antisymmetric subspace of the dimensions $\frac{1}{2}d(d+1)$ and $\frac{1}{2}d(d-1)$ spanned by Hermitian and skew-Hermitian matrices, respectively. Furthermore, the symmetric subspace can be decomposed into one-dimensional subspace $\mathcal{H}^{(1)}$ spanned by the identity matrix $\mathbb{1}$ (invariant under $\Delta^d \otimes \Delta^d$, hence belongs to the trivial subspace) and its orthogonal complement $\mathcal{H}^{(S)}$ of dimension $\frac{1}{2}d(d+1) - 1 = \frac{1}{2}(d-1)(d+2)$. This induces the decomposition $\mathbb{R}^d \otimes \mathbb{R}^d = \mathcal{H}^{(1)} \oplus \mathcal{H}^{(S)} \oplus \mathcal{H}^{(AS)}$. Each subspace is irreducible for $\Delta^d \otimes \Delta^d$ (see Appendix 4). Therefore, the CG series is given by:

$$\Delta^d \otimes \Delta^d = \Delta^1 \oplus \Delta^S \oplus \Delta^{AS}, \tag{32}$$

where Δ^S and Δ^{AS} are the corresponding IR representations of $\mathrm{SO}(d)$ of the dimensions $AS = \frac{1}{2}d(d-1)$ and $S = \frac{1}{2}(d-1)(d+2)$, respectively. Since the trivial representation appears once only, the solution to (28) is one-dimensional and spanned by identity matrix $a_{ij} = a\delta_{ij}$. By applying condition (24) we get $a_{ij} = 0$.

The degeneracy of the solution of Eq. (29) can be found from the decomposition

$$\begin{aligned} \Delta^d \otimes \Delta^d \otimes \Delta^d &= \Delta^d \otimes (\Delta^d \otimes \Delta^d) \\ &= \Delta^d \otimes \left(\Delta^1 \oplus \Delta^S \oplus \Delta^{AS}\right) \\ &= \Delta^d \otimes \Delta^1 \oplus \Delta^d \otimes \Delta^S \oplus \Delta^d \otimes \Delta^{AS}. \end{aligned} \tag{33}$$

Let us apply Lemma 1. The decomposition of the first term $\Delta^d \otimes \Delta^1$ in last equation does not contain the trivial representation because $d > 1$. The second term $\Delta^d \otimes \Delta^S$ contains the trivial representation, if Δ^d and Δ^S are equivalent, which is possible only if $d = S = \frac{1}{2}(d-1)(d+2)$. There is no solution to this equation among d odd numbers. Similarly, the last term contains the trivial representation, only if $d = \frac{1}{2}d(d-1)$, that has solution $d = 3$. Furthermore for $d = 3$ the solution is one-dimensional and is represented by the completely antisymmetric (Levi-Civita) tensor $\mu_{ijk} = \epsilon_{ijk}$, where $\epsilon_{ijk} = +1$ for (ijk) being an even permutation. The solution $d = 3$ is worked out in details in Appendix 5.

6.3 $d = 7$ Case

Here the minimal transitive group on S^6 is the exceptional Lie group G_2. The generators span a 14-dimensional Lie algebra:

$$H(\mathbf{x}) = \begin{pmatrix} 0 & x_1 & -x_2 & x_3 & -x_4 & -x_5 & x_9-x_7 \\ -x_1 & 0 & x_6 & x_7 & x_8-x_5 & x_4-x_{11} & x_3+x_{10} \\ x_2 & -x_6 & 0 & -x_8 & x_9 & x_{10} & x_{11} \\ -x_3 & -x_7 & x_8 & 0 & x_{13}-x_6 & x_{14}-x_2 & x_{12}-x_1 \\ x_4 & x_5-x_8 & -x_9 & x_6-x_{13} & 0 & x_{12} & -x_{14} \\ x_5 & x_{11}-x_4 & -x_{10} & x_2-x_{14} & -x_{12} & 0 & x_{13} \\ x_7-x_9 & -x_3-x_{10} & -x_{11} & x_1-x_{12} & x_{14} & -x_{13} & 0 \end{pmatrix}. \tag{34}$$

We will next show that this generator, in general is not of the form (22), i.e. $H_{ij} = a_{ij} + \mu_{ijk}B_k$. This means that the dynamics generated by macroscopic field B_k exceeds the group G_2. On the other hand, there is no group transitive on S^6 other than G_2 and SO(7) (see Appendix 2). Since the group of transformations exceeds G_2, it has to be SO(7). But the case of SO(7) has been studied in the previous section, where it was shown that no nontrivial solution to Eq. (29) exists in this case.

We apply the analysis from the last section in the present case. We label 7-dimensional IR representation of G_2 as Δ^7. According to Behrends et al. [61] CG series is given by

$$\Delta^7 \otimes \Delta^7 = \Delta^1 \oplus \Delta^7 \oplus \Delta^{14} \oplus \Delta^{27}, \tag{35}$$

hence the trivial representation appears once only. Consequently the solution to (28) is spanned by the identity matrix $a_{ij} = a\delta_{ij}$. Constraint (24) gives $a_{ij} = 0$. Next, in the decomposition

$$\begin{aligned} \Delta^7 \otimes \Delta^7 \otimes \Delta^7 &= \Delta^7 \otimes (\Delta^7 \otimes \Delta^7) \\ &= \Delta^7 \otimes \left(\Delta^1 \oplus \Delta^7 \oplus \Delta^{14} \oplus \Delta^{27}\right) \\ &= \Delta^7 \otimes \Delta^1 \oplus \Delta^7 \otimes \Delta^7 \oplus \Delta^7 \otimes \Delta^{14} \oplus \Delta^7 \otimes \Delta^{27}, \end{aligned} \tag{36}$$

the trivial representation appears once only due to the term $\Delta^7 \otimes \Delta^7$. Therefore, the solution of Eq. (29) is unique (up to a constant) and is given by completely antisymmetric tensor ψ_{ijk} taking the non-zero value of $+1$ for $ijk = 123, 145, 176, 246, 257, 347, 365$. Incidentally, note that ψ_{ijk} is the tensor involved in the definition of the multiplication rule of octonions and seven-dimensional cross product [62]:

$$(\mathbf{a} \times \mathbf{b})_i = \psi_{ijk} a_j b_k, \tag{37}$$

where **a** and **b** are two octonions.

Let us set the macroscopic field of (22) to $B_k^{(1)} = B\delta_{1k}$. The corresponding generator $g_{ij} = \psi_{ijk}B_k^{(1)} = B\psi_{ij1}$ has six nonzero elements $g_{23} = g_{45} = g_{76} = -g_{32} = -g_{54} = -g_{67} = +1$:

$$G = B \begin{pmatrix} 0 & 0 & 0 & 0 & 0 & 0 & 0 \\ 0 & 0 & 1 & 0 & 0 & 0 & 0 \\ 0 & -1 & 0 & 0 & 0 & 0 & 0 \\ 0 & 0 & 0 & 0 & 1 & 0 & 0 \\ 0 & 0 & 0 & -1 & 0 & 0 & 0 \\ 0 & 0 & 0 & 0 & 0 & 0 & -1 \\ 0 & 0 & 0 & 0 & 0 & 1 & 0 \end{pmatrix}. \tag{38}$$

This generator is not of the form (34); therefore the dynamics generated by $B_k^{(1)}$ goes beyond the G_2 group. Since the only transitive groups on S^6 are G_2 and SO(7), and since we already excluded SO(7) in previous section, the Eq. (29) has no solution.

6.4 $d = 4k$ Case ($k = \frac{1}{2}, 1, 2, \ldots$)

In this case the minimal group transitive on $\mathcal{S}^{d-1}$ contains total inversion $E\mathbf{x} = -\mathbf{x}$. Now, we set $R = E$ in the Eq. (29), hence $-\mu_{ijk} = \mu_{ijk}$. This gives only trivial solution $\mu_{ijk} = 0$.

6.5 $d = 4k + 2$ Case ($k = 1, 2, 3, \ldots$)

In this case the minimal transitive group is $\mathrm{SU}(2k + 1)$. For some complex unitary $u \in \mathrm{SU}(2k + 1)$ its representation in ($d = 4k + 2$)-dimensional real space is given by the following matrix:

$$D(u) = \begin{pmatrix} \mathrm{Re}\, u & -\mathrm{Im}\, u \\ \mathrm{Im}\, u & \mathrm{Re}\, u \end{pmatrix}. \tag{39}$$

Note that this representation commutes with the symplectic form $J = \begin{pmatrix} 0 & \mathbb{1} \\ -\mathbb{1} & 0 \end{pmatrix}$, i.e.

$$[D(u), J] = 0, \tag{40}$$

for every u.

Let us analyze the case where u is a real matrix, that is $u \in \mathrm{SO}(2k+1) \lhd \mathrm{SU}(2k+1)$. Then $D(u) = \mathbb{1}_2 \otimes u$, where $\mathbb{1}_2$ is a 2×2 identity matrix. The Eq. (29) can be written in a tensor form:

$$(\mathbb{1}_2 \otimes u) \otimes (\mathbb{1}_2 \otimes u) \otimes (\mathbb{1}_2 \otimes u)|\mu\rangle = |\mu\rangle, \tag{41}$$

or equivalently

$$(\mathbb{1}_2 \otimes \mathbb{1}_2 \otimes \mathbb{1}_2) \otimes (u \otimes u \otimes u)|\widetilde{\mu}\rangle = |\widetilde{\mu}\rangle, \tag{42}$$

where $|\widetilde{\mu}\rangle$ and $|\mu\rangle$ are the ket vectors that correspond to the tensors μ_{ijk} and $\widetilde{\mu}_{ijk}$ and are connected by a suitable transformation. The solution to the last equation can be found in a product form $|\widetilde{\mu}\rangle = |\chi\rangle|\phi\rangle$, where

$$(u \otimes u \otimes u)|\phi\rangle = |\phi\rangle, \tag{43}$$

holds for every $u \in \mathrm{SO}(2k + 1)$. This equation has been analyzed earlier and it has nontrivial solution only if $2k + 1 = 3$ or $d = 6$. In that case, solution $|\phi\rangle$ has components ϕ_{ijk} that are the Levi-Civita tensor ϵ_{ijk}, hence we write the solution as $|\widetilde{\mu}\rangle = |\chi\rangle|\epsilon\rangle$.

We have found non-trivial solution for the case $d = 6$ and the corresponding group is SU(3). The group generators span 8-dimensional Lie algebra and the corresponding real representation reads:

$$H(\mathbf{x}) = \begin{pmatrix} 0 & -x_4 & -x_5 & x_7 & x_1 & x_2 \\ x_4 & 0 & -x_6 & x_1 & x_8 - x_7 & x_3 \\ x_5 & x_6 & 0 & x_2 & x_3 & -x_8 \\ -x_7 & -x_1 & -x_2 & 0 & -x_4 & -x_5 \\ -x_1 & x_7 - x_8 & -x_3 & x_4 & 0 & -x_6 \\ -x_2 & -x_3 & x_8 & x_5 & x_6 & 0 \end{pmatrix}. \tag{44}$$

We set the notation $H_i = H(\mathbf{e}^{(i)})$, where $e_k^{(i)} = \delta_{ik}$ is the kth component of $e_k^{(i)}$.

Similarly to the previous section our goal is to show that $a_{ij} + \mu_{ijk}B_k$ generate transformations that go beyond the SU(3) group. In such a case, the group of transformations exceeds the minimal transitive group. Since there is no group transitive on S^5 other than SU(3) that do not contain the total inversion, one concludes that there is no nontrivial solution to Eq. (29).

Note that the solution to Eq. (28) is twofold $a_{ij} = \alpha\delta_{ij} + \beta J_{ij}$, where J_{ij} is the symplectic form. However, since $a_{ij} = -a_{ji}$ we have $\alpha = 0$. Furthermore, symplectic form J_{ij} does not belong to the set of generators $H(\mathbf{x})$ therefore $\beta = 0$ and finally $a_{ij} = 0$.

Recall that the solution to (42) can be found in the product form $|\chi\rangle|\epsilon\rangle$ where $|\epsilon\rangle$ is the tensor Levi-Civita. Let us set the macroscopic field of (22) to $B_k^{(1)} = B\delta_{1k}$. In that case the generator becomes $G = B\chi \otimes E_1$, where χ_{ab} is some symmetric 2×2 matrix and $[E_1]_{ij} = \epsilon_{ij1}$. One has

$$G = B \begin{pmatrix} 0 & 0 & 0 & 0 & 0 & 0 \\ 0 & 0 & \chi_{11} & 0 & 0 & \chi_{12} \\ 0 & -\chi_{11} & 0 & 0 & -\chi_{12} & 0 \\ 0 & 0 & 0 & 0 & 0 & 0 \\ 0 & 0 & \chi_{21} & 0 & 0 & \chi_{22} \\ 0 & -\chi_{21} & 0 & 0 & -\chi_{22} & 0 \end{pmatrix}. \tag{45}$$

This can be generator of the form (44) if $\chi_{12} = \chi_{21} = 0$ and $\chi_{11} = \chi_{22} = \chi_0$. Therefore $G = B\chi_0 H_6$. On the other hand, the dynamics generated by $B_k^{(1)} = \delta_{k1}$ is invariant under all transformations that keep $B_k^{(1)}$ invariant. In this particular case, it means that G has to commute with the generators H_6 and H_3. This gives only trivial solution $\chi_0 = 0$, hence $G = 0$. Similarly, one can draw the same conclusion for any other $B_k^{(s)} = B\delta_{ks}$. Therefore, the dynamics generated by arbitrary B_k goes beyond the SU(3) group.

7 Going Beyond Three Dimensions

In this section we shall argue that higher-dimensional macroscopic limit may arise as a consequence of a multi-partite invariant interaction among elementary spins. We construct an explicit model of dynamics in analogy to the quantum case and three dimensions (see Appendix 5 for details). However, it remains as an open question if such an ansatz leads to a proper probabilistic theory, in the sense that positivity of probabilities is not guaranteed.

Note that it as an artefact of the three dimensions that the evolution Eq. (15) can be written in the form

$$\frac{d\mathbf{x}}{dt} = \mathbf{B} \times \mathbf{x}, \tag{46}$$

where the vector $\mathbf{B}$ generates evolution with the generator matrix $g_{ij} = \epsilon_{ijk} B_k$. This expression for $d > 3$ is no longer possible. The evolution cannot be generated by a single vector, but a tensor. We will show that such a situation arises in the macroscopic limit if elementary interactions were multi-particle.

Let us start with the dimension $d = 4$. We consider three generalized spins described by a state

$$\psi = \{\mathbf{x}, \mathbf{y}, \mathbf{z}, T^{(12)}, T^{(13)}, T^{(23)}, T^{(123)}, \Lambda\}. \tag{47}$$

Let the spins interact via genuine three-particle, rotationally invariant interaction (see Fig. 2, right) In analogy with the quantum case discussed above, we can consider the dynamical equation for, say the first spin, as follows:

$$\frac{dx_i}{dt} = a\epsilon_{ijkl} T_{jkl}^{(123)} + L_{in}^{(1)} \lambda_n. \tag{48}$$

Here, a is a constant and ϵ_{ijkl} is the completely antisymmetric tensor of four indices, with $\epsilon_{1234} = +1$. It is well know that this tensor is invariant under SO(4) rotations. Analogously, one can write the equations for the other two local Bloch vectors y_i and z_i, as well as for correlations, both bipartite and tripartite and the global parameter.

Next we consider an ensemble of a large number N of spins. Let a single spin interact with each of the N spins via three-partite interaction defined above. In the macroscopic limit the dynamics should factorize and the state of the large system of N spins should not evolve in time. Therefore, all the correlations between single spin and large system factorize:

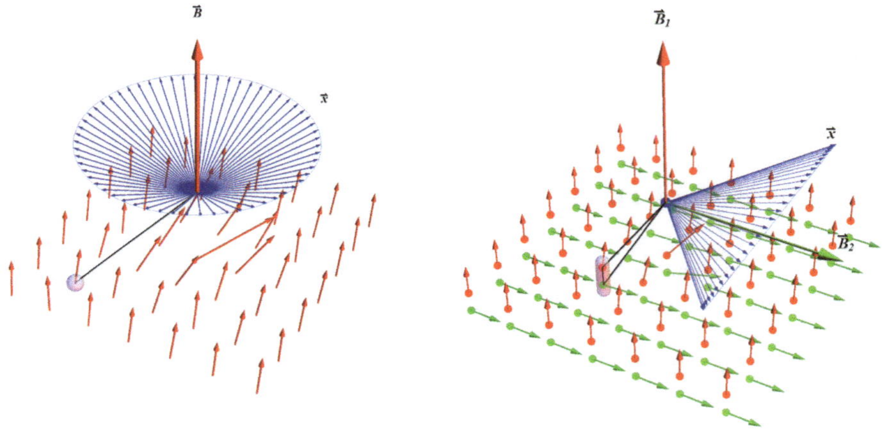

Fig. 2 Dynamics of the generalized spin as generated by its interaction with a single coherent state in three-dimensional space (*left*) or with a pair of coherent states in four-dimensional space (*right*). The coherent state is a collection of a large number of equally prepared constituent spins which are distributed here on a regular lattice. With no pre-existing reference direction all interactions are assumed to be rotationally invariant. In the macroscopic limit of an infinite large coherent states the effect of the spin on the coherent state is negligible and the dynamics becomes separable (i.e. the spin evolves according to the unitary evolution and the coherent state remains unchanged). (*Left*) Rotation of the quantum spin in three dimensions. The spin $\vec{x}$ interacts pairwise with each constituent spin of the coherent state $\vec{B}$. In the macroscopic limit this results in an effective precession of spin $\vec{x}$ around the classical macroscopic field generated by the spin coherent state $\vec{B}$. (*Right*) Rotation of the generalized spin in four dimensions. The spin $\vec{x}$ interacts via a three-particle interaction with each spin-pair, where one spin (*red*) of the pair belongs to coherent state $\vec{B}_1$ and the other one (*green*) to coherent state $\vec{B}_2$. In the macroscopic limit the effective dynamics of the generalized spin is rotation in the plane orthogonal to two macroscopic fields which are represented by the two coherent states $\vec{B}_1$ and $\vec{B}_2$. In the figure it is shown a projection of the dynamics in three dimensions

$$\begin{aligned} T_{ijk}^{(0nm)}(t) &= x_i(t)T_{jk}^{(nm)}(0), \\ T_{ij}^{(0m)}(t) &= x_i(t)y_j^{(m)}(0), \\ \Lambda(t) &= 0, \end{aligned} \tag{49}$$

where index 0 labels the single spin, whereas n labels the nth spins of the large system ($n = 1 \dots N$). The Λ labels the set of all global parameters between the single spin and the large system.

The equation of motion for the single spin reads:

$$\frac{dx_i}{dt} = a\epsilon_{ijkl}x_j \sum_{n,m=1}^{N} J_{nm}T_{kl}^{(nm)}(0), \tag{50}$$

where J_{nm} is the coupling constant between the single spin and spins n and m of the large system. Taking $B_{ij} = a\sum_{n,m=1}^{N} J_{nm}T_{kl}^{(nm)}(0)$ one obtains a reversible dynamics of a single spin:

$$\frac{dx_i}{dt} = \epsilon_{ijkl} B_{kl} x_j , \tag{51}$$

The dynamics is then generated by a covariant tensor field B_{ij}. We can further assume the situation as described in Fig. 2, right. The spins of the large system are arranged in a (regular) lattice such that each cell consist of two spins prepared along orthogonal directions $\vec{n}_1$ and $\vec{n}_2$. The two arrays of spins define two spin-coherent states. If we assume that the small spin interacts with two spins of a single cell, we obtain $B_{ij} = N\langle J\rangle n_{1i} n_{2j}$, where $\langle J\rangle = \frac{1}{N}\sum_{n=1}^{N} J_n$. We can say that dynamics is generated by two spin-coherent states defined by directions $\vec{n}_1$ and $\vec{n}_2$.

The present analysis for $d = 4$ can be generalized to higher-dimensions in a straightforward way. The dynamics of a generalized spin in d dimensions can be obtained from the SO(d) invariant dynamics that is generated by a genuine $(d-1)$-particle interaction. Of course, it is an open question if the set of Eq. (48) leads to a proper physical solution, in the sense that positivity of probabilities is not violated. We leave this question open for future investigation.

8 Conclusions

Physicist study models with extra dimensions. This research appears to be justified as we do not know of convincing arguments why we should necessarily live in three-dimensional space (or 3+1 space-time). In this paper we put a "closeness" requirement on every physical theory, which restricts the possible dimensions. The theory is closed if macroscopic field—which, via interaction with a microscopic system generates its dynamics—itself is described by the theory in the classical limit.

In the operational approach to a physical theory, one expects that the dimension and the symmetry of the state space of the "elementary system" are the same as those of the space in which "laboratory devices" are embedded. This is for the simple reason that the parameters describing the state operationally have no other meaning than that of the parameters that specify the configuration of macroscopic instruments by which the states are prepared, transformed or measured. On the other hand, the states of the macroscopic instruments can be obtained from within the theory in the classical limit; for example, in quantum mechanics, the "magnetic field" is represented by the coherent state of a very large number of equally prepared spins. Arbitrary unitary transformation of the elementary quantum spin (spin-1/2) can be generated by a (group invariant) bipartite interaction between the spin and the "magnetic field" (i.e. between the spin and each of the spins constituting the coherent state that represents the "field"). Therefore quantum theory is closed according to our requirement.

We showed that in no probabilistic theory of spin (where the spin has d components), other than quantum mechanics ($d = 3$), an invariant *pairwise* interaction can generate the group of transformation of the spin. However, if one considers three- or more-spin interactions this possibility might be realized. This opens up a possibility of having higher-dimensional spaces ($d > 3$) and "laboratory devices" embedded in it, which could generate the group of transformation of spin with the state space dimension $d > 3$. We hope that our work will be useful for physicists considering the existence of extra dimensions or other modifications of space-time.

Appendix 1: Dynamics of Spin in Presence of Spin-Coherent State

Here we justify the approximation made in Sect. 4. Namely, we show that Eq. (22) can be realized within quantum mechanics. We follow the idea given in the work by Poulin [46]. Let the large system be a ferromagnet composed of N spin-1/2 particles with the Hamiltonian H_0. We assume that H_0 is rotationally invariant $U^{\otimes N} H_0 U^{\dagger \otimes N} = H_0$ for all single particle rotations $U \in SU(2)$. One particular example of such a system is a Heisenberg ferromagnet with the Hamiltonian:

$$H_0 = -\sum_{n,m=1}^{N} J_{nm} \vec{\sigma^{(n)}} \vec{\sigma^{(m)}}, \tag{52}$$

with $J_{nm} \geq 0$ are the coupling constants. The rotational invariance is an important assumption because there is no external reference direction. The large system itself can be used to define preferred direction in space. Referring to the well known result in solid state physics [63] such a system, although rotationally invariant, can still exhibit spontaneous magnetization bellow the critical temperature. At zero temperature all the spins are aligned along some direction, that we choose to be the $\mathbf{e}_z$-direction. Hence the ground state is $|\psi_0\rangle = |0\rangle^{\otimes N}$ with the energy set to zero $E_0 = 0$ (this is always possible by changing the energy reference point). Let the small system be prepared in a state $|\phi\rangle = \alpha|0\rangle + \beta|1\rangle$ and assume $\sigma_3|0\rangle = |0\rangle$. The system interacts with the large system via Heisenberg interaction, therefore the total Hamiltonian reads

$$H = \sum_{n=1}^{N} J_n \vec{\sigma^{(0)}} \vec{\sigma^{(n)}} + H_0, \tag{53}$$

where J_n is the coupling constant for the interaction between the small spin and nth spin of the large system. Our goal is to show that in macroscopic limit $N \to \infty$, the dynamics becomes separable:

$$e^{itH}|\phi\rangle|\psi_0\rangle = (e^{itH_{eff}}|\phi\rangle)|\psi_0\rangle, \tag{54}$$

where H_{eff} is an effective Hamiltonian.

Firstly, let us compute the following

$$\begin{aligned}
H|\phi\rangle|\psi_0\rangle &= \sum_{n=1}^{N}(J_n\vec{\sigma^{(0)}}\vec{\sigma^{(n)}} + H_0)|\phi\rangle|0\rangle^{\otimes N} \\
&= \sum_{n=1}^{N} J_n\vec{\sigma^{(0)}}\vec{\sigma^{(n)}}|\phi\rangle|0\rangle^{\otimes N} \\
&= \left(\sum_{n=1}^{N} J_n\right)(\sigma_3|\phi\rangle)(\sigma_3^{(n)}|0\rangle^{\otimes N}) \\
&+ \sum_{n=1}^{N} J_n \sum_{i=1}^{2} \sigma_i^{(0)}\sigma_i^{(n)}|\psi\rangle|0\rangle^{\otimes N} \\
&= \left(\sum_{n=1}^{N} J_n\right)(\sigma_3|\phi\rangle)|0\rangle^{\otimes N} + \sum_{n=1}^{N} J_n \sum_{i=1}^{2} \sigma_i^{(0)}\sigma_i^{(n)}|\psi\rangle|0\rangle^{\otimes N} \\
&= |\chi\rangle + |\mu\rangle,
\end{aligned} \tag{55}$$

where

$$|\chi\rangle = \left(\sum_{n=1}^{N} J_n\right)(\sigma_3|\phi\rangle)|0\rangle^{\otimes N}, \tag{56}$$

$$|\mu\rangle = \sum_{n=1}^{N} J_n \sum_{i=1}^{2} \sigma_i^{(0)}\sigma_i^{(n)}|\psi\rangle|0\rangle^{\otimes N}. \tag{57}$$

The norm of $|\chi\rangle$ is easy to compute $\langle\chi|\chi\rangle = (\sum_{n=1}^{N} J_n)^2$. On the other hand, we have:

$$\begin{aligned}
|\mu\rangle &= (\sigma_1|\psi\rangle)(J_1|1\rangle|0\rangle|0\rangle \cdots + J_2|0\rangle|1\rangle|0\rangle \ldots) \\
&+ (i\sigma_2|\psi\rangle)(J_1|1\rangle|0\rangle|0\rangle \cdots + J_2|0\rangle|1\rangle|0\rangle \ldots) \\
&= ((\sigma_1 + i\sigma_2)|\psi\rangle)\,(J_1|1\rangle|0\rangle|0\rangle \cdots + J_2|0\rangle|1\rangle|0\rangle \ldots).
\end{aligned} \tag{58}$$

The norm of $|\mu\rangle$ is given by $\langle\mu|\mu\rangle = \sum_{n=1}^{N} J_n^2$. Let us define the averages

$$\langle J\rangle_N = \frac{1}{N}\sum_{n=1}^{N} J_n, \tag{59}$$

$$\langle J^2\rangle_N = \frac{1}{N}\sum_{n=1}^{N} J_n^2. \tag{60}$$

We assume that $\langle J \rangle_N$ and $\langle J^2 \rangle_N$ have finite values in macroscopic limit. Furthermore, we assume that $\lim_{N\to\infty}\langle J \rangle_N = \langle J \rangle \neq 0$. We can express the norms of $|\chi\rangle$ and $|\mu\rangle$ in terms of these quantities

$$\langle \chi|\chi\rangle = N^2\langle J \rangle_N, \tag{61}$$

$$\langle \mu|\mu\rangle = N\langle J^2 \rangle_N. \tag{62}$$

Now, it is clear that $|\mu\rangle$ is a vector of short length as compared to $|\chi\rangle$ when N is large. Furthermore, in the macroscopic limit, we have $\lim_{N\to\infty}\frac{\langle \mu|\mu\rangle}{\langle \chi|\chi\rangle} = 0$, therefore one can safely remove $|\mu\rangle$ from Eq. (55) when $N \to \infty$:

$$H|\phi\rangle|\psi_0\rangle = (H_{eff}|\phi\rangle)|\psi_0\rangle, \tag{63}$$

where $H_{eff} = N\langle J \rangle\sigma_3$. Now we can prove (54):

$$\begin{aligned} e^{itH}|\phi\rangle|\psi_0\rangle &= \sum_{k=0}^{+\infty}\frac{t^k}{k!}H^k|\phi\rangle|\psi_0\rangle \\ &= \sum_{k=0}^{+\infty}\frac{t^k}{k!}(H_{eff}^k|\phi\rangle)|\psi_0\rangle \\ &= (e^{itH_{eff}}|\phi\rangle)|\psi_0\rangle. \end{aligned} \tag{64}$$

In general, if the large system exhibits the ground state $|\psi_0\rangle = |\vec{n}\rangle^{\otimes N}$ (spin coherent state) magnetized along the direction $\vec{n}$, it will generate an effective Hamiltonian $H_{eff}(\vec{n}) = N\langle J \rangle\vec{n}\vec{\sigma}$.

Appendix 2: Groups Transitive on Spheres

The groups that are transitive on spheres are summarized in Table 1.

For simplicity reasons, we shall study only the minimal group (therefore certainly within the set of physical transformations) that is transitive on a sphere $\mathcal{S}^{d-1}$. If d is odd, the minimal transitive group is the special orthogonal group SO(d) unless $d = 7$. For $d = 7$ the minimal group is the exceptional Lie group G_2. If d is even, there are several options. We distinguish the cases whether the group contains the total inversion $E\mathbf{x} = -\mathbf{x}$ or not. The groups U($d/2$), Sp($d/4$), Sp($d/4$)×U(1), Sp($d/4$)× SU(2), Spin(7) and Spin(9) contain E as well as the group SU($d/2$), if d is multiple of four $d = 4k$ (Ref. [29], page 18). The only d-even groups that do not contain total inversion are SU($d/2$) for $d = 4k + 2$, where $k = 1, 2, 3, \ldots$

Table 1 Table taken from the Ref. [29]

Abstract group	d
SO(d)	$3, 4, 5, \ldots$
SU(d/2)	$4, 6, 8, \ldots$
U(d/2)	$2, 4, 6, 8, \ldots$
Sp(d/4)	$8, 12, 16, \ldots$
Sp($d/4$) × U(1)	$8, 12, 16, \ldots$
Sp($d/4$) × SU(2)	$4, 8, 12, \ldots$
G_2	7
Spin(7)	8
Spin(9)	16

We assume $d > 1$ always. First column shows the abstract group transitive on sphere $\mathcal{S}^{d-1}$, whereas the second column shows the possible value of d. Here SO(2) ≅ U(1) and Sp(1) ≅ SU(2). For a complex matrix U, the real representation is generated by following real matrix $\begin{pmatrix} \mathrm{Re}U & -\mathrm{Im}U \\ \mathrm{Im}U & \mathrm{Re}U \end{pmatrix}$

Appendix 3: Kronecker Product of Irreducible Representations

Here we provide the proof of Lemma 1:

Lemma 1 *CG series of the product $\Delta^{(\mu)} \otimes \Delta^{(\nu)}$, where $\Delta^{(\mu)}, \Delta^{(\nu)}$ are real and irreducible, contains the trivial representation if and only if $\mu = \nu$ and then the trivial representation appears once, only.*

Proof Note that for a real, orthogonal representation $D(g)$ we have $D(g^{-1}) = D^{\mathrm{T}}(g)$, hence $\chi(g^{-1}) = \mathrm{Tr}D(g^{-1}) = \mathrm{Tr}D^{\mathrm{T}}(g) = \chi(g)$. We set $\mu = 1$ with $\Delta^{(1)}(g) = 1$ (trivial representation) and $D(g) = \Delta^{(\mu)}(g) \otimes \Delta^{(\nu)}(g)$. We have the characters $\chi^{(1)}(g) = 1$ and $\chi(g) = \chi^{(\mu)}(g)\chi^{(\nu)}(g)$. The frequency is computed using Eq. (31)

$$
\begin{aligned}
a_1 &= (\chi^{(1)}, \chi) && (65)\\
&= \frac{1}{|\mathcal{G}|} \sum_{g \in \mathcal{G}} \chi^{(1)}(g^{-1})\chi^{(\mu)}(g)\chi^{(\nu)}(g) && (66)\\
&= \frac{1}{|\mathcal{G}|} \sum_{g \in \mathcal{G}} \chi^{(\mu)}(g)\chi^{(\nu)}(g) && (67)\\
&= \frac{1}{|\mathcal{G}|} \sum_{g \in \mathcal{G}} \chi^{(\mu)}(g^{-1})\chi^{(\nu)}(g) && (68)\\
&= (\chi^{(\mu)}, \chi^{(\nu)}) && (69)\\
&= \delta_{\mu\nu}. && (70)
\end{aligned}
$$

QED

Appendix 4: Irreducible Decomposition of the Two-Fold Tensor Representation of SO(d)

Here we show that the decomposition (32)

$$\Delta^d \otimes \Delta^d = \Delta^1 \oplus \Delta^S \oplus \Delta^{AS}, \tag{71}$$

is irreducible unless $d = 4$.

Let the representation $D(\mathcal{G})$ of $\mathcal{G}$ acts on a vector space $\mathcal{V}$. By definition, $D(\mathcal{G})$ is irreducible on $\mathcal{V}$ if $\text{span}\{D(g)\mathbf{x} \mid \forall g \in \mathcal{G}\} = \mathcal{V}$ for every non-zero vector $\mathbf{x} \in \mathcal{V}$.

Firstly, let us analyze the symmetric subspace of all $d \times d$ symmetric, traceless matrices

$$\mathcal{V}_S = \{H \mid H^{\mathrm{T}} = H \ \wedge \ \mathrm{Tr}H = 0\}. \tag{72}$$

This is an invariant subspace under the action of SO(d), because $(RHR^{\mathrm{T}})^{\mathrm{T}} = RHR^{\mathrm{T}}$ for every $R \in \mathrm{SO}(d)$ and $H \in \mathcal{V}_S$. Our goal is to show that the action of SO(d) is irreducible on $\mathcal{V}_S$. Therefore, we have to prove that the set

$$\mathcal{W}(H) = \text{span}\{RHR^{\mathrm{T}} \mid R \in \mathrm{SO}(d)\} = \mathcal{V}_S, \tag{73}$$

for every non-zero $H \in \mathcal{V}_S$. Let us write H in diagonal form $H = \sum_{i=1}^d h_i |i\rangle\langle i|$, where $H = \sum_{i=1}^d h_i = 0$. Since $\mathrm{Tr}H = 0$, the largest and lowest eigenvalue satisfy $h_{\max} > 0$ and $h_{\min} < 0$. For convenience we set $h_{\max} = h_1$ and $h_{\min} = h_2$. Consider the orthogonal matrix $F_{12} \in \mathrm{SO}(d)$ swapping the basis vectors $|1\rangle$ and $|2\rangle$ (swap-rotation in 12-subspace):

$$F_{12} = \text{diag}\left[\begin{pmatrix} 0 & -1 \\ 1 & 0 \end{pmatrix}, 1, 1, 1, \dots\right]. \tag{74}$$

We have $H' = \frac{1}{h_1 - h_2}(H - F_{12}HF_{12}^{\mathrm{T}}) = |1\rangle\langle 1| - |2\rangle\langle 2|$, where $h_1 - h_2 > 0$. If we further rotate in 12-subspace for 45° we obtain

$$R_{45^\circ} H' R_{45^\circ}^{\mathrm{T}} = |1\rangle\langle 2| + |2\rangle\langle 1| = E_{12}. \tag{75}$$

The matrix E_{12} is the element of a standard basis in $\mathcal{V}_S$. Other basis elements E_{ij} can be obtained from E_{12} by suitable rotations. Therefore we have completed the space $\mathcal{V}_S$ starting from an arbitrary element H, hence $\mathcal{W}_S(H) = \mathcal{V}_S$.

In the case of antisymmetric subspace we define

$$\mathcal{V}_{AS} = \text{span}\{H \mid H^{\mathrm{T}} = -H\}. \tag{76}$$

Our goal is to show $\mathcal{W}(A) = \mathcal{V}_{AS}$ for arbitrary $A \in \mathcal{V}_{AS}$. Let $A_{ij} = |i\rangle\langle j| - |j\rangle\langle i|$, for $j > i$ be the standard basis in $\mathcal{V}_{AS}$. It is sufficient to show that $A_{12} \in \mathcal{W}(A)$, and the other basis elements can be obtained from A_{12} by suitable rotations. For an arbitrary antisymmetric matrix $A \in \mathcal{V}_{AS}$ we can find the canonical form by applying suitable rotation $T \in \mathrm{SO}(d)$:

$$\begin{aligned} A' = TAT^{\mathrm{T}} &= \mathrm{diag}\left[\begin{pmatrix} 0 & -a_1 \\ a_1 & 0 \end{pmatrix}, \begin{pmatrix} 0 & -a_2 \\ a_2 & 0 \end{pmatrix}, \ldots, 0, 0, \ldots\right] \\ &= a_1 A_{12} + a_2 A_{34} + \ldots. \end{aligned} \tag{77}$$

If only $a_1 \neq 0$, than $A = a_1 A_{12}$ and we can generate the full basis $\{A_{ij}\}$ in $\mathcal{V}_{AS}$ by applying suitable rotations. Otherwise, we assume that at least two elements a_i are non-zero, and for convenience we set $a_1 \neq 0$ and $a_2 \neq 0$. Let R_{ij} be the rotation that flips ith and jth coordinate only, i.e. $R_{ij}|k\rangle = s|k\rangle$, where $s = -1$ if $k = i$ or $k = j$, otherwise $s = 1$. We get the following

$$A'' = A' - R_{13} A' R_{13}^{\mathrm{T}} = 2a_1 A_{12} + 2a_2 A_{34}. \tag{78}$$

Now if $d > 4$ we further apply R_{15} to A'' and obtain the following

$$A'' - R_{15} A'' R_{15}^{\mathrm{T}} = 4a_1 A_{12}. \tag{79}$$

From here we can generate the full basis A_{ij}, hence $\mathcal{W}(H) = \mathcal{V}_{AS}$. If $d = 4$ the construction above is no longer possible (R_{15} does not exist). In this case the antisymmetric space is reduced to two three-dimensional irreducible subspaces as follows

$$\Delta^4 \otimes \Delta^4 = \Delta^1 \oplus \Delta^9 \oplus \Delta^3_+ \oplus \Delta^3_-. \tag{80}$$

We leave the proof to the curious reader.

Appendix 5: $d = 3$ Solution

We begin with analyzing the fourth tensor power of Δ^d representation of $\mathrm{SO}(d)$ group, as defined in the main text. We have

$$\begin{aligned} \Delta^d \otimes \Delta^d \otimes \Delta^d \otimes \Delta^d &= (\Delta^d \otimes \Delta^d) \otimes (\Delta^d \otimes \Delta^d) \\ &= \left(\Delta^1 \oplus \Delta^{AS} \oplus \Delta^S\right) \otimes \left(\Delta^1 \oplus \Delta^{AS} \oplus \Delta^S\right). \end{aligned} \tag{81}$$

Since $S \neq AS$ for $d > 1$ and $d \neq 4$ (see Appendix 4), according to Lemma 1 the only contributing terms to the trivial representation are $\Delta^1 \otimes \Delta^1$, $\Delta^{AS} \otimes \Delta^{AS}$ and $\Delta^S \otimes \Delta^S$, each of which appears once. Therefore, the tensor K_{ijkl} that is invariant under $\mathrm{SO}(d)$ belongs to the three dimensional IR subspace. We can form a basis in it by combining Kronecker delta tensors δ_{ij}. There are three different ways to combine them into a four-fold tensor, therefore:

$$K_{ijkl} = \alpha\delta_{ij}\delta_{kl} + \beta\delta_{ik}\delta_{jl} + \gamma\delta_{il}\delta_{jk}. \tag{82}$$

From the analysis given in the main text, only $d = 3$ case exhibits non-trivial invariant dynamics. The most general dynamical law for the global state $\psi = (\mathbf{x}, \mathbf{y}, T, \Lambda)$ is given by:

$$\frac{dx_i}{dt} = a\epsilon_{ijk}T_{jk} + L_{in}^{(1)}\lambda_n, \tag{83}$$

$$\frac{dy_i}{dt} = b\epsilon_{ijk}T_{jk} + L_{in}^{(2)}\lambda_n, \tag{84}$$

$$\frac{dT_{ij}}{dt} = -a\epsilon_{ijk}x_k - b\epsilon_{ijk}y_k + L_{ijn}^{(12)}\lambda_n + K_{ijkl}T_{kl}, \tag{85}$$

$$\frac{d\lambda_n}{dt} = Q_{nm}\lambda_m - L_{in}^{(1)}x_i - L_{in}^{(2)}y_i - L_{ijn}^{(12)}T_{ij}. \tag{86}$$

Note that the reversibility requires $K_{ijkl} = -K_{klij}$. If we apply this constraint to the Eq. (82), we obtain $K = 0$.

Next we will find the consistent values for the constants $a, b, L_{in}^{(1)}, L_{in}^{(2)}, L_{ijn}^{(12)}$ such that the solutions to the dynamical Eqs. (83)–(86) above always lead to non-negative probabilities in Eq. (13). We look at the simplest case where all the couplings to global parameters are zero $L_{in}^{(1)} = L_{in}^{(2)} = L_{ijn}^{(12)} = 0$. If our initial state is a product state, than the global parameters remain zero during the evolution and we can safely neglect them from the analysis. In other words, the solution to the dynamical equations admits local tomography ($\Lambda = 0$) and it can be found by solving the following set of equations:

$$\frac{dx_i}{dt} = a\epsilon_{ijk}T_{jk}, \tag{87}$$

$$\frac{dy_i}{dt} = b\epsilon_{ijk}T_{jk}, \tag{88}$$

$$\frac{dT_{ij}}{dt} = -a\epsilon_{ijk}x_k - b\epsilon_{ijk}y_k. \tag{89}$$

Let us find the solution for the initial conditions $\vec{\psi}^{\pm}(0) = \{\mathbf{e}_3, \pm\mathbf{e}_3, \pm\mathbf{e}_3\mathbf{e}_3^{\mathrm{T}}\}$, where $\mathbf{e}_3 = (0, 0, 1)^{\mathrm{T}}$. The only components that evolve in time are $x_3(t)$, $y_3(t)$ and $T_{12}(t) = -T_{12}(t)$, hence the solution has the form:

$$\psi^{\pm}(t) = \left\{ \begin{pmatrix} 0 \\ 0 \\ x^{\pm}(t) \end{pmatrix}, \begin{pmatrix} 0 \\ 0 \\ y^{\pm}(t) \end{pmatrix}, \begin{pmatrix} 0 & \tau(t) & 0 \\ -\tau(t) & 0 & 0 \\ 0 & 0 & \pm 1 \end{pmatrix} \right\}, \tag{90}$$

where $x^{\pm}(t)$, $y^{\pm}(t)$ and $\tau(t)$ are the solutions to:

$$\frac{dx^{\pm}}{dt} = 2a\tau, \tag{91}$$

$$\frac{dy^\pm}{dt} = 2b\tau, \tag{92}$$

$$\frac{d\tau}{dt} = -ax^\pm - by^\pm. \tag{93}$$

Note that the state $\vec{\psi}^\pm(t)$ has to be physical state, that is, probability of Eq. (13) is non-negative $P_{12}(\vec{\psi}|\ \mathbf{a}, \mathbf{b}) \geq 0$ for arbitrary choice of local measurements $\mathbf{a}$ and $\mathbf{b}$. If we set $\mathbf{a} = \mathbf{e}_3$ and $\mathbf{b} = -\mathbf{e}_3$, the positivity condition reads $\frac{1}{4}(x^\pm(t) - y^\pm(t)) \geq 0$. Similarly for $\mathbf{a} = -\mathbf{e}_3$ and $\mathbf{b} = \mathbf{e}_3$ we have $\frac{1}{4}(-x^\pm(t) + y^\pm(t)) \geq 0$. This is possible only if $x^\pm(t) = y^\pm(t)$.

In order to eliminate $\tau(t)$ from the dynamical equations we find the second derivatives in time of $x^\pm$ and $y^\pm$. We obtain:

$$\frac{d^2x^\pm}{dt^2} = -2a^2x^\pm - 2aby^\pm, \tag{94}$$

$$\frac{d^2y^\pm}{dt^2} = -2abx_i - 2b^2y^\pm. \tag{95}$$

This set of equation leads to the symmetric solution $x^\pm(t) = y^\pm(t)$ only if $a^2 = b^2$ or equivalently $b = \pm a$. Note that $a = -b$ case brings new symmetry to the set of dynamical equations, the invariance under particle swap. If one requires such a symmetry, the case $a = b$ can be safely eliminated. However, we will use another argument that has been used in the work of Ref. [26]. We distinguish two cases, and label different solution as $\vec{\psi}^\pm_{\mathrm{MQM}}(t)$ and $\vec{\psi}^\pm_{\mathrm{QM}}(t)$, for $a = b$ and $a = -b$ respectively. The label QM and MQM stands for *quantum mechanics* and *mirror quantum mechanics* and the meaning of notation we explain shortly.

It is straightforward to evaluate the solution of dynamical equations:

$$\psi^+_{\mathrm{MQM}}(t) = \left\{ \begin{pmatrix} 0 \\ 0 \\ \cos 2at \end{pmatrix}, \begin{pmatrix} 0 \\ 0 \\ \cos 2at \end{pmatrix}, \begin{pmatrix} 0 & -\sin 2at & 0 \\ \sin 2at & 0 & 0 \\ 0 & 0 & 1 \end{pmatrix} \right\},$$

$$\psi^-_{\mathrm{MQM}}(t) = \left\{ \begin{pmatrix} 0 \\ 0 \\ 1 \end{pmatrix}, \begin{pmatrix} 0 \\ 0 \\ -1 \end{pmatrix}, \begin{pmatrix} 0 & 0 & 0 \\ 0 & 0 & 0 \\ 0 & 0 & -1 \end{pmatrix} \right\}, \tag{96}$$

$$\psi^-_{\mathrm{QM}}(t) = \left\{ \begin{pmatrix} 0 \\ 0 \\ \cos 2at \end{pmatrix}, \begin{pmatrix} 0 \\ 0 \\ -\cos 2at \end{pmatrix}, \begin{pmatrix} 0 & \sin 2at & 0 \\ -\sin 2at & 0 & 0 \\ 0 & 0 & -1 \end{pmatrix} \right\},$$

$$\psi^+_{\mathrm{QM}}(t) = \left\{ \begin{pmatrix} 0 \\ 0 \\ 1 \end{pmatrix}, \begin{pmatrix} 0 \\ 0 \\ -1 \end{pmatrix}, \begin{pmatrix} 0 & 0 & 0 \\ 0 & 0 & 0 \\ 0 & 0 & -1 \end{pmatrix} \right\}. \tag{97}$$

Our goal is to show that ψ_{QM} and the associate dynamics corresponds to quantum mechanics for two qubits, whereas ψ_{MQM} belongs to so called *mirror quantum mechanics* [26]. The later case has the set of states obtained by partial transpose of two-qubit states. We introduce the matrix representation of $\vec{\psi} = (\mathbf{x}, \mathbf{y}, T)$:

$$\rho(\vec{\psi}) = \frac{1}{4}(\mathbb{1} \otimes \mathbb{1} + x_i\sigma_i \otimes \mathbb{1} + y_i \mathbb{1} \otimes \sigma_i + T_{ij}\sigma_i \otimes \sigma_j), \tag{98}$$

where σ_i, $i = 1, 2, 3$ are the Pauli matrices. Straightforward calculation shows that $\rho(\psi^-_{\mathrm{QM}}(t)) = |\psi(t)\rangle\langle\psi(t)|$ is a density matrix, furthermore, it is a pure quantum state, where $|\psi(t)\rangle = \cos at|0\rangle|1\rangle + i \sin at|1\rangle|0\rangle$. Similarly, one can show that the matrix representation of mirror state $\psi^+_{\mathrm{MQM}}(t)$ is a non-quantum state (unless $\psi^+_{\mathrm{MQM}}(t)$ is product state) that can be obtained from $\psi^-_{\mathrm{QM}}(t)$ by applying total inversion $\mathbf{y} \mapsto -\mathbf{y}$ on the second spin. Note, that is a non-quantum operation. Mirror quantum mechanics is shown to be mathematically inconsistent theory for the tripartite case [26]. Therefore we will adopt only quantum solution.

The set of dynamical Eq. (87) has the corresponding matrix form:

$$\frac{d\rho(\vec{\psi})}{dt} = i[H_{12}, \rho(\vec{\psi})], \tag{99}$$

where H_{12} is the Heisenberg spin-spin interaction $H_{12} = \frac{a}{2}\vec{\sigma}_1\vec{\sigma}_2 = \frac{a}{2}\sum_{i=1}^{3} \sigma_i \otimes \sigma_i$.

References

1. I. Bengtsson, Why is space three-dimensional?, http://www.physto.se/~ingemar/fyra.pdf
2. I.M. Freeman, Why is space three-dimensional? Based on W. Büchel: "Warum hat der Raum drei Dimensionen?," Physikalische Blätter, Vol. 19(12), pp. 547–549 (December 1963). Am. J. Phys. **37**, 1222 (1969)
3. P. Ehrenfest, Proc. Amst. Acad. **20**, 200 (1917)
4. I.F. Herbut, Majorana mass, time reversal symmetry, and the dimension of space. Phys. Rev. D **87**, 085002 (2013)
5. T. Kaluza, Zum Unitätsproblem in der Physik, Akad. Wiss. Berlin. (Math. Phys.), 966–972 (1921)
6. O. Klein, Quantentheorie und fünfdimensionale Relativitätstheorie. Zeitschrift für Physik A **37**(12), 895–906 (1926)
7. L. Randal, *Warped Passages: Unraveling the Mysteries of the Universe's Hidden Dimensions* (Harper Perennial, New York, 2006)
8. I. Antoniadis, A possible new dimension at a few TeV. Phys. Lett. B **246**, 377–384 (1990)
9. N. Arkani-Hamed, S. Dimopoulos, G. Dvali, The Hierarchy problem and new dimensions at a millimeter. Phys. Lett. B **429**(3–4), 263–272 (1998)
10. K. Agashe, A. Pomarol, Focus on extra space dimensions. New J. Phys. **12**, 075010 (2010)
11. J. Barrett, Information processing in general probabilistic theories. Phys. Rev. A. **75**, 032304 (2007)
12. H. Barnum, A. Wilce, Information processing in convex operational theories. Electron. Notes Theor. Comput. Sci. **270**(1), 3–15 (2011)
13. L. Hardy, Quantum theory from five reasonable axioms (2001). arXiv:quant-ph/0101012
14. H. Barnum, J. Barrett, M. Leifer, A. Wilce, A general no-cloning theorem. Phys. Rev. Lett. **99**, 240501 (2007)

15. J.S. Bell, On the Einstein-Podolsky-Rosen paradox, Physics **1**, 195–200 (1964); reprinted in J.S. Bell, *Speakable and Unspeakable in Quantum Mechanics* (Cambridge University Press, Cambridge, 1987)
16. S. Popescu, D. Rohrlich, Quantum nonlocality as an axiom. Found. Phys. **24**, 379 (1994)
17. D.I. Fivel, How interference effects in mixtures determine the rules of quantum mechanics. Phys. Rev. A **59**, 2108 (1994)
18. C.A. Fuchs, Quantum mechanics as quantum information (and only a little more), in *Quantum Theory: Reconstruction of Foundations*, ed. by A. Khrenikov (Växjo University Press, Växjo, 2002)
19. R. Clifton, J. Bub, H. Halvorson, Characterizing quantum theory in terms of information-theoretic constraints. Found. Phys. **33**(11), 1561 (2003)
20. Č. Brukner, A. Zeilinger, Information and fundamental elements of the structure of quantum theory, in *Time, Quantum, Information*, ed. by L. Castell, O. Ischebeck (Springer, Berlin, 2003)
21. A. Grinbaum, Elements of information-theoretic derivation of the formalism of quantum theory. Int. J. Quant. Inf. **1**(3), 289 (2003)
22. G.M. D'Ariano, Operational axioms for quantum mechanics. AIP Conf. Proc. **889**, 79–105 (2006)
23. A. Grinbaum, Reconstruction of quantum theory. Br. J. Philos. Sci. **8**, 387 (2007)
24. P. Goyal, Information-geometric reconstruction of quantum theory. Phys. Rev. A **78**, 052120 (2008)
25. Č. Brukner, A. Zeilinger, Information invariance and quantum probabilities. Found. Phys. **39**, 677 (2009)
26. B. Dakić, Č. Brukner, Quantum theory and beyond: is entanglement special, in *Deep Beauty: Understanding the Quantum World through Mathematical Innovation*, ed. by H. Halvorson (Cambridge University Press, Cambridge, 2011)
27. L. Masanes, M. Müller, A derivation of quantum theory from physical requirements. New J. Phys. **13**, 063001 (2011)
28. L. Hardy, Reformulating and reconstructing quantum theory (2011). arXiv:1104.2066
29. L. Masanes, M.P. Müller, D.P. Garcia, R. Augusiak, Entangling dynamics beyond quantum theory (2011). arXiv:1111.4060
30. J. Rau, Measurement-based quantum foundations. Found. Phys. **41**(3), 380–388 (2011)
31. G. Chiribella, G.M. D'Ariano, P. Perinotti, Informational derivation of quantum theory. Phys. Rev. A **84**, 012311 (2011)
32. H. Barnum, Quantum knowledge, quantum belief, quantum reality: notes of a QBist fellow traveler (2010). arXiv:1003.4555v1
33. C.F. von Weizsäcker, in Quantum theory and the structures of time and space, Eds. L. Castell, M. Drieschner, C.F. von Weizsäcker (Hanser, München, 1975). Papers presented at a conference held in Feldafing, July (1974)
34. R. Penrose, Angular momentum: an approach to combinatorial space-time, in *Quantum Theory and Beyond*, ed. by T. Bastin (Cambridge University Press, Cambridge, 1971)
35. W.K. Wootters, The acquisition of information from quantum measurements, Ph.D. thesis, University of Texas at Austin (1980)
36. A. Einstein, W.J. de Haas, Experimenteller Nachweis des Ampéreschen Molekularströme. Naturwissenschaften **3**, 237–238 (1915)
37. S.J. Barnett, Magnetization by rotation. Phys. Rev. **6**, 239–270 (1915)
38. As noted by A. Peres, in *Quantum Theory: Conpcepts and Methods* (Kluwer Academic Publishers, 2002): "Even if quantum theory is universal, it is not closed. A distinction must be made between endophysical systems—those which are described by the theory—and exophysical ones, which lie outside the domain of the theory (for example, the telescopes and photographic plates used by astronomers for verifying the laws of celestial mechanics). While quantum theory can in principle describe anything, a quantum description cannot include everything. In every physical situation something must remain unanalyzed."
39. P.W. Atkins, J.C. Dobson, Angular momentum coherent states. Proc. R. Soc. A **321**, 321 (1971)
40. J.M. Radcliffe, Some properties of coherent spin states. J. Phys. A: Gen. Phys. **4**, 313 (1971)

41. J. Kofler, Č. Brukner, Classical world arising out of quantum physics under the restriction of coarse-grained measurements. Phys. Rev. Lett. **99**, 180403 (2007)
42. S.D. Bartlett, T. Rudolph, R.W. Spekkens, Reference frames, superselection rules, and quantum information. Rev. Mod. Phys. **79**, 555–606 (2007)
43. M. Dickson, A view from nowhere: quantum reference frames and uncertainty. Stud. Hist. Philos. Mod. Phys. **35**, 195–220 (2004)
44. Y. Aharonov, T. Kaufherr, Quantum frames of reference. Phys. Rev. D **30**, 368 (1984)
45. D. Poulin, J. Yard, Dynamics of a quantum reference frame. New J. Phys. **9**, 156 (2007)
46. D. Poulin, Toy model for a relational formulation of quantum theory (2005). arXiv:0505081v2
47. Č. Brukner, In the Kreisgang between classical and quantum physics, UniMolti modi della filosofia 2008/2, arXiv:0905.3363
48. D.C. Brody, E.M. Graefe, Six-dimensional space-time from quaternionic quantum mechanics. Phys. Rev. D **84**, 125016 (2011)
49. T. Paterek, B. Dakić, Č. Brukner, Theories of systems with limited information content. New J. Phys. **12**, 053037 (2010)
50. G.V. Steeg, S. Wehner, Relaxed uncertainty relations and information processing. Quantum Inf. Comput. **9**(9–10), 0801–0832 (2009)
51. M.P. Müller, L. Masanes, Three-dimensionality of space and the quantum bit: how to derive both from information-theoretic postulates (2012). arXiv:1206.0630
52. H. Araki, On a characterization of the state space of quantum mechanics. Commun. Math. Phys. **75**, 1–24 (1980)
53. S. Bergia, F. Cannata, A. Cornia, R. Livi, On the actual measurability of the density matrix of a decaying system by means of measurements on the decay products. Found. Phys. **10**, 723–730 (1980)
54. W.K. Wootters, Local accessibility of quantum states, in *Complexity, Entropy and the Physics of Information*, ed. by W.H. Zurek (Addison-Wesley, Boston, 1990)
55. G. Chiribella, G.M. D'Ariano, P. Perinotti, Informational derivation of quantum theory. Phys. Rev. A **81**, 062348 (2010)
56. L. Hardy, W.K. Wootters, Limited Holism and real-vector-space quantum theory (2010). arXiv:1005.4870
57. E.C.G. Stueckelberg, Quantum theory in real hilbert space. Helv. Phys. Acta **33**, 727–752 (1960)
58. M. Pawlowski, A. Winter, Hyperbits: the information quasiparticles. Phys. Rev. A **85**, 022331 (2012)
59. D. Montgomery, H. Samelson, Transformation groups of spheres. Ann. Math. **44**, 454–470 (1943)
60. A. Borel, Some remarks about Lie groups transitive on spheres and tori. Bull. A.M.S. **55**, 580–587 (1949)
61. R.E. Behrends, J. Dreitlein, C. Fronsdal, W. Lee, Simple groups and strong interaction symmetries. Rev. Mod. Phys. **34**, 1–40 (1962)
62. J.C. Baez, The octonions. Bull. Am. Math. Soc. **39**, 145–205 (2002)
63. N.W. Ashcroft, N.D. Mermin, *Solid State Physics* (Harcourt College Publishers, San Diego, 1976)

Some Negative Remarks on Operational Approaches to Quantum Theory

Christopher A. Fuchs and Blake C. Stacey

I always like to start with a joke, but due to the Hollywood special effects of Charles H. Bennett, I've got something to share with you. I'll let Charles tell the joke. Many years ago, Asher Peres and I wrote an article called "Quantum Theory Needs No 'Interpretation' " [1], or as Charles would have it:

OPINION

Quantum Theory Needs No 'Interpretation' **besides Ours**

Christopher A. Fuchs and Asher Peres

Recently there has been a spate of articles, reviews, and letters in PHYSICS TODAY promoting various "interpretations" of quantum theory of our experimental activity, then we must be prepared for that, too.

The thread common to all the non-standard "interpretations" is the carry an umbrella. Probability theory is simply the quantitative formulation of how to make rational decisions in the face of uncertainty.

But the paper ended with these words!

> All this said, we would be the last to claim that the foundations of quantum theory are not worth further scrutiny. For instance, it is interesting to search for minimal sets of *physical* assumptions that give rise to the theory.

So, some negative remarks on operational approaches! (I'm awfully loud, aren't I?) [Someone in the audience: "You always are!"] I've been happy to see that this old slide of mine, with the traditional axioms of quantum theory, has gotten such airplay this week.

C.A. Fuchs (✉)
Raytheon BBN Technologies, 10 Moulton Street, Cambridge, MA 02138, USA
e-mail: qbism.fuchs@gmail.com

B.C. Stacey
Martin A. Fisher School of Physics, Brandeis University, Waltham, MA 02453, USA

G. Chiribella and R.W. Spekkens (eds.), *Quantum Theory: Informational Foundations and Foils*, Fundamental Theories of Physics 181,
DOI 10.1007/978-94-017-7303-4_9

Axioms: Quantum

0) Systems exist.

1) Associated with each is a complex vector space $\mathcal{H}$.

2) Measurements correspond to orthonormal bases $|e_i\rangle$ on $\mathcal{H}$.

3) States correspond to density operators ρ on $\mathcal{H}$.

4) Systems combine by tensor producting their vector spaces, $\mathcal{H}_{AB} = \mathcal{H}_A \otimes \mathcal{H}_B$.

5) When no measurement is performed, states evolve by unitary maps $\mathcal{U}$.

I pulled it out of the bin and I haven't used it in quite a long time. In fact, I think it was around 12 years ago, at some meeting in Maryland when I first put the thing up. At the time, it was mostly as an excuse to make a joke at Max Tegmark's expense, who was going around to meetings taking polls of which interpretation was the most popular at which meeting.

Tegmark Poll

Interpretation	Votes
Copenhagen	13
Many Worlds	8
Bohm	4
Consistent Histories	4
Modified Dynamics (GRW)	1
None of the above/undecided	18

And there was a hidden agenda … well it wasn't so hidden … it was a pretty obvious agenda that he wanted to see Many Worlds climb in numbers as time went by, so that he could say that the Many Worlds Interpretation was the eminently reasonable one, and that's decided by democratic vote.

All that caused me to reflect upon what our mission … well, at the time I didn't call myself quantum foundations, and I probably won't be calling myself quantum foundations after this meeting … it caused me to reflect upon what exactly we needed to do as a community to get in a position that we would disband. That there wouldn't be any more quantum foundational meetings—how could we make this sort of stale debate on interpretation, where numbers fluctuated from year to year but otherwise there was no great progress—how might we make that end? And the thing that struck me was that this slide actually said it all. This is one of our great physical theories, one of the great theories of physics, one of the two or three great theories of physics, and look how it's posed!

Associated with each system is a complex vector space. Vectors, tensor products, all of these things. Compare that to one of our other great physical theories, special relativity. One could make the statement of it in terms of some very crisp and clear physical principles: The speed of light is constant in all inertial frames, and the laws of physics are the same in all inertial frames. And it struck me that if we couldn't take the structure of quantum theory and change it from this very overt mathematical speak—something that didn't look to have much physical content at all, in a way that anyone could identify with some kind of physical principle—if we couldn't turn that into something like this, then the debate would go on forever and ever. And it seemed like a worthwhile exercise to try to reduce the mathematical structure of quantum mechanics to some crisp physical statements.

Now the reason I went in a direction like this, where I said, "What would be a good methodology for doing that?," and I landed upon going to each and every one of the axioms and trying to give it an information-theoretic reason *if possible*, was that by that time I had already become pretty convinced that most of the structure of quantum theory was about information [2]. I had convinced myself that quantum states represented information of some sort, or Bayesian degrees of belief, or some might say knowledge... But the question on my mind was, "How much of quantum theory is about information?" Just because some parts of it are about information, it didn't mean that *all* of it had to be about information. And I threw my money on the idea that there would be something about quantum theory that was *not* information-theoretic.

And so, when I wrote the paper [3] that I took this little image from, I wrote of all these axioms, "Give an information-theoretic reason *if possible*!"

> The distillate that remains—the piece of quantum theory with no information theoretic significance—will be our first unadorned glimpse of "quantum reality." Far from being the end of the journey, placing this conception of nature in open view will be the start of a great adventure.

So that was really what I was seeking: to try and tear away all the underbrush that was about information, and find the one piece, or some small number of pieces, of quantum theory that were actually statements about the world independent of information-processing agents, independent of measuring observers and so forth. What I was really after was, I wanted to know what made quantum systems go? What made them interesting and behave in some peculiar way? Could we pinpoint that one thing on a principle that wasn't written in information-theoretic terms or operational terms in any way?

So, some time has passed, and there's been all of this fantastic work! And that's why I'm at this conference, because this has surprised me: the number of people that really threw their hearts and souls and have made all this great progress that we've seen here. We've seen some of these axiom systems—here's the one of Chiribella, D'Ariano and Perinotti [4, 5], their five axioms and one postulate, where they lay down principles of causality, perfect distinguishability, ideal compression, local distinguishability, pure conditioning and purification. And out of that, all written in English, no mathematical equations—I like that!—one pulls together the mathematical structure of quantum theory, *after a lot of work*. But they nail it. And, particularly, one thing I like about this system is the way they distinguish the purification postulate from the other axioms. This now approaches something that I think is along the lines of trying to find a crisp physical principle.

Chiribella, D'Ariano, and Perinotti's "5 Axioms and 1 Postulate"

- **Causality.** The probability of a measurement outcome at a certain time does not depend on the choice of measurements that will be performed later.
- **Perfect Distinguishability.** If a state is not completely mixed, then there exists at least one state that can be perfectly distinguished from it.
- **Ideal Compression.** Every source of information can be encoded in a suitable physical system in a lossless and maximally efficient fashion. Here lossless means that the information can be decoded without errors and maximally efficient means that every state of the encoding system represents a state in the information source.
- **Local Distinguishability.** If two states of a composite system are different, then we can distinguish between them from the statistics of local measurements on the component systems.
- **Pure Conditioning.** If a pure state of system AB undergoes an atomic measurement on system A, then each outcome of the measurement induces a pure state on system B.
- **Purification.** Every state has a purification. For fixed purifying system, every two purifications of the same state are connected by a reversible transformation on the purifying system.

Dakić and Brukner's Axioms

- **Information capacity.** An elementary system has the information carrying capacity of at most one bit. All systems of the same information carrying capacity are equivalent.
- **Locality.** The state of a composite system is completely determined by local measurements on its subsystems and their correlations.
- **Continuity.** Between any two pure states there exists a continuous reversible transformation.

Hardy's New Axioms

- **Definiteness.** Associated with any given pure state is a unique maximal effect giving probability equal to one. This maximal effect does not give probability equal to one for any other pure state.
- **Information Locality.** A maximal measurement on a composite system is effected if we perform maximal measurements on each of the components.
- **Tomographic Locality.** The state of a composite system can be determined from the statistics collected by making measurements on the components.
- **Compound Permutatability.** There exists a compound reversible transformation on any system effecting any given permutation of any given maximal set of distinguishable states for that system.
- **Preparability.** Filters are non-mixing and non-flattening.

Masanes and Müller's "Physical Requirements"

- **Finiteness.** In systems that carry one bit of information, each state is characterized by a finite set of outcome probabilities.
- **Local Tomography.** The state of a composite system is characterized by the statistics of measurements on the individual components.
- **Equivalence of Subspaces.** All systems that effectively carry the same amount of information have equivalent state spaces.
- **Symmetry.** Any pure state of a system can be reversibly transformed into any other.
- **All Measurements Allowed.** In systems that carry one bit of information, all mathematically well-defined measurements are allowed by the theory.

Local Distinguishability, Locality, Tomographic Locality, Local Tomography . . .

Wilce's "Four and a Half Axioms"

Let a physical system be modeled by a pair $(\mathcal{U}, \Omega)$ where $\mathcal{U}$ is a test space with outcome-space X and Ω is a closed, convex, outcome-separating set of continuous states thereon.

- **Symmetry.** There is a compact group G acting continuously on $(\mathcal{U}, \Omega)$, in such a way that (i) G acts fully symmetrically on $\mathcal{U}$, and (ii) G acts transitively on Ω_{ext}.
- **Minimization.** There exists a minimizing G-invariant, positive inner product on V^*.
- **Sharpness.** To every outcome $x \in X$, there corresponds a unique state $\epsilon_x \in \Omega$, $\epsilon_x(x) = 1$.
- **Correlation.** Every state is the marginal of a correlating non-signaling state.
- **Filtering.** For every test E and every $f : E \to (0, 1]$, $\exists$ an order-isomorphism $\phi : V^* \to V^*$ with $\phi(x) = f(x)x$.

I've put all you guys in alphabetical order, so not to offend anyone. Dakić and Brukner's axioms: information capacity, locality, continuity [6]. (They haven't presented on this yet; I guess you will later in the conference.) Hardy's new axioms—we saw his old axioms [7, 8] before, ten years ago, and his new axioms this week—definiteness, information locality, tomographic locality, compound permutability and preparability [9]. Masanes and Müller's axioms, we've already seen this as well: finiteness, local tomography, equivalence of subspaces, symmetry, all measurements are allowed [10]. And we saw Alex Wilce's this morning, written in a little bit more mathematical language—there are some equations in there! [11].

Well, when this whole business—or at least my discussions with Gilles Brassard and Charlie Bennett and many of you started up—I remember Charlie Bennett saying, "How will you know when you get to this physical distillate of quantum theory? How will you know that you've reached the end of the process?" And he sort of jokingly said, "Will it be like pornography?" You know, "the only way you know it's pornography is if you see it." And I think he was right! Not in the way he wanted to be. But I think what I am having a difficulty with, and what I want to try to express is that, of all the progress that's been made, I haven't been able to look at these systems and see something that stands out at me as the essential core of quantum mechanics. Something that's written in physical, nonoperational, noninformation-theoretic terms. So, there's been great progress in putting everything in operational terms, and making that very clear, and one can do that. Can one do the opposite thing? Can one find some nonoperational terms and have most of it in operational or information-theoretic terms but perhaps not all of it?

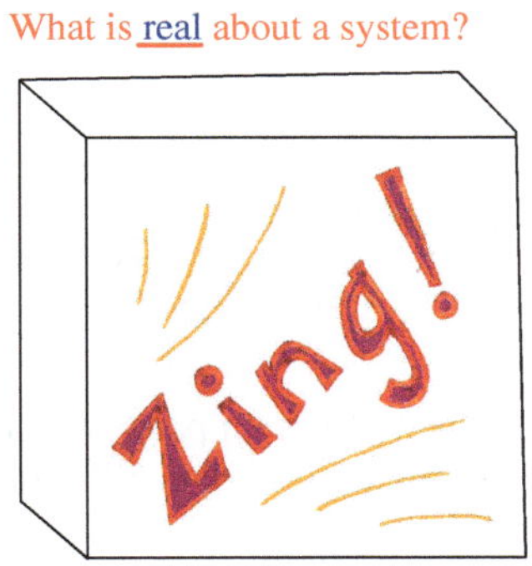

So, I look at these systems and I say, "What is the distillate that's left behind? That I can say this is the Zing, this is the thing that makes quantum systems go?"

And I don't see it. So that's my basic point.

How would I know it if I saw it? Back to Charlie's question. I'm not completely sure, but I think there's a distinction between the two principles for relativity—which of course are girded up by some mathematics, as Lucien Hardy made clear—and the ones that we've seen in the present efforts. And it's this: There's a certain amount of shock value in seeing the two things side-by-side. They're not just sort of inert principles that are lying there and are equivalent to the structure of the theory. But instead, one of them says that the speed of light should be constant in all inertial frames. And if you're accustomed to thinking the speed of light is this measurable quantity, how could it possibly be the same in all frames? Putting these two things together caused a certain amount of shock. So, I wonder whether quantum theory can be written in terms where the physical principle is identified as having a certain amount of shock value.

I guess another thing I'm trying to say is that of the existing axiomatic systems, it doesn't seem, to me at least, that any have quite gone for the jugular vein of quantum theory. You know, when a werewolf attacks a person he'll jump at this vein, and the person will bleed to death, and it'll all be over with. I would like to see an axiomatic system that goes for the weirdest part of quantum theory of all. For instance, most of the ones in the list above are built on the idea of local tomography. Is this a feature of quantum theory that's really weird? No, it doesn't seem to me that it's a feature that's really weird. And similarly with so many of the other ones. Can we find some axiomatic system that really goes after the *weird* part of quantum theory?

Well, what is the weird part? What is the toy that one might want to go after for axiomatizing?

Is it nonlocality? We hear this word all the time in quantum information and quantum foundations. My own sympathy, however, lies with something that Albert Einstein said I believe in the 40s [12]. It's in small print, I'll read it. "Einstein on Locality":

> If one asks what is characteristic of the realm of physical ideas independently of the quantum-theory, then above all the following attracts our attention: the concepts of physics refer to a real external world.

I'm okay with that!

> ...i.e., ideas are posited of things that claim a "real existence" independent of the perceiving subject (bodies, fields, etc.), ... Moreover, it is characteristic of these physical things that they are conceived of as being arranged in a space-time continuum. Further, it appears to be essential for this arrangement of the things introduced in physics that, at a specific time, these things claim an existence independent of one another, insofar as these things "lie in different parts of space."

He puts scare quotes around "lie in different parts of space" I presume because he's meaning this is a tautology: we say things are in different parts of space if they can't directly influence each other.

> Without such an assumption of the mutually independent existence ...

This is the important part now.

> Without such an assumption of the mutually independent existence (the "being-thus") of spatially distant things, an assumption which originates in everyday thought, physical thought in the sense familiar to us would not be possible. Nor does one see how physical laws could be formulated and tested without such a clean separation.

I think this is maybe a debatable point in terms of detail, but I think the idea is fundamentally sound. If you first posit two systems, and then you say, "Oops! Made a mistake, they were really one after all, because any one system can influence any other one," it would be hard to imagine how we come across the usual sort of reasoning that we do. So, my money is not on taking nonlocality as the kind of shock-value principle I'm looking for.

Instead, I'm much more sympathetic to something Asher Peres would have said, or did say!, around 1978: "Unperformed experiments have no results" [13]. If I were looking for a shock-value principle to blame quantum mechanics on, my feeling is that it's something along these lines. What is really at issue here? It's the question of whether quantum measurements reveal some pre-existing value for something that's unknown, or whether in some sense they go toward creating that very value, from the process of measurement. A good way to see how to pose this in more technical terms comes from—I guess I could call it a version of the Free Will theorem [14], but it's much older than Conway and Kochen [15]—has to do with first looking at the EPR criterion of reality [16]. (The bracketed part is my addition.)

> If, without in any way disturbing a system one can [gather the information required to] predict with certainty (i.e., with probability equal to unity) the value of a physical quantity, then there exists an element of physical reality corresponding to this physical quantity.

So, what's the content of that? Is it right?

You can consider a little variation of the EPR thought experiment. Let me take two qutrits and ascribe to them a maximally entangled state, and let me assume locality, the principle that Einstein used in the quote I took from him. And now, consider making a measurement on the left-hand particle in some basis—let's say this purple basis, or alternatively this green basis. I'm going to choose either the green one or the purple one. If I get, let's say, outcome number 2 on the left side, then I can predict that if I were to make the measurement on the right side, I would get outcome number 2. (I've omitted a transpose in drawing this picture.) So, under the assumption of locality, EPR would say, "Aha! It must be the case that there is an element of reality on this side corresponding to outcome number 2 of that measurement. It's something inherent in that body." But similarly, we could talk about the other measurement. And if I were to get outcome 2 for that measurement on one particle, I would predict with certainty that I would get outcome 2 for that measurement on the other particle.

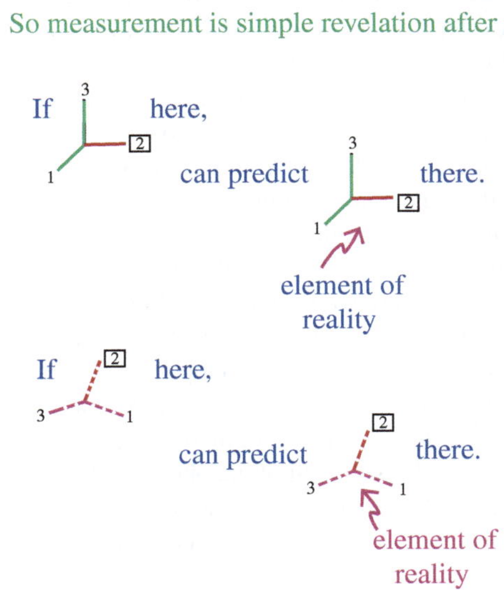

We're talking about noncommuting variables here; by the EPR criterion and locality, they would say it must be the case that there were elements of reality associated with those noncommuting observables. But we could make this more extreme and consider not just two observables but a whole set of them, quite a lot of them, corresponding to one of the Kochen–Specker constructions, for instance the one that Asher Peres found [17–19]. By taking a sufficient number of orthogonal bases, and interlocking them in some interesting way, we can construct a noncolorable set. For each of these, we would have said, by making a measurement here, I draw an inference about the element of reality over here. I can do it for one, I can do it for another, I can do it for another ... I can't do it for all of them without running into trouble!

Cannot be colored:

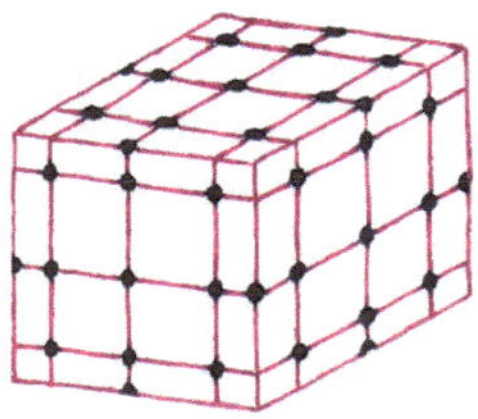

33 rays, Peres

(When completed into full triads, consists of 40 triads made from 57 rays)

I would say that the thing that this teaches me is that there's something bankrupt not in locality, but bankrupt in the EPR criterion of reality. There's something wrong about that. And that's what I'm calling Peres's principle that unperformed experiments have no outcomes. Particularly, quantum probabilities are not probabilities for some pre-existing reality that we're finding when we perform a measurement, but instead for something that's produced in the process of measurement.

So, what would excite me in an axiomatic approach to quantum reconstruction? I think it would be if someone could find a set of axioms that pulled this idea to the forefront and really made it the core of things. Now I don't know how to do that! At all! But I have a little toy approach that I've been playing with for some time that pleases me at least. And that is to recognize that these considerations of Kochen–Specker, including this case where we have the locality assumption, rest on thinking about different experiments in terms of each other. We think about what happens if we perform one experiment, we think about what happens if we were to perform another one—and it tells us that contextuality is at the core of our considerations. What I would like as a goal is a way to push quantum theory's specific form of contextuality all into one corner. If I could show that all of the phenomena of Kochen–Specker and Bell inequality violations and everything else, the whole formal structure, comes out of one corner to do with something with contextuality, I think I would be pleased.

So, let me give you a progress report on that kind of idea [20–22].

To give you the report, I have to make use of a little device I've told so many of you about so many times: a particularly interesting measuring device that I would like to elevate to a standard quantum measurement. It would be an informationally complete measurement. In other words, if I knew the statistics of the outcomes of this measurement, I would be able to reconstruct the quantum state that gave rise to them. That is what I mean by this: I'm getting rid of this symbol ρ which represents a quantum state and putting in its place a probability distribution. This probability distribution refers to the probabilities of the outcomes of this measurement up in the sky, and if this is an informationally complete measurement, we can completely reconstruct the quantum state ρ from the probabilities. This means that we can really cross out ρ completely and use the probabilities instead.

Particularly, to make everything that I say have a pretty form, this device should have the following properties. Suppose you could find d^2 rank-one projection operators, with this nice symmetry condition:

$$\operatorname{tr} \Pi_i \Pi_j = \frac{1}{d+1}, \quad i \neq j. \tag{1}$$

Namely, the Hilbert–Schmidt inner product between any two of the projectors is equal to a constant value determined by the dimension. If you can find d^2 projection operators satisfying this symmetry, you can prove that they have to be linearly independent—they don't have a choice! Moreover, if you renormalize by $1/d$, then they'll sum up to the identity. So, these form the elements of a POVM, and they can be thought of as the outcomes of a quantum measurement. We call it a SIC measurement, where the acronym stands for "Symmetric Informationally Complete" [22–31].

Since this measurement is informationally complete, one can completely determine the quantum state ρ in terms of the probabilities given by the Born Rule:

$$p(i) = \frac{1}{d}\operatorname{tr}(\rho \Pi_i). \tag{2}$$

And because of this great symmetry, there's a lovely reconstruction formula which says that the initial quantum state is just a linear combination of the projection operators, where the expansion coefficients are determined by the probabilities in a really simple way:

$$\rho = \sum_i \left[(d+1)p(i) - \frac{1}{d} \right] \Pi_i. \tag{3}$$

It's just a nice little affine transformation.

We can now explore what state space looks like in terms of probabilities. If we were talking about density operators, we could say, "We know what state space looks like: It's the set of positive semidefinite matrices with trace one." But what does the state space look like in terms of the probabilities themselves?

If you put an arbitrary probability distribution into the reconstruction formula, you have to get a Hermitian matrix, because this is a real combination of projection operators. So you will always get a Hermitian matrix. But for certain choices of the probabilities, you won't get a positive semidefinite one. What this tells you is if you want to specify the set of rank-one positive semidefinite operators, all you have to add to the condition that you have a Hermitian operator is that the trace of the operator squared and the trace of the operator cubed both equal 1:

$$\rho^\dagger = \rho, \ \operatorname{tr} \rho^2 = \operatorname{tr} \rho^3 = 1. \tag{4}$$

And if you translate that into a condition on the probabilities, you get two equations. One says that the probabilities should lie on the surface of a sphere of a certain radius:

$$\sum_i p(i)^2 = \frac{2}{d(d+1)}. \tag{5}$$

And the other one says that they should satisfy a certain cubic condition which isn't necessarily very pretty:

$$\sum_{jkl} c_{jkl}\, p(j) p(k) p(l) = \frac{d+7}{(d+1)^3}, \tag{6}$$

where

$$c_{jkl} = \operatorname{Re} \operatorname{tr}(\Pi_j \Pi_k \Pi_l). \tag{7}$$

But what you can glean from this is that the extreme points form the intersection of two sets, one of which is a sphere and one of which is some cubic curve. So, the pure states can't be the whole sphere, unless the latter condition is trivial: They're going to be some subset of a sphere.

What does all this have to do with contextuality, and maybe a reconstruction of quantum theory by some means associated with contextuality? You can start to get a sense of what that has to do with this from the following little thought experiment.

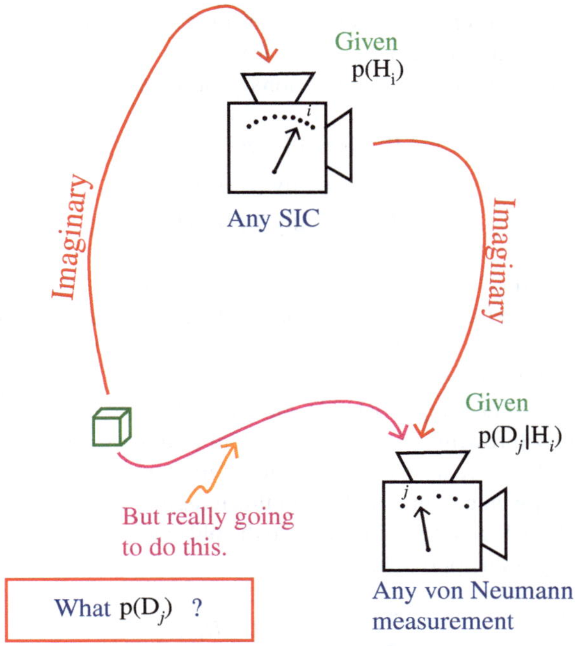

Here's what I'm going to do. I've got my quantum system, and I'm going to throw it into some measuring device—I'll call it "the measuring device on the ground." Let's say it's a von Neumann measuring device. So I'm going to throw this quantum system into here, and when I do, it will generate an outcome. Let me call it D_j. Here, I'm using the letter D to evoke the idea of "data". What I'd like to know is, what are the probabilities for the outcomes D_j?

$p(D_j)$—that's what I would like to calculate. I'm going to throw the system in here; that's what I'm really going to do. Throw it in there and get an outcome. But suppose I only know the probabilities for the outcomes of this very special SIC measurement, and I know the conditional probabilities for the outcomes of the measurement on the ground in terms of the measurement in the sky. In other words, I have conditional probabilities $p(D_j|H_i)$. Do I have enough information to reconstruct the probability I really want?

Well, you might be able to reconstruct it, but you can't use regular probability theory. You can't just say that the probabilities for the bottom path will be given by the classical formula to do with the conditional probabilities for the upper path.

$$p(D_j) \neq \sum_i p(H_i)p(D_j|H_i), \tag{8}$$

because in physics terms one of them is a coherent process, and the other one is an incoherent process. If you speak like a probabilist (or like Leslie Ballentine, when he was here and I made this slide for him) you might say there are extra conditions. The probability on the left-hand side refers to one experiment, and the probabilities on the right-hand side refer to a different experiment. So there's an extra conditional in these expressions, and it's no wonder that you don't get equality in the classical expression because there are extra conditions floating about:

$$\begin{aligned} p(D_j) &\text{ is really } p(D_j|C_1), \\ p(H_i) &\text{ is really } p(H_i|C_2), \\ p(D_j|H_i) &\text{ is really } p(D_j|H_i, C_2). \end{aligned} \tag{9}$$

What is so interesting in this case, however, is that though all of that is true—it's true you can't just use the classical formula for the upper path to calculate the lower path—quantum mechanics nonetheless provides the tools for making the calculation. *It's just a different calculation.*

In usual "physicist language," what is going on is that the probability I really want is one that I get by throwing my quantum system into the second measuring device via a coherent path. There's no measurement on it; there's no decoherence. Yet, the probabilities I am given are ones to do with this imaginary path. An incoherent path.

In any case, the calculation is this: You do the old classical calculation—first form this, the usual Bayesian approach,

$$\sum_i p(H_i)p(D_j|H_i), \tag{10}$$

and then you simply stretch the answer a little bit by a factor that depends on the dimension, and renormalize:

$$p(D_j) = (d+1)\sum_i p(H_i)p(D_j|H_i) - 1. \tag{11}$$

So whereas raw probabilistic considerations say there is an inequality, or at least that there can be an inequality in this expression, quantum mechanics restores equality by changing the formula a little bit. Quantum mechanics adds something, gives some extra structure to probability theory that just raw, plain probability theory itself does not have.

Again, just to emphasize. To get this answer: you just do the classical calculation, and then to get the Born rule you simply modify the classical calculation ever so little!

$$\begin{aligned} p(D_j) &= \mathrm{tr}(\rho \hat{D}_j) \\ &= (d+1)\sum_i p(H_i)p(D_j|H_i) - 1 \\ &= (d+1)[\text{classical calculation}] - 1. \end{aligned} \tag{12}$$

This causes me to wonder whether this is the kind of corner where I can push all of contextuality. The contextuality here is that I've got the calculation for the probabilities directly for the ground path in terms of the probabilities for the sky path, and the formula is modified. Can we take this diagram and this addition to probability theory as a fundamental postulate of quantum mechanics?

Well, what is one wanting to get out of it? The thing you'd like to get out of it is a specification of this convex set that we were talking about earlier. Can it be that this formula somehow implies the convex set that we were talking about, where we had a sphere intersecting a cubic curve? It might imply some features, and you can already see that it might because of the following: Suppose the term under the sum here were a number near zero, or zero in fact. Well, if the number were zero, then we would have $0 - 1$, and that would lead to a negative number on the left-hand side. So we wouldn't actually have a probability on the left-hand side! We might have something that you would call negative probability, but negative probability isn't probability. Alternatively, suppose the sum were close to 1, or in fact 1 itself, then the right-hand side would be $(d + 1) - 1$. That would be d, and that would be larger than 1. So again, this term under the sum can't be too large; otherwise, we'll end up with something that is not a proper probability either. So it gives some hope hope that making this formula hold will give us something like a convex structure, and something along the right variety of one.

OK, so what I *really* want to do—that was all tricksterism, I don't really want to take this one statement (Eq. (12)) as a postulate of quantum mechanics. My starting point should be to consider the most general case, not just where I have a von Neumann measurement on the ground, but where I have any kind of measurement whatsoever on the ground. You can prove that the rule gets a little modified when we consider completely general positive operator valued measures on the ground: The thing that changes is that there is an extra conditional term underneath the sum:

$$\boxed{q(j) = \sum_{i=1}^{d^2} \left[(d+1)p(i) - \frac{1}{d} \right] r(j|i).} \tag{13}$$

So in the case where we have a von Neumann measurement, then all of the rightmost terms under the sum—they sum up to 1 in fact. We thus get this little extra term that is outside of the sum (in Eq. (12)).

I am going to change my notation slightly, because I have more equations to show you. I'm calling the quantum probability, the thing the Born Rule calculates for us, $q(j)$; I'll call the prior probabilities for the measurement in the sky $p(i)$; and I'll call the conditional probabilities for the ground outcomes given the sky outcomes $r(j|i)$.

Here's what I want to show you. Nearly the consistency of this equation alone (Eq. (13))—when I say "nearly," you have to take that a little bit with a little grain of salt, but for me, it's nearly—nearly the consistency of this equation alone implies a significant, nontrivial convex structure. You don't just end up with a sphere, you don't end up with a cube, you don't end up with any polytope—you end up with something that you don't see discussed much in these circles, at least as far as I've seen.

Here's the "nearly." Here's a property I'm going to add to that axiom. It's a property that quantum mechanics has.

Suppose your initial state is the complete garbage state, $\rho = \frac{1}{d}I$, and you actually follow the path in the sky—so in our diagram, we throw the garbage state up into the sky, it goes through the SIC-POVM, and it comes down to the ground for some new POVM, which I'll call $\{G_j\}$. So those are the elements of it. Suppose I look at the outcome j and try to make an inference to what happened up in the sky. Well, then I'd just use the normal Bayes' Rule to calculate the probability for the outcome i in the sky given the outcome j on the ground. And it just works out to be this:

$$\text{Prob}(i|j) = \frac{p(i)r(j|i)}{\sum_k p(k)r(j|k)} \tag{14}$$

$$= \frac{\text{tr}(G_j\Pi_i)}{d \cdot \text{tr}\, G_j}. \tag{15}$$

Now let me redefine this $G_j/\text{tr}\, G_j$ and call it a density operator.

But then look at this probability. The probability for getting i in the sky given j on the ground is just one of these SIC representations of the quantum state ρ_j:

$$\text{Prob}(i|j) = \frac{1}{d}\text{tr}(\rho_j\Pi_i), \;\; \rho_j = \frac{G_j}{\text{tr}\, G_j}. \tag{16}$$

So this is generated by a hidden quantum state ρ_j. Moreover, any ρ_j can be gotten in this way by choosing the measurement on the ground to be an appropriate POVM, because that is a property of quantum mechanics: That for any POVM we can generate a quantum state this way, and vice versa.

Well, all that was really a statement of was the combination of using Bayes' Rule and using this operator representation. So, I'm going to promote it to an axiom:

> Starting from a state of maximal uncertainty for the sky, one can use the posterior state supplied by Bayes' Rule,
>
> $$\text{Prob}(i|j) = \frac{r(j|i)}{\sum_k r(j|k)}, \tag{17}$$
>
> as a valid prior state. Moreover all valid priors can be generated in this way.

We do this all the time: Whenever you gather data, you take your prior, you turn that into a posterior, and then you use that later on for your next prior. So that's all that this rule is telling you, and I have particularized it to having complete ignorance of the sky.

That leads to an immediate consequence for our formula. Because if we just write it out, now supposing the $r(j|i)$ derive from one of these posteriors, the formula becomes this

$$q(j) = \left(\sum_k r(j|k)\right)\left[(d+1)\sum_i p(i)\text{Prob}(i|j) - \frac{1}{d}\right]. \tag{18}$$

But we demand that the number on the left be nonnegative. So that tells us that the the thing in the square brackets here has to be nonnegative, and by the postulate above, this Prob$(i\,|\,j)$ is just some other quantum state. And if we want this to be nonnegative, then it says that the inner product of these two quantum states is bounded below.

Within the probability simplex, if I have a point that is a valid state and I have another point that is a valid state—in other words one that doesn't violate the condition here—their inner product can't be too small. For any two valid priors $\vec{p}$ and $\vec{s}$,

$$\vec{p}\cdot\vec{s} = \sum_i p(i)s(i) \geq \frac{1}{d(d+1)}. \tag{19}$$

That's a nontrivial condition, I would say.

Let me now point out another thing. If we're going to be consistent—we have the measurement in the sky and we have the measurement on the ground—then it has to be the case that we can do the calculation for a measurement on the ground that is exactly the same as the measurement in the sky. Then, self-consistency requires that for any valid $\vec{p}$,

$$p(j) = (d+1)\sum_i p(i)r(j|i) - \frac{1}{d}. \tag{20}$$

If this is going to hold, then the only conditional probabilities $r(j|i)$ that can be allowed are ones of this sort:

$$r(j|i) = \frac{1}{d+1}\left(\delta_{ij} + \frac{1}{d}\right). \tag{21}$$

Now we use the Reciprocity Axiom, the one about Bayes' Rule, to turn these into actual states: All $\vec{p}$ of the form

$$\vec{e}_k = \left[\frac{1}{d(d+1)}, \ldots \frac{1}{d}, \ldots, \frac{1}{d(d+1)}\right] \tag{22}$$

must be valid priors. It says that among our set of states, just because of the existence of these two measuring devices, we have to have states which are flat except for one point, with this particular normalization. We'll call these *basis distributions.*

Something to notice about these states: If we take the inner product of any one of them with itself,

$$\vec{e}_k\cdot\vec{e}_k = \frac{2}{d(d+1)}. \tag{23}$$

Since these correspond to our very special measuring device, let's think of these as being among the extreme points of the set of valid distributions. "Extreme," in the sense means that the norm takes its largest possible value.

This leads to the following notion. I'll call a set S within the probability simplex Δ_{d^2}, which contains these points $\vec{e}_k$ it must contain, *consistent* if for any two points $\vec{p}, \vec{q} \in S$, we have

$$\frac{1}{d(d+1)} \leq \vec{p} \cdot \vec{q} \leq \frac{2}{d(d+1)}. \tag{24}$$

And I will call S *maximal* if adding any further point $\vec{p} \in \Delta_{d^2}$ makes it inconsistent.

So, if I have a set for which every pair of points satisfies Eq. (24) and I add just one more point to it from the simplex, if all of a sudden Eq. (24) is violated, then I say the set is maximal.

Here's an example of a maximal consistent set: if S is the set of quantum states itself, it is a consistent and maximal set. I'll show you that momentarily. As a general problem, it would be nice to characterize *all* such sets. We know that quantum state spaces are among them, but what else is among them?

Let me show you that quantum state space is a maximal consistent set. Suppose I use the SIC representation to turn two probability distributions into operators:

$$\rho = \sum_i \left[(d+1)p(i) - \frac{1}{d} \right] \Pi_i, \tag{25}$$

$$\sigma = \sum_i \left[(d+1)q(i) - \frac{1}{d} \right] \Pi_i. \tag{26}$$

It works out that the Hilbert–Schmidt inner product of these operators can be written in terms of the normal Euclidean inner product of the probability distributions:

$$\operatorname{tr} \rho\sigma = d(d+1)\vec{p} \cdot \vec{q} - 1. \tag{27}$$

Suppose $\vec{p}$ does not correspond to a quantum state. Well, ρ is automatically Hermitian, so if it's not going to correspond to a quantum state, then it has to have a negative eigenvalue. If ρ has a negative eigenvalue, let me choose σ to be the projection onto the direction which gives that negative eigenvalue. I see that I generate a Hilbert–Schmidt inner product that is less than zero:

$$\operatorname{tr} \rho\sigma < 0. \tag{28}$$

Consequently, the inner product of probabilities has to be smaller than our bound:

$$\vec{p} \cdot \vec{q} < \frac{1}{d(d+1)}. \tag{29}$$

There's nothing you can add to quantum state space and still satisfy the consistency condition!

OK. So, quantum state space is in there. But what other properties do these maximal consistent sets have in general which are along the lines of quantum mechanics?

For starters, we can show that maximal consistent sets have to be convex. Let S be a consistent set. If $\vec{p}, \vec{q} \in S$, then for any $\vec{r} \in S$ and $0 \leq x \leq 1$,

$$\frac{1}{d(d+1)} \leq [x\vec{p} + (1-x)\vec{q}\,] \cdot \vec{r} \leq \frac{2}{d(d+1)}. \tag{30}$$

Therefore, maximal consistent sets have to be convex sets.

Let $\vec{p}$ belong to the closure of S. Then there must be a sequence $\vec{p}_t \in S$ converging to $\vec{p}$. But for any $\vec{q} \in S$,

$$\frac{1}{d(d+1)} \leq \vec{p}_t \cdot \vec{q} \leq \frac{2}{d(d+1)}. \tag{31}$$

Therefore, if S is a maximal consistent set, then $\vec{p}$ belongs to S. So maximal consistent sets are closed.

So we have closed and convex sets just from the two conditions for maximal consistent sets. In what sense do we get *nontrivial* sets? There's already a sense in which we get a pretty nontrivial set, just from the lower bound of the consistency condition. For any point that's allowed in my set, there has to be some point on the opposite side of the sphere, around the antipode, that can't be allowed. There has to be a nonincluded region. For any point, there's an excluded spot on the opposite side of the sphere, which is pretty strange.

Moreover, we've got this property: The sphere we're talking about actually reaches outside of the simplex for sufficiently high-dimensional faces. Whatever the object is, not only does it have the property that antipodal regions are excluded, but also it is larger than the simplex it's sitting in. So, it has to be the intersection of a sphere and a simplex.

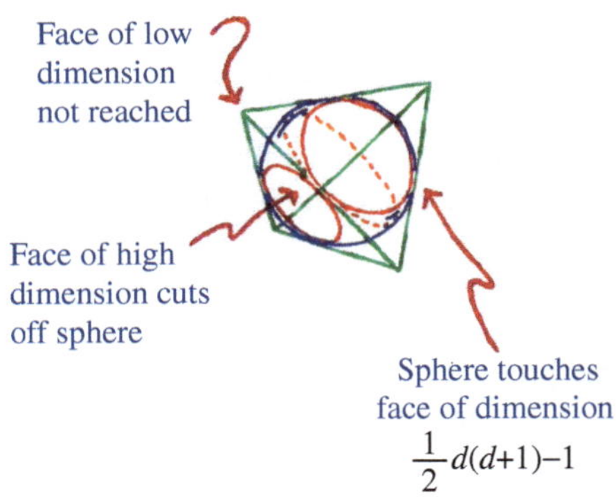

In more detail, the argument goes like this. Re-reference points to the center $\vec{c}$ of the probability simplex:

$$\vec{p}\,' = \vec{p} - \vec{c}. \tag{32}$$

The consistency condition becomes

$$-\frac{1}{d^2(d+1)} \leq \vec{p}\,' \cdot \vec{q}\,' \leq \frac{d-1}{d^2(d+1)}. \tag{33}$$

The sphere is too big for the simplex!

Here's another interesting property. You might ask yourself, "Is there some bound on the number of zeros which can be in a probability vector?" You can use the Schwarz inequality to show that indeed there is. Suppose $\vec{p}$ has n zero values, $p(i) = 0$. Then

$$1 = \left(\sum_{p(i)\neq 0} p(i) \right)^2 \leq (d^2 - n) \sum_{p(i)\neq 0} p(i)^2 \leq (d^2 - n) \frac{2}{d(d+1)}. \tag{34}$$

So,

$$n \leq \frac{1}{2} d(d-1). \tag{35}$$

This automatically implies that there are certain asymmetries to this set. So it started off in a very symmetric way, but suppose you achieve this bound on the number of zeros: We have some $\vec{p}_1$ that has a lot of zeros and then some other values.

$$\vec{p}_1 = (0, 0, 0, 0, p_5, p_6, \ldots, p_n). \tag{36}$$

Let's say that this is a good point. Alternatively, consider another one in which I've rotated around and moved my zeros one spot to the right:

$$\vec{p}_2 = (p_1, 0, 0, 0, 0, p_6, \ldots, p_n). \tag{37}$$

That might still be a good point. But generally, it'll be the case that if I take all of my zeros and shuffle them to the other side of the vector,

$$\vec{p}_3 = (p_1, p_2, \ldots, p_k, 0, 0, 0, 0), \tag{38}$$

all of a sudden the inner product bound will be violated: $\vec{p}_1 \cdot \vec{p}_3$ is too small. So it's not the case that all permutations of the vectors' components lead to allowed points within the set.

All of these things are picking up features that the cubic equation in the quantum state space specification implies. It's a nice, simple condition, but it has lots of consequences, and all these things are consequences of the cubic equation, not the quadratic condition.

Here's one last one. Suppose we have a set of vectors within one of these maximally consistent sets, and they're all of the largest length possible, and moreover, they all have the smallest inner product allowed by the bound. That is, we have $\vec{p}_k{}' \in S$, with $k = 1, \ldots, m$, that saturate both bounds:

$$\vec{p}_k{}' \cdot \vec{p}_k{}' = \frac{d-1}{d^2(d+1)} \;\forall k, \tag{39}$$

$$\vec{p}_k{}' \cdot \vec{p}_l{}' = -\frac{1}{d^2(d+1)} \;\forall k \neq l. \tag{40}$$

Here is the question I would like to ask. For vectors of this variety, is there some maximum number of vectors within this set. How many can be there? That is, how large can we make m?

You can figure this out by forming this combination

$$\vec{G} = \sum_k \vec{p}_k{}', \tag{41}$$

and observing that

$$0 \leq \vec{G} \cdot \vec{G} = \sum_{k,l} \vec{p}_k{}' \cdot \vec{p}_l{}' = \frac{m(d-m)}{d^2(d+1)}. \tag{42}$$

This expression can only be nonnegative if

$$m \leq d. \tag{43}$$

So it says we've started with a simplex of size d^2, and if we demand that we have a set of points which are maximally distant from one another, we can't have more than d of them. Again, this mimics quantum state space! We started with a set of dimension d^2 and found a kind of underlying set of size d.

Moreover, if

$$\vec{G} \cdot \vec{G} = 0, \tag{44}$$

then

$$\sum_{k=1}^{d} \frac{1}{d} \vec{p}_k{}' = 0 \Rightarrow \sum_{k=1}^{d} \frac{1}{d} \vec{p}_k = \vec{c}. \tag{45}$$

This is also the same as in quantum mechanics.

So, challenge: What further postulates must be made to recover quantum state space precisely? That is, how do we recover the convex hull of

$$\sum_i p(i)^2 = \frac{2}{d(d+1)}, \tag{46}$$

$$\sum_{ijk} c_{ijk}\, p(i)p(j)p(k) = \frac{d+7}{(d+1)^3}, \tag{47}$$

with the c_{ijk} possessing the correct properties? I don't know, maybe they'll get really horrible. Maybe it won't be the pretty principle that I want at all. But I feel that at least this is moving in the direction of focusing on contextuality as a primitive notion within quantum theory, and the right particular flavor of contextuality. Of course, what we really want is this thing John Wheeler said:

> If one really understood the central point and its necessity in the construction of the world, one ought to be able to state it in one clear, simple sentence.

I hope you guys keep trying to do that. Thank you!

During the Q&A, there was a comment from Lucien Hardy.

HARDY: Maybe first a comment, and then a more specific question. The way I see it, in terms of this program of finding some postulates or axioms or whatever, is that we're looking at some strange object, and the original mathematical axioms of quantum theory sort of give us a bunch (five or however many there are) of strange vantage points on this strange object. Maybe we're looking at it from odd directions. And as we get better postulates, we're finding better ways to look at it from. So, rather than looking at it from some direction where it doesn't make sense, perhaps we can see it face-on, and we see it from several different directions. So we're moving towards some more reasonable set of vantage points on the given object, so then it wouldn't disturb us of course whether these axioms are unique or not—at least they should somehow be a sensible way of looking at that object. Just sort a general comment. But then the particular point was you said that you thought this equation could be a possible postulate, but you also said that you liked it when postulates were expressed in words. So do you have some words for that equation?

FUCHS: No, that's my big failure at this point. But I liked your analogy at the beginning, and I guess maybe it gives me the tool to express that: It seems that the light has been shining in ways that it kind of takes the corners off of the edge of quantum theory. So we've got this sharp, rather jagged object called quantum theory, and the lights that have been shined on it have made it like a little distinction from classical theory in usual ways. For instance in Giulio and Mauro and Paulo's postulates, they can pinpoint it to one spot, and you can as well. And some of the other axiom systems. So it feels to me that it is kind of dulling … the projection of it is smoother than the real object itself is.[1]

[1]David Mermin says it much better in the paper where he originated the phrase 'shut up and calculate' [33]: "I would rather celebrate the strangeness of quantum theory than deny it, because

But for your question, yeah it's a shortcoming. I don't have any explanation of this equation that is giving some structure other than to know that it works. I've tried to play games to do with Dutch book arguments and some kind of picture of the world that ... uh ... You know I have these ... Like Gilles, you see, Gilles was my teacher. He has conversations with God. So I've had conversations with God where I imagine that God says ... You know, I ask him, "I want the ability to write messages upon the world." And God says,"You know ... I can give you that but that means that the world is going to have to have some loose play in it, because if it were a rigid thing then you wouldn't be able to write messages in it." Then he says, "Ah but the moment I give you some loose play in the world, then I may not be able to predict all the floods and fires for you anymore, because it's got all this loose play in it." So anyway I've played little games like that associated with Dutch book, and I've gotten nowhere.

Conversation with God

> Adam said to God, "I want the ability to write messages onto the world."
>
> God replied, "You ask much of me. If you want to write upon the world, it cannot be so rigid a thing as I had originally intended. The world would have to have some malleability, with enough looseness for you to write upon its properties. It will make your world more unpredictable than it would have been. Because of this looseness, I may not be able to warn you about impending dangers like droughts and hurricanes as effectively as I might have otherwise, but I can make it such if you want."
>
> With that Adam brought all host of uncertainties to his life, but he gained a world where his deeds and actions mattered.

(Footnote 1 continued)
I believe it still has interesting things to teach us about how certain powerful but flawed verbal and mental tools we once took for granted continue to infect our thinking in subtly hidden ways .... [T]he problem with the second generation's iron-fistedly soothing attitude is that by striving to make quantum mechanics appear so ordinary, so sedately practical, so benignly humdrum, they deprive us of the stimulus for exploring some very intriguing questions about the limitations in how we think and how we are capable of apprehending the world."

References

1. C.A. Fuchs, A. Peres, Quantum theory needs no 'interpretation'. Phys. Today **53**(3), 70 (2000)
2. C.A. Fuchs, *Coming of Age with Quantum Information* (Cambridge University Press, Cambridge, 2010)
3. C.A. Fuchs, Quantum mechanics as quantum information (and only a little more) (2002). arXiv:quant-ph/0205039v1. Abridged version, in *Quantum Theory: Reconsideration of Foundations*, ed. by A. Khrennikov (Växjö University Press, Växjö, 2002), pp. 463–543
4. G. Chiribella, G.M. D'Ariano, P. Perinotti, Informational derivation of quantum theory. Phys. Rev. A **84**, 012311 (2011). arXiv:1011.6451
5. Č. Brukner, Questioning the rules of the game. Physics **4**, 55 (2011)
6. B. Dakić, Č. Brukner, Quantum theory and beyond: is entanglement special?, in *Deep Beauty: Understanding the Quantum World Through Mathematical Innovation*, ed. by H. Halvorson (Cambridge University Press, Cambridge, 2011), pp. 365–392. arXiv:0911.0695
7. L. Hardy, Quantum theory from five reasonable axioms (2001). arXiv:quant-ph/0101012v4
8. R. Schack, Quantum theory from four of Hardy's axioms. Found. Phys. **33**, 1461 (2003). arXiv:quant-ph/0210017
9. L. Hardy, Reformulating and reconstructing quantum theory (2011). arXiv:1104.2066
10. M.P. Müller, L. Masanes, Information-theoretic postulates for quantum theory (2012). arXiv:1203.4516
11. A. Wilce, Four and a half axioms for finite dimensional quantum mechanics (2009). arXiv:0912.5530
12. A. Einstein, Quantenmechanik und Wirklichkeit. Dialectica 2, 320–24 (1948). The English translation used here is from D. Howard, Einstein on Locality and Separability, Stud. Hist. Phil. Sci. Part A 16, 171–201 (1985). Another translation can be found in *The Born–Einstein Letters*, ed. by M. Born (Macmillan, 1971), p. 170
13. A. Peres, Unperformed experiments have no results. Am. J. Phys. **46**, 745 (1978)
14. J. Conway, S. Kochen, The Free Will Theorem. Found. Phys. **36**, 1441–73 (2006). arXiv:quant-ph/0604079
15. C.M. Caves, C.A. Fuchs, R. Schack, Subjective probability and quantum certainty, Stud. Hist. Phil. Mod. Phys. **38**, 255–274 (2007). arXiv:quant-ph/0608190
16. A. Einstein, B. Podolsky, N. Rosen, Can quantum-mechanical description of physical reality be considered complete? Phys. Rev. **47**, 777–780 (1935)
17. A. Peres, Two simple proofs of the Kochen-Specker theorem. J. Phys. A **24**, L175 (1991)
18. A. Peres, *Quantum Theory: Concepts and Methods*, Chapter 7 (Kluwer, Dordrecht, 1993)
19. A. Peres, Generalized Kochen–Specker theorem. Found. Phys. **26**, 807–12 (1996), arXiv:quant-ph/9510018
20. C.A. Fuchs, R. Schack, A quantum-Bayesian route to quantum-state space. Found. Phys. **41**, 345–356 (2011). arXiv:0912.4252
21. D.M. Appleby, Å. Ericsson, C.A. Fuchs, Properties of QBist state spaces. Found. Phys. **41**, 564–579 (2011). arXiv:0910.2750
22. C.A. Fuchs, R. Schack, Quantum-Bayesian coherence. Rev. Mod. Phys. **85**, 1693–1715 (2013). arXiv:1301.3274
23. G. Zauner, Quantum designs—foundations of a noncommutative theory of designs. Ph.D. thesis, University of Vienna (1999). http://www.gerhardzauner.at/qdmye.html
24. J.M. Renes, R. Blume-Kohout, A.J. Scott, C.M. Caves, Symmetric informationally complete quantum measurements. J. Math. Phys. **45**(6), 2171 (2004). arXiv:quant-ph/0310075
25. I. Bengtsson, K. Blanchfield, A. Cabello, A Kochen–Specker inequality from a SIC. Phys. Lett. A 376, 374–76 (2010). arXiv:quant-ph/1109.6514
26. A.J. Scott, M. Grassl, SIC-POVMs: A new computer study. J. Math. Phys. **51**, 042203 (2010). arXiv:quant-ph/0910.5784
27. D.M. Appleby, S.T. Flammia, C.A. Fuchs, The Lie algebraic significance of symmetric informationally complete measurements. J. Math. Phys. **52**, 02202 (2011). arXiv:quant-ph/1001.0004

28. J.I. Rosado, Representation of quantum states as points in a probability simplex associated to a SIC-POVM. Found. Phys. **41** 1200–13 (2011). arXiv:quant-ph/1007.0715
29. G.N.M. Tabia, Experimental scheme for qubit and qutrit SIC-POVMs using multiport devices. Phys. Rev. A **86**, 062107 (2012). arXiv:quant-ph/1207.6035
30. G.N.M. Tabia, D.M. Appleby, Exploring the geometry of qutrit state space using symmetric informationally complete probabilities. Phys. Rev. A **88**, 012131 (2013). arXiv:quant-ph/1304.8075
31. D.M. Appleby, C.A. Fuchs, H. Zhu, Group theoretic, Lie algebraic and Jordan algebraic formulations of the SIC existence problem (2013). arXiv:quant-ph/1312.0555
32. J.A. Wheeler, The quantum and the universe, in *Relativity, Quanta and Cosmology,* vol. 2, ed. by M. Pantaleo, F. de Finis (Johnson Reprint Corp., New York, 1979)
33. N.D. Mermin, What's wrong with this pillow?. Phys. Today **42**(4), 9, 11 (1989)

Part III
Categories and Convex Sets

Generalised Compositional Theories and Diagrammatic Reasoning

Bob Coecke, Ross Duncan, Aleks Kissinger and Quanlong Wang

This chapter provides an introduction to the use of *diagrammatic language*, or perhaps more accurately, *diagrammatic calculus*, in quantum information and quantum foundations. We illustrate the use of diagrammatic calculus in one particular case, namely the study of *complementarity* and *non-locality*, two fundamental concepts of quantum theory whose relationship we explore in later part of this chapter.

The diagrammatic calculus that we are concerned with here is not merely an illustrative tool, but it has both (i) a conceptual physical backbone, which allows it to act as a foundation for diverse physical theories, and (ii) a genuine mathematical underpinning, permitting one to relate it to standard mathematical structures.

(i) The conceptual physical backbone concerns *compositionality*. Given two systems, there is also a composite system. This notion of composition is a primitive ingredient of the diagrammatic language. Moreover, the basic elements of the diagrammatic language are *processes*, and states are identified with preparation processes. This paves the way for a framework of *generalised compositional theories* (GCTs), named in analogy to generalised probabilistic theories [1]. The latter

B. Coecke (✉) · A. Kissinger
Department of Computer Science, University of Oxford, Wolfson Building, Parks Road, Oxford OX1 3QD, UK
e-mail: Bob.Coecke@comlab.ox.ac.uk

A. Kissinger
e-mail: alek@cs.ox.ac.uk

R. Duncan
Department of Computer and Information Sciences, University of Strathclyde, Livingston Tower, 26 Richmond Street, Glasgow G1 1XH, UK
e-mail: ross.duncan@strath.ac.uk

Q. Wang
School of Mathematics and System Sciences, Beihang University, XueYuan Road No.37, HaiDian District, Beijing, China
e-mail: qlwang@buaa.edu.cn

G. Chiribella and R.W. Spekkens (eds.), *Quantum Theory: Informational Foundations and Foils*, Fundamental Theories of Physics 181,
DOI 10.1007/978-94-017-7303-4_10

have recently received much attention because one can better understand a theory—quantum theory in particular—by studying it as merely a member of a broader class of theories. Notably, the study of non-locality within this framework has provided important new insights [2, 3]. Whereas generalised probabilistic theories discard everything except the convex probabilistic structure, in contrast, GCTs focus on composition. This approach is informed by techniques used in computer science, logic, and the branch of mathematics called category theory, however its roots can be traced to Schrödinger's conviction that the essential characteristic of quantum theory is the manner in which systems compose [4].

(ii) On the other hand, the diagrammatic language has a well-defined mathematical meaning, which permits any diagram to be interpreted as a definite object in various other concrete mathematical models, for example in Hilbert spaces. This translation can be carried out in a formally precise manner, so that reasoning in the diagrammatic calculus produces true equations in the chosen model. At the same time, the relationship between what is provable in the calculus and what is provable in concrete models can be described to a high degree of precision.

We won't discuss this mathematical basis in detail here, however it may be summarised as follows: the diagrammatic calculus is itself a GCT, and GCTs form a certain class of *monoidal categories*, also known as *tensor categories*. The use of diagrammatic languages for tensors traces back to Penrose in the early 1970s [5], but was only placed on a formal mathematical basis in the late 1980s [6, 7]. Their use in quantum foundations and quantum information began with an abstract (partial) axiomatisation of Hilbert spaces in terms of these categories [8], eventually resulting in so-called *quantum picturalism* [9]. Meanwhile, the diagrammatic compositional language has been adopted by several researchers in quantum foundations [10, 11]. The particular developments related here been used to solve problems in quantum foundations [12, 13] and quantum computation [14–16].

1 Introduction to Quantum Picturalism

1.1 Theories and Diagrams

A generalised compositional theory consists of *systems*, or more accurately *types of systems*, and *processes* which transform systems. A process f which transforms systems of type A into systems of type B is written $f : A \to B$. At the highest level of generality we do not need to give any details as to what A, B, or f are: it is enough to know that that f accepts systems of type A as inputs and produces systems of type B as outputs. The important thing is how systems and processes are combined.

Mathematically speaking, general compositional theories are *strict symmetric monoidal categories*, and a full exposition of their properties would require a lengthy detour into category theory. The interested reader can refer to Mac Lane's classic text

[17] for a thorough treatment. However, we can avoid reading Mac Lane's book[1] by adopting a diagrammatic notation, which absorbs all of the relevant equations into the syntax. This notation is the subject of the first section of this paper.

We will represent processes by *diagrams*, consisting of boxes and wires. The wires are labelled by systems, and the boxes by basic processes.[2] Wires join boxes at the top and bottom; the wires below correspond to the input systems of the process, and those at the top correspond to the output systems. For example:

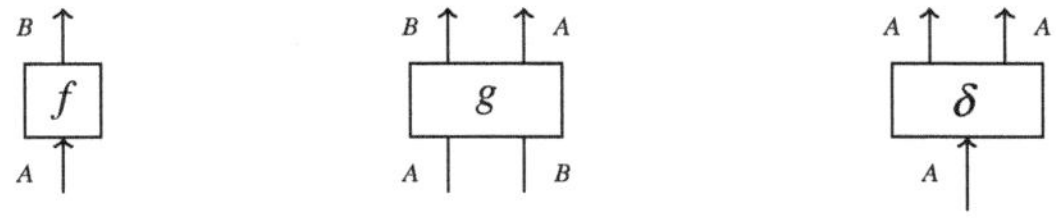

$f : A \to B$ $\quad$ $g : A \otimes B \to B \otimes A$ $\quad$ $\delta : A \to A \otimes A$

The same is true for the diagram as a whole: the wires entering the bottom of the diagram are its input systems, and those leaving from the top are its outputs.

Given processes $f : A \to B$ and $g : B \to C$, it seems obvious that doing f then g is again a process, and we write $g \circ f : A \to C$ to denote this process. In other words, processes admit *sequential combination*; we will usually call this operation *composition*.

Similarly, a pair of systems, say A and B, can be taken together and viewed as a single system, $A \otimes B$. Now, given a pair of processes $f : A \to B$ and $g : A' \to B'$, a new process is obtained by placing them in parallel. We denote the combined process $f \otimes g : A \otimes A' \to B \otimes B'$. This operation of parallel combination is called *tensor*.

In the diagrammatic notation, composition is expressed by plugging the outputs of one box into the inputs of another, and the tensor is given by juxtaposition.

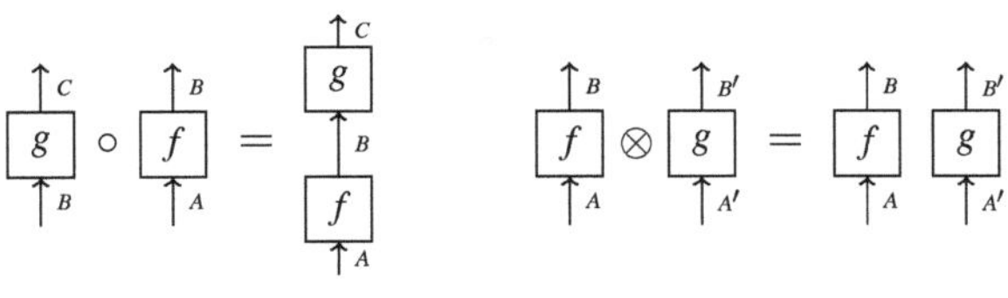

We require that both operations, composition and tensor, are associative and obey the interchange law,

$$(f \otimes g) \circ (h \otimes k) = (f \circ h) \otimes (g \circ k). \tag{1}$$

In the graphical notation, all of these equations become trivial: they boil down the statement that the three diagrams below are unambiguous.

---

[1] We jest; reading Mac Lane's book is eventually unavoidable, however the paper [18] is an easy introduction to the subject of monoidal categories.

[2] The term "basic" simply means a process whose internal structure is of no interest. Typically we construct diagrams from some given set of basic processes.

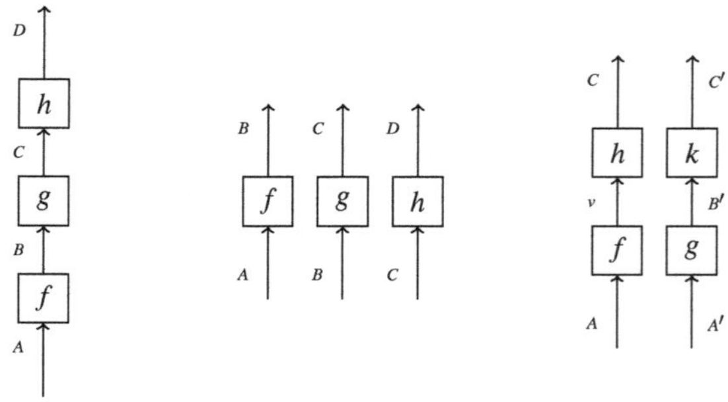

While it is easy to translate these diagrams back into conventional notation, to do so we must make a *choice* of where to put the brackets, even though the theory tells us this choice does not matter. This highlights a key advantage of working with diagrams, namely that the objects which are *equal* in the theory produce the *same diagram*.

In addition to the two operations, composition and tensor, every generalised compositional theory is equipped with certain primitive processes. The simplest process is the process which doesn't do anything at all, simply returning unchanged the system given to it. We assume that for every system A such a null process, called the *identity* and written $1_A : A \to A$, exists. The fact that it does nothing is expressed by the equations

$$1_B \circ f = f = f \circ 1_A$$

for all processes $f : A \to B$. The identity process $1_A : A \to A$ is drawn as a wire without any box on it, while the identity for $A \otimes A'$ is simply the tensor product $1_A \otimes 1_{A'}$, i.e. two wires.

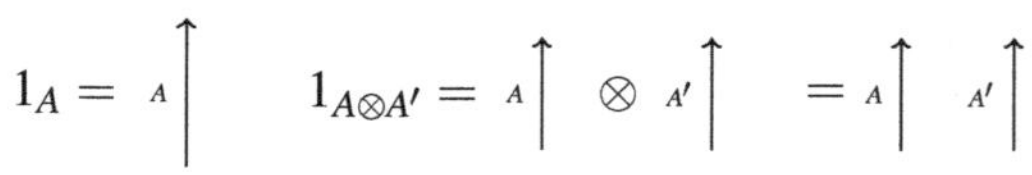

Once again we see an equation absorbed into the notation: since the identity has no effect on a process, the *length* of the wires attached to a box makes no difference.

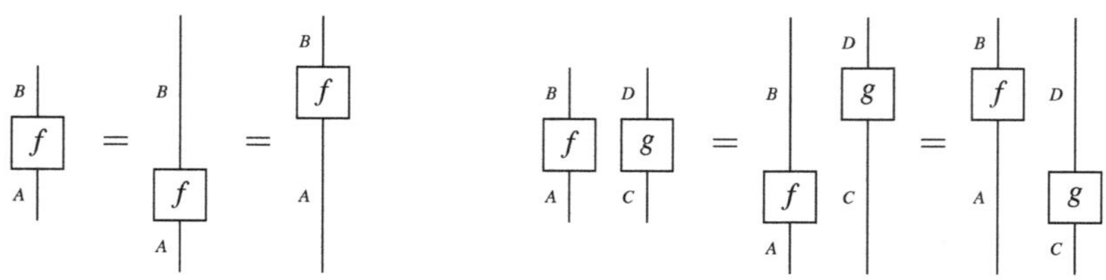

In addition, for every pair of systems A and B there is a process $\sigma_{A,B} : A \otimes B \to B \otimes A$ which exchanges the two systems. The class of theories we consider here are *symmetric*: swapping two systems twice has no effect, hence the equation

$$\sigma_{B,A} \circ \sigma_{A,B} = 1_{A \otimes B}$$

holds for all systems A and B Graphically, the swap is just the crossing of two wires:

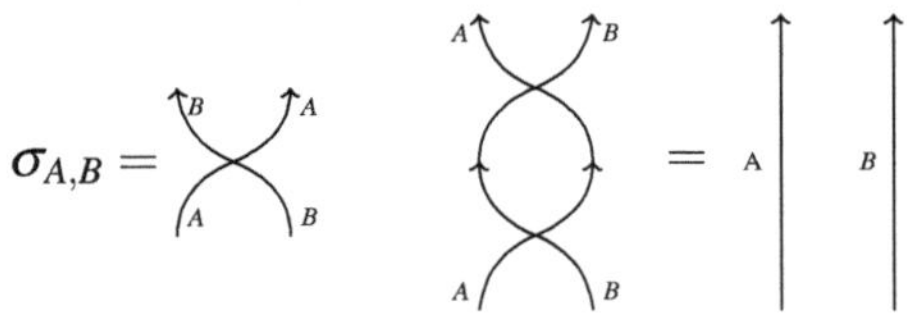

In fact, the swap should satisfy some further *coherence equations*, the details of which can be found in [17]. However, we can again make the graphical notation do the work by allowing wires to cross freely in the diagrams, and saying that only the connectivity of the wires matters, and not their configuration in the page. For example, the following diagrams are equal:

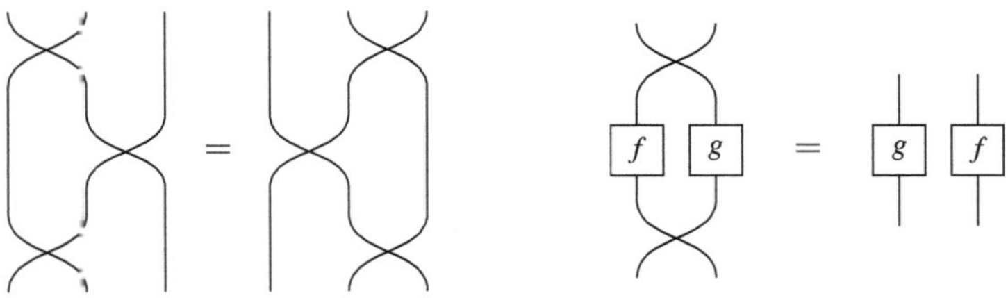

Note that we do not distinguish between wires crossing over and crossing under.

A processes may produce an output without having to consume an input first, or vice versa. Therefore we introduce a null system, or empty system, which we denote I. Hence a process that produces an A from nothing would be written $p : I \to A$. Like the identity process, the null system obeys some equations:

$$A \otimes I = A = I \otimes A \qquad \text{and} \qquad 1_I \otimes f = f = f \otimes 1_I,$$

for all systems A and all processes f. As suggested by the preceding equations, I is represented as empty space in the diagram, and its identity process $1_I : I \to I$ is represented by the empty diagram.

A process of type $s : I \to I$ is called a *scalar*; this name will be justified later. It is clear from the diagrammatic notation that given scalars s and s' we have $s \circ s = s \otimes s' = s' \otimes s = s' \circ s$; i.e. the scalars form a commutative monoid.[3]

In the preceding text we have introduced various transformations of diagrams which, we claim, do not change anything. It is reasonable to ask: when are two

[3] This is actually true even for non-symmetric monoidal categories; see [7].

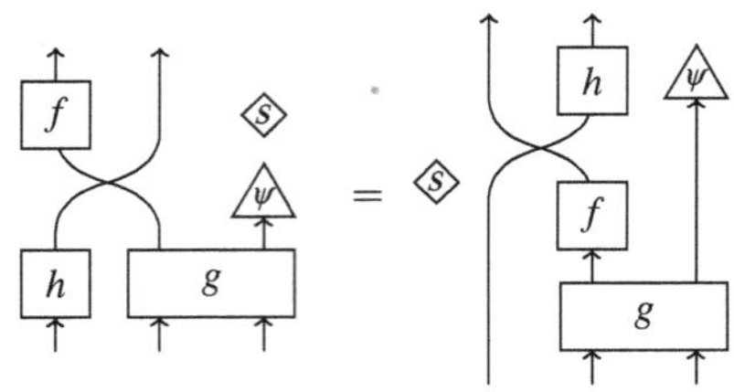

Fig. 1 Examples of topologically equivalent diagrams

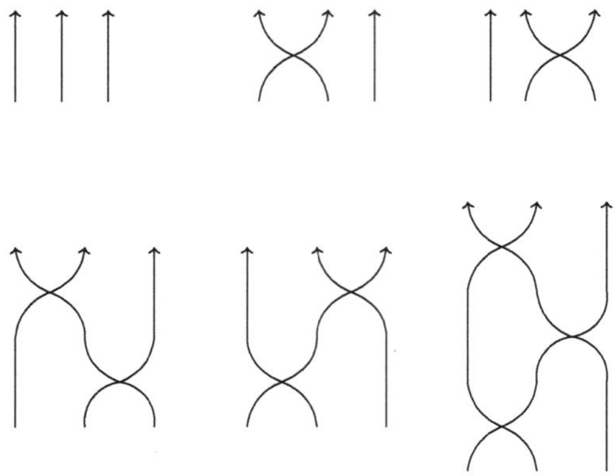

Fig. 2 Example: the symmetric group S_3 presented as diagrams

diagrams considered to be equal? We use a very intuitive notion here: **Two diagrams are considered equal when, keeping the inputs and outputs fixed, one may be transformed to the other by purely topological transformations.** In other words, if starting from one diagram we—by crossing or uncrossing wires, stretching wires, moving boxes along wires, translating boxes in the plane (while maintaining their connections), etc—arrive at the other, then they are equal. In particular, since scalars are not connected to the inputs or outputs of the diagram, they may be placed anywhere in the diagram without altering its meaning (Fig. 1).

Example 1 The simplest non-trivial example is the theory with one primitive system, denoted u, and whose processes are generated by the identity and swap. We call this theory **SymGrp**. Since there is only one basic system, every other system is just an n-fold tensor power of u, hence the systems of the theory can be identified with the natural numbers. In this theory, a process $p : n \to n$ is nothing more than a sequence of swaps; i.e. a permutation on the n-element set. Hence **SymGrp** is exactly the theory of the symmetric groups (Fig. 2).

Example 2 (*Finite-dimensional Hilbert spaces*) The theory called **FHilb** has as its systems all finite-dimensional complex Hilbert spaces. The processes of this theory are all linear maps $f : A \to B$. The sequential composition of processes is the usual composition of linear maps, and the tensor is the usual Kronecker product of vector spaces and maps. The identity process is the identity map, the swap is the evident permutation map, and the null system is the base field, $\mathbb{C}$. Since a linear map $\mathbb{C} \to \mathbb{C}$

is totally determined by its value at 1, we see that the scalars of **FHilb** are nothing more than the complex numbers themselves.

We write $\mathbf{FHilb}_D$ to denote the subtheory **FHilb** restricted to Hilbert spaces of dimension D^n and linear maps between them, for some fixed D. For convenience, we refer to $\mathbf{FHilb}_2$ as **Qubit**. Notice that the systems of **Qubit** are all tensor powers of $\mathbb{C}^2$, and its processes include all quantum circuits, state preparations, and post-selected measurements, justifying the name.

Remark 1 Note that we must specify what the tensor product is to specify what the theory is. For example, another equally valid theory is the collection of finite dimensional Hilbert spaces and linear maps, but with the direct sum as the tensor. This is again a general compositional theory, although since it lacks certain other features we will require later, it will play no role in this presentation.

Example 3 (*Sets and Relations*) An example with very different flavour, but most of the same structure is **FRel**. The systems of **FRel** are all finite sets (considered up to isomorphism[4]), and the processes $r : X \to Y$ are relations between X and Y, that is subsets of $X \times Y$. The composition of relations is given by

$$s \circ r = \{(x, z) \mid \exists y \text{ s.t. } (x, y) \in r \text{ and } (y, z) \in s\}.$$

The identity process is the diagonal relation,

$$1_X = \{(x, x) \mid x \in X\}.$$

The tensor product in **FRel** is the cartesian product $X \otimes Y = X \times Y$, which takes the form

$$r \otimes r' = \{((x, x'), (y, y')) \mid (x, y) \in r \text{ and } (x', y') \in r'\}$$

on processes. The null system is the singleton set $\{*\}$, for which we have $\{*\} \times X \cong X$ for all sets X. There are exactly two relations from $\{*\}$ to itself, namely the total relation and the empty relation. Hence, the scalars of **FRel** are the Boolean monoid, i.e. $\mathbb{Z}_2$ with the usual multiplication.

An important subtheory of **FRel** is **FSet**, obtained by restricting the to relations which are functions: that is, relations $r : X \to Y$ where each x is related to exactly one y. Just as in the case of **FHilb**, we can consider restrictions of **FRel** to systems generated by a set of size D, which we call $\mathbf{FRel}_D$. For example, $\mathbf{FRel}_2$ contains all the Boolean functions. The intersection of $\mathbf{FRel}_2$ and **FSet** consists of precisely the Boolean functions; this theory we denote **Bool**. Many other interesting theories are subtheories of **FRel**; we'll meet some more later.

Since generalised compositional theories all share certain basic structure, it is natural to consider maps between them. Given two such theories **C** and **D**, a map

[4]Since we identify sets of the same cardinality, we can equivalently say that the systems of **FRel** are just the natural numbers.

$F : \mathbf{C} \to \mathbf{D}$ consists of an assignment of each system A in $\mathbf{C}$ to a system FA in $\mathbf{D}$, and an assignment of each process $f : A \to B$ in $\mathbf{C}$ to a process $Ff : FA \to FB$ in $\mathbf{D}$, obeying the following equations:

$$F(A \otimes B) = FA \otimes FB \qquad FI = I$$
$$F(g \circ f) = Fg \circ Ff \qquad F(f \otimes g) = Ff \otimes Fg$$
$$F1_A = 1_{FA} \qquad F\sigma_{A,B} = \sigma_{FA,FB}$$

In the mathematics literature, such a map is called a *strict symmetric monoidal functor*; again, see Mac Lane [17] for the details. The important point to note is that the mapping F sends wires to wires. Therefore, to specify such a mapping it is enough to specify the image of the boxes in a diagram, ensuring that composition and tensor are respected.

Example 4 We can define a map R_D : **SymGrp** $\to$ **FHilb** by setting $R_D(u) = \mathbb{C}^D$ and then everything else is defined by the requirement that R_D is a strict symmetric monoidal functor. Thus we have a D^n dimensional representation of the symmetric group S_n for every D.

In fact, this construction applies equally well to any generalised compositional theory $\mathbf{C}$: all that is required is an assignment of the unique primitive system u to some system of $\mathbf{C}$. Therefore every generalised compositional theory contains all the symmetric groups.

Given a mapping between theories it is easy to calculate the image of a given diagram. One must recursively partition the diagram into tensors and compositions of smaller diagrams until each partition contains exactly one element—that is, either a single wire, a crossing of wires, or a box. The interchange law (Eq. 1) guarantees that the result does not depend on the partition chosen.

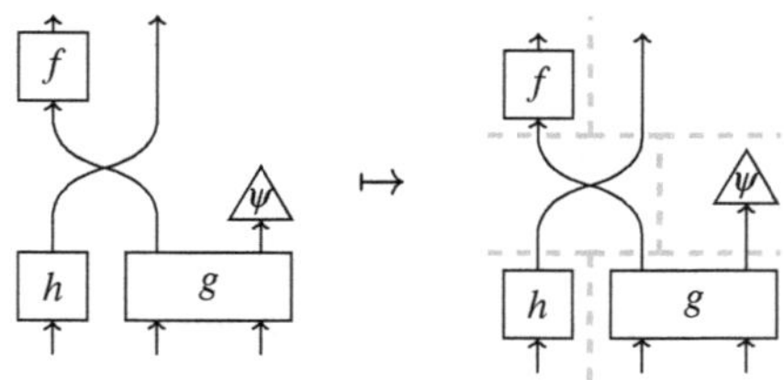

We may now state:

Theorem 1 (Fundamental Theorem of Diagrams) *Given any two generalised compositional theories* $\mathbf{C}$ *and* $\mathbf{D}$*, and a map* $F : \mathbf{C} \to \mathbf{D}$*, for any two diagrams* d *and* d' *in* $\mathbf{C}$*, if* $d = d'$ as diagrams *then* $Fd = Fd'$ *in* $\mathbf{D}$.

This theorem has many variations, and we refer the reader to Selinger's survey article [19] for the full details.

Remark 2 In the diagrams to come, we will often use horizontal separation to indicate separation in space and vertical separation to indicate separation in time. For example,

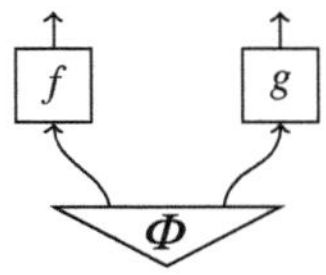

depicts the creation of two systems by the process Φ, which then become spatially separated over some time and are acted upon by processes f and g respectively. Since, as we already know, topologically equivalent diagrams are equal, these separations have no formal status and are purely illustrative.

1.2 Rewrites and Models

Since we wish to generalise over many concrete mathematical structures, we are particularly interested in theories which can be specified *axiomatically*. That is, to specify the theory we state (i) the list of basic systems—typically we'll only have one basic system, the rest being generated by the tensor product—and (ii) the basic processes. The processes of the theory are then *all* the diagrams which can be constructed from these processes and nothing else.

Example 5 (*Boolean Circuits*) A simple example of a compositional theory is **BoolCirc**, the theory of boolean circuits. This theory has only one basic system, the bit b, and the basic processes are the logic gates:

$\wedge : b \otimes b \to b$ $\quad$ $\vee : b \otimes b \to b$ $\quad$ $\neg : b \to b$ $\quad$ $\mathbf{FAN} : b \to b \otimes b$

A process in this theory is a circuit for computing some boolean function, built up from these basic gates (Fig. 3).

It is tempting to assume that **BoolCirc** is related to the theory of Boolean functions, and we can make this precise by specifying a mapping B : **BoolCirc** $\to$ **Bool**. We assign $B(b) = \{0, 1\}$ and define B on the basic processes as follows:

$$B(\wedge) = a : \begin{cases} 00 \mapsto 0 \\ 01 \mapsto 0 \\ 10 \mapsto 0 \\ 11 \mapsto 1 \end{cases} \qquad B(\vee) = o : \begin{cases} 00 \mapsto 0 \\ 01 \mapsto 1 \\ 10 \mapsto 1 \\ 11 \mapsto 1 \end{cases}$$

$$B(\neg) = n : \begin{cases} 0 \mapsto 1 \\ 1 \mapsto 0 \end{cases} \qquad B(\mathbf{FAN}) = \delta : \begin{cases} 0 \mapsto 00 \\ 1 \mapsto 11 \end{cases}$$

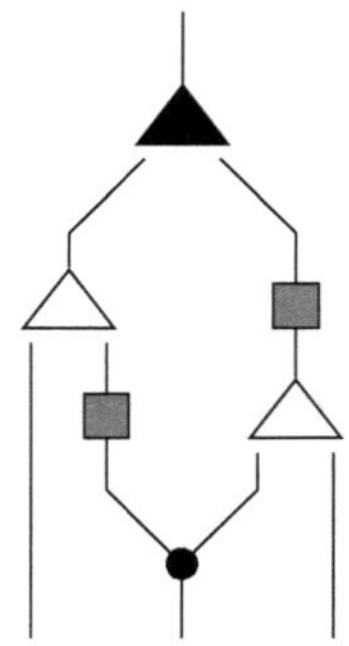

Fig. 3 A Boolean circuit to compute $(x \wedge \neg y) \vee \neg(y \wedge z)$

The mapping B assigns to each diagram the boolean function normally associated with it. However this is not the only possibility. Consider the following mapping, P : **BoolCirc** $\to$ **Bool**. Once again $P(b) = \{0, 1\}$, but now we have the following assignment of processes:

$$P(\wedge) = a : \begin{cases} 00 \mapsto 0 \\ 01 \mapsto 0 \\ 10 \mapsto 0 \\ 11 \mapsto 1 \end{cases} \qquad P(\vee) = p : \begin{cases} 00 \mapsto 0 \\ 01 \mapsto 1 \\ 10 \mapsto 1 \\ 11 \mapsto 0 \end{cases}$$

$$P(\neg) = i : \begin{cases} 0 \mapsto 0 \\ 1 \mapsto 1 \end{cases} \qquad P(\mathbf{FAN}) = \delta : \begin{cases} 0 \mapsto 00 \\ 1 \mapsto 11 \end{cases}$$

The mapping P assigns to each $d : b^n \to b$ in **BoolCirc** an n-variable polynomial over the ring $\mathbb{Z}_2$. (More generally a circuit with multiple outputs produces a list of polynomials, one for each output.)

In fact, as the example of P suggests, the diagrams of **BoolCirc** admit an interpretation in any setting with two binary operations and one unary operation. This is not entirely satisfactory. In order to capture more than the bare syntax of any given theory we need to impose some additional equations on the class of diagrams. We do this via *rewrite rules*.

A rewrite rule consists of a pair of diagrams of the same type, for example $d : A \to B$ and $d' : A \to B$. If this rule is called r then we write $r : d \Rightarrow d'$, or diagrammatically

Whenever d occurs as a subdiagram of a larger diagram e then we can replace d with d' in e, written $e[d] \stackrel{r}{\Rightarrow} e[d']$, or in diagrams:

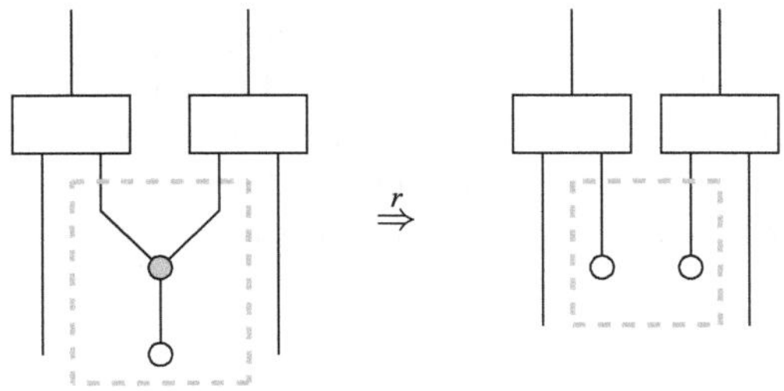

Rewrite rules allow us to define a notion of equality in addition to the basic equality of diagrams Given a collection of rewrite rules $\mathscr{R}$ we write $d \overset{\mathscr{R}}{\Rightarrow} d'$ if there is some rewrite in $\mathscr{R}$ taking d to d'. Evidently $\overset{\mathscr{R}}{\Rightarrow}$ is a transitive relation; let $\overset{\mathscr{R}}{=}$ be its symmetric, reflexive closure. Then we say that two processes are equal according to $\mathscr{R}$ if their corresponding diagrams satisfy $d \overset{\mathscr{R}}{=} d'$. Typically we'll exhibit this equivalence as a sequence of rewrites.

Example 6 (*Boolean circuits*) Consider the following two rewrite rules for **BoolCirc**, expressing respectively the distributivity of AND over OR, and (one half of) De Morgan's law.

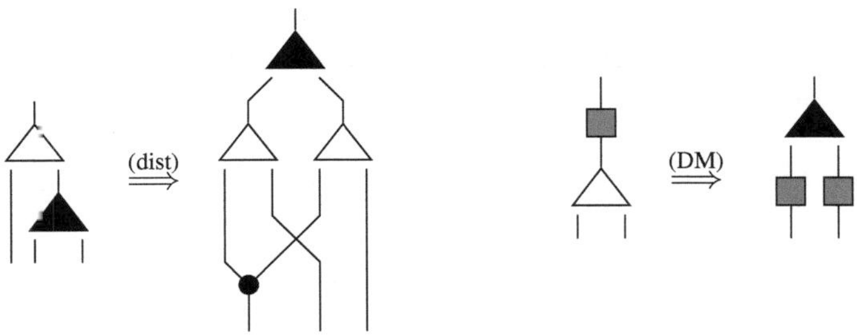

Now we can show that a certain Boolean circuit can be transformed into its disjunctive normal form:

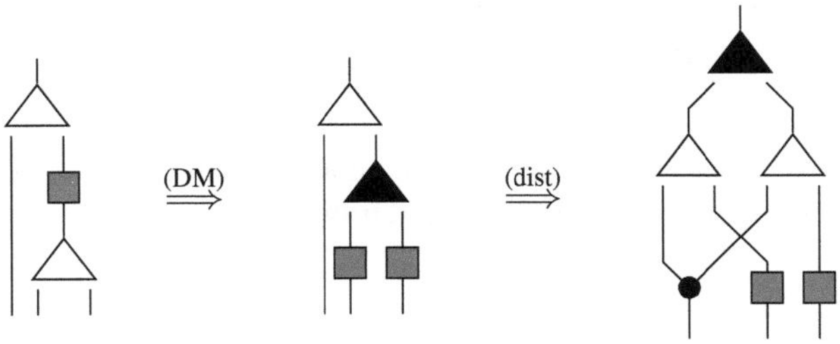

Given a theory **C**, a set of rewrite rules $\mathscr{R}$, and a mapping $F : \mathbf{C} \to \mathbf{D}$, we can ask the following question: if $d \overset{\mathscr{R}}{=} d'$ in **C**, is it the case that $Fd = Fd'$ in **D**?

This property is called *soundness*. A sound mapping $F : \mathbf{C} \to \mathbf{D}$ is called an *interpretation* of $\mathbf{C}$ in $\mathbf{D}$, and the image of $\mathbf{C}$ in $\mathbf{D}$ is called a *model*. In the example above, the mapping B is sound, hence it provides an interpretation of **BoolCirc** (and $\mathscr{R}$) in **Bool**; on the other hand P does not, due to the failure of De Morgan's law. Generally speaking we will always work with a given set of rewrite rules and a given interpretation map, so we will usually say "*the* **D** interpretation of **C**", although in principle there could be many.

Remark 3 The converse property to soundness, $Fd = Fd'$ implies $d = d'$, is called *completeness*. An interpretation which is both sound and complete provides an isomorphism between the formally presented theory and its model. While checking soundness is straightforward, showing completeness is often much more difficult.[5] On the other hand, not having completeness means there are multiple models of a given theory, and the study of the differences between such models is often informative.

Before moving on, we'll introduce an important example, and its standard model.

Example 7 (*Quantum Circuits*) Similar to the example of Boolean circuits, we can also view (post-selected) quantum circuits as generalised compositional theory, called **QuCirc**. Again we have a single basic system, the qubit Q, and the basic processes are a collection of unitary gates, state preparations, and projections from which we construct the other quantum circuits.

$|0\rangle : I \to Q$ $\quad$ $|1\rangle : I \to Q$ $\quad$ $\langle 0| : Q \to I$ $\quad$ $\langle 1| : Q \to I$

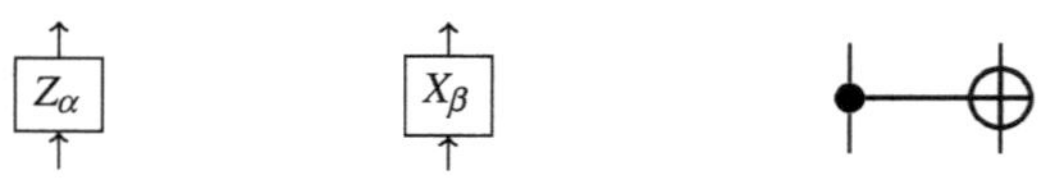

$Z_\alpha : Q \to Q$ $\quad$ $X_\beta : Q \to Q$ $\quad$ $\wedge X : Q \otimes Q \to Q \otimes Q$

From these basic elements we can write down any quantum circuit. We now define the standard interpretation of **QuCirc** into **Qubit**.

[5]To show completeness for a rewrite theory it is typically necessary, but rarely sufficient, to check that the rewrite rules are *confluent*; that is, whenever two rewrites simultaneously apply to a given diagram, then the choice between then (eventually) does not matter. Since this property must hold for every diagram and every pair of rewrites, even a simple rewrite system can produce an extremely large number of cases, necessitating a computer-assisted proof. For example see the work of Lafont on Boolean circuits [20].

$$[\![Q]\!] = \mathbb{C}^2$$

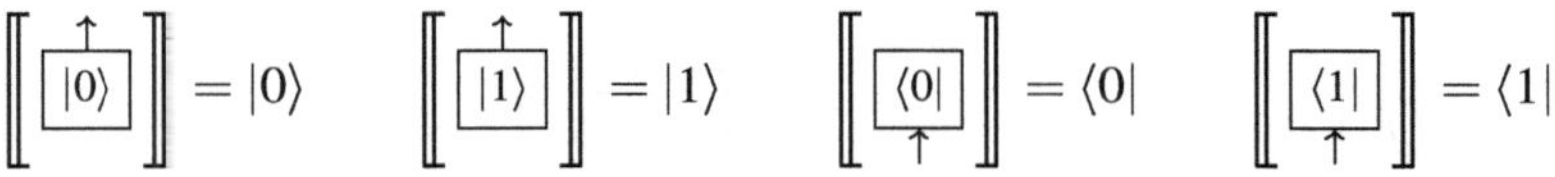

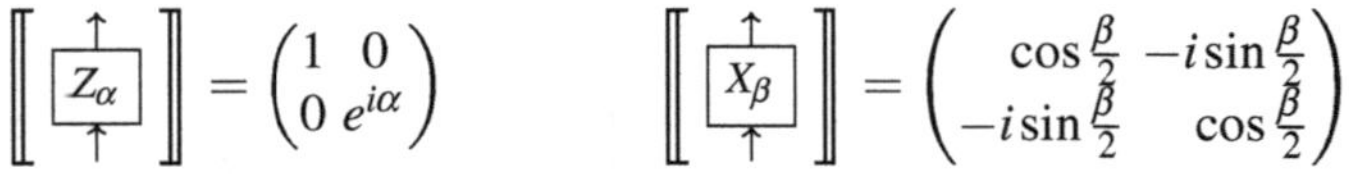

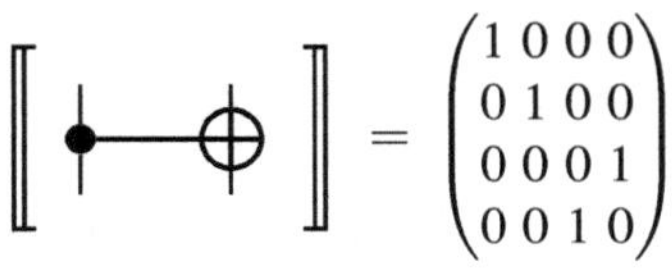

Thanks to the well-known universality result [21] this interpretation demonstrates that **QuCirc** can represent all unitary maps between qubits. In fact, since we have the projections $\langle 0|$, $\langle 1|$, all linear maps can be represented. Note, however, that although all quantum circuits can be represented, without a set of rewrite rules **QuCirc** cannot express any non-trivial equalities between them. We could propose various sound equations here, but there is no known collection of rewrite rules which makes **QuCirc** complete with respect to this interpretation into **Qubit**. If such a set of rewrites did exist, it would constitute provide a presentation of the unitary group by generators and relations.

1.3 The Dagger

Now we introduce the *dagger*. This is simply an operation on the processes of a theory, sending every process $f : A \to B$ to another process $f^\dagger : B \to A$. We call $f^\dagger$ the *adjoint* of f. In the graphical calculus, we represent the dagger by a flip in the horizontal axis:

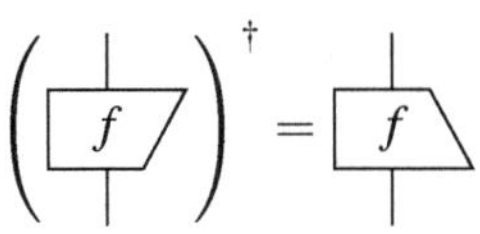

Note that we have made the box asymmetric to make this flipping evident. For more general diagrams, the dagger flips a diagram upside down, preserving all the internal structure. Taking this claim at face value, we can derive the key properties of the dagger:

$(f^\dagger)^\dagger = f$

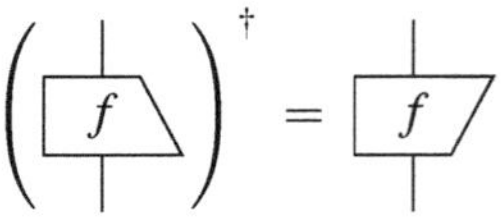

$(g \circ f)^\dagger = f^\dagger \circ g^\dagger$

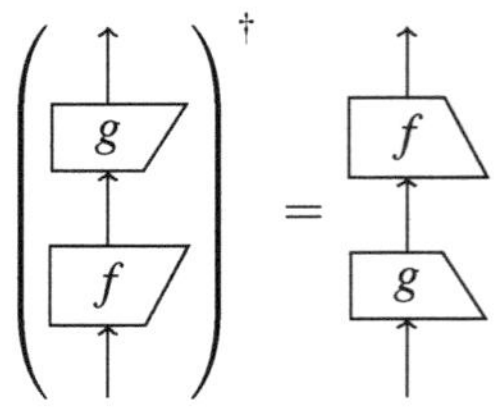

$(f \otimes g)^\dagger = f^\dagger \otimes g^\dagger$

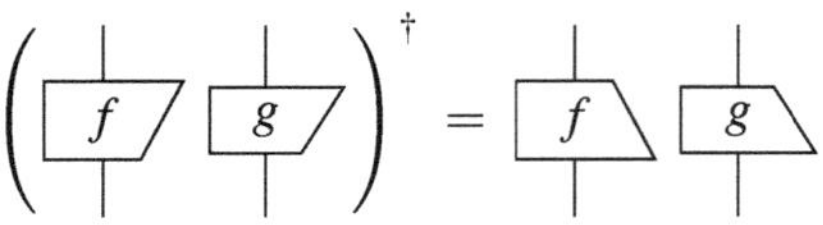

$1_A^\dagger = 1_A$

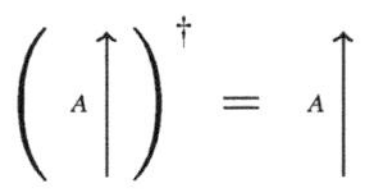

$\sigma_{A,B}^\dagger = \sigma_{B,A}$

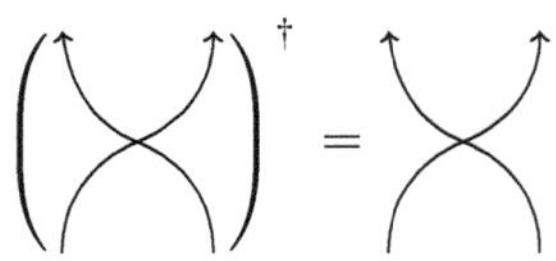

The dagger allows two important concepts to be defined.

Definition 1 A process $f : A \to B$ is called *unitary* if $f \circ f^\dagger = 1_B$ and $f^\dagger \circ f = 1_A$. A process is called *self-adjoint* when $f^\dagger = f$.

Example 8 (*Finite-dimensional Hilbert spaces*) The theory **FHilb** admits a dagger: it is the usual adjoint of a linear map. In this theory, the abstract definitions of unitarity and self-adjointness coincide with the usual one.

Example 9 In the theory **FRel**, the dagger of a relation $r : X \to Y$ is defined by the converse relation, i.e.

$$r^\dagger = \{(y, x) \mid (x, y) \in r\}$$

Here, unitary processes are exactly those relations which encode permutations. A relation is self-adjoint whenever it is symmetric. Hence the self-adjoint unitaries in **FRel** are exactly the permutations of order 2.

We extend the definition of mapping to demand that it also preserves the dagger. That is, given two theories with dagger, we require that a map $F : \mathbf{C} \to \mathbf{D}$ satisfies

$$F(f^{\dagger}) = (Ff)^{\dagger}$$

Example 10 (*Quantum Circuits*) We define a dagger on **QuCirc** as follows:

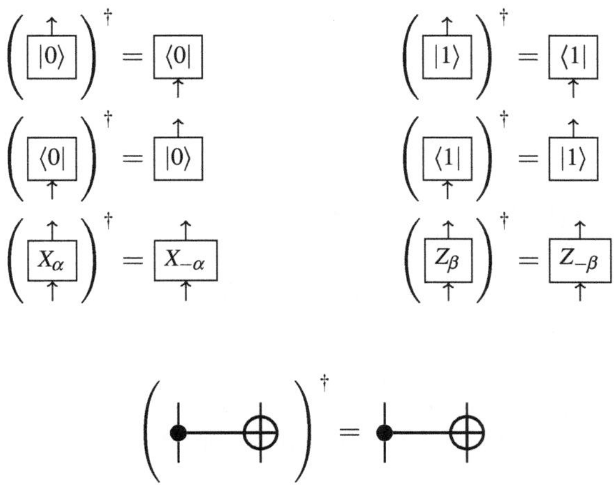

It's now easy to check that the interpretation map introduced earlier, $[\![\cdot]\!] : \mathbf{QuCirc} \to \mathbf{Qubit}$ preserves the dagger as required.

Remark 4 The theory of Boolean circuits, **BoolCirc**, does not admit a dagger. However, we could formally add new basic processes corresponding to the adjoints of the basic processes of **BoolCirc** and thus define a new theory, $\mathbf{BoolCirc}^{\dagger}$. Since the converse of a function is not in general a function, the interpretation $B : \mathbf{BoolCirc} \to \mathbf{Bool}$ no longer makes sense. Instead we must interpret $\mathbf{BoolCirc}^{\dagger}$ over $\mathbf{FRel}_2$, that is as Boolean relations rather than functions. In this case B again defines a valid interpretation $\mathbf{BoolCirc}^{\dagger} \to \mathbf{FRel}_2$. The resulting theory is a model of non-deterministic computation.

In any theory, a process of type $p : I \to A$ is called a *point*, or sometimes a *state*, of A. Dually, a process of type $e : A \to I$ is called a *co-point*, or sometimes an *effect* on A. For example, in **FHilb** the points $\psi : I \to A$ are in one-to-one correspondence with the vectors of A, while in **FRel** a point $s : I \to X$ is precisely a subset of X.

In a theory with a dagger the set of points is isomorphic to the set of copoints (or in other language, for every state there is a corresponding effect and vice versa). This allows us to define another important concept.

Definition 2 Given two points $\psi, \phi : I \to A$ we define their *inner product* as $\phi^{\dagger} \circ \psi$. Dually, the outer product is defined as $\phi \circ \psi^{\dagger}$.

As one may expect, the inner product is always a scalar. The diagrammatic language automatically allows the same tricks—and more—as Dirac notation does in Hilbert spaces. Indeed one can view the diagrammatic language as a 2-dimensional generalisation of Dirac notation.

Example 11 (*Finite-dimensional Hilbert spaces*) In **FHilb** the inner product defined by the dagger, is exactly the usual inner product $\langle \phi | \psi \rangle$.

Example 12 (*Sets and Relations*) In **FRel** the inner product $r^\dagger \circ s$ is 0 if the r and s are disjoint as subsets, and 1 otherwise.

2 Pure State Quantum Mechanics

2.1 The Elements of an Operational Theory

It is remarkable that the the basic language of quantum mechanics—states, effects, unitarity, self-adjointness, inner products, tensor products—can all be defined in the abstract setting of generalised compositional theories. We now have enough material to describe a formal operational framework for pure state quantum mechanics in purely diagrammatic terms.

- A *preparation* is any process which produces a state; that is to say it is process of type $p : I \to A$.

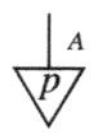

  Preparations are not restricted to producing single systems; a preparation process of type $I \to A_1 \otimes \cdots A_n$ is called *multipartite*. Of course, multipartite preparations need not be separable.

  When interpreted in **FHilb** each preparation process yields a ray in some Hilbert space, which, ignoring global phase, we may identify with a specific quantum state. It may happen, depending on the equations of the formal theory, that different preparation processes produce the same state.
- A *transformation* is any process which acts on states and produces new states, and which is unitary:

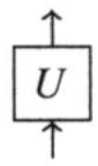

Once again, transformations may act on one or many systems at the same time.

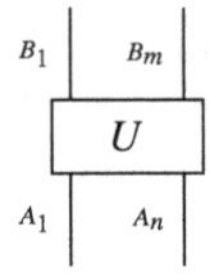

- Measurements are processes which accept quantum inputs and produce classical information about the state which was input. Since, for now, our theory only has pure states, we will work with *non-degenerate post-selected* measurements[6]; i.e. we know that a definite outcome has occurred, and that outcome corresponds to a definite quantum state. Therefore, measurements are one-dimensional effects, represented as co-points:

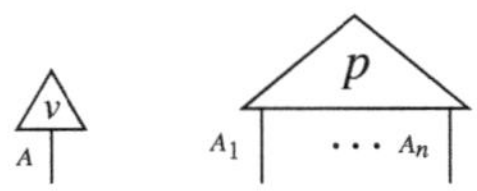

The classical information is implicit in the choice of copoint, and hence not represented. Since copoints do not have quantum outputs, these processes correspond to *demolition measurements*, where the original system is consumed by the measurement process. However, by combining an effect with the corresponding state preparation we can also represent non-demolition measurements:

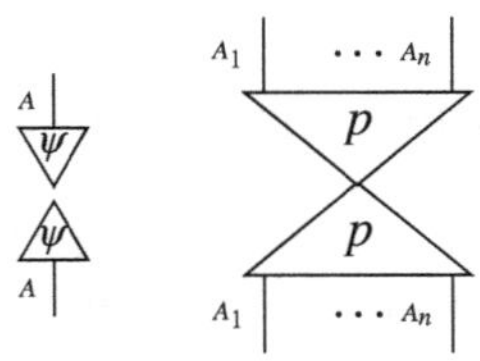

To properly represent the non-determinism of quantum measurements we need to consider mixed states; this is dealt with in Sect. 4. More general measurements can be represented within the theory, however they will not be described here.

This basic recipe—preparations, transformations, and measurements—allows any experimental setup to be described in terms of the processes which realise it. More precisely, since we use post-selected measurements, the diagram really represents a *run* of the experiment where a certain outcome occurred. We call an experiment *closed* when it has no external inputs or outputs. Any closed experiment is necessarily described by a process of type $x : I \to I$; that is, a scalar. This scalar is the abstract

[6]In other words, rank 1 projectors.

counterpart to the probability amplitude for performing the process and observing the specified result. Indeed, when such a diagram is interpreted in **FHilb**, the result is exactly the probability amplitude.

Example 13 The theory **QuCirc** has the structure described above, and we can use it to define a simple experiment. For example, the diagram below corresponds to preparing a qubit in the $|0\rangle$ state, applying a unitary gate to it, and, upon measuring in the computational basis, finding that the qubit is in the state $|1\rangle$.

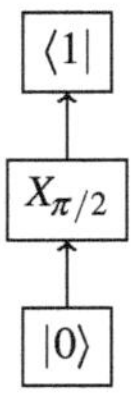

Using the interpretation map $[\![\cdot]\!]$: **QuCirc** → **Qubit** we can calculate the amplitude for this experimental result.

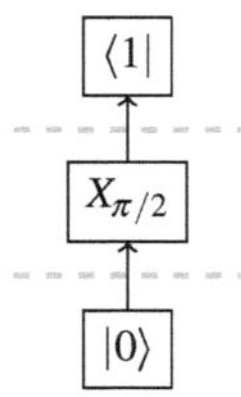

$$\mapsto \quad \langle 1| \circ \frac{1}{\sqrt{2}} \begin{pmatrix} 1 & -i \\ -i & 1 \end{pmatrix} \circ |0\rangle \quad = \quad \frac{-i}{\sqrt{2}}$$

To summarise the elements of the framework, a formal generalised compositional theory consists of:

- A collection of basic systems and processes, corresponding to the available "lab equipment".
- The collection of all diagrams constructed from the basic processes, corresponding to every possible experiment that could be built from the given equipment. We consider diagrams modulo topological equivalence: equivalent diagrams correspond to the same experiment.
- A (possibly empty) collection of axioms, presented as rewrite rules over diagrams, which specify behavioural equivalence of processes. These rules tell us when a piece of the experimental setup can safely be replaced by another without changing the result of the experiment.
- Finally, given the above, we'll usually consider (sound) *interpretation maps* of the formal theory into some concrete mathematical structure, such as Hilbert spaces.

So far we have been operating at an extremely high level of generality. To focus our attention on quantum systems we will now gradually introduce more structure to our theories. We identify certain structural features of the Hilbert space presentation of quantum mechanics, and provide an abstract realisation of those features in terms of basic processes and equations, whose behaviour reproduces various quantum phenomena in the abstract setting of generalised compositional theories.

The rest of this section will layout which basic processes and equations we will need to realise. As we do so, we'll say goodbye to some of the models introduced earlier, but the two most important ones, **FHilb** and **FRel**, will still be applicable.

2.2 Duals

The next piece of structure that will be required is the existence of *duals*.[7]

Definition 3 A system A has a *dual* if there exists a system A^* and processes

$$e_A : I \to A^* \otimes A \quad \text{and} \quad d_A : A \otimes A^* \to I$$

such that we have the following equations:

$$(d_A \otimes 1_A) \circ (1_A \otimes e_A) = 1_A \qquad (1_{A^*} \otimes d_A) \circ (e_A \otimes 1_{A^*}) = 1_{A^*}$$

Since this definition is rather hard to parse we will immediately move to its diagrammatic form. We indicate the dual system A^* by a wire labelled by A but directed in the opposite direction. The maps d_A and e_A are represented by wires with half turns, henceforth "caps" and "cups". The equations above then take the form of "straightening wires":

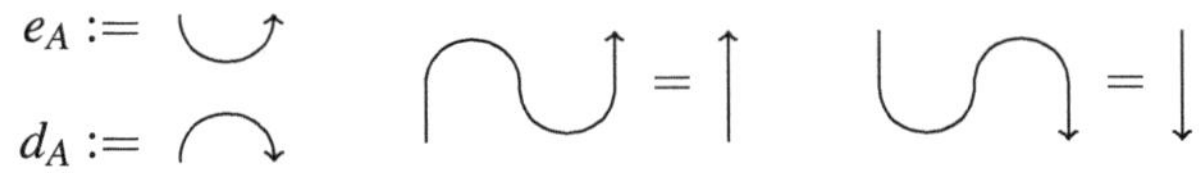

In general a system might have more than one dual, but they are all guaranteed to be isomorphic. We'll assume that every system has a given dual, and in particular $(A \otimes B)^* = B^* \otimes A^*$, in which case $d_{A\otimes B}$ and $e_{A\otimes B}$ take the form of nested caps and cups. Furthermore, we'll assume that the double dual $A^{**} = A$. These simplifications automatically hold in any theory presented diagrammatically; taking them as the general case saves a lot of bureaucracy.

Example 14 (*Finite dimensional Hilbert spaces*) Let A be a Hilbert space of dimension d, then A^* is the usual dual space; that is, the space of linear functionals from A to the complex numbers. Supposing that $\{|a_i\rangle\}$ is a basis for the space A, then the cup and cap are given by the linear maps

$$e_A : 1 \mapsto \sum_i \langle a_i| \otimes |a_i\rangle\,, \qquad d_A : \sum_i |a_i\rangle \otimes \langle a_i| \mapsto 1.$$

[7] For the experts in category theory, this additional structure can be summed up by saying we operate in a dagger-compact category, rather than just a symmetric monoidal category.

However since we are in a finite-dimensional setting we could also choose $A^* = A$; specialising to the case of qubits, we can now view the cup and cap as the preparation and projection onto a Bell state:

$$e_Q = |00\rangle + |11\rangle \qquad d_Q = \langle 00| + \langle 11|$$

Recall the quantum teleportations protocol: Alice has some unknown state that she wishes to send to Bob, but they do not share a quantum channel. However they have a classical channel, and have previously shared a Bell pair. In order to send her qubit to Bob, Alice measures her two qubits in the Bell basis, and transmits the result to Bob. Now Bob simply applies some unitary map (depending on Alice's outcome) to his half of the Bell pair to recover the qubit that Alice wanted to send. Since, for the moment, we are operating in a post-selected setting, we'll assume that Alice observes the outcome corresponding to the state $\Phi_+ = (|00\rangle + |11\rangle)/\sqrt{2})$ at her measurement. In this case Bob need do nothing to his qubit. The whole set up is shown below:

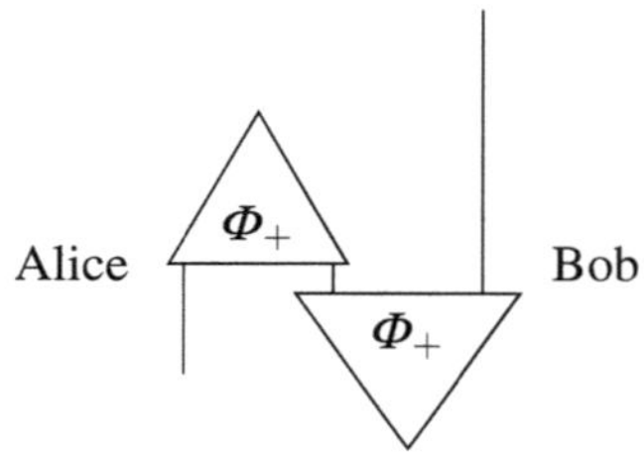

Knowing that the projection onto Φ_+ is just the effect d_Q, we can rewrite the protocol as shown and demonstrate the protocol purely diagrammatically:

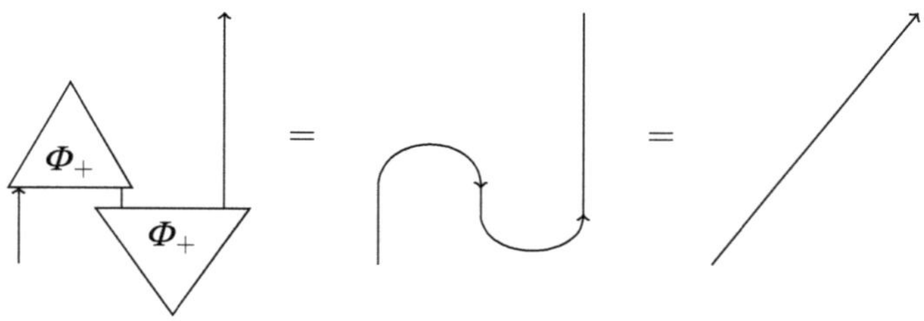

Example 15 (*Sets and Relations*) In **FRel**, the dual of a set X is just the same set X again. The cup is given by the "name" of the identity:

$$e_X = \{(*, (x, x)) \mid x \in X\}$$

while the cap, d_X is just the converse of e_X.

Using caps and cups, we can turn any process $f : A \to B$ into a process on the dual objects going in the opposite direction: $f^* : B^* \to A^*$.

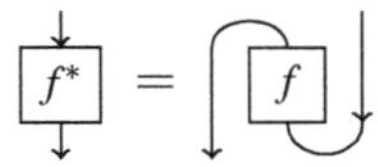

This is sometimes called the transpose of f, but this terminology can be misleading. In **FHilb**, f^* is the map that takes a linear form $\langle \xi | \in B^*$ to $\langle \xi | f \in A^*$. We refer to this map simply as the *upper-star* of f. Clearly, we have $f^{**} = f$.

It is also required that the dagger and the duals interact nicely. More precisely we have the equations:

In any theory with both a dagger and duals, we can define a third operation, the *lower-star* of f as $f_* := (f^\dagger)^* = (f^*)^\dagger$. Again this is involutive, i.e. $f_{**} = f$. We'll return to the uses of the upper and lower stars in Sect. 4.

Finally, the cup and cap can be used to define a *trace* in purely diagrammatic terms:

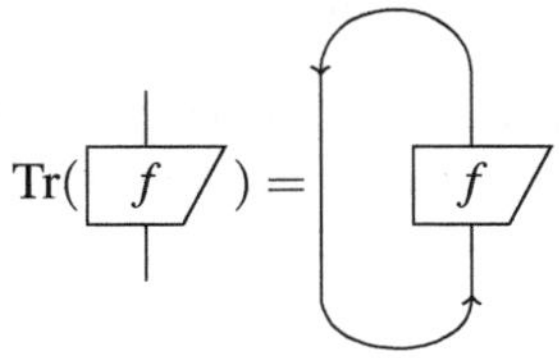

Checking the Hilbert space interpretation, it is easy to see that this coincides with the usual definition.

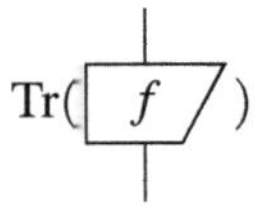

$$= \Big(\sum_i \langle ii|\Big) \circ (1_A \otimes f) \circ \Big(\sum_j |jj\rangle\Big) = \sum_i \langle i| f |i\rangle = \sum_i f_{ii}$$

In the diagrammatic form it is trivial to prove that trace is invariant under cyclic permutation:

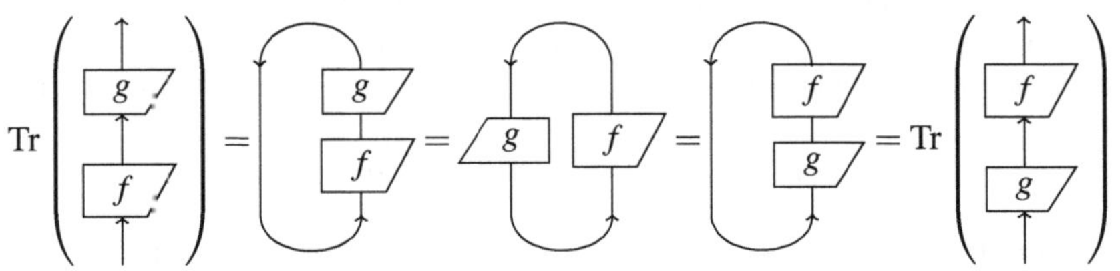

The partial trace can be defined analogously.

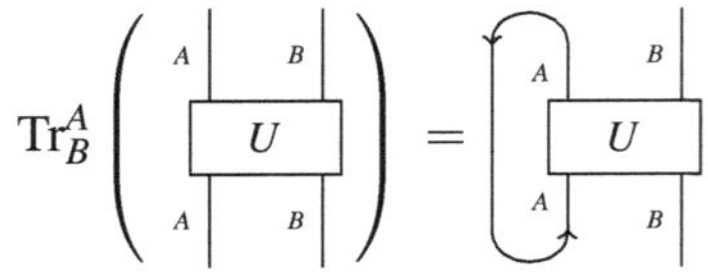

By adding duals we have enlarged the class of possible diagrams, since wires may now loop back from inputs to outputs and vice versa, but the basic principle of diagram equality does not change: **Two diagrams are considered equal if one can be smoothly transformed to another, by bending, stretching, or crossing wires, and moving boxes around**. With this in mind we can update the key theorem.

Theorem 2 (Fundamental Theorem of Diagrams with Daggers and Duals) *Given any two generalised compositional theories* **C** *and* **D** *with daggers and duals, and a map* $F : \mathbf{C} \to \mathbf{D}$, *for any two diagrams* d *and* d' *in* **C**, *if* $d = d'$ as diagrams *then* $Fd = Fd'$ *in* **D**.

Once again, the full details are found in [19].

Remark 5 We need not demand any additional conditions on the class of mappings to guarantee the preservation of duals; since they are defined in terms of processes, the structure is automatically preserved.

2.3 Observable Structures

An *observable* yields classical data from a physical system [22]. The key distinction between classical and quantum data is that classical data may be freely copied and deleted, while this is impossible for quantum data, due to the no-cloning [23, 24] and no-deleting [25] theorems.

In quantum mechanics, an observable is represented by a self-adjoint operator. This (non-degenerate) operator encodes certain classical data as its orthonormal basis of eigenstates, the possible outcomes of the corresponding measurement. Note that if a quantum state is known to be a member of a given orthonormal basis, such as the eigenbasis $\{|a_i\rangle\}$ of some observable, then it *can* be copied and deleted via the maps

$$\delta : |a_i\rangle \mapsto |a_i\rangle \otimes |a_i\rangle \quad \text{and} \quad \varepsilon : |a_i\rangle \mapsto 1.$$

Hence we can view the classical content of a quantum measurement as the possibility to copy and delete its set of outcomes. We will axiomatise quantum observables by describing the copying and deleting operations as algebraic structures inside a general compositional theory. The relevant structure is called a †-*special commutative Frobenius algebra*, and we will now build up its definition one piece at a time.

Definition 4 A *commutative monoid* in $\mathbf{C}$ is a triple (X, μ, η), where μ and η are maps

$$\mu : X \otimes X \to X \qquad\qquad \eta : I \to X$$

which we write graphically as $\mu =$ , $\eta =$ . These operations satisfy the following equations:

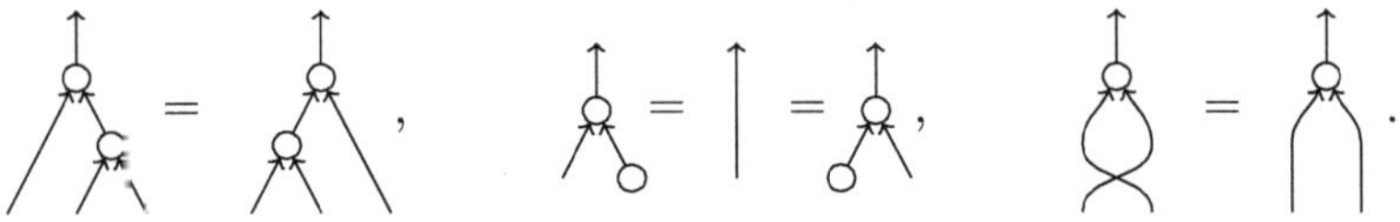

Remark 6 The process μ can be understood as a *multiplication* for systems of type X; the first and last equations assert that this operation is associative and commutative respectively. The process η is the *unit* for this multiplication: the second equation asserts that multiplication by the unit is simply the identity.

The dual to a monoid is a *comonoid.*

Definition 5 A *comonoid* in a theory $\mathbf{C}$ consists of a triple (X, δ, ε) where δ and ε are processes

$$\delta : X \to X \otimes X \qquad\qquad \varepsilon : X \to I$$

satisfying the equations of Definition 4 but in reverse, viz:

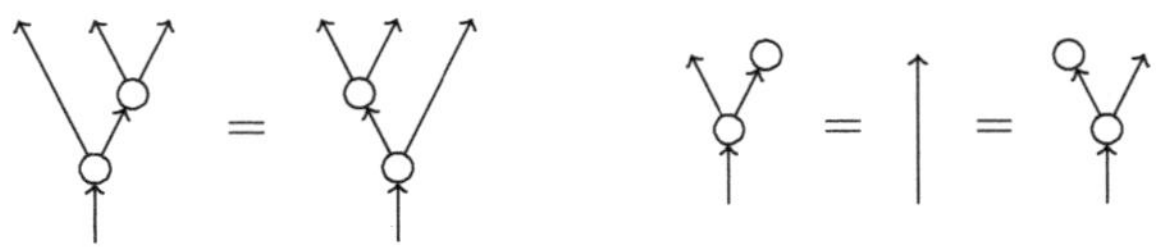

A comonoid is *cocommutative* if it satisfies:

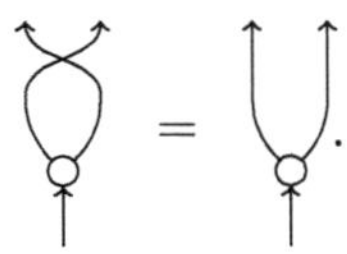

The processes δ and ε are called the *comultiplication* and *counit* respectively.

Example 16 We have already met the basic example of a comonoid: in **FHilb**, for any orthonormal basis $\{x_i\}_i$ of a space X we obtain a comonoid via 'copying' and 'erasing' processes mentioned above:

$$\delta : x_i \mapsto x_i \otimes x_i \qquad \varepsilon : x_i \mapsto 1$$

Remark 7 Thanks to the dagger, if (X, δ, ε) is a comonoid then $(X, \delta^\dagger, \varepsilon^\dagger)$ is automatically a monoid, and vice versa.

Generally speaking, a process is called a *homomorphism* if it preserves some algebraic structure. In the context of GCTs, such preservation is usually expressed by a process commuting with another which reifies that structure. For example:

Definition 6 Given two comonoids (X, δ, ε) and $(X', \delta', \varepsilon')$, a *comonoid homomorphism* is a process $f : X \to X'$ such that

$$\delta' \circ f = (f \otimes f) \circ \delta \quad \text{and} \quad \varepsilon' \circ f = \varepsilon.$$

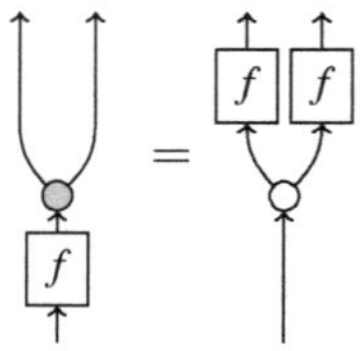

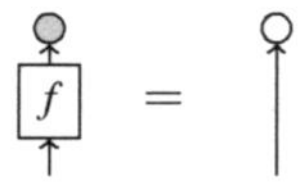

Monoid homomorphisms are defined similarly.

Remark 8 The definition above is the most general, but we will frequently encounter cases where $f : X \to X$ is homomorphism between two comonoids defined on the same object, or from a single comonoid to itself.

The structures of greatest interest for this paper are algebras containing both monoids and comonoids.

Definition 7 A *commutative Frobenius algebra* is a 5-tuple $(X, \delta, \varepsilon, \mu, \eta)$ where

1. (X, δ, ε) is a cocommutative comonoid;
2. (X, μ, η) is a commutative monoid; and,
3. δ and μ satisfy the following equations:

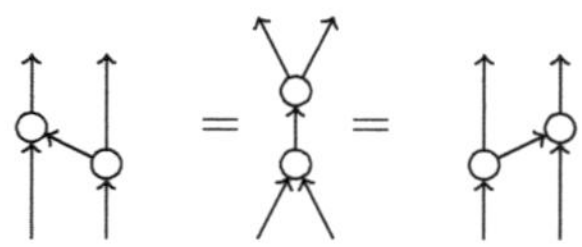

Finally, we can define:

Definition 8 A †*-special Frobenius algebra* (†-SCFA) is a commutative Frobenius algebra

$$\mathscr{O}_\circ = (\mu_\circ : X \otimes X \to X,\ \eta_\circ : I \to X,$$
$$\delta_\circ : X \to X \otimes X,\ \varepsilon_\circ : X \to I)$$

such that $\delta_\circ = (\mu_\circ)^\dagger$, $\varepsilon_\circ = (\eta_\circ)^\dagger$ and 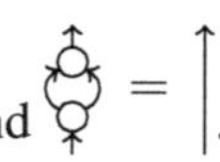.

The preceding definitions may seem rather opaque, and not fully justified by the intuition that a quantum observable is somehow encoded by the maps which copy and delete its eigenstates. However complex it may appear (and we shall shortly simplify it), the importance of Definition 8 rests on the fact [26] that in **FHilb** *every* †-SCFA arises from a comonoid defined by copying an orthonormal basis as described above. Since orthonormal bases define non-degenerate quantum observables, †-SCFAs are also called *observable structures*.

Concretely, given an orthonormal basis $\{|i\rangle\}_i$ then $\delta_{\circ} :: |i\rangle \mapsto |ii\rangle$ defines an observable, and all observables are of this form for some orthonormal basis. The resulting intuition is that $\delta_{\circ}$is an operation that 'copies' basis vectors, and that $\varepsilon_{\circ}$ 'erases' them [22]. We will use the symbolic representation $(\mu_{\circ}, \eta_{\circ}, \delta_{\circ}, \varepsilon_{\circ})$ and the pictorial one  interchangeably.

Example 17 (*Sets and Relations*) Perhaps surprisingly, **FRel** also has many distinct observable structures, which have been classified by Pavlovic [27]. Even on the two element set there are two, namely

$$\delta_{\circ} : i \mapsto \{(i,i)\}$$

$$\delta_{\bullet} : \begin{cases} 0 \mapsto \{(0,0),(1,1)\} \\ 1 \mapsto \{(0,1),(1,0)\} \end{cases}$$

In fact, this pair is *strongly complementary* in the sense to be explained in Sect. 3.

Manipulating observable structures in the graphical language is extremely convenient since they obey a remarkable normal form law. Let $\delta_n : X \to X^{\otimes n}$ be defined by the recursion

$$\delta_0 := \varepsilon \qquad \delta_{n+1} := (\delta_n \otimes 1_A) \circ \delta$$

and define μ_m analogously. Now we have the following important theorem:

Theorem 3 *Given an SCFA* $(X, \delta, \varepsilon, \mu, \eta)$ *let* $f : X^{\otimes m} \to X^{\otimes n}$ *be a map constructed from* δ, ε, μ *and* η *whose graphical form is connected. Then* $f = \delta_n \circ \mu_m$.

Proposition 1 *Given an observable structure* $\mathcal{O}_{\circ}$ *on* X, *let* $(\circ)_n^m$ *denote the '*(n, m)*-legged spider':*

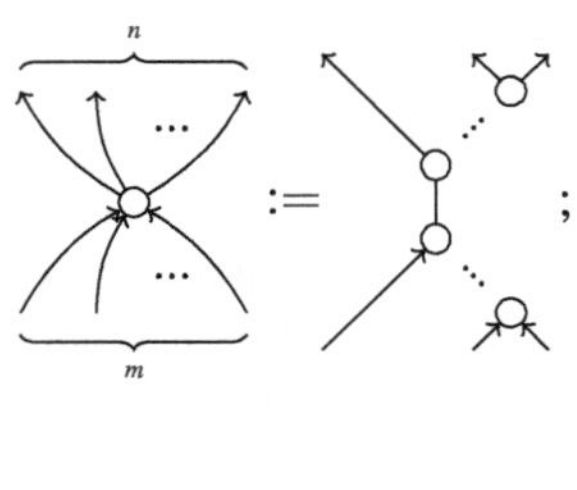

then any morphism $X^{\otimes n} \to X^{\otimes m}$ built from $\mu_\circ, \eta_\circ, \delta_\circ$ and $\varepsilon_\circ$ via †-SMC structure which has a connected graph is equal to the $(\circ)_n^m$. Hence, spiders compose as follows:

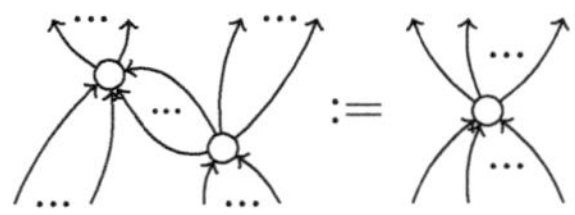

(2)

Thanks to the spider rule (2), every observable structure on X makes X dual to itself (in the sense of Definition 3), via the cup and cap:

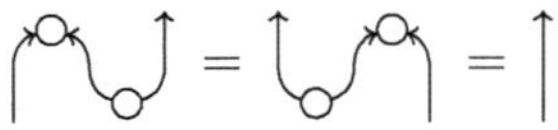

The upper-star with respect to this cup and cap corresponds in **FHilb** to transposition in the given basis. For that reason, we call this the ○-transpose $f^{\circ\mathrm{T}}$. The lower star corresponds to complex conjugation in the basis of $\mathcal{O}_\circ$, so we call it the ○-conjugate $f_{\circledast} := (f^{\circ\mathrm{T}})^\dagger$.

Recall that a process $k : I \to X$ is called a *point of* X. In **FHilb** the points of X are simply vectors in the Hilbert space X. The abstract analogue of the eigenvectors of an observable in **FHilb** are the *classical points* of an observable structure.

Definition 9 A *classical point* for an observable structure is a state that is copied by the comultiplication and deleted by the counit:

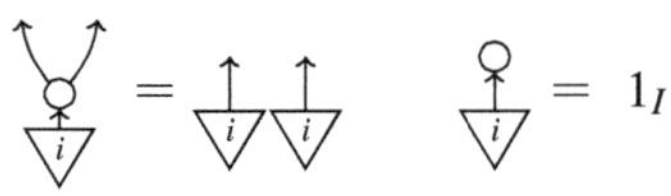

(3)

We will depict classical points as triangles of the same colour as their observable structure.

Remark 9 Another way to say the same thing, is to define classical points as comonoid homomorphisms from the trivial comonoid $(I, 1_I, 1_I)$ to (X, δ, ε).

In quantum computing, it is common to think of elements of a product basis as strings of some kind. E.g. for qubits:

$$|010011\rangle \leftrightarrow |0\rangle \otimes |1\rangle \otimes |0\rangle \otimes |0\rangle \otimes |1\rangle \otimes |1\rangle$$

Such product bases are precisely the classical points of *products of observable structures*. Given an observable structure $\mathcal{O}_\circ$ on X, and another $\mathcal{O}_\circ$ on Y, we form a new observable structure on $X \otimes Y$ by taking their product:

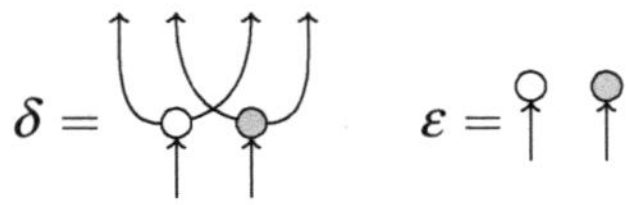

Evidently any pair of classical points for the constituent observable structures will be copied.

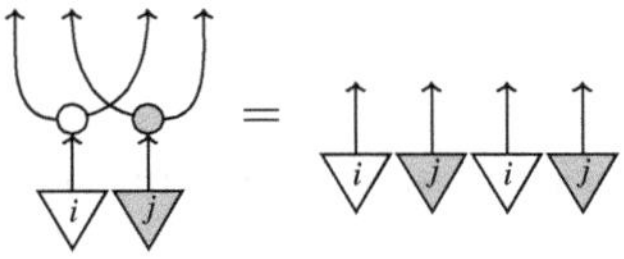

Generalising, the classical points of any n-ary product of observable structures are nothing more than lists of classical points, one for each factor.

Working concretely in Hilbert space, one can use the linear structure to give another set of equations for observable structures. Consider some basis vector $|i\rangle$, then the map $|ii\rangle\langle i|$ has the diagrammatic form:

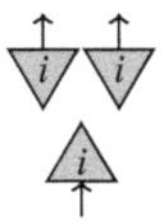

But notice that the sum $\sum_i |ii\rangle\langle i|$ is nothing more than the the map $\delta : |i\rangle \to |ii\rangle$. A similar statement can be made for the counit ε. Hence given the complete set of classical points for an observable structure $\mathcal{O}_\circ$ we have the following equations:

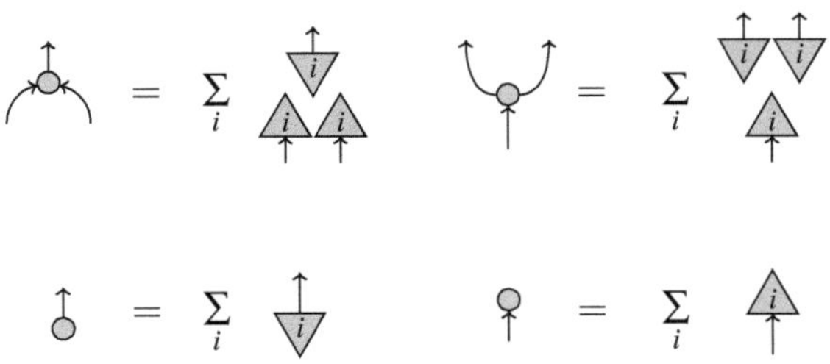

Indeed these can be generalised to arbitrary spiders:

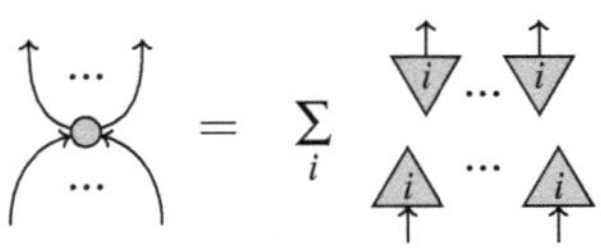

Note that generalised compositional theories do not necessarily admit addition of diagrams: we introduce these equations as way of generalising from concepts defined in Hilbert space to the abstract setting where there need not be any linear structure.

Linear maps have the property that if two maps are equal on every element of a basis, the maps themselves are equal. In analogy to this we define the following:

Definition 10 Let $\mathcal{O}_\circ$ be an observable structure on X, with classical points $\{k_i\}_i$; we say that $\mathcal{O}_\circ$ has *enough classical points* if, for every system Y, and every pair of processes $f, g : X \to Y$, we have

$$(\forall i : f \circ k_i = g \circ k_i) \Rightarrow f = g.$$

This property does not necessarily hold in an arbitrary GCT (although obviously it does in **FHilb**) however when it does hold certain statements can be made stronger. For example, many implications described in the subsequent sections are equivalences if the underlying object has enough classical points.

2.4 Phase Group for an Observable Structure

Let ψ and ϕ be two points of X. Given an observable structure $\mathcal{O}_\circ$ on X, applying the multiplication $\mu_\circ$ to ψ and ϕ yields another point of X:

$$\psi +_\circ \phi := \mu_\circ(\psi \otimes \phi)$$

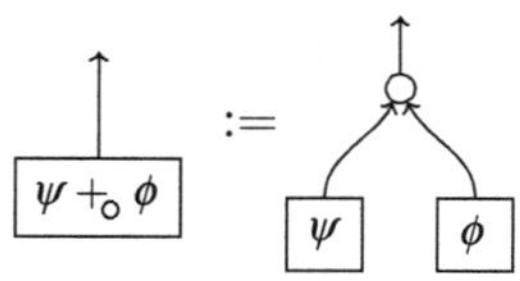

Since $\mu_\circ$ is commutative and associative, and it has a unit point (namely $\eta_\circ$), the operation $+_\circ$ gives the points of X the structure of a commutative monoid.

If we restrict to those points $\psi : I \to X$ which satisfy

$$\psi +_\circ \psi_\circledast = \eta_\circ$$

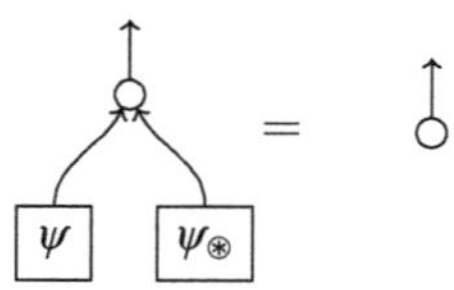

we obtain an abelian group $\Phi_\circ$, called the *phase group* of $\mathcal{O}_\circ$ [28, 29]. The elements of this group are called *phases*. Note the phase group is non-empty, since the unit $\eta_\circ$ satisfies the required equation. We let $-\alpha := \alpha_\circledast$ for phases, and represent these points as circles with one output, labelled by a phase.

Example 18 In **FHilb**, let $\mathcal{O}_\circ$ be defined by some orthonormal basis $\{|i\rangle\}_i$. One can verify by direct calculation that a vector $|\psi\rangle$ lies in the phase group $\Phi_\circ$ if and only if we have $|\langle i|\psi\rangle|^2 = 1/D$, for all i, where D is the dimension of the underlying space. Such vectors are called *unbiased* for the basis $\{|i\rangle\}_i$. The multiplication is then the convolution (pointwise) product.

Concretely, for a qubit observable given by $\mu_\circ = |0\rangle\langle 00| + |1\rangle\langle 11|$, the phases are the unbiased states, which are all of the form:

$$|\alpha\rangle = \begin{pmatrix} 1 \\ e^{i\alpha} \end{pmatrix},$$

with the multiplication:

$$\mu_\circ\left(\begin{pmatrix} 1 \\ e^{i\alpha} \end{pmatrix} \otimes \begin{pmatrix} 1 \\ e^{i\beta} \end{pmatrix}\right) = \begin{pmatrix} 1 \\ e^{i(\alpha+\beta)} \end{pmatrix}.$$

We therefore see that the phase group for a qubit observable is the circle group. It is an easy exercise to check that for a D-dimensional Hilbert space the phase group for any observable is isomorphic to the $(D-1)$-dimensional torus.

The name 'phase group' comes from fact that the elements of the $\Phi_\circ$ correspond to unitary maps which preserve the basis defining $\mathcal{O}_\circ$. We can map any point $\psi : I \to X$ onto an operation on X via the *left action*, $\Lambda_\circ(\psi) = \mu_\circ \circ (\psi \otimes 1_X)$, or in pictures:

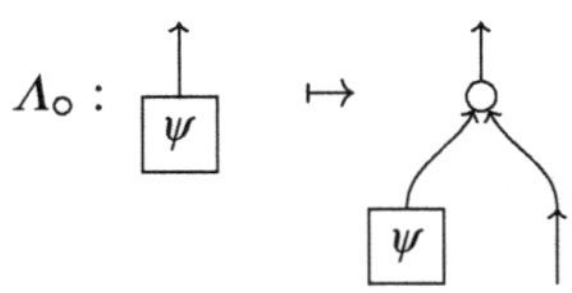

Then we have the following facts:

Proposition 2 *Let $\phi, \psi \in \Phi_\circ$; then*

1. *$\Lambda_\circ(\phi)$ is unitary;*
2. *$\Lambda_\circ(\phi) \circ \Lambda_\circ(\psi) = \Lambda_\circ(\phi + \psi) = \Lambda_\circ(\psi) \circ \Lambda_\circ(\phi)$*
3. *If k is a classical point for $\mathcal{O}_\circ$ then $\Lambda_\circ(\phi) \circ k = k \otimes s$ for some scalar s.*

Proof 1. We show that $(\psi)^\dagger \circ \Lambda_\circ(\psi) = 1$:

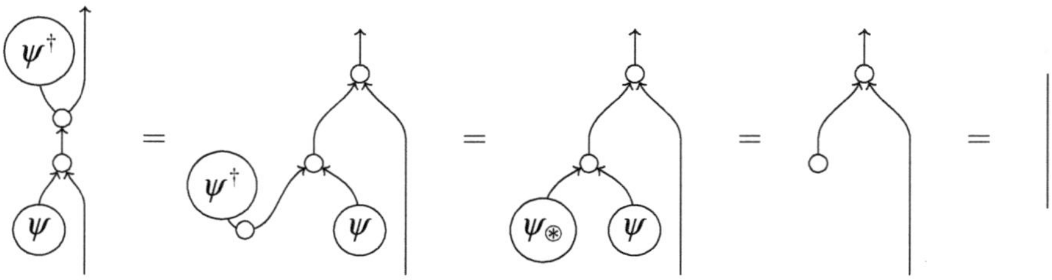

The first equation is the spider rule while the second is the definition of $\psi_\circledast$. The case $\Lambda_\circ(\psi) \circ (\Lambda_\circ(\psi))^\dagger = 1$ is similar.

2. This follows immediately from the associativity and commutativity of $\mu_\circ$:

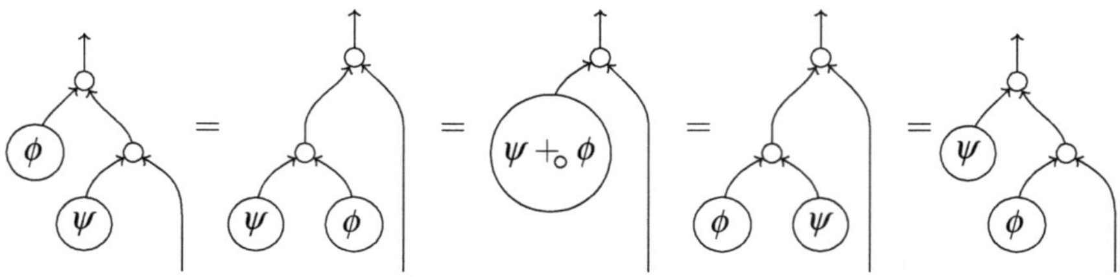

3. This follow from the definition of classical points.

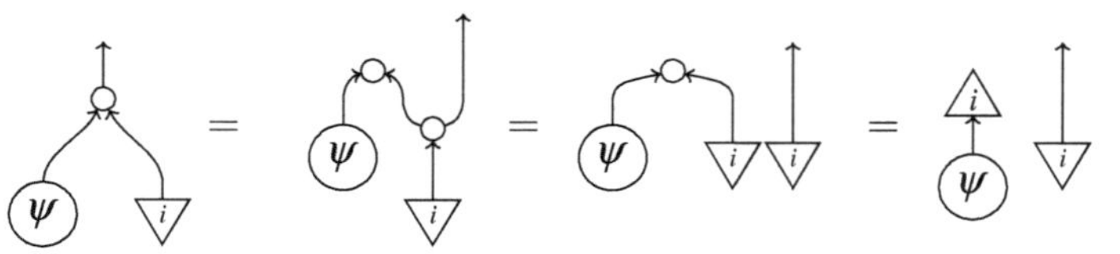

The image $\Lambda_\circ(\Phi_\circ)$ is therefore an abelian subgroup of the unitaries on X, which is isomorphic to $\Phi_\circ$. We refer to these as *phase maps*. If we reinterpret the third part of the preceding proposition in terms of linear algebra, we see that every classical point of $\mathscr{O}_\circ$ is an eigenvector of every phase map in $\Lambda_\circ(\Phi_\circ)$. This in turn "explains" why they commute with each other.

Example 19 Let $\mathscr{O}_\circ$ be defined by $\mu_\circ = |0\rangle\langle 00| + |1\rangle\langle 11|$ as above. Now for $\alpha \in \Phi_\circ$ we have

$$\Lambda_\circ(\alpha) = \begin{pmatrix} 1 & 0 \\ 0 & e^{i\alpha} \end{pmatrix}$$

Hence the phase group in Hilbert spaces is exactly the group of phase shifts relative to the given basis.

Generalising from the preceding discussion, we can now introduce 'spiders decorated with phases':

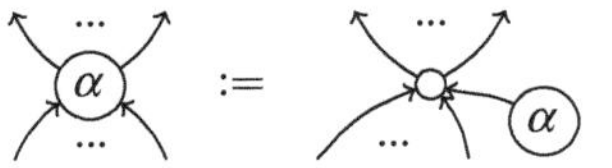

(4)

which compose as follows:

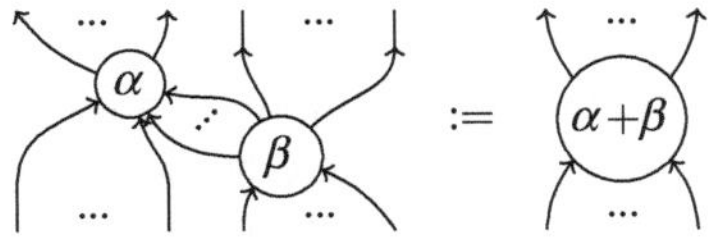

(5)

In the following sections we will refer to this generalised composition rule for phased spiders as *the spider law*.

2.5 *Two Toy Models*

In this section we'll introduce two "toy" models of quantum mechanics. The first is the restriction of quantum mechanics to stabilizer states; this theory we call **Stab**. The second is the toy model due to Spekkens [30], which we refer to as **Spek**. While the first of these is indeed a true subtheory of quantum mechanics, **Spek** is a local hidden variable model. By casting both of these in the language of generalised compositional theories we can see that the difference between is in fact very slight.

Before discussing these concrete models, we'll introduce a formal precursor theory. Let **Toy** be the general compositional theory built from the formal generators:

- one basic system, which we denote T;
- six points $z_0, z_1, x_0, x_1, y_0, y_1 : I \to T$ and their corresponding copoints;
- 24 unitary maps $T \to T$ which form a group isomorphic to the symmetric group S_4;
- one observable structure $\mathcal{O}_0$, whose classical points are z_0 and z_1, and whose phase group comprises the remaining four points.

Note that **Toy** is not fully specified: to do so we ought to say which group the phase group is, and how the corresponding unitaries embed into the endomorphisms of T. Since Φ_0 is a four-element group we have only two choices here: $\mathbb{Z}_4$, or $\mathbb{Z}_2 \times \mathbb{Z}_2$. As we will see this choice will make the difference between stabilizer quantum mechanics and the quantum-like local hidden variable theory.

Let **Stab** be the subtheory of **FHilb** generated by the following elements:

- One basic system $\mathbb{C}^2$, which we call Q.
- Six points $I \to Q$:

$$z_0 = |0\rangle \qquad x_0 = \frac{1}{\sqrt{2}}(|0\rangle + |1\rangle) \qquad y_0 = \frac{1}{\sqrt{2}}(|0\rangle + i\,|1\rangle)$$
$$z_1 = |1\rangle \qquad x_1 = \frac{1}{\sqrt{2}}(|0\rangle - |1\rangle) \qquad y_1 = \frac{1}{\sqrt{2}}(|0\rangle - i\,|1\rangle)$$

- The group of unitaries generated by the matrices:

$$Z_{\pi/2} = \begin{pmatrix} 1 & 0 \\ 0 & i \end{pmatrix} \qquad X_{\pi/2} = \frac{1}{\sqrt{-2i}} \begin{pmatrix} 1 & -i \\ -i & 1 \end{pmatrix}$$

  This group is known in the quantum computation literature as the Clifford group for one qubit; it is isomorphic to S_4. The other key property of this group is that it acts as a permutation on the states defined above, so we cannot generate new states via unitaries.
- An observable structure $\mathscr{O}_\circ$ defined by the basis $|0\rangle$, $|1\rangle$.

Evidently the classical points of $\mathscr{O}_\circ$ are indeed z_0 and z_1 and the remaining points are unbiased for this basis, hence part of $\Phi_\circ$. One can check that

$$\Lambda_\circ(y_0) = Z_{\pi/2}$$

and which generates a four element cyclic subgroup, hence the phase group $\Phi_\circ$ is $\mathbb{Z}_4$.

We now introduce Spekkens' toy theory. The toy theory is a local hidden variable theory, based on epistemic restrictions. There is a single basic system, the toy bit, which can have one of four possible states. We formalise the state space simply as a four-element set. However, we now impose the epistemic restriction that any state preparation (and, dually, measurement) may only narrow down the state to two of the possible four. Hence the "states" of the toy theory are two-element subsets. Although Spekkens' original presentation [30] was informal, the toy theory is ideally studied as subtheory of **FRel**. Following [31], let **Spek** be the subtheory of **FRel** generated by the following elements:

- One basic system, the four element set $\mathbf{4} = \{0, 1, 2, 3\}$.
- Six points:

$$z_0 = \{0, 1\} \qquad x_0 = \{0, 2\} \qquad y_0 = \{0, 3\}$$
$$z_1 = \{2, 3\} \qquad x_1 = \{1, 3\} \qquad y_1 = \{1, 2\}$$

- The full group of permutations on **4**;
- An observable structure $\mathscr{O}_\circ$ defined by

$$\mu_\circ : \begin{array}{l} \{00,11\} \sim 0 \\ \{01,10\} \sim 1 \\ \{22,33\} \sim 2 \\ \{23,32\} \sim 3 \end{array} \qquad \eta_\circ : * \sim \{0,2\}$$

  where we write the tensor as juxtaposition, i.e. $00 = (0, 0)$.

Once again we easily check that the classical points for $\mathcal{O}_{\circ}$ are z_0 and z_1, and the other four form the phase group $\Phi_{\circ}$. The phase group in this case is generated by the transpositions (0 1) and (2 3); hence $\Phi_{\circ} \cong \mathbb{Z}_2 \times \mathbb{Z}_2$.

As should be evident by this point both **Stab** and **Spek** are realisations of the incomplete theory **Toy**. The only notable difference between them is the group structure of $\Phi_{\circ}$. This highlights the importance of the phase group for understanding non-locality in generalised compositional theories.

Remark 10 In the description above the group of unitaries was given *a priori*. This is not necessary. If we include a second observable structure $\mathcal{O}_{\bullet}$, corresponding to the classical points x_0 and x_1, the the union of the two phase groups $\Phi_{\circ}$ and $\mathcal{O}_{\bullet}$ yields *all* unitaries described above. These two observables are complementary in the sense described below. Hence these two theories are in a sense minimal.

3 Complementarity and Strong Complementarity

In the Hilbert space presentation of quantum mechanics, two observables are *complementary* if their bases of eigenstates are mutually unbiased. That is, for any i, j, $|\langle v_i | v'_j \rangle|^2 = 1/D$. In the graphical notation:

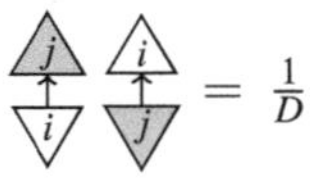

A question posed by Coecke and Duncan [28, 29] was, "Can we represent complementarity purely in terms of interacting observable structures?" It turns out that complementarity is equivalent to a simple diagrammatic equation. First, we can move $1/D$ in the above equation to the other side and express it as a circle, as the trace of the identity always equals D. Then, replace 1 on the RHS with "deleted points".

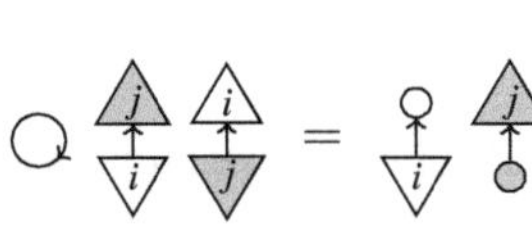

(6)

As we saw in Sect. 2.3, observable structures fix an isomorphism of a space with its dual space, via the transpose. While it is not true in general that $|\psi\rangle^{\circ T} = \langle\psi|$, the transpose does take classical points for a particular observable structure to their adjoints:

$$|v_i\rangle^{\circ T} = \langle v_i| \quad \text{and} \quad |v'_j\rangle^{\bullet T} = \langle v'_j| \ .$$

Graphically:

(7)

Exercise Prove this.

We can rewrite the left hand side of Eq. (6) using this fact.

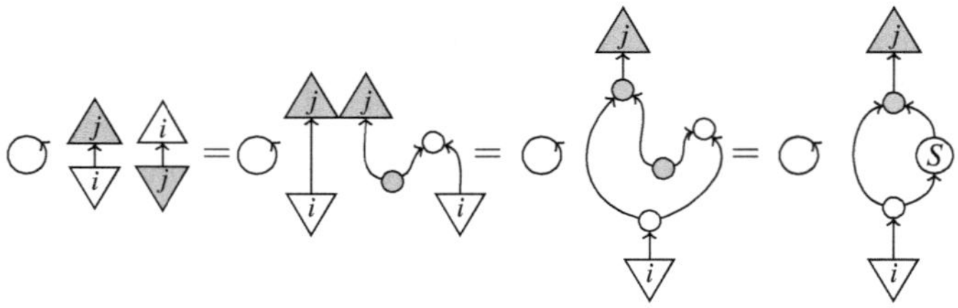

(8)

The last equation follows by substituting the symbol S for its definition, viz:

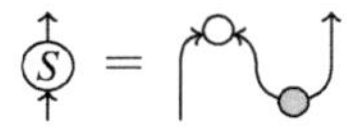

(9)

Unifying Eqs. (6) and (8) we have:

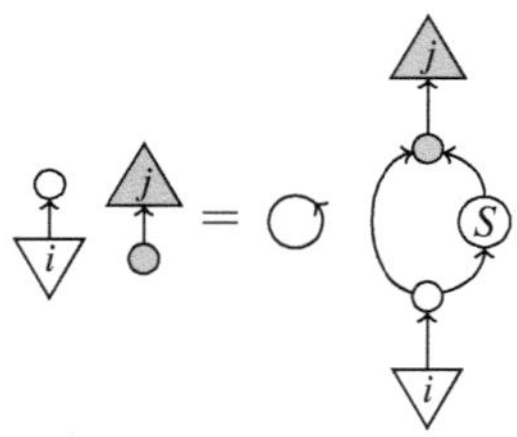

(10)

Since this equation holds for all classical points i and j, if we now appeal to the fact that **FHilb** has enough classical points (cf. Definition 10), we can conclude that identity holds without points:

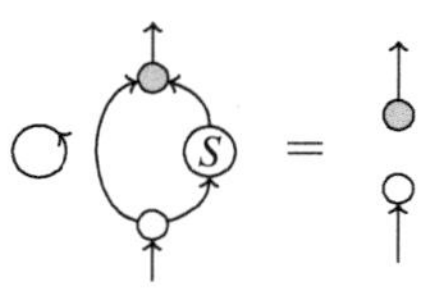

(11)

Remark 11 The above equation is (up to a scalar factor) one of the defining equations a *Hopf algebra*, in which case the map S is called the *antipode*. For that reason, we refer to (11) as the *Hopf law*. As we will see in the next section, subject to some additional assumptions, pairs of complementarity observables do indeed form Hopf algebras with the antipode defined as in Eq. (9).

Notice if we assume Eq. (11) we can derive Eq. (6) without any additional assumptions. In other words, if $\mathcal{O}_{\circ}$ and $\mathcal{O}_{\bullet}$ satisfy the Hopf law their classical points are mutually unbiased. Thus, we take the Hopf law to be the defining equation for our *abstract* notion of complementarity.

Definition 11 A pair $(\mathcal{O}_{\circ}, \mathcal{O}_{\bullet})$ of observables on the same object X is *complementary* iff:

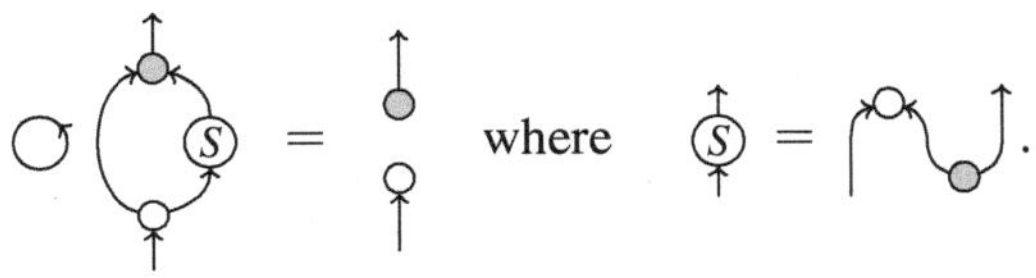

Since every observable in **FHilb** has enough classical points, Definition 11 is equivalent[8] to saying that observables are complementary if their eigenbases are mutually unbiased with respect to the other. (See [28] for more details). Hence, we reclaim the usual notion of quantum complementarity, and extend it to a more general setting.

Definition 12 A pair $(\mathcal{O}_{\circ}, \mathcal{O}_{\bullet})$ of observables on the same object X is *coherent* iff:

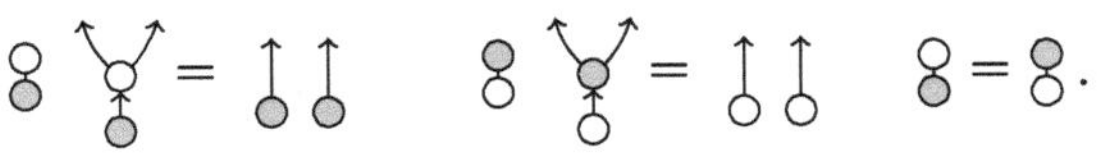

In other words, the unit point $\eta_{\circ}$ () is, modulo a scalar factor, a classical point for $\mathcal{O}_{\bullet}$, and vice versa.

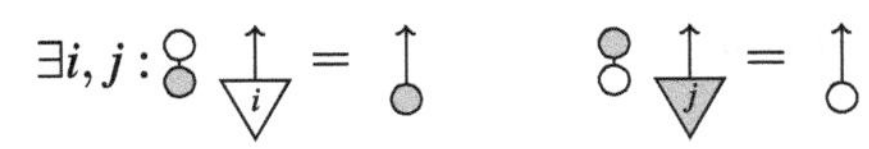

We will assume that the scalar  is always cancellable.

Example 20 Consider the two observables on the Hilbert space $\mathbb{C}^2$ corresponding to the Z and X spins:

[8]Indeed Eqs. (6) and (11) are equivalent in any theory wherever at least one of the observable structures has enough classical points.

$$\delta_{\circ}: \begin{matrix} |0\rangle \mapsto |00\rangle \\ |1\rangle \mapsto |11\rangle \end{matrix} \qquad\qquad \delta_{\bullet}: \begin{matrix} |+\rangle \mapsto |++\rangle \\ |-\rangle \mapsto |--\rangle \end{matrix}$$

$$\eta_{\circ}: \begin{matrix} |0\rangle \mapsto 1 \\ |1\rangle \mapsto 1 \end{matrix} \qquad\qquad \eta_{\bullet}: \begin{matrix} |+\rangle \mapsto 1 \\ |-\rangle \mapsto 1 \end{matrix}$$

Computing $\eta_{\circ}$ we obtain:

$$\eta_{\circ} = (\varepsilon_{\circ})^{\dagger} = \Big[1 \mapsto (|0\rangle + |1\rangle)\Big] = \sqrt{2}|+\rangle$$

which is indeed proportional to a classical point for $\delta_{\bullet}$. By a similar computation we obtain $\eta_{\bullet} = \sqrt{2}\,|0\rangle$, from which the value of their inner product $\begin{smallmatrix}\bullet\\ \circ\end{smallmatrix} = \sqrt{2}$ follows. The equations of Definition 12 can easily be verified from here, demonstrating that $\mathcal{O}_{\circ}$ and $\mathcal{O}_{\bullet}$ are coherent.

Proposition 3 *In* **FHilb** *if $O_{\circ}$ and $O_{\bullet}$ are self-adjoint operators corresponding to complementary observables, one can always choose a pair of coherent observable structures $(\mathcal{O}_{\circ}, \mathcal{O}_{\bullet})$ whose classical points correspond to the eigenbases of $O_{\circ}$ and $O_{\bullet}$.*

Proof Let $\{\,|a_i\rangle\,\}_{i=1}^{n}$ and $\{\,|b_j\rangle\,\}_{j=1}^{n}$ be orthonormal eigenbases for $O_{\circ}$ and $O_{\bullet}$ respectively. Since the bases are mutually unbiased we have

$$|b_j\rangle = \frac{1}{\sqrt{n}}\big[\alpha_{1j}\,|a_1\rangle + \cdots \alpha_{nj}\,|a_n\rangle\,\big]$$

where the α_{ij} are scalars satisfying $|\alpha_{ij}| = 1$. Setting $|a_i'\rangle = \alpha_{i1}\,|a_i\rangle$, we see that $\{|a_i'\rangle\}_i$ is also an orthonormal eigenbasis for $O_{\circ}$, which is still mutually unbiased with respect to $\{|b_j\rangle\}_j$. Now define:

$$\delta_{\circ}: |a_i'\rangle \mapsto |a_i'\rangle \otimes |a_i'\rangle$$
$$\varepsilon_{\circ}: |a_i'\rangle \mapsto 1$$

This choice yields $\eta_{\circ} = (\varepsilon_{\circ})^{\dagger} = \sum_i |a_i'\rangle = \sqrt{n}\,|b_1\rangle$.

In a similar fashion we can choose an eigenbasis $|b_1'\rangle, \ldots, |b_n'\rangle$ for $O_{\bullet}$ such that the resulting $\delta_{\bullet}$ and $\varepsilon_{\bullet}$ satisfy $(\varepsilon_{\bullet})^{\dagger} = \sqrt{n}|a_1'\rangle$. It is straightforward to verify that this can be done such that $|b_1'\rangle = |b_1\rangle$, ensuring the coherence of $\mathcal{O}_{\circ}$ and $\mathcal{O}_{\bullet}$.

For this reason we will from now on assume that pairs of complementary observables are always coherent.

3.1 Strongly Complementary Observables

Many familiar observables, when expressed in terms of algebras, turn out to have useful additional properties. These are called *strongly complementary*; before describing them we will require some preliminary definitions.

Definition 13 A *(commutative) bialgebra* on X is a 4-tuple $(\mu, \eta, \delta, \varepsilon)$ of maps,

$$\mu : X \otimes X \to X \qquad \delta : X \to X \otimes X$$
$$\eta : I \to X \qquad \varepsilon : X \to I,$$

which we write graphically as 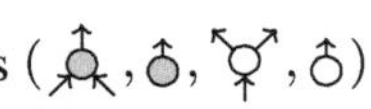, such that:

- (X, μ, η) is a (commutative) monoid;
- (X, δ, ε) is (cocommutative) comonoid; and,
- the following additional equations are satisfied:

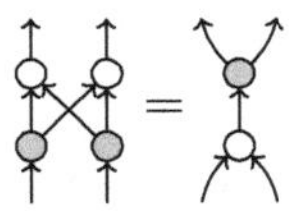 (12)

 (13)

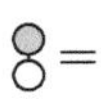 (14)

Remark 12 Note that Eqs. (13) and (14) are very similar to the equations required for the coherence of two observables, per Definition 12. The only difference there is that the scalar  is not assumed to be trivial.

Definition 14 A *(commutative) Hopf algebra* on X is a (commutative) bialgebra on X, augmented with a map $s : X \to X$, called the *antipode*, which satisfies:

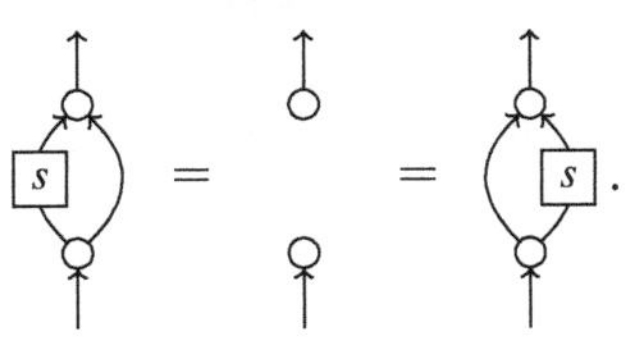 (15)

Again, note the similarity to Eq. (11): the difference is only by a scalar factor.

Definition 15 A pair $(\mathcal{O}_\circ, \mathcal{O}_\bullet)$ of observables on the same object X is *strongly complementary* iff they are coherent and:

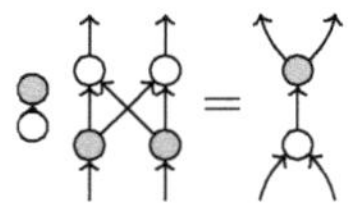

(16)

To expand on this definition slightly, given a pair of strongly complementary observables, if we consider just the monoid part of one and the comonoid part of the other then the resulting structure is, up to a scalar factor, a bialgebra. Note that thanks to the up-down symmetry induced by the † it doesn't matter which is the monoid and which the comonoid. For obvious reasons, the we say that a pair of strongly complementary observables forms a *scaled bialgebra*, and we refer to Eq. (16) as the *bialgebra law*. Notice that we have not, as yet, established any connection between complementarity (Definition 11) and strong complementarity. The following theorem links the two definitions.

Theorem 4 *Let* $\mathcal{O}_\circ$ *and* $\mathcal{O}_\bullet$ *be strongly complementarity observables; then they are complementary.*

Proof Let s be defined by

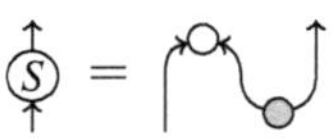

as per Eq. (9). Using the bialgebra law we reason:

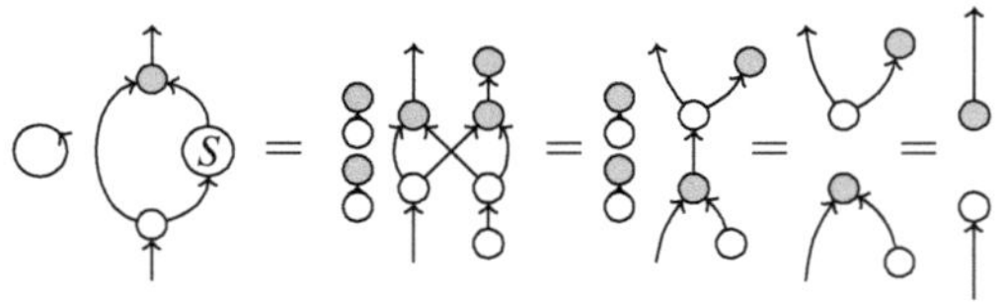

The last equation relies on the fact that $\eta_\circ$ is classical for $\mathcal{O}_\bullet$ (and $\eta_\bullet$ for $\mathcal{O}_\circ$), and (7).

As a consequence, strongly complementary observables always form a *scaled Hopf algebra*. Note that Theorem 4 relies on the fact that $\mathcal{O}_\circ$ and $\mathcal{O}_\bullet$ are Frobenius algebras; it is certainly not the case that every bialgebra is a Hopf algebra.

The converse to Theorem 4 does not hold: it is possible to find coherent complementary observables in **FHilb** which are not strongly complementary. See [29] for a counterexample.

The following lemma about the antipode for a strongly complementary pair was shown in [32].

Lemma 1 *If $(\mathcal{O}_{\circ}, \mathcal{O}_{\bullet})$ are strongly complementary, and have enough classical points then the antipode s is:*

- *self-adjoint;*
- *a monoid homomorphism from $\mathcal{O}_{\circ}$ to itself, and similarly for $\mathcal{O}_{\bullet}$; and*
- *a comonoid homomorphism from $\mathcal{O}_{\circ}$ to itself, and similarly for $\mathcal{O}_{\bullet}$.*

3.2 Strong Complementarity and Phase Groups

For complementary observables, classical points of one observable are always included in the phase group of the other observable, up to a normalizing scalar. Strong complementarity strengthens this property to inclusion as a subgroup. Let $\mathcal{K}_{\bullet}$ be the set of classical points of $\mathcal{O}_{\bullet}$ multiplied by the scalar factor .

Theorem 5 *Let $(\mathcal{O}_{\circ}, \mathcal{O}_{\bullet})$ be strongly complementary observables and let $\mathcal{O}_{\bullet}$ have finitely many classical points. Then $\mathcal{K}_{\bullet}$ forms a subgroup of the phase group $\Phi_{\circ}$ of $\mathcal{O}_{\circ}$. The converse also holds when $\mathcal{O}_{\bullet}$ has 'enough classical points'.*

Proof By strong complementarity it straightforwardly follows that, up to a scalar, $\mu_{\circ}$ applied to two classical points of $\mathcal{O}_{\bullet}$ yields again a classical point of $\mathcal{O}_{\bullet}$:

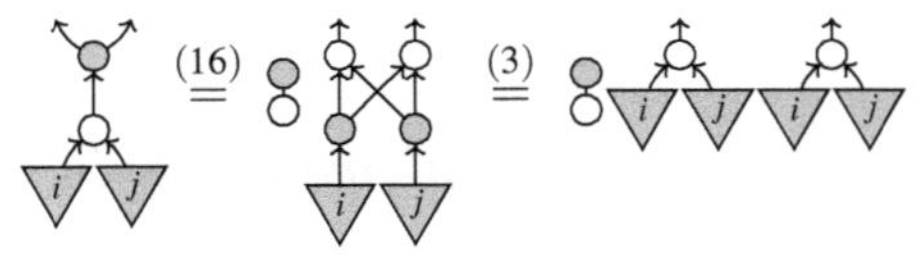

The unit of $\Phi_{\circ}$ is, up to a scalar, also a classical point of $\mathcal{O}_{\bullet}$ by coherence. Hence, $\mathcal{K}_{\bullet}$ is a submonoid of $\Phi_{\circ}$ and any finite submonoid is a subgroup. The converse follows from:

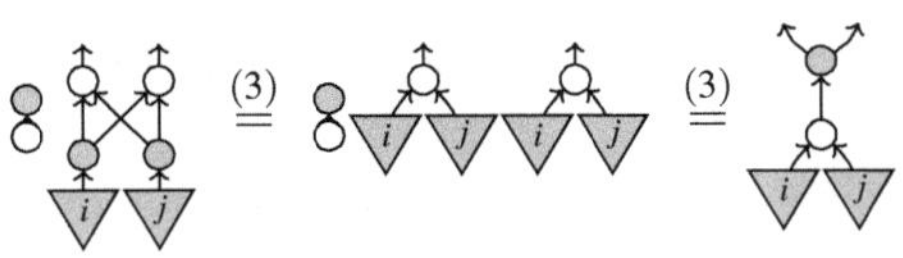

together with the 'enough classical points' assumption.

Recall that the exponent of a group G is the maximum order of any element of that group: $\exp(G) = \max\{|g| : g \in G\}$.

Corollary 1 *For any pair of strongly complementary observables, let $k = \exp(\mathcal{K}_{\bullet})$. Then, assuming $\mathcal{O}_{\bullet}$ has 'enough classical points':*

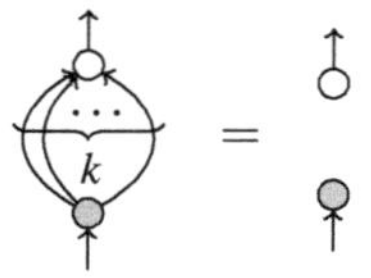

(17)

Proof In a finite abelian group, the order of any element divides $\exp(\mathscr{K}_\circ)$. The result then follows by:

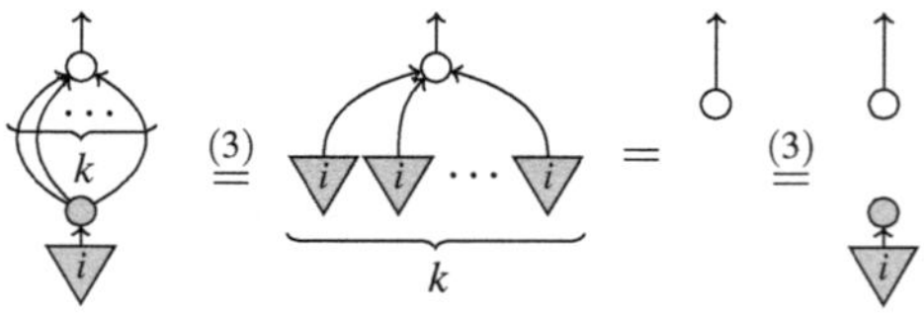

together with the 'enough classical points' assumption.

Proposition 4 *For a pair of strongly complementary observables* $\textcircled{i}$ *is a* $\mathscr{O}_\circ$-*homomorphism for all* $\textcircled{i} \in \mathscr{K}_\circ$. *Conversely, this property defines strong complementarity provided* $\delta_\circ$ *has 'enough classical points'.*

Proof Similar to the proof of Theorem 5.

3.3 Classification of Strong Complementarity in FHilb

Corollary 2 *Every pair of strongly complementary observables in* **FHilb** *is of the following form:*

$$\begin{cases} \delta_\circ :: |g\rangle \mapsto |g\rangle \otimes |g\rangle \\ \varepsilon_\circ :: |g\rangle \mapsto 1 \end{cases} \qquad \begin{cases} \delta_\circ^\dagger :: |g\rangle \otimes |h\rangle \mapsto \frac{1}{\sqrt{D}} |g+h\rangle \\ \varepsilon_\circ^\dagger :: 1 \mapsto \sqrt{D}\,|0\rangle \end{cases}$$

where $(G = \{g, h, \ldots\}, +, 0)$ *is a finite Abelian group. Conversely, each such pair is always strongly complementary.*

Proof By Theorem 5 it follows that the classical points of one observable (here $\mathscr{O}_\circ$) form a group for the multiplication of the other observable (here $\delta_\circ^\dagger$), and in **FHilb** this characterises strong complementarity.

One of the longest-standing open problems in quantum information is the characterisation of the number of pairwise complementary observables in a Hilbert space of dimension D. In all known cases this is $D + 1$, and the smallest unknown case is $D = 6$. We now show that in the case of strong complementarity this number is always 2 for $D \geq 2$.

Theorem 6 *In a Hilbert space with $D \geq 2$ the largest set of pairwise strongly complementary observables has size at most 2.*

Proof Assume that both $(\mathcal{O}_\circ, \mathcal{O}_\circ)$ and $(\mathcal{O}_\circ, \mathcal{O}_\bullet)$ are strongly complementary pairs. By coherence and must be proportional to classical points of $\mathcal{O}_\circ$. If $(\mathcal{O}_\circ, \mathcal{O}_\bullet)$ were to be strongly complementary observables, it is easily shown that so and are proportional to the same classical point. Hence, up to a non-zero scalar:

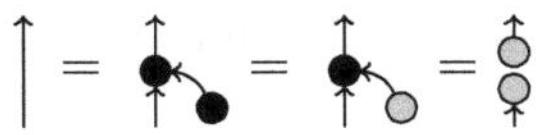

i.e. the identity has rank 1, which fails for $D \geq 2$. By Corollary 2 a strongly complementary pair exists for any $D \geq 2$.

4 Mixed States, Measurements, and "Abstract Probabilities"

For some ket $|\psi\rangle$ in a Hilbert space, there are (at least) four distinct ways to represent $|\psi\rangle$ as a linear map.

It is possible to represent a ket $|\psi\rangle \in H$ as a map $|\psi\rangle : \mathbb{C} \to H$, sending $1 \in \mathbb{C}$ to $|\psi\rangle$. We call such a map a "point" of H, because it does nothing more than picking out a specific element. The second map is the associated bra $\langle\psi| : H \to \mathbb{C}$. This kind of map is called a "co-point". We can also regard such a map as an element of the dual space H^*. But then, H^* is just another Hilbert space, so we could just as well represent $\langle\psi|$ as a point of H^*. That is, a linear map $\langle\psi|^* : \mathbb{C} \to H^*$. There is yet a fourth way to represent $|\psi\rangle$, namely as a linear map $|\psi\rangle^* : H^* \to \mathbb{C}$, sending a bra $\langle\phi| \in H^*$ to the inner product $\langle\phi|\psi\rangle \in \mathbb{C}$.

So, for a given ket $|\psi\rangle$, there are four ways to write it as points or copoints.

The difference in these four pictures is largely notational: the data they represent is the same. However, its a very useful piece of notation, especially when representing functionals between spaces of maps. Firstly, we note that we can represent a map $M: A \to B$ as a special kind of point, $|\Psi_M\rangle \in A^* \otimes B$.

$$M = \sum a_i^j \,|j\rangle \langle i| \qquad \mapsto \qquad |\Phi_M\rangle = \sum a_i^j \,\langle i| \otimes |j\rangle$$

These two objects clearly represent the same data. In fact, this mapping is essentially the Choi-Jamiołkowski isomorphism. However instead of fixing a basis, we rely on the dual space A^*. Thus the value on the right does not depend on a choice of orthonormal basis. By fixing a basis $\mathscr{B} = \{|i\rangle\}$, we can define a transposition map $T_{\mathscr{B}}(|i\rangle) = \langle i|$. Then the usual C-J isomorphism is recovered as $(T_{\mathscr{B}} \otimes 1)\,|\Phi_M\rangle$. However, this *does* depend on a choice of basis, since $T_{\mathscr{B}}$ does.

In [8], the authors refer to $|\psi_M\rangle$ as the "name" of a map. We shall often find this representation more useful than the usual C-J representation, especially in instances involving several distinct orthonormal bases. Using the caps and cups from before, we can isomorphically relate maps and their associated names.

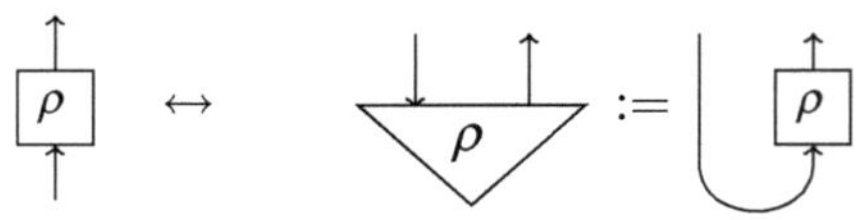

The benefit of working with names of maps, as opposed to the maps themselves becomes clear when we start considering higher-order functionals. For a finite-dimensional Hilbert space H, let $L(H)$ be the space of linear maps from H to itself. When operating on density matrices, we often want to consider maps of the form $\Phi\colon L(H) \to L(K)$. We can either treat this as a genuine, higher-order map, or we can treat it as a first-order map from names to names.

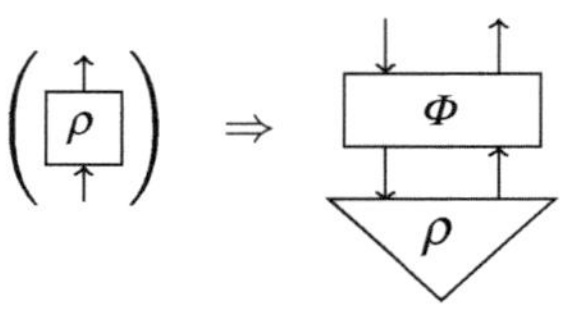

Since, in finite dimensions, we have an isomorphism $L(H) \cong H^* \otimes H$, we know that all maps $\Phi\colon L(H) \to L(K)$ can be represented this way.

In ordinary quantum theory, mixed quantum states are represented as positive operators and operations as completely positive maps, or CPMs. These are maps that take positive operators to positive operators. A general CPM can be written in terms of a set of linear maps $\{B_i : H \to K\}$ called its *Kraus maps*.

$$\Theta(\rho) = \sum_i B_i \rho B_i^\dagger$$

As before, we can collapse the higher-order map Θ to a first- order map by translating the positive operator ρ to its associated name.

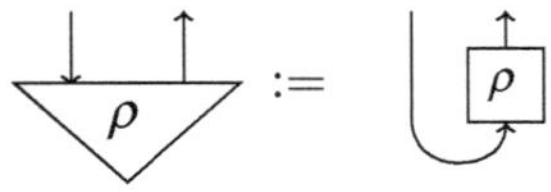

Then, we can encode the Kraus vectors of Θ in a map $B' = \sum |i\rangle \otimes B_i$ and represent Θ as:

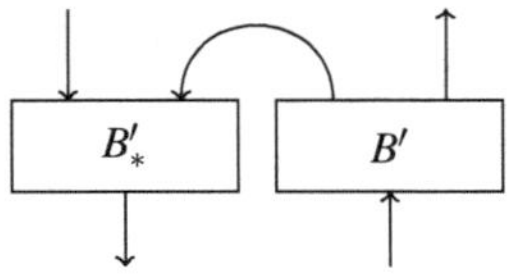

(18)

When we take the elements in Eq. (19) to be morphisms in an arbitrary †-compact category, this gives us an abstract definition of a completely positive map. This is Selinger's representation of CPMs [33].

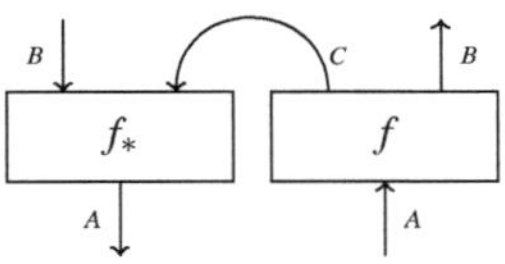

(19)

Important special cases are *states* where $A \cong I$, *effects* where $B \cong I$, and 'pure' maps, where $C \cong I$.

4.1 Measurements and Born Vectors

Returning to quantum mechanics, we can see how a quantum measurement would look in this language. A (projective) quantum measurement $M_{\circ}$ is a CPM that sends trace 1 positive operators (in this case quantum states) to trace 1 positive operators that are diagonal in some ONB (encoding a probability distribution of outcomes). Suppose we wish to measure with respect to an observable $\mathcal{O}_{\circ}$, whose classical points form an ONB $\{|x_i\rangle\}$. The probability of getting the ith measurement outcome is computed using the Born rule.

$$\text{Prob}(i, \rho) = \text{Tr}(|x_i\rangle\langle x_i| \rho)$$

We can write this probability distribution as a vector in the basis $\{|x_i\rangle\}$. That is, a vector whose ith entry is the probability of the ith outcome:

$$M_{\circ}(\rho) = \sum \text{Tr}(|x_i\rangle\langle x_i|\rho)\, |x_i\rangle$$

So, M defines a linear map from density matrices to probability distributions. Expanding this graphically, we have:

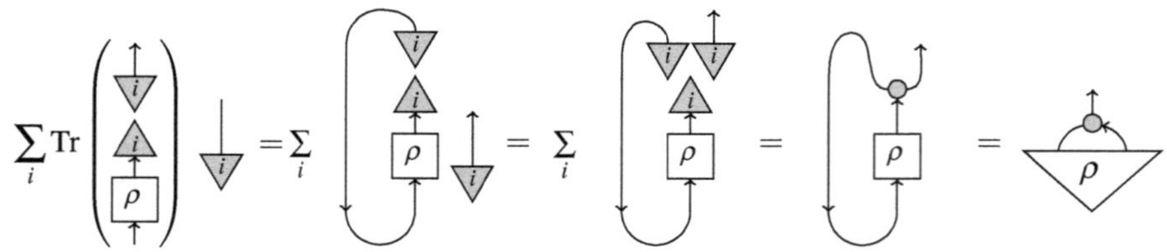

We are now ready to make definitions for abstract measurements and abstract probability distributions, which we shall call Born vector.

Definition 16 For an observable structure $\mathcal{O}_{\circ}$, a measurement is defined as the following map:

$$m_{\circ} :=$$
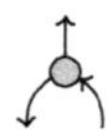

Any point $|\Lambda)_{\circ} : I \to X$ of the following form is called a *Born vector*, with respect to $\mathcal{O}_{\circ}$:

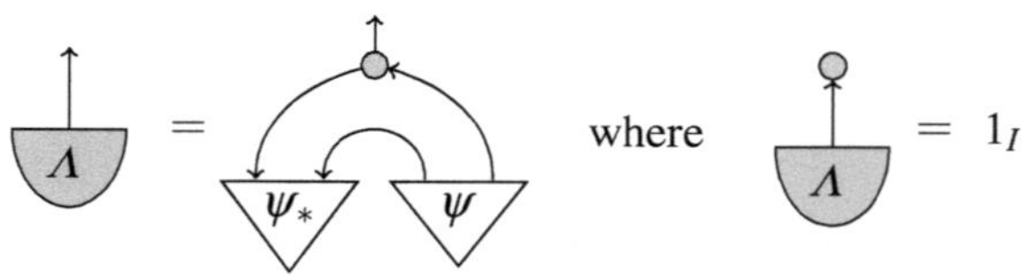

Theorem 7 *In* **FHilb**, *Born vectors for an observable $\mathcal{O}_{\circ}$ are precisely those vectors whose entries are positive and sum to* 1, *when written in the basis of $\mathcal{O}_{\circ}$.*

Proof Let $|\Lambda)_{\circ}$ be a Born vector. Its ith coefficient in the $\mathcal{O}_{\circ}$-basis is given by:

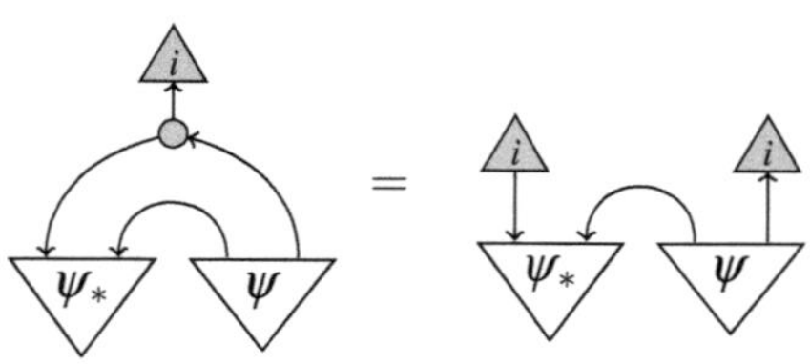

We can see that these coefficients sum to 1 by using the definition of the deleting point:

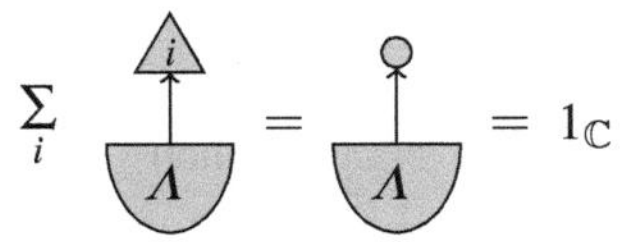

This is a completely positive map from $\mathbb{C}$ to $\mathbb{C}$. In other words, it is a positive scalar. For the converse, assume $|\Lambda)_{\circ}$ is a probability distribution whose ith coefficient in the $\mathcal{O}_{\circ}$-basis is $p_i \in \mathbb{R}_{+}$. Then, letting $\Psi = \Sigma \sqrt{p_i}$ (point i):

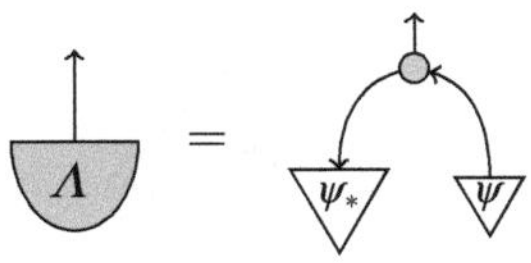

Post-composing with the deleting point yields $\sum(\sqrt{p_i})^2 = \sum p_i = 1$.

Thus Born vectors in **FHilb** correspond precisely to *probability distributions* over classical points.

We can naturally extend the definition above to points of the form $|\Lambda)_{\circ} : I \to X \otimes \ldots \otimes X$ by requiring that they be Born vectors with respect to the product Frobenius algebra $\mathcal{O}_{\circ} \otimes \ldots \otimes \mathcal{O}_{\circ}$.

The adjoint of the measurement map $m_{\circ}^{\dagger}$ is a preparation operation. In **FHilb**, it takes a Born vector $|\Lambda)_{\circ}$ with respect to $\mathcal{O}_{\circ}$ and produces a probabilistic mixture of the (pure) outcome states of $\mathcal{O}_{\circ}$ with probabilities given by $|\Lambda)_{\circ}$.

This leads to a simple classical versus quantum diagrammatic paradigm that applies in all of the models we consider [22]: *classical systems are encoded as a single wire and quantum systems as a double wire*. The same applies to operations, and $m_{\circ}$ and $m_{\circ}^{\dagger}$ allow passage between these types.

Note that the classical data will 'remember' to which observable it relates, cf. the encoding $\sum_i p_i \, |x_i\rangle$. This is physically meaningful since, for example, when one measures position the resulting value will carry specification of the length unit in which it is expressed. If one wishes to avoid interconversion of this 'classical data with memory', one could fix one observable, and unitarily transform the quantum data before measuring. Indeed, if

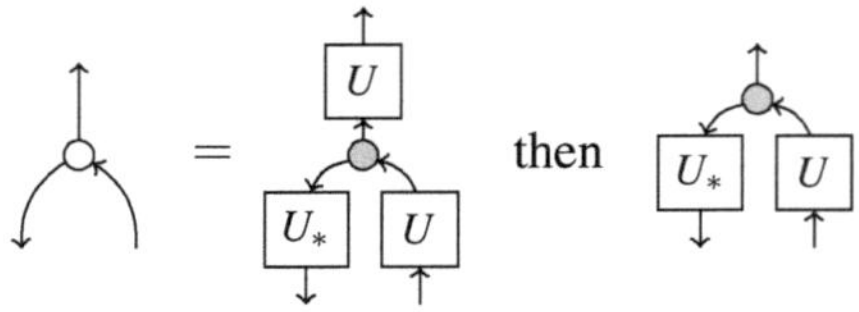

measures the $\mathcal{O}_\circ$-observable but produces $\mathcal{O}_\bullet$-data. In **FHilb**, all observable structures are unitarily isomorphic, so any projective measurement can be obtained in this way. A particularly relevant example is when these unitaries are phases with respect the another observable structure $\mathcal{O}_\circ$.

$m_\bullet^\alpha :=$ 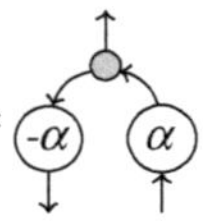

(20)

When $\mathcal{O}_\circ$ is induced by the Pauli spin-Z observable and $\mathcal{O}_\bullet$ by the Pauli spin-X observable, then $m_\bullet = m_\bullet^0$ is an X measurement and $m_\bullet^{\pi/2}$ is a Y measurement. Note however, that both produce Born vectors of outcome probabilities with respect to the $\mathcal{O}_\bullet$ basis. This will be useful in the next section.

5 Example: Non-locality of QM

In 1989 Greenburger, Horne, and Zeilinger provided an analysis [34] of quantum theory which improves on Bell's theorem in one crucial way. Whereas Bell demonstrated a *probabilistic* argument that quantum theory is incompatible with the assumption of local realism (i.e. quantum theory generates correlations that are too high for a classical local hidden variable model), the GHZ/Mermin theorem illustrates a situation that rules out a locally realistic model *possibilistically*. That is, they described a series of experiments for which quantum theory predicts a single, definite outcome that is impossible under the assumption of locality.

Here, we reproduce Mermin's version of this argument [35] using diagrammatic techniques. There are two key ingredients for this translation. The first is a graphical notion of locality. For our purposes, it will suffice to treat locality as the fact that global probability can be traced down to hidden states that determine all measurement outcomes, since we shall show that no hidden state can ever be compatible with the predictions of quantum theory. Hence, there is no point in even considering crafting a local hidden variable representation.

The second key ingredient is *parity*. The GHZ/Mermin trick for producing definite outcomes is to look not at individual measurement outcomes, but the overall parity of those outcomes, i.e. "Did the experiment produce and even or an odd number of clicks?". Considering a single outcome (click or no-click) as an element of the abelian group $\mathbb{Z}_2$, the parity of a set of outcomes is given by their sum in the group.

We already saw in Sect. 3.3 that strongly complementary observables are classified by abelian groups. In two dimensions, there is only one such strongly complementary pair, namely the one corresponding to $\mathbb{Z}_2$. When we prepare a GHZ state with respect to a certain observable (e.g. spin-Z) and conduct measurements using a strongly complementary observable (e.g. spin-X), we will see this $\mathbb{Z}_2$ structure arise.

By combining these two elements (the topological picture of locality and the encoding of abelian groups as strongly complementary observables) we will derive a contradiction. Furthermore in Sect. 5.4, we shall see how strong complementary was embedded in the pre-conditions of a GHZ/Mermin-style argument in the first place.

5.1 A Local Hidden Variable Model

For a particular n-party state $|\psi\rangle$ in some theory, a *local hidden variable (LHV) model* for that state consists of:

- a family of hidden states $|\lambda\rangle$, each of which assigns to any measurement on each subsystem a definite outcome; and,
- a probability distribution on these hidden states,

which simulates the probabilities of that theory. We say that a theory is *local* if each state admits a LHV model.

Consider three systems and four possible (compound) measurement settings, *XXX*, *XYY*, *YXY*, and *YYX*. A hidden state of an underlying LHV model stores one measurement outcome for each setting on each system:

$$|\lambda'\rangle = |\underbrace{\overbrace{+}^{X}\ \overbrace{-}^{Y}}_{\textit{system 1}}\ \underbrace{\overbrace{-}^{X}\ \overbrace{+}^{Y}}_{\textit{system 2}}\ \underbrace{\overbrace{-}^{X}\ \overbrace{+}^{Y}}_{\textit{system 3}}\rangle$$

We can represent this diagrammatically as follows:

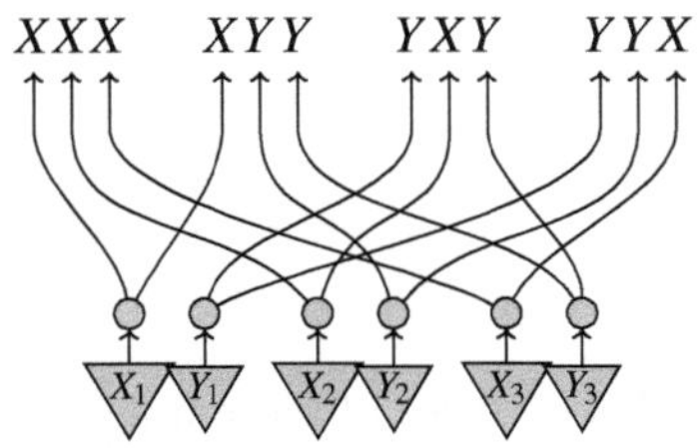

that is, we simply copy those values to each of the four measurement settings.

5.2 Encoding the GHZ State and Computing Correlations, Diagrammatically

To present Mermin/GHZ style argument graphically, we first show how to compute measurement outcomes for an n-party GHZ state graphically. This computation relies on a standard theorem about bialgebras, which relates a graph-theoretic property of diagrams to equality of bialgebra expressions.

Definition 17 Let (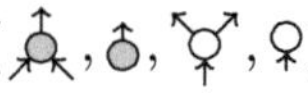) be a commutative, cocommutative bialgebra, and let $\mathscr{D}$ be a diagram consisting only of 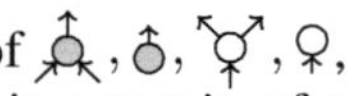, identity maps, and swaps. Then, the characteristic matrix χ of $\mathscr{D}$ is a matrix of natural numbers where the (i, j)th entry represents the number of forward-directed paths connecting the ith input to the jth output.

Example 21 The following terms have characteristic matrix

$$\begin{pmatrix} 0 & 1 \\ 0 & 0 \end{pmatrix} :$$

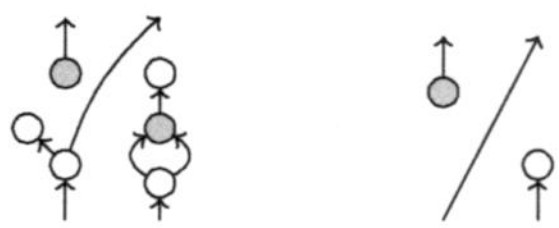

The following terms have characteristic matrix

$$\begin{pmatrix} 1 & 1 \\ 1 & 1 \\ 1 & 1 \end{pmatrix} :$$

Theorem 8 *If two diagrams generated by the same bialgebra have the same characteristic matrix, they are equal as maps.*

Proof (sketch) It is possible to show by case analysis that the three bialgebra equations can be used to move all of the gray dots to the top all the white dots to the bottom.

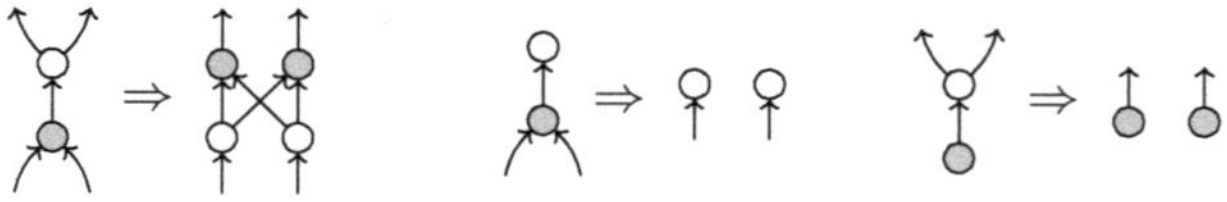

We can furthermore show that all three of these transformations preserve the characteristic matrix of $\mathscr{D}$. Once this is done, we obtain a diagram in normal form:

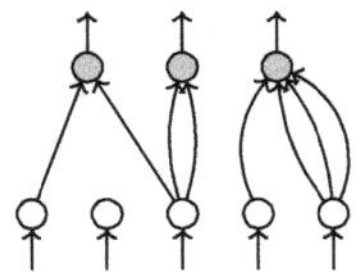

$$\leftrightarrow \quad \begin{pmatrix} 1 & 0 & 0 \\ 0 & 0 & 0 \\ 1 & 2 & 0 \\ 0 & 0 & 1 \\ 0 & 0 & 3 \end{pmatrix}$$

Then, it is possible to show there is *exactly one* such normal form for each characteristic matrix. In fact, it is straightforward to read off the matrix by counting edges in the normal form. Since every diagram can be put into normal form using equations that preserve the characteristic matrix, and normal forms are in 1-to-1 correspondence with characteristic matrices, this completes the proof.[9]

We can now apply the theorem to prove the following corollary.

Corollary 3 *The following equation holds for any connected bipartite graph with directions as shown.*

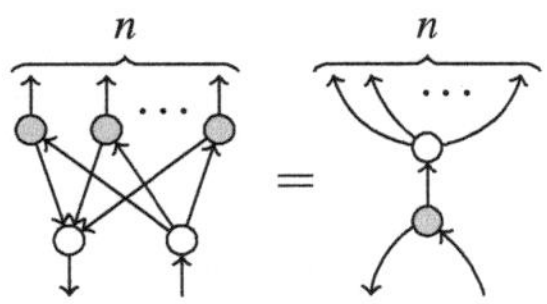

(21)

Proof For the theorem on bialgebras to apply, all of the edges need to be directed upward. For a strongly complementary observable, the edge direction between two different colours can be reversed by applying the antipode S. Then, we use the fact that S is a Frobenius algebra endomorphism to move it down.

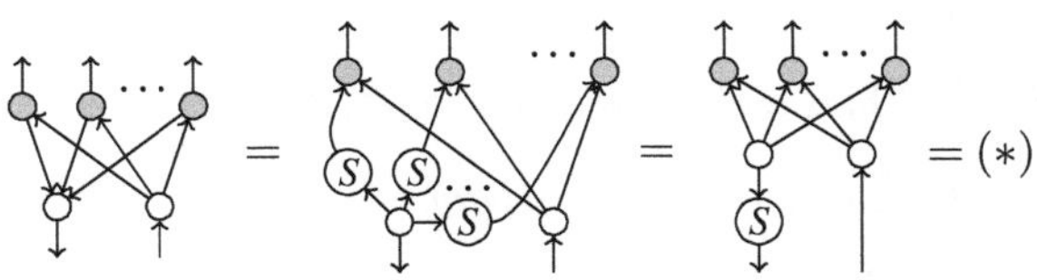

We apply Theorem 8 and the spider theorem to complete the proof.

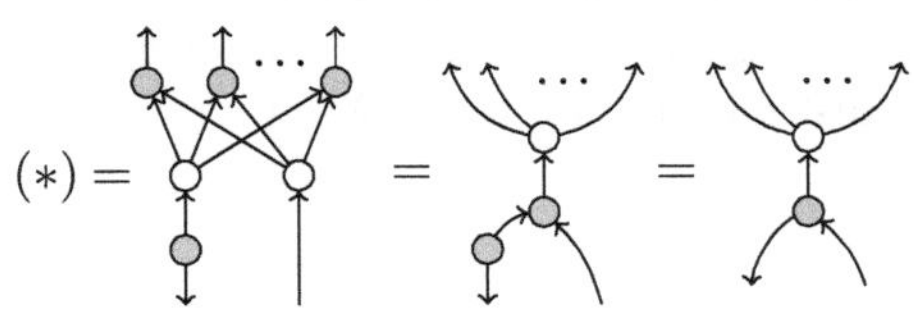

[9]For a formal statement and proof of this theorem, in terms of factorisation systems see [36].

We compute the classical probability distributions ($= \mathcal{O}_\circ$-data) for n measurements against arbitrary phases $\alpha_i \in \Phi_\circ$ on n systems of any type in a generalised $GHZ^n_\circ$-state:

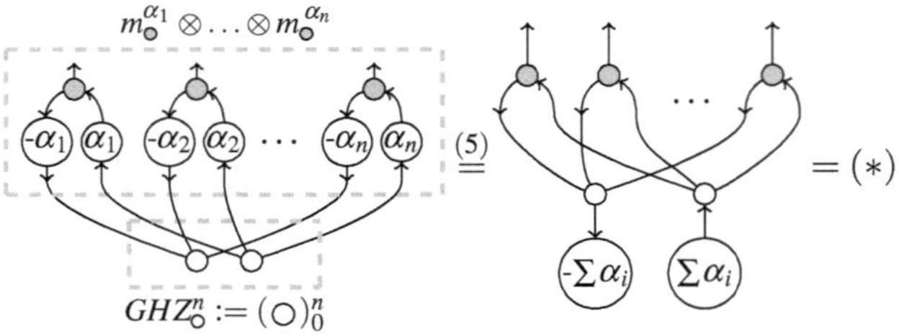

Applying Corollary 3, we note that this is a probability distribution followed by a $\circ$-copy.

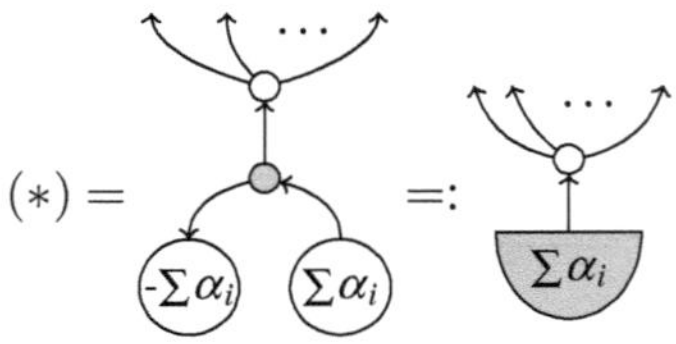

(22)

The following is an immediate consequence.

Theorem 9 *When measuring each system of a GHZ^n_A-state against an arbitrary angle then the resulting classical probability distribution over outcomes is symmetric.*

Theorem 10 *The classical probability distributions for $^{\alpha_1}_\circ \otimes \ldots \otimes m_\circ^{\alpha_n}$-measurements on a GHZ^n_A-state is:*

- *uncorrelated if $|\Sigma\alpha_i)_\circ$ is a classical point for $\mathcal{O}_\circ$ and,*
- *parity-correlated if $|\Sigma\alpha_i)_\circ$ is a classical point i for $\mathcal{O}_\circ$ (i.e. contains precisely those outcomes $i_1 \otimes \ldots \otimes i_n$ such that the sum of group elements $\sum i_k$ is equal to i).*

Example 22 We can compute the outcome distributions for *XXX*, *XYY*, *YXY*, and *YYX* measurements on three qubits in a GHZ-state using the technique described above. First, outcome distribution $|A)_\circ$ for *XXX*:

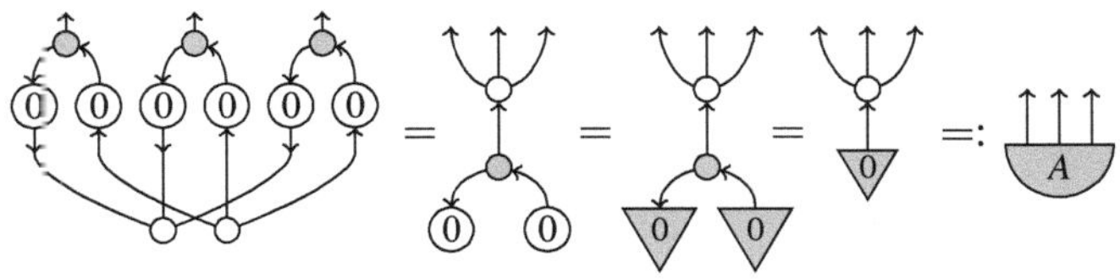

Next, we compute outcome distribution $|B_1)_{\circ}$ for *XYY*:

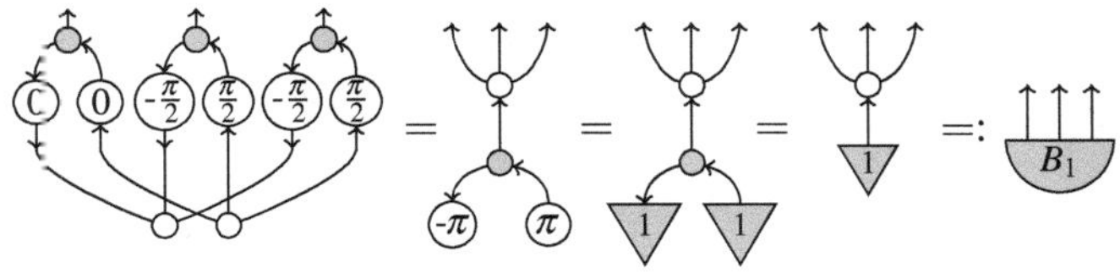

Computing correlations as in Fig. (22) is symmetric in the choice of measurement angle for each of the systems. Thus, for the other two cases (*YXY* and *YYX*), we get the same result: $|B_1)_{\circ} = |B_2)_{\circ} = |B_3)_{\circ}$.

5.3 Deriving the Contradiction

Consider the function:

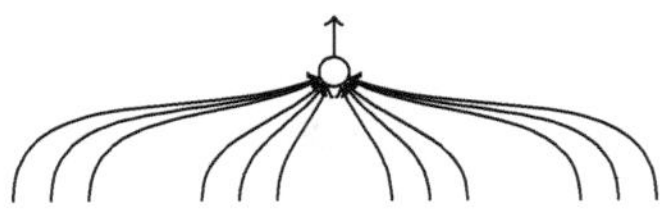

(23)

We have already seen that strongly complementary observables correspond to group algebras. That is,  defines a group algebra over the classical points of $\mathcal{O}_{\circ}$. For qubits there is only one choice: $\mathbb{Z}_2$. Thus, this function computes the parity (i.e. the $\mathbb{Z}_2$-sum) of all outcomes.

Measuring the parity for any local hidden state we obtain:

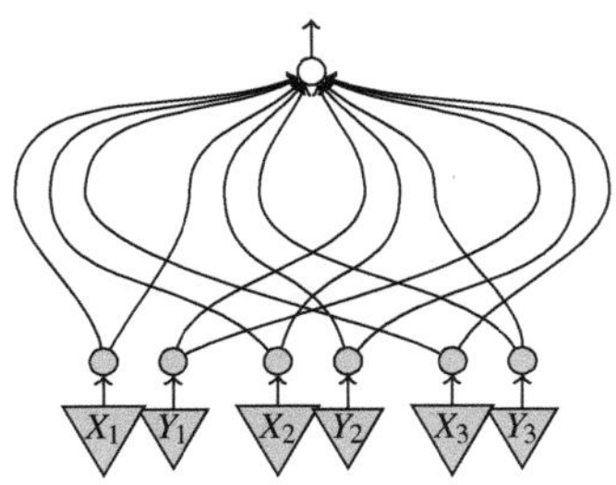

that is, by (17):

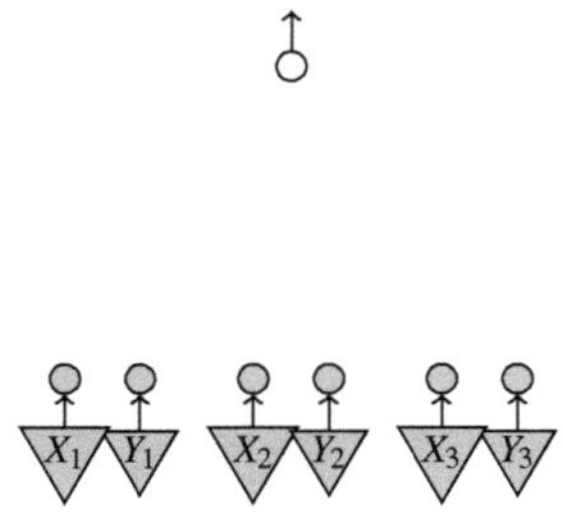

and hence:

and measuring the parity in quantum theory we obtain:

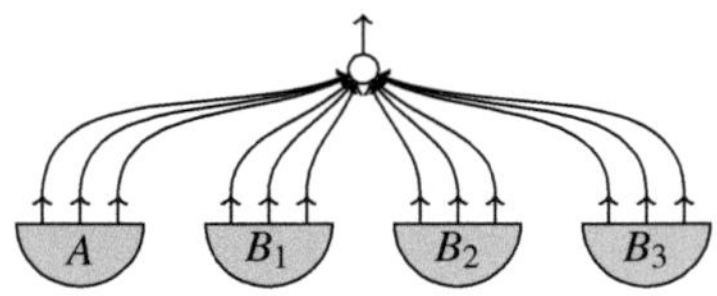

that is, by the previous section:

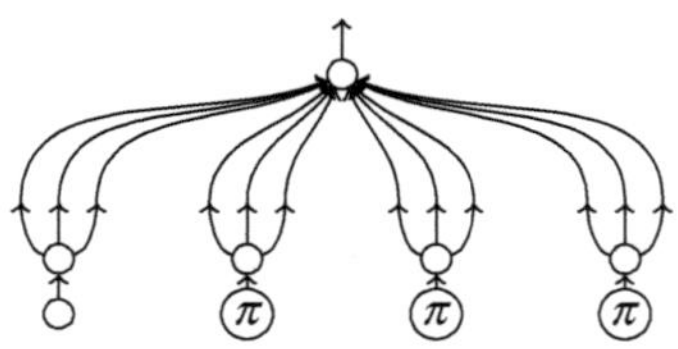

and hence:

which yields a contradiction.

5.4 *GHZ/Mermin Assumptions and the Necessity of Strong Complementarity*

We shall examine two assumptions that play a key role in a GHZ/Mermin style non-locality argument, and show that the presence of a strongly complementary observable arises as a consequence.

The original argument due to Greenburger, Horne, and Zeilinger [34] and later simplifications [35, 37] focus on a state defined in terms of correlated (or anti-correlated) Z-spins and local spin measurements in the XY-plane. We generalise this assumption as follows.

Assumption 1 (*Coherence*) We will use a GHZ state defined with respect to some observable structure $\mathcal{O}_\circ$. Measurements are all conducted within a $\mathcal{O}_\circ$-phase of some coherent observable $\mathcal{O}_\circ$.

In **FHilb**, all observables containing at least one unbiased classical point, w.r.t. $\mathcal{O}_\circ$, are within a $\mathcal{O}_\circ$-phase of a coherent observable, so we could weaken this assumption further. That is, if $\mathcal{O}_\circ$ contains an unbiased classical point, we might as well assume it is coherent, since Assumption 1 allows us to freely choose phases.

By Assumption 1, the correlations associated with each experiment are computed from this diagram:

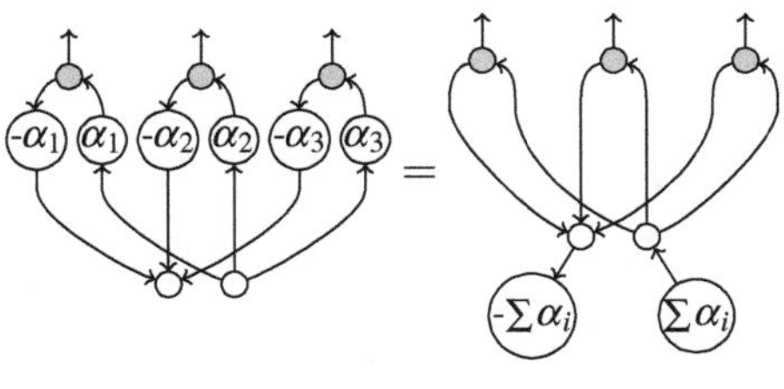

The second assumption is what [34] refers to as "super-classicality". We shall refer to it as *sharpness*.

Assumption 2 (Sharpness) After all subsystems except one are measured, the final measurement outcome is fixed.

The map  is called the *decoherence map* for $\mathcal{O}_\circ$. It projects from the space of all quantum mixed states to the the space of classical mixtures of eigenstates of $\mathcal{O}_\circ$. To assert sharpness, we require that, once two of the three systems are measured, the third is invariant under this map:

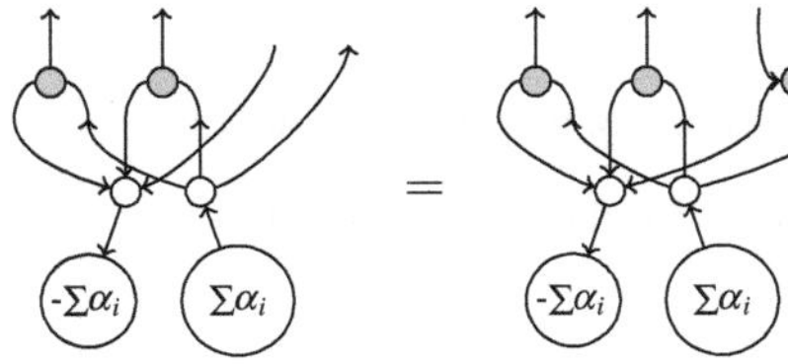

(24)

Plugging the unit of $\mathcal{O}_\circ$ in the 2nd system both for LHS and RHS, and using coherence we obtain:

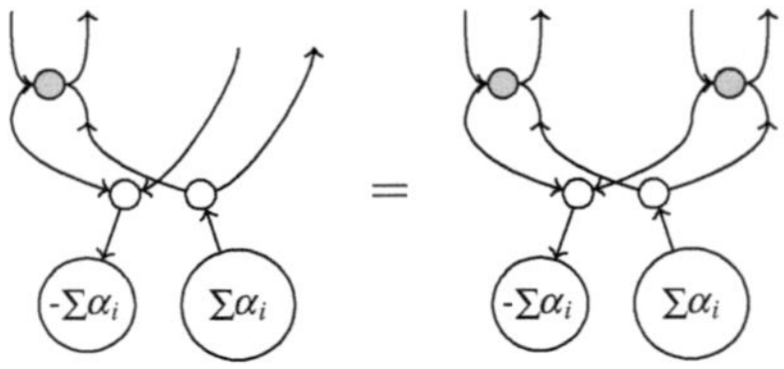

(25)

and by exploiting symmetry we have:

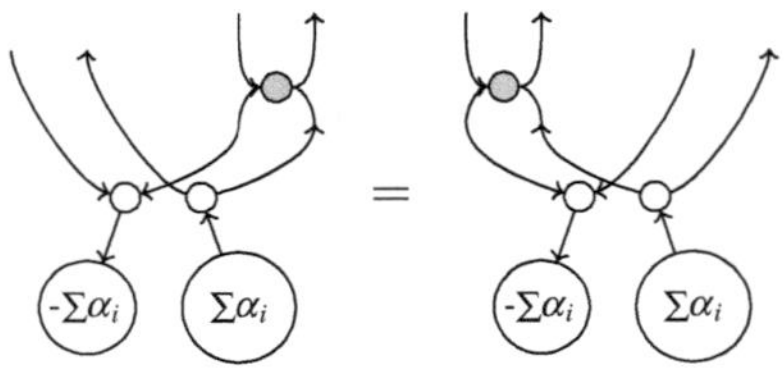

(26)

Hence we obtain:

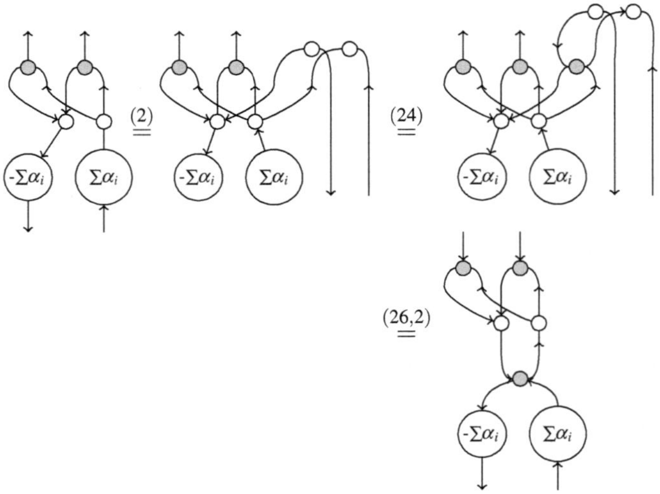

Since $\delta_\circ^\dagger \circ (1_X \otimes \sum_i \alpha_i)$ is unitary it cancels. Thus our assumptions lead us to conclude the following equation for the observable structures $(\mathscr{O}_\circ, \mathscr{O}_\bullet)$:

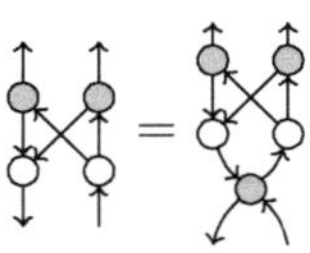

(27)

Proposition 5 *A pair* $(\mathscr{O}_\circ, \mathscr{O}_\bullet)$ *of coherent observables satisfying Eq. (27) are strongly complementary.*

Proof First, we show that Eq. (27) implies the following, for any pair of coherent observables:

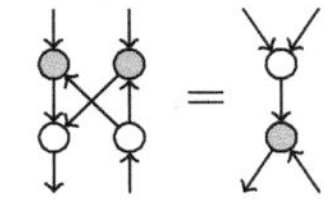 (28)

The proof goes as follows:

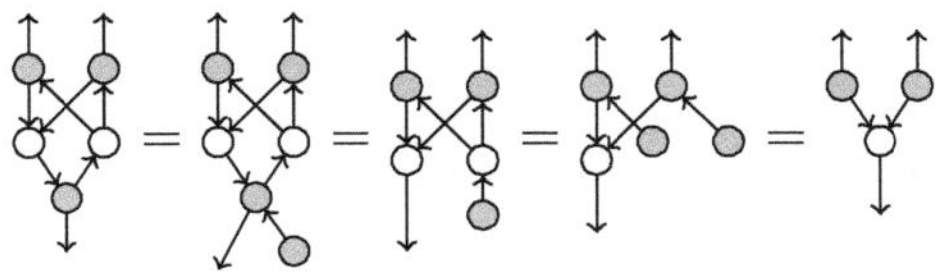

...which implies:

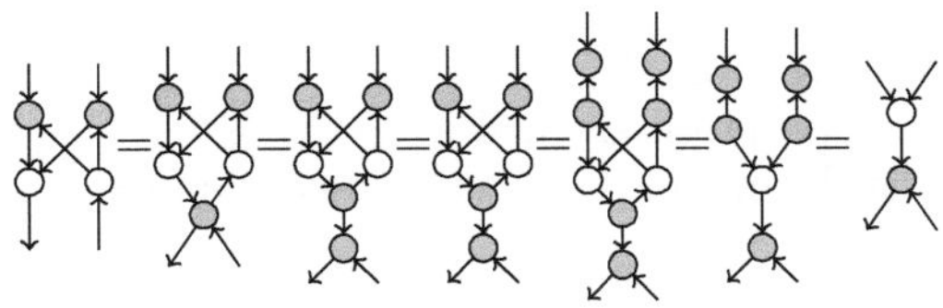

Equation (28) is very nearly the required equation for strong complementarity, but the directions are wrong. However, we can correct this by first showing the following equations, using coherence and (28):

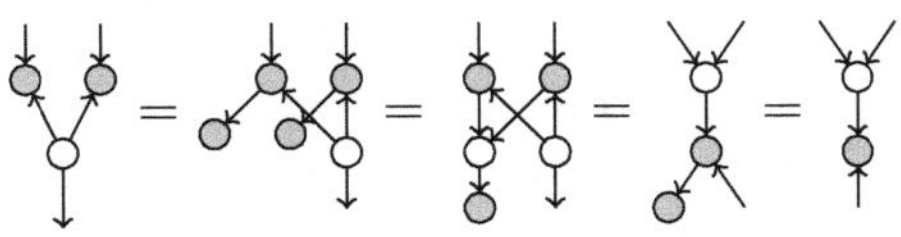

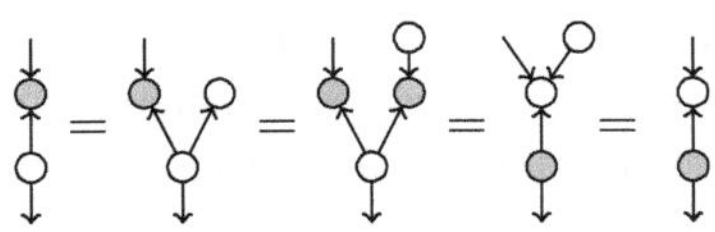

Then, we complete the proof by using the equations above to change the directions of the arrows on the inside:

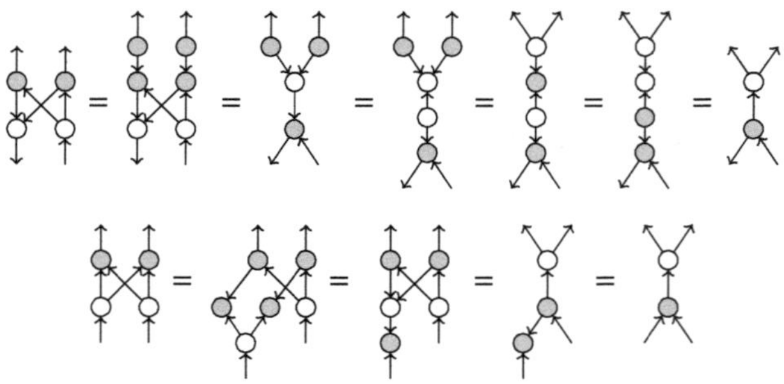

Thus any coherent pair of observable structures satisfying Eq. (27) is a strongly complementary pair.

6 Summary and Further Reading

In this chapter, we developed the notion of a *generalised compositional theory*, a new approach to studying quantum mechanics and constructing foil theories with quantum-like properties. The main building blocks for a GCT are:

- a collection of systems $A, B, C, \ldots,$
- a collection of *primitive* processes, and
- a notion of horizontal composition $\otimes$ and vertical composition $\circ$.

From this sparse setting, we began to add extra pieces of structure.

- symmetry maps $\Rightarrow$ "permutibility of systems"
- dagger $\Rightarrow$ "time-reversed processes"
- duals $\Rightarrow$ "map/state duality"

This structure and its diagrammatic presentation give a rich language for talking about composed processes. We then went on to define various concepts in this framework, often by analogy to their Hilbert space counterparts: pure states, reversible dynamics, quantum observables, complementarity, mixed states, and measurements. Using these ingredients, we worked through a complete example, following Mermin's illustration of a possibilistic locality violation, as predicted by quantum mechanics.

The interested reader can find many papers related to, or extending the formalism introduced in this chapter. One example is the *ZX-calculus*, which is a graphical calculus for the interaction of the Pauli-Z and Pauli-X observable structures. In addition to the usual rules (complementarity, strong complementarity), several other rules are added, which turn out to be complete for stabiliser quantum mechanics [38]. This calculus has been applied to the study of measurement-based quantum computing [15], topological MBQC [16], and quantum protocols [39].

The ideas developed in Sect. 4 originated in [22]. In [40], a simplified formalism for interacting classical and quantum data was developed, and can be viewed as an abstraction of the C*-algebraic approach to the study of quantum information.

References

1. J. Barrett, Phys. Rev. A **75**(3), 032304 (2007)
2. M. Pawlowski, T. Paterek, D. Kazlikowski, V. Scarani, A. Winter, M. Zukowski, Nature **461**, 1101 (2009). arXiv:0905.2292
3. H. Barnum, J. Barrett, L.O. Clark, M. Leifer, R.W. Spekkens, N. Stepanik, A. Wilce, R. Wilke, New J. Phys. **12**, 033024 (2009). arXiv:0909.5075
4. E. Schrödinger, in *Proceedings of the Cambridge Philosophical Society*, vol. 31 (Academic Press, New York, 1935), pp. 555–563
5. R. Penrose, in *Combinatorial Mathematics and Its Applications* (Academic Press, New York, 1971)
6. G.M. Kelly, M.L. Laplaza, J. Pure Appl. Algebra **19**, 193 (1980)
7. A. Joyal, R. Street, Adv. Math. **102**, 20 (1993)
8. S. Abramsky, B. Coecke, in *Proceedings of 19th IEEE Conference on Logic in Computer Science, LiCS'04* (IEEE Press, 2004), pp. 415–425
9. B. Coecke, in *Quantum Theory: Reconsiderations of the Foundations III* (AIP, Press, New York, 2005), pp. 81–98
10. G.M. D'Ariano, G. Chiribella, P. Perinotti, Phys. Rev. A **84**, 012311 (2010).arXiv:1011.6451
11. L. Hardy, in *Deep Beauty: Understanding the Quantum World Through Mathematical Innovation* (Cambridge University Press, Cambridge, 2011), pp. 409–442. arXiv:0912.4740
12. B. Coecke B. Edwards, R.W. Spekkens, Electron. Notes Theor. Comput. Sci. **270**(2), 15 (2011)
13. B. Edwards, Phase groups and local hidden variables. Technical report RR-10-15, Department of Computer Science, University of Oxford (2010)
14. B. Coecke, A. Kissinger, in *Proceedings of ICALP Automata, Languages, and Programming*. Lecture Notes in Computer Science, vol. 6199 (Springer, Heidelberg, 2010), pp. 297–308
15. R. Duncan, S. Perdrix, in *Proceedings of the 37th International Colloquium Conference on Automata, Languages and Programming, ICALP'10: Part II* (Springer, Berlin, 2010), pp. 285–296. http://dl.acm.org/citation.cfm?id=1880999.1881030
16. C. Horsman, New J. Phys. **13**, 095011 (2011). arXiv:1101.4722
17. S. Mac Lane, *Categories for the Working Mathematician*, 2nd edn. (Springer, New York, 1997)
18. B. Coecke, E.O. Paquette, in *New Structures for Physics*. Springer Lecture Notes in Physics, vol. 813 (2011), pp. 173–286
19. P. Selinger, in *New Structures for Physics*. Springer Lecture Notes in Physics, vol. 813 (2011), pp. 289–355
20. Y. Lafont, J. Pure Appl. Algebra **184**(2–3), 257 (2003)
21. A. Barenco, C.H. Bennett, R. Cleve, D.P. DiVincenzo, N. Margolus, P. Shor, T. Sleator, J.A. Smolin, H. Weinfurter, Phys. Rev. A **52**, 3457 (1995). doi:10.1103/PhysRevA.52.3457
22. B. Coecke, E.O. Paquette, D. Pavlovic, in *Semantic Techniques for Quantum Computation* (Cambridge University Press, Cambridge, 2009), pp. 29–69
23. D.G.B.J. Dieks, Phys. Lett. A **92**, 271 (1982)
24. W.K. Wootters, W. Zurek, Nature **299**, 802 (1982)
25. A.K. Pati, S.L. Braunstein, Nature **404**, 164 (2000). arXiv:quant-ph/9911090
26. B. Coecke, D. Pavlovic, J. Vicary, Math. Struct. Comput. Sci. **23**, 555 (2013)
27. D. Pavlovic, in *Proceedings of the Symposium on Quantum Interaction*. Lecture Notes in Computer Science, vol. 5494 (Springer, New York, 2009), pp. 143–157. arXiv:0812.2266
28. B. Coecke R. Duncan, in *Proceedings of ICALP 2008 Automata, Languages, and Programming* Lecture Notes in Computer Science, vol. 5126 (Springer, New York, 2008), pp. 298–310

29. B. Coecke, R. Duncan, New J. Phys. **13**, 043016 (2011). arXiv:0906.4725
30. R.W. Spekkens, Phys. Rev. A **75**, 032110 (2007). arXiv:quant-ph/0401052
31. B. Coecke, B. Edwards, Spekkens's toy theory as a category of processes (2011). arXiv:1108.1978v1[quant-ph]
32. A. Kissinger, Pictures of processes: automated graph rewriting for monoidal categories and applications to quantum computing. Ph.D. thesis, University of Oxford (2012)
33. P. Selinger, Electron. Notes Theor. Comput. Sci. **170**, 139 (2007)
34. D.M. Greenberger, M.A. Horne, A. Zeilinger, in *Bell's Theorem, Quantum Theory, and Conceptions of the Universe*, ed. by M. Kafatos (Sringer, New York, 1989), pp. 69–72
35. N.D. Mermin, Am. J. Phys. **58**, 731 (1990)
36. S. Lack, Theory Appl. Categ. **13**(9), 147 (2004)
37. M.A. Horne, A. Shimony, D.M. Greenberger, A. Zeilinger, Am. J. Phys. **58**, 1131 (1990)
38. M. Backens, in *Proceedings of Quantum Physics and Logic* (2012), pp. 15–27
39. A. Hillebrand, Quantum protocols involving multiparticle entanglement and their representations in the ZX-calculus. Master's thesis, University of Oxford (2011)
40. B. Coecke, C. Heunen, A. Kissinger, in *Proceedings of Quantum Physics and Logic* (2012), pp. 87–100

Post-Classical Probability Theory

Howard Barnum and Alexander Wilce

1 Introduction

This chapter offers a brief introduction to what is often called the *convex-operational approach* to the foundations of quantum mechanics, and reviews selected results, mostly by ourselves and collaborators, obtained using that approach. Broadly speaking, the goal of research in this vein is to locate quantum mechanics (henceforth: QM) within a very much more general, but conceptually very straightforward, generalization of classical probability theory. The hope is that, by regarding QM from the outside, so to say, we shall be able to understand it more clearly. And, in fact, this proves to be the case.

The phrase "convex-operational" deserves some comment. The approach discussed here is "convex" in that it takes the space of states of a physical system to be a convex set (to accommodate the formation of probabilistic mixtures), and draws conclusions from the geometry of this set. It is "operational" in its acceptance of measurements and their outcomes as part of its primitive conceptual apparatus, and in its identification of states with probability weights on measurement outcomes. In this sense, it is conceptually very conservative, differing from classical probability only in that it is *not* assumed that all measurements can be made simultaneously.

H. Barnum (✉)
Department of Physics and Astronomy, University of New Mexico,
Albuquerque, Mexico
e-mail: hnbarnum@aol.com; hbarnum@unm.edu

H. Barnum
Stellenbosch Institute for Advanced Study (STIAS), Wallenberg Research Centre
at Stellenbosch University, Marais Street, Stellenbosch 7600, South Africa

A. Wilce
Department of Mathematics, Susquehanna University, Selinsgrove, PA, USA
e-mail: wilce@susqu.edu

G. Chiribella and R.W. Spekkens (eds.), *Quantum Theory: Informational Foundations and Foils*, Fundamental Theories of Physics 181,
DOI 10.1007/978-94-017-7303-4_11

From this starting point, one is led very naturally to a mathematical framework for a *post-classical* probability theory, which, while varying idiomatically from author to author [9, 29, 31, 34, 41, 42, 48, 51], is more or less canonical. About the first third of what follows is devoted to a detailed discussion of the structure of individual probabilistic models in this framework. Here we exhibit a range of simple non-classical examples, many of them quite different from *either* classical or quantum probabilistic models. At the same time, we try to bring some order to this diversity, by showing that essentially any probabilistic model can be represented in a natural way in terms of an ordered real vector space and its dual, and that processes operating on and between models can be represented by positive linear maps between these associated spaces.

Starting in Sect. 3, we focus on *composites* of probabilistic models, subject to a natural non-signaling constraint. As we shall see, the phenomenon of *entanglement*, often regarded as a hallmark of quantum mechanics, is actually a rather generic feature of non-signaling composites of non-classical state spaces, and thus, more a marker of non-classicality than of "quantumness" per se. Since quantum information theory treats entanglement as a resource, the question then arises of which quantum-information theoretic results can be made to work in a more general probabilistic setting. Section 4 reviews some work in this direction, particularly the generalization of the no-cloning and no-broadcasting theorems of [9, 10], and an analysis of teleportation and entanglement-swapping protocols in terms of conditional states, following [11].

If many non-classical features of QM are not so much quantum as generically non-classical, what *does* single out QM? The question of how to characterize QM in operational or probabilistic terms is a very old one. After many decades of hard-won partial results in this direction (e.g., [4, 5, 23, 39, 56, 65, 80]), the past decade has produced a slew of novel derivations of finite-dimensional QM from fairly simple, transparent, and plausible assumptions [24, 28, 41, 50, 57] (to cite just a few). In Sect. 5, we outline one of these, which recovers the Jordan structure of finite-dimensional quantum theory from symmetry considerations; the specific C^*-algebraic machinery of standard quantum mechanics is then singled out by considerations involving the formation of composite systems. The key tools here are a classical representation theorem for homogeneous, self-dual cones, due to M. Koecher and E. Vinberg [44, 69], and a theorem about tensor products of Jordan algebras due to H. Hanche Olsen [40].

Since the aim of this paper is to provide a brief and accessible introduction to this material, we make some simplifying assumptions. The most important is that we focus *entirely* on finite-dimensional models, even though large parts of the apparatus developed here work perfectly well (and were first developed) in an infinite-dimensional setting. Further assumptions will be spelled out as we go.

Notational conventions Real vector spaces are indicated generically by bold capitals $\boldsymbol{E}$, $\boldsymbol{F}$, etc. The space of linear mappings $\boldsymbol{E} \to \boldsymbol{F}$ is denoted by $\mathcal{L}(\boldsymbol{E}, \boldsymbol{F})$; $\boldsymbol{E}^*$ denotes the dual space of $\boldsymbol{E}$. If $\mathcal{H}$ is a real or complex Hilbert space, $\mathcal{L}_h(\mathcal{H})$ stands for the space of bounded Hermitian operators on $\mathcal{H}$. If X is a set, $\mathbb{R}^X$ denotes the vector space of all real-valued functions on X. We write $\langle x, y\rangle$ for the inner product of two vectors in a real or complex Hilbert space; in the complex case, we take this to be conjugate linear in the *second* argument. (Thus, our $\langle x, y\rangle$ would be $\langle y|x\rangle$, in Dirac notation.)

2 Elementary Probability Theory, Classical and Otherwise

If $\mathcal{H}$ is a Hilbert space, representing a quantum-mechanical system, then each state of that system is represented by a density operator ρ. A possible measurement outcome is represented by an *effect*, i.e., a positive hermitian operator a with $\mathbf{0} \leq a \leq \mathbf{1}$; $\mathrm{Tr}(\rho a)$ gives the *probability* that a will occur (if measured) when the state ρ obtains. This probabilistic apparatus generalizes that of classical probability theory, in that if we fix an *observable*, that is, a set $\{a_1, \ldots, a_n\}$ of effects summing to $\mathbf{1}$, we can understand this as a model of a single, discrete, classical statistical experiment, on which each state ρ defines a probability weight $p(i) := \mathrm{Tr}(\rho a_i)$. The novelty here is that, in general, a pair of observables $\{a_1, \ldots, a_n\}$ and $\{b_1, \ldots b_k\}$ is not *co-measurable*. In classical probability theory, it is always assumed (if often tacitly) that any pair of outcome-sets E_1 and E_2 admit a simultaneous refinement, that is, both can be represented as partitions or "coarse-grainings" of some third outcome-set F. In quantum probability theory, this is not the case. Unless the operators a_i and b_j all commute, there will be no third observable of which E_1 and E_2 are both coarse-grainings.

So, quantum probability theory foregoes the assumption of co-measurability, which is a tenet of classical probability theory. And, indeed, in retrospect, the latter is surely a *contingent* matter, so it is not so very radical a step to renounce it. On the other hand, quantum probability theory replaces the simple axiom of co-measurability with the elaborate apparatus of the Hilbert space $\mathcal{H}$ and its associated space of Hermitian operators. As a framework for an autonomous probability calculus, this seems less than perfectly well motivated, and one can wonder whether, and why, it is necessary. A sensible way to approach this question is simply to drop the co-measurability assumption, without making any special assumptions to replace it. The resulting *post-classical* probability theory, while in some respects rather conservative (probabilities are still real numbers between 0 and 1, summing to unity over complete sets of mutually exclusive possible outcomes), presents a vast, poorly explored, and rather wild landscape, within which even quantum probability theory seems rather tame.

2.1 Test Spaces and Probabilistic Models

There are many more or less equivalent, but stylistically diverse, ways of formulating a post-classical probability theory of the kind indicated above. The approach we take here (due originally to C.H. Randall and D.J. Foulis [33, 34]) begins with a very minimum of raw material.

Definition 1 A *test space* is a pair $(X, \mathcal{M})$ where X is a set of *outcomes* and $\mathcal{M}$ is a covering of X by non-empty sets called *tests*. A *probablity weight* on $(X, \mathcal{M})$ is a function $\alpha : X \to [0, 1]$ with

$$\sum_{x \in E} \alpha(x) = 1$$

for every $E \in \mathcal{M}$.

The intended interpretation is that each $E \in \mathcal{M}$ is the set of mutually exclusive outcomes associated with some probabilistic experiment—anything from rolling a die to asking a question to making a measurement (via some well-defined procedure) of some physical quantity. It is permitted that distinct tests may overlap, that is, that distinct experiments may share some outcomes. The definition of a probability weight requires that, when this is the case, the probability of a given outcome be independent of the measurement used to secure it. In other words, probability weights are *non-contextual.*[1]

Notation: It will often be convenient to let X stand for the pair $(X, \mathcal{M})$, writing $\mathcal{M}(X)$ for $\mathcal{M}$. Also, we'll write $\mathbf{\Omega}(X, \mathcal{M})$ or simply $\Omega(X)$, for the set of all probability weights on a test space $(X, \mathcal{M})$. This is a *convex* subset of $[0, 1]^X \subseteq \mathbb{R}^X$, i.e.,

$$\alpha, \beta \in \mathbf{\Omega}(X) \Rightarrow t\alpha + (1-t)\beta \in \mathbf{\Omega}(X)$$

for all $0 \leq t \leq 1$. Where X is *locally finite*, meaning that every test $E \in \mathcal{M}(X)$ is a finite set, it is not hard to see that $\mathbf{\Omega}(X)$ is closed, and hence compact, with respect to the product topology on $[0, 1]^X$. It follows that $\mathbf{\Omega}(X)$ is the closed convex hull of its extreme points.

Models In constructing a model for a probabilistic system, we may wish to single out certain probability weights as corresponding to possible *states* of the system. It is reasonable to form probability-weighted averages of such states, in order to represent ensembles of systems in different states. It is also reasonable to idealize the situation slightly by assuming that the limit of a sequence of possible states should again count as a possible state. In the same spirit, we shall assume in what follows that X carries a Hausdorff topology, with respect to which states are continuous. This is harmless, since we can always use the discrete topology as a default.[2] Indeed, given that our focus here is exclusively on finite-dimensional models, it is not unreasonable to assume that X is even compact.

[1]The formalism easily accommodates contextual probability assignments, however: simply define $\widetilde{X}$ to be the disjoint union of the test in $\mathcal{M}$—say, to be concrete, $\widetilde{X} = \{(x, E) | x \in E \in \mathcal{M}\}$. In effect, each outcome of $\widetilde{X}$ consists of an outcome of X, plus a *record* of which test was used to secure it. For each test $E \in \mathcal{M}$, let $\widetilde{E} = \{(x, E) | x \in E\}$, and let $\widetilde{\mathcal{M}} = \{\widetilde{E} | E \in \mathcal{M}\}$. Probability weights on $(\widetilde{X}, \widetilde{\mathcal{M}})$ are exactly what one means by *contextual* probability weights on $(X, \mathcal{M})$. There is a natural surjection $\widetilde{X} \to X$ that simply forgets these records; probability weights on $(X, \mathcal{M})$ pull back along this surjection to give us weights on $(\widetilde{X}, \widetilde{\mathcal{M}})$.

[2]A more detailed discussion of test spaces with topological structure can be found in [72].

To make all of this official:

Definition 2 A *probabilistic model*—or, for purposes of this paper, just a *model*—is a structure $(X, \mathcal{M}, \Omega)$, where $(X, \mathcal{M})$ is a test space with X a compact Hausdorff space and Ω is a pointwise-closed (hence, compact), convex set of continuous probability weights on $\mathfrak{Q}(X, \mathcal{M})$. The extreme points of Ω are the *pure states* of the model.

Notation: We henceforth use capital letters A, B, etc. to denote models, writing, e.g., $(X(A), \mathcal{M}(A))$ for the test space belonging to model A, and $\Omega(A)$ for A's state space.

Example 1 (*Classical Models*) (a) The simplest classical models have the structure $(E, \{E\}, \Delta(E))$, where E is a single test (so that $\mathcal{M} = \{E\}$), and where $\Delta(E)$ is the simplex of all probability weights thereon. We might also deem "classical" a broader set of models: those of the form $(E, \{E\}, \Omega)$ where $\Omega \subseteq \Delta(E)$ is any closed, convex set of probability weights sufficiently large to statistically separate different outcomes[3] of the single test E.

(b) A more sophisticated classical model begins with a measurable space S, and identifies statistical experiments with finite or countably infinite partitions of S by measurable subsets. The collection of all such experiments is a test space: let $X(S)$ be the set of non-empty measurable subsets of S (say, with the discrete topology), and let $\mathcal{D}(S)$ be the set of finite partitions of S into measurable subsets. We call $(X(S), \mathcal{D}(S))$ the *partition test space* associated with S. Probability weights on $(X(S), \mathcal{D}(S))$ correspond exactly to finitely additive measures on S. By varying $\mathcal{D}(S)$, we can change the character of the probability weights that are allowed. For example, if $\mathcal{D}_\sigma(S)$ is the set of countable partitions, probability weights on $(X(S), \mathcal{D}_\sigma(S))$ correspond to countably-additive probability measures on S.

Example 2 (*Quantum Models*) (a) The most basic quantum-mechanical model begins with a complex Hilbert space $\mathcal{H}$. The corresponding *quantum test space* is $(X(\mathcal{H}), \mathcal{M}(\mathcal{H}))$ where the outcome space $X(\mathcal{H})$ is the unit sphere of $\mathcal{H}$ (with its usual topology) and where the space $\mathcal{M}(\mathcal{H})$ of tests is the set of unordered orthonormal bases or *frames* of $\mathcal{H}$. Every unit vector $v \in \mathcal{H}$ determines a probability weight α_v on $\mathcal{M}(\mathcal{H})$, defined for all $x \in X(\mathcal{H})$ by

$$\alpha_v(x) = |\langle v, x\rangle|^2 = \mathrm{Tr}(P_v P_x),$$

where P_v and P_x are the rank-one projection operators corresponding to v and x. Accordingly, if W is a density operator on $\mathcal{H}$—a positive semidefinite hermitian operator of trace one, or, equivalently, a convex combination of rank-one projections—then $\alpha_W(x) := \langle Wx, x\rangle = \mathrm{Tr}(WP_x)$ defines a probability weight on $X(\mathcal{H})$. If $\dim(\mathcal{H}) \geq 3$, then Gleason's theorem tells us that every probability weight on $X(\mathcal{H})$ is of this form, but for $\dim(\mathcal{H}) = 2$, there are many others, which one regards as

[3]That is, given any pair of distinct outcomes, there exists a state assigning them different probabilities.

non-physical. In either case, letting $\Omega(\mathcal{H})$ denote the convex set of density operators on $\mathcal{H}$, we obtain the *quantum model* $A(\mathcal{H}) = (X(\mathcal{H}), \mathcal{M}(\mathcal{H}), \Omega(\mathcal{H}))$.

(b) A slightly different model, which we'll call the *projective* quantum model, and which we denote by $A(\mathbb{P}\mathcal{H})$, replaces each outcome $x \in X(\mathbb{P}\mathcal{H})$ by the corresponding rank-one projection operator P_x; tests in $\mathbf{M}(\mathbb{P}\mathcal{H})$ are maximal pairwise orthogonal families of such projections. Again, states correspond to density operators via the recipe $\alpha_W(P_x) = \text{Tr}(WP_x)$ where $P_x \in X(\mathbb{P}\mathcal{H})$. For many purposes, the choice between $A(\mathcal{H})$ and $A(\mathbb{P}\mathcal{H})$ is one of convenience. However, notice that in passing from $A(\mathcal{H})$ to $A(\mathbb{P}\mathcal{H})$ we lose information about phase relations between the unit vectors representing outcomes of $X(\mathcal{H})$, which are important in describing sequential experiments. We won't pursue this here. The paper [79] contains some relevant discussion.

(c) A more sophisticated quantum model might begin with a W^*-algebra $\mathcal{A}$, and take for $\mathcal{M}$, the collection of all (say, finite) sets of projections summing to the identity in $\mathcal{A}$. If $\mathcal{M}$ has no I_2 summand, the Christensen-Yeadon extension of Gleason's theorem [30] identifies the probability weights on $\mathcal{M}$ with states on $\mathcal{A}$. Again, if there are I_2 factors (copies of $M_2(\mathbb{C})$), then one must explicitly limit the states to the quantum-mechanical ones.

By the *dimension* of a model A, we mean the dimension of the span of $\Omega(A)$ in $\mathbb{R}^{X(A)}$. Of course, this will generally be infinite. However, as mentioned in the introduction, our focus in this paper is on finite-dimensional models. In this context, it is also reasonable to concentrate on locally finite models. Indeed, making this official, we make it a standing assumption that, from this point forward,

all models are locally finite and finite-dimensional.

In particular, all *quantum* models $A(\mathcal{H})$ and $A(\mathbb{P}\mathcal{H})$ involve only finite-dimensional Hilbert spaces $\mathcal{H}$.

Dispersion-Free States and Distinguishability One very striking difference between classical and quantum models has to do with the existence of (globally) *dispersion-free*, that is, zero-or-one valued, states. In both of the classical models considered above, all pure states are dispersion-free. Quantum models, in contrast, have *no* dispersion-free states: a pure quantum state still makes only uncertain predictions about the results of most measurements.

Definition 3 A set Ω of probability weights on a test space X is *unital* iff, for every $x \in X$, there exists at least one $\alpha \in \Omega$ with $\alpha(x) = 1$. If there is a *unique* such state, we say that Ω is *sharp*. We say that a model A is unital or sharp if its state space $\Omega(A)$ is a unital, respectively sharp, set of probability weights on the test space $X(A)$.

Like the classical examples, the quantum quantum models $A(\mathcal{H})$ and $A(\mathbb{P}\mathcal{H})$ are sharp; indeed, the unique state α assigning probability one to a given outcome $x \in X(\mathcal{H})$, or to the corresponding outcome $P_x \in X(\mathbb{P}\mathcal{H})$, is the one corresponding to the density operator P_x.

Definition 4 A set Ω of probability weights on a test space X *separates outcomes*, or is *separating*, iff, for all outcomes $x, y \in X$, $\alpha(x) = \alpha(y)$ for all $\alpha \in \Omega$ implies $x = y$. A model A is *separated* iff $\Omega(A)$ separates outcomes of $X(A)$.

The state space of a standard quantum model $A(\mathcal{H})$ is not separating; that of the corresponding projective quantum model $A(\mathbb{P}\mathcal{H})$ *is* separating. As this example illustrates, given a non-separated model A, one *can* always replace $X(A)$ by an obvious quotient test space, in which probabilistically indistinguishable outcomes are identified, to obtain a separated model having the same states. One may or may not *wish* to do so.

A *partition space* is a test space that is isomorphic[4] to a sub-test space of the test space $\mathcal{D}(S)$ of finite partitions of some set S (see Example 1(b)). Any such space supports a separating set of dispersion-free probability weights, namely, the point-masses associated with the points of S. The following is straightforward:

Lemma 1 *If test space has a unital, separating set of dispersion-free states, then it is a partition test space. If it has a sharp separating set of, dispersion-free states, then it is classical.*

In anticipation of later results, we'll write $x \perp y$ to mean that outcomes $x, y \in X(A)$ are *distinguishable* by means of some test $E \in \mathcal{M}(A)$—that is, that $x, y \in E$ and $x \neq y$. At present, there is no linear structure in view, let alone an inner product, so the notation is only suggestive. Later, we'll see that one can often embed X in an inner product space in such a way that the notation can be taken literally.

It will also be useful to introduce the following notion of distinguishability for *states*.

Definition 5 Two states, $\alpha, \beta \in \Omega(A)$ are *sharply distinguishable* iff there exist outcomes $x, y \in X(A)$ with $x \perp y$ such that $\alpha(x) = \beta(y) = 1$. More generally, states $\alpha_1, \ldots, \alpha_n$ are *jointly* sharply distinguishable iff there exists a test $E \in \mathcal{M}(A)$ and outcomes $x_1, \ldots, x_n \in E$ with $\alpha_i(x_j) = \delta_{i,j}$.

The idea is that, if the system is known to be in one of the states $\alpha_1, \ldots, \alpha_n$, then by performing the measurement E we will learn—with probability one—which of these states was the actual one.[5]

---

[4] An isomorphism of test spaces is a bijection from outcomes to outcomes, preserving tests in both directions.

[5] A weaker notion would require only that $\alpha_i(x_i) > 0 = \alpha_i(x_j)$ for each i, j, so that with *some* nonzero probability we learn which state is actual. Notice, too, that the condition of joint sharp distinguishability is a priori much stronger than pairwise sharp distinguishability.

2.2 Further Examples

Classical and quantum examples hardly exhaust the possibilities, of course: a major point of the present framework is to provide us with a maximum of flexibility in constructing *ac hoc* models.

Example 3 (*The Square Bit*) The very simplest non-classical model starts with a test space X be a test space containing just two tests $E = \{x, x'\}$ and $F = \{y, y'\}$, each having two outcomes—as, say, two coins, or a stern-Gerlach apparatus with two angular settings. The convex set $\Omega(X)$ of all probability weights on X is affinely isomorphic to the unit square, under the mapping $\alpha \mapsto (\alpha(x), \alpha(y))$. The model (X, Ω) has, accordingly, been called the *square bit* or *squit* [12]. As $\Omega(X)$ is not a simplex, this model is not entirely classical. On the other hand, as its pure states are all dispersion-free, it is very far from being "quantum".

Remark: A test space in which distinct tests are disjoint—as in the square bit—is said to be *semiclassical*. Such a much test space clearly supports a separating set of dispersion-free states, hence, by Lemma 1, can be represented as a partition test space. However, as the square bit illustrates, such a test space will typically support many more states than are permitted classically.

Greechie Diagrams A useful graphical device for representing small test spaces (those involving only a few outcomes) is to represent each outcome as a dot, and to join outcomes belonging to a test by a straight line or other smooth arc, with arcs corresponding to distinct tetst intersecting, if at all, at a sharp angle, so as to be easily distinguished. Such a representation (first used in the quantum-logical literature) is called a *Greechie diagram* [38]. For example, we might represent a three-outcome classical test by the diagram in Fig. 1a, and the square-bit test space by that in Fig. 1b. The test space pictured in Fig. 1c with two three-outcome tests (the top and bottom rows) and three two-outcome tests (the vertical lines), makes the point that a test space need not have any states at all —in this case, because a state can not at the same time sum to one across the two rows *and* across the three columns.

The following whimsical example (due to D.J. Foulis) is useful as an antidote to several too-comfortable intuitions.

Example 4 (*The Firefly Box*) Suppose a sealed triangular box is divided into three interior chambers, as in the top-down view in Fig. 2a, below. The walls of the box are translucent, while the top, the bottom, and the interior partitions are opaque. In the box is a firefly, free to move about between the chambers (for which purpose, the interior partitions contain small tunnels). Viewed from one side, we might see

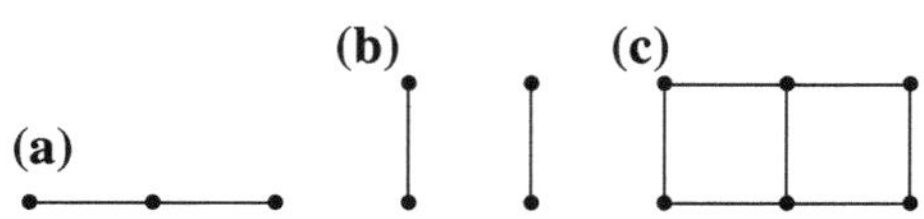

Fig. 1 Various Greechie diagrams

Fig. 2 The firefly box

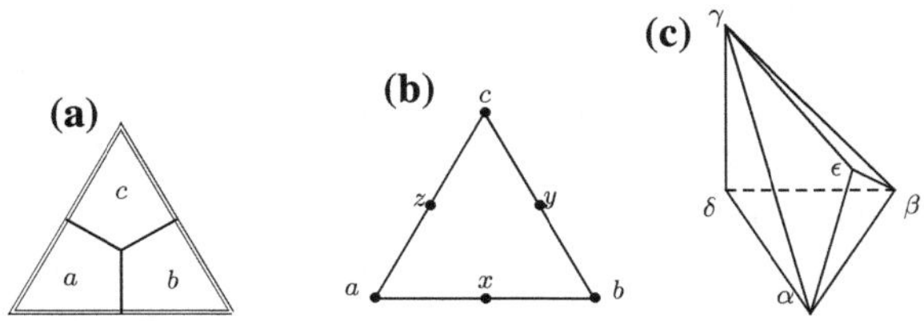

the firefly flashing in chamber a or chamber b, or we might see nothing—the firefly might not be flashing, or might be in chamber c. Thus, we have three experiments, corresponding to the three walls of the box: $\{a, x, b\}$, $\{b, y, c\}$ and $\{c, z, a\}$, where x, y and z are the (distinct) "no-light" outcomes associated with each experiment. The resulting test space $\mathfrak{A} = \{\{a, x, b\}, \{b, y, c\}, \{c, z, a\}\}$ has the Greechie diagram pictured in Fig. 2b.

We can identify several pure states on this test space with concrete situations involving the location, and the internal state (lit or unlit) of the firefly. For example,

$$\alpha(a) = \alpha(y) = 1;\ \alpha(b) = \alpha(c) = \alpha(x) = \alpha(z) = 0$$

corresponds to the firefly's flashing in chamber a. We can define similar states β and γ corresponding to chambers b and c. All of these states are dispersion-free. A fourth dispersion-free pure state, δ, assigns probability 1 to the outcomes x, y and z. This corresponds to the firefly not flashing. These four dispersion-free states separate the six outcomes, and thus allow us, by Lemma 1, to represent the firefly box as a partition test space over a classical state space. *However*, there is also a fifth, *non*-dispersion free pure state, ϵ, given by

$$\epsilon(a) = \epsilon(b) = \epsilon(c) = 1/2;\ \epsilon(x) = \epsilon(y) = \epsilon(z) = 0.$$

This last state is difficult to interpret in any way but to imagine that the firefly *responds* to being observed through a given window by entering (with equal probability) one of the two corresponding chambers. Since any state on this test space is determined by its values at the outcomes x, y and z, the convex set of all probabilty weights for the firefly box is a non-simplicial set in $\mathbb{R}^3$: the pure states α, β and γ correspond to the standard basis vectors $(1, 0, 0)$, $(0, 1, 0)$ and $(0, 0, 1)$, δ corresponds to the origin, and ϵ, to the vector $\frac{1}{2}(1, 1, 1)$. Thus, Ω is affinely isomorpic to a triangular diprism, as pictured in Fig. 2c.

Example 5 (*Grids and Graphs*) Let E be a finite set—for definiteness, say $\{0, 1, \ldots, n-1\}$, with $n \geq 2$. We define two test spaces associated with E:

(a) The *grid test space*, $\mathfrak{Gri}(E)$, consists of all rows and columns of the $n \times n$ array $E \times E$, that is, all sets of the form $\{x\} \times E$ or $E \times \{y\}$.
(b) The *graph test space*, $\mathfrak{Gra}(E)$ consists of the *graphs* of permutations $f : E \to E$, that is, subsets of $E \times E$ of the form $\{(i, f(i)) | i \in E\}$.

Both of these test spaces have outcome-set $X = E \times E$, so a state on either test space can be regarded as an $n \times n$ real matrix with non-negative entries. In the case of $\mathcal{G}ri(E)$, these entries must sum to unity along each row and column; that is, the states on $\mathcal{G}ri(E)$ are exactly the *doubly stochastic matrices*. By the Birkhoff-von Neumann theorem, these all arise as convex combinations of permutation matrices—that is, of the dispersion-free states corresponding to elements of $\mathcal{G}ra(E)$. Similarly, one can show that, for $n \geq 3$, every state of $\mathcal{G}ra(E)$ is an average of *row states*, α^k, given by $\alpha^k(i, j) = \delta_{i,k}$ and *column states* α_k, given by $\alpha_k(i, j) = \delta_{k,j}$.

Every pair of pure states on either $\mathcal{G}ri(E)$ or $\mathcal{G}ra(E)$ is distinguishable by a test in that space. Nevertheless, neither state space is a simplex for $n \geq 3$. The space of doubly-stochastic matrices has $n!$ pure states, which, for $n \geq 4$, exceeds the $n^2 + 1$ states permissible for a simplex in $\mathbb{R}^{n^2}$. For $n \geq 3$, $\mathcal{G}ri(E)$ has only $2n$ pure states; however, the maximally mixed state $\alpha(i, j) \equiv 1/n$, can be represented as a uniform average over just the row states, or over just the column states; similarly, on $\mathcal{G}ri(E)$, it can be represented as a uniform average over any set of permutations the graphs of which partition $E \times E$. By a curious coincidence, the test spaces $\mathcal{G}ri(3)$ and $\mathcal{G}ra(3)$ are isomorphic, so the state space of $\mathcal{G}ra(3)$ is isomorphic to that of $\mathcal{G}ri(3)$, and again, not a simplex.

Remark: We've seen that a variety of convex geometries can arise more or less naturally as the (full) state spaces of test spaces. A natural question is whether *every* possible convex geometry arises in this way. A theorem of F. Shultz [59] shows that in fact, every compact convex set can be represented as the space of probability measures on an orthomodular lattice. The set of decompositions of the unit element in such a lattice is a test space, the probability weights on which correspond precisely to the probability measures on the lattice. Thus, Shultz' theorem implies that every compact convex set can be realized as the full state space of a test space.

Models from Symmetry A *symmetry* of a test space X is a bijection $g : X \to X$ such that both g and g^{-1} preserve tests—in other words, such that for all $E \subseteq X$, we have $gE \in \mathcal{M}(X)$ iff $E \in \mathcal{M}(X)$. (In other words, it is an isomorphism from the test space X to itself.) The set of all symmetries of X is evidently a group, which we'll denote by $G(X)$. There is a natural dual action of $G(X)$ on probability weights on X, given by $g\alpha := \alpha \circ g^{-1}$; a symmetry of a model $A = (X, \Omega)$ is a symmetry of A that also preserves Ω. Again, the symmetries of a model form a group, $G(A) \leq G(X(A))$.

Both classical and quantum test spaces are marked by very strong symmetry properties. In particular, the symmetry group of either kind of system acts transitively on pure states, and also on the set of tests; moreover, any permutation of the outcomes of any given test can be implemented by a symmetry of the entire system. (This is more or less trivial in the case of a classical system; for a quantum system, it amounts to the observation that any permutation of an orthonormal basis for a Hilbert space $\mathcal{H}$ extends to a unitary operator on $\mathcal{H}$.) In contrast, no symmetry of the "firefly box" test space of Example 4 will exchange one of the outcomes a, b, c with one of x, y, z, since each of the former belongs to two tests, while each of the latter belongs only to one.

An *action* of a group G on a test space $(X, \mathcal{M})$ is an action of G on X, with $x \mapsto gx$ a symmetry of $(X, \mathcal{M})$ for every $g \in G$. Recall that an action of G on X is *transitive* iff it has only a single orbit, i.e., for every $x, y \in X$, there exists some $g \in G$ with $y = gx$. The action is *doubly transitive* iff, for any pairs $(x, u), (y, v) \in X^2$, there is some $g \in G$ with $gx = y$ and $gu = v$—i.e., the obvious action of G on X^2 is transitive.

Definition 6 Let G be a group acting by symmetries on a test space X. We say X is *symmetric* under G, or *G-symmetric*, iff G acts transitively on $\mathcal{M}(X)$, and the stabilizer G_E of a test $E \in \mathcal{M}(A)$ acts transitively on E. If X is G-symmetric and G_E acts doubly transitively on E, then X is *2-symmetric* under G. If G_E acts as the full permutation group of E, we say that X is *fully* G-symmetric.

In fact, test spaces with these symmetry properties can be constructed very naturally [76]. Suppose one has a simple measuring device, which can be applied to a system of some sort to produce outcomes in a set E. One might be able to apply this device *in different ways*—for example, by changing the orientation of the apparatus with respect to the system, or by adjusting some controllable physical parameters associated with the system. This suggests that we might be able to build a larger family of experiments—a test space, in other words—starting with the basic measurement E, and adding parameters that keep track of the various ways in which we might deploy it. In many cases, there will be a group G of "physical symmetries" acting on these parameters, and we can often reconstruct the desired test space simply from a knowledge of this group and its relationship to the test E. Specifically, there will be some subgroup H of G that acts to permute the outcomes of E. Let us suppose that H acts transitively on E, so that, for any reference outcome $x_o \in E$, every other outcome $x \in E$ has the form hx_o for some $h \in H$. Let K be any subgroup of G such that $K \cap H = H_0$, where H_0 is the stabilizer in H of a chosen reference outcome $x_o \in E$, and set $X = G/K$. Then there is a well-defined canonical H-equivariant injection $j : E \to X$ given by $j(x) = hK$ where $x = hx_o$. Let us identify E with its image under j, so that $E \subseteq X$. Let $\mathfrak{G}$ be the orbit of E under G, i.e.,

$$\mathfrak{G} := \{gE | g \in G\}.$$

The test space $(X, \mathfrak{G})$ will automatically be symmetric, and will be 2-symmetric or fully symmetric under G as H acts doubly or fully transitively on E. We obtain a G-symmetric *model* by choosing any G-invariant, closed, convex set of probability weights on X.

The choice of the group K extending the stabilizer H_o has a large effect on the combinatorial structure of $(X, \mathfrak{G})$. For example, if $K = H_o$, then $\mathcal{M}$ is a semiclassical test space consisting of disjoint copies of E; in general, a larger choice of K will enforce non-trivial intersections among the tests gE with $g \in G$.

Example 6 As an illustration of this construction, let $E = \{0, 1, \ldots, n-1\}$, and let U be the group of all unitary $n \times n$ matrices, acting in the usual way on $\mathcal{H} = \mathbb{C}^E$. Let $H \leq U$ be the subgroup consisting of permutation matrices, and K, the group of unitaries fixing e_0, the column vector corresponding to $0 \in E$. Then $K \cap H$ is exactly the set of permutation matrices corresponding to permutations fixing 0, i.e., $K \cap H = H_0$. Now $X = G/K$ is the (projective) unit sphere of $\mathcal{H}$, and $\mathcal{M}$ is the set of (projective) frames of $\mathcal{H}$. For another example, let H be the full permutation group $S(E)$ of E and set $G = S(E) \times S(E)$. Embedding H in G by $h \mapsto (h, e)$, the construction above produces the "grid" test space $\mathcal{G}ri(E)$ of Example 6. Using instead the diagonal embedding $h \mapsto (h, h)$ yields the "graph" test space $\mathcal{G}ra(E)$.

2.3 Models Linearized

In many situations, the outcomes of a test space are naturally represented as elements of a vector space. This is obviously the case for the quantum-mechanical examples discussed above, where outcomes are directly identified with unit vectors in $\mathcal{H}$ or with rank-one projections in $\mathcal{L}(\mathcal{H})$. One can also formulate classical probability theory in this way, by considering the vector space of random variables associated with a given measurable space, and identifying measurement outcomes (that is, measurable sets) with the corresponding indicator random variables.

Subject to mild restrictions, such representations are always available. Attached to every probabilistic model A there is a canonical *ordered* real vector space $\boldsymbol{E}(A)$, generated by A's outcomes, with every state of the model represented by a positive linear functional on $\boldsymbol{E}(A)$. There is also a canonical ordered vector space $\mathbf{V}(A)$, generated by A's state-space, such that A's measurement outcomes are represented by elements of $\mathbf{V}(A)^*$. In nice cases, we have $\mathbf{V}(A) \simeq \boldsymbol{E}(A)^*$, whence, in finite-dimensional cases (our focus here), $\mathbf{V}(A)^* = \boldsymbol{E}(A)$, and the two representations coincide. Before discussing these constructions, we pause briefly to review basic facts about ordered vector spaces. For further details, see [2, 32].

Ordered Linear Spaces By a *cone* in a real vector space $\boldsymbol{E}$, we mean a convex subset closed under multiplication by non-negative scalars. K is *pointed*, if $K \cap -K = \{0\}$, and *generating* iff it spans $\boldsymbol{E}$. it spans $\boldsymbol{E}$. An *ordered linear space* is a real vector space $\boldsymbol{E}$, equipped with a closed, pointed, generating cone $\boldsymbol{E}_+$. Such a cone determines a (partial) ordering, invariant under translation and under positive scalar multiplication, on $\boldsymbol{E}$, namely $a \leq b$ iff $b - a \in \boldsymbol{E}_+$.[6] Noticing that $a \geq 0$ iff $a \in \boldsymbol{E}_+$, we refer to $\boldsymbol{E}_+$ as the *positive cone* of $\boldsymbol{E}$.

The basic example is the space $\mathbb{R}^X$ of all real-valued functions on a set X, ordered pointwise. Thus,

$$(\mathbb{R}^X)_+ = \{f \in \mathbb{R}^X \mid f(x) \geq 0 \, \forall x \in X\}.$$

Another example, central to our concerns here, is the space $\mathcal{L}_h(\mathcal{H})$ of bounded *hermitian* operators on a Hilbert space $\mathcal{H}$ (over either $\mathbb{R}$ or $\mathbb{C}$). This space has a

[6] Many authors define ordered linear spaces without requiring that the positive cone be either closed or generating. For our purposes, the present definition is more useful.

standard ordering, induced by the cone $\mathcal{L}_+(\mathcal{H})$ of positive semi-definite operators—that is, $a \in \mathcal{L}_+(\mathcal{H})$ iff $\langle ax, x\rangle \geq 0$ for all vectors $x \in \mathcal{H}$. More generally, the real vector space of self-adjoint elements of a C^*-algebra $\mathcal{A}$ is ordered by the cone of elements of the form aa^*, $a \in \mathcal{A}$.

If $\boldsymbol{E}$ and $\boldsymbol{F}$ are ordered linear spaces, a linear mapping $f : \boldsymbol{E} \to \boldsymbol{F}$ is *positive* iff $f(\boldsymbol{E}_+) \subseteq \boldsymbol{F}_+$, i.e., $f(a) \geq 0$ whenever $a \geq 0$. An *order-isomorphism* between $\boldsymbol{E}$ and $\boldsymbol{F}$ is a positive, invertible linear mapping having a positive inverse. We'll denote the set of positive linear mappings $\boldsymbol{E} \to \boldsymbol{F}$ by $\mathcal{L}_+(\boldsymbol{E}, \boldsymbol{F})$. This is a cone in the space $\mathcal{L}(\boldsymbol{E}, \boldsymbol{F})$. As a special case, the dual space of an ordered vector space $\boldsymbol{E}$ has a natural *dual cone*, $\boldsymbol{E}^*_+ = \mathcal{L}_+(\boldsymbol{E}, \mathbb{R})$. In our present finite-dimensional setting, this is generating, so $\boldsymbol{E}^*$ becomes an ordered vector space in a natural way.

Order-unit spaces An *order-unit* in an ordered vector space $\boldsymbol{E}$ is an element $u \in \boldsymbol{E}_+$ such that, for every $a \in \boldsymbol{E}$, there exists some $\lambda > 0$ with $a \leq \lambda u$. When $\boldsymbol{E}$ is finite-dimensional, this is equivalent to u's belonging to the interior of $\boldsymbol{E}_+$, or to the condition that $\alpha(u) > 0$ for every non-zero $\alpha \in \boldsymbol{E}^*_+$. An *order-unit space* is an ordered linear space equipped with a distinguished order-unit.[7] The key example to bear in mind is the space $\mathcal{L}_h(\mathcal{H})$, ordered as described above, and with the identity operator as order-unit.

An order unit space already provides enough structure to support probabilistic ideas. A *state* on an order-unit space $\boldsymbol{E}$ is a linear functional $\alpha \in \boldsymbol{E}^*$ with $\alpha(u) = 1$. We write $\Omega(\boldsymbol{E})$ for the set of all states on $\boldsymbol{E}$. This is easily seen to be a compact convex subset of $\boldsymbol{E}^*$. An *effect* in $\boldsymbol{E}$ is an element $a \in \boldsymbol{E}$ with $0 \leq a \leq u$, so that $0 \leq \alpha(a) \leq 1$ for every state α. The set of all effects is denoted $[0, u]$. A discrete *observable* on $\boldsymbol{E}$ is a finite set $E = \{a_1, \ldots, a_k\}$ of non-zero effects with $a_1 + \cdots + a_k = u$. Evidently, any state on $\boldsymbol{E}$ restricts to a probability weight on every observable on $\boldsymbol{E}$.

In the special case where $\boldsymbol{E} = \mathcal{L}_h(\mathcal{H})$, the space of Hermitian operators on a Hilbert space $\mathcal{H}$, an effect is a positive operator a with $0 \leq a \leq \mathbf{1}$; all states have the form $\alpha(a) = \mathrm{Tr}(Wa)$ where W is a density operator on $\mathcal{H}$, and an observable is essentially a (discrete) positive-operator valued measure.

Remark: We can regard the set $\mathcal{O}(\boldsymbol{E})$ of observables of $\boldsymbol{E}$ as forming a test space, the outcomes of which are the non-zero effects. This gives us a probabilistic model

$$A_{\max}(\boldsymbol{E}) := ((0, u], \mathcal{O}(\boldsymbol{E}), \Omega(\boldsymbol{E})),$$

which we call the *maximal model* associated with $\boldsymbol{E}$.

As noted above, the state space $\Omega(\boldsymbol{E})$ of an order-unit space is a compact convex set. Dually, let K be a compact convex subset of a finite-dimensional real vector space $\mathbf{W}$. Let $\mathbf{V}(K)$ denote K's span in $\mathbf{W}$, and let $\mathrm{Aff}(K)$ denote the space of affine real-valued functionals $f : K \to \mathbb{R}$. This last is an order-unit space, with order-unit given by $u(\alpha) \equiv 1$ for all $\alpha \in K$. One can show that any affine mapping $T : K \to \mathbf{M}_+$,

[7] In general, one must add the requirement that $\boldsymbol{E}$'s ordering be *archimedean*, meaning that if $x, y \in \boldsymbol{E}$ with $0 \leq nx \leq y$ for all $n \in \mathbb{N}$, then $x = 0$. However, in our finite-dimensional setting, any closed cone induces an archimedean ordering.

where $\mathbf{M}$ is any ordered vector space, extends uniquely to a positive linear mapping $T : \mathbf{V}(K) \to \mathbf{M}$. In particular, every element of $\text{Aff}(K)_+$ extends uniquely to positive linear functional on $\mathbf{V}(K)$. This gives us a canonical isomorphism $\mathbf{V}(K)^* \simeq \text{Aff}(K)$, and hence (in our finite-dimensional setting), an isomorphism $\mathbf{V}(K) \simeq \text{Aff}(K)^*$. In fact, K is naturally embedded in $\text{Aff}(K)^*$ by evaluation, so we can treat K as a subset of $\text{Aff}(K)^*$ (rather than $\mathbf{W}$) and identify $\mathbf{V}(K)$ with $\mathbf{V}(K)^*$.

Linear representations of a model Let $(X, \mathcal{M})$ be a test space and $(\boldsymbol{E}, u)$ an order-unit space. A *representation* of $(X, \mathcal{M})$ on $\boldsymbol{E}$ is a mapping $\phi : X \to \boldsymbol{E}_+$ such that $\sum_{x \in E} \phi(x) = u$ for all tests $E \in \mathcal{M}$. Evidently, any state $\alpha \in \boldsymbol{E}^*$ pulls back to a probability weight $\alpha \circ \phi$ on $(X, \mathcal{M})$. The set of all probability weights arising in this way is compact (since ϕ is linear, hence continuous) and convex, and thus, defines a probabilistic model. By a representation of a probabilistic model A, we mean a representation of $(X(A), \mathcal{M}(A))$ such that every state of $\Omega(A)$ arises in this way from a state on $\boldsymbol{E}$.

In effect, a representation allows us to interpret the test space $(X(A), \mathcal{M}(A))$ as a collection of observables on $\boldsymbol{E}$. As we'll now see, every probabilistic model has a canonical representation in this sense. Let $\mathbf{V}(A) = \mathbf{V}(\Omega(A))$ be the span of $\Omega(A)$ in $\mathbb{R}^{X(A)}$, ordered pointwise on $X(A)$. Then, as discussed above, $\mathbf{V}(A)^* \simeq \text{Aff}(\Omega(A))$ is an order-unit space, with order-unit u_A defined by $u_A(\alpha) \equiv 1$ for all states $\alpha \in \Omega(A)$, and we can identify $\Omega(A)$ with the state-space of $(\mathbf{V}^*(A), u_A)$. Every outcome $x \in X(A)$ defines an effect $\hat{x} \in \mathbf{V}(A)^*$ by evaluation, i.e., $\hat{x}(\alpha) = \alpha(x)$ for all $\alpha \in \mathbf{V}(A)$; moreover, if $E \in \mathcal{M}(A)$, we have $(\sum_{x \in E} \hat{x})(\alpha) = \sum_{x \in E} \alpha(x) = 1$, for all $\alpha \in \Omega(A)$, so $\sum_{x \in E} \hat{x} = u_A$, so $\hat{\cdot} : X(A) \to \mathbf{V}(A)$ is in fact a representation.

Given our standing assumption that models are locally finite, we can define a smaller representing space. Let $\boldsymbol{E}(A)$ be the span of $X(A)$ in $\mathbf{V}(A)^* = \text{Aff}(\Omega(A))$, ordered by the closure of the cone consisting of linear combinations with non-negative coefficients of evaluation functionals $\hat{x}$, $x \in X(A)$:

$$\boldsymbol{E}(A)_+ := \text{cl}\left(\left\{ \sum_i t_i \hat{x}_i \; | x_i \in X, \; t_i \geq 0 \right\}\right).$$

Since A is locally finite, we see that $u_A = \sum_{x \in E} \hat{x}$, where E is any test in $\mathcal{M}(A)$. Hence, u_A belongs to $\boldsymbol{E}_+(A)$, where it continues to function as an order-unit. (To see this, note that if $\alpha \in \boldsymbol{E}(A)^*_+$ and $\alpha(u) = 0$, then $\alpha(\hat{x}) = 0$ for every $x \in X(A)$, whence, $\alpha = 0$.) The space $\boldsymbol{E}(A)$, together with the representation $\hat{\cdot} : X(A) \to \boldsymbol{E}(A)$, is called the *linear hull* of the model A.

Since $\boldsymbol{E}(A) \leq \mathbf{V}(A)^*$ are both spanned by $X(A)$, they have the same dimension, hence, coincide *as vector spaces*. However, the cone of $\boldsymbol{E}(A)$ is generally smaller than that of $\mathbf{V}(A)^*$. Thus, we have canonical positive linear bijections $\boldsymbol{E}(A) \to \mathbf{V}(A)^*$ and, dually, $\mathbf{V}(A) \to \boldsymbol{E}(A)^*$.

Example 7 Let $A = A(\mathcal{H})$ be a finite-dimensional quantum model, as discussed in Example 2. Let us write $\mathbf{V}(A(\mathcal{H}))$ and $\boldsymbol{E}(A(\mathcal{H}))$ as $\mathbf{V}(\mathcal{H})$ and $\boldsymbol{E}(\mathcal{H})$, for short. Remarkably, here all three of the ordered vector spaces $\mathbf{V}(\mathcal{H})$, $\mathbf{V}(\mathcal{H})^*$, and $\boldsymbol{E}(\mathcal{H})$

coincide, all being isomorphic to $\mathcal{L}_h(\mathcal{H})$. Since every hermitian operator in $\mathcal{H}$ is a difference of positive operators, each of which normalizes (in finite dimensions) to a density operator, we have $\mathcal{L}_h(\mathcal{H}) \simeq \mathbf{V}(\mathcal{H})$. The dual of $\mathcal{L}_h(\mathcal{H})$, as an ordered vector space, is isomorphic to $\mathcal{L}_h(\mathcal{H})$: every positive linear functional on $\mathcal{L}_h(\mathcal{H})$ has the form $a \mapsto \text{Tr}(ba)$ for a unique positive hermitian operator b. Hence, $\mathbf{V}(A)^* \simeq \mathcal{L}_h(\mathcal{H})$ as well. Finally, the Spectral Theorem tells us that every positive hermitian operator on $\mathcal{H}$ is a positive linear combination of rank-one projections. It follows that the space $\boldsymbol{E}(\mathcal{H})$ can also be identified with $\mathcal{L}_h(\mathcal{H})$.

State-Completeness The state space of $\mathbf{V}(A)^*$ is exactly $\Omega(A)$. However $\boldsymbol{E}(A)$, having a smaller positive cone than $\mathbf{V}(A)^*$, has a larger state space. Still, as noted above, any state $\alpha \in \boldsymbol{E}(A)^*$ defines a probability weight on $(X(A), \mathcal{M}(A))$. Letting $\widehat{\Omega}(A)$ denote the set of such states, we have $\Omega(A) \subseteq \widehat{\Omega}(A)$. We may regard $\widehat{\Omega}(A)$ as the set of probability weights that are consistent with all of the linear relations among outcomes that are satisfied by the given state space $\Omega(A)$. Call a model A *state-complete* iff $\Omega(A) = \widehat{\Omega}(A)$.

Lemma 2 *Let* $A = (X, \Omega)$ *be a finite-dimensional probabilistic model. Then the following are equivalent:*

(a) A *is state-complete*
(b) $\boldsymbol{E}(A)_+ = \boldsymbol{E}(A) \cap \mathit{Aff}_+(\Omega) \simeq \boldsymbol{E}(A) \cap \mathbf{V}(A)^*_+$;
(c) The canonical mapping $\mathbf{V}(A)_+ \to \boldsymbol{E}(A)^*_+$ *is surjective, so that* $\mathbf{V}(A)$ *and* $\boldsymbol{E}(A)^*$ *are order-isomorphic.*

Proof Clearly, (c) implies (a). To see that (a) implies (b), suppose $f \in \text{Aff}(\Omega)_+ \setminus \boldsymbol{E}(A)_+$. Then (by the finite-dimensional version of the Hahn-Banach separation theorem) there exists some $\alpha \in \boldsymbol{E}(A)^*$ with $\alpha(a) \geq 0$ for all $a \in \boldsymbol{E}(A)_+$ but $\alpha(f) < 0$. We can normalize α so that $\alpha(u) = 1$, in which case $\alpha \in \widehat{\Omega}$. Since f is non-negative on Ω, it follows that $\alpha \notin \Omega$, whence, $\widehat{\Omega} \neq \Omega$, and A is not state-complete. To see that (b) implies (c), suppose $\alpha \in \widehat{\Omega} \setminus \Omega$: then we can find some $f \in \boldsymbol{E}(A)^{**} = \boldsymbol{E}(A)$ with $f(\alpha) < 0$ but $f(\beta) \geq 0$ for all $\beta \in \Omega$. But now $f \in \boldsymbol{E} \cap \text{Aff}_+(\Omega)$, and yet—as $a(\alpha) \geq 0$ for all $a \in \boldsymbol{E}(A)_+$—we have $f \notin \boldsymbol{E}(A)_+$. □

Standing Assumption: *Henceforth,* ***all models are state-complete***.

Accordingly, we can, and shall, write $\boldsymbol{E}(A)$ for $\mathbf{V}(A)^*$.

Remarks: The idea of treating order-unit spaces—or rather, their intervals of effects—as a model for the set of events of an abstract probabilistic (or physical) system, with normalized positive linear functionals playing the role of states, goes back at least to the work of Ludwig and his school [48], and was also developed by Davies and Lewis [29], Edwards [31], and many others. Evidently, this approach—often referred to as *operational* or *convex*—is recovered in the present framework. Indeed, it is possible to regard the additional test-space structure of a probabilistic model as largely dispensible: a kind of builders' scaffolding, to be discarded once the spaces $\mathbf{V}(A)$ and $\mathbf{V}(A)^*$ have been obtained. For many applications, this works perfectly well. However, this additional test-space structure turns out to be useful in many ways, so we prefer to retain it.[8]

Direct Sums of Models A *face* of a convex set K is a convex subset $J \subseteq K$ such that, for all $a, b \in K$ and all $0 \leq t \leq 1$,

$$ta + (1 - t)b \in J \Rightarrow a \in J \text{ and } b \in J.$$

If J and K are cones, then this is equivalent to the condition that $a+b \in J \Rightarrow a \in J$ and $b \in J$. A minimal face of a cone is in fact a ray; we more usually speak of an *extremal ray*. An element of a cone is *ray-extremal*, or simply *extremal*, iff it generates an extremal ray. In finite dimensions, every (closed) cone is the convex hull of its extremal elements.

The *direct sum* of two ordered vector spaces $\boldsymbol{E}$ and $\boldsymbol{F}$ is their vector-space direct sum, $\boldsymbol{E} \oplus \boldsymbol{F}$, equipped with the cone $\boldsymbol{E}_+ \oplus \boldsymbol{F}_+$ consisting of all sums of positive elements from each. This is the smallest cone in $\boldsymbol{E} \oplus \boldsymbol{F}$ making the standard embeddings $\boldsymbol{E}, \boldsymbol{F} \to \boldsymbol{E} \oplus \boldsymbol{F}$ given by $a \mapsto (a, 0)$ and $b \mapsto (0, b)$ (for $a \in \boldsymbol{E}$ and $b \in \boldsymbol{F}$) positive. In this case, $\boldsymbol{E}_+$ and $\boldsymbol{F}_+$ are both faces of $\boldsymbol{E}_+ \oplus \boldsymbol{F}_+$. $\boldsymbol{E}$ is *irreducible* iff not a direct sum.

If X and Y are sets, we write $X \oplus Y$ for their coproduct (or disjointified union),

$$X \oplus Y = \{1\} \times X \cup \{2\} \times Y.$$

If X and Y are test spaces, we make $X \oplus Y$ into a test space by letting $\mathcal{M}(X \oplus Y)$ equal the set $\{E \oplus F | E \in \mathcal{M}(X), F \in \mathcal{M}(Y)\}$. We can understand a test of the form $E \oplus F$ as a two-stage test: first, perform the classical two-outcome test $\{1, 2\}$ (by flipping a coin, say); if the result is 1, measure E, if the result is 2, measure F. A probability weight ω on $X \times Y$ corresponds to an arbitrary choice of a probability weight p on $\{1, 2\}$ and probability weights $\alpha \in \Omega(X)$ and $\beta \in \Omega(Y)$, by

$$\omega(1, x) = p(1)\alpha(x) \text{ and } \omega(2, y) = p(2)\beta(y).$$

The weights p, α and β are uniquely determined by ω, so we can unambiguously write

$$\omega = p\alpha + (1 - p)\beta$$

In other words, $\Omega(X \oplus Y) = \Omega(X) \oplus \Omega(Y)$, whence, $\boldsymbol{E}(X \oplus Y) = \boldsymbol{E}(X) \oplus \boldsymbol{E}(Y)$.

Every discrete classical probabilistic model $(E, \Delta(E))$ is a direct convex sum of trivial models $(\{x\}, \delta_x)$ where $x \in E$ and $\delta_x(x) = 1$. In contrast, the basic quantum

[8]One of many uses for the test space structure is to privilege certain classes of observables on an order-unit space having special order-theoretic properties—for example, the set of observables the outcomes of which lie on extremal rays of $\boldsymbol{E}_+$ forms a test space, or those whose outcomes are atomic effects, i.e., those that lie on extremal rays of $\boldsymbol{E}_+$ *and* are extreme points of $[0, u]$.

model $(X(\mathcal{H}), \Omega(\mathcal{H}))$ is irreducible. More general quantum models associated with matrix algebras (representing quantum systems with superselection rules) arise as direct sums of irreducible quantum models.

2.4 Processes and Categories

In very broad terms, a *probabilistic theory* might be nothing more than a class of probabilistic models. But this usage is really much too broad. Part of the job of a theory is to tell us, not only which models represent "actual" systems, but also something about how such systems can change. In order to speak about systems changing, we need to introduce into the preceding formalism a notion of *process*. A natural place to start is with the idea of a mapping $\phi : \alpha \mapsto \phi(\alpha)$ taking states α of an initial (or input) system A to states of a final (output) system B. To allow for "lossy" processes or processes conditioned on some event, we should permit $\phi(\alpha)$ be be a sub-normalized state of B when α is a normalized state of A. Finally, since randomizing the input state should randomize the output state in the same way, we should expect this ϕ be an affine mapping. Thus, we model a process from A to B by an affine mapping $\phi : \Omega(A) \to \mathbf{V}(B)$ with $u_B(\phi(\alpha)) \leq 1$; or, what is the same thing, by a positive linear mapping $\phi : \mathbf{V}(A) \to \mathbf{V}(B)$ with $u_B \circ \phi \leq u_A$. We can interpret $u_B(\phi(\alpha))$ as the *probability* that ϕ occurs when the initial state is α—or, perhaps more accurately, as the probability that the process occurs, *if* initiated.

To every process $\phi : \mathbf{V}(A) \to \mathbf{V}(B)$, there corresponds a *dual process* $\tau = \phi^* : \boldsymbol{E}(B) \to \boldsymbol{E}(A)$, given by $\phi^*(b) = b \circ \phi$ for any $b \in \boldsymbol{E}(B)$. Operationally, to measure $\phi^*(b)$ on a state α, one first subjects the state α to the process ϕ, and then makes a measurement of the effect b. Note that $\phi^*(u_B)(\alpha) = u_B(\phi(\alpha))$ is the probability that the process ϕ occurs if the initial state is α. In what follows, it will often be more convenient mathematically to deal with these dual processes. In other words, to use physicists' lingo, we'll often work with the "Heisenberg" rather than the "Schrödinger" picture of processes.

Not every positive linear mapping $\mathbf{V}(A) \to \mathbf{V}(B)$ will generally count as a process. As remarked above, it is part of the job of a probabilistic theory to specify those that do. However, it seems reasonable to require that convex combinations of processes and composites of (composable) processes also count as processes. It will also be convenient to assume that, for every pair of systems A and B, there is a *null process* that takes every state $\alpha \in \Omega(A)$ to the *zero state* $0 \in \boldsymbol{E}(B)$. It seems reasonable, also, that there exist a canonical *trivial* system I, corresponding to a test space with only a single outcome, 1, and a single test $\{1\}$. We then have $\boldsymbol{E}(I) = \boldsymbol{E}(I)^* = \mathbb{R}$. We can then require that, for every normalized state $\alpha \in \mathbf{V}(A)$, there exist a process $\mathbb{R} \to \mathbf{V}(A)$ of *preparation*, given by $1 \mapsto \alpha$, and, for every outcome $x \in X(A)$, a process $\mathbf{V}(A) \to \mathbb{R}$ of *registration*, sending $\alpha \in \mathbf{V}(A)$ to $\alpha(x)$. The dual process corresponding to the preparation of α is simply the state α itself, while the process dual to the registration of x is the linear mapping $\mathbb{R} \to \boldsymbol{E}(A)$ sending 1 to x. All of this suggests the following

Definition 7 A (state-complete) **probabilistic theory**[9] is a category $\mathcal{C}$ such that

(1) Every object $A \in \mathcal{C}$ is a probabilistic model;
(2) For all $A, B \in \mathcal{C}$, the set $\mathcal{C}(A, B)$ of morphisms $A \to B$ is a closed, convex subset of $\mathcal{L}_+(\boldsymbol{E}(A), \boldsymbol{E}(B))$, containing the zero mapping, and with $\tau(u_A) \leq u_B$ for all $\tau \in \mathcal{C}(A, B)$;
(3) There is a distinguished *trivial system* I with $\boldsymbol{E}(I) = \mathbb{R}$ and $X = \{1\}$, such that for every $A \in \mathcal{C}$, $X(A) \subseteq \mathcal{C}(I, A)$ and $\Omega(A) \subseteq \mathcal{C}(A, I)$.
(4) The order unit $u_A \in \boldsymbol{E}(A)$ belongs to $\mathcal{C}(I, A)$.

From now on, we work in a fixed probabilistic theory $\mathcal{C}$ ***of this kind.*** We write $\mathcal{C}^*$ for the category having the same objects, but with morphisms $\mathcal{C}^*(A, B)$ the set of mappings $\phi = \tau^* : \mathbf{V}(A) \to \mathbf{V}(B)$ with $\tau \in \mathcal{C}(B, A)$. In effect, $\mathcal{C}$ and $\mathcal{C}^*$ offer, respectively, the "Heisenberg" and the "Schrödinger" picture of the same theory. Depending on context, we shall understand the word "process" to refer either to a morphism $\tau \in \mathcal{C}(A, B)$ for some $A, B \in \mathcal{C}$, or to the dual mapping $\phi = \tau^* : \mathbf{V}(B) \to \mathbf{V}(A)$.

Example 8 By a *standard finite-dimensional quantum theory*, we mean a category $\mathcal{C}$ of quantum-probabilistic models associated with hermitian parts of finite-dimensional complex matrix algebras (equivalently, direct sum of algebras of the form $\mathcal{L}(\mathcal{H})$), with trace-nonincreasing completely positive mappings as morphisms. In this formulation, classical probabilistic theories arise as the degenerate case in which all of the matrix algebras associated with systems in $\mathcal{C}$ are commutative.

Reversible and Probabilistically Reversible Processes A process $\tau \in \mathcal{C}(A, B)$ is *reversible* iff it is invertible as a morphism in $\mathcal{C}$, i.e., there exists an inverse process $\tau^{-1} \in \mathcal{C}(B, A)$ with $\tau^{-1} \circ \tau = \mathrm{id}_A$ and $\tau \circ \tau^{-1} = \mathrm{id}_B$. In this case, $\tau : \boldsymbol{E}(A) \to \boldsymbol{E}(B)$ is an order-automorphism and $\tau^{-1} : \boldsymbol{E}(B) \to \boldsymbol{E}(A)$ is the inverse isomorphism. Moreover, for such a process, we have $\tau(u_A) = u_B$: by assumption, $\tau(u_A) \leq u_B$, and also $\tau^{-1}(u_B) \leq u_A$, whence, as τ preserves order, $u_B \leq \tau(u_A)$. Dually, a process $\phi \in \mathcal{C}^*(A, B)$ is reversible iff it has an inverse in $\mathcal{C}^*(B, A)$; equivalently, ϕ is reversible iff the dual process $\tau = \phi^*$ is reversible. In this case, we have $u_B\phi(\alpha) = 1$ for every normalized state $\alpha \in \Omega(A)$.

There is a weaker but very useful notion, which we shall call *probabilistic reversibility*. This is slightly easier to describe in terms of processes acting on states, rather than effects:

Definition 8 A process $\phi \in \mathcal{C}^*(A, B)$, is *probabilistically reversible* iff it is invertible as a linear mapping $\mathbf{V}(A) \to \mathbf{V}(B)$, with a positive inverse *and* if the inverse mapping ϕ^{-1} is a positive multiple of a process $\phi_o \in \mathcal{C}^*(B, A)$—say, $\phi^{-1} = c\phi_o$ with $c > 0$. (in which case, notice, $c > 1$!)

[9]This definition differs from that of [17], most obviously in that objects are associated with effect spaces, rather than state spaces, but also in taking the test space $X(A)$ to be part of the structure of $A \in \mathcal{C}$.

Operationally, this means that there is some non-zero probability that $\phi_o \circ \phi$ will return the system to its original state. Indeed,

$$u_A(\phi_o(\phi(\alpha))) = c^{-1}u_A(\phi^{-1}(\phi(\alpha))) = c^{-1}u_A(\alpha) = c^{-1},$$

so this probability is exactly $1/c$. In particular, ϕ is reversible with probability one iff $c = 1$, so that ϕ^{-1} is a process in $\mathcal{C}^*(B, A)$—in other words, ϕ is a reversible process. Obviously, the set of probabilistically reversible processes, in either $\mathcal{C}(A, A)$ or $\mathcal{C}^*(A, A)$, is a group, containing, but larger than, the group of all reversible processes on A.

Historical remarks: The representation of what we are calling probabilistic models in terms of an order-unit space and its dual goes back at least to the work of Davies and Lewis [29] and Edwards [31]. A good survey of the relevant functional analysis can be found in [2]. Test spaces—originally called "manuals"—were the basis for a generalized probability theory (and an associated "empirical logic") developed in the 1970s and 1980s by C. H. Randall and D. J. Foulis and their students. See [75] for a survey. Mathematically, of course, a test space is just a hypergraph; the current terminology serves only to reinforce the intended probabilistic interpretation of the nodes as outcomes and the hyper-edges as outcome-sets for various tests.

3 Composition and Entanglement

Consider two systems, A and B, which are not interacting in any obvious, causal sense—for example, systems occupying space-like separated regions of space-time. In this situation, it seems reasonable to assume that what that can be *happen* to each system idividually—the preparation of a state, the making of a measurement, etc.—can happen together, independently.

Another natural (albeit more contingent) requirement is a *no-signaling* condition, forbidding the transmission of information from A to B, or vice versa, by the mere decision to make one measurement rather than another on A, or on B. As we'll see, the phenomenon of *entanglement*, one of the supposed hallmarks of quantum theory, is actually a rather generic feature of composite systems in non-classical probabilistic theories, whether "quantum" or otherwise. (Indeed, the phenomenon even arises in otherwise quite classical theories involving a restricted set of probability weights.)

3.1 Composites of Models

Suppose two parties—Alice and Bob, say—control, respectively, systems A and B, which occur as components of some composite system AB, but are still sufficiently isolated to be prepared and measured separately. At a very minimum, we would expect Alice's making a measurement, E, on here part of the composite system, and Bob's making a measurement, F, on his part, *constitutes* the making of a measurement on the combined system. We would also expect that states of the two component systems can be prepared independently. Formalizing these requirements, we arrive at the following:

Definition 9 A *composite* of two probabilistic models A and B is a model AB, together with a mapping

$$X(A) \times X(B) \to X(AB) : \ (x, y) \mapsto xy$$

such that

(i) for all tests $E \in \mathcal{M}$ and $F \in B$, the *product test* $EF := \{xy | x \in E, y \in F\}$ belongs to $\mathcal{M}(AB)$; and
(ii) for all states $\alpha \in \Omega(A)$ and $\beta \in \Omega(B)$, there exists a *product state* $\alpha \otimes \beta \in \Omega(AB)$ with $(\alpha \otimes \beta)(xy) = \alpha(x)\beta(y)$.

Remarks: There are several ways in which we might plausibly weaken this definition. For instance, we might require only that the product outcome xy be an *effect* in $\boldsymbol{E}(AB)_+$, and the set EF, an observable, but not necessarily a test, of AB.[10] Such possibilities are worth bearing in mind. However, for the purposes of this survey, it seems reasonable to use the more restrictive, but therefore simpler, definition above. Note in (ii) we require only the existence, but not the uniqueness of product states (where a product state for α and β is defined as a state γ with $\gamma(xy) = \alpha(x)\beta(y)$.

The injectivity of the mapping $x, y \mapsto xy$ in condition (i) allows us to identify $X(A) \times X(B)$ with the set of *product outcomes* in Z. Let us write

$$X(A)X(B) := \{xy | x \in X, y \in Y\}.$$

With a slight abuse of notation, we may write $\mathcal{M}(A) \times \mathcal{M}(B)$ for the test space consisting of product tests EF. Condition (i) asserts that $\mathcal{M}(A) \times \mathcal{M}(B)$ is contained in $\mathcal{M}(AB)$, so every state in $\Omega(AB)$ restricts to a state ω_o on the former. Where the restricted state ω_o *determines* the global state ω—that is, where the set $X(A)X(B)$ of product outcomes is state-separating—we say that the composite is *locally tomographic*. In this setting, the joint probabilities of outcomes of measurements on the component systems A and B, completely determine the state of the composite.[11] This is a reasonable, but also a rather strong, restriction. Indeed, while composites in standard complex QM are locally tomographic, this is not the case for real or quaternionic QM. We'll return to this matter below.

---

[10] More radically, one might consider models of systems interacting in such a way that the making of a particular measurement, or the preparation of a particular state, on one component, *precludes* the making of certain measurements, or the preparation of certain states, on the other component. Mathematically, such situations are certainly possible.

[11] Barrett [19] calls this the *global state hypothesis*; the term *locally tomographic* seems to have become more standard.

Example 9 (*Composite quantum models*) If $A(\mathcal{H})$ and $A(\boldsymbol{K})$ are two quantum-mechanical models, associated with finite-dimensional Hilbert spaces $\mathcal{H}$ and $\boldsymbol{K}$, respectively, let

$$A(\mathcal{H})A(\boldsymbol{K}) \; = \; A(\mathcal{H} \otimes \boldsymbol{K})$$

the model associated with $\mathcal{H} \otimes \boldsymbol{K}$. That is, $\mathcal{M}(\mathcal{H} \otimes \boldsymbol{K})$ consists of orthonormal bases for $\mathcal{H} \otimes \boldsymbol{K}$, while $\Omega(\mathcal{H} \otimes \boldsymbol{K})$ consists of density operators on $\mathcal{H} \otimes \boldsymbol{K}$. If $x \in \mathcal{H}$ and $y \in \boldsymbol{K}$ are unit vectors, then $x \otimes y$ is a unit vector in $\mathcal{H} \otimes \boldsymbol{K}$. It is easy to check that $x, y \mapsto x \otimes y$ makes $A(\mathcal{H} \otimes \boldsymbol{K})$ into a composite in the sense of the preceding definition.

3.2 Non-Signaling Composites and Entanglement

The very broad definition of a composite system given above leaves room for situations in which the probability of Bob's obtaining an outcome y will depend on which test $E \in \mathcal{M}(A)$ Alice chooses to measure. This is plausible only in scenarios in which Alice's measurements are able physically to disturb Bob's system. If we wish to model composites in which the two systems A and B are sufficiently isolated from one another that this kind of remote disturbance is ruled out—the obvious situation being one in which A and B are spacelike separated—then we must impose a further constraint.

Definition 10 A probability weight ω on $\mathcal{M}(A) \times \mathcal{M}(B)$ is *non-signaling* iff it has well-defined *marginal* (or *reduced*) states, in the sense that

$$\omega_1(x) := \sum_{y \in F} \omega(xy) \;\text{ and }\; \omega_2(y) := \sum_{x \in E} \omega(xy)$$

are independent of the choice of tests $E \in \mathcal{M}(A)$, $F \in \mathcal{M}(B)$.

If $\omega \in \Omega(AB)$ is non-signaling, then for every $y \in X(B)$ and $x \in X(A)$, we can define the *conditional states* $\omega_{1|y}$ and $\omega_{2|x}$ on A and B, respectively, by

$$\omega_{1|y}(x) := \frac{\omega(xy)}{\omega_2(y)} \text{ and } \omega_{2|x}(y) := \frac{\omega(xy)}{\omega_1(x)}.$$

These are well-defined probability weights on $\mathcal{M}(A)$ and $\mathcal{M}(B)$, respectively. It would seem reasonable to include them in the state spaces of A and B. Therefore, we adopt the following language:

Definition 11 A *non-signaling composite* of A and B is a composite AB in which all states are non-signaling, *and* all conditional states belong to the designated state spaces of A and B—that is, $\omega_{2|x} \in \Omega(B)$ and $\omega_{1|y} \in \Omega(A)$ for all $x \in X(A)$ and $y \in X(B)$.

This has a strong consequence [71]:

Lemma 3 (Bi-Linearization) *Let AB be a non-signaling composite of A and B. Then every state $\omega \in \Omega(AB)$ extends uniquely to a bilinear form on $\boldsymbol{E}(A) \times \boldsymbol{E}(B)$.*

Proof For every $x \in X(A)$, define $\hat{\omega}(x) \in \mathbb{R}^{X(B)}$ by $\hat{\omega}(x)(y) = \omega(x, y)$. Notice that $\omega_{2|x} = \hat{\omega}(x)/\omega_1(x)$. Since the conditional state $\omega_{2|x}$ belongs to $\Omega(B)$, we have $\hat{\omega}(x) \in \mathbf{V}(B) = \boldsymbol{E}(B)^*$, with $\sum_{x\in E} \hat{\omega}(x) = \omega_2$. Dualizing (and remembering that $\boldsymbol{E}(A)$ is finite-dimensional), we have a linear mapping $\hat{\omega}^* : \boldsymbol{E}(B) \to \mathbb{R}^{X(A)}$. Now, $\hat{\omega}^*(y) = \omega_2(y)\omega_{1|y}$; the latter belongs to $\Omega(A)$, so $\hat{\omega}^*(y) \in \mathbf{V}(A) = \boldsymbol{E}(A)^*$ for every $y \in X(B)$. Since $X(B)$ spans $\boldsymbol{E}(B)$, it follows that the range of $\hat{\omega}^*$ lies in $\mathbf{V}(A)$, i.e., we can regard $\hat{\omega}^*$ as a linear mapping $\boldsymbol{E}(A) \to \mathbf{V}(B) = \boldsymbol{E}(B)^*$. Equivalently, we have a bilinear form $\mathcal{B}_\omega(a, b) = \hat{\omega}^*(b)(a)$, which evidently satisfies $B_\omega(x, y) = \omega(xy)$ for all $x \in X(A)$, $y \in X(B)$. Since $X(A)$ and $X(B)$ span $\boldsymbol{E}(A)$ and $\boldsymbol{E}(B)$, the form $\mathcal{B}_\omega$ is uniquely determined by this property. □

It follows that, for a non-signaling composite, the mapping $X(A) \times X(B) \to X(AB) : x, y \mapsto xy$ gives rise to a linear mapping $\otimes : \boldsymbol{E}(A) \otimes \boldsymbol{E}(B) \to \boldsymbol{E}(AB)$, with $\omega(x \otimes y) = \mathcal{B}_\omega(x, y) = \omega(xy)$ for every $\omega \in \boldsymbol{E}(AB)^*$. The composite AB is locally tomographic iff this mapping is surjective.

Corollary 1 *A non-signaling composite AB of models A and B is locally tomographic iff $\boldsymbol{E}(AB) \simeq \boldsymbol{E}(A) \otimes \boldsymbol{E}(B)$, that is,* $\dim(\boldsymbol{E}(AB)) = \dim(\boldsymbol{E}(A)) \dim(\boldsymbol{E}(B))$.

Lemma 3 allow us to extend the definition of conditional states to arbitrary effects, setting

$$\omega_{1|b}(a) = \omega(a \otimes b)/\omega(u \otimes b) \text{ and } \omega_{2|a}(b) = \omega(a \otimes b)/\omega(a \otimes u)$$

for arbitrary effects $a \in \boldsymbol{E}(A)$ and $b \in \boldsymbol{E}(B)$ (with the usual proviso about division by zero). The following bipartite version of the law of total probability is easily verified:

Lemma 4 (Law of Total Probability) *Let AB be a non-signaling composite of A and B; let ω be any state on AB, and let E and F be any two observables on $\boldsymbol{E}(A)$ and $\boldsymbol{E}(B)$, respectively, then*

$$\omega_2 = \sum_{a\in E} \omega_1(a)\omega_{2|a} \text{ and } \omega_1 = \sum_{b\in F} \omega_2(b)\omega_{1|b}$$

Corollary 2 *Let AB be a non-signaling composite of A and B, and let ω be a pure state of AB. If the marginal state ω_2 is pure, then ω_1 is also pure, and $\omega = \omega_1 \otimes \omega_2$.*

Proof It is easy to see that, if a product state $\omega = \omega_1 \otimes \omega_2$ is pure, then both marginals must be pure. Now suppose that one marginal state—say, ω_2—is pure. Since $\omega_2 = \sum_{x \in E} \omega_1(x)\omega_{2|x}$, and the conditional states $\omega_{2|x}$ belong to $\mathbf{V}(B)$, it follows that for every $x \in E$ with $\omega_1(x) > 0$, we must have $\omega_{2|x} = \omega_2$, so that $\omega(xy) = \omega_1(x)\omega_2(y)$ for every such x. The same result holds trivially if $\omega_1(x) = 0$, so we have $\omega(xy) = \omega_1(x)\omega_2(y)$ for all choices of x and y. It follows that $\omega = \omega_1 \otimes \omega_2$. □

Definition 12 A state ω on AB is *separable* iff it is a mixture of product states, that is, $\omega = \sum_i t_i \alpha_i \otimes \beta_i$ where $t_i \geq 0$ and $\sum_i t_i = 1$ A state *not* of this form is said to be *entangled*.

Using this language, the preceding Corollary gives us

Corollary 3 *If AB is a non-signaling composite of models A and B, and ω is an entangled state of AB, then both ω_1 and ω_2 are mixed.*

This is often regarded as the hallmark of entangled *quantum* states; but, as we see, it is really a quite general possibilty arising in any non-classical probabilistic setting. Of course, one can still ask at this point whether entangled states *exist* in any generality, once one leaves the confines of quantum theory. However, as we'll see in Sect. 3.4 below, there is a sense in which *most* non-signaling composites of non-classical models admit entangled states.

The CHSH Inequality Let AA be a non-signaling composite of two copies of A. For any $a, b \in \boldsymbol{E}(A)$ with $-u_A \leq a, b \leq u_A$, let $a' = u_A - a$ and $b' = u_A - b$. For any state ω in AA, define

$$S(\omega; a, b) = \omega(a, b) + \omega(a, b') + \omega(a', b) - \omega(a', b').$$

This is called the CHSH (Clauser-Horn-Shimony-Holt) parameter associated with ω, a and b. of a bipartite If ω is a product state, then $S \leq 2$ for all choices of a and b; as S is affine in ω, it follows that $S \leq 2$ for all separable states. For entangled states it can be larger A priori, the upper bound for S is 4, and this is achieved, for example, if A is the "square bit" of Example 3. However, for bipartite quantum states, the upper bound is much lower. As pointed out by Tsirel'son [66], $S \leq 2\sqrt{2}$ for any quantum bipartite state and any effects a and b. A great deal of work has gone into trying to find a deeper explanation for this bound. [3, 55]. In Sect. 4, we will return to this matter.

Conditioning Maps and Isomorphism States If ω is any non-signaling state on AB, then the associated bilinear form $\mathcal{B}_\omega$ on $\boldsymbol{E}(A) \times \boldsymbol{E}(B)$ gives us a positive linear mapping

$$\widehat{\omega} : \boldsymbol{E}(A) \to \boldsymbol{E}(B)^*$$

defined by

$$\widehat{\omega}(a)(b) = \omega(a \otimes b)$$

for all $a \in \boldsymbol{E}(A)$ and $b \in \boldsymbol{E}(B)$. Notice that $\widehat{\omega}(a) = \omega_1(a)\omega_{2|a}$. Accordingly, we think of $\widehat{\omega}(a)$ as an *un-normalized conditional state* of B given the effect $a \in \boldsymbol{E}(A)$, and refer to $\widehat{\omega}$ as the *conditioning map* associated with ω. Of course, there is also a conditioning map running in the opposite direction. In fact, this is just the adjoint of $\widehat{\omega}$; that is, $\widehat{\omega}^*(b)(a) = \widehat{\omega}(a)(b) = \omega(a, b)$ for all effects $a \in \boldsymbol{E}(A)$ and $b \in \boldsymbol{E}(B)$.

There is a dual construction for effects. An effect $f \in \boldsymbol{E}(AB)$ defines a positive bilinear form on $\mathbf{V}(A) \times \mathbf{V}(B)$ by $(\alpha, \beta) \mapsto f(\alpha \otimes \beta)$. This, in turn, yields a positive linear mapping

$$\hat{f} : \mathbf{V}(A) \to \mathbf{V}(B)^* = \boldsymbol{E}(B)$$

given by $\hat{f}(\alpha)(\beta) = f(\alpha \otimes \beta)$. We call $\hat{f}$ the *co-conditioning map* associated with f.

Definition 13 Let AB be a non-signaling composite of A and B. An *isomorphism state* on AB is a state $\omega \in \Omega(AB)$ such that the conditioning map $\hat{\omega} : \boldsymbol{E}(A) \to \mathbf{V}(B)$ is an order-isomorphism. Dually, an *isomorphism effect* is an effect $f \in \boldsymbol{E}(AB)$ such that the co-conditioning map $\hat{f} : \mathbf{V}(A) \to \boldsymbol{E}(B)$ is an order-isomorphism.

If there exists an isomorphism state on a composite AA of A with itself, then we have $\boldsymbol{E}(A) \simeq \mathbf{V}(A) = \boldsymbol{E}(A)^*$.[12] More generally, we shall say that A is *weakly self-dual* iff there exists an order-isomorhism $\boldsymbol{E}(A) \simeq \mathbf{V}(A)$. Although this is a strong constraint on the structure of a probabilistic model, it is nevertheless satisfied by many examples that are neither quantum nor classical. For example, the models associated with state spaces that are regular 2-dimensional polytopes—that is, regular n-gons—are weakly self-dual.

As we'll discuss further in Sect. 5, quantum models satisfy a much stronger form of self-duality: not only does there exist an order-isomorphism $\mathbf{V}(\mathcal{H}) \simeq \boldsymbol{E}(\mathcal{H})$, but this is given by an inner product on $\boldsymbol{E}(\mathcal{H}) = \mathcal{L}(\mathcal{H})$, namely, $a \mapsto \mathrm{Tr}(a \cdot)$.

Theorem 1 ([14]) *Let A and B be irreducible, and let AB be any locally-tomographic, non-signaling composite of A with B. Then any isomorphism state in AB is pure in $\Omega(AB)$, and any isomorphism effect is extremal in $\boldsymbol{E}(AB)_+$.*

If A and B are not irreducible, an isomorphism state on AB need not be pure. For example, if $A = B = (E, \Delta(E))$, then any state uniformly correlating A and B—say $\omega(x, x) = 1/|E|$ and $\omega(x, y) = 0$ for $x \neq y$—is an isomorphism state, but will be pure only if $|E| = 1$.

[12] The converse is not quite true: an order-isomorphism $\boldsymbol{E}(A) \simeq \mathbf{V}(A)$ defines a non-signaling state on $A \otimes_{\max} B$ (see definition 14 below), but need not correspond to a state of AB.

3.3 Quantum Composites

This is a good place at which to pause for a second and more detailed look at quantum-mechanical composites. As noted earlier in Example 2, the mapping $X(\mathcal{H}) \times X(\boldsymbol{K}) \mapsto X(\mathcal{H} \otimes \boldsymbol{K})$ given by $x, y \mapsto x \otimes y$ turns $A(\mathcal{H} \otimes \boldsymbol{K})$ into a composite of the models $A(\mathcal{H})$ and $A(\boldsymbol{K})$. This mapping extends to the bilinear mapping

$$\boldsymbol{E}(\mathcal{H}) \times \boldsymbol{E}(\boldsymbol{K}) = \mathcal{L}_h(\mathcal{H}) \times \mathcal{L}_h(\boldsymbol{K}) \to \mathcal{L}_h(\mathcal{H} \otimes \boldsymbol{K}) = \boldsymbol{E}(\mathcal{H} \otimes \boldsymbol{K}),$$

that sends $a, b \in \mathcal{L}_h(\mathcal{H}) \times \mathcal{L}_h(\boldsymbol{K})$ to the operator $a \otimes b$ on $\mathcal{H} \otimes \boldsymbol{K}$ (given by $(a \otimes b)(x \otimes y) = ax \otimes by$ for all $x \in \mathcal{H}$, $y \in \boldsymbol{K}$). Hence, by Lemma 3, $A(\mathcal{H} \otimes \boldsymbol{K})$ is a *non-signaling* product of $A(\mathcal{H})$ and $A(\boldsymbol{K})$.

Conditioning Let $\mathcal{H}$ be a complex Hilbert space. For any vectors $x, y \in \mathcal{H}$, let $x \odot y$ denote the rank-one operator on $\mathcal{H}$ given by $(x \odot y)z = \langle z, y\rangle x$. (In Dirac notation, this is $|x\rangle\langle y|$.) If x is a unit vector, then $x \odot x = P_x$, the orthogonal projection operator associated with x.

The mapping $x, y \mapsto x \odot y$ is sesquilinear, that is, linear in its first, and conjugate linear in its second, argument; it therefore extends to a linear mapping $\mathcal{H} \otimes \bar{\mathcal{H}} \to \mathcal{L}(\mathcal{H})$, where $\bar{\mathcal{H}}$ is the conjugate space of $\mathcal{H}$, taking any vector $v = \sum_i t_i x_i \otimes \bar{y}_i$ to the corresponding operator $\hat{v} := \sum_i t_i x_i \odot y_i$. It is easy to see that this is injective and hence, on dimensional grounds, an isomorphism. It is useful to note that

$$\langle \hat{v}(x), y\rangle = \langle v, y \otimes \bar{x}\rangle$$

for all $x, y \in \mathcal{H}$. Hence, if v is any unit vector in $\mathcal{H} \otimes \bar{\mathcal{H}}$, the corresponding pure state $\omega = \alpha_v$ of $A(\mathcal{H} \otimes \bar{\mathcal{H}})$ assigns joint probabilities to outcomes $x \in X(\mathcal{H})$ and $\bar{y} \in X(\bar{\mathcal{H}})$ by

$$\omega(x, \bar{y}) = |\langle v, x \otimes \bar{y}\rangle|^2 = \left|\langle \hat{v}(y), x\rangle\right|^2$$

so that the conditional state $\omega_{2|\bar{y}}$ is exactly the pure state associated with the unit vector $\hat{v(y)}/\|\hat{v(y)}\|$. (The rather special fact that conditioning a pure bipartite quantum state by a measurement outcome always leads to a pure state—the *pure conditioning property*—has been exploited in [24, 74].)

Purification and Correlation Suppose now that α is a state on $A(\mathcal{H})$, represented by a density operator W on $\mathcal{H}$ with spectral resolution

$$W = \sum_{x \in E} \lambda_x P_x = \sum_{x \in E} \lambda_x x \odot x$$

where E is an orthonormal basis for $\mathcal{H}$ and $\sum_{x \in E} \lambda_x = \mathrm{Tr}(W) = 1$. Functional calculus gives us $W^{1/2} = \sum_{x \in E} \lambda_x^{1/2} x \odot x$. We can interpret this as a unit vector in $\mathcal{H} \otimes \bar{\mathcal{H}}$, namely

$$\Psi_W := \sum_{x \in E} \lambda_x^{1/2} x \otimes \bar{x}. \tag{1}$$

This, in turn, defines a bipartite state on the composite quantum system $A\bar{A} := A(\mathcal{H} \otimes \overline{\mathcal{H}})$. The marginal, or reduced, state of the first component system is given by

$$\omega_1(a) = \mathrm{Tr}(P_{\Psi_W}(a \otimes \mathbf{1}_{\overline{\mathcal{H}}}) = \langle (a \otimes \mathbf{1}_{\bar{\mathcal{H}}})\Psi_W, \Psi_W\rangle = \mathrm{Tr}(Wa)$$

so the pure state corresponding to Ψ_W is a *dilation* (or *purification*) of the given mixed state W. Now observe that if $u, v \in X(\mathcal{H})$ with $u \perp v$, then we have

$$\langle \Psi_W, u \otimes \bar{v}\rangle \ = \sum_{x \in E} \lambda_x^{1/2} \langle x, u\rangle \langle \bar{x}, v\rangle = 0.$$

Evidently, the pure state ω corresponding to Ψ_W sets up a *perfect correlation* between $E \in \mathcal{M}(\mathcal{H})$ and the corresponding test $\bar{E} = \{\bar{x} | x \in \boldsymbol{E}\} \in \mathcal{M}(\bar{\mathcal{H}})$, with

$$\omega(x, \bar{x}) = |\langle \Psi_W, x \otimes \bar{x}\rangle|^2 = |\lambda_x^{1/2}|^2 = \lambda_x.$$

An especially interesting case arises when α is the maximally mixed state, i.e., when $W = \frac{1}{n}\mathbf{1}$ (where $n = \dim(\mathcal{H})$). In this case, we have

$$\Psi_W = \frac{1}{\sqrt{n}} \sum_{x \in E} x \otimes \bar{x}.$$

This is the analogue, on $\mathcal{H} \otimes \bar{\mathcal{H}}$, of the *maximally entangled*, or *EPR*, state on $\mathcal{H} \otimes \mathcal{H}$. Notice that Ψ_W is *independent* of the choice of E (since every orthonormal basis of $\mathcal{H}$ is an eigenbasis for $\mathbf{1}$). Hence, Ψ_W simultaneously correlates *every* test $E \in \mathcal{M}(\mathcal{H})$ with its counterpart in $\mathcal{M}(\overline{\mathcal{H}})$. Moreover, the correlation is *uniform*, in that the probabilities of correlated pairs $x \otimes \overline{x}$ of outcomes is uniformly $1/n$. As we'll see later, the existence of such a uniformly correlating state between two isomorphic systems has interesting consequences.

Local Tomography If $\mathcal{H}$ and $\boldsymbol{K}$ are real or complex Hilbert spaces of dimensions m and n, respectively, As was remarked above, $A(\mathcal{H} \otimes \boldsymbol{K})$ is a non-signaling composite of $A(\mathcal{H})$ and $A(\boldsymbol{K})$. It is easily checked that $\dim \boldsymbol{E}(A) = \dim \mathcal{L}_h(\mathcal{H}) = m^2$ if $\mathcal{H}$ is complex and $(m^2 + m)/2$ if $\mathcal{H}$ is real. Hence, the dimension of the real vector space $\boldsymbol{E}(\mathcal{H} \otimes \boldsymbol{K}) = \mathcal{L}_h(\mathcal{H} \otimes \boldsymbol{K})$ of Hermitian operators is $(mn)^2 = m^2n^2$, so in fact $\mathcal{L}_h(\mathcal{H} \otimes \boldsymbol{K}) = \mathcal{L}_h(\mathcal{H}) \otimes \mathcal{L}_h(\boldsymbol{K})$, and the composite system is locally tomographic. On the other hand, if $\mathcal{H}$ and $\boldsymbol{K}$ are real, the dimension of $\mathcal{L}_h(\mathcal{H} \otimes \boldsymbol{K})$ is $((mn)^2 - mn)/2 + mn = ((mn)^2 + mn)/2$, while the product of the dimensions of $\mathcal{L}_h(\mathcal{H})$ and $\mathcal{L}_h(\boldsymbol{K})$ is

$$\frac{(m^2 + m)}{2} \cdot \frac{(n^2 + n)}{2} = \frac{m^2n^2 + m^2n + mn^2 + mn}{4}.$$

This is strictly less than $(m^2n^2 + mn)/2$, which in turn is less then $(mn)^2$, so in this case, $\boldsymbol{E}(AB)$ is strictly larger than $\boldsymbol{E}(A) \otimes \boldsymbol{E}(B)$. Thus, for real Hilbert spaces $\mathcal{H}$ and $\boldsymbol{K}$, the standard composite $\mathcal{M}(\mathcal{H} \otimes \boldsymbol{K})$ is *not* locally tomographic. (Neither do we have local tomography for quaternionic Hilbert spaces, though here, one needs to be more careful about the formulation of the relevant tensor products. See [6] and [46] for more details.)

3.4 Maximal and Minimal Tensor Products

Let AB be a non-signaling composite of two systems A and B. As noted above, if AB is locally tomographic, then $\boldsymbol{E}(AB) \simeq \boldsymbol{E}(A) \otimes \boldsymbol{E}(B)$ as vector spaces. In this section, we consider more closely the possibilities for such a composite.

As we saw earlier, any non-signaling state ω on any composite system AB induces a bilinear form on $\boldsymbol{E}(A) \times \boldsymbol{E}(B)$. If AB is locally tomographic, then we can identify ω with this form. We then see that there are two extreme possibilities for the set of states on a locally tomographic composite AB: maximally, we may include *all* positive, normalized bilinear forms on $\boldsymbol{E}(A) \times \boldsymbol{E}(B)$; minimally, we may restrict our attention to the closed convex hull of the product states.

It will simplify further discussion to put these ideas into a broader context [54]:

Definition 14 Let $\boldsymbol{E}$ and $\boldsymbol{F}$ be any two finite-dimensional ordered vector spaces. The *minimal tensor cone* on $\boldsymbol{E} \otimes \boldsymbol{F}$ is the cone generated by pure tensors $a \otimes b$ with $a \in \boldsymbol{E}_+$ and $b \in \boldsymbol{F}_+$. The *maximal tensor cone* is the cone consists of all tensors τ such that $(f \otimes g)(\tau) \geq 0$ for all $(f, g) \in \boldsymbol{E}_+^* \times \boldsymbol{F}_+^*$—in other words, the cone of tensors corresponding to positive bilinear forms on $\boldsymbol{E}^* \times \boldsymbol{F}^*$.

Evidently, $(\boldsymbol{E} \otimes_{\min} \boldsymbol{F})_+ \subseteq (\boldsymbol{E} \otimes_{\max} \boldsymbol{F})_+$; in general, the inclusion is proper. These two cones on $\boldsymbol{E} \otimes \boldsymbol{F}$ give us two different *ordered* tensor products, which we denote by $\boldsymbol{E} \otimes_{\min} \boldsymbol{F}$ and $\boldsymbol{E} \otimes_{\max} \boldsymbol{F}$, respectively. It is not difficult to see that (in finite dimensions) we have

$$(\boldsymbol{E} \otimes_{\min} \boldsymbol{F})^* = \boldsymbol{E}^* \otimes_{\max} \boldsymbol{F}^* \text{ and } (\boldsymbol{E} \otimes_{\max} \boldsymbol{F})^* = \boldsymbol{E}^* \otimes_{\min} \boldsymbol{F}^*.$$

Let $\mathcal{B}(\boldsymbol{E}, \boldsymbol{F})$ stand for the space of bilinear forms on $\boldsymbol{E} \times \boldsymbol{F}$ (so that $(\boldsymbol{E} \otimes \boldsymbol{F})^* \simeq \mathcal{B}(\boldsymbol{E}, \boldsymbol{F})$), and $\mathcal{B}_+(\boldsymbol{E}, \boldsymbol{F})$, for the cone of all bilinear forms that are non-negative on $\boldsymbol{E}_+ \times \boldsymbol{F}_+$. Then we also have

$$(\boldsymbol{E}^* \otimes_{\max} \boldsymbol{F}^*)_+ \simeq \mathcal{B}(\boldsymbol{E}, \boldsymbol{F})_+ \text{ and } (\boldsymbol{E} \otimes_{\max} \boldsymbol{F})_+ \simeq \mathcal{B}_+(\boldsymbol{E}^*, \boldsymbol{F}^*).$$

If $\boldsymbol{E}$ and $\boldsymbol{F}$ are order-unit spaces, with order-units u and v, say, then $u \otimes v$ is an order unit for both $\boldsymbol{E} \otimes_{\max} \boldsymbol{F}$ and $\boldsymbol{E} \otimes_{\min} \boldsymbol{F}$.

Let AB be any locally tomographic composite of models A and B. As noted above, every state on AB induces a positive bilinear form on $\boldsymbol{E}(A) \times \boldsymbol{E}(B)$. This gives us a positive linear mapping

$$\mathbf{V}(AB) \mapsto \mathcal{B}(\boldsymbol{E}(A), \boldsymbol{E}(B)) = \mathbf{V}(A) \otimes_{\max} \mathbf{V}(B).$$

The existence of product states (condition (ii) in definition 9) tells us that the image of $\mathbf{V}(AB)_+$ under this mapping contains $(\mathbf{V}(A) \otimes_{\min} \mathbf{V}(B))_+$. Thus, if AB is locally tomographic, we have order-embeddings

$$\mathbf{V}(A) \otimes_{\min} \mathbf{V}(B) \leq \mathbf{V}(AB) \leq \mathbf{V}(A) \otimes_{\max} \mathbf{V}(B).$$

This raises the question of whether we can define, in a canonical way, locally tomographic composites—call them $A \otimes_{\max} B$ and $A \otimes_{\min} B$—in such a way that

$$\mathbf{V}(A \otimes_{\min} B) = \mathbf{V}(A) \otimes_{\min} \mathbf{V}(B) \text{ and } \mathbf{V}(A \otimes_{\max} B) = \mathbf{V}(A) \otimes_{\max} \mathbf{V}(B).$$

To this end, we define the *maximal composite* of models A and B, $A \otimes_{\min} B$ by setting $X(A \otimes_{\max} B) = X(A) \times X(B)$ and $\mathcal{M}(A \otimes_{\min} B) = \mathcal{M}(A) \times \mathcal{M}(B)$, and taking $\Omega(A \otimes_{\max} B)$ to be the state space of $\boldsymbol{E}(A) \otimes_{\min} \boldsymbol{E}(B)$. Then $\mathbf{V}(A \otimes_{\max} B) = \mathbf{V}(A) \otimes_{\max} \mathbf{V}(B)$. Define the *minimal composite* of A and B, $A \otimes_{\min} B$, to be the maximal model associated with the order-unit space $\boldsymbol{E}(A) \otimes_{\max} \boldsymbol{E}(B)$ (see the remark on page 13). Then $\mathbf{V}(A \otimes_{\min} B) = \mathbf{V}(A) \otimes_{\min} \mathbf{V}(B)$.

Thus, $A \otimes_{\max} B$ is the smallest possible locally tomographic composite of A and B, in the sense of having the fewest possible tests, and the largest, in the sense of having the most states; dually, $A \otimes_{\min} B$ has the largest possible set of tests, and hence, the smallest possible state space. (One might say, roughly speaking, that $A \otimes_{\min} B$ admits *no* entanglement between effects, and, consequently, admits all possible entangled states; at the other extreme, $A \otimes_{\max} B$ admits every possible entangled bipartite effect and, in consequence, admits no entanglement of states.)

If $\Omega(A)$ or $\Omega(B)$ is a simplex, then it is easy to show that $\mathbf{V}(A) \otimes_{\max} \mathbf{V}(B) \simeq \mathbf{V}(A) \otimes_{\min} \mathbf{V}(B)$ and $\boldsymbol{E}(A) \otimes_{\max} \boldsymbol{E}(B) \simeq \boldsymbol{E}(A) \otimes_{\min} \boldsymbol{E}(B)$. Thus, a classical system admits no entangled states or effects in any locally tomographic non-signaling composite with another system. There is a partial converse:

Theorem 2 ([54]) *The following are equivalent:*

(a) $\Omega(A \otimes_{max} B)$ *contains no entangled state for any model* B,
(b) $\Omega(A \otimes_{max} B)$ *contains no entangled state,* B *the square bit (Example 3),*
(c) $\Omega(A)$ *is a simplex.*

It follows that any *non*-classical system A—one with a non-simplicial state space—will admit *some* locally tomographic, non-signaling composite AB that admits entangled states. In this sense, entanglement is a highly generic phenomenon in non-classical probability theory.

3.5 Monoidal Probabilistic Theories

Earlier, we decided to represent a probabilistic *theory* as a category of probabilistic models with positive mappings as morphisms. It is not unreasonable to require that, if A, B and C are three systems, we should be able to form tripartite composites $(AB)C$ and $A(BC)$. We'd perhaps like to require that these be the same, i.e., that we have an *associative* rule of composition. This is not a trivial requirement—one can readily imagine situations in which the composition of systems might not be associative[13]—but it is a natural one, especially if we think of AB as simply modelling the two systems *considered together*, but not necessarily interacting in any way.

A *symmetric monoidal category* is a category $\mathcal{C}$, equipped with a bi-functor $\mathcal{C} \times \mathcal{C} \stackrel{\otimes}{\longrightarrow} \mathcal{C}$, such that for all $A, B, C, D \in \mathcal{C}$,

$$A \otimes (B \otimes C) \simeq (A \otimes B) \otimes C \quad \text{and} \quad A \otimes B \simeq B \otimes A$$

by means of natural isomorphsism $\alpha_{A;BC}$ and $\sigma_{A,B}$ belonging to $\mathcal{C}$; and also equipped with a *tensor unit*, I, and natural isomorphisms

$$I \otimes A \simeq A \simeq A \otimes I$$

This point of view has been extensively developed in the the categorical semantics for quantum theory developed by Abramsky and Coecke [1, 26] and Selinger [61], and also in the work of Baez and his students [6, 7].

Definition 15 A *monoidal probabilistic theory* is a probabilistic theory $\mathcal{C}$, equipped with a rule of composition $A, B \mapsto AB$ assigning, to each pair of models $A, B \in \mathcal{C}$, a composite AB in the sense of Definition 9, and making $\mathcal{C}$ a symmetric monoidal category. We shall say that $\mathcal{C}$ is *non-signaling*, respectively *locally tomographic*, iff AB is non-signaling or locally tomographic for every pair $A, B \in \mathcal{C}$.

This definition implies that, for all $A, B \in \mathcal{C}$ and all states $\alpha \in \Omega(A)$, $\beta \in \Omega(B)$, there is a distinguished product state $\alpha \otimes \beta$ with $(\alpha \otimes \beta)(xy) = \alpha(x)\beta(y)$ for all $x \in X(A)$, $y \in X(B)$. Similarly, for any (dual) processes $\tau_1 \in \mathcal{C}(A)$ and $\tau_2 \in \mathcal{C}(B)$, there exists a process $(\tau_1 \otimes \tau_2) \in \mathcal{C}(AB)$ with $(\tau_1 \otimes \tau_2)(a \otimes b) = \tau_1(a) \otimes \tau_2(b)$ for all effects $a \in \mathbf{E}(A)$ and $b \in \mathbf{E}(B)$.

Finite-dimensional classical and quantum probability theory are both monoidal with respect to their usual rules of composition. The minimal and maximal tensor products are each naturally associative, and hence make the category of *all* probabilistic models into a monoidal probabilistic theory; but neither is entirely satisfactory: the former provides for entangled states, but does not permit entangled effects, while the latter provides for entanglement between effects, but allows none between states.

[13] Consider, for instance, the case of (Farmer ⊗ Hen) ⊗ Fox *vs.* Farmer ⊗ (Hen ⊗ Fox).

That a probabilistic theory support a *single* "tensor product" that accommodates entanglement of both states and effects, is a non-trivial constraint. To be sure, one might consider probabilistic theories equipped with more than one rule of composition; however, the interactions among different non-signaling compositions on a given theory can be very delicate. It therefore seems reasonable to begin by investigating the simpler possibilities for a theory equipped with a single privileged, monoidal rule of composition. Accordingly, ***in the balance of this paper, we work in a monoidal probabilistic theory*** $\mathfrak{C}$.

Historical Remarks Tensor products of compact convex sets or of order-unit spaces were studied in a number of papers in the late 1960s, notably that of Namioka and Phelps [54]. The fact that the marginal of an entangled pure state must be a mixed state already appears there, albeit not in these terms, as do the definitions of what we are calling the maximal and minimal tensor products. Our treatment composite systems derives from that of Foulis and Randall [36, 46]. Some first attempts to understand probabilistic theories as symmetric monoidal categories of probabilistic models can be found in [15, 17]; work in this direction is ongoing.

4 Post-Classical Information Processing

As we've seen, entangled bipartite states and effects arise very naturally, not only in quantum theory, but in almost any context in which we form non-signaling composites of non-classical systems. While this observation goes back at least to [45, 46] in the late 1980s, it remained unexploited. Entanglement lies at the heart of quantum information theory, so it natural to wonder to what extent quantum information-theoretic results carry over to other non-classical settings. It turns out that a great many such results do have analogues for probabilistic theories that are far more general than quantum mechanics. While the exploration of this post-classical information theory is still in its infancy, it has already shed considerable light on the scope and meaning of several key quantum-informational results.

In this section, we review in some detail two of these. The first is the no-cloning theorem, and its generalization, the no-broadcasting theorem. These hold in *any* finite-dimensional theory having a state space that is not a simplex. The second is the existence of a teleportation protocol, or, a bit more generally, of an entanglement-swapping protocol. Here, some restrictions need to be made, but they are of moderate strength. For example, any monoidal probabilistic theory in which individual systems are *weakly self-dual*, and composites include states corresponding to isomorphisms witnessing the weak self-duality, supports a certain kind of teleportation. Moreover, when viewed in this generality, teleportation loses most of its mystery: it is simply a form of classical conditioning, one which appears startling only owing to the appearance of isomorphism states.

4.1 Cloning and Broadcasting

To *clone* a state of a system A means, very broadly, to produce two *independent* copies of that state by means of some physical process. In the present formalism, if the initial state belongs to a system A, this would require a positive linear mapping

$$\phi : \mathbf{V}(A) \to \mathbf{V}(AA)$$

such that $\phi(\alpha) = \alpha \otimes \alpha$. There is no difficulty in producing a mapping that clones a *particular* state: indeed, the constant mapping $\Omega(A) \to \Omega(AA)$ given by $\beta \mapsto \alpha \otimes \alpha$ for all $\beta \in \Omega(A)$ is affine, and hence, extends uniquely to a positive linear mapping $\mathbf{V}(A) \to \mathbf{V}(AA)$. However, this mapping is (highly!) state-dependent. One might ask whether one could *jointly* clone a collection of states, say, $\alpha_1, \ldots, \alpha_n$. That is: given such a set of states, can one find a *single*, norm-nonincreasing, positive linear mapping $\mathbf{V}(A) \to \mathbf{V}(AA)$ that clones them all, in the sense that $\phi(\alpha_i) = \alpha_i \otimes \alpha_i$ for all i?

If the states α_i are jointly distinguishable, the answer is yes. If $\{a_i\}$ is an observable on A with $\alpha_i(a_i) = 1$ for all i, then the mapping ϕ defined by

$$\phi(\beta) = \sum_i \beta(a_i)\, \alpha_i \otimes \alpha_i$$

does the trick. The *no-cloning theorem* is essentially the converse: if there exists a single process that will clone all of the states $\alpha_1, \ldots, \alpha_n$, then there exists an observable that distinguishes them. In the case of a discrete classical model, where all pure states are jointly distinguishable, this is no restriction on the clonability of pure states; but quantum pure states, which are not jointly distinguishable, are in general not jointly clonable.

The quantum no-cloning theorem was first proved, independently, by Wootters and Zurek [78] and by Dieks [27]. That the same result holds for arbitrary probabilistic theories is proved in [9]. We omit the proof here, but the idea is simple: if we can clone each of the states $\alpha_1, \ldots, \alpha_n$ with a single mapping, then by iterating this process, we can create arbitrarily large ensembles of independent copies of an unknown state $\alpha \in \{\alpha_1, \ldots, \alpha_n\}$ and, by making measurements on this ensemble, we can use statistics to distinguish among them.

We say that a state $\rho \in \Omega$ is *broadcast* by an affine mapping $\phi : \Omega \to \Omega \otimes \Omega$ iff the bipartite state $\phi(\rho)$ has marginal states $\phi(\rho)_1$ and $\phi(\rho)_2$ both equal to ρ. If ρ can be expressed as a mixture of distinguishable—hence, clonable—states $\alpha_1, \ldots, \alpha_n$, say $\rho = \sum_i t_i \alpha_i$, then one can broadcast ρ using a cloning map ϕ for the states $\alpha_1, \ldots, \alpha_n$: the state $\phi(\rho) = \sum_i t_i \alpha_i \otimes \alpha_i$ has both marginal states equal to ρ, as required. The quantum *no-broadcasting theorem* of Barnum et al. [8] tells us that, conversely, two *quantum* states are jointly broadcastable iff, regarded as density operators, they commute—which, by the Spectral Theorem, is equivalent to requiring that all are convex combinations of some single set of distinguishable pure states. In fact, this is a corollary of a more general result:

Theorem 3 ([9, 10]) *Let Γ be the set of states broadcast by an affine mapping $\phi : \Omega \to \Omega \otimes \Omega$. Then Γ is the simplex generated by a set of distinguishable states in Ω, which are cloned by ϕ.*

(Although we omit the proof here, it is not especially difficult. This is in contrast to earlier proofs of the quantum no-broadcasting result [8, 47], which were not especially easy.)

4.2 Remote Evaluation

Suppose $\mathcal{C}$ is a locally tomographic, monoidal probabilistic theory. Consider two parties, Alice and Bob, occupying arbitrarily distant sites. Suppose that Alice controls a pair of systems, say $A_0, A_1 \in \mathcal{C}$, while Bob controls a system $B \in \mathcal{C}$. Since $\mathcal{C}$ is monoidal, we can represent Alice's two systems together as a single bipartite system $A = A_0A_1$, and the entire Alice-Bob system, by the tripartite composite $AB = (A_0A_1)B \simeq A_0(A_1B)$.

Now suppose that the composite system A_1B is in a state ω, while Alice's system A_0 is in a state α, independent of the A_1B sub-system. Then the total state of the system $AB = A_0(A_1B)$ is $\alpha \otimes \omega$. Now let Alice make a measurement on her system $A = A_0A_1$, obtaining a result represented by an effect $f \in \boldsymbol{E}(A)$; suppose Bob also makes a measurement on his system, B, obtaining a result represented by an effect $b \in \boldsymbol{E}(B)$, so that the joint outcome of these two measurements is $f \otimes b$.

Lemma 5 (Remote Evaluation) *With notation as above, let $\widehat{\omega} : \boldsymbol{E}(A_1) \to \mathbf{V}(B)$ and $\widehat{f} : \mathbf{V}(A_0) \to \boldsymbol{E}(A_1)$ be the conditioning and co-conditioning maps associated with the state ω and the effect f. Then, for all $\alpha \in \mathbf{V}(A_0)$ and all $b \in \boldsymbol{E}(B)$,*

$$(f \otimes b)(\alpha \otimes \omega) = \hat{f}(\hat{\omega}(\alpha))(b). \tag{2}$$

The proof is easy: one simply checks that the formula is correct when ω is a product state and f is a product effect. Since we are working with locally tomographic composites, product states and product effects span $\boldsymbol{E}(A_1B)^*$ and $\boldsymbol{E}(A_0A_1)$, respectively, so (2) holds for all choices of ω and f. Nevertheless, the result is somewhat surprising, for it asserts that the mapping

$$\tau := \hat{\omega} \circ \hat{f} : \mathbf{V}(A_0) \to \mathbf{V}(B)$$

can be implemented, probabilistically, by means of a preparation of A_1B in the joint state ω and a (successful) observation of f on A_0A_1. In particular, when Alice observes the effect f, the corresponding un-normalized conditional state of Bob's system is

$$(f \otimes -)(\alpha \otimes \omega) = \tau(\alpha).$$

Note that the probability of the process τ occurring in state α is $u_B(\tau(\alpha))$, which is exactly the marginal probability $(\alpha \otimes \omega)_1(f)$ of Alice's obtaining f. In what follows, we refer to the pair (f, ω) as a *remote evalution* protocol for the process $\tau = \hat{f} \circ \hat{\omega}$.

We can reformulate the notion of conditioning and co-conditioning map, and the remote evaluation Lemma (Lemma 5), in purely categorical terms. In fact, both make sense in any symmetric monoidal category $\mathcal{C}$. Given objects $A, B \in C$ and a morphism $\omega : A \otimes B \to I$, there is a canonical mapping $\hat{\omega} : \mathcal{C}(I, A) \to \mathcal{C}(B, I)$ given by

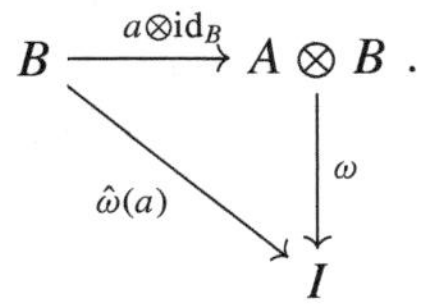

(3)

Dually, if $f \in \mathcal{C}(I, A \otimes B)$, there is a natural mapping $\hat{f} : \mathcal{C}(A, I) \to \mathcal{C}(I, B)$ given by

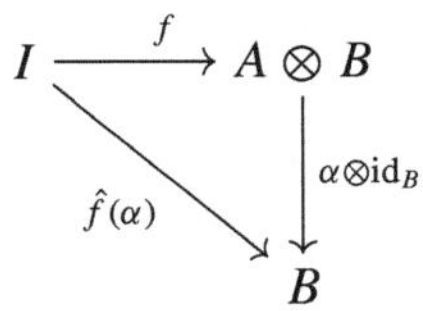

(4)

If $\mathcal{C}$ is a monoidal probabilistic theory, then $\hat{\omega}$ and $\hat{f}$, defined in this way, correspond exactly to the conditioning and co-conditioning maps associated with the bipartite state $\omega : A \otimes B \to I$ and effect $f : I \to A \otimes B$. Combining diagrams (3) and (4), and taking advantage of the monoidal structure of $\mathcal{C}$—in particular, the fact that $\alpha \otimes \omega = (I \otimes \omega) \circ (\alpha \otimes \mathrm{id}_{A_0 \otimes A_1})$—we have

$$\hat{\omega}(\hat{f}(\alpha) \otimes \mathrm{id}_B) = \omega \circ (\alpha \otimes \mathrm{id}_{A_1 B}) \circ (f \circ \mathrm{id}_B) = (\alpha \otimes \omega) \circ (f \otimes \mathrm{id}_B) \quad (5)$$

which precisely expresses Lemma 5.

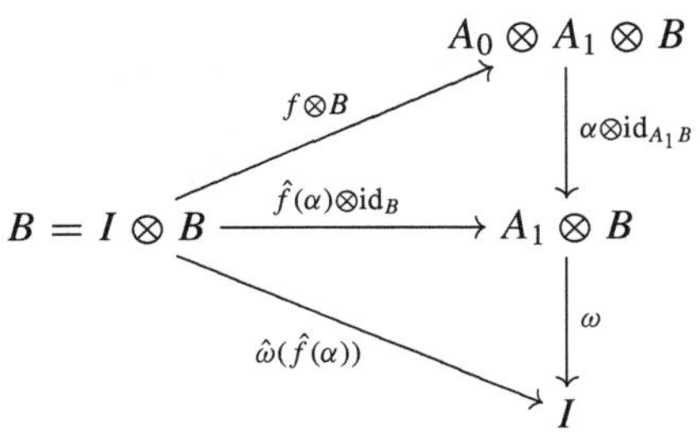

(6)

This has an important corollary. Since $\omega \circ (\alpha \otimes \mathrm{id}_{A_1 B}) = \alpha \otimes (\mathrm{id}_{A_0} \circ \omega)$, we can re-write (6) as

$$\hat{\omega}(\hat{f}(\alpha) \otimes \mathrm{id}_B) = \alpha \circ (\mathrm{id}_{A_0} \otimes \omega) \circ (f \otimes \mathrm{id}_B)$$

Thus, the dual process $\tau^* : \boldsymbol{E}(B) \to \boldsymbol{E}(A)$ corresponding to the process $\tau = \hat{\omega} \circ \hat{f}$ arising in the remote evaluation protocol, is in fact a morphism in $\mathcal{C}(A_0, B)$.

Conclusive Teleportation In the special case in which the models A_0, A_1 and B are isomorphic and weakly self-dual, we can consider a remote evaluation protocol in which both the effect $f \in \boldsymbol{E}(A)$ and the state $\omega \in \Omega(A_1 B)$ correspond to order isomorphisms $\hat{f} : \mathbf{V}(E_0) \simeq \boldsymbol{E}(A_1)$ and $\hat{\omega} : \boldsymbol{E}(A_1) \simeq \mathbf{V}(B)$. In this case, the process $\tau = \hat{\omega} \circ \hat{f}$ is again an order-isomorphism. If this scenario is repeated many times, Bob can perform sufficiently many measurements to determine $\tau(\alpha)$ with reasonable confidence, and then *compute* the value of α. On the other hand, if τ is probabilistically reversible, in a single run of the scenario Bob can actually correct his state, with non-zero probability, so that it agrees with α. In this case, we may say that the state α has been *teleported* from Alice's system A_0 to Bob's system B, and refer to (f, ω) as a *conclusive teleportation protocol*. If τ is reversible with probability 1, we shall say that (f, ω) is a *strong* conclusive teleportation protocol.

Deterministic Teleportation Suppose now that Alice has access to an observable $\{f_i\}$ on $A = A_0 A_1$, with each of the effects f_i an isomorphism effect. Each of these effects, in combination with the isomorphism state ω, gives rise to a conclusive teleportation protocol, implementing the order-isomorphism $\tau_i = \hat{\omega} \circ \hat{f}_i : \mathbf{V}(A_0) \simeq \mathbf{V}(B)$. If Alice is permitted to communicate (classically) with Bob, then upon observing outcome f_i, she can instruct Bob to implement the inverse process τ^{-1}, which he can do with probability $c_i := u_B \tau_i^{-1}(\alpha)$. It follows that the post-measurement state of Bob's system will be $\sum_i c_i \alpha = \alpha$. In particular, $\sum_i c_i = \sum_i u_B \tau_i^{-1}(\alpha)$. Say that A_0 *supports a deterministic teleportation protocol* iff there exist such an effect f and such a state ω on suitable composites $A_0 A_1$ and $A_1 B$, with $B \simeq A$. In fact, one might as well take these comosites to be $A_o \otimes_{\min} A_1$ and $A_1 \otimes_{\max} B$ (in order to have as many effects as possible on the former, and as many states as possible on the latter.)

Theorem 4 ([11]) *Suppose there exist a finite group G acting transitively on A's pure states, and a G-equivariant order-isomorphism $\boldsymbol{E}(A) \simeq \boldsymbol{E}(A)^*$. Then A supports a deterministic teleportation protocol.*

Entanglement Swapping Suppose that, like Alice, Bob controls a bipartite system $B = B_1 B_2$. Assume here that A_0, A_1, B_1 and B_0 are all isomorphic to one another. Given an entangled state ω between A_1 and B_1, and isomorphism effects f on $A = A_0 A_1$ and g on $B = B_1 B_0$, we find that, for any state μ on $A_0 A_0$, we have (up to the obvious symmetrizers and associators)

$$(f \otimes g)(\mu \otimes \omega) = g(\hat{\omega} \circ \hat{f} \circ \hat{\mu}^*).$$

Since this holds for any choice of $g \in \boldsymbol{E}(B)$, we have

$$(\mu \otimes \omega)_{B|f} = \hat{\omega} \circ \hat{f} \circ \hat{\mu}^*$$

If $\tau = \hat{\omega} \circ \hat{f}$ is probabilistically reversible, then upon Bob's executing the reverse process, the state μ has been transferred from $A_0 B_2$ to $B = B_1 B_0$.

Teleportation and Compact Closure Let $\mathcal{C}$ be any symmetric monoidal category. A *dual* for an object $A \in \mathcal{C}$ is an object $B \in \mathcal{C}$, together with two morphisms, $\eta : I \to B \otimes A$ and $\epsilon : A \otimes B \to I$—called the *unit* and *co-unit*, respectively—such that

$$(\epsilon \otimes \mathrm{id}_A) \circ (\mathrm{id}_A \otimes \eta) = \mathrm{id}_A \quad \text{and} \quad (\mathrm{id}_B \otimes \epsilon) \circ (\eta \otimes \mathrm{id}_A) = \mathrm{id}_B \tag{7}$$

In view of the discussion above, if $\mathcal{C}$ is a monoidal probabilistic theory and f, ω is a conclusive teleportation protocol for a pair of systems $A, B \in \mathcal{C}$, then the remote evaluation lemma tells us that f and ω function as a unit and co-unit, respectively, for A and B. A symmetric monoidal category in which every object has a dual is said to be *compact closed*. A *compact structure* on a compact closed category is a specification, for every object $A \in \mathcal{C}$, of a distinguished dual $A' \in \mathcal{C}$. Where $A = A'$ for every $A \in \mathcal{C}$, this structure is *degenerate*.[14]

Theorem 5 ([15]) *Let $\mathcal{C}$ be a monoidal probabilistic theory. The following are equivalent.*

(a) $\mathcal{C}$ admits a compact closed structure.
(b) Every $A \in \mathcal{C}$ can be teleported through some $B \in \mathcal{C}$;
(c) Every morphism in $\mathcal{C}$ has the form $\hat{\omega} \circ \hat{f}$ for some bipartite state ω and bipartite effect f in $\mathcal{C}$.

Proof The equivalence of (a) and (b) is clear from the preceding discussion. To see that these are in turn equivalent to (c), suppose first that (a) and (b) hold. Choose for each $A \in \mathcal{C}$ a dual system A', a state $\omega_A \in \mathcal{C}(A \otimes A', I)$, and an effect $f_A \in \mathcal{C}(I, A' \otimes A)$ with $\hat{\omega_A} = \hat{f_A}^{-1}$. Then for any morphism $\tau \in \mathcal{C}(A, B)$, let $f_\tau \in \mathcal{C}(I, A \otimes B)$ be the effect $f_A \circ (A' \otimes \tau)$. It is easily checked that then $\hat{f_\tau} = \tau \circ \hat{f_A}$, so that $\tau = \hat{f_\tau} \circ \hat{\omega_A}$. Conversely, if (c) holds, then for each A, the identity mapping id_A factors as $\hat{\omega}_A \circ \hat{f}_A$ for some $\omega_A \in \mathcal{C}(B \otimes A, I)$ and some $f \in A \otimes B$. It follows that $\hat{\omega}_A = \hat{f}_A^{-1}$, so this gives us a compact closed structure. □

[14] Duals, where they exist, are canonically isomorphic. Hence, for most purposes, the choice of one rather than another object as "the" dual is irrelevant. The existence of a degenerate compact structure is, however, a real constraint [15, 62].

4.3 Ensemble Steering

Let B be a probabilistic model. An *ensemble* for a state $\beta \in \Omega(B)$ is a finite set of of states $\beta_i \in \mathbf{V}(B)_+$ such that $\sum_i \beta_i = \beta$. We can understand such an ensemble as representing one possible way of *preparing* the state β, namely, to choose one of the normalized states $\hat{\beta}_i := \beta_i/u(\beta_i)$ with probability $p_i = u_B(\beta_i)$.

One way to do *this* is to begin with a bipartite state ω on a non-signaling composite AB, with marginal $\omega_2 = \beta$. Then for any observable $E = \{a_i\}$ on A, the un-normalized conditional states $\beta_i := \hat{\omega}(a_i)$ are an ensemble for β. That is: by measuring E, we prepare not only the marginal state ω_B, but *a particular ensemble* for this state. By choosing to measure a different observable, we will typically obtain a different ensemble for β. If A and B are *quantum* systems, and if ω is a pure entangled state of AB, then *any* ensemble for ω_2 can be obtained in this way from a suitable choice of measurement on A. This phenomenon was first observed by Schrödinger [60], who called it *steering*. The concept extends readily to the setting of an arbitrary non-signaling composite.

Definition 16 Let AB be a non-signaling composite of probabilistic models A and B. A bipartite state $\omega \in AB$ is *steering for its B marginal*, or *B-steering*, for short, iff, for every ensemble (convex decomposition) $\omega_2 = \sum_i \beta_i$, where β_i are un-normalized states of B, there exists an observable $E = \{a_i\}$ on A with $\beta_i = \hat{\omega}(a_i)$. We say that ω is *bi-steering* iff it's steering for both marginals.

The relevance of steering to information processing became evident when Bennett and Brassard [21], in the same paper that introduced quantum key distribution, considered a natural quantum scheme for another important cryptographic primitive, bit commitment, and showed that ensemble steering can be used to break it. In the proposed scheme, the two possible values to which Alice can commit are represented by two distinct ensembles for the same density matrix. She is to send samples from the ensemble to Bob in order to commit, and later reveal which states she drew so that Bob can check that she used the claimed ensemble. However, by sending to Bob, not a draw from the ensemble, but one of two systems in an entangled pure bipartite state with the specified density matrix as its marginal, and keeping the other system, she can realize either ensemble after she has already sent the systems to Bob by making measurements on her entangled system, enabling her to perfectly mimic commitment to either bit.

Later Mayers, and Lo and Chau, showed that *no* information-theoretically secure quantum bit commitment protocol can exist. The techniques they used to defeat putative protocols do not literally use steering, but are closely related to the Bennett-Brassard steering attack, in particular in Alice's retention of a system *purifying* the systems she sends to Bob in the course of the protocol.

The paper [14] studies steering in the context of general probabilistic theories. If α is any state on A and β is a *pure* state on B, then $\omega = \alpha \otimes \beta$ is trivially steering for $\omega_2 = \beta$ since the latter has no non-trivial ensembles. In particular, any pure product state will be steering for both of its marginals. Any isomorphism state $\omega \in \mathbf{V}(AB)$ will also be steering.

It follows almost immediately from the definition, that if ω is steering for its B-marginal, then the image, $\hat{\omega}(\boldsymbol{E}(A)_+)$, of the positive cone in $\boldsymbol{E}(A)$, is a face of $\mathbf{V}(B)_+$. Indeed, we have

Lemma 6 *If ω is steering for its B-marginal, then $\hat{\omega}(\boldsymbol{E}(A)_+) = \text{Face}(\omega_2)$.*

Here Face(ω_2) refers to the face generated by ω_2, i.e., the smallest face of $\mathbf{V}(B)_+$ containing ω_2. The converse of Lemma (6) is false.

A probabilistic theory $\mathcal{C}$ *supports uniform universal steering* if, for every system $B \in \mathcal{C}$, there exists a system $A_B \in \mathcal{C}$ such that every state $\beta \in A$ is the marginal of some B-steering state $\omega \in A_B B$. If one can always take $A_B = A$, we say that $\mathcal{C}$ *supports universal self-steering.*

Theorem 6 *Let $\omega \in \Omega(AB)$ be steering for ω_2, where ω_2 is interior to $\mathbf{V}(B)_+$, so that* Face$(\omega_2) = \mathbf{V}(B)_+$. *If $\hat{\omega}$ is injective (non-singular), then $\hat{\omega}$ is an order isomorphism. If $\mathbf{V}(B)$ is irreducible, therefore, by Proposition 1, $\hat{\omega}$ it is pure.*

In other words, if A and B have the same dimension, then the states that are steering for an *interior marginal* are precisely the isomorphism states (and hence, are steering for both marginals).

Steering is closely related to an important property of quantum theory called *homogeneity*.

Definition 17 Let $\mathcal{G}$ be a group of order-automorphisms of an ordered vector space $\boldsymbol{E}$. We say that $\boldsymbol{E}$ is *homogeneous with respect to* $\mathcal{G}$ if $\mathcal{G}$ acts transitively on the *interior* of the positive cone $\boldsymbol{E}_+$. That is, for every pair of interior points a, b of $\boldsymbol{E}_+$, there exists an element $g \in \mathcal{G}$ with $ga = b$. We say $\boldsymbol{E}$ is *homogeneous* if it is homogeneous with respect to some group of order-automorphisms, or, equivalently, if it is homogeneous with respect to the group Aut$(\boldsymbol{E})$ of all order-automorphisms.

It can be shown that the cone $\mathcal{L}_+(\mathcal{H})$ of positive operators on a finite-dimensional Hilbert space $\mathcal{H}$ is homogeneous with respect to the group of order-automorphisms of $\mathcal{L}(\mathcal{H})$. As we discuss below in Sect. 5, the combination of homogeneity and strong self-duality comes close to characterizing finite-dimensional quantum theory among probabilistic theories generally. More precisely, the *Koecher-Vinberg Theorem* asserts that if $\boldsymbol{E}$ is an ordered linear space whose positive cone $\boldsymbol{E}_+$ is both homogeneous and self-dual, then $\boldsymbol{E}$ can be given the structure of a euclidean Jordan algebra. With this in mind, the following result is particularly intriguing:

Theorem 7 *For a model with irreducible state space $\mathbf{V}(A)$ the following are equivalent:*
(a) A is homogeneous;
(b) Every normalized state in the interior of $\Omega(A)$ is the A-marginal of an isomorphism state in $B \otimes_{max} A$, where B is any (fixed) model with state space order-isomorphic to $\mathbf{V}(A)^*$.

From this we obtain:

Corollary 4 *For any model with irreducible state space A, the following are equivalent:*
(a) $\mathbf{V}(A)_+$ *is weakly self-dual and homogeneous;*
(b) Every normalized state in the interior of $\Omega(A)$ *is the marginal of an isomorphism state in* $A \otimes_{max} A$.

Corollary 4, combined with Theorem 7, gives

Theorem 8 *In any theory that supports universal uniform steering, every irreducible, finite-dimensional state space in the theory is homogeneous.*

In light of Corollary 5, we also have

Theorem 9 *In any theory that supports universal self-steering, every irreducible, finite-dimensional state space in the theory is homogeneous and weakly self-dual.*

Therefore, the distance between probabilistic theories allowing universal self-steering, and those whose state-spaces are Jordan-algebraic is just that between weak and strong self-duality.

In [13] it was shown that an asymptotically exponentially secure bit commitment protocol, based (like the original Bennett-Brassard one-qubit protocol) on the nonuniqueness of convex decomposition in nonclassical state spaces, exists in any theory containing some nonclassical state spaces, coupled only by the minimal tensor product (so that there is no entanglement between them). In a nonclassical theory in which all states can be steered, by contrast, this type of bit commitment protocol can always be defeated.

4.4 Entropy and Information Causality

Classical information theory begins with the Gibbs-Shannon entropy $H(p) = -\sum_i p_i \log(p_i)$ of a discrete probabiilty weight $p_1, \ldots, p_n$. Analogously, in quantum theory the *von Neumann entropy* of the state corresponding to a density operator ρ is given by $S(\rho) := \mathrm{Tr}\rho \log \rho$. This is related to the classical Gibbs-Shannon entropy in two important ways. On one hand, $S(\rho)$ is the minimum of the Gibbs-Shannon entropies $-\sum_i p_i \log p_i$ of the probability weights $p_i = \mathrm{Tr}(\rho e_i)$ that ρ induces on quantum tests $\{e_i\}$. (This turns out to be achieved when the measurement is in a diagonalizing basis). Alternatively, $S(\rho)$ is the minimum Gibbs-Shannon entropy of the probabilities p_i arising in representations of ρ as a mixture $\rho = \sum_i p_i \rho_i$ of pure states ρ_i. (This again turns out to be achieved for an ensemble whose states are the rank-one projectors corresponding to a diagonalizing basis).

Both of these characterizations make sense in the context of an arbitrary probabilistic model, but in general, they are not equivalent.

Definition 18 Let α be a state on A. For each test $E \in \mathcal{M}(A)$, define the *local measurement entropy* of α at E, $H_E(\alpha)$, to be the classical (Shannon) entropy of $\alpha|_E$, i.e.,

$$H_E(\alpha) := -\sum_{x \in E} \alpha(x) \log(\alpha(x)).$$

The *measurement entropy* of α, $H(\alpha)$, is the infimum of $H_E(\alpha)$ as E ranges over $\mathcal{M}(A)$, i.e.,

$$H(\alpha) := \inf_{E \in \mathcal{M}(A)} H_E(\alpha).$$

Note that the measurement entropy of a state $\alpha \in \Omega(A)$ depends entirely on the structure of the test space $\mathbf{M}(A)$, and not on the geometry of the state space Ω.

We shall assume in what follows that the measurement entropy of a state is actually achieved on some test, i.e., that $H(\alpha) = H_E(\alpha)$ for some $E \in \mathcal{M}(A)$. This is the case in quantum theory, and can be shown to hold much more generally, given some rather weak analytic requirements on the model A ([12], Appendix B.) It follows that $H(\alpha) = 0$ if and only if there is a test such that α assigns probability 1 to one of its outcomes.

Notation: It will often be convenient to write $H(\alpha)$ as $H(A)$, where context makes clear which state is being considered. If AB is a non-signaling composite, and $H(AB)$ represents $H(\omega)$, we shall write $H(A)$ and $H(B)$ for the marginal entropies $H(\omega_1)$ and $H(\omega_2)$. It is easily checked that the measurement entropy is *subadditive*, i.e.,

$$H(AB) \leq H(A) + H(B).$$

Definition 19 Let α be a state on A. The *mixing* (or *preparation*) entropy for α, denoted $S(\alpha)$, is the infimum of the classical (Shannon) entropy $H(p_1, \ldots, p_n)$ over all finite convex decompositions $\alpha = \sum_i p_i \alpha_i$ with α_i pure states in $\Omega(A)$.

Again, we write $S(A)$ for $S(\alpha)$ where α belongs to the state space Ω of a system $A = (\mathcal{M}, \Omega)$. In contrast to measurement entropy, the mixing entropy of a state depends only on the geometry of the state space Ω, and is independent of the choice of test space $\mathcal{M}(A)$. The mixing entropy is essentially the same as the entropy defined for elements of compact convex sets by A. Uhlmann in [67].

We call a theory *monoentropic* if mixing entropy equals measurement entropy, for every state of every model in the theory. Appendix B of [12] considers some implications of monoentropicity. For instance, it is shown that any monoentropic model A in which the set of pure states is closed in $\Omega(A)$ is sharp.

We define conditional and mutual information in terms of measurement entropy via formulas that also hold classically:

Definition 20 The *conditional measurement entropy* between A and B is defined to be

$$H(A|B) := H(AB) - H(B). \tag{8}$$

The (measurement-based) *mutual information* is defined to be:

$$I(A:B) := H(A) + H(B) - H(AB). \tag{9}$$

Intuitively, one might expect that $I(A:B)$ should not *decrease* if we recognize that B is a part of some larger composite system BC—i.e., we might expect that $I(A:B) \leq I(A:BC)$. Simple algebraic manipulations (using Eqs. (8) and (9)) allow us to reformulate this condition in various ways.

Lemma 7 *The following are equivalent:*

(a) $I(A:BC) \geq I(A:B)$
(b) $H(A|BC) \leq H(A|B)$
(c) $H(AB) + H(BC) - H(B) \leq H(ABC)$
(d) $I(A:B|C) \geq 0$, *where* $I(A:B|C) = H(A|C) + H(B|C) - H(AB|C)$.

The measurement entropy is said to be *strongly subadditive* if it satisfies the equivalent conditions (a)–(d). (Condition (c) is what is usually termed "strong subadditivity" (SSA).) A probabilistic theory in which conditions (a)–(d) are satisfied for all systems A, B and C will also be called *strongly subadditive*. Despite the intution mentioned above, strong subadditivity can fail in general theories, which is perhaps a signal that mutual information as defined above should not be interpreted in general as "the information each system contains about the other".

The Holevo Bound and the Data Processing Inequality The strong subadditivity inequality is crucial to deriving bounds on many quantum information-transmission protocols, and the conditions under which it is satisfied with equality are also of great importance. Another extremely important inequality—derivable, in the quantum setting, from strong subadditivity—is the *Holevo bound*, which figures in an expresssion for the highest achievable rate of classical information transmission through a noisy quantum channel.

The standard formulation of the Holevo bound can apply to a general theory, if the entropies are interpreted as measurement entropies: it asserts that if Alice prepares a state $\rho = \sum_{x \in E} p_x \rho_x$ for Bob, then, for any measurement F that Bob can make on his system,

$$I(E:F) \leq \chi,$$

where $\chi := H(\rho) - \sum_{x \in E} p_x H(\rho_x)$ (often called the *Holevo quantity*).

Suppose that Alice has a classical system $A = (\{E\}, \Delta(E))$ and Bob a general system B. Alice's system is to serve as a record of which state of B she prepared. The situation above is modeled by the joint state $\omega^{AB} = \sum_{x \in E} p_x \delta_x \otimes \beta_x$, where

δ_x is a deterministic state of Alice's system with $\delta_x(x) = 1$. Bob's marginal state is $\omega_2 = \sum_{x \in E} p_x \beta_x$. It is easily shown that, $H(\omega^{AB}) = H(A) + \sum_{x \in E} p_x H(\beta_x)$. Hence,

$$\begin{aligned} I(A:B) &= H(A) + H(B) - H(AB) \\ &= H(A) + H(B) - \left(H(A) + \sum_{x \in E} p_x H(\beta_x) \right) \\ &= H(\omega_B) - \sum_{x \in E} p_x H(\beta_x) = \chi. \end{aligned}$$

So the content of the Holevo bound is simply that the mutual information between the measurement of Alice's classical system and any measurement on Bob's system is no greater than $I(A : B)$,

$$I(E : F) \leq I(A : B).$$

(While this is certainly natural, in general theories it does not always hold.)

Both strong subadditivity and the Holevo bound are instances of a more basic principle. The *data processing inequality* (DPI) asserts that, for any systems A, B and C, and any physical process $\mathcal{E} : B \to C$,

$$I(A : \mathcal{E}(B)) \leq I(A : B)$$

where $I(A : \mathcal{E}(B))$ refers to the mutual information of the state resulting from applying $\mathrm{id}_A \otimes \mathcal{E})$ to the state of AB. The strong subadditivity of entropy amounts to the DPI for the process that simply discards a system (the *marginalization map* $BC \to C$). The Holevo bound is the DPI for the special case of measurements, which can be understood as processes taking a system into a classical system which records the outcome.

Information Causality In a widely discussed paper [55], M. Pawlowski et al. introduced a constraint on a non-signaling probabilistic theory, which they called *information causality*, in terms of the following protocol. Two parties, Alice and Bob, share a joint non-signaling state, known to both of them. Alice receives a random bit string $\boldsymbol{e}$ of length N; after making measurements, she sends Bob a message, $\boldsymbol{f}$, a bit-string of length length m or less. Bob receives a random variable G, encoding a number, $k = 1, \ldots, N$, which he takes as the instruction to measure Alice's k-th bit. After making a suitable measurement, and taking into account both its outcome and Alice's message, Bob produces his guess, b_k. Information causality is the requirement that

$$\sum_{k=1}^{N} I(e_k : b_k | G = k) \leq m. \tag{10}$$

The main result of [55] is that if a theory contains states that violate the CHSH inequality by more than the Tsirel'son bound, then it violates information causality. In particular, if Alice and Bob can share PR boxes, then using a protocol due to van Dam [68], they can violate information causality maximally, meaning that Bob's guess is correct with certainty, and the left hand side of Eq. (10) is N. Pawlowski et al. also give a proof, using fairly standard manipulations of quantum mutual information, that quantum theory *does* satisfy information causality.

One of the principal results of [12] is a sufficient condition for a general probabilistic theory to be information-causal. The following is a strengthening of that result:

Theorem 10 *Suppose that a theory is strongly subadditive, and satisfies the Holevo bound. Then the theory satisfies information causality. It follows that any theory satisfying these conditions cannot violate Tsirel'son's bound.*

Since strong subadditivity and the Holevo bound follow from the data processing inequality, we have the following:

Corollary 5 *Any theory in which measurement-based mutual information satisfies the data processing inequality satisfies information causality.*

In [12], monoentropicity was assumed in addition to SSA and Holevo. As noted there, it was only used to derive that $H(A|B) \geq 0$ when A is classical. However, this follows easily from strong subadditivity in the equivalent (cf. Lemma 7) form $I(A:B|C) \geq 0$, when we let A and B be identical perfectly correlated classical systems. We have ([12], Addendum):

$$I(A:B|C) = H(A|C) + H(B|C) - H(AB|C) \tag{11}$$

$$= H(AC) - H(C) + H(BC) - H(C) - H(ABC) + H(C) \tag{12}$$

$$= H(AC) + H(BC) - H(ABC) - H(C). \tag{13}$$

Since A, B are perfectly correlated classical systems, $H(AC) = H(BC) = H(ABC)$. Consequently, in this case $I(A:B|C) = H(AC) - H(C) \equiv H(A|C)$. By SSA, this is ≥ 0.

4.5 Other Developments

There is much more to say about information processing in general probabilistic theories than we have room to discuss here. We remark in particular on [20], in which a version of the de Finetti theorem is proved for states on test spaces.

5 Characterizing Quantum Theory

As we've seen, a great number of information-processing phenomena first discovered in association with quantum theory, are actually rather more generally *post-classical*, as opposed to quantum, in character. This brings us back to the question of how to *characterize* quantum theory in operational or probabilistic terms. The idea is to identify one or more features of quantum theory that can be expressed in purely operational-probabilistic terms—roughly, without any special reference to the Hilbert space structure, but only in terms of primitive concepts such as states, effects, tests, processes, etc.—and that, taken together, *uniquely* specify quantum (or quantum-plus-classical) models. This is an old problem, and also a somewhat vague one, since what counts as a satisfactory solution will be, to some extent, a matter of taste. Even so, striking progress has been made in the past several years, leading to several different, more-or-less satisfactory characterizations of quantum mechanics as a probability theory [24, 28, 50, 57]. In this section, we review one of these [17, 18, 74, 76], which makes use of the equivalence between homogeneous self-dual cones and Euclidean Jordan algebras.

It will be convenient (and largely harmless) to assume in this section that every model A is *outcome-closed*, meaning that the image of $X(A)$ in $\boldsymbol{E}(A)$ is closed. This is a weak condition, virtually always satisfied in practice.

5.1 Homogeneity and Self-Duality

Let $\boldsymbol{E}$ be (for the moment) any finite-dimensional ordered linear space. Given a bilinear form $\mathcal{B} : \boldsymbol{E} \times \boldsymbol{E} \to \mathbb{R}$, we define the *internal dual* (with respect to $\mathcal{B}$) of the cone $\boldsymbol{E}_+$ to be the cone

$$\boldsymbol{E}^+ := \{a \in \boldsymbol{E} | \forall x \in \boldsymbol{E}_+,\ \mathcal{B}(a, x) \geq 0\}.$$

We say that $\mathcal{B}$ is *positive on* $\boldsymbol{E}_+$, or simply *positive*, iff $\boldsymbol{E}_+ \subseteq \boldsymbol{E}^+$—in other words, if the linear mapping $\beta : \boldsymbol{E} \to \boldsymbol{E}^*$ given by $\beta(a)(x) = \mathcal{B}(a, x)$ is positive.

Definition 21 $\boldsymbol{E}$ is *self-dual with respect to* $\mathcal{B}$ iff $\boldsymbol{E}^+ = \boldsymbol{E}_+$. We shall say that $\boldsymbol{E}$ is *weakly self-dual* iff there exists a bilinear form $\mathcal{B}$ with respect to which $\boldsymbol{E}$ is self-dual, and *strongly* self-dual, if there exists an *inner product* on $\boldsymbol{E}$ having this feature.

Weak self-duality is equivalent to the existence of an isomorphism state in $A \otimes_{\max} A$. As discussed above, this is equivalent to the requirement that there exist *some* composite of three copies of A that supports a conclusive teleportation protocol, and to the requirement that states on A arise as marginals of steering states in a composite of A with itself [14]. Strong self-duality is much less easy to motivate, but we will discuss several ways in which it can be justified in the next section.

Recall that $\boldsymbol{E}$ is *homogeneous* with respect to a group $\mathcal{G}$ of order-automorphisms if $\mathcal{G}$ acts transitively on the *interior* of the positive cone $\boldsymbol{E}_+$, so that for every pair of interior points a, b of $\boldsymbol{E}_+$, there exists an element $g \in \mathcal{G}$ with $ga = b$.

Classical and quantum probabilistic models are both homogeneous and self-dual. Somewhat more generally, let $\boldsymbol{E}$ be a euclidean Jordan algebra. This is a finite-dimensional real vector space $\boldsymbol{E}$ equipped with a commutative bilinear operation $\bullet$ satisfying the *Jordan identity* $a^2 \bullet (b \bullet a) = (a^2 \bullet b) \bullet a$ for all $a, b \in \boldsymbol{E}$, and equipped with a canonical trace such that $\langle a, b \rangle := \mathrm{Tr}(a \bullet b)$ is an inner product, with $\langle a \bullet b, c \rangle = \langle a, b \bullet c \rangle$ for all $a, b, c \in \boldsymbol{E}$. The set $\boldsymbol{E}_+ = \{a^2 | a \in \boldsymbol{E}\}$ (where $a^2 = a \bullet a$) is a cone in $\boldsymbol{E}_+$, and one can show is homogeneous with respect to the group of order-automorphisms of $\boldsymbol{E}$, and self-dual with respect to the tracial inner product. Remarkably, there is a converse, to be found in work of M. Koecher [44] and E. Vinberg [69].

If G be any closed subgroup of $\mathrm{Aut}(\boldsymbol{E})$, acting transitively on the interior of $\boldsymbol{E}_+$, then G is a Lie subgroup of $GL(\boldsymbol{E})$. Let $\mathfrak{g}$ denote its Lie algebra, and let $\mathfrak{g}_u$ denote the Lie algebra of the stabilizer $G_u \leq G$ of the order-unit. The following formulation of the Koecher-Vinberg Theorem summarizes the construction of the Jordan product on $\boldsymbol{E}$. See [32] for a proof (also, the Appendix to [18] contains a fairly detailed outline of the proof and some additional remarks pertinent to the precise version given below):

Theorem 11 (Koecher-Vinberg) *Let $\boldsymbol{E}_+$ be self-dual with respect to some inner product on $\boldsymbol{E}$, and let G be a closed, connected subgroup of $\mathrm{Aut}(\boldsymbol{E})$, acting transitively on the interior of $\boldsymbol{E}_+$. Then*

(a) It is possible to choose a self-dualizing inner product on $\boldsymbol{E}_+$ in such a way that $G_u = G \cap \mathcal{O}(\boldsymbol{E})$ (where $\mathcal{O}(\boldsymbol{E})$ is the orthogonal group with respect to the inner product);
(b) If $G = G^{\dagger}$ with respect to this inner product, then $\mathfrak{g}_u = \{X \in \mathfrak{g} | X^{\dagger} = -X\} = \{X \in \mathfrak{g} | Xu = 0\}$, and $\mathfrak{g} = \mathfrak{g}_u \oplus P$, where $P = \{X \in \mathfrak{g} | X^{\dagger} = X\}$;
(c) In this case the mapping $P \to \boldsymbol{E}$, given by $X \mapsto Xu$, is a vector-space isomorphism. Letting L_a be the unique element of P with $L_a u = a$, define

$$a \bullet b = L_a b$$

for all $a, b \in \boldsymbol{E}$. Then $\bullet$ makes $\boldsymbol{E}$ a formally real Jordan algebra, with identity element u.

In [43], Jordan, von Neumann and Wigner classified Euclidean Jordan algebras as belonging to one of two broad types, plus one exceptional example. These are

(a) ***Hermitian parts of matrix algebras*** over $\mathbb{R}$, $\mathbb{C}$ or $\mathbb{H}$, ordered as usual;
(b) ***Spin factors***, in which the normalized state space is a ball of dimension n; and
(c) ***The Exceptional Jordan Algebra*** of positive semidefinite 3×3 hermitian matrices over the Octonions.

Thus, it would seem that if we can motivate both homogeneity and self-duality in operational terms, we will go a great way towards obtaining an operational characterization of finite-dimensional QM. This problem is taken up in the next section.

We then discuss the consequences of assuming that a monoidal probabilistic theory consisting of Jordan models has locally tomographic composites. Here a theorem of H. Hanche-Olsen [40] can be invoked to show that, so long as the theory contains even a single instance of the simplest quantum-mechanical system—a qubit—every system allowed by the theory must be quantum.

5.2 Motivating Homogeneity and Self-Duality

Let us call a model A *HSD* (Homogeneous and self-dual) iff its linear hull $\mathbf{E}(A)$—or, equivalently, its dual, $\mathbf{V}(A)$—is homogeneous and self-dual. Why should this be the case? In this section, we discuss several possible answers.

Homogeneity Call a state $\alpha \in \Omega(A)$ *non-singular* iff $\alpha(x) > 0$ for all $x \in X(A)$. Since $\mathbf{V}(A)$ is finite-dimensional, it is easy to see that α is non-singular iff it lies in the interior of the cone $\mathbf{V}(A)_+$. If α and β are two states and τ is a probabilistically reversible process such that $\tau(\alpha) = t\beta$ for some $0 < t \leq 1$, then $t^{-1}\tau$ is an order-automorphism of $\mathbf{V}(A)$ taking α to β. This gives us a perfectly serviceable operational interpretation of homogeneity: any non-singular state can be prepared from any other, up to normalization, by a probabilistically reversible process.[15] As noted above, homogeneity is also implied by either of the following conditions:

(a) Every interior state is the marginal of an isomorphism state
(b) Every state is the marginal of a steering state.

Yet another way of arriving at the homogeneity of $\mathbf{V}(A)$ can be found in [74].

Self-Duality Self-duality seems less clear-cut, but can be obtained as a consequence of certain symmetry assumptions. Perhaps the simplest and most dramatic is the following beautiful result due to M. Mueller and C. Ududec. Call two states $\alpha, \beta \in \Omega(A)$ *sharply distinguishable by effects* iff there exists an effect a such that $\alpha(a) = 1$ and $\beta(a) = 0$. Mueller and Ududec call a system *bit-symmetric* iff every such pair of states can be mapped to any other such pair by a symmetry of the state cone, that is, an affine symmetry of Ω. They then prove:

Theorem 12 ([52]) *If* $\Omega(A)$ *is bit-symmetric, then* $\mathbf{V}(A)$ *(and hence,* $\mathbf{E}(A)$*) is self-dual.*

It is worth noting that not every self-dual model is bit-symmetric. For instance, if Ω is a 2-dimensional regular $2n + 1$-gon, then $\mathbf{V}(\Omega)$ is self-dual, but Ω is not bit-symmetric. Bit-symmetry is thus a very restrictive, yet very plausible, and operationally meaningful, constraint.

---

[15] One might raise the *aesthetic* objection that it is awkward to make special reference to the interior state. But it is difficult to see how this is any worse aesthetically than making special reference to, say, pure states.

A more involved condition having a somewhat similar flavor, but dealing with the test space structure $X(A)$ rather than the pure states of A, is worth mentioning. Call A *bi-symmetric* iff it is 2-symmetric under $G(A)$ and if $G(A)$ acts transitively on pure states. As discussed in Sect. 2.2, it is quite easy to construct such models one at a time. Recall that A is *sharp* iff for every outcome x, there is a unique state α with $\alpha(x) = 1$.

Theorem 13 ([76]) *Let $\mathcal{C}$ be a monoidal probabilistic theory in which every model is bi-symmetric. If $A \in \mathcal{C}$ is irreducible and sharp, then $\mathbf{E}(A)$ is self-dual.*

Another way of obtaining self-duality from bi-symmetry involves the notion of a conjugate system:

Definition 22 A *conjugate* for a model A is a structure $(\overline{A}, \gamma_A, \eta_A)$, where $\overline{A}$ is a model, $\gamma_A : A \to \overline{A}$ is an isomorphism, and η_A is a bipartite state (on some non-signaling composite) $A\overline{A}$ such that

$$\eta_A(x, \gamma_A(x)) = 1/n$$

for every $x \in X(A)$. We'll call γ_A the *conjugation map* and η_A, the *correlator* for the given conjugate.

Notice that this implies that every test $E \in \mathcal{M}(A)$ has cardinality n.

Example 10 Let $A = A(\mathcal{H})$ be the quantum model associated with a complex Hilbert space $\mathcal{H}$, and $\overline{A} = A(\bar{\mathcal{H}})$ associated with the conjugate Hilbert space. Define a mapping $\gamma_A : X(\mathcal{H}) \to X(\bar{\mathcal{H}})$ by $\gamma_A : x \mapsto \bar{x}$ (strictly speaking, the identity map!). Then, as discussed in Sect. 3.3, $\eta_A(x, \gamma_A(y)) = |\langle \Psi, x \otimes y \rangle|^2 = \mathrm{Tr}(P_\Psi P_{x\otimes y})$ is a correlator, where Ψ is the 'EPR' state.

If A has a conjugate, then it has a conjugate for which the correlator η_A is symmetric, in the sense that $\eta(x, \gamma_A(y)) = \eta(y, \gamma_A(x))$, and invariant, in the sense that $\eta_A(gx, \gamma_A(gy)) = \eta(x, \gamma_A(y))$. Indeed, $\eta^T(x, \gamma_A(y)) := \eta(y, \gamma_A(x))$ is again a correlator; averaging η and η^T gives us a symmetric correlator. If η is symmetric, then for all symmetries $g \in G(A)$, $\eta^g(x, y) = \eta(gx, gy)$ is again a symmetric correlator; averaging over G yields an invariant symmetric correlator. Henceforth, we assume that correlators are symmetric and invariant. It follows that the bilinear form

$$\mathcal{B}(a, b) := \eta(a, \gamma_A(b))$$

is *orthogonalizing*, meaning that $\mathcal{B}(x, y) = 0$ for all $x \perp y$ in $X(A)$. For the following, see [76]:

Theorem 14 *Let A be irreducible, bi-symmetric, and have a conjugate $(\bar{A}, \gamma_A, \eta_A)$. Then (a) $\mathcal{B}$ is an inner product on $\mathbf{E}$, and (b) A is self-dual with respect to $\mathcal{B}$ iff η_A is an isomorphism state iff A is sharp.*

5.3 *HSD and Jordan Models*

Call a model *A HSD* (Homogeneous and self-dual) iff the cone $\boldsymbol{E}_+$ is homogeneous under *some* group $\mathcal{G}(A)$ of order-automorphisms, and self-dual with respect to *some* inner product. If A is an HSD model, then by the Koecher-Vinberg theorem, $\boldsymbol{E}(A)$ carries a unique euclidean Jordan structure with respect to which the order unit, u, is the identity and $\langle a, u\rangle = \mathrm{Tr}(a)$.

An *idempotent* in a Jordan algebra $\boldsymbol{E}$ is an element $e \in \boldsymbol{E}_+$ with $e^2 = e \bullet e = e$. Idempotents in the special Jordan algebra $\mathcal{L}_h(\mathcal{H})$ are precisely orthogonal projection operators. A *primitive* idempotent is an idempotent that is not a sum of other non-zero idempotents; thus, in the context of $\mathcal{L}_h(\mathcal{H})$, a primitive idempotent is a rank-one projection operator. Any Euclidean Jordan algebra $\boldsymbol{E}$ carries a canonical trace functional, with $\mathrm{Tr}(ab) = \langle a, b\rangle$, and one can show that $\mathrm{Tr}(e) = 1$ for any primitive idempotent. A *Jordan frame* in a Euclidean Jordan algebra $\boldsymbol{E}$ is a set$e_1, \ldots, e_n$ of primitive idempotents summing to u. The Spectral Theorem for Euclidean Jordan algebras asserts that every $a \in \boldsymbol{E}$ has a unique representation as a sum of the form $\sum_{e\in E} t_e e$ over a Jordan frame E, where $\{t_e | e \in E\}$ are non-negative real coefficients. It follows that the extremal elements of the cone $\boldsymbol{E}_+$ are exactly the primitive idempotents. The group of order-automorphisms of $\boldsymbol{E}$ fixing the unit u acts transitively on the set of Jordan frames, so all Jordan frames have the same size, the *rank* of $\boldsymbol{E}$. (Indeed, regarding the set of Jordan frames as a test space, this group acts fully transitively, i.e., any permutation of a Jordan frame can be implemented by an order-automorphism of $\boldsymbol{E}$.)

Definition 23 A probabilistic model A is *uniform* iff its test have a uniform cardinality n, and the uniformly mixed probability weight $\mu(x) \equiv 1/n$ belongs to $\Omega(A)$.

If A is an HSD model, then every primitive idempotent e in $\boldsymbol{E}(A)$ defines a pure state, $\langle\epsilon|$, and this is the unique pure state assigning probability 1 to the effect corresponding to e. By a *Jordan model*, we mean an HSD model A such that every outcome in $X(A)$ is a primitive idempotent in $\boldsymbol{E}(A)$, or, equivalently, every test is a Jordan frame. Evidently, such a model is unital, indeed, sharp, and uniform.

In fact, these properties characterize Jordan models. Suppose that A is HSD. By an easy extension of the converse to the Krein-Mil'man theorem, any closed, generating subset of $\mathbf{V}(A)_+$ contains a point on every extremal ray of $\mathbf{V}(A)_+$. By our standing assumption of outcome-closure, $X(A)$ is closed in $\boldsymbol{E}(A) \simeq \mathbf{V}(A)$. By construction, it is generating. Therefore, every extremal ray of $\boldsymbol{E}(A)_+$ consists of multiples of an outcome. Giving $\boldsymbol{E}(A)$ its standard Jordan structure, primitive idempotents generate extremal rays of $\boldsymbol{E}(A)_+$, so every primitive idempotent in $\boldsymbol{E}(A)$ is a positive multiple of an outcome in $X(A)$.

Call an outcome $x \in X(A)$ *unital* iff there exists at least one state $\alpha \in \Omega(A)$ with $\alpha(x) = 1$.

Lemma 8 *Let A be HSD, and let* $\mathbf{E}(A)$ *have its canonical Jordan structure. Then:*

(a) Every extremal unital outcome $x \in X(A)$ *is a primitive idempotent.*
(b) If A is uniform, then every unital outcome is extremal, hence, a primitive idempotent.
(c) If A is both unital and uniform, it is a Jordan model.

Proof (a) Let $x \in X(A)$ be extremal. As discussed above, there then exists some $t > 0$ such that $tx =: e$, a primitive idempotent. Now suppose f is a primitive idempotent representing a pure state of $\mathbf{E}$, with $\langle f, x\rangle = 1$. Then

$$t = t\langle f, x\rangle = \langle f, tx\rangle = \langle f, e\rangle \leq 1,$$

by the Cauchy-Schwarz inequality. Now notice that

$$t^2\langle x, x\rangle = \langle e, e\rangle = 1$$

so $\langle x, x\rangle = 1/t^2$. Choosing any $E \in \mathcal{M}(A)$ with $x \in E$, we now have

$$1 = \langle e, u\rangle = t\langle x, u\rangle = t\left(\langle x, x\rangle + \sum_{y \in E\setminus\{x\}} \langle x, y\rangle\right) \geq t\langle x, x\rangle = t/t^2 = 1/t,$$

so that $t \geq 1$. Thus, $t = 1$, and $x = e$, a primitive idempotent.

(b) Let $x = \sum_i s_i x_i$ where the x_i are extremal outcomes and $s_i \geq 0$. Let μ be the uniform state on A. Then

$$\frac{1}{m} = \mu(x) = \sum_i s_i \mu(x_i) = \sum_i s_i \frac{1}{m}$$

so $\sum_i s_i = 1$. If x is unital, therefore, there exists a primitive idempotent f with

$$1 = \langle f, x\rangle = \sum_i s_i \langle f, x_i\rangle.$$

Since the coefficients s_i are convex, we have $\langle f, x_i\rangle = 1$ for every i with $s_i \neq 0$. But then, every x_i is a unital extremal outcome and so, by part (a), a primitive idempotent. It follows (again by the Cauchy-Schwarz inequality) that $s_i \neq 0$ implies $x_i = f$, whence, $x = f$ is again a primitive idempotent. (c) now follows at once from (a) and (b). □

5.4 Composites of Jordan Models

Suppose a probabilistic theory $\mathcal{C}$ consists entirely of Jordan models. Under what conditions can one equip $\mathcal{C}$ with an associative compositional structure so as to obtain a *monoidal* probabilistic theory? Subject to two further requirements, this this is possible *only* if $\mathcal{C}$ is in fact a standard quantum theory:

Theorem 15 ([18]) *Let $\mathcal{C}$ be a symmetric monoidal category of Jordan probabilistic models such that (i) for every $A, B \in \mathcal{C}$, the composite AB is locally tomographic, and (ii) at least one system in $\mathcal{C}$ has the structure of a qubit. Then every model in $\mathcal{C}$ is the hermitian part of a complex matrix algebra.*

The proof of this result exploits the following theorem due to H. Hanche-Olsen [40]. A *JB algebra* is a Jordan algebra $\boldsymbol{E}$ equipped with a norm making it a Banach space, and satisfying $\|a^2\| = \|a\|^2$ and $\|a^2\| \leq \|a^2 + b^2\|$ for all $a, b \in \boldsymbol{E}$. In finite dimensions, this is the same thing as a euclidean Jordan algebra.

Theorem 16 (Hanche-Olsen) *If $\boldsymbol{E}$ is a JB algebra and $\mathbf{M}_2$ is the Jordan algebra of 2×2 hermitian matrices over $\mathbb{C}$, then $\boldsymbol{E}$ is the Hermitian part of a complex matrix algebra iff there exists a Jordan product on $\boldsymbol{E} \otimes \mathbf{M}_2$ such that*

$$(a \otimes \mathbf{1}) \bullet (b \otimes \mathbf{1}) = ab \otimes \mathbf{1} \text{ and } (\mathbf{1} \otimes x) \bullet (\mathbf{1} \otimes y) = \mathbf{1} \otimes xy \tag{14}$$

for all $a, b \in \boldsymbol{E}$ and all $x, y \in \mathbf{M}_2$.

Essentially, [18] shows that if AB is a non-signaling HSD composite of HSD models A and B, then local tomography forces the Jordan product on $\boldsymbol{E}(AB)$ to satisfy (14). A key step is the following observation.

Lemma 9 *Suppose A is a Jordan model. Let AA be a non-signaling composite of A with itself. If AA is Jordan, then the trace form on $\boldsymbol{E}(AA)$ factors.*

Proof By definition of a composite, if $x, y \in X(A)$, then $x \otimes y$ is an outcome in $X(AA)$. Since x and y are unital in A, $x \otimes y$ is unital in $X(AA)$. Indeed, the pure product state $\langle x| \otimes \langle y|$ assigns $x \otimes y$ probability 1 (again, by definition of a composite). Hence, by part (b) of Lemma 8, $x \otimes y$ is a primitive idempotent in $\boldsymbol{E}(AA)$. But then we also have $\langle x \otimes y | x \otimes y \rangle = 1$, and this is the unique pure state with this property. Hence, $\langle x| \otimes \langle y| = \langle x \otimes y|$, so that

$$\langle x \otimes y | a \otimes b \rangle = \langle x | a \rangle \langle y | b \rangle$$

for all $a, b \in \boldsymbol{E}(A)$. Since $X(A)$ spans $\boldsymbol{E}(A)$, the same holds with arbitrary elements of $\boldsymbol{E}(A)$ in place of x and y, i.e., the inner product factors. □

Local tomography is a strong constraint on a probabilistic theory. The fact that real and quaternionic quantum mechanics are not locally tomographic should at least slightly temper our willingness to adopt it. A classification of non-locally tomographic non-signaling composites of Jordan models is the subject of on-going work.

6 Conclusion

The framework we have sketched here for a post-classical probability theory has several virtues. It is conceptually conservative, mathematically straightforward, and easily accommodates free mathematical constructions, as well as the introduction of further structure (for example, one can readily topologize the concept of a test space; see [72, 73]). Still, at present, what we have is indeed just the sketch of a framework. Its further development offers many interesting opportunities. We close by mentioning five areas for further work.

Quantum Axiomatics. As long as we restrict our attention to finite-dimensional probabilistic models, it seems that there are many different axiomatic packages—that is, many different clusters of plausible constraints—that locate orthodox QM, or its near environs, within the wild landscape of general post-classical probabilistic theories. In addition to the approach via homogeneity and self-duality, sketched in Sect. 4, there are various derivations of finite-dimensional QM in the spirit of Hardy's axioms [41], including work by Rau [57], Dakic and Brukner [28], Masanes and Mueller [50] and Chiribella, D'Ariano and Perinotti [24]. A different approach [37] exploits information geometry. There is also the completeness theorem of Selinger [63] for dagger-compact categories. This is not even to mention the various axiomatic treatments of quantum theory given in the older quantum-logical literature.[16] It would be of great interest to know how all of these various axiomatizations (most of which share at least a few assumptions), are related to one another. The mathematical framework developed here seems ideal for this task.

Infinite-Dimensional Models Of even greater interest would be to extend the results of these efforts to infinite-dimensional settings. Individually, infinite-dimensional probabilistic models have been well-studied [29, 31], and tools are available for dealing with composites in this setting, too [71]. However, the line of argument developed in Sect. 5, depending as it does on the Koecher-Vinberg Theorem, does not generalize easily to the infinite-dimensional setting. Efforts in this direction are just getting under way, and there is a great deal more work to be done.

Quantum Field Theory Algebraic quantum field theory associates an algebra of observables to each open subset of spacetime. An obvious project would be to consider a probabilistic theory in which each such region is associated with a probabilistic model, subject to the constraint that the model associated with a union of spacelike separated regions be a non-signaling composite of the models associated with the regions individually.

Applications; Post-Quantum Information Theory The notion of a probabilistic model is very broad. It would likely be a fruitful exercise to look for concrete information-theoretic applications in which models that are neither classical nor quantum arise. In anticipation of this, it would be reasonable to further develop the post-classical

[16] This last has sometimes been criticized for being "too mathematical"—that is, insufficiently operational and at the same time, too technically involved. It's worth pointing out, however, that much of it becomes significantly simpler when specialized to the finite-dimensional case.

information theory sketched in [12, 64], especially by investigating in some detail such ideas as *channel capacity* in this setting.

The Measurement Problem. Even though we take measurements and measurement-outcomes as primitives, nothing prevents us from asking whether these can be modeled dynamically *within* the formal framework presented here. Certain versions of the measurement problem can be formulated as theorems in this framework, leading one to wonder whether various strategies for resolving the *quantum* measurement problem—e.g., some version of "many worlds" interpretations, or the apparatus of decoherence—have analogues in the setting of a general probabilistic theory. If so, this would shed some light on *how* these interpretive moves work; if not, then the existence of such an analogue could be regarded as another constraint on a probabilistic theory, taking us closer to orthodox QM. A further discussion of these matters can be found in [75].

References

1. S. Abramsky, B. Coecke, A categorical semantics of quantum protocols, in *Proceedings of the 19th Annual IEEE Symposium on Logic in Computer Science (LICS'04)* (2004), pp. 415–425
2. E. Alfsen, F.W. Shultz, *Geometry of State Spaces of Operator Algebras* (Birkhäuser, Boston, 2003)
3. J. Allcock, N. Brunner, M.Pawlowski, V. Scarani, Recovering part of the quantum boundary from information causality (2009). arXiv:0906.3464.v3
4. H. Araki, On a characterization of the state space of quantum mechanics. Commun. Math. Phys. **75**, 1–24 (1980)
5. I. Amemiya, H. Araki, A remark on Piron's paper. Publ. Res. Inst. Math. Sci. Kyoto Univ. Ser. A **2**, 423–429 (1967)
6. J. Baez, Quantum quandaries: a category-theoretic perspective, in *The Structural Foundations of Quantum Gravity*, ed. by D. Rickles, S. French, J. Saatsi (Oxford University Press, Oxford, 2006) (arXiv:0404040.v2, 2004)
7. J. Baez, M. Stay, Physics, topology, logic and computation: a Rosetta stone, in *New Structures for Physics*. Lecture Notes in Physics, vol. 813, ed. by B. Coecke (Springer, Berlin, 2011) (arXiv:0903.0340, 2009)
8. H. Barnum, C. Caves, C. Fuchs, R. Josza, B. Schumacher, Noncommuting mixed states cannot be broadcast. Phys. Rev. Lett. **76**, 2818–2821 (1996)
9. H. Barnum, J. Barrett, M. Leifer, A. Wilce, Cloning and broadcasting in generic probabilistic theories (2006). arXiv:quant-ph/0611295
10. H. Barnum, J. Barrett, M. Leifer, A. Wilce, A generalized no-broadcasting theorem. Phys. Rev. Lett. **99**, 240501–240504 (2007). arXiv:0707.0620
11. H. Barnum, J. Barrett, M. Leifer, A. Wilce, Teleportation in general probabilistic theories, in *Mathematical Foundations of Information Flow*. AMS Proceedings of Symposia in Applied Mathematics, ed. by S. Abramsky, M. Mislove (American Mathematical Society, Providence, 2012). arXiv:0805.3553
12. H. Barnum, J. Barrett, L. Clark, M. Leifer, R. Spekkens, N. Stepanik, A. Wilce, R. Wilke, Entropy and information causality in general probabilistic theories. New J. Phys. **12**, 033024 (2010) (N J. Addendum, Phys. **12**, 129401 (2012)) (arXiv:0909.5075)
13. H. Barnum, O. Dahlsten, M. Leifer, B. Toner, Nonclassicality without entanglement enables bit-commitment (2008). arXiv:0803.1264
14. H. Barnum, P. Gaebler, A. Wilce, Ensemble steering, weak self-duality, and the structure of probabilistic theories (2009). arXiv:0912.5532

15. H. Barnum, R. Duncan, A. Wilce, Symmetry, compact closure, and dagger compactness for categories of convex operational models. J. Philos. Logic **42**, 501–523 (2013). (arxiv:1004.2920)
16. H. Barnum, C. Fuchs, J. Renes, A. Wilce, Influence-free states on compound quantum systems, arXiv:quant-ph/0507108v1 (2005)
17. H. Barnum, A. Wilce, Ordered linear spaces and categories as frameworks for information-processing characterizations of quantum theory (2009). arXiv:0908.2354
18. H. Barnum, R. Duncan, A. Wilce, Symmetry, compact closure, and dagger compactness for categories of convex operational models. J. Philos. Logic **42**, 501–523 (2013). arXiv:1004.2920
19. J. Barrett, Information processing in generalized probabilistic theories (2005). arXiv:quant-ph/0508211v3
20. J. Barrett, M. Leifer, The deFinetti theorem for test spaces. New J. Phys. **11** (2009). arXiv:0712.2265
21. C.H. Bennett, G. Brassard, Quantum cryptography: public key distribution and coin tossing, in *Proceedings of the IEEE International Conference on Computers, Systems, and Signal Processing*, Bangalore (1984), p. 175
22. C.H. Bennett, G. Brassard, C. Crepeau, R. Jozsa, A. Peres, W.K. Wootters, Teleporting an unknown quantum state via dual classical and Einstein-Podolsky-Rosen channels. Phys. Rev. Lett. **70**, 1895–1899 (1993)
23. G. Birkhoff, J. von Neumann, The logic of quantum mechanics. Ann. Math. **37**, 823–843 (1936)
24. G. Chiribella, G.M. D'Ariano, P. Perinotti, Probabilistic theories with purification. Phys. Rev. A **81**, 062348 (2010). arXiv:0908.1583
25. G. Chiribella, G.M. D'Ariano, P. Perinotti, Information-theoretic derivation of quantum theory. Phys. Rev. A **84**, 012311 (2011). arXiv:1011.6451
26. B. Coecke, A universe of interacting processes and some of its guises, in *Deep Beauty: Understanding the Quantum World Through Mathematical Innovation*, ed. by H. Halvorsen (Cambridge University Press, Cambridge, 2011)
27. D. Dieks, Communication by EPR devices. Phys. Lett. A **92**, 271172 (1982)
28. B. Dakic, C. Brukner, Quantum theory and beyond: is entanglement special? (2009). arXiv:0911.0695
29. L. Davies, An operational approach to quantum probability, Commun. Math. Phys. **17**, 239–260 (1970)
30. A. Dvurecenskij, *Gleason's Theorem and Its Applications* (Kluwer, Dordrecht, 1993)
31. C.M. Edwards, The operational approach to algebraic quantum theory. Commun. Math. Phys. **16**, 207–230 (1970)
32. J. Faraut, A. Koranyi, *Analysis on Symmetric Cones* (Oxford University Press, Oxford, 1994)
33. D. Foulis, C. Randall, An approach to empirical logic. Am. Math. Mon. **77**, 363–374 (1970)
34. D. Foulis, C. Randall, The empirical logic approach to the physical sciences, in *Foundations of Quantum Mechanics and Ordered Linear Spaces*, ed. by A. Hartkämper, H. von Neumann (Springer, Berlin, 1974)
35. D. Foulis, C. Randall, What are quantum logics and what ought they to be?, in *Current Issues in Quantum Logic*, ed. by E.G. Beltrametti, B.C. van Fraassen (Plenum, New York, 1981)
36. D. Foulis, C. Randall, Empirical logic and tensor products, in *Interpretations and Foundations of quantum Theory*, ed. by H. Neumann (B. I. Wisssenschaft, Mannheim, 1981)
37. P. Goyal, From information geometry to to quantum theory. New J. Phys. **12**, 023012 (2010). arXiv:0805.2770
38. R. Greechie, Orthomodular lattices admitting no states. J. Comb. Theory **10**, 119–132 (1971)
39. J. Gunson, On the algebraic structure of quantum mechanics. Commun. Math. Phys. **6**, 262–285 (1967)
40. H. Hanche-Olsen, JB algebras with tensor products are C^*-algebras, in *Operator Algebras and their Connections with Topology and Ergodic Theory*, ed. by H. Araki et al., Lecture Notes in Mathematics, vol. 1132 (Springer, Berlin,1985)
41. L. Hardy, Quantum theory from five reasonable axioms (2000). arXiv:quant-ph/00101012
42. A. Holevo, *Probabilistic and Statistical Aspects of Quantum Mechanics* (North-Holland, Amsterdam, 1982) (Second edition published by Edizioni della Normale, Pisa, 2011)

43. P. Jordan, J. von Neumann, E.P. Wigner, On an algebraic generalization of the quantum-mechanical formalism. Ann. Math. **35**, 29–64 (1934)
44. M. Koecher, Die geodätischen von positivitätsbereichen. Mathematische Annalen **135**, 192–202 (1958)
45. M. Kläy, Einstein-Podolsky-Rosen experiments: the structure of the sample space I, II. Found. Phys. Lett. **1**, 205–244 (1988)
46. M. Kläy, C.H. Randall, D.J. Foulis, Tensor products and probability weights. Int. J. Theor. Phys. **26**, 199–219 (1987)
47. G. Lindblad, A general no-cloning theorem. Lett. Math. Phys. **47**, 189–196 (1999)
48. G. Ludwig, *Foundations of Quantum Mechanics* (Springer, New York, 1985)
49. G. Mackey, *Mathematical Foundations of Quantum Mechanics* (Addison Wesley, 1963)
50. L. Masanes, M. Müller, A derivation of quantum theory from physical requirements. New J. Phys. **13**, 063001 (2011). arXiv:1004.1483
51. B. Mielnik, Geometry of quantum states. Commun. Math. Phys. **9**, 55–80 (1968)
52. M. Mueller, C. Ududec, The computational power of quantum mechanics determines its self-duality (2011). arXiv:1110.3516
53. G. de la Torre, Ll. Masanes, A. Short, M. Mueller, Deriving quantum theory from its local structure and reversibility (2011). arXiv:1110:5482
54. I. Namioka, R. Phelps, Tensor products of compact convex sets. Pac. J. Math. **31**, 469–480 (1969)
55. M. Pawlowski, T. Paterek, D. Kaszlikowski, V. Scarani, A. Winter, M. Zukowski, Information causality as a physical principle. Nature **461**, 1101 (2009)
56. C. Piron, *Foundations of Quantum Physics* (W.A. Benjamin, Reading, 1976)
57. J. Rau, On quantum versus classical probability. Ann. Phys. **324**, 2622–2637 (2009). arXiv:0710.2119
58. I. Satake, *Algebraic Structures of Symmetric Domains* (Publications of the Mathematical Society of Japan, no. 14) (Princeton University Press, Iwanami Shoten, 1980)
59. F. Shultz, A characterization of state spaces of orthomodular lattices. J. Comb. Theory **17**, 317–328 (1974)
60. E. Schrödinger, Probability relations between separated systems. Proc. Camb. Philos. Soc. **32**, 446–452 (1936)
61. P. Selinger, Towards a semantics for higher-order quantum computation, in *Proceedings of the 2nd International Workshop on Quantum Programming Languages, Turku, Finland*. Turku Center for Computer Science, Publication No. 33 (2004), pp. 127–143
62. P. Selinger, Autonomous categories in which A is isomorphic to A^* extended abstract, in *Proceedings of the 7th International Workshop on Quantum Physics and Logic (QPL 2010)* (Oxford, 2010), pp. 151–160
63. P. Selinger, Finite dimensional Hilbert spaces are complete for dagger-compact categories (extended abstract), in *Proceedings of the 5th International Workshop on Quantum Physics and Logic (QPL 2008), Reykjavik*, ENTCS, vol. 270, pp. 113–119 (2011)
64. A. Short, S. Wehner, Entropy in general physical theories. New J. Phys. **12**, 033023 (2010). arXiv:0909.4801
65. M.P. Soler, A characterization of Hilbert spaces by orthomoular spaces. Commun. Algebra **23**, 219–243 (1995)
66. B. Tsirel'son, Quantum generalizations of Bell's inequality. Lett. Math. Phys. **4**, 93–100 (1980)
67. A. Uhlmann, On the Shannon entropy and related functionals on convex sets. Lett. Math. Phys. **4**, 93–100 (1980)
68. W. van Dam, Implausible consequences of superstrong nonlocality. Nat. Comput. **12**, 9–12 (2013). arXiv:quant-ph/0501159
69. E.B. Vinberg, Homogeneous cones, Dokl. Acad. Nauk. SSSR **141**, 270–273 (1960). English translation: Soviet Math. Dokl. **2**, 1416–1619 (1961)
70. J. von Neumann, *Mathematical Foundations of Quantum Mechanics* (Princeton University Press, Princeton, 1955)

71. A. Wilce, The tensor product in generalized measure theory. Int. J. Theor. Phys. **31**, 1915–1928 (1992)
72. A. Wilce, Topological test spaces. Int. J. Theor. Phys. **44**, 1227–1238 (2005). arXiv:quant-ph/0405178
73. A. Wilce, Symmetry and topology in quantum logic. Int. J. Theor. Phys. **44**, 2303–2316 (2005)
74. A. Wilce, Four and a half axioms for finite-dimensional quantum theory, in *Probability in Physics: Essays in Honor of Itamar Pitowsky*, ed. by Y. Ben-Menahem, M. Hemmo (2012) (arXiv:0912.5530, 2009)
75. A. Wilce, Formalism and interpretation in quantum theory. Found. Phys. **40**, 434–462 (2010). http://philsci-archive.pitt.edu/3794/1/bub6d-post.pdf
76. A. Wilce, Symmetry, self-duality, and the Jordan structure of quantum theory (2011). arXiv:1110.6607
77. G. Wittstock, Ordered normed tensor products, in *Foundations of Quantum Mechanics and Ordered Linear Spaces*. Springer Lecture Notes in Physics, vol. 29, ed. by A. Härtkamper, H. Neumann (Springer, Berlin, 1974)
78. W.K. Wootters, W.H. Zurek, A single quantum cannot be cloned. Nature **299**, 802–803 (1982)
79. R. Wright, Spin manuals: empirical logic talks quantum mechanics, in *Mathematical Foundations of Quantum Theory*, ed. by A.R. Marlowe (Academic Press, New York, 1977)
80. N. Zierler, Axioms for non-relativistic quantum mechanics. Pac. J. Math. **11**, 1151–1169 (1961)

Part IV
Quantum Versus Super-Quantum Correlations

Information Causality

Marcin Pawłowski and Valerio Scarani

1 Certain Things Should Not Happen

Like many people working in quantum information science, Bob had spent a few weeks in the Centre for Quantum Technologies in Singapore, collaborating with Alice. Some time after he left, Alice finished preparing ten tutorials for her module on quantum biology. She thought of sharing them with Bob, who was preparing to teach a similar module in his university. However, the latest policies allow only 1 Mb attachment per year to an e-mail,[1] and each tutorial alone amounts at 1 Mb. Alice is in a dilemma: which tutorial will be the best for Bob? It would be much simpler to let Bob choose. But this means that the information about all the tutorials must be made available in Bob's location. How can that happen by sending only a much smaller amount of information?

Alice remembers having shared with Bob, when he was in Singapore, a one-time pad key and even several qubits maximally entangled with hers. Quantum channels

[1]As the reader may expect, this restriction is *not* really implemented in Singapore at the time of writing: we may have to wait for the next generation of managers.

M. Pawłowski (✉)
Faculty of Mathematics, Physics and Informatics,
Gdańsk University, 80-952 Gdańsk, Poland
e-mail: dokmpa@univ.gd

M. Pawłowski
Department of Mathematics, University of Bristol, Bristol BS8 1TW, UK

V. Scarani
Centre for Quantum Technologies, National University of Singapore,
3 Science Drive 2, Singapore 117543, Singapore
e-mail: physv@nus.edu.sg

V. Scarani
Department of Physics, National University of Singapore,
2 Science Drive 3, Singapore 117542, Singapore

G. Chiribella and R.W. Spekkens (eds.), *Quantum Theory: Informational Foundations and Foils*, Fundamental Theories of Physics 181,
DOI 10.1007/978-94-017-7303-4_12

can perform tasks that appear incredible to the classically-minded. Can then these shared resources be helpful for this specific task? Alice does not believe it: she knows that shared randomness and entanglement are no-signaling resources. So, she argues, how could they help in sending new information, like the tutorials, which did not even exist at the time of the sharing?

In this text, we show that Alice's argument is wrong: no-signaling resources could in principle solve that task. Her final conclusion is nevertheless correct: the no-signaling resources that exist in our world cannot solve that task. Why? It is probably beyond physics to answer this question. Maybe simply because certain things should not happen?

2 The Context

2.1 *Defining Quantum Physics*

Definire means to *find the boundary*. In order to define quantum physics, therefore, one can't invoke the "typically quantum" notions of coherence and entanglement: if anything, these notions fix the boundaries of classical physics. One really needs to go at the quantum *finis terræ*. However, all known natural phenomena can be made to fit in the quantum framework. So, are there any boundaries to be found at all?

We leave the question open regarding boundaries in nature. But there are certainly boundaries in the world of physical theories. In quantum *theory*: (i) physical systems must be described by Hilbert spaces, their pure states by one-dimensional projectors, with the rule that orthogonal vectors describe fully distinguishable states; and (ii) the evolution in time must be reversible. As well known by now, pretty much all the formalism stems from these two requirements: a clear boundary, a sharp definition, and a very successful one. However, curiosity is not assuaged: recipes (i) and (ii) define a boundary *with what*? What is there *outside*? How would physics be if (i) and (ii) would not be true?

2.2 *No-Signaling Is Not Enough*

2.2.1 No-Signaling as a Principle

It is far from easy to invent decent, consistent answers to the previous questions. Even the anarchical freedom of science fiction has ultimately produced a single creative alternative: *signaling*, in all its possible variations (faster-than-light travel, teleportation of matter between distant locations, etc.). No-signaling is certainly a boundary, and a very constraining one at that: just think how tiny is the portion of the universe that the human kind may hope to visit, unless a family of kind wormholes

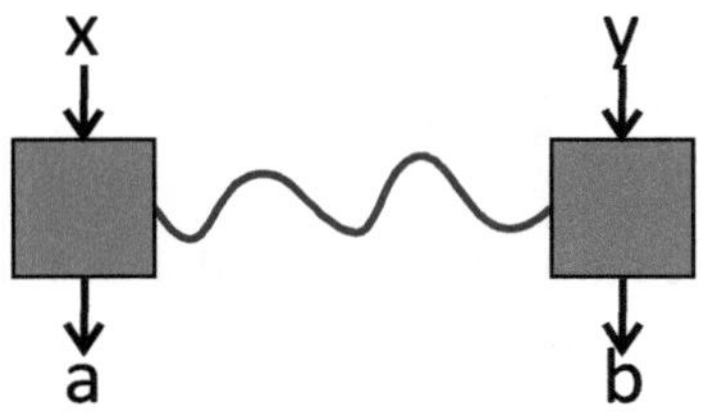

Fig. 1 The representation of a bipartite no-signaling probability distribution, or "no-signaling box", used in this text. The *wavy line* is not meant as a material connection, but only as a reminder of the existence of correlations. The PR-box is defined by $x, y, a, b \in \{0, 1\}$, random marginals i.e. $P(a|x) = P(b|y) = \frac{1}{2}$, and perfect correlations satisfying $a \oplus b = xy$

comes to rescue. So let us take this single suggestion seriously: *is no-signaling the physical principle that defines our (quantum) universe?*

Popescu and Rohrlich were the first to raise this question explicitly, and to find a *negative answer* [1]. The counter-example uses a simple mathematical object that had been described some years earlier by Rastall [2]; nowadays it is customarily referred to as *the PR-box*.[2]

2.2.2 The PR-Box and the CHSH Game

The PR-box is a specific bipartite no-signaling probability distribution with both binary input and output (Fig. 1). Alice can input a bit x and receives a bit a as output; and similarly Bob can input a bit y and receives a bit b as output. The PR-box is specified by the rule

$$P_{PR}(a, b|x, y) = \frac{1}{2}\,\delta_{a\oplus b=xy}, \tag{1}$$

where the symbol $\oplus$ indicates sum modulo 2. In other words, a and b are always locally random; they are equal in the three cases $(x, y) = (0, 0)$, $(0, 1)$ and $(1, 0)$, while they are different when $(x, y) = (1, 1)$.

The PR-box is tailored to violate maximally the Clauser-Horne-Shimony-Holt (CHSH) Bell inequality [3]. For the purpose of this paper, we present this criterion as *the CHSH game*. Alice and Bob are given two binary inputs and must produce, without communication, binary outcomes satisfying (1). If the inputs are distributed randomly, the probability of success is

[2]In the remainder of this section, we introduce notions and tools that are pretty basic for people working in the field, in order to address a more general readership and also to have a consistent discourse in the text. The reader in need of a more tutorial introduction can refer to Sect. 5 of: V. Scarani, *Quantum information: primitive notions and quantum correlations*, in: C. Miniatura et al. (eds), *Ultracold Gases and Quantum Information—Les Houches 2009 session XCI* (Oxford University Press, Oxford, 2011). Preprint available as http://arXiv.org/abs/0910.4222

$$p_{\text{CHSH}} = \frac{1}{4}\sum_{x,y=0}^{1} P(a \oplus b = xy|x, y)\,. \tag{2}$$

If Alice and Bob are allowed to use only classical shared randomness, their winning probability is bounded as $p_{\text{CHSH}} \leq p_C = \frac{3}{4}$. If they can share entanglement, their winning probability is increased up to the *Tsirelson bound* [4]

$$p_{\text{CHSH}} \leq p_Q = \frac{2+\sqrt{2}}{4} \approx 85\,\% \tag{3}$$

which is still smaller than one. By construction, the PR-box reaches $p_{\text{CHSH}} = 1$.

This simple argument proves that no-signaling cannot be the only physical principle that defines our quantum world. At least another constraint is in place, that limits the probability of success of the CHSH game. We can thus rephrase the questions of our curiosity: *given that we live in a world, in which Bell's inequalities are violated, why are they then not violated as much as no-signaling would allow?* Any physical principle (or collection thereof) claiming to come close to a definition of quantum physics should be able to deal with the riddle of the Tsirelson bound.

2.3 Mathematical Framework

We focus on an operational generalization of quantum kinematics (states and measurement, without dynamics). The measurement process is defined as "choosing an input and getting an output". The information about the state of the system is contained in the observed probability distributions of the outputs, for each input. Since we focus on bipartite systems, let us fix the notations: the inputs of Alice and Bob are written $x \in \mathcal{X}$ and $y \in \mathcal{Y}$, respectively; the outputs (we assume that every input leads to the same number of possible outcomes) are written $a \in \mathcal{A}$ and $b \in \mathcal{B}$ respectively. So, for each x, y, Alice and Bob can reconstruct the probability distribution $P_{xy} = \{P(a, b|x, y)\big|a \in \mathcal{A}, b \in \mathcal{B}\}$. All that Alice and Bob know about the system and the measurements is captured by the *probability point*

$$\mathcal{P} = \{P_{xy}\big|x \in \mathcal{X}, y \in \mathcal{Y}\}\,. \tag{4}$$

A priori, each P_{xy} is specified by $|\mathcal{A}||\mathcal{B}|-1$ values because of normalization; therefore $\mathcal{P}$ is generically specified by $|\mathcal{X}||\mathcal{Y}|\big(|\mathcal{A}||\mathcal{B}| - 1\big)$ values.

For the following, it is important to classify probability points as follows:

- $\mathcal{P}$ belongs to the *classical set* if it can be written as a convex combination of local deterministic points, i.e. points of the form $P(a, b|x, y) = \delta_{a=f(x)}\delta_{b=g(y)}$. These points are the extremal points of the classical set; since there are finitely many of them, namely $|\mathcal{A}|^{|\mathcal{X}|}\,|\mathcal{B}|^{|\mathcal{Y}|}$, the classical set is a polytope. In summary,

the classical polytope contains all the $\mathcal{P}$ that can be obtained from "local (hidden, or not hidden) variables".
- $\mathcal{P}$ belongs to the *quantum set* if there exist a state ρ and projectors $\{\Pi_a^x, \Pi_b^y\}$ such that

$$P(a,b|x,y) = \mathrm{Tr}\left(\rho \Pi_a^x \Pi_b^y\right), \tag{5}$$

where the projectors must satisfy $[\Pi_a^x, \Pi_b^y] = [\Pi_a^x, \Pi_{a'}^x] = [\Pi_b^y, \Pi_{b'}^y] = 0$ for all a, b, x, y. There is no loss of generality in considering only projective measurements, since the dimensionality of ρ is not restricted. For finite-dimensional Hilbert spaces, these relations between projectors are fulfilled if and only if there is a tensor product representation $\Pi_a^x = \pi_a^x \otimes \mathbb{1}$ and $\Pi_b^y = \mathbb{1} \otimes \pi_b^y$ [5].
- $\mathcal{P}$ belongs to the *no-signaling set* if $P(a|x,y) = P(a|x)$ and $P(b|x,y) = P(b|y)$ for all a, b, x, y. This set is also a polytope. Clearly the classical set is included in the quantum set, which is included in the no-signaling set. Notice also that the no-signaling constraints reduce the number of values required to specify a probability point $\mathcal{P}$ down to $|\mathcal{X}||\mathcal{Y}|\big(|\mathcal{A}|-1\big)\big(|\mathcal{B}|-1\big) + |\mathcal{X}|\big(|\mathcal{A}|-1\big) + |\mathcal{Y}|\big(|\mathcal{B}|-1\big)$.

In this framework, *we are looking for a physical principle, which would single out the quantum set within the no-signaling polytope.*

Before continuing, we want to stress a difference with other operational approaches, in particular with the line of research on axiomatics [6]. There, a lot is built on the assumption of tomography: it is supposed that some given $\mathcal{P}$'s are known to carry all the information the system. This is physically possible if the degrees of freedom under study and the measurements that are being performed on it have been characterized. Here, on the contrary, we work in a *completely black-box scenario*, ultimately the same as in Bell's theorem and in device-independent assessments [7]. In such a scenario, the point $\mathcal{P}$ can never be claimed to be "the state", with the idea of complete information that this term conveys. Rather, $\mathcal{P}$ encodes just the information that can be gathered from the black boxes. This is also one of the reasons why we start out with bipartite systems: in a black-box scenario, the behavior of a single system can always be described in terms of hidden variables.

3 Information Causality: The Task

The statement of "no-signaling" is the impossibility of a task, namely, sending any amount of information by sampling a bipartite probability distribution. Every device independent principle must have a task (an information processing protocol) and a statement about it. In this section we aim at explaining the choice of the task and the statement of Information Causality. We start by asking the question: in what sense the PR-box is to powerful?

3.1 The Power of the PR-Box

The first device independent principle that put some bounds on the winning probability of the CHSH game was that of nontrivial communication complexity [8]. It has been shown that the access to perfect PR-boxes allows two parties to solve any communication complexity problem with the transmission of a single bit. Later this result has been improved in [9] where it was shown that the same happens even if the boxes are a little noisy, i.e. they allow for the success probability in the CHSH game greater than $\frac{3+\sqrt{6}}{6} \approx 0.908$. The question whether this principle can be used to derive even stronger limits is still open.

The simple idea behind taking this approach to study nonlocality is that if nothing seems to be wrong with the PR-boxes if the parties are not communicating (and no communication must be the case if we would like to use the no-signaling principle) then maybe there is something wrong with them when the communication takes place. To see why this should be the case let us put ourselves in the place of Bob, the owner of one part of the PR-box. When we choose our setting to be $y = 0$ we know that the outcome of our part of the box is going to be equal to the outcome of Alice $b = a$. If we choose $y = 1$ instead then we can expect $b = a \oplus x$. We see that we can choose to learn any one of the two independent bits a or $a \oplus x$ by choosing different settings. Granted that these two bits are perfectly random, but their randomness is the same. What we mean by that is that both of them are generated by XORing something deterministic (i.e. 0) or controlled by Alice (i.e. x) with the same random bit a. This is important because it allows, by transmitting later only a single bit form Alice to Bob, to erase the randomness in any of the bits that Bob might want to get regardless of his choice.

This property of the PR-box has been exploited in [10] in the context of oblivious transfer (Fig. 2). Imagine that Alice has two bits x_0 and x_1. She can send only one bit of classical communication to Bob who is interested in one of the bits (Alice does not know in which). Let the index of the bit that Bob is interested in be k. If they have access to a PR-box they can do this. Alice inputs $x = x_0 \oplus x_1$ in her part of the box and, after reading a, sends the one bit message $m = x_0 \oplus a$ to Bob. Bob inputs $y = k$, reads b and computes $C = m \oplus b = x_0 \oplus a \oplus b$. It is easy to see that $C = x_k$. Indeed, if $k = 0$ then $a = b$ and $C = x_0$; if $k = 1$, then $b = a \oplus x$ and $C = x_0 \oplus x = x_0 \oplus x_0 \oplus x_1 = x_1$.

Earlier we have promised that this analysis will show us what goes wrong if we consider the protocols with PR-boxes and communication. We are almost there. Look

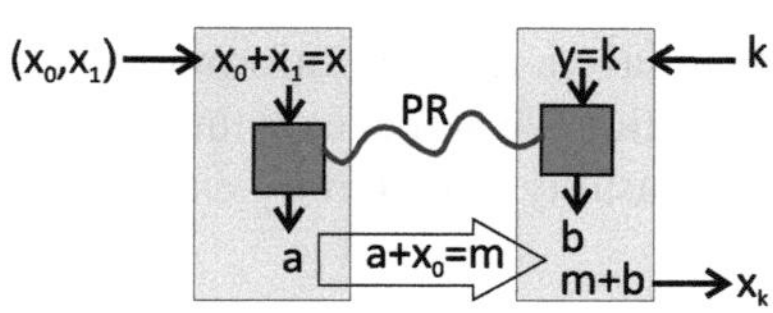

Fig. 2 Implementation of perfect oblivious transfer using the PR-box and one bit of communication

at the situation in the Bob's laboratory when he has already received Alice's message but he has not yet chosen which bit to decode. Considered as a black box his lab now has, in some sense, two bits. True that the extraction of one will destroy the other but, since any can be decoded, they both must be there. But we have transmitted only a single bit and the PR-boxes are supposed to be no-signaling so they cannot be used to transmit the other. Somehow the amount of information that the lab of Bob has is larger than the amount it received. Things like this should not happen.

3.2 *Random Access Codes*

The protocol that we have just described is called (2,1,1) Random Access Code (RAC) [11]. It allows Alice to encode two bits x_0 and x_1 into a single bit message m in such a way that Bob can decode any bit he chooses to. The notion generalizes to that of (N, M, p) RAC, which allows Alice to encode N bits into M bit message in such a way that the worst case probability of Bob decoding any of these bits correctly is p.[3] We can talk here as well about the average success probability instead of the worst case since Yao's principle [12] applied to RACs allows, with the use of shared randomness, to make these two equal [13]. There are many different types of RACs with slightly different properties which depend on the resources that we allow to be used. The most important distinction among the known codes lies in what is being communicated (classical bits or qubits).

In the code presented above the bits are decoded correctly as long as the correlations $a = b$ for $y = 0$ and $a = b \oplus x$ for $y = 1$ are always true. If they occur with probability p then the box can win the CHSH game with this probability and, at the same time, the average success probability of $(2, 1, p)$ RAC is also p. Therefore, we see that finding a way to bound the success probability of the RAC is equivalent to finding the bound on the probability to win the CHSH game.

3.3 *Task and Statement of Information Causality*

We are now in a position to define the *task*, to which the principle of Information Causality is going to apply. It is the same as a (N, M, p) Random Access Code, where N and M are classical bits (Fig. 3). Notice that it does not matter how this information is encoded: when we refer to "sending the M bit message", it should be understood as a single use of a channel with classical communication capacity M.

The *statement* of Information Causality requests that, in the task just defined, *the amount of information potentially available to Bob about Alice's input cannot*

[3] Earlier we have mentioned that the protocol that we have described is for oblivious transfer. It might puzzle the reader that we are now referring to it as RAC. The difference between these two is that in the oblivious transfer there is one more requirement: Bob after choosing to decode one bit cannot learn anything about the other. In RAC there is no such assumption although in the optimal RACs it is always the case.

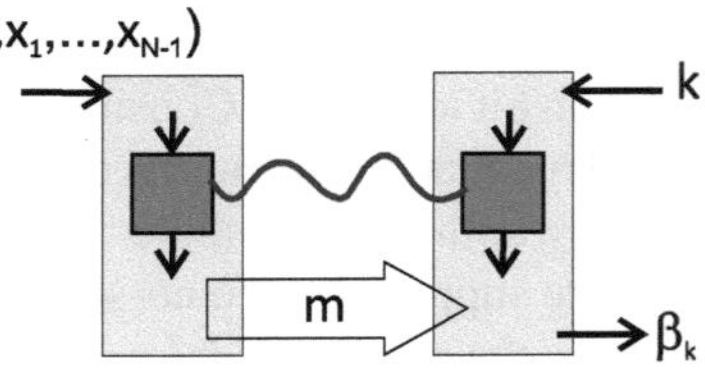

Fig. 3 The task of Information Causality is the same as the one that defines a Random Access Code. Alice receives N bits, and Bob is asked to guess one of them. Alice is allowed to send a message m that carries M bits, where $M < N$ to avoid trivialities. Moreover, Alice and Bob can share a no-signaling resource—and in fact, in all this study the goal is to compare the power of such resources. The usual figure of merit is the success probability $p = \sum_{k=0}^{N-1} \text{Prob}(\beta_k = x_k|k)$; Information Causality rather quantifies the amount of information that is potentially available in Bob's location

exceed M bits. This potentiality is the key to the Information Causality's success. If we would consider only the information that Bob *actually* gets, then this principle would be equivalent to no-signaling (indeed, imposing that Bob can actually receive only M bits is equivalent to stating that any additional resource is no-signaling). However, this little tweak makes all the difference as we will see in the next section.

3.4 The Reason for the Name

But before we get to it, we would like to take this opportunity and make a short comment on the choice of the name for our principle. We do this mainly because several people have asked us for the justification of our choice.

Let us reiterate that *Information Causality is about forbidding more information to be potentially available to the receiver than has been sent by the sender*. We hope that expressing our principle in that form makes the choice of the name clearer. Causality is the ability to change something over space-time. In the task we are considering, what gets changed is the information that Bob has about the particular bits of Alice. Before the protocol is run it is, by definition, zero. The cause is the transmission of the message, which increases the information. The statement about the task is that this increase in information is limited. In other words, we are putting a limit on the effect that the cause can have in the terms of information. Hence the name.

4 Mathematics

4.1 The Figure of Merit

Now we are ready to present the principle of Information Causality (shortened IC from now on) in its formal version. There are many different measures of information to choose from but in our case the choice is quite obvious. Since the task is about communicating over a channel with a specified classical communication capacity and because Shannon's celebrated single letter formula relates it to mutual information we take this measure. Therefore, the amount of information that Bob can potentially have about the variable x_i of Alice is given by $I(x_i : \beta_i)$ where β_i is the random variable that he generates when using his optimal procedure for maximizing the amount of information about this particular x_i. The statement of IC is that

$$\sum_{i=1}^{N} I(x_i : \beta_i) \leq M. \tag{6}$$

Note that we the variables x_i do not have to be binary. We do not make any assumptions about their alphabets. The definition of IC that we have given here is slightly stronger than the one given in the original paper [14]. There we have assumed that the communication from Alice to Bob is over a noiseless classical channel. This assumption can be lifted and, as we show in the next section, our principle will still hold in the quantum theory.

4.2 *Information Causality Holds for Quantum No-Signaling Resources*

IC sounds like a reasonable thing to expect from the universe but so does locality, determinism or the notion of absolute time. Therefore, in the presentation of a new principle, there should always be a proof that it is not violated by nature. Now we present a proof that IC holds in the classical and quantum information theory. We focus on quantum correlations because classical correlations form a subset of quantum correlations.

Let us denote by ρ_B Bob's part of the shared quantum state and $\vec{x}$ the set of all Alice's variables x_i. We begin by showing that after receiving the message $\vec{m}$, which was communicated over the channel with the classical communication capacity M, from Alice all the classical and quantum information he has does not have more than M bits of information about $\vec{x}$:

$$I(\vec{x} : \vec{m}, \rho_B) \leq M. \tag{7}$$

For the proof we use the chain rule for mutual information, $I(\vec{x} : \vec{m}, \rho_B) = I(\vec{x} : \rho_B) + I(\vec{x} : \vec{m}|\rho_B)$. Since at the beginning of the protocol Bob knows nothing about the variables of Alice $I(\vec{x} : \rho_B) = 0$, and the second term $I(\vec{x} : \vec{m}|\rho_B) = I(\vec{x}, \rho_B : \vec{m}) - I(\rho_B : \vec{m})$ is bounded by M due to the positivity of the mutual information and the fact that $\vec{m}$ is a message sent over the channel with the classical communication capacity M.

In the case of independent Alice's input bits condition (7) limits the information gain about the individual bits as well because

$$I(\vec{x} : \vec{m}, \rho_B) \geq \sum_{i=1}^{N} I(x_i : \vec{m}, \rho_B). \tag{8}$$

This inequality is also proved using the chain rule. Finally, we observe that Bob's output bit β_i is obtained at the end from $\vec{m}$ and ρ_B. Hence, the data processing inequality implies $I(x_i : \vec{m}, \vec{B}) \geq I(x_i : \beta_i)$ which gives us (6).

4.3 *Information-Theoretical Derivation of the Tsirelson Bound*

Here we show that any theory which allows for the violation of the Tsirelson bound violates also IC. To this end we consider a concatenated RAC. Let us explain what we mean by this.

Previously we have presented a code which encodes two classical bits into a single one and gives the average probability of correct decoding equal to the winning probability of the CHSH game. We may think about it as a pair of black boxes. Alice puts two bits into hers and it returns a single bit which she sends to Bob. Bob then puts this message into his box, makes a choice which bit he wants to learn and gets a value which with the probability p is equal to the bit he is interested in. Now imagine that Alice gets four bits instead of two and she is still limited to one bit of communication. She and Bob can construct a RAC for this task with the pairs of the same boxes they used previously with the help of concatenation procedure. It works like this: The parties need three pairs of boxes. Alice puts two of her bits into her first box and the remaining two into the second. The boxes have produced two messages which she does not send to Bob but puts into her third box, instead. It is the output of this final box that she sends to Bob. He inputs it to his box from the third pair and chooses to learn the message generated by the first or the second box of Alice. He inputs this message into one of his other boxes - the one paired with the box of Alice that generated this message, and then he can retrieve the bit. The overall success probability is now $p^2 + (1-p)^2$ if the success probability for each pair of boxes is p.

The generalization of this procedure is quite straightforward. If the parties use n levels of concatenation (using just a single pair of boxes corresponds to $n = 1$) they can encode 2^n bits using $2^n - 1$ pairs of boxes. The overall success probability of decoding the desired bit correctly is $p_n = \frac{1+E^2}{2}$, where E is the bias of the probability p (i.e. $p = \frac{1+E}{2}$).

If β_i is Bob's best guess of x_i and they are equal with the probability p_n then $I(x_i : \beta_i) = 1 - h(p_n)$, where $h(.)$ is Shannon's binary entropy. By expanding it into the Taylor series one gets that

$$1 - h\left(\frac{1+E^n}{2}\right) \geq \frac{E^{2n}}{2\ln 2}. \tag{9}$$

Since only one bit has been communicated, IC implies

$$1 \geq \sum_{i=1}^{2^n} I(x_i : \beta_i) \geq 2^n \frac{E^{2n}}{2\ln 2} = \frac{1}{2\ln 2}\left(2E^2\right)^n \tag{10}$$

for any n. This is going to be true only if $2E^2 \leq 1$ or, equivalently, $E \leq \frac{1}{\sqrt{2}}$. This puts a bound on the winning probability of the CHSH game $p \leq \frac{1}{2}\left(1 + \frac{1}{\sqrt{2}}\right)$ which is exactly the Tsirelson bound.

Quite straightforward generalization of this method can be employed if the probabilities of guessing different bits are different. In [15] it has been used to derive the bound on the efficiency of the RAC's

$$\sum_{i=1}^{N} E_i^2 \leq 1, \tag{11}$$

where E_i is the bias of the guessing probability for the i'th bit.

4.4 Entropic Approach

In order to prove that IC holds in quantum mechanics we have used the properties of mutual information. This means that something must go wrong with entropy measures for superstrong nonlocal boxes, as indeed was discussed shortly after the first IC paper [16]. In the latest development [17], it has been shown that all the properties necessary for the derivation of IC are consequences of only two conditions. This means that even if we choose a measure of information different than the mutual information, the objects exhibiting more nonlocality than the quantum theory allows will violate at least one of these conditions.

The conditions proposed in [17] are for the entropies $H(.)$. The information that object A has about B can be defined as for the von Neumann entropies as $I(A : B) = H(A) + H(B) - H(A, B)$. The first of the conditions is consistency: if A is a classical random variable, then $H(X)$ is equal to the Shannon entropy of X. The second is evolution with an ancilla: for any two systems A and B, whenever a transformation is performed on B alone, one must have $\Delta H(A, B) \geq \Delta H(B)$. It can be understood as saying that local transformations can only destroy correlations not create them.

Since the consistency condition is nothing more than the normalization of the entropy, it must be the second one which is violated by the superstrong nonlocality. This provides another characterization of what is wrong with no-signaling theories that violate Tsirelson bound: even though they cannot instantaneously send information at a distance, they can create correlations which is just as unacceptable.

Recently a slightly generalized version of IC has been proposed [18]. It keeps all its reasonable appeal and leads to entropic inequalities that are strictly stronger than in the original version. Recall that the reasoning that lead us to stating IC included two steps. In the first step, we argued that if the Bob's part of the system together with the message are treated as a single black box, then the information it has about the settings of Alice cannot exceed the classical communication capacity of the channel. If we associate random variable e with this black box we can express this statement formally as

$$H(\vec{m}) \geq I(\vec{x} : e). \tag{12}$$

In the second step, we argued that the random variable β_i is obtained locally from e, therefore the data processing inequality implies

$$\forall_i \quad H(x_i|\beta_i) \geq H(x_i|e). \tag{13}$$

If we sum up all the inequalities (12) and (13) and use the subadditivity of the entropy we obtain

$$H(\vec{m}) + \sum_i H(a_i|\beta_i) \geq H(\vec{x}), \tag{14}$$

which is equivalent to (6) in the case when the x_i are independent. But nothing forces us to sum up all the terms with the same weight. In fact, we can use a different one for each of the inequalities and get that, for all $w_i \geq 0$ and every $p(e|\vec{x})$, it holds

$$w_0 H(\vec{m}) + \sum_i w_i H(a_i|\beta_i) \geq w_0 I(\vec{x} : e) + \sum_i w_i H(a_i|e), \tag{15}$$

which is strictly stronger than the original IC. It remains to be seen if this new version of the principle leads to tighter bounds on what is possible in our world and what is not.

5 (Un?)expected Complexity

The fact that IC solves the riddle of the Tsirelson bound has been considered as a remarkable success. But of course, the ultimate goal is far more ambitious: is IC *the* physical principle that defines our quantum universe? In other words, does IC define exactly the quantum set within the no-signaling polytope, in any scenario? In the following, we refer to this scientific quest as to *the IC program.*

Several subsequent studies have witnessed partial success and lead to a wealth of unanswered questions—which are of course also an asset for research, at least as long as their complexity does not suffocate the driving motivation. In this last section, we review the status of the IC program.

5.1 Non-isotropic Correlations

The recovery of the Tsirelson bound proves that IC defines the quantum set if one considers the single-parameter family of "isotropic correlations", that is, the probability points that can be written as a convex combination of the PR-box and the white noise. In the first extension of the basic result, the authors considered whether IC defines the whole quantum set in the CHSH scenario [19]. The conclusion is that we don't know yet. Specifically, the paper focused on two-parameter families (recall that the no-signaling polytope lives in an eight-dimensional space). The violation of IC is assessed using the same explicit protocol described above, which is not guaranteed to be optimal *a priori*. For some families, IC is found to be violated as soon as one leaves the quantum set; in other cases, a finite gap is left. Similar results have been obtained by studying the probability points that admit a Hardy's paradox [20].

Adopting an optimistic view on the IC program, one may surmise that the gap is only due to the specific protocol using concatenated RAC. Indeed, a subsequent paper showed that this protocol is provably not optimal for some points [21]. Indeed, some points, which do not exhibit a violation of IC under that protocol, can be "distilled" to points which do violate IC under the same protocol. In other words, if the process of "distillation" is added to the protocol, the gap shrinks. However, it is not yet fully closed. Notice that, apart from the fact itself of belonging to the quantum set, we know don't know any sufficient condition for IC to be respected.[4]

The scary part of it all comes when one realizes that we are still speaking of the elementary CHSH scenario: two parties, two inputs and two outputs! Quantum physics is certainly more than that. What can one say for more general scenarios?

5.2 Comparison with "Macroscopic Locality"

The first natural generalization consists in keeping the bipartite scenario and enlarging the alphabets of the inputs and/or the outputs of the no-signaling resource. Obviously, this can in principle be done by keeping the task as a RAC involving bits. For simplicity, though, the only larger-alphabet study published so far [23] generalized also the task to a RAC in which Alice receives N classical dits and send $M = 1$ classical dit to Bob. The underlying no-signaling resources are such that $|\mathcal{X}| = |\mathcal{A}| = |\mathcal{B}| = d$, while $|\mathcal{Y}| = 2$.

The main result of this paper is the observation that IC comes closer to defining the quantum set than does *macroscopic locality* (ML). The latter is another criterion proposed with a similar scope [24]. It basically says that, in an experiment with many independent sources, the coarse-grained statistics should not violate Bell's inequalities. For instance, imagine a down-conversion experiment in which one would not

[4]A sufficient condition for IC to hold has been given [22], but for a fixed protocol (how to use the no-signaling resource, coding of the signal bit etc.); it is therefore of limited scope.

be able to count photons and had to rely on proportional counting: then the observed currents and their fluctuations could be compatible with a classical source.

The correlations that satisfy ML have been characterized completely: they form a set which is close, but not identical, to the quantum set. Therefore, it is a necessary condition for the IC program to succeed, that IC can rule out more correlations than ML does. Reference [23] provides examples of correlations for which it is indeed the case.

5.3 IC and Multi-partite Correlations

Complexity is further increased if one moves from bipartite to multipartite situations. Even in the simplest tripartite scenario (two inputs and two outputs per party), the structure of the no-signaling polytope is appalling [25].

One can certainly take multipartite boxes and use them as underlying no-signaling resource in a bipartite scenario: for instance, in the tripartite case, Alice may hold two of the input-output ports and even wire them together, while Bob keeps the third port. This has been tried, and the result is somehow expected: bipartite IC is powerful enough rule out many examples of non-quantum points [26], but not all. In fact, two different examples have been reported of tripartite probability points, which are definitely not quantum but which exhibit classical behavior in any bipartite scenario [27, 28].

In themselves, these results do not vanify the IC program: it is not surprising, after all, that the quantum set of multipartite scenarios can't be captured by a bipartite criterion. However, to find a natural generalization of the IC task to more parties has proved daunting: the several attempts we are aware of through private communications have not lead to any interesting development. There has been no systematic attempt of classifying those failures, but loosely speaking, the obstacle seems to be that most (if not all) multipartite communication task can ultimately be broken down into a succession of pairwise communications.

6 Conclusion

Formulated just five years ago, Information Causality has immediately attracted the attention of the scientific community. The reason for this success may be purely sociological: the idea that physics may be defined in terms of information processing has been lingering for many years and IC came to fill in the expectation. But we prefer to think in more "objective" terms: as we were trying to argue all along this text, IC is a very sensible thing to assume about the universe.

Improvement on the initial study have proved technically challenging: often restricted to extremely specific examples, they have nevertheless provided interesting information about the power of the notion of IC and unraveled some of its complex features. A few more of these specific studies will certainly be welcome; but if the IC program has to succeed, one will have to find a much more comprehensive approach. It is our sincere wish that this short review be outdated soon.

We acknowledge the support of TEAM programme of foundation for Polish Science (FNP), the NCN grant 2013/08/M/ST2/00626, the Singapore Ministry of Education and the National Research Foundation of Singapore.

Note Added in Proof

We want to highlight two papers that appeared in the period between the completion of this short review and its final printing.

Navascués and co-authors [29] conjectured that none of the information-theoretical principles presented so far will single out the quantum set, but at best a specific "almost quantum" set that is known to be strictly larger than it. The conjecture is actually proved for all the other principles but IC; the authors also show that all the results on IC published so far do not contain any counter-example to their conjecture.

Chaves and co-authors [30] managed to derive hitherto unknown entropic inequalities using the formalism of causal sets. As a consequence of one of these new bounds, they were able to extend slightly the set of correlations that provably violate IC.

References

1. S. Popescu, D. Rohrlich, Found. Phys. **24**, 379 (1994)
2. P. Rastall, Found. Phys. **15**, 963 (1985)
3. J.F. Clauser, M. Horne, A. Shimony, R.A. Holt, Phys. Rev. Lett. **23**, 880 (1969)
4. B.S. Cirelson, Lett. Math. Phys. **4**, 93 (1980)
5. B.S. Tsirelson, Bell inequalities and operator algebras (2006). http://www.imaph.tu-bs.de/qi/problems/33.html
6. For instance: L. Hardy, arXiv:quant-ph/0101012v4; B. Dakić, Č. Brukner, in Deep Beauty: Understanding the Quantum World Through Mathematical Innovation, ed. by H. Halvorson (Cambridge University Press, 2011), pp. 365–392; G. Chiribella, G.M. D'Ariano, P. Perinotti, Phys. Rev. A 84, 012311 (2011); L. Masanes, M. Müller, New J. Phys. 13, 063001 (2011)
7. For instance: A. Acín et al., Phys. Rev. Lett. 98, 230501 (2007); S. Pironio et al., Nature 464, 1021 (2010); R. Colbeck, A. Kent, J. Phys. A: Math. Theor. 44, 095305 (2011); R. Rabelo et al., Phys. Rev. Lett. 107, 050502 (2011)
8. W. van Dam, arXiv:quant-ph/0501159
9. G. Brassard, H. Buhrman, N. Linden, A.A. Méthot, A. Tapp, F. Unger, Phys. Rev. Lett. **96**, 250401 (2006)
10. S. Wolf, J. Wullschleger, arXiv:quant-ph/0502030
11. A. Ambainis, A. Nayak, A. Ta-Shma, U. Vazirani, J. ACM **49**(4), 496 (2002)
12. A.C. Yao, Proceedings of 11th STOC 14, 209 (1979)
13. A. Ambainis, D. Leung, L. Mancinska, M. Ozols, arXiv:0810.2937
14. M. Pawłowski, T. Paterek, D. Kaszlikowski, V. Scarani, A. Winter, M. Żukowski, Nature **461**, 1101 (2009)
15. M. Pawłowski, M. Żukowski, Phys. Rev. A **81**, 042326 (2010)
16. A.J. Short, S. Wehner, New J. Phys. 12, 033023 (2010); H. Barnum, J. Barrett, L. Orloff Clark, M. Leifer, R. Spekkens, N. Stepanik, A. Wilce, R. Wilke. New J. Phys. **12**, 033024 (2010)
17. S.W. Al-Safi, A.J. Short, Phys. Rev. A **84**, 042323 (2011)

18. S. Beigi, A. Gohari, arXiv:1111.3151v1
19. J. Allcock, N. Brunner, M. Pawłowski, V. Scarani, Phys. Rev. A **80**, 040103 (2009)
20. A. Ahanj, S. Kunkri, A. Rai, R. Rahaman, P.S. Joag, Phys. Rev. A **81**, 032103 (2010); MD.R. Gazi, A. Rai, S. Kunkri, R. Rahaman. J. Phys. A: Math. Theor. **43**, 452001 (2010)
21. J. Allcock, N. Brunner, N. Linden, S. Popescu, P. Skrzypczyk, T. Vértesi, Phys. Rev. A **80**, 062107 (2009)
22. L.-Y. Hsu, I.-C. Yu, F.-L. Lin, arXiv:1010.3419
23. D. Cavalcanti, A. Sallés, V. Scarani, Nat. Commun. **1**, 136 (2010)
24. M. Navascués, H. Wunderlich, Proc. R. Soc. Lond. A **466**, 881–890 (2009)
25. S. Pironio, J.-D. Bancal, V. Scarani, J. Phys. A: Math. Theor. **44**, 065303 (2011)
26. Y. Xiang, W. Ren, Quantum Inf. Comput. **11**, 948 (2011)
27. R. Gallego, L. Würflinger, A. Acín, M. Navascués, Phys. Rev. Lett. **107**, 210403 (2011)
28. T.H. Yang, D. Cavalcanti, M.L. Almeida, C. Teo, V. Scarani, New J. Phys. **14**, 013061 (2012)
29. M. Navascués, Y. Guryanova, M.J. Hoban, A. Acín, Nature Commun **6**, 6288 (2015)
30. R. Chaves, C. Majenz, D. Gross, Nat. Commun. **6**, 5766 (2015)

Macroscopic Locality

Miguel Navascués

1 Introduction

From the beginning of the 20th century, we have had at our disposal three alternative models of reality—newtonian, einstenian and quantum—each of which appears logically consistent. These theories, though, are not independent: there are inclusion relations between them. Indeed, note that the framework needed to describe physical systems in classical physics can be recovered as a low energy limit of the framework used in general relativity. Likewise, quantum theory allows recovering classical dynamics when measurements are sufficiently coarse-grained [1], or in the limit $\hbar \rightarrow 0$. Moreover, the fact that both quantum mechanics and general relativity provide accurate descriptions of reality in their respective domains implies that these two models emerge as limits of a third (yet unknown) physical theory.

This third theory is expected to reveal its nature in particle experiments dealing with energies of the order of the Planck mass $m_P \approx 1.2 \times 10^{19}$ GeV/c^2, inaccessible with current technology. In order to infer properties of this mysterious model that contains Quantum Physics and General Relativity as particular cases, we are thus bound to rely on logical reasoning and physical intuition rather than experimental feedback.

One approach to the problem, initiated by Hardy [2] and further developed in [3–7], is to find new formulations of Quantum Mechanics in terms of physically compelling axioms. The idea stems from the fact that while it is fairly easy to propose extensions of General Relativity, any small modification of Quantum Mechanics will most likely lead to inconsistencies. The hope here is that reducing Quantum Mechanics to a set of physical properties should point out new ways to generalize it.

M. Navascués (✉)
H.H. Wills Physics Laboratory, University of Bristol,
Tyndall Avenue, Bristol BS8 1TL, UK
e-mail: M.Navascues@bristol.ac.uk

G. Chiribella and R.W. Spekkens (eds.), *Quantum Theory: Informational Foundations and Foils*, Fundamental Theories of Physics 181,
DOI 10.1007/978-94-017-7303-4_13

Up to now, this line of research has proven very fruitful, allowing to recover finite dimensional Quantum Mechanics from first principles [3–7].

Another, more operational, approach was introduced in [8], where Rohrlich and Popescu proposed to classify physical theories regarding their ability to generate correlations between distant points in space.

For the sake of clarity, suppose, for instance, that two space-like separated parties (call them Alice and Bob) share a given bipartite physical system. Alice and Bob are allowed to measure certain observables X and Y (out of a finite set) on their subsystems, and these measurements will report them some outcomes a and b, respectively. Assuming a complete ignorance about the Physics involved in these processes, Alice and Bob could regard their system as a black box where they input a pair of symbols X, Y and obtain a pair of outputs a, b.

Popescu and Rorlich proposed that the choice of Alice's interaction should not affect Bob's statistics and viceversa. This principle, known as the *no-signaling condition*, translates at the level of probabilities as

$$\sum_a P(a,b|X,Y) = \sum_a P(a,b|X',Y) \equiv P(b|Y)$$
$$\sum_b P(a,b|X,Y) = \sum_b P(a,b|X,Y') \equiv P(a|X), \qquad (1)$$

for whatever interactions X, X' (Y, Y') available to Alice (Bob).

Despite its simplicity, the no-signaling condition imposes a strong constraint on the set of possible correlations present in a given physical theory. When represented in a real space, the set of all no-signaling distributions forms a polytope, i.e., a convex set defined by a finite number of linear inequalities.

Unfortunately, the no-signaling constraint is not strong enough. Indeed, the *no-signaling polytope* contains probability distributions that are so weird that soon people started to think that they could not be present in any reasonable physical theory. This motivated different works that ruled out some of the correlations of the no-signaling polytope on the grounds that they would make communication or computation trivial [9, 10], or violate the principle of information causality [11], see the corresponding chapter in this volume.

In [12], the authors proposed reduction to Classical Physics in the macroscopic limit as a fundamental axiom to be satisfied by any reasonable physical theory. Note that the theory that describes our universe has to recover Quantum Theory and General Relativity in some limits, and both these theories allow recovering the framework of Classical Physics. It thus seems inevitable that any reasonable physical theory must reduce to Classical Physics in suitable limits.

Notice also that a connection with Classical Physics may come together with a Correspondence Principle to derive the dynamics of the theory from classical models of the physical system at stake. A classical macroscopic limit is thus desirable from a practical point of view: since Classical Physics is the only known theory that relates Quantum Mechanics and General Relativity, it seems natural to resort to it in order to find a consistent dynamics for a deeper theory.

Finally, the notion that any theory has to recover Classical Physics is somehow implicit in Rorlich and Popescu's formalism. Expressions like $P(a, b|X, Y)$ assume that Alice is bound to apply only one out of the set of possible interactions, and not, for example, a linear combination of interactions X_1, X_2. The observer itself is thus regarded as classical in this scenario, and so the world in which it lives should also have some notion of classicality.

In this chapter we will show that the existence of a classical limit bounds the strength of the correlations measured by space-like separated observers in a non-trivial way. In a nutshell, the fact that there exists a classical limit implies that 'natural' macroscopic experiments involving distant parties measuring many microscopic systems in a coarse-grained way admit a local hidden variable model. This property, which we will call *macroscopic locality* [12], in turn imposes strong restrictions on the correlations generated by such microscopic systems. The main goal of this chapter is to characterize such restrictions.

The structure of this chapter is as follows: first, we will illustrate the meaning of macroscopic locality (ML) by means of a specific example. Later, in Sects. 2.2, 2.3, we will define and characterize Macroscopic Locality in a more general framework where experimentalists are allowed to apply sequential interactions. In Sect. 2.4, we will attain one of the main goals of this chapter, that is, to characterize the set Q^{ml} of all multipartite correlations *compatible* with ML. We will next prove that quantum correlations are contained in Q^{ml}, and, in Sect. 4, we will study the differences and similarities between the two sets. Finally, we will present our conclusions.

2 Macroscopic Locality

2.1 Some Preliminary Thoughts

Think of the following bipartite scenario (Fig. 1): a particle pair is produced and two experimentalists, call them Alice and Bob, receive one particle each. Within the given setup, Alice (Bob) can interact with her/his particle in two different ways $X = 0, 1$ ($Y = 0, 1$). As a result of each interaction, Alice's (Bob's) particle will follow one of two possible paths, the upper or the lower, and eventually will impinge on one of Alice's (Bob's) two detectors, as shown in Fig. 1. If Alice and Bob repeat the experiment many times, they will be able to estimate the probabilities $P(a, b|X, Y)$, i.e., the probability that Alice's and Bob's particles impinge on detectors $a, b = 0, 1$ when they apply the interactions X, Y.

This is the schema of a bipartite experiment of non-locality. We say that $P(a, b|X, Y)$ is *local* (sometimes called *classical*) if it can be expressed as

$$P(a, b|X, Y) = \sum_{\lambda} P(\lambda) P_A(a|X, \lambda) P_B(b|Y, \lambda), \tag{2}$$

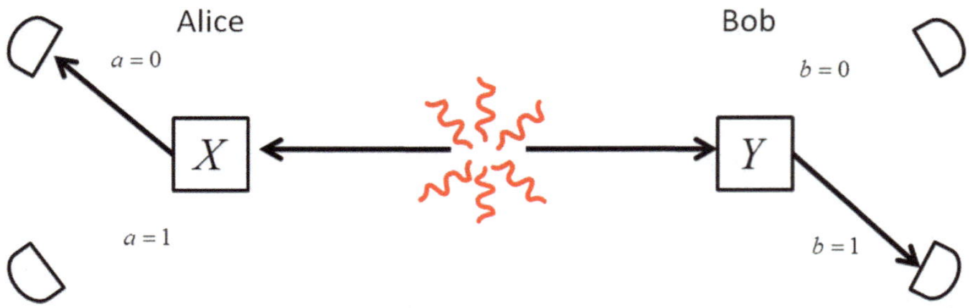

Fig. 1 A microscopic experiment of non-locality. A particle pair is produced. The interaction that Alice (Bob) subjects her (his) particle will make it take the path $a = 0$ or $a = 1$ ($b = 0$ or $b = 1$). Clicks on the detectors at the end of each path are associated to measurement outcomes. In the figure, the outcomes are $a = 0, b = 1$

for some probability distributions $P(\lambda)$, P_A, P_B. Equivalently, a distribution is local if there exists a global probability distribution for the variables $\{a_X, b_Y : X, Y\}$ with marginals $P(a_X, b_Y) = P(a, b|X, Y)$. Otherwise, we say that $P(a, b|X, Y)$ is *non-local*. There is plenty of evidence that our world is non-local (or non-classical) at the microscopic level, so it should not surprise us that an experiment like the one described above produces a non-local distribution.

Suppose now that $P(a, b|X, Y)$ is of the form:

$$P(a, b|X, Y) = \frac{\epsilon}{2}\delta_{a\oplus b, XY} + \frac{1-\epsilon}{4}, \tag{3}$$

where $0 \leq \epsilon \leq 1$. Such a probability distribution is known as an *isotropic Popescu-Rorlich (PR) box* [13]. It is local for $\epsilon \leq \frac{1}{2}$, and quantum for $\epsilon \leq \frac{1}{\sqrt{2}}$.

If there existed a microscopic event producing a particle pair following a distribution of the form (3), one would expect to find natural 'macroscopic' sources of $N \gg 1$ independent identical pairs. According to our guiding principle, namely, the existence of a classical limit, the observations performed by two classical experimentalists situated at a distance from this source should admit a classical description.

And what would two classical observers see in the vicinity of such sources? That will depend on Alice and Bob's 'classical' measurement devices, which we will model through their microscopic counterparts. Correspondingly, we will assume that Alice's (Bob's) interactions over her (his) beam will affect each particle individually, and so the net effect of such interactions will be to split the incident beam into two sub-beams of lower intensity, see Fig. 2. Also, since $N \gg 1$, Alice and Bob's detectors will measure sums of clicks or intensities rather than individual clicks. That is, as opposed to noticing a click in detector a, Alice will measure the intensity $I^a_{A|X}$.

Now, it is very unlikely to recover Classical Physics in a macroscopic experiment if we allow Alice's and Bob's detectors to have an arbitrary (microscopic) precision. For instance, if Alice and Bob realized that all their intensities are multiples of a smaller quantity, they could postulate that their beams are composed of pairs of

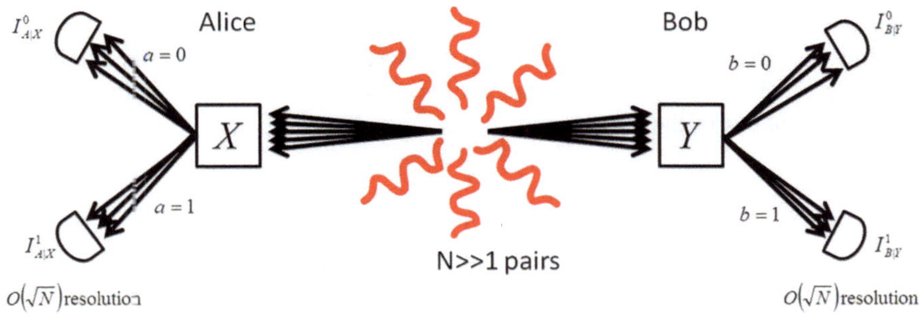

Fig. 2 A macroscopic experiment of non-locality. N independent particle pairs are produced. Alice's and Bob's interactions apply to all particles on each beam. This time, the intensities measured at each detector (with precision $O(\sqrt{N})$) are the outcomes of the experiment

correlated elementary particles, and derive the (in general, non-local) microscopic distribution $P(a, b|X, Y)$ from their classical data. This consideration leads to the extra assumption that the resolution of such detectors should just be able to detect intensity fluctuations of order $O(\sqrt{N})$. According to the Central Limit Theorem, such will be the expected size of the intensity fluctuations every time they repeat the experiment, so the assumption seems quite natural.[1]

We thus conclude that, under the above conditions (where two parties are conducting coarse-grained extensive measurements over a natural source of particle pairs), any macroscopic experiment should be describable in terms of a classical physical model. A necessary condition for this is the existence of a local hidden variable model (LHVM) for the distributions[2]

$$P(I_A^0, I_B^0|X, Y). \tag{4}$$

That is actually the original definition of Macroscopic locality [12]: namely, the requirement that coarse-grained ($O(\sqrt{N})$) extensive observations of macroscopic sources of N independent particle pairs admit a LHVM in the limit $N \to \infty$.

By the central limit theorem, when $N \to \infty$, $P(I_A^0, I_B^0|X, Y)$ become bivariate gaussian distributions with covariance matrix[3] proportional to

$$\gamma_{XY} = \begin{pmatrix} \frac{1}{4} & \frac{\epsilon}{2}\delta_{X\cdot Y,0} - \frac{\epsilon}{4} \\ \frac{\epsilon}{2}\delta_{X\cdot Y,0} - \frac{\epsilon}{4} & \frac{1}{4} \end{pmatrix}, \tag{5}$$

[1] A similar coarse-graining was required in [1] to prove the emergence of macroscopic realism from quantum mechanical systems. Note, however, that the resolution Δm considered there satisfies $\Delta m \gg O(\sqrt{N})$.

[2] Note that, more generally, we could have demanded the existence of a LHVM for $P(I_A^0, I_A^1, I_B^0, I_B^1|X, Y)$. However, in this class of experiments, it can be observed experimentally that $I_A^0 + I_A^1 = I_B^0 + I_B^1 = N$, and so the locality condition reduces to (4).

[3] The covariance matrix of a set of random variables $\xi_1, \xi_2, \ldots$ is defined as $\gamma_{ij} \equiv \langle\xi_i\xi_j\rangle - \langle\xi_i\rangle\langle\xi_j\rangle$. It can be verified that any covariance matrix must be positive semidefinite, see Appendix 1.

see Appendix 2.

Now, suppose that there exists a global measure $d\rho$ (or LVHM) for the intensities $\{I^0_{A|X}, I^0_{B|Y} : X, Y = 0, 1\}$. Then one could use such a measure to define a *global* covariance matrix of the form

$$\Gamma = \begin{pmatrix} \frac{1}{4} & \lambda_1 & \frac{\epsilon}{4} & \frac{\epsilon}{4} \\ \lambda_1 & \frac{1}{4} & \frac{\epsilon}{4} & -\frac{\epsilon}{4} \\ \frac{\epsilon}{4} & \frac{\epsilon}{4} & \frac{1}{4} & \lambda_2 \\ \frac{\epsilon}{4} & -\frac{\epsilon}{4} & \lambda_2 & \frac{1}{4} \end{pmatrix}. \quad (6)$$

Here the rows and columns of the matrix correspond to the intensities $I^0_{A|X=0}$, $I^0_{A|X=1}$, $I^0_{B|Y=0}$, $I^0_{B|Y=1}$, and $\lambda_1, \lambda_2 \in \mathbb{R}$ resp. represent the values $\langle I^0_{A|X=0} I^0_{A|X=1}\rangle - \langle I^0_{A|X=0}\rangle\langle I^0_{A|X=1}\rangle$, $\langle I^0_{B|Y=0} I^0_{B|Y=1}\rangle - \langle I^0_{B|Y=0}\rangle\langle I^0_{B|Y=1}\rangle$ as calculated via the measure $d\rho$. Note that λ_1, λ_2 are not observable, and thus can only be computed with the extra knowledge $d\rho$.

At this moment there comes a crucial observation: in order for $\Gamma(\lambda_1, \lambda_2)$ to be a covariance matrix, it must be positive semidefinite. This implies that there must exist a choice of λ_1, λ_2 such that $\Gamma(\lambda_1, \lambda_2) \geq 0$.

By symmetry under the exchange of Alice and Bob, it is easy to see that we can take $\lambda_1 = \lambda_2 = \lambda$. Since the minimum eigenvalue of $\Gamma(\lambda, \lambda)$ is $1/4 - (1/4)\sqrt{2\epsilon^2 + 8\epsilon|\lambda| + 16\lambda^2}$, the condition for positivity is thus equivalent to the existence of $\lambda \in \mathbb{R}$ such that

$$2\epsilon^2 + 8\epsilon|\lambda| + 16\lambda^2 \leq 1. \quad (7)$$

It is easy to see that the above equation can only hold if $\epsilon \leq \frac{1}{\sqrt{2}}$, i.e., if the isotropic box belongs to the quantum region.

We have just shown that post-quantum isotropic PR boxes are incompatible with the principle of Macroscopic Locality (ML).

2.2 The Macroscopic Scenario

In this section, we will consider the transition from microscopic to macroscopic experiments in complex multipartite scenarios where each party is allowed to interact sequentially with its particle beam. Before starting, though, some comments about basic notation are in order. Despite its popularity in nonlocality research, denoting probabilities by $P(a, b|X, Y)$ and intensities by $I^a_{A|X}$ soon becomes messy when we have to deal with macroscopically local models. For this reason, along this article we will adopt the representation introduced by Tsirelson [14]. In this notation, any possible outcome we may measure after the application of an interaction X is to be denoted by a symbol a that allows identifying X. That way, interactions X can be regarded as disjoint sets of possible outcomes a. For any pair of interactions

X, Y, available at Alice's and Bob's lab, respectively, and any pair of outcomes $a \in X, b \in Y$, the expression $P(a, b|X, Y)$ thus becomes redundant, and can be substituted by $P(a, b)$.

In a generic multipartite microscopic experiment, each of the space-like separated sites has access to a set of local interactions $X, Y, Z, \ldots$. Given a possible outcome a, the mappings $X(a) = X, O(a) = i$ will return, respectively, the measurement setting and site i where such a measurement is performed.

An *experimental setting* S is any arrangement of interactions that an experimentalist at lab i can prepare in order to measure intensities in a macroscopic experiment. For example, in Fig. 3, the experimental setting on site 1 consists on applying interaction X over the main beam and then interactions Z and Z' to the particles following a trajectory a or a', respectively. Given an experimental setting at site i, we will call *arc* to the trajectory followed by a particle since its arrival at lab i until it impinges on a detector. Any arc s can thus be completely specified by an ordered sequence of outcomes $s \equiv a_1 \rightarrow a_2 \rightarrow a_3 \rightarrow \cdots \rightarrow a_m$, with $O(a_j) = i$, and therefore any experimental setting can be identified with the set of all arcs it generates. Interactions applied in different arcs will be regarded as different, i.e., the specification of each interaction must make reference to all prior interactions. Coming back to Alice's experimental setting S in Fig. 3, we hence have that $S = \{a \rightarrow c, a \rightarrow c', a' \rightarrow d, a' \rightarrow d'\}$.

We say that two different arcs s, s' are *locally orthogonal* if they both appear in the same experimental setting, that is, if $s = s_1 \rightarrow a \rightarrow s_2$ and $s' = s_1 \rightarrow a' \rightarrow s_2'$, with $X(a) = X(a'), a \neq a'$. Also, two arcs s, s' are *space-like separated* if they correspond to experimental settings on different sites, i.e., iff $O(s) \neq O(s')$. Given two space-like separated arcs $s \equiv a_1 \rightarrow \cdots \rightarrow a_m, t \equiv b_1 \rightarrow \cdots \rightarrow b_{m'}$, $P(s, t)$ will represent the probability that the first particle of a pair returns the sequence of outcomes $(a_1, \ldots, a_m)$ when interactions $X(a_1), \ldots, X(a_m)$ are sequentially applied, and the second particle outputs $(b_1, \ldots, b_{m'})$ when $X(b_1), \ldots, X(b_{m'})$ are effected.

In this scenario, the no-signaling condition translates as

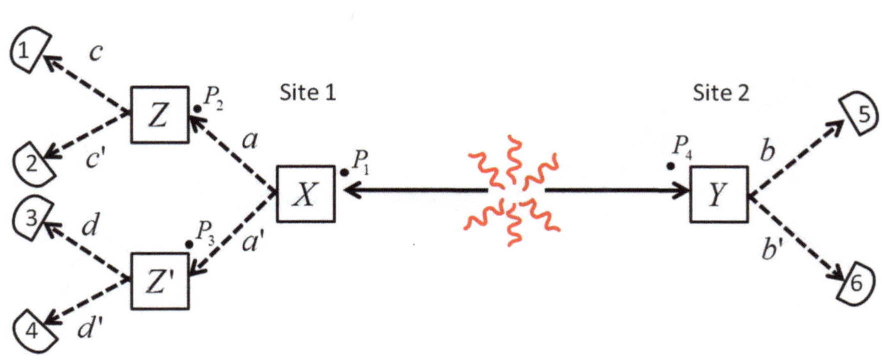

Fig. 3 A microscopic experiment with sequential measurements

Definition 1 No-signaling condition
Let $\{s_j\}_{j=1}^{K}$ be a collection of space-like separated arcs. Then, for any $k \in \{1, 2, \ldots, K\}$ and any interaction X, with $O(X) = k$,

$$\sum_{a \in X} P(s_1, \ldots, s_k \to a, \ldots s_K) = P(s_1, \ldots s_k, \ldots s_K). \tag{8}$$

When we bring a microscopic experiment of non-locality to the macroscopic scale, we end up in the scenario depicted in Fig. 4. Here the experimental outcomes are coarse-grained intensity measurements conducted at the end of each arc. The intensity measured at site i at the end of the arc s will be denoted by I^s, and by $\bar{I}^s$ we will refer to the measured intensity fluctuation, i.e., $\bar{I}^s = I^s - \langle I^s \rangle$. Notice that intensities corresponding to the same arc s, but belonging to different experimental settings, are identified with this notation. The reason is that the different interactions effected on each measurement setting can be space-like separated from arc s. For example, in the scenario depicted in Fig. 4 the regions where interactions Z, Z' are applied can be arbitrarily far away from each other, and so the intensities $I^{a\to c}$, $I^{a\to c'}$ ($I^{a'\to d}$, $I^{a'\to d'}$) would be regarded as independent of Z' (Z) by classical observers.

In a realistic situation, we cannot expect the K parties to be able to realize any possible experimental setting. Any experimentalist at site i will have to work under space and budget constraints. Consequently, the length of the available arcs will have to be limited, and so will be the (finite) set of accessible experimental settings $\mathcal{S}_{\text{acc}}$.

For any particular choice of experimental settings $\{S_i\}_{i=1}^{K} \subset \mathcal{S}_{\text{acc}}$, the K parties performing a macroscopic experiment can estimate the marginal probability distributions

$$P\left(\{I^s : s \in \cup_{i=1}^{K} S_i\}\right), \tag{9}$$

corresponding to all the intensities measured in a single experiment with space-like separated measurement settings $S_1, S_2, \ldots$ with precision $O(\sqrt{N})$.

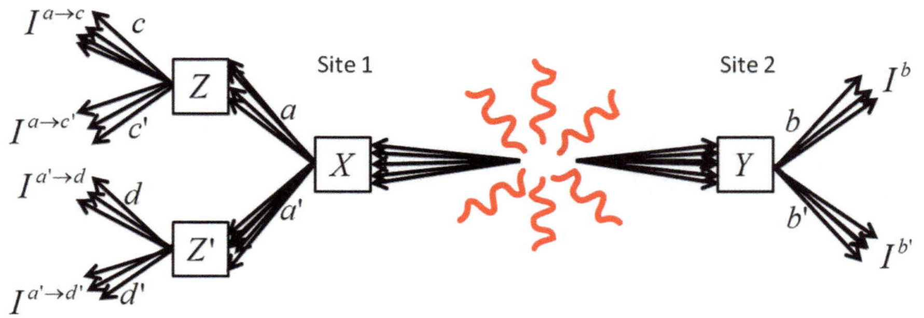

Fig. 4 A macroscopic experiment with sequential measurements

Following Sect. 2.1, the experiments performed by the K parties will satisfy ML iff there exists a joint measure $P(\{I^s : s \in S \in \mathcal{S}_{\text{acc}}\})$ for all the intensities I^s accessible in the family of experiments denoted by $\mathcal{S}_{\text{acc}}$, compatible with the marginal distributions (9) in the limit $N \rightarrow \infty$.

2.3 Characterization of Macroscopic Locality

By Appendices 1, 2, a necessary and sufficient condition for ML in a given set of accessible experimental settings $\mathcal{S}_{\text{acc}}$ is the existence of a positive semidefinite matrix γ, with rows and columns labeled by arcs, and satisfying

$$\gamma_{ss} = P(s) - P(s)^2, \tag{10}$$

$$\gamma_{ss'} = P(s, s') - P(s)P(s') \text{ for } s, s' \text{ space-like separated.} \tag{11}$$

$$\gamma_{ss'} = -P(s)P(s') \text{ for } s, s' \text{ locally orthogonal.} \tag{12}$$

for all $s \in S$, $s' \in S'$, with $S, S' \in \mathcal{S}_{\text{acc}}$.

Of course, if there is a LHVM behind all possible experiments that the K parties can perform without any limitation on the size of the settings, then there will exist a LHVM for all experiments involving a finite set of experimental settings $\mathcal{S}_{\text{acc}}$. However, in principle there could be weird K-partite microscopic correlations such that, for any finite set of available experimental settings $\mathcal{S}_{\text{acc}}$ there is a $\mathcal{S}_{\text{acc}}$-dependent LHVM that describes the observed intensity fluctuations, but nevertheless there is not a single classical model independent of $\mathcal{S}_{\text{acc}}$ that is compatible with all experimental data!

This possibility is ruled out by the next result.

Lemma 2 *Let P be a set of K-partite microscopic correlations. If for any finite set of available experimental settings $\mathcal{S}$ there exists a LHVM $P_{\mathcal{S}}(\{\bar{I}^s, s \in S \in \mathcal{S}\})$, then there exists a setting-independent LHVM compatible with all possible macroscopic experimental observations. That is, for any finite set of arcs $\vec{o}$ there exists a measure $P_{\vec{o}}$ for the intensity fluctuations $\{\bar{I}^s : s \in \vec{o}\}$ in agreement with experimental data, such that, for any two finite sets $\vec{o}, \vec{o}'$,*

$$\mathcal{P}_{\vec{o}}\left(\{\bar{I}^s : s \in \vec{o} \cap \vec{o}'\}\right) = \mathcal{P}_{\vec{o}'}\left(\{\bar{I}^s : s \in \vec{o} \cap \vec{o}'\}\right). \tag{13}$$

Proof Let $(\mathcal{S}_n)$ be a sequence of finite sets of experimental settings such that, for all n, $\mathcal{S}_n \subset \mathcal{S}_{n+1}$ and such that, for any possible setting S, there exists an n such that $S \in \mathcal{S}_n$. According to the previous remarks, this implies that there exists a sequence of positive semidefinite matrices (γ^n) of increasing size that satisfy conditions (10), (11) and (12) for all $s, s' \in \mathcal{S}_n$. Note as well that all the entries of γ^n are bounded by $1/4$. If we extend to the infinity all the rows and columns of these matrices by adding zeros, we will end up with a sequence of vectors γ^n in the ball l_∞. It is not difficult to see that there exists an entry-wise convergent subsequence $(\gamma^{n_k})_{n_k}$, call $\bar{\gamma}$ its limit. The infinite dimensional matrix $\bar{\gamma}$ hence satisfies

1. Any finite submatrix of $\bar{\gamma}$ is positive semidefinite.
2. $\bar{\gamma}$ satisfies conditions (10), (11) and (12) for all $s, s' \in \vec{O}$.

Let $\mathcal{P}_{\vec{o}}$ be the gaussian probability distribution for the variables $\{\bar{I}^s : s \in \vec{o}\}$ with covariance matrix $\{\langle \bar{I}^s \bar{I}^{s'} \rangle = \bar{\gamma}_{s,s'}\}$ and zero displacement vector. Clearly, $P_{\vec{o}}$ satisfies the conditions of the lemma. □

Remark 3 The proof of the previous lemma shows that a microscopic distribution is macroscopically local iff there exists a matrix $\bar{\gamma}$ satisfying points 1 and 2. Now, consider the (infinite) matrix Γ, whose rows are columns are numbered by the symbol $\mathbb{I}$[4] and any arc $s \in \vec{O}$, and is defined by the following relation

$$\Gamma = \begin{pmatrix} 1 & \vec{p}^T \\ \vec{p} & \tilde{\gamma} \end{pmatrix}. \tag{14}$$

Here $\vec{p}$ is a vector whose components are numbered by arcs s and such that $p_s = P(s)$, and $\tilde{\gamma}_{ss'} = \bar{\gamma}_{ss'} + p_s p_{s'}$. Given $P(s)$, we can easily switch from one matrix to the other. Note also that, for whatever finite set of arcs $\vec{o}$, the submatrix $\{\Gamma_{\alpha\beta} : \alpha, \beta \in \vec{o} \cup \{\mathbb{I}\}\}$ is positive semidefinite iff $\{\bar{\gamma}_{\alpha\beta} : \alpha, \beta \in \vec{o}\}$ is positive semidefinite. Indeed, by Schur's theorem [15], $\Gamma_{\vec{o}\cup\{\mathbb{I}\}}$ is positive semidefinite iff $\tilde{\gamma}_{\vec{o}} - \vec{p}_{\vec{o}}\vec{p}_{\vec{o}}^T = \bar{\gamma}_{\vec{o}}$ is positive semidefinite.

Conditions (10), (11) and (12), together with the definition of Γ, translate into the following rules

1. $\Gamma_{\mathbb{I}\mathbb{I}} = 1$.
2. $\Gamma_{\mathbb{I}s} = P(s)$.
3. $\Gamma_{ss} = P(s)$.
4. $\Gamma_{ss'} = P(s, s')$, for s, s' space-like separated.
5. $\Gamma_{ss'} = 0$, for s, s' locally orthogonal.

The remark above shows that the existence of a LHVM for all possible macroscopic experiments is equivalent to the existence of an object Γ satisfying conditions 1–5 and such that any finite submatrix of it is positive semidefinite. The advantage with respect to the previous formulation is that Γ only depends *linearly* on the original microscopic probabilities. In Sect. 4, this will allow us to compute maximal violations of linear Bell inequalities in general macroscopically local theories via semidefinite programming [16].

Remark 4 Remark 3, in combination with Lemma 2, suggests an operational hierarchy of constraints to be satisfied by any K-partite distribution $P(s_1, s_2, \ldots, s_K)$ in order to be macroscopically local (in the line of [17]). Given an increasing sequence of sets of experimental settings $(\mathcal{S}_i)$ such that no setting is left out, it is thus enough

[4] Intuitively, if Γ were a quantum moment matrix, "$\mathbb{I}$" would correspond to the identity operator.

to check that, for each i, there exists a positive semidefinite matrix Γ satisfying conditions 1–5 in Remark 3 for any $s, s' \in S \in \mathcal{S}_i$. From Lemma 2, it follows that such a hierarchy is complete.

2.4 The Set of Correlations Compatible with ML

Let P_1, P_2 be two independent distributions held by K parties. In principle, an experimentalist at site i conducting a microscopic experiment on the *composed distribution* $P_1 \otimes P_2$ could measure X_1 on P_1 and, depending on the outcome, measure Y or Y' on P_2. In this sort of experiments, a generic arc s_{12} at site i is thus decomposed as the interlacing of two arcs s_1, s_2 associated to measurements on the boxes $1, 2$, respectively.[5] Since systems P and P' are independent, the probability that the K particles follow the arcs $\{s_{12}^i\}_{i=1}^K$ is therefore given by $P(s_{12}^1, \ldots, s_{12}^K) = P(s_1^1, \ldots, s_1^K)P(s_2^1, \ldots, s_2^K)$.

The joint use of two or more independent distributions to generate new statistics is known as "wiring" [18], and in some circumstances it can be used to distill Bell inequality violations [13]. This observation raises the possibility of the existence of macroscopically local distribution P with the property that $P^{\otimes n}$ or some other allocation of many copies of the distribution P allows generating non-local macroscopic intensities. Such a distribution P, though macroscopically local, would not be *compatible* with the principle of macroscopic locality.

If such were the case, we would be in a conundrum. On one hand, it may be that distributions like P' never appear naturally at the macroscopic scale, and so such non-local intensities are never observed. This would not contradict the fact that Nature seems to be local at big scales, but would lead to a restriction on the dynamics of this Universe, that allows for the existence of macroscopic sources of P, but not of P'. On the other hand, we could simply postulate that *any* physical system can be brought to the macroscopic scale, and so we could ban the existence of P on the grounds that it allows to engineer P', which, in turn, generates non-local macroscopic correlations. This is the approach we will stick to along this chapter.

Once this point has been clarified, the next question to address is how to determine if a given probability distribution P is compatible with ML. In general, one would expect the answer to depend on the rest of available correlations $\{P'\}$, since it could well be that P and P' alone only lead to macroscopically local experiments, but nonetheless allow to distill macroscopic non-locality when they belong to the same space of physical states.

In this section we will show that such is not the case: any set of physical systems unable to produce non-local intensities by themselves cannot be wired into a macro-

[5] For instance, if a_1, a_2 are outcomes corresponding to P_1; and b, to P_2, the arc $s_{12} = a_1 \to b \to a_2$ corresponds to the event of measuring $X(a_1)$ on the first box, then $X(b)$ on the second and then $X(a_2)$ on the first, and obtaining the sequence of outcomes (a_1, b, a_2). In this case, the arc s_{12} corresponds to the interlacing of $s_1 = a_1 \to a_2$ and $s_2 = b$.

scopically non-local system when they are brought together. Ergo, there exists a maximal set of correlations Q^{ml} compatible with macroscopic locality that is closed under wirings. We will characterize this set at the end of the section.

2.4.1 A Closure Result

We will begin by showing that K-partite macroscopically local distributions are closed under local wirings. That is, once a number of such correlations has been distributed between the parties and those are not allowed to communicate classically, any wiring they may perform on their systems will not allow them to generate macroscopically non-local correlations. That $\otimes_i P_i$ cannot be used to distill macroscopic non-locality when each P_i is macroscopically local follows by induction from the next theorem.

Theorem 5 *Let P_1, P_2 be K-partite macroscopically local distributions. Then, $P_1 \otimes P_2$ is also macroscopically local.*

Proof Call $\vec{O}^1$ ($\vec{O}^2$) the set of all arcs s (s') pertaining to system P_1 (P_2); $\vec{O}^{12}$ will denote the set of all arcs generated by interlacing arcs from boxes P_1 and P_2. If P_1 and P_2 are ML, then from the last remark, there must exist two infinite matrices Γ_1, Γ_2 that satisfy conditions 1–5 for all $s, s' \in \vec{O}^1$ and $\vec{O}^2$, respectively, and such that any finite submatrix of them is positive semidefinite. Following the lines of [17], we have that there must exist two sets of vectors $V_1 = \{|s\rangle_1 : s \in \vec{O}^1 \cup \{\mathbb{I}_1\}\}$, $V_2 = \{|s\rangle_2 : s \in \vec{O}^2 \cup \{\mathbb{I}_2\}\}$, with $\langle s|s'\rangle_1 = \Gamma^1_{ss'}$ ($\langle s|s'\rangle_2 = \Gamma^2_{ss'}$) for all $s, s' \in \vec{O}^1 \cup \{\mathbb{I}_1\}(\vec{O}^2 \cup \{\mathbb{I}_2\})$.

Now, define the vectors[6]

$$
\begin{aligned}
|\mathbb{I}\rangle_{12} &\equiv |\mathbb{I}\rangle_1 \otimes |\mathbb{I}\rangle_2, \\
|s_{12}\rangle_{12} &\equiv |s_1\rangle_1 \otimes |s_2\rangle_2,
\end{aligned} \tag{15}
$$

where the arc $s_{12} \in \vec{O}^{12}$ is understood to arise by interlacing the arcs $s_1 \in \vec{O}^1$, $s_2 \in \vec{O}^2$ (without altering the order in which the outcomes of $\vec{O}^1$ and $\vec{O}^2$ appear).

Then we can construct the matrix Γ^{12} as

$$
\Gamma^{12}_{ss'} = \langle s|s'\rangle, \tag{16}
$$

for $s, s' \in \vec{O}^{12} \cup \{\mathbb{I}\}$. Clearly, any finite submatrix of Γ will be positive semidefinite.

We will now see that Γ^{12} satisfies conditions 1–5 of Remark 3 for the new set of correlations $P_1 \otimes P_2$.

[6]In case and $s_1 = \emptyset$ ($s_2 = \emptyset$), take $|s_1\rangle_1 = |\mathbb{I}\rangle_1$ ($|s_2\rangle_2 = |\mathbb{I}\rangle_2$).

1. $\Gamma^{12}_{\mathbb{I}\mathbb{I}} = \langle \mathbb{I}|\mathbb{I}\rangle_{12} = \langle \mathbb{I}|\mathbb{I}\rangle_1 \langle \mathbb{I}|\mathbb{I}\rangle_2 = 1$.
2. $\Gamma^{12}_{\mathbb{I}s} = \langle \mathbb{I}|s\rangle_{12} = \langle \mathbb{I}|s_1\rangle_1 \langle \mathbb{I}|s_2\rangle_2 = P(s_1)P(s_2) = P(s)$.
3. $\Gamma^{12}_{ss} = \langle s|s\rangle_{12} = \langle s_1|s_1\rangle_1 \langle s_2|s_2\rangle_2 = P(s_1)P(s_2) = P(s)$.
4. If $s, s' \in \vec{O}^{12}$ are space-like separated, then s_1 and s'_1 (s_2 and s'_2) are space-like separated. It follows that $\Gamma^{12}_{s,s'} = \langle s|s'\rangle_{12} = \langle s_1|s'_1\rangle_1 \langle s_2|s'_2\rangle_2 = P(s_1, s'_1)P(s_2, s'_2) = P(s, s')$.
5. Let s, s' be locally orthogonal. Then, $s = t_{12} \to a \to t'_{12}$, $s' = t_{12} \to a' \to t''_{12}$, where $a \neq a'$ but $X(a) = X(a')$. Suppose w.l.o.g. that $a \in O^1$. Then, $|s\rangle = |s_1\rangle \otimes |s_2\rangle$, and $|s'\rangle = |s'_1\rangle \otimes |s'_2\rangle$, with $s_1 = t_1 \to a \to t'_1$ and $s'_1 = t_1 \to a' \to t''_1$. That is, s_1 and s'_1 are locally orthogonal. This implies that $\langle s_1|s'_1\rangle = 0$, and so $\Gamma^{12}_{s,s'} = \langle s|s'\rangle_{12} = \langle s_1|s'_1\rangle_1 \langle s_2|s'_2\rangle_2 = 0$.

Therefore, $P_1 \otimes P_2$ is macroscopically local. □

We have just proven that ML cannot be activated by local wirings without communication. The next section shows, however, that one can distill macroscopic non-locality from ML boxes via a prior non-local engineering and/or local postselections.

2.4.2 Activation of ML

Consider a tripartite scenario where Alice, Bob and Charlie have each a pair of measurement settings with two possible outcomes. For clarity, let us return momentarily to the standard notation in non-locality, where probabilities are denoted as $P(a, b, c|X, Y, Z)$, and a, b, c, X, Y, Z take values in $\{0, 1\}$. Then one can check that the tripartite set of correlations

$$P \equiv P(a, b, c|X, Y, Z) = \frac{1}{4}\delta_{a\oplus b\oplus c = XYZ} \tag{17}$$

generates local intensities. Moreover, if several copies of P were distributed to Alice, Bob and Charlie, together with any amount of shared randomness, and the parties were not allowed to communicate, any wiring they performed on their subsystems would not allow them to distill macroscopic non-locality. The reason is that the (gaussian) marginal distributions of the macroscopic intensity fluctuations observed by the three parties are completely determined by the bipartite correlations between different intensities. These, in turn, just depend on the bipartite probability distributions $P(a, b|X, Y)$, $P(a, c|X, Z)$, $P(b, c|Y, Z)$. Since such bipartite distributions also arise from the local tripartite distribution $L(a, b, c|X, Y, Z) = 1/8, \forall a, b, c$, any wiring $\mathcal{W}$ of m copies of P will be macroscopically indistinguishable from $\mathcal{W}(L^{\otimes m})$, and thus macroscopically local.

However, if Charlie measures $Z = 1$ and announces his outcome, then Alice and Bob would be sharing a perfect PR box [8], which, as we saw in Sect. 2, is macroscopically non-local [12].

Also, suppose that Alice's and Charlie's *separate degrees of freedom* are integrated into just one particle, call it AC, and imagine a macroscopic experiment

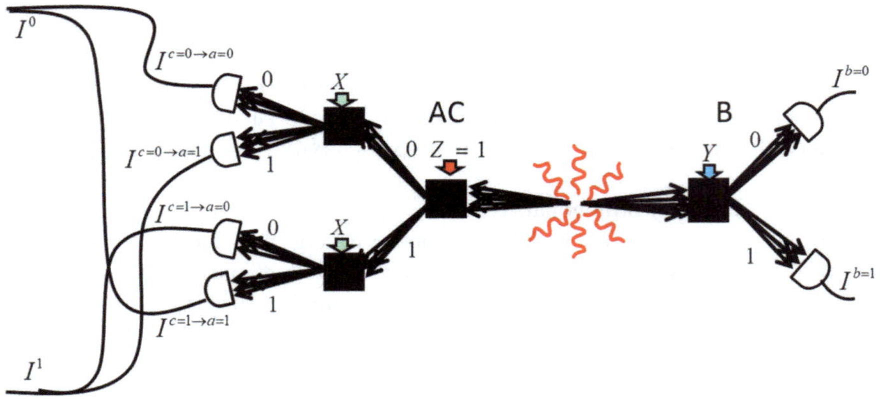

Fig. 5 Activation of macroscopic non-locality. By summing pairs of intensities, Alice and Bob can reproduce the macroscopic correlations generated by perfect PR boxes

where several independent pairs of AC/B particles are generated and sent to Alice and Bob. Then Alice could apply two consecutive interactions over her particle beam, as shown in Fig. 5. The first of such interactions, $Z = 1$, would address Charlie's degree of freedom, and split the particle beam into two different sub-beams. The subsequent application of an arbitrary interaction X to Alice's degree of freedom in AC, would subsequently split each sub-beam, thus ending up with four intensities on Alice's lab, each with mean value proportional to $NP(a, c|X, Z)$. Defining $\vec{I}_A \equiv (I^{c=0\to a=0}, I^{c=0\to a=1}, I^{c=1\to a=0}, I^{c=1\to a=1})$ and $\vec{I}_B \equiv (I^{b=0}, I^{b=1})$, it can be verified that the observed macroscopic distributions $P(\vec{I}_A, \vec{I}_B|X, Y)$ do not admit a LVHM, see Fig. 5.

P thus contains some *hidden macroscopic non-locality*, that can be activated either with one bit of communication or by joining two separate degrees of freedom into one.

2.4.3 The Set Q^{ml}

Theorem 5 shows that, once several macroscopically local systems have been distributed, the separated parties are not able to distill macroscopic non-locality. Now, a *general* wiring of a finite set of distributions $\{P_i\}_{i=1}^m$ will start (or not) with some post-selective measurements[7] (e.g.: in the previous section, in order to activate macroscopic non-locality, Charlie had to measure his subsystem and announce his measurement outcome). Afterwards, the remaining separate degrees of freedom will be distributed between the different parties, wirings will be made and a macroscopic

[7]Here, by a post-selective measurement, we understand the generation of a new state or set of correlations by preparing a (non-local) experimental setting and conditioning the final state on a specific arc.

experiment will take place. Invoking Theorem 5, it follows that, if such post-selected systems are already macroscopically local, then any wiring of them will be macroscopically local as well. On the other hand, any distribution P compatible with macroscopic locality has to remain macroscopically local under postselection. These considerations lead us to the following set:

Definition 6 A probability distribution P belongs to Q^{ml} iff, for any previous post-selective measurement and subsequent distribution in space of its separate degrees of freedom, the corresponding K-partite system is macroscopically local.

Following Remark 3, the necessary and sufficient conditions for a (conditional) set of correlations to be macroscopically local amount to the possibility to complete a sequence of growing matrices (whose determined entries are *linear*[8] in the microscopic probabilities and whose undetermined entries satisfy certain linear relations) in such a way that all of them are positive semidefinite. It follows that Q^{ml} is a convex set.

From the previous observations, it is clear that Q^{ml} is closed under wirings when the postselection part is deterministic (i.e., when the state distributed to the parties has the form $\otimes_i P_{i|a_i}$, where $P_{i|a_i}$ is the distribution P_i conditioned on the outcome(s) a). That Q^{ml} is closed under wirings in general is due to the fact that, after a generic postselection phase, the parties will be distributed convex combinations of boxes of the form $\otimes_{i,a} P_{i|a}$. Any experimental setting S_i (or wiring) on each side i will then only produce a convex combination of that same setting applied to the boxes $\otimes_{i,a} P_{i|a}$. Closure under wirings follows then from the convexity of the set Q^{ml}.

In sum, Q^{ml} is the maximal set of macroscopically local correlations that is closed under wirings.

Note that, in order to determine if a given set of probabilities $\{P(s)\}_s$ belongs to Q^{ml}, one would have to check for the existence of infinite dimensional covariance matrices for any possible postselection $\{P(s|s') : s\}$. That amounts to look for an infinite number of infinitely-sized matrices, not an easy task! In Sect. 4, however, we will see that in standard scenarios one just has to consider a finite set of finite-dimensional matrices.

3 Quantum Mechanics Satisfies Macroscopic Locality

In the last sections, we have defined ML and provided a semidefinite programming characterization of the set of microscopic correlations compatible with this principle. It is now time to prove that Quantum Mechanics satisfies ML.

---

[8] In principle, according to Remark 3, for a set of correlations conditioned on $\tilde{s}$ the determined entries of Γ should be of the form $P(s|\tilde{s}) = P(s, \tilde{s})/P(\tilde{s})$ and thus highly non-linear in P. Note, though, that we can always redefine such a matrix Γ as $\Gamma' = P(\tilde{s})\Gamma$. Obviously, as long as $P(\tilde{s}) \neq 0$, the positivity of Γ' is equivalent to that of Γ, but Γ' depends linearly on P.

Let $\mathcal{S}_{\text{acc}}$ be any set of experimentally accessible settings, and let $|\psi\rangle \in \mathcal{H}$ be the joint state of the corresponding quantum microscopic experiment (w.l.o.g., we can assume it to be pure). For any interaction X, call E_a the projector operator corresponding to outcome $a \in X$. Clearly, $E_a E_{a'} = \delta_{aa'} E_a$, for $a, a' \in X$ and $\sum_{a \in X} E_a = \mathbb{I}$. Also, if $O(a) \neq O(b)$, $[E_a, E_b] = 0$, i.e., observables corresponding to different parties commute. One can argue that during the course of the experiment the quantum system could experience some evolution U, perhaps depending on the sequence of past interactions effected on the particle. We will solve this issue by switching to the Heisenberg picture and redefining the measurement operators at each point of each arc via $E \to UEU^\dagger$. Since the experiment is assumed to be performed under space-like separation, operators belonging to different parties will still commute.

Associate to each measurement X an auxiliary Hilbert space $\mathbb{C}^{|X|}$. We will call such systems *registers*, and use $\mathcal{H}_r = \bigotimes_X \mathbb{C}^{|X|}$ to denote the space of all of them. Intuitively, the registers are going to hold a record of the outcomes we observe when we interact with particle p_k sequentially. Let $\{|j\rangle\}_{j=0}^{|X|-1}$ be an orthonormal basis of $\mathbb{C}^{|X|}$; and ϕ, a function which maps any outcome $a \in X$ to a natural number between 0 and $|X| - 1$ in such a way that $\phi(a) \neq \phi(a')$, for $a \neq a', a, a' \in X$. Now, consider the unitary $U_X \in B(\mathcal{H}_r \otimes \mathcal{H})$ given by

$$U_X = \sum_{a \in X} V_X^{\phi(a)} \otimes E_a, \tag{18}$$

where V_X is a unitary which acts non-trivially only over the register X as $V|j\rangle = |j + 1 \pmod{|X|}\rangle$. For any measurement outcome $a \in X$, call $\bar{\Pi}_a \in B(\mathcal{H}_r \otimes \mathcal{H})$ the projector that acts non-trivially over register X as $|\phi(a)\rangle\langle\phi(a)|$. For any *fragment* of an arc $s = a_1 \to a_2 \to \ldots \to a_m$, $\bar{\Pi}_s$ will denote the projector

$$\bar{\Pi}_s = \bar{\Pi}_{a_m} \bar{\Pi}_{a_{m-1}} \ldots \bar{\Pi}_{a_1}. \tag{19}$$

Analogously, U_s and E_s will represent the unitary operator $U_{X(a_m)} U_{X(a_{m-1})} \ldots U_{X(a_1)}$ and the non-hermitian operator $E_{a_m} \ldots E_{a_1}$, respectively. Now, define the projector

$$\Pi_s = U_s^\dagger \bar{\Pi}_s U_s, \tag{20}$$

for $s \neq \mathbb{I}$ and $\mathbb{I}_{\mathcal{H}_r} \otimes \mathbb{I}_{\mathcal{H}}$ otherwise, and denote the state $\bigotimes_X |0\rangle_X \in \mathcal{H}_r$ by $|\vec{0}\rangle$. We claim that the positive semidefinite matrix

$$\Gamma_{ss'} = \langle\vec{0}|\langle\psi|\Pi_s \Pi_{s'}|\vec{0}\rangle|\psi\rangle. \tag{21}$$

satisfies the conditions (3).

Indeed:

1. $\Gamma_{\mathbb{I}\mathbb{I}} = \langle\vec{0}|\vec{0}\rangle\langle\psi|\psi\rangle = 1$
2. $\Gamma_{\mathbb{I}s} = \Gamma_{\mathbb{I}s} = \Gamma_{ss} = \langle\vec{0}|\langle\psi|\Pi_s|\vec{0}\rangle|\psi\rangle = \langle\psi|E_s^\dagger E_s|\psi\rangle = P(s)$.

3. Let s, s' be space-like separated. Then, the operators $\bar{\Pi}_s, U_s$ commute with $\bar{\Pi}_{s'}, U_{s'}$. It follows that

$$\begin{aligned}\Gamma_{ss'} = & \langle\vec{0}|\langle\psi|\Pi_s\Pi_{s'}|\vec{0}\rangle|\psi\rangle = \langle\vec{0}|\langle\psi|U_s^\dagger U_{s'}^\dagger\bar{\Pi}_s\bar{\Pi}_{s'}U_sU_{s'}|\vec{0}\rangle|\psi\rangle = \\ & = \langle\psi|E_{s'}^\dagger E_s^\dagger E_sE_{s'}|\psi\rangle = P(s,s').\end{aligned} \tag{22}$$

4. Let s, s' be locally orthogonal. Then, $s = s_1 \to a \to s_2$, $s' = s_1 \to a' \to s_2'$, with $a, a' \in X(a)$, $a \neq a'$, and so,

$$\begin{aligned}\Pi_s &= U_{s_1\to a}^\dagger U_{s_2}^\dagger\bar{\Pi}_{s_1\to a}\bar{\Pi}_{s_2}U_{s_2}U_{s_1\to a},\\ \Pi_{s'} &= U_{s_1\to a'}^\dagger U_{s_2'}^\dagger\bar{\Pi}_{s_1\to a'}\bar{\Pi}_{s_2'}U_{s_2'}U_{s_1\to a'}.\end{aligned} \tag{23}$$

Note that $U_{s_1\to a} = U_{s_1\to a'}$. Also, $\bar{\Pi}_{s_1\to a}, \bar{\Pi}_{s_1\to a'}$ commute with $U_{s_2}, U_{s_2'}$ (because they act over different subsystems). This, together with the relation $\bar{\Pi}_{s_1\to a}\bar{\Pi}_{s_1\to a'} = 0$, implies that $\Pi_s\Pi_{s'} = 0$, and, consequently, $\Gamma_{ss'} = \langle\vec{0}|\langle\psi|\Pi_s\Pi_{s'}|\vec{0}\rangle|\psi\rangle = 0$.

4 Predictions of ML

Note that up to now we have not discussed the dynamics of theories respecting ML. This is so because the formalism of black boxes only allows to speak about correlations between distant parties, independently of how those correlations originated. Consequently, the only predictions we can expect from the ML axiom are limits to the non-locality exhibited by the physical theories subject to them. We already saw, in Sect. 2, that supra-quantum isotropic PR boxes are not compatible with ML. In this section, we will explore further how ML constrains bipartite and tripartite correlations. We will see how these results compare to the no-signaling, quantum and classical cases. But first we have to point out a practical observation.

Currently, the state of the art in non-locality research is to consider scenarios where each party interacts with its subsystem only once, i.e., the length of all accessible arcs is 1. We will call this kind of scenario the *standard picture*. In the standard picture only probabilities of the type $P(a_1, \ldots, a_K)$ are considered. To determine if such distributions are ML, we just have to check the existence of LHVMs for a finite set of intensities $\{I^a\}$ in a finite set of experiments, i.e., ML can be certified in a finite number of steps. Along this Section we will always consider standard picture scenarios.

4.1 The Bipartite Case

In order to find out if a set of probabilities $P(a,b)$ is compatible with ML, it is enough to check for the existence of a LHVM for the intensity fluctuations $\{\bar{I}^a, \bar{I}^b : a \in X, b \in Y, \forall X, Y\}$. By Remark 3, we can therefore identify Q^{ml} as Q^1 [12], a set of correlations proposed in [17] as a first approximation to the set Q of quantum correlations. Q^1 is defined as the set of bipartite distributions $P(a,b)$ such that there exists a *positive semidefinite* matrix Γ, whose columns and rows are numbered by the symbol $\mathbb{I}$, Alice's outcomes a and Bob's outcomes b with the structure

$$\Gamma = \begin{pmatrix} 1 & \vec{P}(a)^T & \vec{P}(b)^T \\ \vec{P}(a) & A & P(a,b) \\ \vec{P}(a) & P(b,a) & B \end{pmatrix} \tag{24}$$

with $A_{a,a} = P(a)$, $B_{b,b} = P(b)$.
From Sect. 3, we know that $Q \subset Q^1$. Moreover, this inclusion is strict [12, 17]. However, in a sense, the two sets are quite close.

Consider, for instance, a scenario where both Alice and Bob perform s dichotomic measurements, and, for any pair of measurement settings X_i, Y_j, define the two-point correlators

$$E_{ij} \equiv \sum_{a\in X_i, b\in Y_j} P(a,b)(-1)^{\phi(a)\oplus\phi(b)}, \tag{25}$$

where ϕ is a function that assigns the values 0 and 1 to the two outcomes associated with each measurement. In [12] it was shown that the maximum of any Bell inequality of the form

$$\sum_{i,j} c_{i,j} E_{ij} \tag{26}$$

among all possible sets of correlations $P(a,b)$ compatible with ML is the same as the quantum optimum. This implies, as shown in Sect. 2.1, that the maximum violation of the Clauser-Horn-Shimony-Holt (CHSH) inequality [19]

$$CHSH \equiv E_{00} + E_{10} + E_{01} - E_{11} \le 2. \tag{27}$$

allowed in ML theories is the Tsirelson bound, $2\sqrt{2}$ [20].

In this respect, a much more powerful and general result is derived in [21]: consider a bipartite non-locality scenario involving dichotomic observables, and let

$$f(E_i^A, E_j^B, E_{ij}) \le R, \tag{28}$$

with $E_i^A = \sum_{a\in X_i}(-1)^{\phi(a)}P(a)$, $E_j^B = \sum_{b\in Y_j}(-1)^{\phi(b)}P(b)$, be a necessary condition for a microscopic distribution to be classical, i.e., let the former expression be a Bell inequality. Then, the relation

$$f\left(0, 0, \frac{2}{\pi}\arcsin\left[\frac{E_{ij} - E_i^A E_j^B}{\sqrt{1-(E_i^A)^2}\sqrt{1-(E_j^B)^2}}\right]\right) \leq R \tag{29}$$

holds for all distributions compatible with ML.

Given expression (28), inequality (29) is proven by considering the macroscopic intensity fluctuations $\bar{I}_A^i = \sum_{a\in X_i}(-1)^{\phi(a)}\bar{I}^a$, $\bar{I}_B^j = \sum_{b\in Y_j}(-1)^{\phi(b)}\bar{I}^b$ generated by many microscopic systems following a ML distribution $P(a, b)$ with two-point correlators $\{E_{ij}\}$ and mean values $\{E_i^A, E_j^B\}$. By hypothesis, $P(\bar{I}_A^i, \bar{I}_B^j)$ is a gaussian local distribution. It follows that the dichotomic distribution $P\left(\text{sgn}(\bar{I}_A^i), \text{sgn}(\bar{I}_B^j)\right)$, with two-point correlators

$$\tilde{E}_{ij} = \langle\text{sgn}(\bar{I}_A^i)\cdot\text{sgn}(\bar{I}_B^j)\rangle = \frac{2}{\pi}\arcsin\left(\frac{E_{ij} - E_i^A E_j^B}{\sqrt{1-(E_i^A)^2}\sqrt{1-(E_j^B)^2}}\right), \tag{30}$$

and average values $\langle\text{sgn}(\bar{I}_A^i)\rangle = \langle\text{sgn}(\bar{I}_B^j)\rangle = 0$, is also local and thus subject to (28).

Applying the former result to the CHSH inequality (27), for example, we deduce that any microscopic distribution compatible with ML must satisfy

$$\frac{2}{\pi}\arcsin\left(\frac{E_{11} - E_1^A E_1^B}{\sqrt{1-(E_1^A)^2}\sqrt{1-(E_1^B)^2}}\right) + \frac{2}{\pi}\arcsin\left(\frac{E_{12} - E_1^A E_2^B}{\sqrt{1-(E_1^A)^2}\sqrt{1-(E_2^B)^2}}\right) -$$
$$\frac{2}{\pi}\arcsin\left(\frac{E_{21} - E_2^A E_1^B}{\sqrt{1-(E_2^A)^2}\sqrt{1-(E_1^B)^2}}\right) - \frac{2}{\pi}\arcsin\left(\frac{E_{22} - E_2^A E_2^B}{\sqrt{1-(E_2^A)^2}\sqrt{1-(E_2^B)^2}}\right) \leq 2. \tag{31}$$

This is a strengthening of the non-linear condition discovered by Landau [30], which can be derived from Eq. (31) by taking $E_i^A = E_j^B = 0$ for all i, j. As shown in [17, 21], for this particular scenario of two settings and two outputs, this condition (and the ones derived by symmetry considerations) is also sufficient to single out all no-signaling correlations compatible with ML.

It turns out that there exist microscopic distributions with biased outcomes (i.e., with some $E_\cdot \neq 0$) attaining the Tsirelson bound which are also compatible with this condition. On the other hand, any set of quantum correlations maximizing the CHSH violation can be shown to have unbiased outcomes [22]. We thus conclude that ML alone is not sufficient to characterize the bipartite quantum set of correlations.

How does ML compare with other physical axioms at the correlation level? Many physical principles have been proposed to constrain the set of all bipartite distributions beyond the non-signalling set, like non-trivial communication complexity [9], Non-local Computation [10] and Information Causality [11]. However, so far only those correlations compatible with Information Causality (IC) have been thoroughly studied [23, 24]. Although a concrete characterization of IC correlations is still missing, current literature suggests that IC imposes weaker constraints than ML when applied to scenarios with a small number of measurement outcomes, like the CHSH scenario. Note, indeed, that, when applied to single out the set of physical two-point correlators, IC does not seem to recover the quantum set [23]. It has been shown, nevertheless, that in setups with a large number of measurement outcomes, there exist distributions compatible with ML which would allow two parties to violate IC [24]. Both principles hence seem to be independent of one another.

4.2 The Tripartite Case

Here we will briefly analyze the scenario with three separate degrees of freedom (i.e., three particles), two settings and two outcomes. The correlations will thus have the form $P(a, b, c)$.

The tripartite case is the simplest scenario where one can study the phenomenon of monogamy of correlations [25]. Consider, for instance, how much Alice and Bob can violate the bipartite CHSH Bell inequality [19], see Eq. (27), for a fixed value $CHSH_{AC}$ of the CHSH parameter with respect to Alice and Charlie. From arguments of extensibility, we know that $CHSH_{AB}$, $CHSH_{AC}$ cannot be both non-local (i.e., greater than 2) at the same time [25]. Moreover, as shown in [26], the no-signaling condition alone implies that

$$|CHSH_{AB}| + |CHSH_{AC}| \leq 4. \tag{32}$$

Toner and Verstraete [27] found that, in quantum theories, this inequality can be replaced by a stronger one, namely,

$$|CHSH_{AB}|^2 + |CHSH_{AC}|^2 \leq 8, \tag{33}$$

that, in particular, allows recovering the original Tsirelson bound [20]. Both inequalities are tight in the no-signaling polytope and the set of quantum correlations, respectively.

It is thus intriguing how these inequalities evolve when we move from one theory to another following the inclusion chain Classical Physics $\subset$ Quantum Physics $\subset$ ML $\subset$ NS.

To find the solution, we had to perform linear optimizations over the set of all tripartite distributions compatible with ML. In order to prove that $P(a, b, c)$ is compatible with ML, it is enough to check that the intensities generated in the three scenarios depicted in Fig. 6 (tripartite case, bipartite with recombination of separate degrees of freedom and bipartite with post-selection) admit a LHVM. This implies checking the positivity of 10 covariance matrices. We performed the corresponding SDP calculations with the MATLAB package *YALMIP* [28] in combination with *SeDuMi* [29].

The results can be seen in Fig. 7, that shows the trade-off between $CHSH_{AB}$ and $CHSH_{AC}$ for different classes of theories. The predictions of ML are disappointing for their simplicity: the no-signaling bound is just complemented with the requirement that ML only allows violations of the CHSH inequality up to $2\sqrt{2}$.

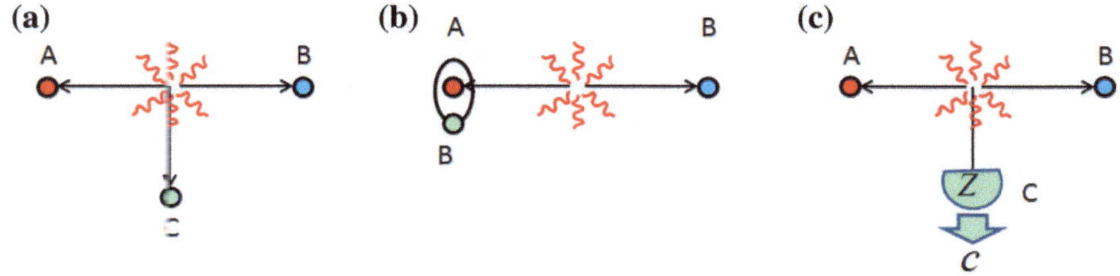

Fig. 6 The GHZ scenario. These are the only three non-trivial ways (modulo permutations of the parties) in which three separate degrees of freedom can be distributed

Fig. 7 Monogamy of bipartite correlations. The plot shows the trade-off between Alice and Bob's and Alice and Charlie's CHSH parameter in different theories. The *yellow regions* corresponds to the accessible points exclusive to the no-signaling polytope (bound (32)). The *green zone* shows the limits compatible with ML. The predictions of QM (Eq. (33)) and classical physics are denoted in *grey* and *brown*, respectively

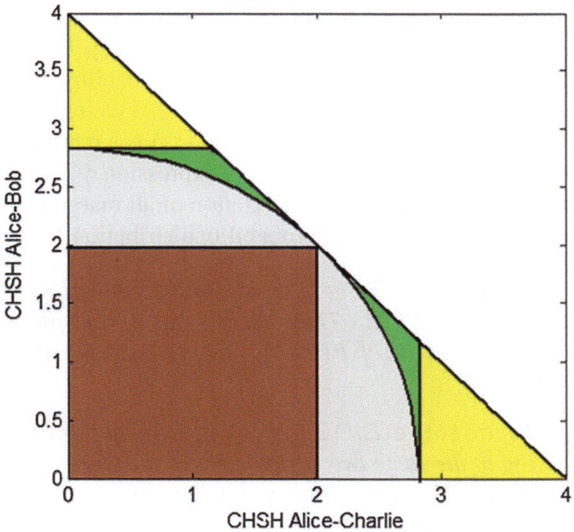

5 Conclusion

In this chapter we have introduced the axiom of Macroscopic Locality as a fundamental principle to be satisfied by future physical theories that aim at describing our Universe. We derived a consistent set Q^{ml} of ML multipartite correlations, which we showed to contain strictly the set Q of quantum correlations. In the process, we noted the phenomenon of macroscopic non-locality activation, whereby K parties sharing a macroscopically local multipartite distribution can generate macroscopic non-locality via clustering or classical communication. We showed how to compute the boundaries of ML in standard nonlocality scenarios and connected our results with previous works on quantum correlations. Our analysis revealed that, in spite of the similarities between Q^{ml} and Q, there exist bipartite correlations compatible with ML which are impossible to approximate by means of quantum systems. This offers some hope to the possibility that quantum mechanics is experimentally falsified in the future via bipartite Bell-type experiments.

Appendix 1: Local Gaussian Distributions

In this appendix, we will show a simple criterion to decide when a set of gaussian marginal distributions admits a local hidden variable model. Let, then, Ξ be a set of variables $\Xi = \{\xi_1, \xi_2, \ldots, \xi_M\}$, and let $\{\Xi_i\}_i$ be a collection of subsets of Ξ such that, for any i, there exists a gaussian probability distribution $P_i(\Xi_i)$ with zero mean and covariance matrix γ^i for all $\xi \in \Xi_i$. We remind the reader that the covariance matrix of a set of variables $(x_1, \ldots, x_n)$ is a matrix whose entries are labeled by the variable indices and given by the expression $\gamma_{ij} \equiv \langle x_i x_j\rangle - \langle x_i\rangle\langle x_j\rangle$. The following theorem provides a characterization of all marginal probability distributions $P_i(\Xi_i)$ that arise from a global probability distribution $P(\Xi)$.

Theorem 7 *Let Ξ_i be sets of continuous variables $\xi_1, \xi_2, \ldots$, as defined above, and let $\Xi = \bigcup_i \Xi_i$. Then, there exists a joint probability density $P(\Xi)$ such that $P(\Xi_i)d\Xi_i = d\Xi_i \int P(\Xi)\prod_{\xi\in\Xi\setminus\Xi_i} d\xi$ holds for all i iff the following conditions are satisfied.*

1. *For all $\xi, \xi' \in \Xi_i \cap \Xi_j$, $\gamma^i_{\xi\xi'} = \gamma^j_{\xi\xi'}$, that is, covariance matrix entries corresponding to the same two variables have the same value.*
2. *There exists a positive semidefinite matrix γ whose entries are labeled by the elements of Ξ and such that, for any i and $\xi, \xi' \in \Xi_i$,*

$$\gamma_{\xi\xi'} = \gamma^i_{\xi\xi'}. \tag{34}$$

Notice that, in case ξ, ξ' do not both belong to one of the sets Ξ_i, the coefficient $\gamma_{\xi\xi'}$ does not appear among the entries of $\{\gamma^j\}_j$.

Proof We will first prove that, if $P(\Xi)$ exists, then conditions 1 and 2 are satisfied. First of all, if there exists a joint probability distribution for the variables in Ξ, then, for any pair of variables $\xi, \xi' \in \Xi$ the mean value $\langle \xi\xi' \rangle$ is uniquely defined, and so, if $\xi, \xi' \in \Xi_i \cap \Xi_j$, then $\gamma^i_{\xi\xi'} = \langle \xi\xi' \rangle = \gamma^j_{\xi\xi'}$. Condition 1 is thus satisfied. To see that condition 2 is also respected define the symmetric real matrix

$$\gamma_{\xi\xi'} \equiv \int \xi\xi' P(\Xi) d\Xi. \tag{35}$$

From previous considerations, it is clear that Eq. (34) applied to γ holds. To see that γ is positive semidefinite, multiply γ on both sides by an arbitrary vector $\vec{v}$. We have that

$$\vec{v}^T \gamma \vec{v} = \sum_{\xi,\xi'} v_\xi v_{\xi'} \gamma_{\xi\xi'} = \int \left(\sum_{\xi} v_\xi \xi \right)^2 P(\Xi) d\Xi \geq 0. \tag{36}$$

Since $\vec{v}$ was an arbitrary vector, it follows that, indeed, $\gamma \geq 0$.

Now we will prove the opposite implication: suppose that there exists a positive semidefinite matrix γ fulfilling Eq. (34). One can then check that the gaussian distribution $P(\Xi) \propto e^{-\vec{\xi}^T \gamma^{-1} \vec{\xi}/2}$ admits $P_i(\Xi_i)$ as marginals, as long as γ is invertible.

In case γ is not invertible, let $r = \text{rank}(\gamma)$, let $\{\vec{u}^i\}_{i=1}^r$ be a basis for its range; and $\{\vec{v}^i\}_{i=r+1}^M$, a basis for its kernel. Now, perform a change of variables $\xi'_i = \sum_i u^i_j \xi_j$, for $i = 1, \ldots, r$ and $\xi'_i = \sum_i v^i_j \xi_j$, for $i = r+1, \ldots, M$. Since, for all $\vec{v}^i$, $(\vec{v}^i)^T \gamma \vec{v}^i = \langle (\xi'_i)^2 \rangle = 0$, it follows that $\xi'_i = 0$, for $i = r+1, \ldots, M$. The distribution of the remaining $\{\xi'_i\}_{i=1}^r$ is thus given by $P(\{\xi'_1, \ldots, \xi'_r\}) \propto e^{-\vec{\xi'}^T (\gamma')^{MP} \vec{\xi'}/2}$. Here the symbol MP denotes the Moore-Penrose inverse. □

Appendix 2: Macroscopic Locality

Here we will study the conditions under which the intensity fluctuations generated by independent sets of multipartite microscopic correlations admit a classical model.

As explained in the main text, a macroscopic experiment will involve a source of N identical and independent K-tuples of particles, all parties are allowed to perform identical microscopic interactions over the particle beams they receive, and their detectors have a resolution that only allows measuring intensity fluctuations of the order $O(\sqrt{N})$.

Given a possible arc $s \in S \subset \vec{O}_i$, define the observable d^s_l as equal to 1 if party i's particle from the lth K-tuple impinges on detector $D(s)$ at the end of the arc s. If we label by $\bar{I}^s$ the intensity fluctuation measured by this party in detector $D(s)$, it is straightforward that

$$\bar{I}^s \propto \sum_{l=1}^{N} [d_l^s - P(s)]. \tag{37}$$

Since the precision of the party's detectors only allow it to detect fluctuations of the order $\sqrt{N}$, the ith experimentalist will be measuring a truncation (in principle, up to an arbitrary number of decimal places) of the variable

$$\bar{I}^s = \frac{\sum_{l=1}^{N}[d_l^s - P(s)]}{\sqrt{N}}. \tag{38}$$

Using the notation $P(s,s) = P(s)$, and $P(s,s') = 0$ if s, s' are locally orthogonal arcs, we have that, for any two space-like separated, locally orthogonal or identical arcs s, s',

$$\langle \bar{I}^s \bar{I}^{s'} \rangle = P(s,s') - P(s)P(s'). \tag{39}$$

By virtue of the Central Limit Theorem [31], in the limit $N \to \infty$, for any collection of local settings $\bar{S} = \{S_i\}_{i=1}^K$, the distribution of the variables $\{\bar{I}^s : s \in S_i, \text{ for some } i\}$ will converge to a multivariate gaussian distribution with zero mean and covariance matrix given by

$$\gamma_{ss'}^{\bar{S}} = P(s,s') - P(s)P(s'). \tag{40}$$

According to Appendix 1, for any finite number of local settings, the set of intensity fluctuations arising from a finite set of accessible local experimental settings $\mathcal{S}_{acc}$ will admit a LHVM iff there exists a positive semidefinite covariance matrix γ that has $\gamma^{\bar{S}}$ as a submatrix for all collections $\bar{S} \in \mathcal{S}_{acc}$.

References

1. J. Kofler, C. Brukner, Phys. Rev. Lett. **99**, 180403 (2007)
2. L. Hardy, arXiv:quant-ph/0101012
3. B. Dakic, C. Brukner, arXiv:0911.0695
4. Ll Masanes, M.P. Mueller, New J. Phys. **13**, 063001 (2011)
5. G. Chiribella, G.M. D'Ariano, P. Perinotti, Phys. Rev. A **84**, 012311 (2011)
6. L. Hardy, arXiv:1104.2066
7. G. de la Torre, Ll. Masanes, A.J. Short, M.P. Mueller, arXiv:1110.5482
8. D. Rohrlich, S. Popescu, Found. Phys. **24**(3), 279 (1995)
9. G. Brassard, H. Buhrman, N. Linden, A.A. Methot, A. Tapp, F. Unger, Limit on nonlocality in any world in which communication complexity is not trivial. Phys. Rev. Lett. **96**, 250401 (2006)
10. N. Linden, S. Popescu, A.J. Short, A. Winter, Phys. Rev. Lett. **99**, 180502 (2007)
11. M. Pawlowski, T. Paterek, D. Kaszlikowski, V. Scarani, A. Winter, M. Zukowski, Nature **461**, 1101 (2009)
12. M. Navascués, H. Wunderlich, Proc. R. Soc. A **466**, 881–890 (2009)

13. N. Brunner, P. Skrzypczyk, Phys. Rev. Lett. **102**, 160403 (2009)
14. B.S. Tsirelson, J. Sov. Math. **36**, 557 (1987)
15. R.A. Horn, C.R. Johnson, *Matrix Analysis* (Cambridge University Press, Cambridge, 1999)
16. L. Vandenberghe, S. Boyd, SIAM Rev. **38**, 49 (1996)
17. M. Navascués, S. Pironio, A. Acín, Phys. Rev. Lett. **98**, 010401 (2007); M. Navascués, S. Pironio, A. Acín, New J. Phys. **10**(7), 073013 (2008)
18. J. Allcock, N. Brunner, N. Linden, S. Popescu, P. Skrzypczyk, T. Vertesi, Phys. Rev. A **80**, 062107 (2009)
19. J.F. Clauser, M.A. Horne, A. Shimony, R.A. Holt, Phys. Rev. Lett. **23**, 880 (1969)
20. B.S. Cirel'son, Lett. Math. Phys. **4**, 93 (1980)
21. T.H. Yang, M. Navascués, L. Sheridan, V. Scarani, Phys. Rev. A **83**, 022105 (2011)
22. R. Werner, M. Wolf, QIC **1**(3), 1 (2001)
23. J. Allcock, N. Brunner, M. Pawlowski, V. Scarani, Phys. Rev. A **80**, 040103(R) (2009)
24. D. Cavalcanti, A. Salles, V. Scarani, Nat. Commun. **1**, 136 (2010)
25. Ll Masanes, A. Acín, N. Gisin, Phys. Rev. A. **73**, 012112 (2006)
26. B. Toner, Proc. R. Soc. A **465**, 59–69 (2009)
27. B. Toner, F. Verstraete, arXiv:quant-ph/0611001
28. J. Löfberg, YALMIP: A toolbox for modeling and optimization in MATLAB, http://control.ee.ethz.ch/joloef/yalmip.php
29. J.F. Sturm, SeDuMi, a MATLAB toolbox for optimization over symmetric cones, http://sedumi.mcmaster.ca
30. L. Landau, Found. Phys. **18**, 449–460 (1988)
31. H. Tijms, *Understanding Probability: Chance Rules in Everyday Life* (Cambridge University Press, Cambridge, 2004)

Guess Your Neighbour's Input: No Quantum Advantage but an Advantage for Quantum Theory

Antonio Acín, Mafalda L. Almeida, Remigiusz Augusiak and Nicolas Brunner

1 Introduction

Quantum theory is arguably the most accurate scientific theory designed so far. However, despite this success, we still lack a deep understanding of the foundations of the theory. An important goal in the foundations of quantum mechanics is therefore to recover quantum theory from alternative sets of axioms, motivated by physical principles rather than mathematical ones [1].

In particular, one aspect of quantum theory that has attracted considerable attention recently is that of quantum nonlocal correlations. Quantum nonlocality [2, 3], a valuable resource for information processing [4–7], is the strongest manifestation of quantum correlations; distant observers performing local measurements on

A. Acín (✉) · R. Augusiak
ICFO–Institut de Ciències Fotòniques, 08860 Castelldefels, Barcelona, Spain
e-mail: antonio.acin@icfo.es

R. Augusiak
e-mail: remigiusz.augusiak@icfo.es

A. Acín
ICREA–Institució Catalana de Recerca I Estudis Avançats, 08010 Castelldefels, Barcelona, Spain

M.L. Almeida
Centre for Quantum Technologies, National University of Singapore, 3 Science Drive 2, Singapore 117543, Singapore
e-mail: mafaldalalmeida@gmail.com

N. Brunner (✉)
Département de Physique Théorique, Université de Genève, 1211 Genève, Switzerland

N. Brunner
H.H. Wills Physics Laboratory, University of Bristol, Bristol BS8 1TL, UK
e-mail: N.Brunner@bristol.ac.uk

G. Chiribella and R.W. Spekkens (eds.), *Quantum Theory: Informational Foundations and Foils*, Fundamental Theories of Physics 181,
DOI 10.1007/978-94-017-7303-4_14

a shared entangled state, may observe correlations between their measurement outcomes which could provably not have been obtained in any local theory.

The strength of quantum correlations appears however to be limited, in a way that cannot be yet explained by any physical principles. Consider for instance the principle stating that information cannot be transmitted instantaneously, the so-called no-signaling principle. Although this principle is satisfied by all quantum correlations, preventing from a direct conflict with relativity, it does not single out quantum correlations. Indeed there exist no-signaling correlations which are stronger than those allowed in quantum mechanics [8], usually referred to as super-quantum correlations. Why such correlations would be unlikely to exist in nature and whether there exist physical principles singling out quantum correlations are important issues in the foundations of quantum mechanics [9–11].

Several approaches have been investigated to discuss this problem. The first consists in investigating the capabilities for information processing of super-quantum correlations, and to compare them with that of quantum correlations. Interestingly it was shown that the availability of certain super-quantum correlations, instead of quantum correlations, would tremendously increase the power of classical communication. In particular, it was shown that some of them would collapse communication complexity [12–14] (hence dramatically reducing the amount of classical communication required to solve a large class of problems [5]) or violate simple and very natural information-theoretic principles (usually satisfied by quantum theory). The latter include the principles of information causality [15, 16] and macroscopic locality [17]. A second approach, perhaps less demanding, starts from assuming 'local quantum mechanics'. In other words the statistics of local measurements are assumed to follow Born's rule. What other principle should then be imposed in order for the global statistics to be quantum? In the bipartite case, it turns out that the no-signaling principle is enough to single out quantum correlations [18, 19]. That is, imposing local quantum mechanics and no-signaling is enough to recover quantum correlations. Importantly, while both of these approaches have proven to be (at least partially) successful in the case of two parties, none of them can tackle the general multipartite scenario [19].

Here we review a simple multipartite game called 'Guess your neighbour's input' (GYNI) [20], the rules of which can be understood intuitively from its name. Despite its innocuous appearance, the game captures crucial features of multipartite quantum correlations. The main aspect of the game is the following. Whereas players sharing quantum resources do not have any advantage over players sharing classical resources, it turns out that players sharing super-quantum correlations can have an advantage over players sharing either classical or quantum resources. In other words, the limitation of quantum resources is here not a mere consequence of the no-signaling principle. Hence, the game of GYNI provides a natural separation between quantum and super-quantum correlations. More generally these results point towards a strengthening of the no-signaling principle, in the general multipartite case, obeyed by quantum mechanics. Therefore, whereas the game of GYNI may seem a priori useless from a quantum perspective, it does in fact bring a novel and fresh perspective on the foundations of quantum theory [21].

Although it is not clear yet what fundamental principle lies behind the quantum limitations for GYNI, several important features of such a principle can already be identified. In particular, this principle must be genuinely multipartite, which can be shown directly from the GYNI game. This is because there exist multipartite super-quantum correlations, that will nevertheless satisfy any bipartite principle [22] (see also [23]). Hence no principle that is inherently bipartite, such as no trivial communication complexity or information causality, can recover quantum correlations.

Moreover, it was shown, using GYNI, that there exist multipartite super-quantum correlations obeying the Born rule locally [19]. Therefore, in the multipartite case, the no-signaling principle is not enough to recover quantum correlations from local quantum mechanics. This result also has fundamental consequences on extensions of Gleason's theorem [24] to composite systems.

Finally, we also review applications of GYNI beyond quantum foundations. In particular, the game turns out to be strongly related [25, 26] to topics of quantum information theory, namely bound entanglement [27] and unextendible product bases [28]. This is surprising since these subjects seem to be completely unconnected at first sight. This connection deepens our understanding of Bell inequalities with no quantum advantage. In particular it allows us to derive such inequalities from unextendible product bases.

This chapter is structured as follows. In Sect. 2, after giving a brief background introduction to nonlocal correlations, we review the GYNI game and derive the winning probabilities for various types of correlations (local, quantum, and no-signaling). Applications of GYNI are reviewed in Sects. 3 and 4. First, in Sect. 3, we discuss results on the extension of Gleason's theorem for composite systems. Then, in Sect. 4, we shall see that any information-theoretic principle capturing quantum correlations must be genuinely multipartite. In Sect. 5, after reviewing in detail the connection between GYNI and unextendible product bases, we will make use of this connection to go beyond GYNI, and to better understand the structure of Bell inequalities with no quantum advantage. Finally, we conclude in Sect. 6.

2 Guess Your Neighbour's Input

2.1 Background: Classical, Quantum and No-Signalling Correlations

The definition of (non)locality was introduced by Bell [2], as a rigorous physical and mathematical framework to test the Einstein-Podolsky-Rosen paradox. Consider two distant observers, Alice and Bob, sharing a physical system, and performing local measurements on their subsystems. Alice and Bob's choice of observables are labeled by x_1 and x_2 respectively, and take outcomes a_1 and a_2. The joint probability distribution of outcomes, conditioned on the choice of observables, is represented by $P(a_1, a_2|x_1, x_2)$. This set of data is described as *local* (or classical) if and only if $P(a_1, a_2|x_1, x_2)$ can be reproduced by a local hidden-variable model, that is, iff it can be written in the form

$$P_L(a_1, a_2|x_1, x_2) = \sum_{\lambda} P(\lambda)P(a_1|x_1, \lambda)P(a_2|x_2, \lambda). \tag{1}$$

Here individual outcomes are completely specified by the choice of local observables and the shared (hidden) random variable λ.[1] Indeed, Alice and Bob's outcomes may be correlated via the variable λ, which is distributed with probability distribution $P(\lambda)$.

The probability distribution $P(a_1, a_2|x_1, x_2)$ is said to be realizable in quantum mechanics (or in short, to be *quantum*) if and only if it can be written in the following form:

$$P_Q(a_1, a_2|x_1, x_2) = \mathrm{tr}(\rho_{AB} M_{a_1}^{x_1} \otimes M_{a_2}^{x_2}), \tag{2}$$

where the state of system ρ_{AB} is defined by a density operator on the joint Hilbert space $\mathscr{H}_A \otimes \mathscr{H}_B$, and $M_{a_1}^{x_1}$, $M_{a_2}^{x_2}$ are local generalized measurements (positive semi-definite operators on the local Hilbert space such that $\sum_{a_j} M_{a_j}^{x_j} = \mathbb{1}$ $(j = 1, 2)$ with $\mathbb{1}$ denoting an identity matrix of the dimension following from the context). Indeed quantum correlations are stronger than classical ones, hence there exist quantum distributions $P_Q(a_1, a_2|x_1, x_2)$ which cannot be written in the form (1).

A crucial feature of both classical and quantum correlations is that they satisfy the no-signalling principle: instantaneous information transmission is impossible. More formally the principle says that Alice's measurement outcome is uncorrelated to Bob's choice of measurement, that is

$$\forall_{a_1, x_1, x_2, x_2'} \quad \sum_{a_2} P_{NS}(a_1, a_2|x_1, x_2) = \sum_{a_2} P_{NS}(a_1, a_2|x_1, x_2') \equiv P_A(a_1|x_1). \tag{3}$$

Similar equations must be obeyed for Bob's marginal distribution. Correlations satisfying this principle, as well as normalization and positivity, are referred to as *nonsignaling* correlations [29]. Interestingly, there exist nonsignaling correlations that are not quantum [8], i.e. cannot be written in the form (2).

The above definitions are naturally generalized to the multipartite case: local correlations between N parties are described by

$$P_L(a_1, \ldots, a_N|x_1, \ldots, x_N) = \sum_{\lambda} P(\lambda)P(a_1|x_1, \lambda)P(a_2|x_2, \lambda)\ldots P(a_N|x_N, \lambda), \tag{4}$$

Quantum correlations are given by

$$P_Q(a_1, \ldots, a_N|x_1, \ldots, x_N) = \mathrm{tr}(\rho M_{a_1}^{x_1} \otimes \cdots \otimes M_{a_N}^{x_N}), \tag{5}$$

[1] By simplicity, we consider λ to be discrete, but all the formulation can be extended to the continuous case.

where ρ denotes the quantum state shared between the parties. Finally nonsignalling correlations are defined such that the marginal distribution $P(a_{i_1}, \ldots, a_{i_k}|x_{i_1}, \ldots, x_{i_k})$ of any subset $\{i_1, \ldots, i_k\}$ of the N parties does not depend on the measurement settings of the remaining parties, that is

$$P_{NS}(a_{i_1}, \ldots, a_{i_k}|x_1, \ldots, x_N) = P_{NS}(a_{i_1}, \ldots, a_{i_k}|x_{i_1}, \ldots, x_{i_k}) \tag{6}$$

This guarantees that any subset of the parties is unable to signal to the remaining parties by their choice of measurement.

In order to distinguish between these three kinds of correlations (local, quantum, and nonsignaling) one devises a Bell test, involving a certain number (usually finite) of parties, observables and outcomes. It is convenient to represent a probability distribution $P(a_1, \ldots, a_N|x_1, \ldots, x_N)$ as a vector of probabilities $\mathbf{P}$, with entries $P(\mathbf{a}|\mathbf{x}) = P(a_1, \ldots, a_N|x_1, \ldots, x_N)$. In this vector space, Bell inequalities are given by linear expressions

$$S = \sum_j \alpha_j P_j \leq \omega_c. \tag{7}$$

where P_j denotes the jth component of $\mathbf{P}$. The coefficients α_j are real. The bound of the inequality, i.e. ω_c, is the largest value of the Bell polynomial S for any local probability distribution, i.e. of the form (4). The set of local correlations defines a convex polytope. Hence it can be described by a finite set of linear inequalities, that are called tight Bell inequalities.

The local set is a strict subset of the set of quantum correlations. The latter is still a convex set, although no longer a polytope. It can, however, be described by an infinite set of quantum Bell inequalities, similar to (7) but replacing the classical bound by a quantum one, ω_q, which may in general exceed the classical bound, i.e. $\omega_q \geq \omega_c$.

Finally, the set of no-signalling correlations is also a convex polytope, which is strictly larger than the quantum set. Its facets are given by positivity inequalities, stating that joint probabilities are positive. The largest value of a Bell polynomial S for any no-signaling probability distribution is denoted ω_{ns}; indeed, in general $\omega_{ns} \geq \omega_q$.

The scene being set, let us bring in the protagonists.

2.2 The GYNI Game

Consider N players disposed on a ring. The game starts with each player receiving a (private) input bit x_i (say from a referee), distributed according to the probability distribution $q(\mathbf{x})$. Now, the name of the game says it all: the goal is that each player makes a correct guess a_i of his (say) right-hand side neighbour's input bit (see Fig. 1), that is

$$\underset{i}{\forall} \quad a_i = x_{i+1}, \tag{8}$$

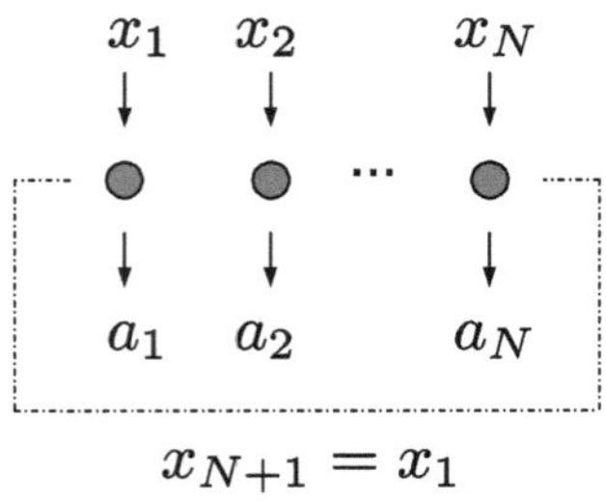

Fig. 1 The GYNI game. The goal is that each party outputs its right-neighbour's input: $a_i = x_{i+1}$

where $x_{N+1} \equiv x_1$. Importantly, the players are successful if and only if all the parties make a correct guess.

The winning probability is then given by

$$\omega = \sum_{\mathbf{x}} q(\mathbf{x}) P(\mathbf{a}_i = \mathbf{x}_{i+1} | \mathbf{x}_i) \tag{9}$$

with $P(\mathbf{a}_i = \mathbf{x}_{i+1}|\mathbf{x}_i) = P(a_1 = x_2, \ldots, a_N = x_1 | x_1, \ldots, x_N)$. Note that no communication between the players is allowed during the game. However, during the preparation stage of the game, the players are informed of the distribution $q(\mathbf{x})$ of the inputs. They are allowed to establish a common strategy, which will consist in utilizing in a judicious way physical resources they are allowed to share. Here our aim will be to find out how good the parties can perform at the game when sharing classical, quantum, or no-signaling correlations. Formally, the game represents a multipartite Bell test, and Eq. (9) has the structure of a multipartite Bell inequality (see (7)). Hence our goal will be to determine the bounds ω_c, ω_q and ω_{ns}, corresponding to the classical, quantum and no-signalling bounds of the GYNI Bell inequality.

2.3 No Quantum Advantage

A central feature of the GYNI game is that the maximum winning probability in the quantum world is exactly the same as in a classical one. In other words, the GYNI inequalities (9) have the same classical and quantum bound, i.e. $\omega_c = \omega_q$, for any distribution of inputs $q(\mathbf{x})$.

Classical bound. Let us start by analyzing the best classical performance. Any probabilistic classical strategy (which includes the use of shared randomness), can be decomposed into a convex sum of deterministic strategies. This means that players can achieve the best winning probability ω_c by making a definite guess a_i for each input bit x_i. Hence it is enough to analyze such cases. Imagine that their deterministic strategy allows them to succeed when receiving some input string $\mathbf{y}$, i.e. $a_i(y_i) = y_{i+1}$, $\forall i$. The input strings have an interesting orthogonality property. Consider any input bit string $\mathbf{x}$ that is different from $\mathbf{y}$ and $\bar{\mathbf{y}}$, where $\bar{\mathbf{y}}$ denotes the negated bit string, i.e. $\bar{y}_i = y_i + 1$. For any such a string $\mathbf{x} \neq \mathbf{y}, \bar{\mathbf{y}}$, there is some player i such that $x_i = y_i$ and $x_{i+1} \neq y_{i+1}$. That is, there is always some player which will make a wrong guess. He will receive the bit $x_i = y_i$, and output $a_i(y_i) = y_{i+1}$ according to the strategy, while the correct output would be $a_i = \bar{y}_{i+1}$. Note that by choosing the strategy such that $a_i(\bar{y}_i) = \bar{y}_{i+1}, \forall i$, it is still possible to score when receiving $\bar{\mathbf{y}}$. The score of this strategy, which is in fact the best classical winning probability (see below), is then

$$\omega_c = \max_{\mathbf{x}}[q(\mathbf{x}) + q(\bar{\mathbf{x}})], \tag{10}$$

achieved by using $\mathbf{y}$ such that $q(\mathbf{y}) + q(\bar{\mathbf{y}}) = \max_{\mathbf{x}}[q(\mathbf{x}) + q(\bar{\mathbf{x}})]$. For more details, we refer the reader to Ref. [20].

Quantum bound. If players have access to quantum systems, the most general protocol involves a quantum state ρ of arbitrary Hilbert space dimension and general quantum measurements $M_{x_i}^{a_i}$ corresponding to a probability distribution

$$P(a_1, \ldots, a_N | x_1, \ldots, x_N) = \mathrm{tr}(\rho M_{x_1}^{a_1} \otimes \cdots \otimes M_{x_N}^{a_N}). \tag{11}$$

The best quantum winning probability is then the maximum expected value of the Bell operator

$$\omega_q = \max_{\psi, \mathrm{meas}} \sum_{\mathbf{x}} q(\mathbf{x}) \langle M_{\mathbf{x}} \rangle. \tag{12}$$

where $M_{\mathbf{x}} \equiv M_{x_1}^{x_2} \otimes \cdots \otimes M_{x_N}^{x_1}$. Notice that it is enough to optimize over pure states $|\psi\rangle$ and projective measurements $M_{a_i}^{x_i} M_{a_i'}^{x_i} = \delta_{a_i = a_i'} M_{a_i}^{x_i}$, since there are no restrictions on the size of local Hilbert spaces. Following a similar reasoning to the classical case, take projectors $M_{\mathbf{y}}$ and $M_{\mathbf{x}}$, where $\mathbf{x} \neq \mathbf{y}, \bar{\mathbf{y}}$. Then there is some local projector i, defined on the same basis $x_i = y_i$, but projecting on orthogonal subspaces $x_{i+1} \neq y_{i+1}$. Consequently, the measurement projectors also obey an orthogonality condition,

$$M_{\mathbf{y}} M_{\mathbf{x}} = 0 \quad \text{if} \quad \mathbf{x} \neq \mathbf{y}, \bar{\mathbf{y}}. \tag{13}$$

This property is sufficient to show that

$$\sum_{\mathbf{x}} q(\mathbf{x}) \langle M_{\mathbf{x}} \rangle \leq \max_{\mathbf{x}}[q(\mathbf{x}) + q(\bar{\mathbf{x}})]\,, \tag{14}$$

which proves that the best quantum winning probability is the same as the classical one

$$\omega_q = \omega_c. \tag{15}$$

Indeed, the derivation of the best winning probabilities, in both the classical and quantum case, relies on a rather natural orthogonality property (either of deterministic local strategies, or of orthogonal measurement projectors). Interestingly such a property is not a consequence of the no-signaling, and does in general not hold for no-signaling correlations, as we shall see in the next section.

2.4 No-Signalling Advantage

We have just seen that quantum resources provide no advantage over classical ones for GYNI. Intuitively, this could be understood as follows. The game of GYNI is in some sense clearly related to the notion of signaling. Indeed, if all players can guess correctly their neighbour's input with a high probability, this implies signaling. As quantum correlations are no-signaling, they cannot help. Following this line of reasoning, it may seem then natural to conjecture that any no-signaling resources will provide no advantage over classical resources for GYNI. Surprisingly, however, this intuition turns out to be wrong. There exist in fact super-quantum no-signaling correlations which provide an advantage, as we shall see now.

2.4.1 Correlated Inputs

Consider a particular version of the GYNI game in which the inputs are correlated in the following way: $q(\mathbf{x})$ is uniform on the set of inputs that satisfy the parity condition:

$$q(\mathbf{x}) = \begin{cases} 1/2^{N-1} & \text{if } x_1 \oplus \cdots \oplus x_{\hat{N}} = 0 \\ 0 & \text{otherwise,} \end{cases} \tag{16}$$

where $\hat{N} = N$ if N is odd and $\hat{N} = N - 1$ if N is even. Using Eq. (10), it is easy to check that in classical or quantum theory, the success probability is limited by $\omega_c = 1/2^{N-1}$. We will see that, allowing for super-quantum correlations, this limit can be beaten: the best winning probability ω_{ns} is lower-bounded by $\omega_{ns}/\omega_c \geq 4/3$. Unlike the previous example, here, although each party still has absolute uncertainty about his neighbour's input, no-signalling correlations are able to exploit a global correlation (the parity of the input-string) to increase the chance of correct guess.

3-player game. Let us first consider the simplest game, featuring three players.[2] The GYNI inequality is then simply given by

$$\omega = \frac{1}{4}\left[P(000|000) + P(110|011) + P(011|101) + P(101|110)\right] \leq \frac{1}{4}, \tag{17}$$

[2]Note that for 2 players, no-signaling correlation provide no advantage.

where the bound holds for any local or quantum strategy.

Let us first derive an upper bound on the no-signaling winning probability. Consider the first three terms in (17). The no-signalling principle implies that

$$\begin{aligned} P(000|000) &\le \sum_{a_3} P(00a_3|000) = \sum_{a_3} P(00a_3|001), \\ P(110|011) &\le \sum_{a_2} P(1a_20|011) = \sum_{a_2} P(1a_20|001), \\ P(011|101) &\le \sum_{a_1} P(a_111|101) = \sum_{a_1} P(a_111|001). \end{aligned} \tag{18}$$

Notice that the terms appearing in the right hand side of the three above equations have no overlap. Hence, from normalization, it follows that $P(000|000) + P(110|011) + P(011|101) \le 1$. We apply a similar reasoning to the remaining combinations of three probability terms of Eq. (17), such that we get

$$3[P_{NS}(000|000) + P_{NS}(110|011) + P_{NS}(011|101) + P_{NS}(101|110)] \le 4. \tag{19}$$

Hence we obtain an upper limit to the no-signalling winning probability: $\omega_{ns} \le 1/3$. From this derivation, we also conclude that it is only possible to reach this limit if every probability term in the GYNI inequality (17) has the value 1/3 (see [20]).

Now, it turns out that this upper bound can be reached by actual no-signaling probability distributions. There exist two (among 45) inequivalent classes of extremal tripartite no-signaling boxes [30], that reach the best winning no signaling probability $\omega_{ns} = 1/3$ (see [20]).

Finally note an interesting feature of inequality (17). It is a tight Bell inequality, that is, it defines a facet of the polytope of local correlations [31]. Hence it identifies a portion of the quantum boundary which is of maximal dimension [20].

N-player game. Next let us consider the general case of N players, using the condition (16) on the inputs. For any N, no-signaling correlations provide an advantage. To show this, we prove that resources that provide a winning probability ω/ω_c, in the game with N players, can provide at least the same ratio ω/ω_c for $N+1$ players. The strategy is very simple: players 1 to N play exactly as in the N-player game, while player $N+1$ outputs his input, $a_{N+1} = x_{N+1}$. This guess is correct when $x_{N+1} = x_1$, which happens with probability 1/2. Since $\omega_c(N+1) = (1/2)\omega_c(N)$, the ratio remains the same:

$$\frac{\omega}{\omega_c}(N) = \frac{\omega}{\omega_c}(N+1). \tag{20}$$

Then, for any $N \ge 3$, the best no-signalling sucess probability is at least as good as $(4/3)\omega_c$. This lower bound is achieved if the first 3 players use the optimal no-signalling strategy for the 3-player game, while the remaining output their inputs.

They can however do better: using linear programming, we obtained that $\omega_{ns}/\omega_c = 4/3$, for $N = 4$; $\omega_{ns}/\omega_c = 16/11$, for $N = 5, 6$; and $\omega_{ns}/\omega_c = 64/42$, for $N = 7, 8$.

Inspired by these three values, we provide the following guess for the non-signalling violation of the GYNI Bell inequality: for odd N one has $\omega_{ns}/\omega_c = 2^{N-1}/\alpha_N$ and

$$\alpha_N = \sum_{k=0}^{\lfloor \frac{N-1}{4} \rfloor} \binom{N}{\frac{N+1}{2} + 2k}, \tag{21}$$

while for even N one has the same value as for $N - 1$, that is $\omega_{ns}/\omega_c = 2^{N-2}/\alpha_{N-1}$. The expression of α_N may seem obscure but has an easy interpretation. Consider all the binomial coefficients $c(N, l) = N!/(l!(N-l)!)$. Now, α_N is nothing but the sum of all the coefficients $c(N, (N+1)/2 + 2k)$, where $0 \leq k \leq (N-1)/4$. It is straightforward to see that this expression reproduces the values obtained using linear programming up to $N = 7$. For instance for $N = 7$ one has $c(7, l) = (1, 7, 21, 35, 35, 21, 7, 1)$ and $\alpha_N = c(7, 4) + c(7, 6) = 42$. Moreover, one can also see that in the limit of $N \to \infty$, $\omega_{ns}/\omega_c \to 2$. As shown in the next section, this is an upper bound for GYNI valid for an arbitrary number of parties. If our guess is correct, the bound becomes tight in the limit of an infinite number of parties.

Remarkably, it turns out that the N-partite GYNI Bell inequalities (with promise (16)), hereafter referred to as GYNI$_N$, are tight for an arbitrary odd N [26] and for $N = 4, 6$ [20]. It is conjectured that they are tight for any N.

2.4.2 Upper Bounds on ω_{ns}

From the winning probability in the classical case (Eq. (10)), we know that $q(\mathbf{x}) \leq \omega_c$ for any $\mathbf{x}$, from which we get the bound $\omega \leq \omega_c \sum_{\mathbf{x}} P(\mathbf{a}_i = \mathbf{x}_{i+1}|\mathbf{x}_i)$. Something more meaningful is obtained if we now assume the distributions to be no-signalling. Take the summation $\sum_{\mathbf{x}} P(\mathbf{a}_i = \mathbf{x}_{i+1}|\mathbf{x})$. Repeatedly applying the no-signalling condition (3), (first to party N, then to $N - 1$ and so on), we get

$$\begin{aligned}\sum_{x_1,\dots,x_N} & P_{NS}(x_2, \dots, x_N, x_1|x_1, \dots, x_{N-1}, x_N) \\ & \leq \sum_{x_1,\dots,x_N} P_{NS}(x_2, \dots, x_N|x_1, \dots, x_{N-1}) \\ & = \sum_{x_1,\dots,x_{N-1}} P_{NS}(x_2, \dots, x_{N-1}|x_1, \dots, x_{N-2}) = \dots = 2. \end{aligned} \tag{22}$$

We conclude that the success probability for no-signalling correlations is bounded by

$$\omega_{ns} \leq 2\omega_c, \tag{23}$$

which means that, in general, no-signalling correlations do not allow deterministic success. As we could predict, for some input distributions, perfect guessing is only possible if players communicate.

2.4.3 Completely Uniform Distributions of Inputs

We have seen that the GYNI game with correlated inputs (16) provides a sharp distinction between classical/quantum and no-signaling best winning strategies. Now we provide a non-trivial example where such difference is not present: the completely uniform distribution over the inputs, i.e. $q(\mathbf{x}) = 1/2^N$. We obtain a tight upper bound on ω_{ns} by noticing that $2q(\mathbf{x}) = \omega_c$, which leads to

$$\omega_{ns} = \frac{\omega_c}{2} \sum_{\mathbf{x}} P_{NS}(\mathbf{a}_i = \mathbf{x}_{i+1}|\mathbf{x}_i) \leq \omega_c. \tag{24}$$

Classical and no-signalling resources provide exactly the same best winning probability, in a situation where each player has, a priori, no information about the input of its neighbour.

Now that we presented the GYNI Bell inequality, we review in the next sections the application of this inequality in two different contexts, both related to the characterization of quantum correlations.

3 Application 1: Gleason's Theorem for Multipartite Systems

Gleason's Theorem [24] is a celebrated theorem in the foundations of quantum mechanics that allows recovering the Born rule for quantum probabilities from the structure of quantum measurements. Recall that a quantum measurement acting on a Hilbert space of dimension d corresponds to a set of k positive operators, $M_i \geq 0$ with $i = 1, \ldots, k$ such that $\sum_i M_i = \mathbb{1}$. Gleason's Theorem aims at characterizing maps from quantum measurements to probability distributions. The maps Λ have to satisfy the following properties:

1. For any positive operator $0 \leq M \leq \mathbb{1}$ one has $\Lambda(M) \geq 0$.
2. Given a quantum measurement, that is, given a set of k positive operators summing up to the identity, one has

$$\sum_{i=1}^{k} \Lambda(M_i) = 1. \quad (25)$$

Note that the considered maps are non-contextual, as the measurement operators are mapped into probabilities independently of the structure of the measurement they belong to.

Gleason's Theorem implies that all maps satisfying the two requirements 1 and 2 can be written as $\Lambda(M) = \mathrm{tr}(\rho M)$ for a given quantum state ρ, that is, ρ a positive operator of trace one. We sketch here the idea of the proof, while its detailed version may be found e.g. in Ref. [32]. Notice, however, that the author of [32] imposes an additional linearity condition on Λ which, as shown in what follows, can be simply inferred from 1 and 2. Indeed, consider two measurement operators M_1, M_2 such that $M_3 = \mathbb{1} - (M_1 + M_2) \geq 0$. Consider now the two different measurements $\{M_1, M_2, M_3\}$ and $\{M_1 + M_2, M_3\}$. The second measurement is simply a coarse-grained version of the first in which the two first outcomes are grouped together. A direct application of property 2 above implies that $\Lambda(M_1)+\Lambda(M_2) = \Lambda(M_1+M_2)$. This together with properties 1 and 2 imply that the map Λ, initially defined for positive operators, can be uniquely extended to a linear map acting on all operators. It immediately follows that it can be written as $\mathrm{tr}(XM)$ for an operator X. But then, the condition 1 implies the positivity of the operator X and its normalization follows from condition 2. On the other hand, one checks by hand that any of these maps satisfies conditions 1 and 2.

This theorem is a seminal result in the foundations of quantum theory. In particular, it implies that Born's rule for the computation of measurement probabilities can be derived from the Hilbert space structure of quantum measurements and the two natural conditions provided above.

3.1 Gleason Correlations

Gleason's Theorem was initially established for single systems. It was later extended to composite systems in Refs. [33, 34]. The scenario consists of N independent observers. To each observer j, with $j = 1, \ldots, N$, one associates a Hilbert space of dimension d_j and a structure of quantum measurements given by sets of positive operators summing up to the identity. For the sake of simplicity, we take in what follows all the local dimensions equal, $d_i = d$, $\forall i$. We denote by $\{M_{i_j}^{(j)}\}$, $i_j = 1, \ldots, k_j$ the sets of positive operators defining a measurement for each observer, that is, $\sum_{i_j} M_{i_j}^{(j)} = \mathbb{1}$. The extension of the theorem then aims at characterizing those maps from measurements by each observer to probability distributions. In what follows, for the ease of notation, we restrict the analysis to the simplest bipartite case, although it can be easily generalized to an arbitrary number of parties. The map is requested to satisfy the following conditions:

1. For pairs of positive operators, $M_{i_1}^{(1)}, M_{i_2}^{(2)}$, where $0 \leq M_{i_1}^{(1)}, M_{i_2}^{(2)} \leq \mathbb{1}$ one has $\Lambda(M_{i_1}^{(1)}, M_{i_2}^{(2)}) \geq 0$.
2. For pairs of measurements, $\{M_{i_1}^{(1)}\}, \{M_{i_2}^{(2)}\}$, where $0 \leq M_{i_1}^{(1)}, M_{i_2}^{(2)} \leq \mathbb{1}$ one has

$$\sum_{i_1,i_2=1}^{k_1,k_2} \Lambda(M_{i_1}^{(1)}, M_{i_2}^{(2)}) = 1. \tag{26}$$

3. Given two complete quantum measurements by one of the observers, say the second, $\{M_{i_2}^{(2)}\}$ and $\{N_{i_2}^{(2)}\}$, the map has to be such that

$$\sum_{i_2=1}^{k_2} \Lambda(M_{i_1}^{(1)}, M_{i_2}^{(2)}) = \sum_{i_2=1}^{k_2'} \Lambda(M_{i_1}^{(1)}, N_{i_2}^{(2)}). \tag{27}$$

The new condition, i.e., the third one, can be understood as the natural formalization of the no-signalling principle in the considered framework: the marginal probability distribution seen by one of the observers cannot depend on the measurement performed by the other observer. The generalization to an arbitrary number of parties of these requirements is straightforward. Now Λ maps tuple of positive operators $M_i^{(1)}, \ldots, M_{i_N}^{(N)}$ into non-signalling probability distributions.

The generalization of the theorem to this scenario, that we call multipartite Gleason's Theorem, states that all such maps can be written as

$$\Lambda(M_{i_1}^{(1)}, \ldots, M_{i_N}^{(N)}) = \mathrm{tr}\left(W M_{i_1}^{(1)} \otimes \ldots \otimes M_{i_N}^{(N)}\right), \tag{28}$$

where W is an operator which is positive on product states $|\psi_1\rangle \ldots |\psi_N\rangle$. These operators can thus be viewed as locally positive quantum states, or equivalently as entanglement witnesses [35] (below we adopt this last terminology). As above, it is clear that maps of the form (28) satisfy the previous three requirements and the non-trivial part of the result is proving the opposite direction.

As the set of entanglement witnesses is larger than the set of quantum states (or, in other words, there exist operators W that are non-positive, but positive on product states) the set of distributions (28), called in what follows *Gleason correlations*, is in principle larger than the quantum set. However, it was shown in [18, 19] that the two sets actually coincide for two parties. Thus, as it happens for single-party systems, imposing the structure of quantum measurements for the observers gives the quantum correlations.

The proof of the equivalence between Gleason and bipartite quantum correlations exploits the Choi-Jamiołkowski (CJ) isomorphism [36] that relates maps to operators. In this case, any witness W can be written as $(I \otimes \Upsilon)(\Phi)$, where Υ is a positive map and Φ is the projector onto the maximally entangled state $|\Phi\rangle = (1/\sqrt{d}) \sum_i |ii\rangle \in \mathbb{C}^d \otimes \mathbb{C}^d$ and I stands for the identity map. With the aid of Ref. [37], one can prove that any normalized witness can also be written as $(I \otimes \Lambda)(\Psi)$, where Λ is now a positive and trace-preserving map, while Ψ is a projector onto some pure bipartite state.[3] It then follows that

$$\begin{aligned}\operatorname{tr}(W M_{a_1}^{x_1} \otimes M_{a_2}^{x_2}) &= \operatorname{tr}[(I \otimes \Lambda)(\Psi) M_{a_1}^{x_1} \otimes M_{a_2}^{x_2}] \\ &= \operatorname{tr}[\Psi M_{a_1}^{x_1} \otimes \Lambda^*(M_{a_2}^{x_2})] \\ &= \operatorname{tr}(\Psi M_{a_1}^{x_1} \otimes \widetilde{M}_{a_2}^{x_2}),\end{aligned} \tag{30}$$

where Λ^* is the dual[4] of Λ and $\widetilde{M}_{a_2}^{x_2} = \Lambda^*(M_{a_2}^{x_2})$ defines a valid quantum measurement because the dual of a positive trace-preserving map is positive and unital, that is, $\Lambda^*(\mathbb{1}) = \mathbb{1}$.

The next natural question is as to whether the equivalence between quantum and Gleason correlations holds for an arbitrary number of parties. Surprisingly, the answer to this question turns out to be negative, as there are local measurements acting on entanglement witnesses that produce supra-quantum correlations [19]. Before reviewing this result, it is worth mentioning that local measurements on entanglement witnesses that can be written as

$$W = \sum_k \left(\Lambda_{A_1}^k \otimes \cdots \otimes \Lambda_{A_N}^k\right)(\rho_k), \tag{31}$$

where ρ_k are N-party quantum states, $\Lambda_{A_i}^k$ are positive trace preserving maps and the number of terms in the sum is arbitrary, do not lead to supra-quantum correlations. This is a rather straightforward generalization of the equivalence proof in the bipartite case.

In order to prove that in the multipartite case the set of Gleason correlations contains quantum correlations as a strict subset, we provide an example of entanglement witness and local measurements giving nonsignalling correlations which violate the three-partite GYNI Bell inequality (17). Let us start by introducing the following set of four fully product vectors from the three-qubit Hilbert space:

---

[3]To see this explicitly let us first notice that for a normalized witness W it holds that $W = (I \otimes \Lambda)(\Phi)$ with trace-preserving Λ iff $W_A = \operatorname{tr}_B W = \mathbb{1}/d$. Then, if $W_A \neq \mathbb{1}/d$ but it is of full rank, one introduces another witness $\widetilde{W} = (1/d)(W_A^{-1/2} \otimes \mathbb{1}) W (W_A^{-1/2} \otimes \mathbb{1})$. Clearly, $\widetilde{W}_A = \mathbb{1}/d$ and thus $\widetilde{W}$ is isomorphic to a trace-preserving positive map $\widetilde{\Lambda}$. Consequently,

$$W = d(\sqrt{W_A} \otimes \mathbb{1}) \widetilde{W} (\sqrt{W_A} \otimes \mathbb{1}) = d(\sqrt{W_A} \otimes \mathbb{1})(I \otimes \widetilde{\Lambda})(\Phi)(\sqrt{W_A} \otimes \mathbb{1}) = (I \otimes \widetilde{\Lambda})(\Psi), \tag{29}$$

where Ψ denotes a projector onto some normalized pure state $|\Psi\rangle = \sqrt{d}(\sqrt{W_A} \otimes \mathbb{1})|\Phi\rangle$ of full Schmidt rank. Finally, if W_A is rank-deficient, one constructs yet another witness $W' = W + \mathscr{P}_A^{\perp} \otimes \mathbb{1}$, where $\mathscr{P}_A^{\perp} = \mathbb{1} - \mathscr{P}_A$ with $\mathscr{P}_A$ denoting a projector onto the support of W_A. Then, W_A' is of full-rank and therefore W' admits the form (29). To complete the proof, it suffices to notice that $W = (\mathscr{P}_A \otimes \mathbb{1}) W' (\mathscr{P}_A \otimes \mathbb{1})$, and hence W also assumes the form (29) with a normalized pure state $|\Psi\rangle = \sqrt{d}[\mathscr{P}_A (W_A')^{1/2} \otimes \mathbb{1}]|\Phi\rangle = \sqrt{d}(W_A^{1/2} \otimes \mathbb{1})|\Phi\rangle$ which is now not of full Schmidt rank.

[4]The dual map Λ^* of Λ is the map such that $\operatorname{tr}(A\Lambda(B)) = \operatorname{tr}[\Lambda^*(A)B]$.

$$|\psi_1\rangle = |000\rangle, \qquad |\psi_2\rangle = |1e^{\perp}e\rangle, \qquad |\psi_3\rangle = |e1e^{\perp}\rangle, \qquad |\psi_4\rangle = |e^{\perp}e1\rangle, \quad (32)$$

where $|e\rangle \in \mathbb{C}^2$ is an arbitrary vector different from $|0\rangle$ and $|1\rangle$, while $|\overline{e}\rangle$ stands for a vector orthogonal to $|e\rangle$. One checks by hand that there is no other three-qubit fully product vector orthogonal to all $|\psi_i\rangle$s; such sets of product vectors are called *unextendible product bases* (UPBs) [28] (see Sect. 5.1 for a detailed discussion on UPBs and more examples).

As noticed in [28], the set (32), called Shifts UPB, can be used for a simple construction of bound entangled state, i.e., an entangled state from which any type of maximally entangled state cannot be distilled [27]. The state is given by $\rho_{\mathrm{UPB}} = (\mathbb{1} - \Pi_{\mathrm{UPB}})/4$, where Π_{UPB} denotes the projector onto span$\{|\psi_i\rangle\}$.

Let us now consider the normalized entanglement witness detecting ρ:

$$W = \frac{1}{4 - 8\varepsilon}(\Pi_{\mathrm{UPB}} - \varepsilon\mathbb{1}), \quad (33)$$

where

$$\varepsilon = \min_{|\alpha\beta\gamma\rangle} \langle\alpha\beta\gamma|\Pi_{\mathrm{UPB}}|\alpha\beta\gamma\rangle. \quad (34)$$

The fact that there is no fully product vector orthogonal to $|\psi_i\rangle$ implies that $\varepsilon > 0$, and, on the other hand, it is fairly easy show that $\varepsilon < 1/2$. One also notices that $\mathrm{Tr}(W\rho_{UPB}) = -\varepsilon/4(1 - 2\varepsilon)$.

Now, one can see that the witness W, when measured along the local bases in the definition of the UPB (32), leads to correlations that produce a value of GYNI game equal to $\omega/\omega_c = (1-\varepsilon)/(1-2\varepsilon)$, which is larger than one for all $0 < \varepsilon \leq 1/2$. Thus, these correlations represent an example of Gleason correlations with no quantum analogue.

4 Application II: Quantum Correlations and Information Principles

As mentioned in the introduction, an intense research effort has recently been devoted to understand why nonlocality appears to be limited in quantum mechanics. Information concepts have been advocated as the key missing ingredient needed to single-out the set of quantum correlations [9–11]. The main idea is to identify 'natural' information principles, satisfied by quantum correlations, but violated by super-quantum correlations. The existence of the latter would then have implausible consequences from an information-theoretic point of view. Celebrated examples of these principles are information causality [15] or non-trivial communication complexity [12]. While the use of these information concepts has been successfully applied to specific scenarios [13, 14, 16, 38, 39], proving, or disproving, the validity of a principle for quantum correlations is extremely challenging. On the one hand, it is rather difficult to derive the Hilbert space structure needed for quantum correlations from information quantities. On the other hand, proving that some super-quantum correlations are fully compatible with an information principle seems out of reach, as one needs to consider all possible protocols using these correlations and show that none of them leads to a violation of the principle. Hence it is still unclear whether this approach is able to fully recover the set of quantum correlations.

Therefore it is relevant to derive general features of a principle that could potentially identify quantum correlations. Using GYNI, it was recently shown that such a principle must be genuinely multipartite. More specifically, no principle that is inherently bipartite (i.e. referring only to correlations between two sets of parties) can characterize the set of quantum correlations when three of more observers are involved [22]. The rest of this section is devoted to this result.

Before reviewing the result, it is worth recalling that, so far, most information-theoretic principles have been formulated in the bipartite scenario. Actually, even the general formulation of the no-signalling principle has a bipartite structure: correlations among N observers are compatible with the no-signalling principle whenever there exists no partition of the N parties into two groups such that the marginal probability distribution of one set of the parties depends on the measurements performed by the other set of parties (see (6)). Moving to information causality, it considers a scenario in which a first party, Alice, has a string of n_A bits. Alice is then allowed to send m classical bits to a second party, Bob. The principle of information causality bounds the amount of information Bob can gain on the n_A bits held by Alice whichever protocol they implement making use of the pre-established bipartite correlations and the message of m bits. Alice and Bob can violate this principle when they have access to some super-quantum correlations [15]. In the case $m = 0$, information causality implies that in absence of a message, pre-established correlations do not allow Bob to gain any information about any of the bits held by Alice, which is nothing but the no-signaling principle. This suggests the following generalization of information causality to an arbitrary number of parties, mimicking what is done for the no-signalling principle: given some correlations $P(a_1, \ldots, a_n|x_1, \ldots, x_N)$, they are said to be compatible with information causality whenever all bipartite correlations constructed from them satisfy this principle. This generalization ensures the correspondence between no-signaling and information causality when $m = 0$ for an arbitrary number of parties. This generalization of information causality has recently been applied to the study of extremal tripartite non-signaling correlations [23].

Regarding non trivial communication complexity, it studies how much communication is needed between two distant parties to compute probabilistically a function of some inputs in a distributed manner. It can also be interpreted as a generalization of the no-signaling principle, as it imposes constraints on correlations when a finite amount of communication is allowed between parties. Different multipartite generalizations of the principle have been studied, see [5]. However, as for information causality, one can always consider the straightforward generalization in which the principle is applied to every partition of the N parties in two groups.

We are now in position to review the proof of the impossibility of characterizing quantum correlations for an arbitrary number of parties using bipartite principles. For simplicity, we restrict the analysis to tripartite correlations.

4.1 Time-Ordered-Bilocal Correlations and GYNI

The first ingredient in the proof is the characterization of multipartite correlations such that any bipartite correlations constructed from them have a classical local model. By definition, correlations satisfying this property do not violate any bipartite principle satisfied by classical correlations.

A priori, one would think that if the correlations $P(a_1, a_2, a_3|x_1, x_2, x_3)$ have a local model along all possible bipartitions, namely $A_1 - A_2A_3$, $A_2 - A_1A_3$ and $A_3 - A_1A_2$, that is,

$$\begin{aligned} P(a_1, a_2, a_3|x_1, x_2, x_3) &= \sum_\lambda P_1(\lambda)P_1(a_1|x_1, \lambda)P_1(a_2, a_3|x_2, x_3, \lambda) \\ &= \sum_\lambda P_2(\lambda)P_2(a_2|x_2, \lambda)P_2(a_1, a_3|x_1, x_3, \lambda) \\ &= \sum_\lambda P_3(\lambda)P_3(a_3|x_3, \lambda)P_1(a_1, a_2|x_1, x_2, \lambda), \end{aligned} \tag{35}$$

then, any bipartite object constructed from it also has a local model. This intuition however has proven to be wrong in [40], where it was shown how non-local bipartite correlations can be derived from correlations having a decomposition of the form of (35). The characterization of multipartite correlations such that a local model exists for any bipartite correlations derived from it is then subtler than expected. Indeed, at the moment, it is unknown what is the largest set of correlations having this property [40]. It has however been shown in [22] that the set of time-ordered-bilocal correlations (TOBL) do fulfill this requirement. Tripartite correlations have a TOBL model whenever they can be written as

$$\begin{aligned} P(a_1, a_2, a_3|x_1, x_2, x_3) &= \sum_\lambda P_\lambda^{i|jk} P(a_i|x_i, \lambda)P_{j\to k}(a_j, a_k|x_j, x_k, \lambda) \\ &= \sum_\lambda P_\lambda^{i|jk} P(a_i|x_i, \lambda)P_{j\leftarrow k}(a_j, a_k|x_j, x_k, \lambda) \end{aligned} \tag{36}$$

for $(i, j, k) = (1, 2, 3), (2, 3, 1), (3, 1, 2)$, with the distributions $P_{j\to k}$ and $P_{j\leftarrow k}$ obeying the conditions

$$P_{j\to k}(a_j|x_j, \lambda) = \sum_{a_k} P_{j\to k}(a_j, a_k|x_j, x_k, \lambda), \tag{37}$$

Table 1 Different examples of deterministic bipartite probability distributions $P_{23}(a_2, a_3|x_2, x_3, \lambda)$ characterized by output assignments to the four possible combination of measurements

x_2	x_3	a_2	a_3
0	0	0	0
0	1	0	1
1	0	1	1
1	1	1	0

x_2	x_3	a_2	a_3
0	0	0	0
0	1	1	1
1	0	0	0
1	1	1	1

x_2	x_3	a_2	a_3
0	0	1	0
0	1	0	1
1	0	0	1
1	1	1	0

Left inputs and outputs corresponding to a point $P_{2\to 3}(a_2, a_3|x_2, x_3, \lambda)$ in the decomposition (36). *Center* inputs and outputs corresponding to a point $P_{2\leftarrow 3}(a_2, a_3|x_2, x_3, \lambda)$ in (36). *Right* inputs and outputs corresponding to a distribution which allows signaling in the two directions

$$P_{j\leftarrow k}(a_k|x_k, \lambda) = \sum_{a_j} P_{j\leftarrow k}(a_j, a_k|x_j, x_k, \lambda). \tag{38}$$

The notion of TOBL correlations first appeared in [30] (see [40] and [41] for a proper introduction and further motivation for such models). As can be seen from the relations (37) and (38) we impose the distributions $P_{j\to k}$ and $P_{j\leftarrow k}$ to allow for signaling at most in one direction, indicated by the arrow (see Table 1).

To understand the operational meaning of these models, consider the bipartition $1|23$ for which systems 2 and 3 act together. In this situation, $P(a_1, a_2, a_3|x_1, x_2, x_3)$ can be simulated if a classical random variable λ with probability distribution $p_\lambda^{1|23}$ is shared by parts 1 and the composite system 2–3, and they implement the following protocol: given λ, 1 generates its output according to the distribution $P(a_1|x_1, \lambda)$; on the other side, and depending on which of the parties 2 and 3 measures first, 2–3 uses either $P_{2\to 3}(a_2, a_3|x_2, x_3, \lambda)$ or $P_{2\leftarrow 3}(a_2, a_3|x_2, x_3, \lambda)$ to produce the two measurement outcomes. Likewise, any other bipartition of the three systems admits a classical simulation.

By construction, the set of tripartite TOBL models is convex and is included (in fact, it is strictly included [40]) in the set of tripartite probability distributions of the form (35). Moreover, TOBL models always produce classical correlations under post-selection: indeed, suppose that we are given a tripartite distribution $P(a_1, a_2, a_3|x_1, x_2, x_3)$ satisfying condition (36), and a postselection is made on the outcome $\tilde{a}_3$ of measurement $\tilde{x}_3$ by party 3. Then, one has

$$P(a_1, a_2|x_1, x_2\tilde{x}_3\tilde{a}_3) = \sum_\lambda P'_\lambda P(a_1|x_1, \lambda) P'(a_2|x_2, \lambda), \tag{39}$$

with

$$P'_\lambda = \frac{p_\lambda^{1|23}}{P(\tilde{a}_3|\tilde{x}_3)} P_{2\leftarrow 3}(\tilde{a}_3|\tilde{x}_3, \lambda), \qquad P'(a_2|x_2, \lambda) = P_{2\leftarrow 3}(a_2|x_2, \tilde{x}_3, \tilde{a}_3, \lambda). \tag{40}$$

Postselected tripartite TOBL boxes can thus be regarded as elements of the TOBL set with trivial outcomes for one of the parties.

We now demonstrate that any possible bipartite correlations derived from many uses of TOBL correlations have a local model and, thus, are compatible with any bipartite principle satisfied by classical (and obviously quantum) correlations. The most general protocol consists in distributing an arbitrary number of boxes described by $P^1, P^2, \ldots, P^N$ among three parties which are split into two groups, A and B. Both groups can process the classical information provided by their share of the N boxes. For instance, outputs generated by some of the boxes can be used as inputs for other boxes (see Fig. 2). This local processing of classical information is usually referred to as *wirings*. Thus, in order to prove our result in full generality, we should consider all possible wirings of tripartite boxes. We show next that if $P^1, P^2, \ldots, P^N$ are in TOBL, then the resulting correlations P_{fin} obtained after any wiring protocol have a local decomposition with respect to the bipartition $A|B$, and therefore fulfill any bipartite information principle.

For simplicity, we illustrate our procedure for the wiring shown in Fig. 2, where boxes P^1, P^2, P^3 are distributed between two parties A and B, and party A only holds one subsystem of each box. The construction is nevertheless general: it applies to any wiring and also covers situations where for some TOBL boxes party A holds two subsystems instead of just one (or even the whole box).

From (36) we have

$$P^i(a_1^i, a_2^i, a_3^i|x_1^i, x_2^i, x_3^i) = \sum_{\lambda^i} P^i_{\lambda^i} P^i_1(a^i_1|x^i_1, \lambda^i) P^i_{2\to 3}(a^i_2, a^i_3|x^i_2, x^i_3, \lambda^i) \quad (41)$$

$$= \sum_{\lambda^i} P^i_{\lambda^i} P^i_1(a^i_1|x^i_1, \lambda^i) P^i_{2\leftarrow 3}(a^i_2, a^i_3|x^i_2, x^i_3, \lambda^i), \quad (42)$$

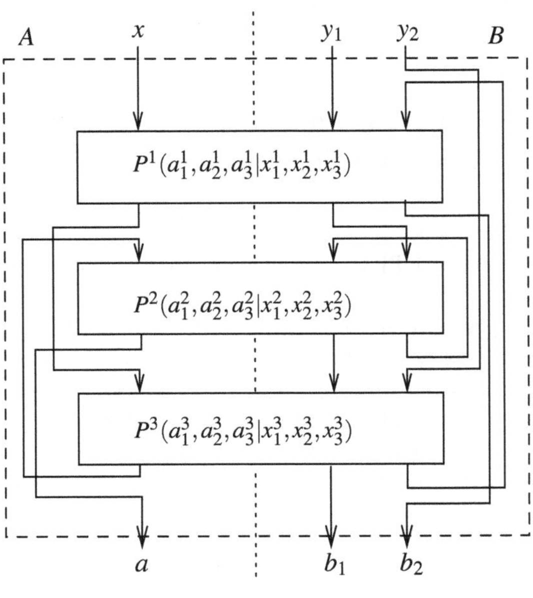

Fig. 2 Wiring of several tripartite correlations distributed among parties A and B. The generated bipartite box accepts a bit x (two bits y_1, y_2) as input on subsystem A (B) and returns a bit a (two bits b_1, b_2) as output. Relations (41, 42) guarantee that the final bipartite distribution $P_{\text{fin}}(a, (b_1, b_2)|x, (y_1, y_2))$ admits a local model

for $i = 1, 2, 3$. Consider the first box that receives an input, in our case subsystem 2 of P^1. The first outcome a_2^1 can be generated by the probability distribution $P^1_{2\to 3}(a_2^1, a_3^1|x_2^1, x_3^1, \lambda^1)$ encoded in the hidden variable λ^1 that models these first correlations. This is possible because for this decomposition a_2^1 is defined independently of x_3^1, the input in subsystem 3. Then, the next input x_3^2, which is equal to a_2^1, generates the output a_3^2 according to the probability distribution $P^2_{2\leftarrow 3}(a_2^2, a_3^2|x_2^2, x_3^2, \lambda^2)$ encoded in λ^2. The subsequent outcomes a_2^i and a_3^i are generated in a similar way. The general idea is that outputs are generated sequentially using the local models according to the structure of the wiring on 2–3. Finally, subsystem 1 can generate its outputs a^i by using the probability distribution $P_1^i(a_1^i|x^i, \lambda^i)$. This probability distribution is independent of the order in which parties 2 and 3 make their measurement choices for any of the boxes. Averaging over all hidden variables one obtains P_{fin}. This construction provides the desired local model for the final probability distribution.

The final step in the proof consists of showing that there exist correlations in the TOBL set that do not have a quantum realization. This was shown by means of the GYNI inequality. More precisely, it can be proven that, contrary to quantum correlations, this inequality is violated by TOBL correlations:

$$\begin{aligned} &\text{maximize} \quad P(000|000) + P(110|011) + P(011|101) + P(101|110) \\ &\text{subject to} \quad P(a_1, a_2, a_3|x_1, x_2, x_3) \in \text{TOBL}. \end{aligned} \tag{43}$$

The maximization yields a value of 7/6, implying the existence of supra-quantum correlations in TOBL. The form of the TOBL correlations leading to this violation can be found in [22]. Later, another example of supra-quantum correlations in TOBL was provided in [23], where the authors proved that an extremal point of the no-signalling polytope for three parties and two two-outcome measurements per party is also in TOBL and has no quantum realization.

5 Generalization of GYNI: Bell Inequalities Without Quantum Violation and Unextendible Product Bases

The relation between GYNI's Bell inequality and the three-qubit unextendible product basis (UPB) was used in the previous section to show that, contrary to the bipartite case [18] (see also [19]), in the three-partite scenario Gleason correlations define a larger set than the quantum ones. In fact, this link can be generalized and used to derive a systematic construction of nontrivial Bell inequalities without quantum violation from UPB, see Ref. [25]. That is, all the Bell inequalities derived using this construction lack a quantum violation, nevertheless, they are nontrivial in the sense that there exist some nonsignalling correlations violating them. They also complement the results of Ref. [19] providing new examples of multipartite scenarios where Gleason correlations are different from the quantum ones. Finally, some UPBs lead to *tight* Bell inequalities with no quantum violation, providing novel examples of this intriguing phenomenon [26].

Our aim in this section is to recall the method from Refs. [25, 26] and then discuss properties of the resulting Bell inequalities. We provide some classes of nontrivial Bell inequalities with no quantum violation associated to UPBs. Finally, we go beyond UPB and show that there are also sets of orthogonal product vectors that are not UPBs but can be associated to nontrivial Bell inequalities. Before that let us recall the notion of unextendible product bases and briefly review their properties.

5.1 Unextendible Product Bases

We start by introducing an N-partite product Hilbert space

$$H = \mathbb{C}^{d_1} \otimes \ldots \otimes \mathbb{C}^{d_N}, \tag{44}$$

where d_i $(i = 1, \ldots, N)$ denote the dimensions of the local Hilbert spaces. In what follows we call an element $|\psi\rangle$ of H *fully product* if it assumes the form $|\psi\rangle = \otimes_{i=1}^{N} |\psi_i\rangle \equiv |\psi_1, \ldots, \psi_N\rangle$ with $|\psi_i\rangle \in \mathbb{C}^{d_i}$.

Then, let us consider a set of orthogonal product vectors

$$S = \left\{ |\Psi_m\rangle = |\psi_m^{(1)}\rangle \otimes \ldots \otimes |\psi_m^{(N)}\rangle \right\}_{m=1}^{|S|}, \tag{45}$$

where $|\psi_m^{(i)}\rangle$ $(m = 1, \ldots, |S|)$ are local vectors belonging to $\mathbb{C}^{d_i}$ and $|S| \leq \dim H$. With this we have the following definition [28].

Definition 1 Let S be a set of orthogonal fully product vectors (45) from H. We call S *unextendible product basis* (UPB) if it spans a proper subspace in H, i.e., $|S| < \dim H$, and there is no product vector $\otimes_{i=1}^{N} |\phi_i\rangle \in H$ orthogonal to span$S$.

The notion of unextendible product bases reflects the peculiar feature of some composite Hilbert spaces H, which can be represented as the direct sum of two orthogonal subspaces, one spanned by product vectors, and another one that does not contain any of them. That is, all the states belonging to this second subspace must be entangled (see Fig. 3). This has interesting consequences from the quantum information point of view. As first observed by Bennett and coworkers [28], UPBs can be used for constructing bound entangled states, i.e., states that are entangled but nevertheless no pure-state entanglement can be distilled from them by means of local operations and classical communication [27].

To be more precise, following [28], let us consider a particular UPB U, and the normalized projector onto the subspace of H orthogonal to U, i.e.,

$$\rho = \frac{1}{\dim H - |U|} \left(\mathbb{1} - \Pi \right). \tag{46}$$

Fig. 3 Schematic definition of a UPB: a set of orthogonal product vectors S spanning a proper subspace $\text{span}S \subset H$ such that there is no fully product vector $\otimes_{i=1}^{N}|\phi_i\rangle \in H$ orthogonal to S. A normalized projector onto $(\text{span}S)^{\perp}$ (46) is a bound entangled state [28]

By Π and $\mathbb{1}$ we denoted, respectively, the projector onto the subspace spanned by U and the identity operator acting on H. Since there is no product vector orthogonal to U, the support of ρ consists only of entangled states, implying that ρ must be entangled. Also, it immediately follows from Eq. (46) that ρ has all partial transpositions positive, which implies that ρ is bound entangled [27].

To illustrate the above definitions we consider the following examples of UPBs.

Example 1 We start from the TILES UPB, one of the first bipartite UPBs introduced in Ref. [28]. It consists of five two-qutrit vectors of the form

$$U_{\text{TILES}} = \{|0\rangle(|0\rangle - |1\rangle), |2\rangle(|1\rangle - |2\rangle), (|0\rangle - |1\rangle)|2\rangle, (|1\rangle - |2\rangle)|0\rangle, \\ (|0\rangle + |1\rangle + |2\rangle)^{\otimes 2}\}. \tag{47}$$

As shown in [28], no product state can be orthogonal to all these vectors. Notice that in the two-qutrit Hilbert space there only exist five-elements UPBs and all of them are known [28, 42–44].

Example 2 Second, let us consider a general class of N-qubit unextendible product bases with odd $N = 2k - 1$ ($k \in \mathbb{N}; k \geq 2$) given by the following $2k$ vectors [42]:

$$U_{\text{GenShifts}} = \{|0\ldots0\rangle, |1e_1\ldots e_{k-1}\overline{e}_{k-1}\ldots\overline{e}_1\rangle, |\overline{e}_1 1 e_1\ldots e_{k-1}\overline{e}_{k-1}\ldots\overline{e}_2\rangle, \ldots, \\ |e_1\ldots e_{k-1}\overline{e}_{k-1}\ldots\overline{e}_1 1\rangle\} \tag{48}$$

with $\{|0\rangle, |1\rangle\}$ and $\{|e_i\rangle, |\overline{e}_i\rangle\}$ ($i = 1, \ldots, k-1$) being k arbitrary but different bases in $\mathbb{C}^2$. The ith ($i \geq 2$) vector in (48), except for the first two ones, is obtained from the vector $i - 1$ by shifting all the local vectors by one to the right, and thus the name *Generalized Shifts*.

Example 3 Third, let us consider the general class of UPBs found by Niset and Cerf [45]. Here we take the Hilbert space $H = (\mathbb{C}^d)^{\otimes N}$, where $N \geq 3$ and $d \geq N - 1$, and the following set of $N(d-1)+1$ vectors:

$$U_{\text{NC}} = \{|e_{d-1}\rangle^{\otimes N}\} \cup \bigcup_{i=0}^{N-1} S_i, \tag{49}$$

where

$$S_0 = \{|0, 1, \ldots, d-1\rangle|e_0\rangle, \ldots, |0, 1, \ldots, d-1\rangle|e_{d-2}\rangle\} \tag{50}$$

and $S_i = V^i S_0$ $(i = 1, \ldots, N-1)$ with V denoting a unitary permutation operator such that $V|x_1\rangle \ldots |x_N\rangle = |x_N\rangle|x_1\rangle \ldots |x_{N-1}\rangle$ for $|x_i\rangle \in \mathbb{C}^d$, and $\{|e_i\rangle\}_{i=0}^{d-1}$ is any orthogonal basis in $\mathbb{C}^d$ different from the standard one. Notice that U_{NC} can straightforwardly be generalized to an arbitrary local dimension $d_i \geq N-1$ $(i = 1, \ldots, N)$ just by adjusting both bases at each site to the respective dimension [45].

Both classes of multipartite UPBs from Examples 2 and 3 (here up to local unitary operations) recover, for $N = 3$, the already introduced Shifts UPB (32), i.e., $U_{\text{Shifts}} = \{|000\rangle, |1\overline{e}e\rangle, |e1\overline{e}\rangle, |\overline{e}e1\rangle\}$ with $\{|e\rangle, |\overline{e}\rangle\}$ being an arbitrary basis of $\mathbb{C}^2$ different from the standard one. Clearly, this set can be slightly generalized by taking the second basis different at each site, that is, $\{|000\rangle, |1\overline{e}_2e_3\rangle, |e_11\overline{e}_3\rangle, |\overline{e}_1e_21\rangle\}$ (the first basis can be fix to the standard one by a local unitary operation). As shown by Bravyi [46], any three-qubit UPB can be written in this canonical form by local unitary operations and permutations of the parties.

5.2 *Constructing Bell Inequalities with No Quantum Violation from Unextendible Product Bases*

We are now ready to recall the method from [25, 26] allowing one to associate a nontrivial Bell inequality with no quantum violation to UPB's. However, as shown below, in order for the construction to work the UPB cannot be generic and has to satisfy a certain property.

5.2.1 The Construction

To begin, consider again the product Hilbert space H and the set of vectors S. For the time being, we do not assume S to be a UPB, keeping, however, the assumption that elements of S are orthogonal product vectors from H. Then, let us collect all different local vectors appearing in all vectors $|\Psi_m\rangle$ at the ith site in the local sets

$$S^{(i)} = \left\{|\psi_m^{(i)}\rangle\right\}_{m=1}^{s_i} \qquad (i = 1, \ldots, N), \tag{51}$$

where $s_i \leq |S|$. Subsequently, among elements of $S^{(i)}$ we search for mutually orthogonal vectors and collect them in separate subsets $S_n^{(i)}$ $(n = 0, \ldots, k_i)$ such that $S_0^{(i)} \cup \ldots \cup S_{k_i}^{(i)} = S^{(i)}$ for any i (see Fig. 4). Notice that these subsets may, but do not have to, span the corresponding Hilbert space $\mathbb{C}^{d_i}$. The idea, as it becomes clearer below, is to associate local measurements to each party from the given UPB.

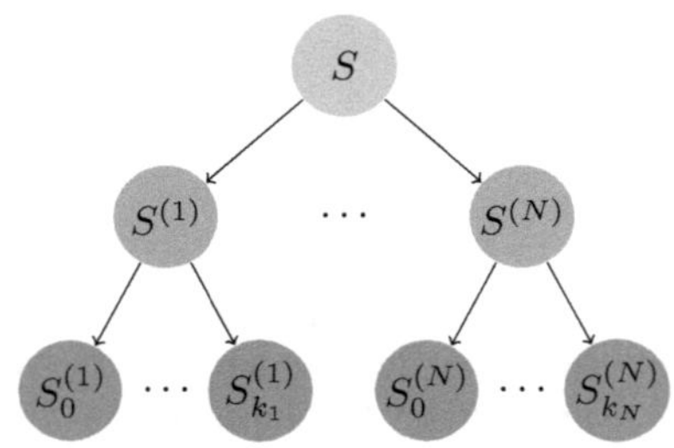

Fig. 4 Schematic description of our construction. From the set S having the local independence property, one constructs the local sets $S^{(i)}$ $(i = 1, \ldots, N)$ by collecting different local vectors in $|\Psi_m\rangle$. Then, one distinguishes local subsets $S_m^{(i)}$ of mutually orthogonal vectors among elements of each local set $S^{(i)}$

It should be emphasized that there exist sets S for which the local subsets cannot be unambiguously defined. This is, for instance, the case for vectors U_{TILES} [cf. Eq. (47)]. At both sites there are five different vectors $|0\rangle$, $|0\rangle - |1\rangle$, $|1\rangle - |2\rangle$, $|2\rangle$, and $|0\rangle + |1\rangle + |2\rangle$. Clearly, the first one is orthogonal to the third and fourth ones, however, the latter are not mutually orthogonal. Then, in order to avoid this ambiguity, we consider only those sets S that have the following property.

Local independence property. *Let S be a set of orthogonal product vectors from H. For each party i, construct the subset of states $S_k^{(i)}$ as described above. The set S is said to satisfy the local independence property whenever, for each party, two vectors belonging to different subsets $S_k^{(i)}$ and $S_l^{(i)}$ $(k \neq l)$ are not orthogonal.*

In other words, what we need is that the local subsets are constructed in such a way that by replacing one of them by another set of orthogonal vectors of the same size, we keep the orthogonality of the product states of the initial UPB S. In yet another words, the above property guarantees that the orthogonality of S is preserved under any unitary rotation of elements of any local subset $S_k^{(i)}$, which, in a sense, makes them independent. This is why we call this property local independence property. This property also implies that given a UPB S, there is a unique way of defining the sets $S_k^{(i)}$ for each party.

A particular example of a set having the above property is the already introduced Shifts UPB (32). At each site there are four different vectors $|0\rangle$, $|1\rangle$, $|e\rangle$, and $|\overline{e}\rangle$, which can be grouped in two distinct sets $S_0 = \{|0\rangle, |1\rangle\}$ and $S_1 = \{|e\rangle, |\overline{e}\rangle\}$. Since, by the very assumption, $|e\rangle \neq |0\rangle, |1\rangle$, none of the vectors from S_0 is orthogonal to none of elements of S_1, and hence U_{Shifts} has the local independence property.

Interestingly, as it can easily be checked, all sets of orthogonal vectors in multi-qubit Hilbert spaces satisfy the local independence property, as all local subsets contain at most two elements. On the other hand, the example of TILES UPB shows that this is in general not the case when local dimensions are larger than two.

Let us now pass to our construction of Bell inequalities. The construction of the different sets $S_k^{(i)}$ for each party defines a set of local measurements. Now, to every vector $|\Psi_m\rangle$ from S [cf. Eq. (45)] we can associate a conditional probability $P(\mathbf{a}_m|\mathbf{x}_m)$, or, strictly speaking, vectors of measurements settings and outcomes

$$\mathbf{a}_m = (a_m^{(1)}, \ldots, a_m^{(N)}) \quad \text{and} \quad \mathbf{x}_m = (x_m^{(1)}, \ldots, x_m^{(N)}) \tag{52}$$

in the following way:

- the measurement setting $x_m^{(i)}$ of the observer i is given by the index k enumerating the subset $S_k^{(i)}$ containing $|\psi_m^{(i)}\rangle$,
- the measurement outcome $a_m^{(i)}$ corresponds to the position of $|\psi_m^{(i)}\rangle$ in the set $S_k^{(i)}$.

Eventually, we simply add the obtained conditional probabilities and maximize the resulting expression over all classical correlations, which leads us to the following Bell inequality

$$\sum_{m=1}^{|S|} P(\mathbf{a}_m|\mathbf{x}_m) \leq 1. \tag{53}$$

The value of the right-hand side of the above, the so-called classical bound, directly follows from the orthogonality of elements of S. In fact, as these events come from a set of orthogonal product vectors with the local independence property, for each pair of vectors $|\Psi_m\rangle, |\Psi_n\rangle$ $(m \neq n)$, there exists a site, say i, such that $|\psi_m^{(i)}\rangle \perp |\psi_n^{(i)}\rangle$. Consequently, the associated conditional probabilities $P(\mathbf{a}_m|\mathbf{x}_m)$ and $P(\mathbf{a}_n|\mathbf{x}_n)$ involve at site i the same measurement setting but different outcomes. This means that for any deterministic local model, if one of these two probabilities is one, the other has to be zero. Let us further call such probabilities orthogonal. Since the above holds for any pair of conditional probabilities, the left-hand side of (53) clearly amounts to at most one.

Notice that, in principle, we can consider more general inequalities by combining the conditional probabilities $P(\mathbf{a}_m|\mathbf{x}_m)$ $(m = 1, \ldots, |S|)$ with arbitrary positive weights q_m. However, the use of different weights leads to Bell inequalities that are weaker and certainly cannot be tight (see below).

5.2.2 Properties

Let us now shortly characterize the obtained Bell inequalities (53). We collect their most important properties in the following theorem [25, 26].

Theorem 1 *Let S be a set of orthogonal product vectors from H having the local independence property. Then the following implications are true:*

(i) the associated Bell inequality (53) is not violated by quantum correlations
(ii) if S is a UPB in H, then the Bell inequality (53) is nontrivial in the sense that it is violated by some nonsignalling correlations,
(iii) if S is a full basis in H or can be completed to one in such a way that it maintains the local independence property, the associated Bell inequality (53) is not violated by any nonsignalling correlations.

Proof (i): Let us, in contrary, assume that indeed the Bell inequality (53) associated to S is violated by a quantum state ρ. Then, there exist local measurement operators and the resulting Bell operator, denoted B, such that $\mathrm{tr}(B\rho) > 1$. This means that at least one of the eigenvalues of B has to exceed one. On the other hand, it is clear that the local measurement operators can be assumed to be projective; if ρ violates (53) with a non-projective measurements, one is able to find another quantum state ρ', possibly acting on a larger Hilbert space, violating the same Bell inequality with projective measurements.

Let then $P_m = \otimes_{i=1}^{N} P_m^{(i)}$ denote a product projective measurement operator corresponding to $P(\mathbf{a}_m|\mathbf{x}_m)$, which, in general, may be different from the corresponding vectors $|\Psi_m\rangle \in S$. Clearly, orthogonality of the conditional probabilities $P(\mathbf{a}_m|\mathbf{x}_m)$ is translated to the orthogonality of the corresponding P_m. Precisely, as already stated, any pair of probabilities $P(\mathbf{a}_m|\mathbf{x}_m)$ and $P(\mathbf{a}_n|\mathbf{x}_n)$ has at some site, say i, the same settings but different outcomes, implying that $P_m^{(i)} \perp P_n^{(i)}$ and hence $P_m \perp P_n$. As a result all P_m ($m = 1, \ldots, |S|$) are orthogonal and the Bell operator $B = \sum_m P_m$ is again a projector contradicting the fact that for some ρ, $\mathrm{tr}(B\rho) > 1$.

(ii): Our proof is constructive, that is, for any Bell inequality associated to a UPB we provide non-signalling correlations violating it. We denote by Π the projector onto $\mathrm{span}S$, and introduce, in a full analogy to (33), the following witness

$$W = \frac{1}{|S| - \dim H}(\Pi - \varepsilon \mathbb{1}) \tag{54}$$

with ε being a positive number defined as

$$\varepsilon = \min\langle x_1, \ldots, x_N|\Pi|x_1, \ldots, x_N\rangle, \tag{55}$$

where the minimum is computed over all fully product vectors from H and, as S is UPB, one has $\varepsilon > 0$. One directly checks that this witness detects entanglement of the state (46) constructed from the UPB S, i.e., $\mathrm{tr}(W\rho) < 0$. This, after substituting the exact form of ρ, see (46), can be rewritten as

$$\mathrm{tr}(W\Pi) > 1. \tag{56}$$

Clearly, Π can be seen as a Bell operator corresponding to the Bell inequality associated to S. To complete the proof of (ii) it suffices to notice that local measurements performed on an entanglement witness, in particular (54), always give nonsignalling correlations (see e.g. [18, 19]).

(iii): Let us start from the case when $|S| < \dim H$ and assume that S can be completed to a basis of H maintaining the local independence property (if $H = (\mathbb{C}^2)^{\otimes N}$ one can always do that provided S is completable). Let then $|\Psi_m\rangle$ ($m = |S| + 1, \ldots, \dim H$) denote product orthogonal vectors completing S, i.e., $\mathrm{span}(S \cup \{|\Psi_m\rangle\}_m) = H$. Consequently, one can associate a Bell inequality (53) to the set

S and conditional probabilities $P(\mathbf{a}_m|\mathbf{x}_m)$ to the new vectors $|\Psi_m\rangle$ ($m = |S| + 1, \ldots, \dim H$) in an unambiguous way. Then

$$\sum_{m=1}^{|S|} P(\mathbf{a}_m|\mathbf{x}_m) \leq \sum_{m=1}^{\dim H} P(\mathbf{a}_m|\mathbf{x}_m) \leq 1, \tag{57}$$

meaning that it suffices to prove that the Bell inequality appearing on the right-hand side (the one constructed from a full basis in H) is trivial. For this purpose, we note that the latter is saturated by the uniform probability distribution $P(\mathbf{a}|\mathbf{x}) = 1/\dim H$ for any $\mathbf{a}$ and $\mathbf{x}$, which is an interior point of the corresponding polytope of classical correlations. Consequently, this Bell inequality is saturated by all vertices of the polytope, and hence by any affine combination thereof, in particular, all nonsignalling correlations. ■

It is illuminating to see how the properties of S determine the properties of the associated Bell inequality. Orthogonality of the elements of S implies that the inequality lacks a quantum violation. If S is additionally a UPB, then the Bell inequality is nontrivial because it detects some nonsignalling correlations. On the other hand, the inequality is trivial if S is or can be completed to a full basis in H, while maintaining the local independence property. In the case of $H = (\mathbb{C}^2)^{\otimes N}$, apart from sets that can only be completed to a UPB, the implication (iii) becomes an equivalence [26]. In the higher-dimensional case, however, there are sets of orthogonal product states that have the local independence property, are not UPBs but cannot be extended maintaining the local independence property (see Sect. 5.3).

The more important and interesting question concerns the tightness of these Bell inequalities. As shown in Refs. [25, 26] there exist example of both tight and nontight Bell inequalities associated to UPBs (see Sect. 5.2.3 for examples). Thus, at the moment, it remains unclear what decides tightness.

5.2.3 Examples

To illustrate the construction, let us apply it to some particular examples of sets S, in particular those presented in Sect. 5.1.

Example 4 Using the already exploited relation between the GYNI Bell inequality (17) and Shifts UPB let us show how the above construction works in practice. As already noticed, U_{Shifts} has two different bases at each site $S_0 = \{|0\rangle, |1\rangle\}$ and $S_1 = \{|e\rangle, |\bar{e}\rangle\}$. The vector $|e\rangle \in \mathbb{C}^2$ is, by assumption, different than $|0\rangle$ and $|1\rangle$, and hence U_{Shifts} has the local independence property. We then associate a conditional probability to every vector in U_{Shifts}:

$$\begin{aligned} |000\rangle &\mapsto P(000|000), & |1\bar{e}e\rangle &\mapsto P(110|011), \\ |e1\bar{e}\rangle &\mapsto P(011|101), & |\bar{e}e1\rangle &\mapsto P(101|110). \end{aligned} \tag{58}$$

By simply adding the above probabilities we get (17). In fact, we can reverse the construction and derive a new family of N-qubit UPB from the Bell inequality GYNI_N [25, 26]. Moving to tightness, recall that GYNI_N were proven to be tight for odd N [26] (and also numerically for some values of even N).

Interestingly, the inequality GYNI_3, which is the only tight tripartite Bell inequality with no quantum violation in the scenario of two dichotomic measurements per site, is associated to the only class of UPB in $(\mathbb{C}^2)^{\otimes 3}$ [46].

Example 5 Second, let us consider the Generalized Shifts UPB (48). The corresponding Hilbert space is $H = (\mathbb{C}^2)^{\otimes N}$ with $N = 2k - 1$ for integer $k \geq 2$. Following the above rules, at each site one can define k local subsets $S_0 = \{|0\rangle, |1\rangle\}$ and $S_i = \{|e_i\rangle, |\bar{e}_i\rangle\}$ $(i = 1, \ldots, k - 1)$, which will later define k observables. We then associate a conditional probability to every element of $U_{\mathrm{GenShifts}}$:

$$
\begin{aligned}
|0\ldots 0\rangle &\mapsto P(0\ldots 0|0\ldots 0)\\
|1e_1\ldots e_{k-1}\bar{e}_{k-1}\ldots\bar{e}_1\rangle &\mapsto P(10\ldots 01\ldots 1|01\ldots k-1, k-1\ldots 1)\\
&\vdots\\
|e_1\ldots e_{k-1}\bar{e}_{k-1}\ldots\bar{e}_1 1\rangle &\mapsto P(0\ldots 01\ldots 11|1\ldots k-1, k-1\ldots 10). \qquad (59)
\end{aligned}
$$

Summing all these probabilities up, we get the N-partite Bell inequality with odd N:

$$
P(0\ldots 0|0\ldots 0) + \sum_{i=1}^{2k-1} D^i P(10\ldots 01\ldots 1|01\ldots k-1, k-1\ldots 1) \leq 1, \qquad (60)
$$

where D denotes an operation shifting the input and output vectors by one to the right, i.e., $D(x_1, \ldots, x_N) = (x_N, x_1, \ldots, x_{N-1})$. Notice that since at each site one has k two-element local subsets S_i, the Bell inequality (60) corresponds to the scenario with k dichotomic observables per site.

Due to Theorem 1, all the Bell inequalities (60) are nontrivial. However, it is unclear whether they are tight. For $N = 3$ the above class recovers the GYNI_3 which is tight, while already for $N = 5$ the corresponding Bell inequality is not tight.

Example 6 Consider now the class of UPBs provided in Ref. [45], i.e., U_{NC} presented in example 3. Here $H = (\mathbb{C}^d)^{\otimes N}$ with $d \geq N-1$. From Eqs. (49) and (50) it follows that at each site one can distinguish two local subsets $S_0 = \{|i\rangle\}_{i=0}^{d-1}$, i.e., the standard basis, and $S_1 = \{|e_i\rangle\}_{i=0}^{d-1}$. Since the elements of U_{NC} are orthogonal irrespectively of the choice of the second basis, U_{NC} has the local independence property. Associating conditional probabilities to elements of U_{NC} and summing them up, one gets the N-partite Bell inequality:

$$
P(d-1, \ldots, d-1|1, \ldots, 1) + \sum_{i=0}^{N-1}\sum_{j=0}^{d-2} D^i P(0, 1, \ldots, d-1, j|0, \ldots, 0, 1) \leq 1, \qquad (61)
$$

where D is defined as before and D^0 is an identity.

Theorem 1 says that all the Bell inequalities (61) are nontrivial. However, it is not clear whether they are tight in general. For $N = 3$ and $d = 2$, this class gives GYNI_3, but for $N = 4$ and $d = 3$ one checks that the resulting Bell inequality is not tight.

The previous examples may lead to the conjecture that GYNI_N is the only situation in which the inequality is tight. However, this is not the case: within our framework, one can also obtain tight Bell inequalities with no quantum violation from UPBs that are independent of GYNI_N. An example is, for instance, the following four-partite Bell inequality

$$\begin{aligned} p(0000|0000) + p(1000|0111) + p(0110|1012) + p(0001|0110) \\ +p(1011|0001) + p(1101|0102) + p(1110|1101) \leq 1 \end{aligned} \tag{62}$$

which was found recently in Ref. [26].

5.3 *Further Generalizations*

In this last section, we present a further generalization of these results and show that it is possible to derive nontrivial Bell inequalities with no quantum violation already from sets of orthogonal product vectors that are not UPB. In fact, if at least a local dimension d_i in H is larger than two, there exist sets of orthogonal product vectors that (i) are not UPBs, in the sense of Definition 1, but (ii) the associated Bell inequalities (53), following the rules from Sect. 5.2.1, lack quantum violation and are violated by non-signalling correlations.

To be more precise, let us again consider a set of orthogonal product vectors S and let us split the local sets $S^{(i)}$ [cf. Eq. (51)] into subsets $S_k^{(i)}$ following the same rules as above. We introduce the definition:

Definition 2 Let S be a set of orthogonal fully product vectors from H having the local independence property. Then, if $|S| < \dim H$ and there does not exist a product vector $\otimes_{i=1}^{N}|\phi_i\rangle \in H$ with $|\phi_i\rangle \in S^{(i)}$ that is orthogonal to all vectors from S, we call S a *weak unextendible product basis* (wUPB).

In this definition, we relax the notion of unextendibility of the set by considering only product vectors made of states from the sets $S^{(i)}$. Clearly, any UPB is also a wUPB. Also if all $d_i = 2$ in Eq. (44), these two notions are equivalent. If, however, at least one of the local dimensions d_i is larger than two, there exist wUPB that are not UPB. As a particular example consider the following.

Example 7 Consider the following set of vectors from $H = \mathbb{C}^2 \otimes \mathbb{C}^2 \otimes \mathbb{C}^3$:

$$S = \{|000\rangle, |1\overline{e}f\rangle, |e1\overline{f}\rangle, |\overline{e}e1\rangle, |\overline{e}e2\rangle, |e1\hat{f}\rangle\}, \tag{63}$$

where $|f\rangle$, $|\overline{f}\rangle$, and $|\hat{f}\rangle$ are three orthogonal vectors from $\mathbb{C}^3$. At the first two sites one distinguishes two local sets $S_0^{(1)} = S_0^{(2)} = \{|0\rangle, |1\rangle\}$ and $S_1^{(1)} = S_1^{(2)} = \{|e\rangle, |\overline{e}\rangle\}$, while at the third site $S_0^{(3)} = \{|0\rangle, |1\rangle, |2\rangle\}$ and $S_1^{(3)} = \{|f\rangle, |\overline{f}\rangle, |\hat{f}\rangle\}$.

The set S has the local independence property because irrespectively of the choice of all these subsets, all its elements are orthogonal. However, it is clearly not a UPB because $|e0g\rangle$ and $|\overline{e}\overline{e}g\rangle$ with $\mathbb{C}^3 \ni |g\rangle \perp |0\rangle, |f\rangle$ are orthogonal to S. Still, S is a wUPB: there is no product vector $|\phi_1\rangle|\phi_2\rangle|\phi_3\rangle \in \mathbb{C}^2 \otimes \mathbb{C}^2 \otimes \mathbb{C}^3$ with $|\phi_i\rangle \in S_j^{(i)}$ $(i = 1, 2, 3;\ j = 1, 2)$, orthogonal to S.

Following the rules given in Sect. 5.2.1, any wUPB can be associated to a Bell inequality (53) with no quantum violation that is violated by nonsignalling correlations. In fact, we have the following theorem.

Theorem 2 *If S is a wUPB, the associated Bell inequality (53) is violated by some nonsignalling correlations.*

Proof The proof goes along the same lines as in point (ii) of Theorem 1. It suffices to consider the same operator as in Eq. (54) with Π denoting now a projector onto the subspace spanned by the wUPB S and the minimum in Eq. (55) taken over product vectors $\otimes_{i=1}^N |\phi_i\rangle \in H$ with local vectors $|\phi_i\rangle \in S^{(i)}$. Notice that if S is a wUPB but not UPB, the operator W is no longer an entanglement witness, but still a Hermitian operator.

One can see that measuring such W along the settings corresponding to the local sets $S_k^{(i)}$ produces non-signalling correlations [19]. The value of the Bell inequality (53) associated to these non-signalling correlations is given by $|S|(1-\varepsilon)/(|S| - \varepsilon \dim H)$. This, due to the fact that $|S| < \dim H$ and $\varepsilon < |S|/\dim H$, is always larger than one. ■

To conclude, let us notice that the Bell inequality corresponding to the set (63):

$$\begin{aligned} &p(000|000) + p(110|011) + p(011|101) + p(101|110) \\ &+p(012|101) + p(102|110) \leq 1, \end{aligned} \tag{64}$$

which has two three-outcome observables at the third site, is tight. This is because it is a lifted tripartite GYNI Bell inequality (17) [47].

6 Conclusions

'Guess your neighbour's input' is a multipartite nonlocal game that, despite its simplicity, captures important features of multipartite correlations. Moreover, it has unexpected connections to topics in quantum foundations and quantum information theory. In particular, it shows that the natural multipartite generalization of Gleason's Theorem fails for more than two parties, that intrinsically multipartite principles are needed to characterize quantum correlations and that there exists a link between unextendible orthogonal product bases and Bell inequalities with no quantum violation.

From a speculative point of view, GYNI suggests that we are lacking an intrinsically multipartite principle in our understanding of correlations. Indeed, the most interesting feature of the game is that it represents a multipartite strengthening of the no-signaling principle, which is by construction a bipartite principle, obeyed by quantum correlations. This naturally raises the question of what physical or information-theoretic principles lie behind GYNI. The principle of Local Orthogonality, proposed in Ref. [48], represents a possible solution to this question.

Acknowledgments Discussions with T. Fritz are acknowledged. This work was supported by the ERC starting grant PERCENT, the EU AQUTE and QCS projects, the Spanish CHIST-ERA DIQIP, FIS2008-00784 and FIS2010-14830 projects, and the UK EPSRC. R. A. is supported by the Spanish MINCIN through the Juan de la Cierva program.

References

1. L. Hardy, arXiv:quant-ph/0101012v4
2. J.S. Bell, Physics **1**, 195 (1964)
3. N. Brunner, D. Cavalcanti, S. Pironio, V. Scarani, S. Wehner, Rev. mod. Phys. **86**, 419 (2014)
4. A.K. Ekert, Phys. Rev. Lett. **67**, 661 (1991)
5. H. Buhrman, R. Cleve, S. Massar, R. de Wolf, Rev. Mod. Phys. **82**, 665 (2010)
6. J. Barrett, L. Hardy, A. Kent, Phys. Rev. Lett. **95**, 010503 (2005); A. Acín, et al., Phys. Rev. Lett. **98**, 230501 (2007); L. Masanes, S. Pironio, A. Acín, Nat. Commun. **2**, 238 (2011)
7. S. Pironio, et al., Nature **464**, 1021 (2010); R. Colbeck, Ph.D. thesis, University of Cambridge; R. Colbeck, A. Kent, J. Phys. A: Math. Theor. **44**(9), 095305 (2011)
8. S. Popescu, R. Rohrlich, Found. Phys. **24**, 379 (1994)
9. G. Brassard, Nat. Phys. **1**, 2 (2005)
10. S. Popescu, Nat. Phys. **2**, 507 (2006)
11. R. Clifton, J. Bub, H. Halvorson, Found. Phys. **33**, 1561 (2003)
12. W. van Dam, Nonlocality & communication complexity. Ph.D. thesis, University of Oxford (2000), arXiv:quant-ph/0501159v1
13. G. Brassard, H. Buhrman, N. Linden, A.A. Méthot, A. Tapp, F. Unger, Phys. Rev. Lett. **96**, 250401 (2006)
14. N. Brunner, P. Skrzypczyk, Phys. Rev. Lett. **102**, 160403 (2009)
15. M. Pawłowski, T. Paterek, D. Kaszlikowski, V. Scarani, A. Winter, M. Żukowski, Nature **461**, 1101 (2009)
16. J. Allcock, N. Brunner, M. Pawłowski, V. Scarani, Phys. Rev. A **80**, 040103(R) (2009)
17. M. Navascues, H. Wunderlich, Proc. R. Soc. Lond. A **466**, 881 (2009)
18. H. Barnum, S. Beigi, S. Boixo, M.B. Elliott, S. Wehner, Phys. Rev. Lett. **104**, 140401 (2010)
19. A. Acín, R. Augusiak, D. Cavalcanti, C. Hadley, J.K. Korbicz, M. Lewenstein, M. Piani, Phys. Rev. Lett. **104**, 140404 (2010)
20. M.L. Almeida, J.-D. Bancal, N. Brunner, A. Acín, N. Gisin, S. Pironio, Phys. Rev. Lett. **104**, 230404 (2010)
21. A. Winter, Nature **466**, 1053 (2010)
22. R. Gallego, L. Würflinger, A. Acín, M. Navascués, Phys. Rev. Lett. **107**, 210403 (2011)
23. T.H. Yang, D. Cavalcanti, M. Almeida, C. Teo, V. Scarani, New J. Phys. **14**, 013061 (2012)
24. A. Gleason, J. Math. Mech. **6**, 885 (1957)

25. R. Augusiak, J. Stasińska, C. Hadley, J.K. Korbicz, M. Lewenstein, A. Acín, Phys. Rev. Lett. **107**, 070401 (2011)
26. R. Augusiak, T. Fritz, M. Kotowski, M. Kotowski, M. Pawłowski, M. Lewenstein, A. Acín, Phys. Rev. A **85**, 042113 (2012)
27. M. Horodecki, P. Horodecki, R. Horodecki, Phys. Rev. Lett. **80**, 5239 (1998)
28. C.H. Bennett, D.P. DiVincenzo, T. Mor, P.W. Shor, J.A. Smolin, B.M. Terhal, Phys. Rev. Lett. **82**, 5385 (1999)
29. J. Barrett, N. Linden, S. Massar, S. Pironio, S. Popescu, D. Roberts, Phys. Rev. A **71**, 022101 (2005)
30. S. Pironio, J.-D. Bancal, V. Scarani, J. Phys. A: Math. Theor. **44**, 065303 (2011)
31. C. Śliwa, Phys. Lett. A **317**, 165 (2003)
32. P. Busch, Phys. Rev. Lett. **91**, 120403 (2003)
33. D. Foulis, C. Randall, Interpret. Found. Quantum Theory **5**, 920 (1979); M. Kläy, C. Randall, D. Foulis, Int. J. Theor. Phys. **26**, 199 (1987); H. Barnum, C.A. Fuchs, J.M. Renes, A. Wilce, arXiv:quant-ph/0507108
34. N.R. Wallach, arXiv:quant-ph/0002058
35. M. Horodecki, P. Horodecki, R. Horodecki, Phys. Lett. A **223**, 1 (1996); B.M. Terhal, Phys. Lett. A **271**, 319 (2000)
36. A. Jamiołkowski, Rep. Math. Phys. **3**, 275 (1972); M.-D. Choi, Linear Algebra Appl. **10**, 285 (1975)
37. M. Horodecki, P. Horodecki, R. Horodecki, Open Syst. Inf. Dyn. **13**, 103 (2006)
38. A. Ahanj, S. Kunkri, A. Rai, R. Rahaman, P.S. Joag, Phys. Rev. A **81**, 032103 (2010)
39. D. Cavalcanti, A. Salles, V. Scarani, Nat. Commun. **1**, 136 (2010)
40. R. Gallego, L. Würflinger, A. Acín, M. Navascués, Phys. Rev. Lett. **109**, 070401 (2012)
41. J. Barrett, S. Pironio, J.-D. Bancal, N. Gisin, Phys. Rev. A **88**, 014102 (2013)
42. D.P. DiVincenzo, T. Mor, P.W. Shor, J.A. Smolin, B.M. Terhal, Commun. Math. Phys. **238**, 379 (2003)
43. J.M. Leinaas, J. Myrheim, P.Ø. Sollid, Phys. Rev. A **81**, 062330 (2010)
44. Ł. Skowronek, J. Math. Phys. **52**, 122202 (2011)
45. J. Niset, N.J. Cerf, Phys. Rev. A **74**, 052103 (2006)
46. S.B. Bravyi, Quantum Inf. Process. **3**, 309 (2004)
47. S. Pironio, J. Math. Phys. **46**, 062112 (2005)
48. T. Fritz, A.B. Sainz, R. Augusiak, J. Bohr Brask, R. Chaves, A. Leverrier, A. Acin, Nat. Commun. **4**, 2263 (2013)

The Completeness of Quantum Theory for Predicting Measurement Outcomes

Roger Colbeck and Renato Renner

1 Introduction

In this chapter we look at the question of whether quantum theory is optimal in terms of the predictions it makes about measurement outcomes, or whether, instead, there could exist an alternative theory with improved predictive power. This was much debated in the early days of quantum theory, when many eminent physicists supported the view that quantum theory will eventually be replaced by a deeper underlying theory. Our aim will be to show that no alternative theory can extend the predictive power of quantum theory, and hence that, in this sense, quantum theory is complete.

Before turning to this question, it is worth reflecting on why one might think that quantum theory may not be optimally predictive. A key factor is that the theory is probabilistic. This is in stark contrast with classical theory, which is deterministic at a fundamental level. Even in classical theory there are scenarios where we may assign probabilities to various events, for example when making a weather forecast. However, this isn't in conflict with our belief in underlying determinism, but, instead, the fact that we assign probabilities simply reflects a lack of knowledge (about the precise value of certain physical quantities) when making the prediction. By analogy, we might imagine that even if we know the quantum state of a system before measurement (i.e., its wave function), we are also in a position of incomplete knowledge, and that additional information might be provided in a higher theory.

A further argument for incompleteness was given by Einstein, Podolsky and Rosen (EPR) [1]. They argued that whenever the outcome of an experiment can be predicted

R. Colbeck (✉)
Department of Mathematics, University of York, York YO10 5DD, UK
e-mail: roger.colbeck@york.ac.uk

R. Renner (✉)
Institute for Theoretical Physics, ETH Zurich, 8093 Zurich, Switzerland
e-mail: renner@phys.ethz.ch

G. Chiribella and R.W. Spekkens (eds.), *Quantum Theory: Informational Foundations and Foils*, Fundamental Theories of Physics 181,
DOI 10.1007/978-94-017-7303-4_15

with certainty, there should be a counterpart in the theory representing its value. They then consider measurements on a maximally entangled pair. In this scenario, the outcome of any measurement on one member of the pair can be perfectly predicted given access to the other member. Since the particles can be far apart, a measurement on one shouldn't, say EPR, affect the other in any way. They hence argue that there should be parts of the theory allowing these perfect predictions and, hence, that the quantum description is incomplete.

Following EPR, one might hope that quantum theory can be explained in terms of an underlying deterministic theory. Such a view was put into doubt by the Bell-Kochen-Specker theorem, independently discovered by Kochen and Specker [2] and by Bell [3], who showed that an underlying deterministic theory is not possible if one demands non-contextuality and freedom of choice. (A non-contextual theory is one in which the probability of a particular measurement outcome occurring depends only on the projector associated with that outcome, and not on the entire set of projectors that specify the measurement according to quantum theory.) Furthermore it was also shown by Bell [4] that there cannot be an underlying theory that is compatible with *local causality* and freedom of choice (we will explain this in more detail in Sect. 5). It is also worth noting that an assumption about locality can be seen as a physical means of justifying certain non-contextuality conditions.

In this chapter, we consider arbitrary alternative theories and ask whether they could have more predictive power than quantum theory. We remark that this question is different from those asked by Kochen and Specker and by Bell, whose goal was to rule out theories with certain specific properties such as non-contextuality or local causality. In this work, we do not demand any of these properties. The only assumption we make about a theory (beyond compatibility with quantum theory) is that it is compatible with a notion of free choice defined with respect to a natural causal order—see later. Roughly, the freedom of choice assumption demands that the theory can be applied to a setting where an experimenter makes certain choices independently of certain pre-existing parameters. It is worth noting that quantum theory is compatible with this assumption, as we would expect, since it is a reasonable theory.

To illustrate how an alternative theory may enable improved predictions over those of quantum theory while remaining probabilistic, one might imagine that the quantum state is supplemented by an additional parameter Z. When measuring one half of a maximally entangled pair of qubits, it could be that if $Z = 0$ the extended theory assigns outcome 0 with probability $3/4$, and outcome 1 with probability $1/4$, while, if $Z = 1$, the extended theory assigns outcome 0 with probability $1/4$, and outcome 1 with probability $3/4$. The extended theory would thus provide more information than quantum theory, which predicts that both outcomes occur with probability $1/2$. Furthermore, if Z is uniformly distributed, the quantum predictions are recovered when Z is unknown (and hence the extended theory is compatible with quantum theory).

This particular example is rather artificial and its purpose is merely to illustrate that—in principle—a theory that is more informative than quantum theory is conceivable. However, there are historical precedents of this type, for instance related to the problem of determining the mass of chemical elements. Take, as an example, the atomic mass of chlorine. Before the discovery of isotopes, its average atomic mass was measured to be 35.5. However, it was later discovered that chlorine in fact naturally occurs as two isotopes with atomic masses 35 and 37 (in approximate ratio 3:1). By introducing isotopes, the theory was extended in such a way that the mass of an individual atom could be better predicted.

Using a more advanced apparatus to measure the masses of individual atoms (rather than averaging), the theory without isotopes would predict that 35 occurs with probability $3/4$, and 37 occurs with probability $1/4$, while knowledge of the isotope would allow the outcome to be predicted perfectly for each atom. Note that the predictions made before the discovery of isotopes were not incorrect, but are simply the natural ones to make without knowledge of the different isotopes (and hence the new theory is compatible with the old one).

Returning to quantum theory, various alternatives, motivated in a more physical way than our earlier example, have been proposed in the past, some of which we will review later (see Sect. 5). Similarly to quantum theory, these alternatives provide rules to compute predictions for future measurement outcomes, based on certain (additional) parameters.

The aim of this chapter is to explain recent results relating the predictive power of quantum theory to that of possible alternative theories [5–7]. For this, we first need to specify what we mean by "quantum theory" and by "alternative theories", and how they can be compared (Sect. 3). The central requirement we impose on any alternative theory is that it be compatible with a notion of "free choice" defined with respect to a natural causal order. This means that the theory can be applied consistently in scenarios where measurements are chosen independently of certain other events (Sect. 4). We then discuss the implications of some existing results to our main question. These impose constraints on any alternative theory that is compatible with quantum theory; for instance, no such theory can be locally deterministic (Sect. 5). The last sections are then devoted to the recent, more general, results. A central claim is that no alternative theory that is compatible with quantum theory can improve the predictions of quantum theory (Sects. 6 and 7). Furthermore, if such an alternative theory is also at least as informative as quantum theory, then it is necessarily equivalent to quantum theory (Sect. 8). In this sense, quantum theory is complete. We conclude with a discussion of how these results relate to known hidden-variable theories, in particular the de Broglie-Bohm theory, and mention some applications (Sect. 9).

2 Preliminaries

2.1 Notation

On a technical level, the main results presented in this chapter are theorems about random variables (RVs) whose (joint) probability distribution satisfies certain assumptions. In the following we introduce our notation for such RVs and their distributions.

We usually denote RVs using upper case letters, and use lower case letters to specify particular values they can take. Thus, $X = x$ means that the RV X takes the value x. We write P_X to denote the probability distribution of the RV X, with $P_X(x)$ being the probability that $X = x$. For two RVs, X and Y, P_{XY} represents their joint distribution. We also use $P_{X|Y}$ to represent the conditional distribution of X given Y. We use $P_{X|Y=y} := P_{X|Y}(\cdot, y)$ to denote the distribution of the RV X conditioned on $Y = y$. We often abbreviate this distribution to $P_{X|y}$. If Y is a discrete random variable, then $P_{X|Y} = P_{XY}/P_Y$. More generally, if X takes values from the set $\mathcal{X}$ and Y takes values from set $\mathcal{Y}$, then $P_{X|Y}$ is defined such that, for all $\tilde{\mathcal{X}} \subseteq \mathcal{X}$ and $\tilde{\mathcal{Y}} \subseteq \mathcal{Y}$,

$$P_{XY}(\tilde{\mathcal{X}}, \tilde{\mathcal{Y}}) = \int_{\tilde{\mathcal{Y}}} \int_{\tilde{\mathcal{X}}} \mathrm{d}P_Y(y)\mathrm{d}P_{X|y}(x).$$

We also use $P(X = Y)$ to denote the probability that the RVs X and Y have equal values. We will only use this for discrete RVs, so we can take $P(X = Y) := \sum_x P_{XY}(x, x)$. Likewise, $P(X \neq Y) := 1 - P(X = Y)$.

2.2 Distance Between Probability Distributions

Our technical argument uses the *variational distance* to quantify the closeness of two probability distributions. For two distributions, P_X and Q_X, over a discrete set, $\mathcal{X}$, it is defined by

$$D(P_X, Q_X) := \frac{1}{2} \sum_{x \in \mathcal{X}} |P_X(x) - Q_X(x)|.$$

This measure is connected to the distinguishability of the two distributions. Specifically, suppose we have a black box that samples either from P_X or Q_X. Then, given one sample, the maximum probability of successfully guessing whether the sample has been generated from P_X or Q_X equals $\frac{1}{2}(1 + D(P_X, Q_X))$. Thus, if two distributions are close in variational distance, they are virtually indistinguishable. Appendix A summarizes some properties of $D(\cdot, \cdot)$ that are used in this work.

2.3 Measuring Correlations

A useful approach towards characterizing alternative theories is to consider the correlations (between the outcomes of two distant measurements) that can be reproduced by a given theory. The strength of these correlations may then, for instance, be compared to those occurring in quantum theory. To quantify correlations, we use an extension of a measure that has been proposed by Pearle [8] and, independently, by Braunstein and Caves [9], based on earlier work by Clauser et al. [10].

The correlation measure is tailored to a specific bipartite setup where measurements are carried out at two separate locations. One of the measurements is specified by a parameter A and has outcome X. The other is specified by a parameter B and has outcome Y. It is furthermore assumed that the outcomes X and Y take values from the set $\{0, 1, \ldots, M-1\}$ and that the parameters A and B are labelled by elements from the sets $\{0, \frac{2}{2N}, \ldots, \frac{2N-2}{2N}\} =: \mathcal{A}_N$ and $\{\frac{1}{2N}, \frac{3}{2N}, \ldots, \frac{2N-1}{2N}\} =: \mathcal{B}_N$, respectively, where $M \geq 2$ and $N \geq 2$ are integers. The correlation measure, in the following denoted by $I_{M,N}$, is then defined by

$$I_{M,N}(P_{XY|AB}) := P(X \oplus 1 \neq Y | A = 0, B = \frac{2N-1}{2N}) + \sum_{\substack{a \in \mathcal{A}_N, b \in \mathcal{B}_N \\ |a-b| = 1/(2N)}} P(X \neq Y | A = a, B = b),$$

where $\oplus$ denotes addition modulo M. This measure is depicted in Fig. 1. Note that the measure only depends on the conditional distribution $P_{XY|AB}$, and that stronger correlations have a lower value of $I_{M,N}$.

We will be particularly interested in the correlations that quantum theory predicts for measurements on two maximally entangled systems. To specify these correlations, define

$$|\psi_M\rangle := \frac{1}{\sqrt{M}} \sum_{i=0}^{M-1} |i\rangle \otimes |i\rangle , \tag{1}$$

where $\{|i\rangle\}$ is an orthonormal basis. We will consider the correlations produced by a particular set of measurements on states of this form. To construct these measure-

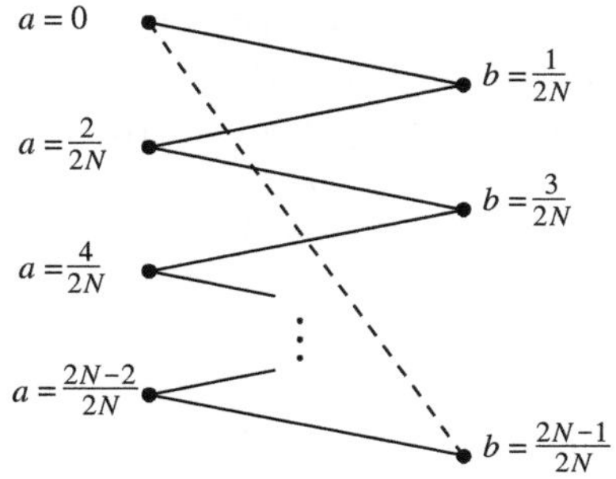

Fig. 1 Illustration of the terms in the correlation measure $I_{M,N}$. This measure is defined as the sum of the probabilities of obtaining different outcomes when measuring two subsystems in neighbouring bases (depicted with the *solid lines*), and of obtaining $X \oplus 1$ different from Y for $a = 0$, $b = \frac{2N-1}{2N}$ (depicted with the *dashed line*)

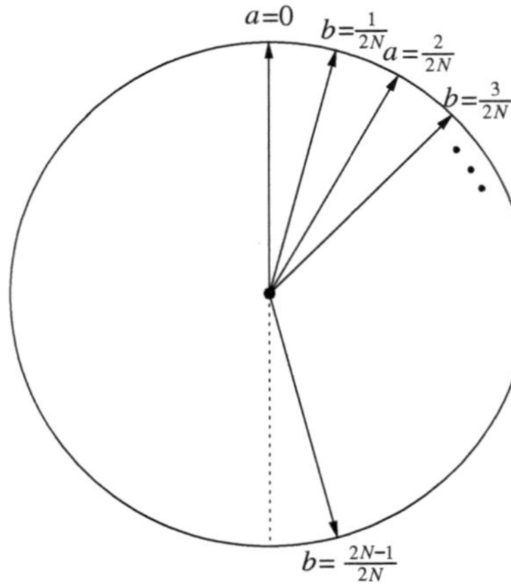

Fig. 2 Depiction of the measurements used to achieve the quantum value of the correlation measure $I_{2,N}$. The circle represents the $\left\{|0\rangle, \frac{1}{\sqrt{2}}(|0\rangle + |1\rangle)\right\}$ plane of the Bloch sphere. The *arrows* depict the Bloch vectors associated with the 0 outcome (i.e. E_0^a or F_0^b are the projectors onto these states). Those for the 1 outcome lie in the opposite direction and are not depicted. In the limit of large N, the measurements for neighbouring bases ($|a-b| = \frac{1}{2N}$) are virtually identical and the outcomes are almost always perfectly correlated. Conversely, for $a = 0, b = \frac{2N-1}{2N}$ and large N, the measurements are virtually opposite of one another and the outcomes are almost always perfectly anti-correlated

ments, consider the generalized Pauli operator $\hat{X}_M \equiv \sum_{l=0}^{M-1} |l\rangle\langle l \oplus 1|$ (where again $\oplus$ denotes addition modulo M), and define for any $a \in \mathcal{A}_N$ and $b \in \mathcal{B}_N$ the POVMs[1] $\{\bar{E}_x^a\}_x$ and $\{\bar{F}_y^b\}_y$ by[2]

$$\bar{E}_x^a \equiv (\hat{X}_M)^a \, |x\rangle\langle x| \, (\hat{X}_M^\dagger)^a \tag{2}$$

$$\bar{F}_y^b \equiv (\hat{X}_M)^b \, |y\rangle\langle y| \, (\hat{X}_M^\dagger)^b . \tag{3}$$

To understand the idea behind the arguments presented here, it is sufficient to consider the case where $M = 2$. Defining $|[\theta]\rangle = \cos\frac{\theta}{2}|0\rangle + \sin\frac{\theta}{2}|1\rangle$, the POVM element $\bar{E}_x^a$ then corresponds to the projector onto $|[(a+x)\pi]\rangle$ and, likewise, $\bar{F}_y^b$ is the projector onto $|[(b+y)\pi]\rangle$ (cf. Fig. 2).

We now define $\bar{P}_{XY|AB}^{M,N}$ as the conditional distribution of the outcomes of two separate quantum measurements, specified by $\{\bar{E}_x^a\}_x$ and $\{\bar{F}_y^b\}_y$, respectively, applied to two separate subsystems with joint state $|\psi_M\rangle$, i.e.,

$$\bar{P}_{XY|ab}^{M,N}(x,y) := \langle\psi_M| \, \bar{E}_x^a \otimes \bar{F}_y^b \, |\psi_M\rangle .$$

---

[1] See Sect. 3.1 for a definition.

[2] Note that $\bar{E}_x^a$ and $\bar{F}_y^b$ depend on M, but we suppress this dependence in our notation for brevity.

One can verify that the correlation strength, quantified with the above correlation measure, $I_{M,N}$, equals

$$I_{M,N}(\bar{P}^{M,N}_{XY|AB}) = 2N\left(1 - \frac{\sin^2\frac{\pi}{2N}}{M^2\sin^2\frac{\pi}{2MN}}\right) \leq \frac{\pi^2}{6N} \tag{4}$$

(see the appendix of [7] for details of this calculation).

As we will discuss later, a key feature of these correlations is that for all M, $I_{M,N}(\bar{P}^{M,N}_{XY|AB})$ tends to zero as N tends to infinity.

3 Quantum and Alternative Theories

The aim of this chapter is to make statements about physical theories, i.e., quantum theory as well as possible alternatives to it. However, in order to derive our result, we do not need to provide a comprehensive mathematical definition for the concept of a "physical theory". Rather, it suffices to focus on one crucial feature that we expect any theory to have, namely that it allows us to compute predictions about values that can be observed (e.g., in an experiment). These predictions, which need not be deterministic, are generally based on certain parameters that characterize the (experimental) setup, i.e., how it has been prepared (its initial state), the evolution it undergoes, and which measurements are going to be applied.

3.1 Predictions of Quantum Theory

In this chapter we consider experimental setups that within quantum theory can be described by taking the Hilbert space, $\mathcal{H}$, of the system to have finite dimension. The state of the system will be described by a RV Ψ, and we will also use RVs A and X to specify the measurement process and the observed outcome respectively. Since A and X refer to experimental parameters, we take them to be finite. Within quantum theory, Born's rule can be used to generate a prediction of X given A and Ψ.

Most generally the state of the system $\Psi = \psi$ can be given in the form of a density operator on $\mathcal{H}$, although we will often only need to consider pure states. Furthermore, any measurement process $A = a$ can be characterized by a *Positive Operator Valued Measure (POVM)* on $\mathcal{H}$, i.e., a family of positive operators $\{E^a_x\}_x$ labelled by the possible measurement outcomes $x \in \mathcal{X}$ such that $\sum_x E^a_x = \mathbb{1}_{\mathcal{H}}$.

For our treatment, we will assume that any evolution of the system prior to the measurement $\{E^a_x\}_x$ is already accounted for by its quantum state, i.e., that $\Psi = \psi$ is the state of the system directly before the measurement is applied.[3] The predic-

[3] Alternatively, one may work in the Heisenberg picture, for instance, and use the POVM to account for the evolution.

tions that quantum theory makes about the measurement outcome X can then be represented by a conditional distribution $\bar{P}_{X|A\Psi}$, which is given by

$$\bar{P}_{X|a\psi}(x) = \mathrm{tr}(E_x^a \psi) \quad \forall\, x \in \mathcal{X}. \tag{5}$$

We note that, by considering an extension of the Hilbert space $\mathcal{H}$, we may describe any quantum-mechanical measurement process equivalently as a *projective measurement*, i.e., one for which the POVM $\{E_x^a\}_x$ consists of orthogonal projectors.[4] Furthermore, we call a set of POVMs $\{E_x^a\}$ on $\mathcal{H}$ *tomographically complete* if the values $\bar{P}_{X|a\psi}(x)$ for all a and x are sufficient to determine ψ on $\mathcal{H}$ uniquely.[5]

For later reference, we also note that, according to quantum theory, any possible evolution of a quantum system, S, corresponds to a unitary mapping on a larger state space (that may include the environment of the system). In the case of a measurement process, this larger state space includes the measurement device, D. Specifically, a projective measurement, say $\{E_x^a\}_x$, would correspond to an isometry of the form

$$|\psi\rangle_S \mapsto \sum_x \sqrt{E_x^a}\, |\psi\rangle_S \otimes |x\rangle_D ,$$

where $\{|x\rangle_D\}$ are orthonormal states of the measurement device (and possibly also its environment) that encode the outcome. The outcome X of the original measurement may then be recovered by a subsequent projective measurement on D in the basis $\{|x\rangle_D\}$.

3.2 *Predictions of Alternative Theories*

In an alternative theory, the measurement process A with outcome X, as described above in terms of the quantum formalism, may admit a different description. This description could involve other parameters, which we denote by Z (one might think of Z as the list of all parameters used by the theory to describe the system's state before the measurement A is chosen).[6] For any values $A = a$ and $Z = z$ of these parameters, the theory specifies a rule for computing the probability distribution, $P_{X|az}$, for the measurement outcome X. Hence, in the following, if we want to make

[4]According to Naimark's theorem, there exists a Hilbert space $\bar{\mathcal{H}}$ that contains $\mathcal{H}$ as a subspace as well as orthogonal projectors $\bar{P}_x^a$ in $\bar{\mathcal{H}}$ such that for each $x \in \mathcal{X}$ the POVM element E_x^a is the projection of $\bar{P}_x^a$ into $\mathcal{H}$.

[5]An example of a tomographically complete set of projective POVMs in the case of a single qubit are the three POVMs whose elements are projectors onto (i) $|0\rangle$ and $|1\rangle$, (ii) $(|0\rangle + |1\rangle)/\sqrt{2}$ and $(|0\rangle - |1\rangle)/\sqrt{2}$, and (iii) $(|0\rangle + i\,|1\rangle)/\sqrt{2}$ and $(|0\rangle - i\,|1\rangle)/\sqrt{2}$.

[6]In [5], Z was modelled more generally as a system with input and output. For simplicity, we ignore this higher level of generality in this work.

a statement about the predictive power of a given theory,[7] it is sufficient to consider the properties of the corresponding distributions $P_{X|az}$.

Since we want to use theories to make predictions, we usually think of Z as (in principle) learnable. However, this is merely an interpretive statement, and none of the conclusions of this work are affected if Z is instead thought of as fundamentally hidden and hence unlearnable in principle. The only thing that changes in the latter case is the interpretation of certain statements. In particular, one may not want to call the condition $P_{XZ|AB} = P_{XZ|A}$, derived in Sect. 7.1, a "no-signalling" condition, or to speak about "predictions" made based on Z if Z is not learnable in principle. Note that although the experimental parameters A and X are finite, no such restriction is placed on the parameters of the alternative theory, which we allow to take values from an arbitrary set.

3.3 Compatibility of Predictions

The predictions computed within two different theories (e.g., quantum theory and an alternative theory) are generally not identical. Nevertheless, they may be *compatible with each other*. In order to say what this means we introduce a definition of when a joint distribution is compatible with another marginal distribution (this is formulated such that it is directly in the form that we need it, but the definition is readily generalized to other situations).

Definition 1 We say that a distribution $P_{AXZZ'}$ is *compatible* with $\bar{P}_{X|AZ'}$ if $P_{X|AZ'} = \bar{P}_{X|AZ'}$.[8] In other words,

$$\bar{P}_{X|az'} = P_{X|az'} = \int_{\mathcal{Z}} \mathrm{d}P_{XZ|az'}(\cdot, z) \ \forall\, a, z',$$

where the conditional distribution on the right-hand-side is derived from $P_{AXZZ'}$ (see Sect. 2.1).

Now suppose $Z \in \mathcal{Z}$ and $Z' \in \mathcal{Z}'$ are the parameters of two different theories, and that their predictions (about the outcome X of a measurement A) are given by conditional probability distributions $\tilde{P}_{X|AZ}$ and $\bar{P}_{X|AZ'}$, respectively.[9] The predictions can be considered *compatible with each other* if there exists a joint distribution $P_{AXZZ'}$ that is compatible with both $\tilde{P}_{X|AZ}$ and $\bar{P}_{X|AZ'}$.

To relate the definition of compatibility back to the earlier example of the isotopes, by way of illustration, we can take Z' to be the chemical element, and imagine that (without knowledge of the isotope) when sending individual chlorine atoms (call

[7]When referring to the predictive power of a theory, we mean predictions based on the value Z.

[8]We require that the distributions on each side of the equality are defined for the same pairs (a, z').

[9]Note that the conditional probability distribution $\tilde{P}_{X|AZ}$ (and, similarly, $\bar{P}_{X|AZ'}$) may in principle be defined only for a restricted set of pairs (a, z).

this $Z' = \text{Cl}$) through a measurement device we observe a mass of $X = 35$ with probability $3/4$ and $X = 37$ with probability $1/4$. However, given knowledge of the isotope, $Z = {}^{35}\text{Cl}$ or $Z = {}^{37}\text{Cl}$, we can make the prediction perfectly. It is easy to see that there exists $P_{XZZ'}$ that is compatible with both predictions:

$$P_{XZZ'}(x, z, z') = \begin{cases} 3/4 \text{ if } x = 35,\ z = {}^{35}\text{Cl and } z' = \text{Cl} \\ 1/4 \text{ if } x = 37,\ z = {}^{37}\text{Cl and } z' = \text{Cl} \\ 0 \text{ otherwise.} \end{cases}$$

3.4 Comparing the Accuracy of Predictions

The predictive powers of different theories can be compared provided the theories are compatible with each other.[10] If one theory has predictions $\tilde{P}_{X|AZ}$ and the other has predictions $\bar{P}_{X|AZ'}$, then we can consider $P_{AXZZ'}$ such that $P_{X|AZ} = \tilde{P}_{X|AZ}$ and $P_{X|AZ'} = \bar{P}_{X|AZ'}$. Roughly speaking, the predictions based on Z improve over those based on Z' if $\bar{P}_{X|AZZ'}$ and $\bar{P}_{X|AZ'}$ differ. However, because Z and Z' may take values over arbitrary sets, we require that the distributions differ on more than a measure zero set. This motivates the following definition.

Definition 2 Let $P_{AXZZ'}$ be a distribution compatible with $\tilde{P}_{X|AZ}$ and $\bar{P}_{X|AZ'}$. $\tilde{P}_{X|AZ}$ is said to give *improved predictions* over $\bar{P}_{X|AZ'}$ with respect to P if for some $A = a$

$$\int_{\mathcal{Z},\mathcal{Z}'} D(P_{X|az'}, P_{X|azz'})\mathrm{d}P_{ZZ'|a}(z, z') > 0.$$

This can again be illustrated using the earlier example of the isotopes. The theory that includes the information Z' about the particular isotope naturally gives improved predictions over the one that only specifies the chemical element Z with respect to the distribution $P_{X|ZZ'}$ given in the previous section.

We remark that quantum-mechanical predictions based on pure states generally give improved predictions over those derived from mixed states. To see this, imagine a system that is prepared in a pure state ψ_C depending on a random bit C, and assume that a measurement with outcome X is performed. If C is unknown, with $C = 0$ and $C = 1$ being equally likely, the distribution of X is, according to quantum theory, given by (5) with ψ substituted by the mixed state $\frac{1}{2}\psi_0 + \frac{1}{2}\psi_1$. However, if we had access to C, we could use (5) with ψ replaced by ψ_C, resulting in a more accurate prediction.

Clearly, when studying the question of whether there can be more informative theories than quantum theory, we need to consider specifications of states and measurement processes that are maximally informative among all predictions that are possible within quantum theory. Hence, following the above remark, we will restrict

[10] If two theories make incompatible predictions, then at least one of the sets of predictions would be falsifiable in principle.

our attention to quantum states that correspond to pure density operators and to projective measurements.

We will use the above notions of compatibility and improved predictions to compare quantum theory to alternative theories. For this, we let $Z' \equiv \Psi$ be the quantum state of a system and consider the conditional distribution $\bar{P}_{X|A\Psi}$ defined by (5), and take the predictions of the alternative theory to be $\tilde{P}_{X|AZ}$ (based on a parameter Z). We require that there exists a distribution $P_{AXZ\Psi}$ compatible with both $\bar{P}_{X|A\Psi}$ and $\tilde{P}_{X|AZ}$. We will often consider the case where Ψ is fixed to some particular state ψ. In this case, we can drop explicit mention of it, so that the condition that $\tilde{P}_{X|AZ}$ gives improved predictions with respect to P can be simplified to

$$\int_{\mathcal{Z}} D(P_{X|a}, P_{X|az}) \mathrm{d}P_{Z|a}(z) > 0,$$

for example.

Note that because we will always take the case where $P_{AXZ\Psi}$ is compatible with the predictions of the alternative theory, $\tilde{P}_{X|AZ}$, we henceforth make this condition implicit in the notation by dropping the tilde on the alternative theory.

4 Freedom of Choice

As explained above, physical theories involve certain parameters, and it is generally assumed (often implicitly) that these can be chosen freely. Quantum mechanics, for instance, allows us to compute the probabilities of a measurement outcome X depending on the system's state Ψ as well as a description of the measurement process, A, and our understanding is that these parameters can in principle be chosen freely (e.g., by an experimenter carrying out a measurement of her choice). In fact, one may argue that a description of nature that does not involve any such choices—thereby not allowing us to compute conclusions for different initial conditions—cannot be reasonably termed a theory [11].

It is worth noting that by assuming free choice, we are not making any metaphysical assertion that the real world contains, say, agents with free will, or anything of that sort. Instead, allowing free choice is a property that we require of a theory. In essence, it means that the theory gives predictions for all possible values of the free parameters, and furthermore, that it does so no matter what happened elsewhere in the theory. Without such an assumption, depending on other events described by the theory, certain values of the 'free' parameters could be unavailable, in the sense that the theory would not be able to predict a response to them.

In this section, we specify what we mean by such free choices. The idea is that, for a given theory, the statement that a parameter of the theory, say A, is considered *free* is equivalent to saying that A is uncorrelated with all values (described by the theory) that are outside the future of A. For this definition to make sense mathematically, we need to establish a notion of *future*. We do this by introducing a *causal order*, i.e., a (partial) ordering of events. We stress, however, that the causal order is only used to define free choice and plays no further part in the argument.[11]

4.1 Causal Order

Let Γ be the set of all parameters required for the description of an experiment within a given theory. In particular, Γ may contain variables that specify the (joint) state in which the relevant physical systems have been prepared (in the following usually denoted by Ψ for quantum theory and by Z for more general theories), the choice of measurements (denoted A and B), as well as the measurement outcomes (denoted X and Y). For any such set of variables Γ, we can define a causal order $\rightsquigarrow$ as follows.

Definition 3 A *causal order* $\rightsquigarrow$ for Γ is a preorder relation[12] on Γ.

If $A \rightsquigarrow X$, we say that X *is in the (causal) future of* A, and if this doesn't hold, we write $A \not\rightsquigarrow X$. These relations can be conveniently specified by a diagram (see Fig. 3 for an example). Note that the causal order $\rightsquigarrow$ should not be interpreted as specifying actual causal dependencies,[13] but instead indicates that such causal dependencies are not precluded (by the theory).

A typical—but for the following considerations not necessary—requirement on a causal order is that it be compatible with relativistic space time. Consider, for example, an experiment where a parameter A is chosen at a given space time point $\mathbf{r}_A$ and where a measurement outcome X is observed at another space time point $\mathbf{r}_X$. One would then naturally demand that $A \rightsquigarrow X$ if and only if $\mathbf{r}_X$ lies in the future light cone of $\mathbf{r}_A$. This captures the idea that the choice A is made at an earlier time than the observation of X, with respect to any reference frame.

4.2 Free Random Variables

To define the notion of a "free choice", we consider a set Γ of RVs equipped with a causal order. (As above, Γ should be thought of as the set of all parameters relevant for the description of an experiment within a given theory.)

Definition 4 With respect to a set of RVs Γ equipped with a causal order and a set $\mathcal{A}$, we say that $A \in \Gamma$ is a *free choice* if $P_A(a) > 0$ for all $a \in \mathcal{A}$ and

---

[11] In particular, we do not assume *local causality* within the specified causal order.

[12] That is, $\rightsquigarrow$ is a binary relation on the set Γ that is reflexive (i.e., $A \rightsquigarrow A$) and transitive (i.e., $Z \rightsquigarrow A$ and $A \rightsquigarrow X$ imply $Z \rightsquigarrow X$).

[13] I.e., $A \rightsquigarrow X$ is not meant to imply that there is necessarily a physical process such that changing A imposes a change of X.

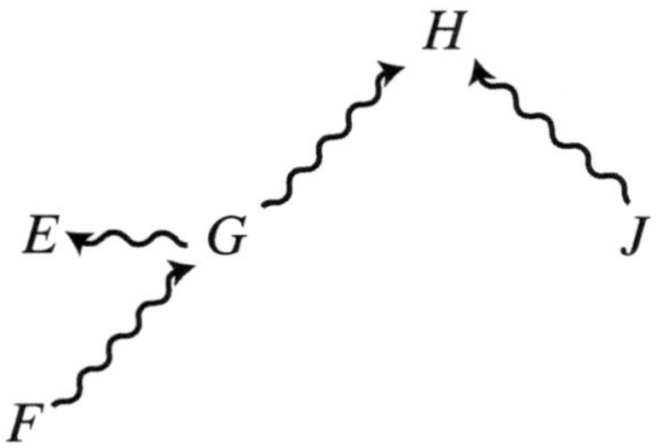

Fig. 3 Free choice and causal order. An arbitrary causal order is depicted for random variables E, F, G, H and J The *arrows* correspond to the relation $\rightsquigarrow$. For example, G lies in the future of F, i.e., $F \rightsquigarrow G$, but not of J, i.e., $J \not\rightsquigarrow G$. Because of the transitivity property, it follows that $F \rightsquigarrow E$, for example. In this setting we would say that, for instance, G is free if it is uncorrelated with F and J, i.e., $P_{GFJ} = P_G \times P_{FJ}$

$$P_{A\Gamma_A} = P_A \times P_{\Gamma_A}$$

holds, where Γ_A is the set of all RVs $X \in \Gamma$ such that $A \not\rightsquigarrow X$.[14]

We stress that this notion of free choice makes sense only with respect to a causal order, and cannot be defined on its own (see [12] for further explanation). Obviously, whether a variable from the set Γ is considered free depends on the causal order that we impose. If the causal order is taken to be the one induced by relativistic space time (see the description above), then this definition coincides with the notion of a *free variable* as used by Bell [11].[15] We remark that both standard quantum theory and classical theory in relativistic space time allow for free choices within such a causal order.

5 Constraints on Theories Compatible with Quantum Theory

We discuss here the implication of some well-known results to our main question, whether an extension of quantum theory can have improved predictive power. Although they were not asking the same question, the works of Bell [4] and Leggett [13] imply constraints on such higher theories, and hence can be seen as special cases of the general theorem presented in Sect. 6, which excludes all alternative theories

[14]By definition, the set Γ_A also excludes A.

[15]In [11], Bell discusses the assumption that the settings of instruments are *free variables*, which he characterizes as follows: "For me this means that the values of such variables have implications only in their future light cones."

whose predictions are more informative than quantum theory. Furthermore, we will show that Bell's theorem and the result of Leggett follow as corollaries of our main theorem.

5.1 Bipartite Setup

The statements described below refer to a bipartite setup which involves two separate measurements, specified by parameters A and B, and with outcomes X and Y, respectively. As before, we consider a theory that allows us to compute predictions about these measurements based on a parameter (or list of parameters) Z, which describes the system's state before the measurement process is started. Furthermore, in order to define free choices, we need to specify a causal order. The technical claims described in this section can be applied to any causal order that satisfies the following conditions:

(i) $A \rightsquigarrow X$ and $B \rightsquigarrow Y$;
(ii) $A \not\rightsquigarrow Z$ and $B \not\rightsquigarrow Z$;
(iii) $A \not\rightsquigarrow Y$ and $B \not\rightsquigarrow X$.

Condition (i) corresponds to the requirement that the measurement is specified before its outcome is obtained. Condition (ii) captures the fact that the parameters of the theory, Z, on which the predictions are based, should not only become available after the measurement process is started. This assumption can be considered necessary in order to reasonably talk about "predictions". Finally, Condition (iii) demands that the arrangement of the two measurements should be such that neither of them lies in the future of the other. (Note that, assuming a relativistic space time structure, this would correspond to a setup where the measurements are space-like separated.) Together, the three conditions imply a causal order in which A is considered free if $P_{ABYZ} = P_A \times P_{BYZ}$, and likewise for B. The causal orders respecting (i)–(iii) are illustrated in Fig. 4.

5.2 Local Deterministic Theories

Within the bipartite setup described above a local deterministic theory is one for which all conditional probabilities $P_{X|az}(x)$ and $P_{Y|bz}(y)$ are equal to either 0 or 1. Such theories were introduced by Bell [4] who proved that no such theory can reproduce the predictions of quantum theory.

Within the terminology of the present paper, Bell's argument implies the following theorem that makes use of the correlations $\bar{P}^{M,N}_{XY|AB}$ that quantum theory predicts for the measurements on the maximally entangled state $|\psi_M\rangle$ defined in Sect. 2.3.

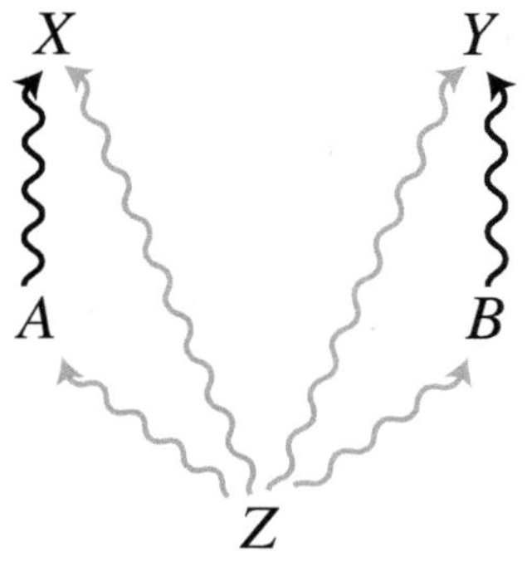

Fig. 4 The causal orders for which our argument applies. We consider a setup with two separate measurements, one depending on a choice A with outcome X, and the other with choice B and outcome Y. Moreover, Z denotes all extra parameters that may be used to make predictions about the outcomes. The figure illustrates all of the causal orders compatible with our requirements (i)–(iii). The *black arrows* originating from A and B are required, while each of the *grey arrows* originating from Z is optional

Theorem 1 (No higher theories are locally deterministic) *For any probability distribution P_{ABXYZ} at least one of the following cannot hold:*

- Freedom of choice:[16] *A and B are free with respect to any of the causal orders depicted in Fig. 4;*
- Compatibility with quantum theory: *P_{ABXYZ} is compatible with the predictions $\bar{P}^{M,N}_{XY|AB}$ of quantum theory (for some $M \geq 2$, $N \geq 2$);*
- Local determinism:

$$P_{X|az}(x) \in \{0, 1\} \quad \forall\, a, z \text{ s.t. } P_{AZ}(a, z) > 0$$
$$P_{Y|bz}(y) \in \{0, 1\} \quad \forall\, b, z \text{ s.t. } P_{BZ}(b, z) > 0.$$

The theorem follows directly from the general non-extendibility theorem described in Sect. 6. One may also prove it directly using the correlation measure $I_{M,N}$ defined in Sect. 2.3 for $M = N = 2$. The central idea is to show that, under the free choice assumption, all correlations explained by a locally deterministic model satisfy the inequality $I_{2,2} \geq 1$, which corresponds to the CHSH inequality [10]. (The free choice assumption ensures that $P_{AB|z}$ has full support for each z, and hence that the conditional distributions $P_{X|az}$ and $P_{Y|bz}$ are well defined for any a, b, and z.) The assertion then follows from the fact that $I_{2,2}(\bar{P}^{2,2}_{XY|AB}) = 2 - \sqrt{2} < 1$ (see Eq. 4).

[16]The freedom of choice assumption is often not mentioned explicitly, but its necessity has been stressed by Bell in later work [11].

5.3 Stochastic Local Causal Theories

In his later work, Bell dropped the assumption of determinism and considered more general stochastic models. He adopted the following definition of locality called *local causality*, which leads to the relation $P_{XY|ABZ} = P_{X|AZ}P_{Y|BZ}$ [14]. Expanding the left hand side using Bayes' rule, this can be broken down into four separate relations, $P_{X|ABZ} = P_{X|AZ}$, $P_{Y|ABZ} = P_{Y|BZ}$, $P_{X|ABYZ} = P_{X|ABZ}$ and $P_{Y|ABXZ} = P_{Y|ABZ}$. The first two of these have sometimes been termed *parameter independence* and imply that, even given access to Z, there cannot be signalling between the two measurement processes.

The last two conditions have been termed *outcome independence*. They do not have an obvious operational significance (such as no-signalling), and do not in general hold for the theories we consider in this work. We note, however, that they are automatically satisfied in any deterministic model, where each of the outcomes X and Y is a function of A, B, and Z. Conversely, as we argue just below, if a theory is locally causal then the predictions it makes about the outcomes of measurements on the entangled state $|\psi_2\rangle = \frac{1}{\sqrt{2}}(|00\rangle + |11\rangle)$ are necessarily deterministic. (This is the essence of the EPR argument [1].)

To see this, note that for any projective measurement (specified by $A = a$) applied to the first part of $|\psi_2\rangle$, there exists another projective measurement (specified by $B = b_a$) on the second part such that the outcomes are perfectly correlated. For example, if $A = a$ corresponds to the POVM $\{|0\rangle\langle 0|, |1\rangle\langle 1|\}$, and if we choose $B = b_a$ such that it corresponds to the same POVM, then $P_{XY|ab_a}(0,0) = P_{XY|ab_a}(1,1) = \frac{1}{2}$. This means that X is determined by Y, i.e., $P_{X|ab_a yz}(x) = \delta_{x,y} \in \{0,1\}$ for all a, x, y and z. Applying now the conditions of local causality, we obtain $P_{X|abyz}(x) = P_{X|az}(x) \in \{0,1\}$, which corresponds to the assumption of local determinism. Hence, there is an analogue of Theorem 1 in which the local determinism condition is weakened to Bell's local causality condition. This modified theorem also follows from our main result via the above argument.

We remark that, as we shall see below (Lemma 1), the freedom of choice assumption implies parameter independence, but is not strong enough to imply local causality, since it doesn't imply outcome independence.

5.4 Leggett-Type Theories

In [13], Leggett introduced what he calls a "non-local hidden variable" model, which attempts to give an explanation of quantum correlations that is partly local and partly non-local. The presence of non-local hidden variables in his model leads to an incompatibility with the free choice assumption. However, since the behaviour of the non-local variables is not specified in Leggett's model, we can consider a slightly modified version in which they are ignored (hence forth, when we speak about Leggett's model, we refer to the local part of it). The model is then compatible with our notion of free choice, and offers improved predictive power for measurements on maximally entangled particles. We note that the model is not a full-fledged theory, as it only specifies how the outcomes of spin measurements are obtained.

Leggett's model is based on the idea of assigning to each spin particle a three-dimensional vector (in addition to its quantum mechanical state). In particular, if we consider two spin particles, each measured on one side within the bipartite setup described above, we need to specify two such vectors, denoted **u** and **v**, respectively. To connect this to our general discussion, we may think of these vectors as part of Z, i.e., Z takes as values pairs $(\mathbf{u}, \mathbf{v})$. As above, we denote the choice of measurement on each side by A and B. Restricting to projective spin measurements, the two choices may be labelled by three-dimensional vectors, denoted **a** and **b**, respectively, indicating their orientation in space (see, for example, [15] for more details). The predictions for the measurement outcomes X and Y, as prescribed by Leggett's model, are then given by

$$P_{X|\mathbf{auv}}(x) = \frac{1}{2}(1 + (-1)^x \mathbf{a} \cdot \mathbf{u}) \tag{6}$$

$$P_{Y|\mathbf{buv}}(y) = \frac{1}{2}(1 + (-1)^y \mathbf{b} \cdot \mathbf{v}). \tag{7}$$

In order to completely define the model, one would also need to assign probabilities to all possible values $Z = (\mathbf{u}, \mathbf{v})$, i.e., specify a probability distribution P_Z (which, in general, depends on the quantum state). However, the following theorem, which is a corollary of results in [13, 15–17], implies that there is no such assignment for which Leggett's model can be made compatible with quantum theory.

Theorem 2 (No higher theories obey the Leggett conditions) *There exists a quantum distribution $\bar{P}_{XY|AB\psi_2}$, generated by measurements on the state $|\psi_2\rangle$ (defined in (1)), such that for any probability distribution P_{ABXYZ} at least one of the following cannot hold:*

- Freedom of choice: *A and B are free with respect to any of the causal orders depicted in Fig. 4;*
- Compatibility with quantum theory: *P_{ABXYZ} is compatible with $\bar{P}_{XY|AB\psi_2}$;*
- Leggett rule: *$P_{XY|ABZ}$ satisfies Eqs. 6 and 7 for all values $A = a$, $B = b$, and $Z = (\mathbf{u}, \mathbf{v})$.*

We will not give a proof of this theorem here, since it follows from the more general results presented in the next section. To see this, it is sufficient to observe that, when measuring the entangled state $|\psi_2\rangle = \frac{1}{\sqrt{2}}(|00\rangle + |11\rangle)$, for instance, quantum theory prescribes that $\bar{P}_{X|\mathbf{a}}(x) = \frac{1}{2}$, independently of the orientation **a** of the measurement. Conversely, for any given $Z = (\mathbf{u}, \mathbf{v})$, Leggett's model predicts a non-uniform distribution whenever the measurement orientation **a** is not orthogonal to the vector **u**. The Leggett model is therefore more informative than quantum theory, and hence excluded by Lemma 3 (as well as the more general Theorem 3) below.

5.5 Other Constraints

Here we summarize a few other known constraints on theories compatible with quantum mechanics. One of the first results in this direction was that the quantum outcomes cannot be predetermined within a non-contextual model [2, 3]. In such a model, one assumes the existence of a map from the set of projectors to the set $\{0, 1\}$ such that for every set of projectors that constitute a POVM, only one member of that set is mapped to 1 (the element that maps to 1 is interpreted as the outcome that will occur if a measurement described by that POVM is carried out). Such a model is *non-contextual* in that whether or not a particular outcome occurs depends only on the individual projectors, and not on the set of projectors making up the POVM. The Bell-Kochen-Specker Theorem [2, 3] implies that no such assignment can exist if the Hilbert space dimension is at least 3.

Hardy [18] later showed that within any extended theory, an infinite number of underlying states are required, even to describe a single qubit, and Montina [19, 20] proved, under the assumption of Markovian dynamics, that the number of real parameters that an extended theory needs to characterize a state in Hilbert space dimension M is at least $2M - 2$ (the same as the number of parameters needed to specify a pure quantum state up to a global phase).

In addition, a claim in the same spirit as our non-extendibility theorem (presented in the next section) has been obtained recently [21] under the assumption of measurement non-contextuality, introduced in [22].

6 The Non-extendibility Theorem

This section is devoted to the key result of this chapter, asserting that quantum theory is maximally informative. Stated informally, we make the following claim, first made in [5].

Claim 1 *No alternative theory that is compatible with quantum theory and allows for free choice (with respect to the discussed causal orders) can give improved predictions.*

The main technical statement is a generalization of the theorems discussed in the previous section. The setup is broadly the same, but instead of the condition that the higher theory remains compatible with quantum theory for measurements on maximally entangled states, we require this for a wider class of states. Furthermore, rather than considering theories that satisfy local determinism or the Leggett rule, the claim is about arbitrary theories that make improved predictions.

The main technical theorem is as follows (this should be read as a purely mathematical statement about bipartite pure states, whose significance to the extendibility of quantum theory will be explained subsequently).

Theorem 3 *Let $\mathcal{H}_S$, $\mathcal{H}_{\tilde{S}}$, $\mathcal{H}_D$ and $\mathcal{H}_{\tilde{D}}$ be finite dimensional Hilbert spaces, $|\phi\rangle_{SD} \in \mathcal{H}_S \otimes \mathcal{H}_D$ be a pure state and $\{|\hat{y}\rangle_D\}$ be a Schmidt basis on D. Then there exists a state $|\Gamma\rangle_{\tilde{S}\tilde{D}} \in \mathcal{H}_{\tilde{S}} \otimes \mathcal{H}_{\tilde{D}}$ and local POVMs $\{E_x^a\}$ and $\{F_y^b\}$ on $S\tilde{S}$ and $D\tilde{D}$, respectively, with $F_y^{b_0} = |\hat{y}\rangle\langle\hat{y}|_D \otimes \mathbb{1}_{\tilde{D}}$ for some $b = b_0$, such that, for any probability distribution P_{ABXYZ} at least one of the following cannot hold:*

- Freedom of choice: *A and B are free with respect to any of the causal orders depicted in Fig. 4;*
- Compatibility with quantum theory: *P_{ABXYZ} is compatible with the predictions $\bar{P}_{XY|AB(\phi\otimes\Gamma)}$ of quantum theory for the measurements $\{E_x^a\}$ and $\{F_y^b\}$ on $|\phi\rangle_{SD} \otimes |\Gamma\rangle_{\tilde{S}\tilde{D}}$.*[17]
- Improved predictions: *$P_{Y|b_0Z}$ gives improved predictions over $\bar{P}_{Y|b_0\phi}$ with respect to P.*

To understand the implications of this theorem, consider a fixed measurement $\hat{a}$ on a system S. Assume that, according to quantum theory, the system (before the measurement) is in a pure state, denoted ψ, and that the measurement corresponds to a projective POVM, $\{\hat{E}_{\hat{x}}^{\hat{a}}\}$. Quantum theory then gives a probabilistic prediction $\bar{P}_{\hat{X}|\hat{a}\psi}$ for the measurement outcome $\hat{X}$, which depends on ψ and $\{\hat{E}_{\hat{x}}^{\hat{a}}\}$ (see Eq. 5). Our aim is to compare this quantum-mechanical prediction with the prediction $P_{\hat{X}|\hat{a}Z}$ that may be obtained by an alternative theory, whose parameters we denote by Z.

We then consider the joint state of the measured system, S, and the measurement device, D, after the measurement $\hat{a}$. Following the discussion in Sect. 3, according to quantum theory, this state can be assumed to have the form

$$|\phi\rangle_{SD} = \sum_{\hat{x}} \sqrt{\hat{E}_{\hat{x}}^{\hat{a}}}\, |\psi\rangle_S \otimes |\hat{x}\rangle_D\,. \tag{8}$$

Note that the POVM $\{F_y^{b_0}\}$ defined by Theorem 3 corresponds to a measurement of D in the basis $\{|\hat{x}\rangle_D\}$. The outcome, Y, of this measurement can therefore be seen as reading out the outcome of the original measurement, specified by $\{\hat{E}_{\hat{x}}^{\hat{a}}\}$, and it is this that we want to predict.

We now apply Theorem 3 to $|\phi\rangle_{SD}$. If we assume that P_{ABXYZ}, in addition to being compatible with quantum theory, satisfies the freedom of choice assumption, then the theorem implies that its third condition, improved predictions, cannot hold. This means that

$$\int_{\mathcal{Z}} D(P_{Y|b_0\phi}, P_{Y|b_0z})\mathrm{d}P_Z(z) = 0.$$

[17]Formally, $\bar{P}_{XY|AB(\phi\otimes\Gamma)}$ is given by

$$P_{XY|ab(\phi\otimes\Gamma)}(x, y) = \mathrm{tr}((E_x^a \otimes F_b^y)\, |\phi \otimes \Gamma\rangle\langle\phi \otimes \Gamma|)\,.$$

[NB: the states and POVM elements do not factor in the same way.].

It hence follows that $\bar{P}_{Y|b_0\phi}$ is equal to $P_{Y|b_0Z}$ (*almost surely* over P_Z), and hence the alternative theory doesn't give better predictions. This is what is stated informally as Claim 1.

7 Proof of Theorem 3

The theorem follows from three statements, which we formulate and prove separately. An overview of the argument is as follows. We consider the previously introduced bipartite scenario and any of the causal orders depicted in Fig. 4. We begin by showing that free choice with respect to this causal order implies that the alternative theory is no-signalling (see Lemma 1). In the second part of the argument, we show that for measurements on maximally entangled states, if quantum theory is correct, no higher theory can give improved predictions about the outcomes (see Lemma 3). In the final part of the argument, we generalize this to measurements on an arbitrary bipartite entangled state. More precisely, we show that for any such state, there exist local measurements that generate correlations arbitrarily close to those generated by r maximally entangled states for some sufficiently large integer r. Hence, from the second part of the argument, these measurements can have no improved predictions.

7.1 Part I: No-signalling From Free Choice

In this part, we show that if A and B are free choices with respect to one of the given causal orders, then there is no signalling within the alternative theory (i.e. no signalling even given access to Z).[18]

Lemma 1 *The freedom of choice assumption implies* $P_{XZ|AB} = P_{XZ|A}$ *and* $P_{YZ|AB} = P_{YZ|B}$.

Proof That A is free within the specified causal order implies $P_{A|BYZ} = P_A$ and hence

$$P_{YZA|B} = P_{YZ|B} \times P_{A|BYZ} = P_A \times P_{YZ|B}, \text{ and}$$
$$P_{YZA|B} = P_{A|B} \times P_{YZ|AB} = P_A \times P_{YZ|AB}.$$

We therefore have $P_{YZ|AB} = P_{YZ|B}$. The relation $P_{XZ|AB} = P_{XZ|A}$ follows by symmetry. □

[18]As explained in Sect. 3, the interpretation of this as "no-signalling" may change if Z is thought of as in principle unlearnable. However, since we only use this condition for an intermediate step, its interpretation is not relevant for our argument.

7.2 Part II: Non-extendibility for Measurements on Maximally Entangled States

In the second part of the argument, we show that the claim holds for particular measurements on maximally entangled pairs of qubits. The proof uses the correlation measure $I_{M,N}$ introduced in Sect. 2.3. The following lemma shows that this measure, applied to a distribution $P_{XY|AB}$, gives a bound on how well any additional information, Z, can be correlated to the outcome X. Note that the lemma is independent of quantum theory and is simply a property of probability distributions.

Lemma 2 *Let $P_{XYZ|AB}$ be a distribution that obeys $P_{XZ|AB} = P_{XZ|A}$ and $P_{YZ|AB} = P_{YZ|B}$. Then, for all a and b, we have*

$$\int_{\mathcal{Z}} D(P_{X|abz}, \hat{P}_X) \mathrm{d}P_{Z|ab} \leq \frac{M}{4} I_{M,N}(P_{XY|AB}), \tag{9}$$

where $\hat{P}_X$ denotes the uniform distribution on X.

The proof is based on an argument given in [7], which develops results of [5, 17, 23, 24].

Proof We first consider the quantity $I_{M,N}$ evaluated for the conditional distribution $P_{XY|ABz} = P_{XY|ABZ}(\cdot, \cdot|\cdot, \cdot, z)$, for any fixed z. The idea is to use this quantity to bound the variational distance between the conditional distribution $P_{X|az}$ and the distribution, $P_{X\oplus 1|az}$, which corresponds to the distribution of X if its values are shifted by 1 modulo M. If this distance is small, it follows that the distribution $P_{X|az}$ is roughly uniform. Because this holds for any $Z = z$, X must be independent of Z.

It is first worth noting that the conditions of the lemma ($P_{XZ|AB} = P_{XZ|A}$ and $P_{YZ|AB} = P_{YZ|B}$) imply $P_{X|ABZ} = P_{X|AZ}$ and $P_{Y|ABZ} = P_{Y|BZ}$ respectively, and together imply $P_{Z|AB} = P_Z$.

Let $\hat{P}_X$ be the uniform distribution on X. For $a_0 := 0$, $b_0 := 2N - 1$, we have

$$\begin{aligned}
& I_{M,N}(P_{XY|ABz}) \\
&= P(X \oplus 1 \neq Y|a_0, b_0, z) + \sum_{\substack{a \in \mathcal{A}_N, b \in \mathcal{B}_N \\ |a-b|=1/(2N)}} P(X \neq Y|a, b, z) \\
&\geq D(P_{X\oplus 1|a_0 b_0 z}, P_{Y|a_0 b_0 z}) + \sum_{\substack{a \in \mathcal{A}_N, b \in \mathcal{B}_N \\ |a-b|=1/(2N)}} D(P_{X|abz}, P_{Y|abz}) \\
&= D(P_{X\oplus 1|a_0 z}, P_{Y|b_0 z}) + \sum_{\substack{a \in \mathcal{A}_N, b \in \mathcal{B}_N \\ |a-b|=1/(2N)}} D(P_{X|az}, P_{Y|bz}) \\
&\geq D(P_{X\oplus 1|a_0 z}, P_{X|a_0 z}) \\
&\geq \frac{4}{M} D(P_{X|a_0 b_0 z}, \hat{P}_X).
\end{aligned} \tag{10}$$

The first inequality follows from the fact that $D(P_{X|\Omega}, P_{Y|\Omega}) \leq P(X \neq Y|\Omega)$ for any event Ω (see Lemma 6 in Appendix A). The final inequality follows from Lemma 7 in Appendix A. Furthermore, we have used the conditions $P_{X|abz} = P_{X|az}$ and $P_{Y|abz} = P_{Y|bz}$, and the triangle inequality for D. By symmetry, this relation holds for all a and b.

We now take the average over z on both sides of (10). The left-hand-side gives

$$\begin{aligned}
&\int_{\mathcal{Z}} I_{M,N}(P_{XY|ABz})\mathrm{d}P_{Z|ab}(z)\\
&= \int_{\mathcal{Z}} I_{M,N}(P_{XY|ABz})\mathrm{d}P_{Z}(z)\\
&= \int_{\mathcal{Z}} P(X \oplus 1 \neq Y|a_0, b_0, z)\mathrm{d}P_{Z|a_0b_0}(z) +\\
&\sum_{\substack{a\in\mathcal{A}_N, b\in\mathcal{B}_N\\ |a-b|=1/(2N)}} \int_{\mathcal{Z}} P(X \neq Y|a, b, z)\mathrm{d}P_{Z|ab}(z)\\
&= P(X \oplus 1 \neq Y|a_0, b_0) + \sum_{\substack{a\in\mathcal{A}_N, b\in\mathcal{B}_N\\ |a-b|=1/(2N)}} P(X \neq Y|a, b, c)\\
&= I_{M,N}(P_{XY|AB}), \qquad (11)
\end{aligned}$$

where we used the condition $P_{Z|ab} = P_Z$ several times. Using

$$\left\langle D(P_{X|abz}, \hat{P}_X)\right\rangle_z := \int_{\mathcal{Z}} D(P_{X|abz}, \hat{P}_X)\mathrm{d}P_{Z|ab}(z),$$

we establish (9). □

We now apply Lemma 2 to the quantum correlations $\bar{P}^{M,N}_{XY|AB}$ arising from measurements on the maximally entangled state ψ_M (cf. Sect. 2.3). In the limit where N tends to infinity, we have $\lim_{N\to\infty} I_{M,N}(\bar{P}^{M,N}_{XY|AB}) = 0$ (for all $M \geq 2$), and hence we can establish that $P_{X|abz} = \bar{P}^{M,N}_{X|ab}$ almost surely over Z (note that $\bar{P}^{M,N}_{X|ab}(x) = \bar{P}^{M,N}_{X|a}(x) = \hat{P}_X(x) = \frac{1}{M}$ for all x). Under the freedom of choice assumption and assuming compatibility with quantum theory this implies $P_{X|az} = \bar{P}^{M,N}_{X|a}$ almost surely over Z, i.e., Z gives no additional information about the measurement outcome, X.

Taking Parts I and II together, we obtain the following lemma, which may be of independent interest.

Lemma 3 (No higher theories give improved predictions for measurements on one half of a pair of particles in the state ψ_M) *For any $\delta > 0$ and any M there exists an N such that for any probability distribution P_{ABXYZ}, at least one of the following three conditions cannot hold:*

- Freedom of choice: *A and B are free with respect to any of the causal orders depicted in Fig. 4;*

- Compatibility with quantum theory: *P_{ABXYZ} is compatible with $\bar{P}^{M,N}_{XY|AB}$*[19]*;*
- Improved predictions: *There exists a value $A = a$ such that $\langle D(P_{X|az}, \bar{P}^{M,N}_{X|a})\rangle_z > \delta$, where $\langle\cdot\rangle_z$ denotes the expectation value over z.*

Hence, if P_{ABXYZ} is compatible with quantum theory and satisfies the freedom of choice assumption then the third condition cannot hold. Since the lemma holds for any $\delta > 0$, under these assumptions, $\langle D(P_{X|az}, \bar{P}^{M,N}_{X|a})\rangle_z = 0$. This implies that the quantum predictions $\bar{P}^{M,N}_{X|a}$ are the same as those of the alternative theory $P_{X|az}$ (almost surely over Z).

7.3 Part III: Generalization to Arbitrary Measurements

The last part of the proof of Theorem 3 consists of generalizing Lemma 3, which applies to specific measurements on a maximally entangled state, to measurements on the general state $|\phi\rangle_{SD}$. The proof relies on the concept of embezzling states [25]. These are entangled states that can be used to extract any desired maximally entangled state locally and without communication. More precisely, we will use the following lemma, which is implicit in [25].

Lemma 4 *For any $\varepsilon > 0$ and for any $k \in \mathbb{N}$ there exists a bipartite state $|\Gamma^k\rangle_{\tilde{S}\tilde{D}}$, the* embezzling state, *such that for any $m \le k$, there exist local isometries, $U_m : \mathcal{H}_{\tilde{S}} \mapsto \mathcal{H}_{\tilde{S}} \otimes \mathcal{H}_{S'}$ and $V_m : \mathcal{H}_{\tilde{D}} \mapsto \mathcal{H}_{\tilde{D}} \otimes \mathcal{H}_{D'}$ that perform the transformation*

$$U_m \otimes V_m : |\Gamma^k\rangle_{\tilde{S}\tilde{D}} \mapsto |\Gamma^k\rangle_{\tilde{S}\tilde{D}} \otimes |\psi_m\rangle_{S'D'}$$

with fidelity[20] *at least $1 - \varepsilon$, where $|\psi_m\rangle_{S'D'} := \frac{1}{\sqrt{m}} \sum_{x=0}^{m-1} |\hat{x}\rangle_{S'} |\hat{x}\rangle_{D'}$ denotes a maximally entangled state of two m dimensional systems.*

Note that the state $|\phi\rangle_{SD}$ considered in Theorem 3 can be represented by its Schmidt decomposition as

$$|\phi\rangle_{SD} = \sum_{\hat{y}} \sqrt{p_{\hat{y}}}\, |\hat{y}\rangle_S \otimes |\hat{y}\rangle_D \, .$$

We now consider an embezzling state on $\tilde{S}\tilde{D}$ and use Lemma 4 to define isometries $\hat{U}$ and $\hat{V}$ on $S\tilde{S}$ and $D\tilde{D}$, respectively, which are controlled by the entry $\hat{y}$ in the registers S or D, and build up entanglement between registers S' and D', i.e.,

[19] Note that this condition is (by definition) only satisfied if P_A and P_B have full support.

[20] The *fidelity* between two pure states $|\psi\rangle$ and $|\phi\rangle$ is $|\langle\psi|\phi\rangle|$.

$$\hat{U} = \sum_{\hat{y}} |\hat{y}\rangle\langle\hat{y}|_S \otimes U_{m(\hat{y})}$$
$$\hat{V} = \sum_{\hat{y}} |\hat{y}\rangle\langle\hat{y}|_D \otimes V_{m(\hat{y})}.$$

The integers $m(\hat{y})$ are chosen such that the state resulting from applying $\hat{U} \otimes \hat{V}$ to $|\phi\rangle_{SD} \otimes |\Gamma^k\rangle_{\bar{S}\bar{D}}$ is close to a state of the form

$$\Big(\frac{1}{\sqrt{M}}\sum_{\hat{y}}\sum_{\hat{y}'=0}^{m(\hat{y})-1} |\hat{y},\hat{y}'\rangle_{SS'} \otimes |\hat{y},\hat{y}'\rangle_{DD'}\Big) \otimes |\Gamma^k\rangle_{\bar{D}\bar{S}},$$

with $\sum_{\hat{y}} m(\hat{y}) = M$, for some integer M. (This can be achieved to arbitrary precision for sufficiently large k and $m(\hat{y})$.) Note that the first part of this state corresponds to a maximally entangled pair, $|\psi_M\rangle$, between the registers SS' and DD'.[21] We now construct the POVMs $\{E_x^a\}$ and $\{F_y^b\}$ by concatenating the operations $\hat{U}$ and $\hat{V}$ with the measurements $\{\bar{E}_x^a\}$ and $\{\bar{F}_y^b\}$ introduced in Sect. 2.3. More precisely, we define

$$E_x^a := \hat{U}^\dagger \cdot \Big[\Big(\bar{E}_x^a\Big)_{DD'} \otimes \mathbb{1}_{\bar{D}}\Big] \cdot \hat{U}$$
$$F_y^b := \hat{V}^\dagger \cdot \Big[\Big(\bar{F}_y^b\Big)_{SS'} \otimes \mathbb{1}_{\bar{S}}\Big] \cdot \hat{V}$$

with $a \in \mathcal{A}_N$ and $b \in \mathcal{B}_N$, for some large N. In addition we define

$$F_y^{b_0} = |\hat{y}\rangle\langle\hat{y}|_S \otimes \mathbb{1}_{\bar{S}}.$$

Assume now that the freedom of choice as well as the compatibility with quantum theory assumption are satisfied. Furthermore, let $X = (\hat{X}, \hat{X}')$ and $Y = \hat{Y}$ be the outcomes of the measurements $A = a_0$ and $B = b_0$, respectively. Note that quantum theory predicts that the outcomes of the measurements of a_0 and b_0 are in agreement, in the sense that $\hat{X} = Y$ holds with probability 1. Hence, together with the no-signalling conditions (cf. Lemma 1) we find that

$$\bar{P}_{Y|b_0(\phi\otimes\Gamma)} = \bar{P}_{\hat{X}|a_0(\phi\otimes\Gamma)}$$
$$P_{Y|b_0 Z} = P_{\hat{X}|a_0 Z}.$$

Now let $\delta > 0$. Lemma 3 implies that by taking N large enough,

---

[21] As a simple example, consider the state $|\phi\rangle_{SD} = \frac{1}{2}\left|\hat{0}\right\rangle_S\left|\hat{0}\right\rangle_D + \frac{\sqrt{3}}{2}\left|\hat{1}\right\rangle_S\left|\hat{1}\right\rangle_D$. In this case we would take $m(0) = 1$ and $m(1) = 3$ to yield a state of the form $\frac{1}{2}(\left|\hat{0}\hat{0}\right\rangle_{SS'}\left|\hat{0}\hat{0}\right\rangle_{DD'} + \left|\hat{1}\hat{0}\right\rangle_{SS'}\left|\hat{1}\hat{0}\right\rangle_{DD'} + \left|\hat{1}\hat{1}\right\rangle_{SS'}\left|\hat{1}\hat{1}\right\rangle_{DD'} + \left|\hat{1}\hat{2}\right\rangle_{SS'}\left|\hat{1}\hat{2}\right\rangle_{DD'})$ after the transformation.

$$\left\langle D(P_{X|a_0 z}, \bar{P}_{X|a_0(\phi\otimes\Gamma)})\right\rangle_z \leq \delta,$$

and hence the same relation holds for the marginals of these distributions, i.e.,

$$\left\langle D(P_{\hat{X}|a_0 z}, \bar{P}_{\hat{X}|a_0(\phi\otimes\Gamma)})\right\rangle_z \leq \delta.$$

Combining this with the above identities and noting that $\bar{P}_{Y|b_0(\phi\otimes\Gamma)} = \bar{P}_{Y|b_0\phi}$, we find that $\left\langle D(P_{Y|b_0 z}, \bar{P}_{Y|b_0\phi})\right\rangle_z \leq \delta$, and since this holds for any $\delta > 0$, we have

$$\left\langle D(P_{Y|b_0 z}, \bar{P}_{Y|b_0\phi})\right\rangle_z = 0\,.$$

In other words, $P_{Y|b_0 Z}$ does not give improved predictions over $\bar{P}_{Y|b_0\phi}$ with respect to P. □

8 Alternative Theories Are Equivalent to Quantum Theory

In this section, we discuss an implication of the non-extendibility theorem (Theorem 3) to a long-standing debate on the nature of the quantum mechanical wave function. The debate centres around whether it should be interpreted as a subjective quantity, for example a state of knowledge about some underlying physical reality, or whether it should instead be interpreted as objective (real).[22] The wave function could be considered subjective if there existed an alternative theory, with predictions based on a parameter Z, whose predictions are the same whether or not the wave function, Ψ, is taken into account,[23] and in which two different wave functions, say ψ and ψ', are compatible with the same value of the parameter, say $Z = z$. Formally, if Ψ takes values from a finite set, this would mean that there exist z, ψ, and $\psi' \neq \psi$ such that $P_{\Psi|Z}(\psi|z) > 0$ and $P_{\Psi|Z}(\psi'|z) > 0$. This is sometimes called a ψ-epistemic view of the wave function and contrasts with the ψ-ontic, or objective, view [27] (we refer to [26, 28, 29] for arguments in favour of the ψ-epistemic view). In the objective view, in any alternative theory whose predictions are the same whether or not Ψ is taken into account, the wave function is uniquely determined by the parameters of the alternative theory. In other words, there exists a (deterministic) function, f such that $\Psi = f(Z)$ holds almost surely (over Z). In this section, we give an argument in support of this.

Our result is based on the following simple lemma, which applies when the RV Ψ takes values from a finite set of wave functions, $\mathcal{P}$.

---

[22]Note that in some subjective interpretations (e.g. [26]) there is no underlying physical reality—the wave function is simply a state of knowledge about future measurement outcomes and nothing more.

[23]If the predictions are about a measurement outcome X based on a setting A, then this condition reads $P_{X|az\psi} = P_{X|az}$.

Lemma 5 *Let $\{E_x^a\}$ form a tomographically complete set of POVMs for $a \in \mathcal{A}$, and let $P_{AXZ\Psi}$ be a probability distribution such that:*

- *A is a free choice with respect to a causal order in which $A \not\rightsquigarrow Z$ and $A \not\rightsquigarrow \Psi$.*
- *The predictions based on Z are at least as informative as those based on Ψ, i.e., $P_{X|az\psi} = P_{X|az}$ whenever $P_{X|az\psi}$ is defined.*[24]
- *The predictions based on Ψ are at least as informative as those based on Z, i.e., $P_{X|az\psi} = P_{X|a\psi}$ whenever $P_{X|az\psi}$ is defined, where $P_{X|a\psi}(x) = \mathrm{tr}(E_x^a\psi)$.*

Then there exists a function, $f : \mathcal{Z} \mapsto \mathcal{P}$, such that $\Psi = f(Z)$.

Proof Combining the second and third conditions of the lemma we have

$$P_{X|a\psi} = P_{X|az}, \tag{12}$$

whenever $P_{X|az\psi}$ is defined. If A is a free choice then $P_{AZ\Psi} = P_A \times P_{Z\Psi}$, hence (12) holds provided that $P_A(a) > 0$.

Suppose now that z, ψ and ψ' are such that $P_{\Psi|z}(\psi) > 0$ and $P_{\Psi|z}(\psi') > 0$. From (12), this implies $P_{X|a\psi} = P_{X|a\psi'}$ for all a such that $P_A(a) > 0$. Since the set of measurements with $P_A(a) > 0$ is tomographically complete, this can only be satisfied if $\psi = \psi'$. Since we can repeat this for any pair of wave functions in $\mathcal{P}$, the existence of the function f such that $\Psi = f(Z)$ follows. □

Combining Theorem 3 with Lemma 5, we can establish the main result of this section, which we state informally as follows.

Claim 2 ([6]) *In any alternative theory that is at least as informative as quantum theory and compatible with free choice (with respect to the discussed causal orders), there is a one-to-one correspondence between the parameters of the alternative theory and the quantum state (up to a possible removable degeneracy*[25] *in the parameters of the alternative theory).*

To establish this, as before, we use Z to denote the parameters of the alternative theory. The statements in the remainder of this paragraph hold almost surely over $P_{Z|\psi}$ for any $\psi \in \mathcal{P}$, where $\mathcal{P}$ is any finite set of wave functions (we omit writing these conditions below for brevity). Theorem 3 shows that under the freedom of choice assumption, quantum theory is at least as informative as any alternative theory. We hence satisfy the conditions of Lemma 5 so find $\Psi = f(Z)$, for some function $f : \mathcal{Z} \mapsto \mathcal{P}$. Furthermore, since Z cannot improve the predictions for any $\Psi = \psi$, any z in $f^{-1}(\psi)$ must give identical predictions. Hence, if $f^{-1}(\psi)$ contains more than one element, this corresponds to a removable degeneracy in the parameters of the alternative theory.

---

[24] Note that if Z takes values from a finite set, this is the same as saying that $P_{X|AZ}$ does not give improved predictions over $P_{X|A\Psi}$ with respect to P.

[25] Any degeneracy is *removable* in the sense that it has no operational effect, i.e., one can define another theory without the degeneracy (but otherwise identical) without affecting the predictive power.

8.1 Related Work

An interpretation of the wave function as a subjective state of knowledge about some underlying theory has also been ruled out by Pusey et al. [30] via a different argument using different assumptions which we now summarize. They consider the preparation of multiple quantum systems, with states Ψ_i, where each system is associated with a particular parameter in the higher theory, Z_i. Pusey et al. assume that the joint distribution of these is product, i.e.

$$P_{Z_1 Z_2 \ldots \Psi_1 \Psi_2 \ldots} = P_{Z_1 \Psi_1} \times P_{Z_2 \Psi_2} \times \ldots . \quad (13)$$

Starting from this assumption, they show that there cannot exist two distinct states, ψ and ψ', such that for each i there exists a value of $Z_i = z_i$ satisfying $P_{\Psi_i|z_i}(\psi) > 0$ and $P_{\Psi_i|z_i}(\psi') > 0$.

We note that the product nature of the joint distribution, Eq. (13), is related to free choice of preparation. In particular, it implies

$$P_{\Psi_1 Z_2 \ldots Z_N \Psi_2 \ldots \Psi_N} = P_{\Psi_1} \times P_{Z_2 \ldots Z_N \Psi_2 \ldots \Psi_N} .$$

If we take the causal order to be such that $\Psi_i \not\rightsquigarrow \Psi_j$ and $\Psi_i \not\rightsquigarrow Z_j$ for $j \neq i$ (as would be natural if we make space-like separated preparations), then this is equivalent to saying that Ψ_1 can be chosen freely. (Note that (13) embodies more than just this.)

It was subsequently noted [31] that the separability assumption can be weakened, in essence to the assumption that there exists a particular set of parameters in the higher theory that are compatible with every product state composed of ψ and ψ', i.e., there exist values of the parameters, $z_1, \ldots, z_N$, such that

$$P_{\Psi_1 \ldots \Psi_N | z_1, \ldots, z_N}(\psi^{(\prime)}, \ldots, \psi^{(\prime)}) > 0,$$

where each $\psi^{(\prime)}$ is independently either ψ or ψ' (so that the above represents 2^N conditions). This condition can be further weakened [32] such that the parameters of the alternative theory for multiple systems need not be made up only of the individual parts, but could be replaced or supplemented with global parameters (provided these are also compatible with all the product state preparations).

An alternative argument against an interpretation of the quantum state as a state of knowledge about an underlying reality can be found in [33].

We remark that some models in which the wave function is subjective have been developed for restricted scenarios. For example, by modifying an earlier model by Bell [3], Lewis et al. constructed a model for a single qubit in which the wave function is subjective, and extended that model to arbitrary dimensions [34]. These models are not in conflict with Claim 2 because they treat only single systems, and cannot be extended to bipartite scenarios while allowing for free choice with respect to one of the causal orders of Fig. 4. However, they do show that the same type of result cannot be established when looking only at single systems.

9 Discussion

The main statements described in this chapter about the completeness of quantum theory are based on two assumptions. One of them is that quantum theory is correct, and is implicit in the question of completeness. The other is that of free choice within a natural causal order. It is worth commenting on the existence of alternative models that are not compatible with this assumption.

A prominent example is the de Broglie-Bohm model [35, 36] which recreates quantum correlations, providing higher explanation in the form of hidden particle positions. These can be thought of as parameters of a higher theory that would allow perfect predictions of the outcomes. However, introducing these parameters comes at a price: it is incompatible with the freedom of choice assumption of our theorems. In fact, for the bipartite setting discussed above, if Z includes the particle positions of the de Broglie-Bohm model, we have some non-local behaviour, so that $P_{X|abz} = P_{X|az}$, for instance, does not hold. Thus, given Lemma 1, it follows that A and B cannot be free choices with respect to any of the causal orders of Fig. 4.

There are at least two ways to avoid our conclusions. The first is to maintain free choice, but assume that the alternative theory has a different causal order (in particular, one in which either $A \not\rightsquigarrow Y$ or $B \not\rightsquigarrow X$ does not hold). The second is to consider alternative theories without free choice, in which the measurement settings A and B may depend on the additional parameters Z (sometimes, this view is argued for by imagining that the additional parameters are permanently hidden).

One may take the view that the freedom of choice assumption, which demands complete independence between the chosen settings and the other variables, is relatively strong, and perhaps contemplate alternative theories where this assumption is weakened. Some results in this direction can be found in [37], where a theorem similar to Lemma 3 is established under a relaxed free choice assumption, and provided there is no signalling at the level of the underlying theory.

Finally, we note that the result presented here has a generic application in quantum cryptography. Standard security proofs for schemes such as quantum key distribution [38, 39] are based on the assumption (usually not stated explicitly) that quantum theory is complete. If this were not the case, it could be that a scheme is proven secure within quantum theory, yet an adversary can break it by exploiting information available in a higher theory. However, our non-extendibility theorem, Theorem 3, implies that it is sufficient to make only the weaker assumption that quantum theory is correct, since this implies completeness.

Acknowledgments We are grateful to Klaas Landsman, Gijs Leegwater and Robert Spekkens for helpful comments on an earlier draft of this chapter.

Appendix: Variational Distance

The following is a list of the main properties of the variational distance $D(\cdot, \cdot)$ used in this work (note that we only use this measure for discrete distributions):

- $D(\cdot, \cdot)$ is a metric on the space of probability distributions.
- $D(\cdot, \cdot)$ is upper bounded by 1.
- The variational distance of marginal distributions cannot be larger than that of the joint distributions: $D(P_X, Q_X) \leq D(P_{XY}, Q_{XY})$ for any P_{XY} and Q_{XY}.
- It is convex: If $\{\alpha_i\}$ satisfy $\alpha_i \geq 0$ and $\sum_i \alpha_i = 1$, and $\{P_X^i\}$ and $\{Q_X^i\}$ are sets of distributions over X, then $D(\sum_i \alpha_i P_X^i, \sum_i \alpha_i Q_X^i) \leq \sum_i \alpha_i D(P_X^i, Q_X^i)$.
- For a joint distribution P_{XY}, the variational distance of the marginal distributions is bounded by the probability that the RVs X and Y have different values: $D(P_X, P_Y) \leq P(X \neq Y)$.

The first four properties follow straightforwardly from the definition. The last is proved in the following.

Lemma 6 *Let X and Y be two discrete random variables jointly distributed according to P_{XY}. Then the variational distance between the marginal distributions P_X and P_Y is bounded by*

$$D(P_X, P_Y) \leq P(X \neq Y).$$

Proof Let $P_{XY}^{\neq} := P_{XY|X \neq Y}$ be the joint distribution of X and Y conditioned on the event that they are not equal. Similarly, define $P_{XY}^{=} := P_{XY|X=Y}$. We then have

$$P_{XY} = p_{\neq} P_{XY}^{\neq} + (1 - p_{\neq}) P_{XY}^{=}$$

where $p_{\neq} := P(X \neq Y)$. By linearity, the marginals of these distributions satisfy the same relation, i.e.,

$$\begin{aligned} P_X &= p_{\neq} P_X^{\neq} + (1 - p_{\neq}) P_X^{=} \\ P_Y &= p_{\neq} P_Y^{\neq} + (1 - p_{\neq}) P_Y^{=}. \end{aligned}$$

Hence, by convexity of the variational distance,

$$\begin{aligned} D(P_X, P_Y) &\leq p_{\neq} D(P_X^{\neq}, P_Y^{\neq}) + (1 - p_{\neq}) D(P_X^{=}, P_Y^{=}) \\ &\leq p_{\neq}, \end{aligned}$$

where the last inequality follows because the variational distance is at most 1, and $D(P_X^{=}, P_Y^{=}) = 0$. □

Lemma 7 *Let X be a random variable taking values in the set $\{0, 1, \ldots, M - 1\}$ that is distributed according to P_X, and $\hat{P}_X$ be the uniform distribution over X. Then*

$$D(P_X, \hat{P}_X) \leq \frac{M}{4} D(P_{X\oplus 1}, P_X) .$$

Proof Since D is convex, we have

$$\begin{aligned} D(P_X, \hat{P}_X) &= D\left(\frac{1}{M}\sum_{i=0}^{M-1} P_X, \frac{1}{M}\sum_{i=0}^{M-1} P_{X\oplus i}\right) \\ &\leq \frac{1}{M}\sum_{i=0}^{M-1} D(P_X, P_{X\oplus i}). \end{aligned}$$

Now, noting that $D(P_{X\oplus(i-1)}, P_{X\oplus i}) = D(P_{X\oplus 1}, P_X)$ for all i we have for $i \leq M/2$

$$\begin{aligned} D(P_X, P_{X\oplus i}) &\leq D(P_X, P_{X\oplus(i-1)}) + D(P_{X\oplus(i-1)}, P_{X\oplus i}) \\ &= D(P_X, P_{X\oplus(i-1)}) + D(P_{X\oplus 1}, P_X) , \end{aligned}$$

from which we obtain $D(P_X, P_{X\oplus i}) \leq i D(P_{X\oplus 1}, P_X)$ by repeated application. A similar analysis in the case $i \geq M/2$ gives $D(P_X, P_{X\oplus i}) \leq (M - i)D(P_{X\oplus 1}, P_X)$. Thus,

$$\begin{aligned} &\sum_{i=0}^{M-1} D(P_X, P_{X\oplus i}) \\ &\quad \leq \left(\sum_{i=0}^{\lfloor M/2\rfloor} i + \sum_{i=\lfloor M/2\rfloor+1}^{M-1} (M - i)\right) D(P_{X\oplus 1}, P_X) \\ &\qquad\qquad \leq \frac{M^2}{4} D(P_{X\oplus 1}, P_X), \end{aligned}$$

from which we obtain the claimed result. □

References

1. A. Einstein, B. Podolsky, N. Rosen, Can quantum-mechanical description of physical reality be considered complete? Phys. Rev. **47**, 777–780 (1935)
2. S. Kochen, E.P. Specker, The problem of hidden variables in quantum mechanics. J. Math. Mech. **17**, 59–87 (1967)
3. J.S. Bell, On the problem of hidden variables in quantum mechanics, in *Speakable and Unspeakable in Quantum Mechanics*, Chap. 1 (Cambridge University Press, Cambridge, 1987)
4. J.S. Bell, On the Einstein-Podolsky-Rosen paradox, in *Speakable and Unspeakable in Quantum Mechanics*, chap. 2 (Cambridge University Press, Cambridge, 1987)
5. R. Colbeck, R. Renner, No extension of quantum theory can have improved predictive power. Nat. Commun. **2**, 411 (2011)
6. R. Colbeck, R. Renner, Is a system's wave function in one-to-one correspondence with its elements of reality? Phys. Rev. Lett. **108**, 150402 (2012)

7. R. Colbeck, R. Renner, A system's wave function is uniquely determined by its underlying physical state. e-print (2013) arXiv:1312.7353
8. P.M. Pearle, Hidden-variable example based upon data rejection. Phys. Rev. D **2**, 1418–1425 (1970)
9. S.L. Braunstein, C.M. Caves, Wringing out better Bell inequalities. Ann. Phys. **202**, 22–56 (1990)
10. J.F. Clauser, M.A. Horne, A. Shimony, R.A. Holt, Proposed experiment to test local hidden-variable theories. Phys. Rev. Lett. **23**, 880–884 (1969)
11. J.S. Bell, Free variables and local causality, in *Speakable and Unspeakable in Quantum Mechanics*, chap. 12 (Cambridge University Press, Cambridge, 1987)
12. R. Colbeck, R. Renner, A short note on the concept of free choice (2013) arXiv:1302.4446
13. A.J. Leggett, Nonlocal hidden-variable theories and quantum mechanics: an incompatibility theorem. Found. Phys. **33**, 1469–1493 (2003)
14. J.S. Bell, La nouvelle cuisine, in *Speakable and Unspeakable in Quantum Mechanics*, 2nd edn chap. 24 (Cambridge University Press, Cambridge, 2004),
15. C. Branciard et al., Testing quantum correlations versus single-particle properties within Leggett's model and beyond. Nat. Phys. **4**, 681–685 (2008)
16. S. Gröblacher et al., An experimental test of non-local realism. Nature **446**, 871–875 (2007)
17. R. Colbeck, R. Renner, Hidden variable models for quantum theory cannot have any local part. Phys. Rev. Lett. **101**, 050403 (2008)
18. L. Hardy, Quantum ontological excess baggage. Stud. Hist. Philos. Mod. Phys. **35**, 267–276 (2004)
19. A. Montina, Exponential complexity and ontological theories of quantum mechanics. Phys. Rev. A **77** 022104 (2008)
20. A. Montina, State-space dimensionality in short-memory hidden-variable theories. Phys. Rev. A **83**, 032107 (2011)
21. Z. Chen, A. Montina, Measurement contextuality is implied by macroscopic realism. Phys. Rev. A **83** 042110 (2011)
22. R.W. Spekkens, Contextuality for preparations, transformations, and unsharp measurements. Phys. Rev A **71**, 052108 (2005)
23. J. Barrett, L. Hardy, A. Kent, No signalling and quantum key distribution. Phys. Rev. Lett. **95**, 010503 (2005)
24. J. Barrett, A. Kent, S. Pironio, Maximally non-local and monogamous quantum correlations. Phys. Rev Lett. **97**, 170409 (2006)
25. W. van Dam, P. Hayden, Universal entanglement transformations without communication. Phys. Rev A **67**, 060302(R) (2003)
26. C.M. Caves, C.A. Fuchs, R. Schack, Quantum probabilities as Bayesian probabilities. Phys. Rev. A **65**, 022305 (2002)
27. N. Harrigan, R.W. Spekkens, Einstein, incompleteness, and the epistemic view of quantum states. Found. Phys. **40**, 125–157 (2010)
28. R.W. Spekkens, Evidence for the epistemic view of quantum states: a toy theory. Phys. Rev. A **75**, 032110 (2007)
29. M.S. Leifer, R.W. Spekkens, Formulating quantum theory as a causally neutral theory of Bayesian inference. e-print (2011) arXiv:1107.5849
30. M.F. Pusey, J. Barrett, T. Rudolph, On the reality of the quantum state. Nat. Phys. **8**, 476–479 (2012)
31. M.J.W. Hall, Generalisations of the recent Pusey-Barrett-Rudolph theorem for statistical models of quantum phenomena. e-print (2011) arXiv:1111.6304
32. M. Schlosshauer, A. Fine, On a recent quantum no-go theorem. e-print (2012) arXiv:1203.4779
33. L. Hardy, Are quantum states real? e-print (2012) arXiv:1205.1439
34. P.G. Lewis, D. Jennings, J. Barrett, T. Rudolph, Distinct quantum states can be compatible with a single state of reality. Phys. Rev. Lett. **109**, 150404 (2012)
35. L. de Broglie, La, mécanique ondulatoire et la structure atomique de la matière et du rayonnement. J. de Physique, Serie VI **VIII**, 225–241, (1927)

36. D. Bohm, A suggested interpretation of the quantum theory in terms of hidden variables. I. Phys. Rev. **85**, 166–179 (1952)
37. R. Colbeck, R. Renner, Free randomness can be amplified. Nat. Phys. **8**, 450–454 (2012)
38. C.H. Bennett, G. Brassard, Quantum cryptography: public key distribution and coin tossing, in *Proceedings of IEEE International Conference on Computers, Systems, and Signal Processing*, (IEEE, New York, 1984), pp. 175–179
39. A.K. Ekert, Quantum cryptography based on Bell's theorem. Phys. Rev. Lett. **67**, 661–663 (1991)

MIX
Papier aus verantwortungsvollen Quellen
Paper from responsible sources
FSC® C105338

If you have any concerns about our products,
you can contact us on
ProductSafety@springernature.com

In case Publisher is established outside the EU,
the EU authorized representative is:
Springer Nature Customer Service Center GmbH
Europaplatz 3, 69115 Heidelberg, Germany

Printed by Libri Plureos GmbH
in Hamburg, Germany